Lecture Notes in Electrical Engineering

Volume 915

The book series *Lecture Notes in Electrical Engineering* (LNEE) publishes the latest developments in Electrical Engineering—quickly, informally and in high quality. While original research reported in proceedings and monographs has traditionally formed the core of LNEE, we also encourage authors to submit books devoted to supporting student education and professional training in the various fields and applications areas of electrical engineering. The series cover classical and emerging topics concerning:

- Communication Engineering, Information Theory and Networks
- Electronics Engineering and Microelectronics
- Signal, Image and Speech Processing
- Wireless and Mobile Communication
- Circuits and Systems
- Energy Systems, Power Electronics and Electrical Machines
- Electro-optical Engineering
- Instrumentation Engineering
- Avionics Engineering
- Control Systems
- Internet-of-Things and Cybersecurity
- Biomedical Devices, MEMS and NEMS

For general information about this book series, comments or suggestions, please contact leontina.dicecco@springer.com.

To submit a proposal or request further information, please contact the Publishing Editor in your country:

China

Jasmine Dou, Editor (jasmine.dou@springer.com)

India, Japan, Rest of Asia

Swati Meherishi, Editorial Director (Swati.Meherishi@springer.com)

Southeast Asia, Australia, New Zealand

Ramesh Nath Premnath, Editor (ramesh.premnath@springernature.com)

USA, Canada:

Michael Luby, Senior Editor (michael.luby@springer.com)

All other Countries:

Leontina Di Cecco, Senior Editor (leontina.dicecco@springer.com)

** **This series is indexed by EI Compendex and Scopus databases.** **

Anuradha Tomar · Hasmat Malik · Pramod Kumar ·
Atif Iqbal
Editors

Proceedings of 3rd International Conference on Machine Learning, Advances in Computing, Renewable Energy and Communication

MARC 2021

Springer

Editors
Anuradha Tomar
Division of Instrumentation and Control Engineering
Netaji Subhas University of Technology
Delhi, India

Hasmat Malik
Division of Electrical Power Engineering
Faculty of Engineering
School of Electrical Engineering
University of Technology Malaysia (UTM)
Johor Bahru, Malaysia

Pramod Kumar
Department of Computer Science and Engineering
Krishna Engineering College
Uttar Pradesh, India

Atif Iqbal
Department of Electrical Engineering
Qatar University
Doha, Qatar

ISSN 1876-1100 ISSN 1876-1119 (electronic)
Lecture Notes in Electrical Engineering
ISBN 978-981-19-2827-7 ISBN 978-981-19-2828-4 (eBook)
https://doi.org/10.1007/978-981-19-2828-4

This Springer imprint is published by the registered company Springer Nature Singapore Pte Ltd.
The registered company address is: 152 Beach Road, #21-01/04 Gateway East, Singapore 189721, Singapore

Preface

The papers presented at the 3rd International Conference on Machine Learning, Advances in Computing, Renewable Energy and Communication (MARC 2021) held at Krishna Engineering College in Ghaziabad, Uttar Pradesh, India, on December 10 and 11, 2021, are compiled in this volume. The International Conference on Machine Learning, Advances in Computing, Renewable Energy and Communication focuses on advanced research in the area of electrical and computer science engineering and will provide a forum for sharing insights, experiences and interaction on various facts of evolving technologies and patterns related to these areas. The objective of MARC 2021 is to provide a platform for leading academic scientists, researchers, scholars and students to get together to share their results and compare notes on their research discovery in the development of electrical engineering and high-performance computing. Numerous participants attended the conference, made technical presentations and indulged in various technical discussions. The number of paper published in this volume and the number of unpublished presentations at the conference indicate the evidence of growing interest among students, researchers and teachers in manufacturing and advanced computing. More than 270 research papers were submitted, out of which 68 were accepted and presented along with a brief report as an editorial.

We thank all the contributors of this book for their valuable effort in producing high-class literature for research community. We are sincere thankful to the reviewers to provide the all reviews/comments/suggestions in a short period of time.

We would like to extend our sincere gratitude to Springer LNEE for giving Krishna Engineering College the opportunity and the platform to organize this conference which helped in reaching out to the eminent scholars and the fellow researchers in the field of electrical and computer science engineering and helping them in widening the areas of the subject.

We express our sincere gratitude to our Patrons Dr. M. K. Singh, Krishna Engineering College, Ghaziabad, Dr. Manoj Goel, Krishna Engineering College, Ghaziabad, Uttar Pradesh, and Prof. Sukumar Mishra, IIT Delhi, for their motivation and support in hosting MARC 2021. Our sincere thanks and appreciation to our General

Chair Prof. Bijaya Ketan Panigrahi, IIT Delhi, Dr. Shailesh Tiwari, Krishna Engineering College, Ghaziabad, Uttar Pradesh, and Dr. Yog Raj Sood, NIT Hamirpur, H.P., for their solid support blended with encouragement and incomparable motivations to achieve the remarkable milestone. We wish to acknowledge our gratitude to Program Chair Dr. Durgesh Pant, USAC & USERC, Dehradun and Dr. Vinay Rishiwal, M. J. P. Rohilkhand University, Bareilly set the tone of the conference at the higher launch. We sincerely acknowledge all the keynote speakers for disseminating your knowledge, experience and thoughts. We also acknowledge Dr. Nuzhat Fatema (UniSZA, Malaysia) for organizing a tutorial on Data-Driven Intelligent Approaches for forecasting and sharing the knowledge, and experience. We express our sincere gratitude to the management of Krishna Engineering College, conference executive chair, publication chair and technical committee members for their kind support and motivation.

We wish to thank our colleagues and friends for their insight and helpful discussion during the production of this book. We would like to highlight the contribution, suggestion and motivation of Prof. Imtiaz Ashraf, Aligarh Muslim University, India; Prof. M. S. Jamil Asghar, Aligarh Muslim University, India; Prof. Salman Hameed, Aligarh Muslim University, India; Prof. A. H. Bhat, NIT Srinagar, India; Prof. Kouzou Abdellah, Djelfa University, Algeria; Prof. Jaroslaw Guzinski, Gdansk University of Technology; Prof. Akhtar Kalam, Victoria University of Technology, Australia; Prof. Mairaj Ud Din Mufti, NIT Srinagar, India; Prof. Majid Jamil, JMI, India; Prof. Majed A. Altotaibi, King Saud University, Saudi Arabia; Prof R. K Jarial, NIT Hamirpur (HP), India; Prof. Rajesh Kumar, GGSIPU, India; Prof. Anand Parey, IIT Indore, India; and Prof Yogesh Pandya, PIEMR, Indore, India.

We would like to express our gratitude and our love and affection to our family members for their intense feeling of deep affection.

New Delhi, India — Dr. Anuradha Tomar
UTM Johor Bahru, Malaysia — Dr. Hasmat Malik
Ghaziabad, India — Dr. Pramod Kumar
Doha, Qatar — Prof. Atif Iqbal

Contents

About the Editors

Dr. Anuradha Tomar has 12 years plus experience in research and academics. She is currently working as Assistant Professor in Instrumentation & Control Engineering Division of Netaji Subhas University, Delhi, India. Dr. Tomar has completed her Postdoctoral research in Electrical Energy Systems Group, from Eindhoven University of Technology (TU/e), the Netherlands and has successfully completed European Commission's Horizon 2020, UNITED GRID and UNICORN TKI Urban Research projects. She has received her B.E. Degree in Electronics Instrumentation & Control with Honours in the year 2007 from University of Rajasthan, India. In the year 2009, she has completed her M.Tech. Degree with Honours in Power System from National Institute of Technology Hamirpur. She has received her Ph.D. in Electrical Engineering, from Indian Institute of Technology Delhi (IITD). Dr. Anuradha Tomar has committed her research work efforts towards the development of sustainable, energy efficient solutions for the empowerment of society, humankind. Her areas of research interest are Operation & Control of Microgrids, Photovoltaic Systems, Renewable Energy based Rural Electrification, Congestion Management in LV Distribution Systems, Artificial Intelligent & Machine Learning Applications in Power System, Energy conservation and Automation. She has authored or co-authored 69 research/review papers in various reputed International, National Journals, and Conferences. She is an Editor for books with International Publication like Springer, Elsevier. She has also filled seven Indian patents on her name. Dr. Tomar is Senior member of IEEE, Life member of ISTE, IETE, IEI, and IAENG.

Dr. Hasmat Malik is a Senior Lecturer at the Division of Electrical Power Engineering, Faculty of Engineering, School of Electrical Engineering, Universiti Teknologi Malaysia (UTM) Johor, Malaysia.

He served as an Assistant Professor for more than five years with the Division of Instrumentation and Control Engineering, Netaji Subhas Institute of Technology (NSIT), Delhi, India; and 3.5 years as a Postdoctoral scholar with the Berkeley Education Alliance for Research in Singapore (A research center of the University of California, Berkeley, USA) and the National University of Singapore (NUS),

Singapore. He is currently a Chartered Engineer (CEng) and a Professional Engineer (PEng). He is a fellow of the Institution of Electronics and Telecommunication Engineering (IETE), a Senior Member of the Institute of Electrical and Electronics Engineers (IEEE) USA, a Life Member of the Indian Society for Technical Education (ISTE), and a member of the Institution of Engineering and Technology (IET), U.K and the Institute of Engineers (India).

He has supervised 25 master's students and 5 Ph.D. are ongoing. He is involved in several large research and development projects. He has authored/co-authored more than 100 research articles, nine books, and 15 chapters in other books, published by IEEE, Springer, and Elsevier. He received the POSOCO Power System Award (PPSA-2017) for his Ph.D. work for research and innovation in the area of power systems. He also received the best research papers awards from IEEE INDICON-2015, and the Full Registration Fee Award from IEEE SSD-2012, Germany.

His research interests include artificial intelligence, machine learning and big-data analytics for renewable energy, smart building and automation, condition monitoring, and online fault detection and diagnosis (FDD).

Prof. (Dr.) Pramod Kumar is working as Professor, Head (CSE) and Dean (CSE and IT) in Krishna Engineering College, Ghaziabad since January 2018. He also served as Director of the Tula's Institute, Dehradun, Uttarakhand. He has more than 22 years of experience in academics. He completed his Ph.D. in Computer Science and Engineering in 2011 and M.Tech. (CSE) in 2006. He is a senior member of IEEE (SMIEEE) and Joint Secretary of IEEE U.P. Section. He has published widely in International Journals and Conferences his research finding related to Computer Network, Internet of things (IoT) and Machine Learning. He has Authored/Co-Authored more than 70 research papers and several chapters in edited books. He has supervised and co-supervised several M.Tech. and Ph.D. students. He has organized more than 10 IEEE International Conferences and all the research paper of these conferences is now available in IEEE Explore. He is the editor of two books. He has conducted more than 15 Faculty Development Program in the collaboration of EICT, IIT Roorkee, EICT, IIT Kanpur and AKTU. He has organised IEEE National workshop on Research Paper writing on 27 March 2017 and IEEE Women Symposium on 21 February 2016.

Prof. Atif Iqbal (Senior Member, IEEE, Fellow IET, Fellow IE) received the B.Sc. and M.Sc. degrees in engineering (power system and drives) from Aligarh Muslim University (AMU), Aligarh, India, in 1991 and 1996, respectively, and Ph.D. degree from Liverpool John Moores University, Liverpool, U.K., in 2006, and the D.Sc. degree (Habilitation) in control, informatics, and electrical engineering from the Gdansk University of Technology, in 2019. He is a Full Professor with the Department of Electrical Engineering, Qatar University, Doha, Qatar, and a former Full Professor with the Department of Electrical Engineering, AMU, Aligarh, India. He has been a Lecturer with the Department of Electrical Engineering, AMU, since 1991, where he has served as a Full Professor, until August 2016. Dr. Iqbal has been listed in top 2% highly cited scientists of the world (data released by Stanford University, USA). The world ranking in 2019 was #649 and the current ranking is #622. He has published widely in international journals and conferences on his research findings related to power electronics, variable speed drives, and renewable energy sources. He has authored or coauthored more than 520 research articles and four books and several chapters in edited books. He has supervised several large research and development projects worth more than multimillion USD. He was a recipient of the Maulana Tufail Ahmad Gold Medal for standing first at the B.Sc.Engg. (Electrical) Exams from AMU, in 1991. He was also a recipient of the Outstanding Faculty Merit Award academic year 2014–2015 and the Research Excellence Awards at Qatar University, in 2015 and 2019. He has received several best research papers awards, e.g., at IEEE ICIT-2013, IET-SEISCON-2013, SIGMA 2018, IEEE CENCON 2019, IEEE ICIOT 2020, ICSTEESD-20, Springer ICRP 2020, IEEE GUCON 2021. He has also received the Gold Medal for his B.Sc. degree. He is the Vice-Chair of the IEEE Qatar Section. He is also an Associate Editor of the IEEE Transactions on Industrial Electronics and IEEE Access and the Editor-in-Chief of the *Journal of Electrical Engineering* (I'manager). He was a former Associate Editor of the *IEEE Transactions on Industry Application* and a former Guest Associate Editor *IEEE Transactions on Power Electronics*. He is head of the design team of Power Electronics and Drives equipment at Powerlab Instruments, Chennai, India. His research interests include smart grid, complex energy transition, active distribution network, electric vehicles drivetrain, sustainable development and energy security, distributed energy generation, and multiphase motor drive systems.

Editorial: Machine Learning, Advances in Computing, Renewable Energy and Communication (MARC)

A. Tomar, H. Malik, P. Kumr, and A. Iqbal

Abstract Machine learning (ML) is the subcategory of artificial intelligence (AI), which has the capability to imitate human behavior intelligently as per the task performed by the human. In the modern time, any organization implements AI by using ML so that system's behavior of interchangeably and ambiguously is updated automatically through the experience without any delay. So, current advances in AI have involved ML. The ML starts with data (i.e., any kind of data starting from primary to secondary data). These data are collected and preprocessed to be used as training and testing the ML models being utilized for different applications such as regression, prediction, forecasting, classification, clustering, management, design, optimization, security, IoTs, health care, digitization, automation, control, privacy protection and e-commerce. In this book, the applications AI, ML and its advancement for different applications have been presented into different chapters, including the state-of-the-art and implementation in the various research domains of engineering and science.

Keywords Machine learning · Artificial intelligence · Data analytics · Renewable energy · Advance computing · Diagnosis · Communication · Smart grid · Fault detection · Prediction · Forecasting · Power system · Optimization · Management · Safety analysis · Smart city · Electric vehicle · Energy storage · Condition monitoring

A. Tomar
ICE Division, NSUT, New Delhi, India

H. Malik (✉)
Division of Electrical Power Engineering, School of Electrical Engineering, Faculty of Engineering, University of Technology Malaysia (UTM), Johor Bahru, Malaysia
e-mail: hasmat@utm.my

P. Kumr
Department of Computer Science & Engineering, KEC, Mumbai, India

A. Iqbal
Department of Electrical Engineering, Qatar University, Doha, Qatar

A. Tomar et al. (eds.), *Proceedings of 3rd International Conference on Machine Learning, Advances in Computing, Renewable Energy and Communication*, Lecture Notes in Electrical Engineering 915, https://doi.org/10.1007/978-981-19-2828-4_1

1 Introduction

The ML is the study of different computer algorithms which can enhance themselves through the experience and utilization of data properties [1–9]. Basically, ML is a part of AI. First time, ML was coined by Arthur Samuel in 1959. Generally, ML algorithms create a model using data/training samples, which may develop the model for different applications (i.e., forecasting, prediction, classification, clustering, regression, optimization, etc., in the engineering, science and management domain of sentiment analysis, natural language understanding, machine learning control, user behavior analytics, syntactic pattern recognition, knowledge graph embedding, agriculture, structural health monitoring, natural language processing, handwriting recognition, adaptive Web site, medical diagnosis, astronomy, online advertising, software engineering, speech recognition, bioinformatics, search engines, cheminformatics, insurance, citizen science, climate science, linguistics, computer networks, computer vision, optimization, credit card fraud detection, data quality, economics, financial market analysis, information retrieval, machine translation, recommender systems, telecommunication, banking, theorem proving, time-series forecasting, behaviorism, anatomy, affective computing, brain-machine interfaces, DNA sequence classification, general game playing, internet fraud detection, marketing, robot locomotion, machine perception and sequence mining) without knowing system information [10–156]. For an example, the broader way of the applications is as follows: (1) Application of AI, machine learning and advances (AIMLA) in condition monitoring, fault detection and diagnosis domain [10–69], (2) Application of AIMLA in prediction and forecasting domain [70–102], (3) Application of AIMLA in power system/smart grid domain [103–138] and (4) Application of AIMLA in other relevant domain [139–156]. The subcategory of ML is known as computational statistics, which is utilized for various applications but not all [157]. In the enhancement and development of ML, a book was written by Nilsson in 1960, which includes the application of pattern classification [158]. ML approaches are conventionally classified into three broader categories (e.g., supervised [159, 160], unsupervised [161, 162] and reinforcement learning [163, 164]), which depend on the nature of the system [1–9].

This edited book focuses on ML, advances in computing, renewable energy and communication. It is a collection of sixty-eight chapters, which represent a few of the state-of-the-art technologies in real-world applications such as: (1) computer knowledge management, (2) VLSI circuit design, (3) IoT network security, (4) healthcare system, (5) mask detection, (6) container culture, (7) speaker recognition/identification, (8) hospital management system, (9) Hadoop system, (10) V2V communication, (11) satellite communication, (12) life cycle models, (13) disease diagnosis, (14) wireless network, (15) PD analysis, (16) big data analytics, (17) review data analysis, (18) solar energy forecasting, (19) image processing, (20) water demand management and prediction, (21) safety analysis, (22) missing value evaluation, (23) Maze problem solution, (24) privacy analysis, (25) smart city, (26) control system, (27) MPPT application, (28) ELD for clean energy, (29) Sentiment analysis, (30) EV applications, (31) ESS area and implementation, (32) PV system, (33) energy management system, (34) fire detection, (35) generation and its management, (36) optimization applications, (37) AI in administration, (38) smart house, (39) forecasting and prediction, (40) container-as-a-service, (41) hybrid renewable resources implementation, (42) converters and inverters design, (43) e-commerce application, (44) robotics and its uses, (45) mining cluster and analysis, (46) design and development, (47) pandemic analysis, (48) human behavior analysis and detection, (49) forgery detection, (50) cloud computing, (51) advancement and (52) renewable energy management and harness.

In this chapter, Mitali Chugh presented a deep drive into a knowledge management system for improving software processes and products. The author explores the knowledge management literature, identifies the knowledge gaps and enhances future research in this domain. Demonstrated results highlight how software process improvement has been supported by knowledge management.

In Chap. 2, Kalivaraprasad, B. et al. presented an area and delayed efficient approximate hybrid adder for VLSI circuit designs. The hybrid adder is implemented with configurable levels of accuracy. The parallel prefix adder is designed and applied in the proposed hybrid n-bit adder to minimize the critical path time.

In Chap. 3, Shilpa B. Sarvaiya et al. proposed an approach for improvement in the security of IoT network using IP binding techniques. The proposed methodology will help to cross-check the sender and receiver node which keep the security up and maintain the variation in data handling.

In Chap. 4, Raveena Yadav and Vinod Kumar presented a healthcare system with IoT wireless sensor network. The main aim of this chapter is to represent a review of the smart healthcare system with researchers' work in this field. Another objective is to provide the role of sensors in this system.

In Chaps. 5 and 6, Arun Chauhan et al. and Avinashwar presented a comparative analysis of face mask detection models. Authors tried to compare the different models for the recognition of mask on the face using different real-world dataset.

In Chap. 7, Priyansh Pathak, Prabhishek Singh presented Kubernetes and docker the star duo of container culture. This chapter focuses on two major stars of the virtualization world Kubernetes and Docker. The paper gives an insight in the beauty of the world of containers, Kubernetes and Docker; it is an exciting journey through the topic of container orchestration and creation.

In Chap. 8, Atul Sharma and Sunil Kumar Singla presented a robust gender identification system for speaker recognition using linear discriminant analysis stepwise dimension reduction using linear discriminant analysis stepwise dimension reduction (LDASDR) which is based on a well-established feature selection algorithm called linear discriminant analysis stepwise feature selection (LDASFS). Three different feature sets, namely acoustic and cepstral, have been separately used for the neural network classifier's input.

In Chap. 9, Rekha S. Dange and Bharati B. Sayankar presented the cloud-integrated hospital allotment system for dynamic patients integrated with IoT. Authors are proposing a review work of the cloud-integrated automatic hospital and patient monitoring system to make life easy for doctors and patients for live tracking of all the individual prospects of critical patient health data.

In Chap. 10, Maithri. Using a MapReduce approach for large datasets, C and Chandramouli proposed an agglomerative hierarchical clustering with parallel implementation. It divides the dataset into data blocks; each is classified into groups on different distributed data nodes of the Hadoop MapReduce with agglomerative hierarchical clustering method. The significance of the approach provides in clustering model over distributed environment for large data analysis.

In Chap. 11, Vinod Kumar et al., reviewed Vasudev et al.'s protocol and find some of the design flaws and security features like forward secrecy, vehicle server impersonation attack, insider attack and others weaknesses.

In Chap. 12, Mukesh Kumar Mishra and Priyanka Ahlawat reviewed different schemes and mentioned that the authentication of a satellite is done and key generation for the satellite and updating of the key and privacy and integrity is maintained.

In Chap. 13, Mitali Chugh presented a survey-based study that was carried out in software engineering organizations in India. Initially, the study introduces the prevalent software development life cycle models followed by the empirical results of the survey that have significant takeaways for the practitioners in software organizations.

In Chap. 14, Sirineni Harshitha et al. focus on early recognition, anticipation and therapies of genuine illnesses like bosom malignancy, Parkinson's and diabetes. The assessment of malignancy by exploring histopathological pictures personates a genuine part in the patient's turn of events, and deep learning strategies are utilized to get a bunch of boundaries from images used to construct convolutional networks.

In Chap. 15, Ashutosh Kumar Choudhary and Surendra Rahamatkar presented the certificate issuing protocols along with their respective predictive and classification engines. The statistical performance comparison between these algorithms to conclude the best available trust-establishment techniques in a given network.

In Chap. 16, Kajol Chaudhari et al. presented the comparative investigation and determination of partial discharge source using a self-organizing map and K-nearest neighbor methods of artificial neural network.

In Chap. 17, Yogesh et al. mainly focused on the different privacy and security issues and probable solution for that. Big healthcare data has substantial ability to enhance and improve the patient's condition for the medical outcomes, expect outbreaks of epidemics, advantage precious insights, keep away from preventable diseases, minimize the price of health care and enhance the high-satisfactory life in general.

In Chap. 19, Neeraj, Pankaj Gupta and Anuradha Tomar compared the capability of the recurrent neural network (RNN) and long short-term memory (LSTM) in the form of accuracy. The short-term solar energy generation can be accurately estimated using RNN and LSTM models. RNN and LSTM models were trained to predict solar energy generation in Airport Depot, Delhi, India.

In Chap. 20, Deepti Vadicherla and Vijay Gadicha presented some of the recent research articles on crack detection and reviewed and checked their feasibility toward the crack detection task.

In Chap. 21, Dr. Priyanka Ahlawat presented an improved algorithm, allowing the network designer to further endure a better number of adversarial attacks. It is done by compromising a link where the keys used in the communication link have been eavesdropped by the attacker during eavesdropping.

In Chap. 22, Shashi Bhushan proposed a hybrid approach-based convolutional neural network (HABCNN) for liver tumor segmentation, which has been mathematically modeled to resolve the issue of cancer detection. The kidney and spleen are segmented initially. After cancer tissue segmentation, GLCM was used to extract features from tumor parts. The modern segmentation approach should ensure that the precise boundary structure of the cancer area is known so that important information can be better diagnosed.

In Chap. 23, Shashi Bhushan presented demand assessment in the water appropriation network, which gives significant information to checking and controlling frameworks. The LSTM forecast is utilized to study consumption and absolute requirement.

In Chap. 24, Dhanvanth and Rohith Rajesh presented the prediction of accidents in an upcoming race, data visualization using plots to draw meaningful inferences and logistic regression to predict the winner of a race.

In Chap. 25, M. Chenna Keshava et al. presented a missing person identification framework is implemented, which combines an effective CNN-hinged deep learning strategy for feature extraction with KNN and SVM classifiers for categorical classification.

In Chap. 26, Harshak Krishnaa et al. proposed a genetic algorithm framework to solve the two-dimensional toy maze problem. The proposed framework is named as GAmaze. The novelty of GAmaze lies in the way it updates the population from generation to generation. The GAmaze uses fitness information of the candidate solutions to optimize the individuals' learning capacity and make them reach the final goal by making them understand how to overcome obstacles and find a path.

In Chap. 27, Vibhor Sharma proposed a novel secure vector product for protecting data privacy in vertically partitioned dataset. The proposed bottom-up approach results in comparison with the following sequential approach.

In Chap. 28, Neeraj Chugh presented the trending and highly cited topics in anomaly detection for IoT-enabled smart cities from 2011 to 2020 and identify the trends in this field. The future works can conduct bibliometric analysis periodically to comprehend dynamics in the studied area and present a comparative account with the result set of the current study.

In Chap. 29, Shafqat Nabi Mughal et al. proposed a method to create an accurate and efficient MPPT algorithm for peak power extraction based on artificial intelligence (AI). In this study, a photovoltaic system is modeled and simulated in MATLAB using Perturb and Observe and the proposed fuzzy logic-based controller.

In Chapter 30, Prashant et al. developed a GA-based multi-objective method for solving load dispatch problem in a cost-effective way. The existing methods involve more cost for both thermal and solar generating units. In order to overcome such issues, genetic algorithm is applied.

In Chap. 31, Sachin Pachauri et al. discussed the notion of WPT and modern innovations that have made human life more meaningful. New wireless power technology scopes are now available, and they are in fierce rivalry with one another.

In Chap. 32, Kiran Deshpande et al. discussed the comprehensive performance evaluation of novel big data log analytic framework. This approach gathers and analyzes raw heterogeneous log data relating to vital IT infrastructure within the educational organization and is addressed in this chapter.

In Chap. 33, Sandeep Kumar et al. reviewed and resolved an issue with the MOOC courses review rating that it has never been addressed previously as per best of our knowledge.

In Chap. 34, Tina et al. proposed neural augmentation with the inbuilt concept of meta-learning which will be helpful in the automatic training of appropriate augmentations in the deep neural network. Meta-learning framework in the deep learning model leverages to self-learn new tasks that embrace a wide spectrum for learning techniques in the neural network.

In Chap. 35, Dogga Raveendhra et al. discussed the effect of the parasitic components of motor on CMV is successfully investigated in three-phase ZSI fed IM drive. For the investigation of various parameters related to common mode voltage effects, simulation model has been established by considering various practical parameters such as parasitic capacitances between various parts of induction motor such as stator to frame, stator to rotor, rotor to shaft and shaft to ground.

In Chap. 36, Harish Pulluri et al. proposed a hybrid technique to solve DED. When used as a search level, the basic genetic algorithm (GA) takes longer to get nearly optimal results. The proposed technique uses a three-parent crossover and diversity operator resulting in increasing the potential for both exploration and exploitation of the algorithm technique.

In Chap. 37, Dogga Raveendhra et al. presented a solar-powered battery-assisted remote area power system (RAPS) that employs a single-stage power conditioning unit. This chapter discusses and involves comprehensive mathematical modeling of power converters and their various modes of operation and design guidelines.

In Chap. 38, Dinanath Prasad et al. discuss and open about the involvement of multi-string solar photovoltaic system (SPPV) in grid-connected systems. Solar-powered plants can be used to supply domestic needs as well as surplus can be fed to utility grid when not in use. Grid-connected systems are controlled solar strings feeding the utility grid.

In Chap. 39, Amisha Srivastava reviewed the various aspects of energy management, viz. home energy management system, building energy management system, advanced metering infrastructure, electric vehicle and demand side energy management system.

In Chap. 40, Malti Gautam Singh and Sharini Rithigaa et al. proposed a system to design low-cost and effective detection and prevention system using GSM modules. This proposed system can quickly detect forest fire at the initial stage.

In Chapter 41, Sonali Vyas et al. discussed the concept of sustainable IT ecologies ensuring need-centered administration of resources at both supply and demand end plus the implementation concept of urban energy management system with the help of resource microgrids.

In Chap. 42, Gagandeep Kaur et al. proposed an optimal and efficient consumption of distributed generation using sift computing techniques.

In Chap. 43, Muskaan nagpal and Richa mishra discussed implementation of various algorithms on CIFAR10 dataset was related to the time taken by the algorithms for training as the size of the dataset is huge.

In Chap. 44, Nishu Bansal and Swimpy Pahuja reviewed different anomaly detection strategies that have been presented in recent years. This overview assists the reader in gaining a fundamental understanding of anomaly detection and the strategies available to identify it.

In Chap. 45, Bandana et al. presented a work that conducts a feasibility assessment and size optimization of a grid-connected solar photovoltaic/biomass/biogas/battery-based hybrid system for a village in India for supplying continuous energy at a minimal cost. The PSO technique has been implemented for reducing the cost of the developed hybrid system.

In Chap. 46, Meet Vasani et al. proposed prediction system that considers various parameters such as English language score, university rank, statement of purpose, letter of recommendation, cumulative grade point average and research experience for predicting the chances of admission in a university using ensemble model.

In Chap. 47, Priyanka Malhotra et al. proposed IoT-based innovative door system that provides detection of the symptoms of COVID-19 or other viral infections and controls the outbreak of infection. The intelligent door system facilitates testing the symptoms like fever or high temperature and helps maintain social distancing. An automatic door opening system helps in contactless operation.

In Chap. 48, Archana Yadav et al. presented a comprehensive survey of work-load prediction approaches in cloud environments. It also highlights the emerging challenges like resource wastage, excess power consumption and quality of service violations, etc.

In Chap. 49, Anshita Malviya et al. presented a survey on containers and its architecture and how containers are better than virtual machines for software development.

In Chap. 50, V. Sowmya Sree and C. Srinivasa Rao presented the performance analysis of DPFC when operating with and without genetic algorithm-based fuzzy logic controller, and the simulation results are validated using MATLAB/Simulink software.

In Chap. 51, Shital Pawar and Suhas Patil presented a machine-to-machine communication of industrial IoT system for MQTT protocol with one case study. To estimate the performance of the IoT applications, we have also represented the delay model for M2M communication of IoT applications.

In Chap. 52, Prashant et al. proposed a viable alternative strategy for battery charging employing a non-isolated bidirectional converter connected with a solar PV system. This chapter concludes that a bidirectional converter can work as an alternative for the charging and discharging of the auxiliary power supply.

In Chap. 53, Tanushree Sanwal and Puja Sareen discussed an endeavor to examine the relationship between stress, organizational performance and social intelligence. This chapter also analyzes the significance of ergonomics in prevention and management of stress.

In Chap. 54, Aashi Pathak et al. presented imputing missing data in electronic health records which shows how the medical data is collected, how it travels through various layers and how to deal with the data that has missing values. All the techniques' accuracy has been mentioned in the chart and table. It is observed that the machine learning approach is better than the statistical approaches.

In Chap. 55, Garima Jaiswal et al. reviewed the work done by the researchers to detect document forgery through source printer identification, verifying identity documents and signature. This study analyzes the machine learning and deep learning approaches, results, datasets, and limitations used to detect document forgery for each domain.

In Chap. 56, Rajesh et al. proposed a hybrid model for deep fake detection (HMDD), which aims to achieve better classification using simple deep learning models and learn a considerable number of features with the same efficiency throughout.

In Chap. 57, Hardik A. Gangadwala et al. presented a comparative study of K-means and fuzzy C-means methods. These methods offer an adequate and precise investigation of proxy server log. The least distance between clusters is calculated with the aid of using the Euclidean distance equation.

In Chap. 58, Deepti Prit Kaur et al. discussed the meta-analysis of students' learning in the theoretical course on "Basics of Electronics Engineering" integrated with virtual lab sessions through remote experimentation.

In Chap. 59, Naveen and Uttam Sharma lighten up some well-known and newly discovered security threats. We have provided a collective package of machine learning algorithms that possibly reduce the vulnerability to some extent.

In Chap. 60, Aylapogu Pramod Kumar proposed level shifter which can convert voltages between 0.5 and 1 V. In 45 nm technology, the result was accomplished with the help of a cadence tool. At 45 nm, the projected idea may run at a reduced fixed energy of 3.26 pw.

In Chap. 61, Aashi Pathak et al. discussed four popular missing data imputation algorithms: expectation–maximization (EM), MICE, KNN imputation and mean imputation based on the available data. As KNN proves to be the most robust and most efficient technique to impute the missing data in health care, it gives the highest accuracy on the dataset that has the missingness ranging from 2 to 82%.

In Chap. 62, Shreya Biswas et al. reviewed the most recent work done in predicting personality using MBTI from 2010 to 2020. This chapter gives a literature review comparing and contrasting all these works based on parameters like dataset used, algorithm used, feature extraction method used and limitations.

In Chap. 63, Anuj Rani et al. discussed multiple image forger detection algorithms are evaluated to check their efficacy using their detection accuracy and execution time. Discrete wavelet transform (DWT) is used for image decomposition at the preprocessing stage. This work aims to suggest a best copy–move forgery detection algorithm among SIFT, SURF and KAZE.

In Chap. 64, Umang Kant and Dr. Vinod Kumar presented the way in which the Internet of things (IoT) based cloud-centric framework for weight management. This cloud-centric proposed framework uses predictive computing to analyze the physical activities and the dietary intake of the registered users.

In Chap. 65, Ashish Tiwari discussed the efficient cloud orchestration services in computing. The research may be increased by using machine learning languages. The use of a genetic algorithm makes the work stronger to get the performance of this research work.

In Chap. 66, Jainendra Singh and Zaheeruddin proposed an algorithm to make gathering and transmitting of data more energy efficient. The proposed protocol, energy efficient cluster-based data aggregation (EECDA), will help to increase network lifetime by ensuring uniform distribution of energy among the nodes.

In Chap. 67, Rishabh Singh discussed the hyperspectral imaging approach to detect document forgery through ink mismatch detection. The proposed system utilizes PCA to handle multiple dimensions in HSIs. The method is applied to the UWA writing ink HSI (WIHSI) database. The results are compared with the state-of-art results that prove the proposed approach's potential.

In Chap. 68, Jitendra Rajpurohit and Tarun K. Sharma proposed a modified variant of JSO called greedy jellyfish search optimizer (GJSO). GJSO changes the current ocean movement of JSO and adds greedy behavior in the search process. The proposed variant has been tested on 20 diverse functions. The results are compared with two other peer algorithms. The results show the superiority of the proposal over unimodal functions and comparative performance over multimodal functions.

In conclusion, editor(s) express their gratitude to many people whose continuous support and coordination have made this book production successful. As editors, we especially thank the authors for their contributions. We are grateful to the reviewers for their valuable comments and prompt responses. We sincerely thank the Springer Nature and staff for their immense support and guidance during this book's complete process.

A. Tomar, H. Malik, P. Kumr and A. Iqbal

Editor(s).

Acknowledgements The editors would like to acknowledge the support from Krishna Engineering College (KEC), Ghaziabad, India to manage the event of 3rd International Conference on Machine Learning, Advances in Computing, Renewable Energy and Communication (MARC-2021) - Virtual Mode. The editors extend their appreciation and acknowledgement to the Intelligent Prognostic Private Limited, India, to provide all types of technical and non-technical facilities, cooperation and support in each stage to make this book real.

References

1. Tomar A et al (2020) Machine learning, advances in computing. Renew Energy Commun (Springer, Berlin, LNEE 768:659. https://doi.org/10.1007/978-981-16-2354-7. (ISBN 978-981-16-2354-7)
2. Iqbal A et al (2020) Renewable power for sustainable growth (Springer, Berlin, LNEE) 723:805. https://doi.org/10.1007/978-981-33-4080-0. (ISBN 978-981-33-4082-4)
3. Ahmad MW et al (2022) Intelligent data-analytics for power and energy systems (Springer, Berlin, LNEE) 802:641. https://doi.org/10.1007/978-981-16-6081-8. (ISBN 978-981-16-6081-8)
4. Fatema N et al (2021) Intelligent data-analytics for condition monitoring: smart grid applications, Elsevier, 268 p https://www.sciencedirect.com/book/9780323855105/intelligent-data-analytics-for-condition-monitoring. (ISBN: 978-0-323-85511-2)
5. Iqbal A et al (2020) Soft computing in condition monitoring and diagnostics of electrical and mechanical systems, 496 p. Springer, Berlin. https://doi.org/10.1007/978-981-15-1532-3. (ISBN 978-981-15-1532-3)
6. Iqbal A et al (2020) Meta heuristic and evolutionary computation: algorithms and applications, 949 p. Springer, Berlin. https://doi.org/10.1007/978-981-15-7571-6. (ISBN 978-981-15-571-6)
7. Jafar A et al (2021) AI and machine learning paradigms for health monitoring system: intelligent data analytics, vol 86, p 513. Springer, Berlin, SBD. https://doi.org/10.1007/978-981-33-4412-9. ISBN 978-981-33-4412-9
8. Srivastava S et al (2019) Applications of artificial intelligence techniques in engineering, SIGMA 2018, vol 1, 698:643. Springer, AISC. https://doi.org/10.1007/978-981-13-1819-1. (ISBN 978-981-13-1818-4)
9. Srivastava S et al (2019) Applications of artificial intelligence techniques in engineering, SIGMA 2018, vol 2 (Springer, AISC, vol 697, 647 p). https://doi.org/10.1007/978-981-13-1822-1. (ISBN 978-981-13-1821-4)
10. Malik H, Mishra S (2016) Application of gene expression programming (GEP) in power transformers fault diagnosis using DGA. IEEE Trans Ind Appl 52(6):4556–4565. https://doi.org/10.1109/TIA.2016.2598677
11. Ahmad MW et al (2020) A fault diagnostic and post-fault reconfiguration scheme for interleaved boost converter in PV-based system. IEEE Trans Power Electron 36(4):3769–3780. https://doi.org/10.1109/TPEL.2020.3018540
12. Malik H, Mishra S (2017) Artificial neural network and empirical mode decomposition based imbalance fault diagnosis of wind turbine using turbsim, FAST and simulink. IET Renew Power Gener 11(6):889–902. https://doi.org/10.1049/iet-rpg.2015.0382
13. Malik H, Mishra S (2018) Application of GEP to investigate the imbalance faults in direct-drive wind turbine using generator current signals. IET Renew Power Gener 12(3):279–291. https://doi.org/10.1049/iet-rpg.2016.0689
14. Malik H, Sharma R (2017) Transmission line fault classification using modified fuzzy Q learning. IET Gen Trans Distrib 11(16):4041–4050. https://doi.org/10.1049/iet-gtd.2017.0331
15. Yadav AK et al (2013) Application of neuro-fuzzy scheme to investigate the winding insulation paper deterioration in oil-immersed power transformer. Electr Power Energy Syst 53:256–271. https://doi.org/10.1016/j.ijepes.2013.04.023
16. Ahmad MW et al (2021) Non-invasive model-based open-circuit switch fault detection of AC-bypass leg switches in transformerless PV inverter. IEEE J Emerg Sel Topics Power Electron. https://doi.org/10.1109/JESTPE.2021.3098195
17. Mishra S et al (2014) Selection of Most Relevant Input Parameters Using Waikato Environment for Knowledge Analysis for Gene Expression Programming Based Power Transformer Fault Diagnosis. International Journal of Electric Power Components and Systems 42(16):1849–1862. https://doi.org/10.1080/15325008.2014.956952

18. Malik H, Mishra S (2017) Selection of most relevant input parameters using principle component analysis for extreme learning machine based power transformer fault diagnosis model. Int J Electric Power Components Syst 45(12):1339–1352. https://doi.org/10.1080/15325008.2017.1338794
19. Sharma R, Malik H (2017) EMD and ANN based intelligent fault diagnosis model for transmission line. J Intel Fuzzy Syst 32(4):3043–3050. https://doi.org/10.3233/JIFS-169247
20. Saad Ahmaduddin S, Malik H (2018) Gene expression programming (GEP) based intelligent model for high performance concrete comprehensive strength analysis. J Intel Fuzzy Syst 35(5):5403–5418. https://doi.org/10.3233/JIFS-169822
21. Shah AK et al (2018) EMD and ANN based intelligent model for bearing fault diagnosis. J Intel Fuzzy Syst 35(5):5391–5402. https://doi.org/10.3233/JIFS-169821
22. Malik H, Mishra S (2017) FAST and simulink based simulation investigation of wind turbine faults. Int J Renew Energy Technol 8(3/4):286–304. https://doi.org/10.1504/IJRET.2017.088970
23. Malik H (2018) Wavelet and Hilbert Huang transform based wind turbine imbalance fault classification model using K-nearest neighbor algorithm (in Press). Int J Renew Energy Technol 9(1/2). https://doi.org/10.1504/IJRET.2018.090105
24. Alotaibi MA et al (2022) Power quality disturbance analysis using data-driven EMD-SVM hybrid approach. J Intel Fuzzy Syst 42(2):669–678. https://doi.org/10.3233/JIFS-189739
25. Kukker A et al Reinforcement learning based genetic fuzzy classifier for transformer faults. IETE J Res 1–12. https://doi.org/10.1080/03772063.2020.1732844
26. Alotaibi MA et al (2022) Cyberattacks identification in IEC 61850 based substation using proximal support vector machine. J Intel Fuzzy Syst 42(2):1213–1222. https://doi.org/10.3233/JIFS-189783
27. Ahmad MW et al (2022) Development of wide area monitoring system for smart grid application. J Intel Fuzzy Syst 42(2):827–839. https://doi.org/10.3233/JIFS-189752
28. Azeem A et al (2022) Real-time harmonics analysis of digital substation equipment based on IEC-61850 using hybrid intelligent approach. J Intel Fuzzy Syst 42(2):741–754. https://doi.org/10.3233/JIFS-189745
29. Sarita K et al (2022) Principal component analysis technique for early fault detection. J Intel Fuzzy Syst 42(2):861–872. https://doi.org/10.3233/JIFS-189755
30. Bisht VS et al (2022) A data-driven intelligent hybrid method for health prognosis of lithium-ion batteries. J Intel Fuzzy Syst 42(2):897–907. https://doi.org/10.3233/JIFS-189758
31. Nageswara Rao P et al (2020) Global sliding mode suspension control and condition monitoring of bearingless switched reluctance motor under eccentric faults. Energies 13(20):5485, 1–38. https://doi.org/10.3390/en13205485
32. Azeem A et al (2021) Design of hardware setup based on IEC 61850 communication protocol for detection & blocking of harmonics in power transformer. Energies 14(24):8284, 1–27. https://doi.org/10.3390/en14248284
33. Bhattacharjee T et al (2022) Hardware development and interoperability testing of a multivendor-IEC-61850-based digital substation. Energies 15(5):1785, 1–19. https://doi.org/10.3390/en15051785
34. Sharma R et al (2020) Fuzzy reinforcement learning based intelligent classifier for power transformer faults (in Press). ISA Trans. https://doi.org/10.1016/j.isatra.2020.01.016
35. Malik H, Almutairi A (2021)Modified fuzzy-Q-learning (MFQL)-based mechanical fault diagnosis for direct-drive wind turbines using electrical signals. IEEE Access 9:52569–52579. https://doi.org/10.1109/ACCESS.2021.3070483
36. Chandra R et al (2020) A survey of failure mechanisms and statistics for critical electrical equipment in buildings. In: IECON 2020 the 46th annual conference of the IEEE industrial electronics society, pp 1955–1961. https://doi.org/10.1109/IECON43393.2020.9254225.63
37. Pandya Y et al (2018) Feature extraction using EMD and classifier through artificial neural networks for gearbox fault diagnosis. Book chapter in Applications of artificial intelligence techniques in engineering, advances in intelligent systems and computing, vol 697, pp 309–317. https://doi.org/10.1007/978-981-13-1822-1_28

38. Kaushal P et al (2018) A hybrid intelligent model for power quality disturbance classification. Book chapter in applications of artificial intelligence techniques in engineering, advances in intelligent systems and computing 697:55–63. https://doi.org/10.1007/978-981-13-1822-1_6
39. Reza MW et al (2018) Wide area monitoring system using integer linear programming. Book chapter in Applications of artificial intelligence techniques in engineering, advances in intelligent systems and computing, vol 698, pp 23–30. https://doi.org/10.1007/978-981-13-1819-1_3
40. Sharma T et al (2018) A novel intelligent bifurcation classification model based on artificial neural network (ANN). Book chapter in Applications of artificial intelligence techniques in engineering, advances in intelligent systems and computing 698:53–61. https://doi.org/10.1007/978-981-13-1819-1_6
41. Chack D et al (2018) A novel intelligent transmission line fault diagnosis model based on EEMD and multiclass PSVM. Book chapter in Applications of artificial intelligence techniques in engineering, advances in intelligent systems and computing 698:85–92. https://doi.org/10.1007/978-981-13-1819-1_9
42. Singh KV et al (2017) Condition monitoring of wind turbine gearbox using electrical signatures. In: IEEE ICMDCS, pp 1–6. https://doi.org/10.1109/ICMDCS.2017.8211718
43. Mishra S, Malik H Application of fuzzy Q learning (FQL) technique to wind turbine imbalance fault identification using generator current signals. In: Proceedings of IEEE PIICON-2016, pp 1–6, 25–27 Nov 2016. https://doi.org/10.1109/POWERI.2016.8077283
44. Mishra S, Malik H Application of gene expression programming (GEP) to investigate the health condition of direct-drive wind turbine using FAST and TurbSim. In: Proceedings IEEE IICPE-2016, pp 1–6, 17–19 Nov 2016. https://doi.org/10.1109/IICPE.2016.8079508
45. Malik H, Aggarwal A, Sharma R (2016) Feature extraction using EMD and classification through probabilistic neural network for fault diagnosis of transmission line. In: Proceedings IEEE ICPEICES-2016, pp 1–6. https://doi.org/10.1109/ICPEICES.2016.7853709
46. Sharma R et al (2016) Selection of most relevant input parameters using weka for artificial neural network based transmission line fault diagnosis model. In: Proceedings of the international conference on nanotechnology for better living, vol 3, No 1, pp 176. https://doi.org/10.3850/978-981-09-7519-7nbl16-rps-176
47. Mishra S, Malik H (2016) Application of extreme learning machine (ELM) in paper insulation deterioration estimation of power transformer. In: Proceedings of the international conference on nanotechnology for better living, vol 3, No 1, pp 209. https://doi.org/10.3850/978-981-09-7519-7nbl16-rps-209
48. Kumar G et al (2016) Learning vector quantization neural network based external fault diagnosis model for three phase induction motor using current signature analysis. Elsevier Procedia Comput Sci 93:1010–1016. https://doi.org/10.1016/j.procs.2016.07.304
49. Mishra S, Malik H Proximal support vector machine (PSVM) based imbalance fault diagnosis of wind turbine using generator current signals. In: Elsevier energy procedia, vol 90, pp 593–603, 15–17 Dec 2015, IIT Bombay. https://doi.org/10.1016/j.egypro.2016.11.228
50. Sharma S et al (2015) External fault classification experienced by three-phase induction motor based on multi-class ELM. Elsevier Procedia Comput Sci 70:814–820. https://doi.org/10.1016/j.procs.2015.10.122
51. Mishra S, Malik H (2015) Application of LVQ network in fault diagnosis of wind turbine using turbsim, FAST and simulink. Michael Faraday IET Int Summit 2015:474–480. https://doi.org/10.1049/cp.2015.1679
52. Khatri A et al (2015) Probabilistic neural network based incipient fault identification using DGA dataset. Elsevier Procedia Comput Sci 58:665–672. https://doi.org/10.1016/j.procs.2015.08.086
53. Mishra S, Malik H (2015) Application of probabilistic neural network in fault diagnosis of wind turbine using FAST, turbsim and simulink. Elsevier Procedia Comput Sci 58:186–193. https://doi.org/10.1016/j.procs.2015.08.052
54. Mittal AP et al External fault identification experienced by 3-phase induction motor using PSVM. In: Proceedings IEEE international conference on power India (PIICON 2014), 5–7 Dec. 2014, New Delhi. https://doi.org/10.1109/POWERI.2014.7117762

55. Mishra S, Malik H Feature selection using rapidminer and classification through probabilistic neural network for fault diagnostics of power transformer. In: Proceedings IEEE international conference on emerging trends and innovation in technology (INDICON 2014), 11–13 Dec 2014, Pune. https://doi.org/10.1109/INDICON.2014.7030427
56. Mishra S, Malik H Fault identification of power transformers using proximal support vector machine (PSVM). In: Proceedings IEEE international conference on power electronics (IICPE 2014), 8–10 Dec 2014, NIT Kurusherta. https://doi.org/10.1109/IICPE.2014.7115842
57. Mishra S, Malik H Application of gene expression programming (GEP) in power transformers fault diagnosis using DGA. In: Proceedings IEEE international conference on power India (PIICON 2014), 5–7 Dec 2014, New Delhi. https://doi.org/10.1109/POWERI.2014.7117782
58. Mishra S, Malik H (2015) Extreme learning machine based fault diagnosis of power transformer using IEC TC10 and its related data. In: Proceedings IEEE India annual conference (INDICON-2015), pp 1–5. https://doi.org/10.1109/INDICON.2015.7443245
59. Singh S et al (2012) UV/VIS response based fuzzy logic for health assessment of transformer oil. In: Elsevier procedia engineering, ISSN: 1877-7058, vol 30, pp 905–912. https://doi.org/10.1016/j.proeng.2012.01.944
60. Mahto T et al (2012) Make use of DGA to carry out the transformer oil-immersed paper deterioration condition estimation with fuzzy-logic. In: Elsevier procedia engineering, ISSN: 1877-7058, vol 30, pp 569–576. https://doi.org/10.1016/j.proeng.2012.01.900
61. Jarial RK et al Application research based on modern technology for transformer health index estimation. In: Proceedings IEEE international multi conference on systems, signals and devices (SSD), pp 1–7, 20–23 March 2012, Chemnitz. https://doi.org/10.1109/SSD.2012.6198012
62. Mahto MT et al (2019) Condition monitoring and fault detection & diagnostics of wind energy conversion system (WECS). In: Springer Nature book: soft computing in condition monitoring and diagnostics of electrical and mechanical systems, pp 121–154. https://doi.org/10.1007/978-981-15-1532-3_5
63. Bakhsh FI et al (2019) Fault analysis of variable frequency transformer (VFT) for power transfer in-between synchronous grids. In: Springer Nature book: soft computing in condition monitoring and diagnostics of electrical and mechanical systems, pp 269–286. https://doi.org/10.1007/978-981-15-1532-3_12
64. Mahto T et al (2011) An expert system for incipient fault diagnosis and condition assessment in transformers. In: Proceedings IEEE international conference on computational intelligence and communication networks, pp 138–142. https://doi.org/10.1109/CICN.2011.27
65. Yadav AK et al (2011) Make use of UV/VIS spectrophotometer to determination of dissolved decay products in mineral insulating oils for transformer remnant life estimation with ANN. In: Proceedings IEEE international conference on engineering sustainable solutions, pp 1–6, INDICON. https://doi.org/10.1109/INDCON.2011.6139574
66. Jarial RK et al Application research based on modern technology to investigating causes and detection of failures in transformers on the bases of importance level. In: Proceedings IEEE international conference on engineering sustainable solutions, pp 1–6, INDICON. https://doi.org/10.1109/INDCON.2011.6139577
67. Kushwaha N et al (2011) Paper insulation deterioration estimation of power transformer using fuzzy-logic: part-2. In: Proceedings IEEE international conference on engineering sustainable solutions, pp 1–5, INDICON. https://doi.org/10.1109/INDCON.2011.6139532
68. Jarial RK et al (2011) Application of modern technology for fault diagnosis in power transformer energy management. In: Proceedings IEEE international conference on communication system's network technologies, pp 376–381. https://doi.org/10.1109/CSNT.2011.84
69. Jarial RK et al (2011) Fuzzy-logic applications in cost analysis of transformer's main material weight. In: Proceedings IEEE international conference on computational intelligence and communication networks, pp 386–389. https://doi.org/10.1109/CICN.2011.81
70. Yadav AK et al (2015) Application of rapid miner In ANN based prediction of solar radiation for assessment of solar energy resource potential of 76 sites in northwestern India. Renew Sustain Energy Rev 52:1093–1106. https://doi.org/10.1016/j.rser.2015.07.156

71. Yadav AK et al (2014) Selection of most relevant input parameters using WEKA for artificial neural network based solar radiation prediction models. Renew Sustain Energy Rev 31:509–519. https://doi.org/10.1016/j.rser.2013.12.008
72. Yadav AK et al (2018) Daily array yield prediction of grid-interactive photovoltaic plant using relief attribute evaluator based radial basis function neural network. Renew Sustain Energy Rev 81, Part 2:2115–2127. https://doi.org/10.1016/j.rser.2017.06.023
73. Azeem A et al (2018) k-NN and ANN based deterministic and probabilistic wind speed forecasting intelligent approach. J Intel Fuzzy Syst 35(5):5021–5031. https://doi.org/10.3233/JIFS-169786
74. Arora P et al (2018) Wind energy forecasting model for northern-western region of India using decision tree and MLP neural network approach. Interdiscip Environ Rev 19(1):13–30. https://doi.org/10.1504/IER.2018.089766
75. Fatema N et al (2022) Hybrid approach combining EMD, ARIMA and Monte Carlo for multi-step ahead medical tourism forecasting. J Intel Fuzzy Syst 42(2):1235–1251. https://doi.org/10.3233/JIFS-189785
76. Khursheed T et al (2022) Multi-step ahead time-series wind speed forecasting for smart-grid application. J Intel Fuzzy Syst 42(2):633–646. https://doi.org/10.3233/JIFS-189736
77. Fatema N et al (2022) Deterministic and probabilistic occupancy detection with a novel heuristic optimization and back-propagation (BP) based algorithm. J Intel Fuzzy Syst 42(2):779–791. https://doi.org/10.3233/JIFS-189748
78. Alotaibi MA et al (2022) A new hybrid model combining EMD and neural network for multi-step ahead load forecasting. J Intel Fuzzy Syst 42(2):1099–1114. https://doi.org/10.3233/JIFS-189775
79. Yadav AK et al (2020) A novel hybrid approach based on relief algorithm and fuzzy reinforcement learning approach for predicting wind speed. Sustain Energy Technol Assess 43. https://doi.org/10.1016/j.seta.2020.100920
80. Yadav AK et al (2021) Case study of grid-connected photovoltaic power system installed at monthly optimum tilt angles for different climatic zones in India. IEEE Access 9:60077–60088. https://doi.org/10.1109/ACCESS.2021.3073136
81. Yadav AK et al (2021) Novel approach to investigate the influence of optimum tilt angle on minimum cost of energy based maximum power generation and sizing of PV systems: a case study of diverse climatic zones in India. IEEE Access 9:110103–110115. https://doi.org/10.1109/ACCESS.2021.3102153
82. Yadav AK et al (2014) Comparison of different artificial neural network techniques in prediction of solar radiation for power generation using different combinations of meterological variables. In: Proceedings IEEE international conference on power electronics, drives and energy systems (PEDES-2014), pp 1–5. https://doi.org/10.1109/PEDES.2014.7042063
83. Kumar G et al (2016) Generalized regression neural network based wind speed prediction model for western region of India. Elsevier Procedia Comput Sci 93:26–32. https://doi.org/10.1016/j.procs.2016.07.177
84. Garg P et al (2016) Infogain attribute evaluator and ANN based wind speed prediction model for Rajasthan, north-west region of India. In: Proceedings of the international conference on nanotechnology for better living, vol 3, No 1, p 233. https://doi.org/10.3850/978-981-09-7519-7nbl16-rps-233
85. Savital et al (2016) Wind speed and power prediction of prominent wind power potential states in India using GRNN. In: Proceedings IEEE ICPEICES-2016, pp 1–6. https://doi.org/10.1109/ICPEICES.2016.7853220
86. Savita et al (2016) Application of artificial neural network for long term wind speed prediction. In: Proceedings IEEE CASP, pp 217–222, 9–11 June 2016. https://doi.org/10.1109/CASP.2016.7746168
87. Yadav AK et al (2018) Short term wind speed forecasting for power generation in Hamirpur, Himachal Pradesh, India, using artificial neural networks. Book chapter in Applications of artificial intelligence techniques in engineering, advances in intelligent systems and computing 697:263–271. https://doi.org/10.1007/978-981-13-1822-1_24

88. Vinoop P et al (2018) PSO-NN-based hybrid model for long-term wind speed prediction: a study on 67 cities of India. Book chapter in Applications of artificial intelligence techniques in engineering, advances in intelligent systems and computing 697:319–327. https://doi.org/10.1007/978-981-13-1822-1_29
89. Singh et al M (2018) Comparative study of different neural networks for 1-year ahead load forecasting. In: Applications of artificial intelligence techniques in engineering, advances in intelligent systems and computing, vol 697, pp 31–42. https://doi.org/10.1007/978-981-13-1822-1_4
90. Yadav V et al (2018) Forecasting of nitrogen dioxide at one day ahead using non-linear autoregressive neural network for environmental applications. Book chapter in Applications of artificial intelligence techniques in engineering, advances in intelligent systems and computing 698:615–623. https://doi.org/10.1007/978-981-13-1819-1_58
91. Yadav AK et al (2018) 10-min ahead forecasting of wind speed for power generation using nonlinear autoregressive neural network. Book chapter in Applications of artificial intelligence techniques in engineering, advances in intelligent systems and computing 698:235–244. https://doi.org/10.1007/978-981-13-1819-1_23
92. Garg S et al (2018) Long-term solar irradiance forecast using artificial neural network: application for performance prediction of Indian cities. Book chapter in Applications of artificial intelligence techniques in engineering, advances in intelligent systems and computing 697:285–293. https://doi.org/10.1007/978-981-13-1822-1_26
93. Azeem A et al Application of Waikato environment for knowledge analysis based artificial neural network models for wind speed forecasting. In: Proceedings IEEE PIICON-2016, pp 1–6, 25–27 Nov 2016. https://doi.org/10.1109/POWERI.2016.8077352
94. Saad S et al Selection of most relevant input parameters using WEKA for artificial neural network based concrete compressive strength prediction model. In: Proceedings IEEE PIICON-2016, pp 1–6, 25–27 Nov 2016. https://doi.org/10.1109/POWERI.2016.8077368
95. Azeem A et al (2016) Artificial neural network based intelligent model for wind power assessment in India. In: Proceedings IEEE PIICON-2016, pp 1–6, 25–27 Nov 2016. https://doi.org/10.1109/POWERI.2016.8077305
96. Yadav AK et al (2015) ANN based prediction of daily global solar radiation for photovoltaics applications. In: Proceedings IEEE India annual conference (INDICON), pp 1–5. https://doi.org/10.1109/INDICON.2015.7443186
97. Reddy Chimmula VK et al (2021) Deep learning and statistical based daily stock price forecasting and monitoring. In: Springer Nature book: AI and machine learning paradigms for health monitoring system: intelligent data analytics, under book series "Studies in Big Data", pp 203–216. https://doi.org/10.1007/978-981-33-4412-9_13
98. Fatema N et al (2020) Metaheurestic algorithm based hybrid model for identification of building sale prices. In: Springer Nature book: metaheuristic and evolutionary computation: algorithms and applications, under book series "studies in computational intelligence", pp 689–704. https://doi.org/10.1007/978-981-15-7571-6_32
99. Fatema N et al (2020) Data-driven occupancy detection hybrid model using particle swarm optimization based artificial neural network. In: Springer Nature book: metaheuristic and evolutionary computation: algorithms and applications, under book series "studies in computational intelligence", pp 283–297. https://doi.org/10.1007/978-981-15-7571-6_13
100. Fatema N et al (2019) Data driven intelligent model for sales prices prediction and monitoring of a building. In: Springer Nature book: soft computing in condition monitoring and diagnostics of electrical and mechanical systems, pp 407–421. https://doi.org/10.1007/978-981-15-1532-3_18
101. Yadav A et al (2011) Application research based on artificial neural network (ANN) to predict no load loss for transformer design. In: Proceedings IEEE international conference on communication system's network technologies, pp 180–183. https://doi.org/10.1109/CSNT.2011.45
102. Mahto T et al (2012) Impact of usage duration on mobile phones EMI characteristics. In: Proceedings IEEE international conference on communication system's network technologies, pp 558–562. https://doi.org/10.1109/CSNT.2012.126

103. Vigya et al (2021) Renewable generation based hybrid power system control using fractional order-fuzzy controller. Energy Rep 7C:641–653. https://doi.org/10.1016/j.egyr.2021.01.022
104. Jadoun VK et al (2021) Optimal scheduling of non-convex cogeneration units using exponentially varying whale optimization algorithm. Energies 14(4):1–30. https://doi.org/10.3390/en14041008
105. Mahto T et al (2021) Fractional order fuzzy based virtual inertia controller design for frequency stability in isolated hybrid power systems. Energies 14(6):1634. https://doi.org/10.3390/en14061634
106. Mahto T et al (2018) Load frequency control of a solar-diesel based isolated hybrid power system by fractional order control using particle swarm optimization. J Intel Fuzzy Syst 35(5):5055–5061. https://doi.org/10.3233/JIFS-169789
107. Nandan NK et al (2018) Solving nonconvex economic thermal power dispatch problem with multiple fuel system and valve point loading effect using fuzzy reinforcement learning. J Intel Fuzzy Syst 35(5):4921–4931. https://doi.org/10.3233/JIFS-169776
108. Devarapalli R et al (2022) An approach to solve OPF problems using a novel hybrid whale and sine cosine optimization algorithm. J Intel Fuzzy Syst 42(2):957–967. https://doi.org/10.3233/JIFS-189763
109. Rao BV et al (2022) Wind integrated power system to reduce emission: an application of Bat algorithm. J Intel Fuzzy Syst 42(2):1041–1049. https://doi.org/10.3233/JIFS-189770
110. Bajaj M et al (2021) Optimal design of passive power filter using multi-objective pareto-based firefly algorithm and analysis under background and load-side's nonlinearity. IEEE Access 9:22724–22744. https://doi.org/10.1109/ACCESS.2021.3055774
111. Gupta S et al (2021) A hybrid Jaya-Powell's pattern search algorithm for multi-objective optimal power flow incorporating distributed generation. Energies 14(10), 2831:2–24. https://doi.org/10.3390/en14102831
112. Singh S et al (2021) Influence of wind power on modeling of bidding strategy in a promising power market with modified gravitational search algorithm. Appl Sci 11(10), 4438:2–16. https://doi.org/10.3390/app11104438
113. Singh S et al (2021) Strategic bidding in the presence of renewable sources for optimizing the profit of the power suppliers. IEEE Access 9:70221–70232. https://doi.org/10.1109/ACCESS.2021.3078288
114. Chankaya M et al (2021) Generalized Normal Distribution Algorithm based control of 3-phase 4-wire grid-tied PV-hybrid energy storage system. Energies 14(14), 4355:1–22. https://doi.org/10.3390/en14144355
115. Singh S et al (2021) Impacts of renewable sources of energy on bid modeling strategy in an emerging electricity market using oppositional gravitational search algorithm. Energies 14(18), 5726:1–22. https://doi.org/10.3390/en14185726
116. Gupta et al S (2021) A robust optimization approach for optimal power flow solutions using Rao algorithms. Energies 14(17), 5449, 1–28. https://doi.org/10.3390/en14175449
117. Chankaya M et al (2021) Multi-objective grasshopper optimization based MPPT and VSC control of grid-tied PV-battery system. Electronics 10(22), 2770:1–24. https://doi.org/10.3390/electronics10222770
118. Prakash P et al (2021) A novel hybrid approach for optimal placement of non-dispatchable distributed generations in radial distribution system. Mathematics 9(24), 3171:1–27. https://doi.org/10.3390/math9243171
119. Nagendra K et al (2021) Novel neural network-based load frequency control scheme: a case study of restructured power system. IEEE Access 9:162231–162242. https://doi.org/10.1109/ACCESS.2021.3133360
120. Shabbiruddin et al (2021) Fuzzy-based investigation of challenges for the deployment of renewable energy power generation. Energies 15(1), 58:1–16
121. Prakash P et al (2022) A Novel analytical approach for optimal integration of renewable energy sources in distribution systems. Energies 15(4), 1341:1–23. https://doi.org/10.3390/en15041341

122. Chankaya M et al (2022) Stability analysis of chaotic grey-wolf optimized grid-tied PV-hybrid storage system during dynamic conditions. Electronics 11(4), 567:1–23. https://doi.org/10.3390/electronics11040567
123. Mohammad K et al (2022) Fuzzy-logic-based comparative analysis of different maximum power point tracking controllers for hybrid renewal energy systems. Mathematics 10(3), 529:1–28. https://doi.org/10.3390/math10030529
124. Kumar N et al (2022) Application of fractional order-PID control scheme in automatic generation control of a deregulated power system in the presence of SMES unit. Mathematics 10(3), 521:1–16. https://doi.org/10.3390/math10030521
125. Minai AF et al (2022) Performance analysis and comparative study of a 467.2 kWp grid-interactive spv system: a case study. Energies, 15(3), 1107:1–19. https://doi.org/10.3390/en15031107
126. Reddy VKC et al (2019) Novel application of relief algorithm in cascade ANN model for prognosis of photovoltaic maximum power under sunny outdoor condition of Sikkim India: a case study. In: Springer Nature book: soft computing in condition monitoring and diagnostics of electrical and mechanical systems, pp 387–405. https://doi.org/10.1007/978-981-15-1532-3_17
127. Yadav AK et al (2020) ANN- and multiple linear regression-based modelling for experimental investigation of photovoltaic module maximum power production under outdoor condition of mountainous region. In: Springer nature book: modern maximum power point tracking techniques for photovoltaic energy systems, pp 229–245. https://doi.org/10.1007/978-3-030-05578-3_8
128. Fatema N et al (2019) Big-data analytics based energy analysis and monitoring for multi-story hospital buildings: case study. In: Springer Nature book: soft computing in condition monitoring and diagnostics of electrical and mechanical systems, pp 325–343. https://doi.org/10.1007/978-981-15-1532-3_14
129. Yadav AK et al (2018) Chapter 11: techno economic feasibility analysis of different combination of PV-wind-diesel-battery hybrid system. In: Elsevier Book: hybrid-renewable energy systems in microgrids, pp 203–218. https://doi.org/10.1016/B978-0-08-102493-5.00011-X
130. Minai AF et al (2020) Metaheuristics paradigms for renewable energy systems: advances in optimization algorithms. In: Springer Nature book: metaheuristic and evolutionary computation: algorithms and applications, under book series "studies in computational intelligence, pp 35–61. https://doi.org/10.1007/978-981-15-7571-6_2
131. Mahto T et al (2020) Traffic signal control to optimize run time for energy saving: a smart city paradigm. In: Springer Nature book: metaheuristic and evolutionary computation: algorithms and applications, under book series "studies in computational intelligence, pp 491–497. https://doi.org/10.1007/978-981-15-7571-6_21
132. Rahi OP et al (2012) Power system voltage stability assessment through artificial neural network. In: Elsevier procedia engineering, ISSN: 1877-7058, vol 30, pp 53–60. https://doi.org/10.1016/j.proeng.2012.01.833
133. Mahto T et al (2018) Fractional order control and simulation of wind-biomass isolated hybrid power system using particle swarm optimization. Book chapter in Applications of artificial intelligence techniques in engineering, advances in intelligent systems and computing 698:277–287. https://doi.org/10.1007/978-981-13-1819-1_28
134. Yadav AK et al (2016) Tilt angle calculation for installation of PV systems for mountainous regions of Himachal Pradesh India. In: Proceedings IEEE ICEPES 2016, pp 205–209, 14–16 Dec. 2016. https://doi.org/10.1109/ICEPES.2016.7915931
135. Yadav AK et al (2015) Optimization of tilt angle for installation of solar photovoltaic system for six sites in India. In: Proceedings IEEE international conference on energy economics and environment (ICEEE-2015), pp 1–4. https://doi.org/10.1109/EnergyEconomics.2015.7235078
136. Fatima K et al (2022) Intelligent approach-based maximum power point tracking for renewable energy system: a review. In: Malik H, Ahmad MW, Kothari D (eds) Intelligent data analytics for power and energy systems. lecture notes in electrical engineering, vol 802, pp 373–405. Springer. https://doi.org/10.1007/978-981-16-6081-8_19

137. Kumar N et al (2022) Modeling and analysis of an intelligent approach for load frequency control in a deregulated power system: a case study based on different control schemes. In: Malik H, Ahmad MW, Kothari D (eds) Intelligent data analytics for power and energy systems. Lecture notes in electrical engineering, vol 802, pp 61–83. Springer. https://doi.org/10.1007/978-981-16-6081-8_4
138. Yadav AK et al (2020) Optimization of tilt angle for intercepting maximum solar radiation for power generation. In: Springer Nature book: optimization of power system problems (Methods, Algorithms and MATLAB Codes), pp 203–232. https://doi.org/10.1007/978-3-030-34050-6_9
139. Gopal C et al (2022) Digital transformation through advances in artificial intelligence and machine learning. J Intel Fuzzy Syst 42(2):615–622. https://doi.org/10.3233/JIFS-189787
140. Shahid A et al (2018) Decrypting wrist movement from MEG signal using SVM classifier. J Intel Fuzzy Syst 35(5):5123–5130. https://doi.org/10.3233/JIFS-169796
141. Fatema N et al (2022) Data driven intelligent model for quality management in healthcare. J Intel Fuzzy Syst 42(2):1155–1169. https://doi.org/10.3233/JIFS-189779
142. Sanaullah A et al (2022) Analyzing impact of relationship benefit and commitment on developing loyalty using machine intelligence approach. J Intel Fuzzy Syst 42(2):699–712. https://doi.org/10.3233/JIFS-189742
143. Smriti S et al (2018) Intelligent tools and techniques for signals, machines and automation. J Intel Fuzzy Syst 35(5):4895–4899. https://doi.org/10.3233/JIFS-169773
144. Zhou L et al (2021) An optimal higher order likelihood distribution based approach for strong edge and high contrast restoration. IEEE Access 9:109012–109024. https://doi.org/10.1109/ACCESS.2021.3101413
145. Kumar D et al (2021) 6D-chaotic system and 2D fractional discrete cosine transform based encryption of biometric templates. IEEE Access 9:103056–103074. https://doi.org/10.1109/ACCESS.2021.3097881
146. Asyraf A et al (2021) Machine learning approach for targeting and recommending a product for project management. Mathematics 9(16), 1958:1–26. https://doi.org/10.3390/math9161958
147. Arvind D et al (2021) Likelihood estimation and wavelet transformation based optimization for minimization of noisy pixels. In: IEEE access, vol 9, pp 132168–132190. https://doi.org/10.1109/ACCESS.2021.3113857
148. Jain H, Fatema N (2018) Layer recurrent neural network based intelligent user activity classification model using smartphone. J Intel Fuzzy Syst (JIFS) 35(5):5085–5097. https://doi.org/10.3233/JIFS-169793
149. Nuzhat F (2018) Application of neuro-fuzzy scheme to improve purchasing process in a hospital. J Intel Fuzzy Syst (JIFS) 35(5):5131–5146. https://doi.org/10.3233/JIFS-169797
150. Anil BK et al (2011) Application research based on fuzzy logic to predict minimum loss for transformer design optimization. In: Proceedings IEEE International conference on computational intelligence and communication networks, pp 207–211. https://doi.org/10.1109/CICN.2011.41
151. Yadav AK et al (2011) Cost analysis of transformer's main material weight with artificial neural network (ANN). In: Proceedings IEEE international conference on communication system's network technologies, pp 184–187. https://doi.org/10.1109/CSNT.2011.46
152. Khatri A et al (2012) Optimal design of power transformer using genetic algorithm. In: Proceedings IEEE international conference on communication system's network technologies, pp 830–833. https://doi.org/10.1109/CSNT.2012.180
153. Goel P et al (2019) Application of evolutionary reinforcement learning (ERL) approach in control domain: a review. Smart Innov Commun Comput Sci Ser 670:273–288. https://doi.org/10.1007/978-981-10-8971-8_25
154. Roy N et al (2018) Extreme learning machine-based image classification model using handwritten digit database. Book chapter in Applications of artificial intelligence techniques in engineering, advances in intelligent systems and computing 697:607–618. https://doi.org/10.1007/978-981-13-1822-1_57

155. Kukker A et al (2016) Foreamrm movements classification of EMG signals using Hilbert Huang transform and artificial neural network. In: Proceedings IEEE PIICON-2016, pp 1–6, 25–27 Nov. 2016. https://doi.org/10.1109/POWERI.2016.8077417
156. Fatema N. Brain health assessment via classification of EEG signals for seizure and non-seizure conditions using extreme learning machine (ELM). In: Malik H, Srivastava S, Sood Y, Ahmad A (eds) Applications of artificial intelligence techniques in engineering. advances in intelligent systems and computing, vol 697. Springer. https://doi.org/10.1007/978-981-13-1822-1_10
157. What is Machine Learning? www.ibm.com. Retrieved 2021-08-15. Available online at https://www.ibm.com/cloud/learn/machine-learning. Accessed on 10 Dec 2021
158. Nilsson N (1965) Learning machines, McGraw Hill
159. Mohri M, Rostamizadeh A, Talwalkar A (2012) Foundations of machine learning, The MIT Press ISBN 9780262018258
160. Russell SJ, Norvig P (2010) Artificial intelligence: a modern approach, 3rd edn, Prentice Hall ISBN 9780136042594
161. Hinton G, Sejnowski T (1999) Unsupervised learning: foundations of neural computation. MIT Press. ISBN 978-0262581684
162. Duda RO, Hart PE; Stork DG (2001) Unsupervised learning and clustering. Pattern classification, 2nd ed. Wiley. ISBN 0-471-05669-3
163. Bishop CM (2006) Pattern recognition and machine learning, Springer, ISBN 978-0-387-31073-2
164. Sutton RS, Barto AG (2018) Reinforcement learning: An introduction, 2nd ed. MIT Press. ISBN 978-0-262-03924-6

A Deep Drive into Knowledge Management for Improving Software Process and Product: Visions and Research Directions

Mitali Chugh

Abstract Software engineering organizations strive to develop the high-quality software. To accomplish the goal of the developing a high-quality software, the software process has to be improved. To facilitate the improvement of the software process, knowledge management plays a significant role, and the comprehensions on knowledge management studies explore the different perspectives to enhance the software process. Up to the present time, little research has revealed how knowledge management initiatives are important to enhance the development process to further improve the software product. This research appraises the literature on software development by centering and inspecting knowledge management visions in software development to address the stated research gap. Specifically, the aim of the present work is to explore the knowledge management literature, identify the knowledge gaps, and enhance future research in this domain. Our findings highlight how software process improvement has been supported by knowledge management. It also put emphasis on the prominence of information technology for knowledge management.

Keywords Knowledge management · Research directions · Software process improvement · Software product

1 Introduction

Nowadays, software development initiatives in software engineering (SE) organizations are highly knowledge intensive. Knowledge sharing in the organization must be carried out in an efficient and cost-effective way for improved software quality in terms of product as well as services. The knowledge-intensive software development efforts pose attention-grabbing challenges for organizations. KM is an enabler in the life cycle for development of software from requirements elicitation, to coding and testing, to software implementation and maintenance; it even extends to the improvement of practices of software development. KM is a team effort and involves

M. Chugh (✉)
UPES, Bidholi, Dehradun, Uttarakhand 248001, India
e-mail: mchugh@upes.ac.in

A. Tomar et al. (eds.), *Proceedings of 3rd International Conference on Machine Learning, Advances in Computing, Renewable Energy and Communication*, Lecture Notes in Electrical Engineering 915, https://doi.org/10.1007/978-981-19-2828-4_2

the coordination and integration of several sources of knowledge; a company needs to keep track of continuously growing knowledge. If organizations do not manage this continuously growing knowledge efficiently, the strength of such knowledge is compromised, and it becomes a problem rather than an advantage. IT-based solutions that support KM must be aligned with software developers' requirements to provide the correct solutions to problems when knowledge seekers require them. The research in the area of KM for improving the software process is significant for the SE organizations to gain the competitive advantage in the dynamic market. The present is intended to guide the research directions in the sphere of KM for enriching the software process. The paper comprises the sections as: Sect. 2 discusses the problems in the SE domain. Section 3 includes the insights on the role of KM in software development process and information technology (IT) a facilitator for KM when employed for enhancing the software process. Section 4 is about the research directions and concludes the study.

2 Problems Related to Software Engineering

The term 'Software Engineering' was introduced in 1968 at a North Atlantic Treaty Organization (NATO) conference in Germany. It mainly focused on exceeding the software project budgets, time, and tarnished functionality—'software crisis'. In effect, SE has proved to be quite dissimilar when compared with other disciplines of engineering because of its unkind nature [1]. The researchers and practitioners are working for a long time to overcome the problems in development; however, there is still a need for more studies to address the stated issues. It is always a challenge to anticipate the concerns for the development of software without practically doing it, and the fast introduction of fresh technologies makes software maintenance difficult.

2.1 Those Who Forget History Are Doomed to Repeat It!

SE is multi-faceted, involving individuals functioning in various phases. Ongoing technology advancements mean that work is no longer static; fresh problems arise and are resolved; diversification takes place, and banks of knowledge are created each day. Organizations face difficulties in keeping track of new knowledge; the danger is that knowledge is rarely captured and stored in order to acquire benefits in new projects to escape postponements and overruns in budgets in the projects. Software engineers and stakeholders working on related or similar projects do not realize that better and faster outcomes could be attained by studying the practices adopted in preceding projects [2]. Software engineers know very well that their efforts are not retained due to a deficiency of knowledge sharing within the organization, and as a result, they have to work on each problem from scratch [3]. Many researchers in literature infer that software engineers do not learn from the knowledge acquired in earlier projects and constantly repeat faults in software development projects [4–6].

2.2 When Employees Leave, What Do We Lose?

Maintaining a high level of organizational knowledge is quite significant. This is challenging for SE organizations given that they are human and knowledge intensive [7]. In SE organizations, software development practices and procedures are dependent on the software engineer's knowledge about these practices. When software team members leave their posts, companies fail to store the institutional knowledge that they carry away with them; in various software companies, there is an absence or shortage of adequate transference sessions to stem the loss. The problem is that knowledge is possessed by the organizational teams and not by the organization. Therefore, knowledge gaps occur when personnel change jobs; the concern becomes more critical when the member leaving the organization is the only expert in a particular field. Moreover, when a team comprises members who belong to different departments, as happens in most SE organizations, the team is dissolved after the project is complete; the acquired experience and knowledge leave the organization without being stored in the organizational repository.

2.3 Cultural Issues and Inadequate Communication

Software development involves the association and interaction of people from diverse cultural backgrounds. Cultures vary in several perspectives, leading to potential difficulties in structural approaches; these may include hierarchy, sense of time, and methods used for communication [8]. Many individuals consider such variances enriching; however, misunderstandings can arise, particularly among those who do not know each other well. Software development, especially during its inception phases, requires a lot of communication [9]. Communications can be formal or informal. Formal communications require a clear interface for crucial software development tasks like project status updating, discussion of project issues, allocation of tasks for different phases, and project deadlines. An interface that is unclear results in time delays and lets the problem fall through the cracks [10].

3 Insights from the KM Literature

Having explored the problems of software engineering, it is evident that KM forms an important practice for SE organizations. Therefore, now, we present insights on the associations for KM with software product and process along with the role of IT

Table 1 Scheme for literature review

S. No.	Study dimension	Areas covered
1	KM and software product	Knowledge evolution and management, knowledge sharing
2	KM and software process improvement	KM in requirement analysis, coding, testing, and maintenance
3	IT as a facilitator for KM and SPI	Role of IT as a catalyst in different empirical studies

for facilitating the KM. To review the literature, the following scheme is followed (Table 1):

3.1 Knowledge Management and Software Product

KM facilitates software organizations by offering creation, storage, retrieval sharing, and application of knowledge through methodical procedures. Scholars in their qualitative and empirical studies have mentioned that use of KM practices and procedures expedite development of unswerving, cost-effective, functional, and maintainable software products.

A study related to knowledge evolution management for software product lines has been carried out by Abbas et al. [11]. They explore how a software product line could get advantage from fresh knowledge emanating from diverse source activities and articles at run time. They present the idea that products self-optimize by sharing the knowledge at run time, hence improving the quality faster as compared to traditional software development.

The research work in software organizations of Iran led by Khosravi et al. [12] suggested a structure to improve the novel software product performance and moderate its budget. KM procedures and approaches constrain knowledge loss and facilitate to share knowledge sharing among software developers for decreasing budgets and tackling faults. The view is supported by Serna et al. in their study [13]. Table 2 presents the summary of KM and software product studies.

3.2 KM and Software Process

In the last decade, SPI has gained momentum in the software development organizations. Accordingly, there has been a proliferation of models and initiatives focused on the growth of SPI initiatives as the IDEAL model, the CMM model, and SPICE.

The SPI literature incorporates proof of successful organizations and SPI program reports, e.g., Motorola [14], Hughes [15], NASA [16]. The SE professionals have always shared the significance of knowledge related to development aids SPI, and

Table 2 Summary of knowledge management and software product

Reference	KM aspect covered	Findings	Conclusion
Abbas et al. [11]	Knowledge evolution management	Software product line could get advantage from fresh knowledge emanating from diverse source activities	The various aspects of knowledge effect/enhance the quality of the software product
Khosarvi et al. [12] and Serna et al. [13]	Knowledge sharing	KM procedures and approaches facilitate developers and enhance software product	

developers acquire knowledge to deal with faults from former development practices [17].

Mitchell and Seaman examined the knowledge usage and recommend that elimination of interruptions to knowledge flow facilitates SPI. This study was conducted with regard to the budgets and schedule overruns that keep on afflicting software engineering [18]. Their results present that knowledge flow limitations once eliminated reduce the period to complete a software engineer's tasks, support in achieving targets, and enhance quality of process. SPI is the outcome. Schneider et al. present that in software development, the requirements are not distinguished nor implemented accurately on the grounds that the process for the most part is influenced by human knowledge [19]. To tackle the issue, the authors distinguished the strategies related to the knowledge creation theory by [20] and examined how they help in dealing these issues.

Serna et al. have proposed KM methodologies in the stage of requirement analysis [13]. De Souza et al. narrate KM advantages in testing phase of software, putting forward that software testing is facilitated by KM and is an imminent research field [21]. Itkonen and Lassenius have put forward the role of tester's knowledge in an empirical study [22].

Other studies converse about the implication of KM in software maintenance and evolution [23–25]. Table 3 presents the summary of KM and SPI.

3.3 IT—An Enhancer in KM Initiatives for Software Process Improvement

Mehta et al. empirically studied IT effect on knowledge exchange. Their results reveal that knowledge exchange is facilitated by IT use as compared to knowledge combination. The reason for this outcome is that knowledge combination involves more application. Hence, the team members come together to talk over strategies, something that is hard to assist through IT [27].

Table 3 Summary of KM and SPI

Reference	Focus of the work	Findings	Conclusion
Schneider et al. [19]	Studying the effect of knowledge creation theory for distinguishing and implementing requirements accurately	The knowledge creation strategies in software projects help to manage risk in KM	
Itkonen and Lassenius [22]. DeSouza et al. [26]	To study KM initiatives facilitate software-testing phase of software development	KM facilitates software-testing phase	
Serna and Serna [23], DeVasconcelos et al. [24]	To study implications of KM in software maintenance and evolution	KM is vital procedure to support software maintainers	
Mitchell and Seaman [18]	To study application of KM at project level in context of budget and schedule overruns	Elimination of knowledge flows reduces time to complete software development tasks and enhances quality of process	
Serna et al. [13]	Methodologies of knowledge management in requirement engineering	Social cooperation practices and artificial intelligence for knowledge creation upgrade the requirement engineering process	
Khalil and Khalil [25]	To present the theoretical model on KM for the agile software development organizations	Present insights on KM for agile software development including continuous communication, iterative development, knowledge repositories, and engineering practices	

Moreno and Cavazotte mention maximum KM initiatives include the application of KMS. KMS is planned and created to upgrade the organizational procedures of knowledge management. This enables employees to access knowledge needed to make decisions and carry out their tasks. The general merits of KMS are typically identified with improvements in adaptability, creativity, responsiveness, decision-making, and efficiency. The model presented in the study involves a systematic empirical investigation and incorporates critical factors backing the success of KM initiatives and efficient usage of KMS [28].

In an empirical investigation, Merlo provides a distinctive comprehension into KM practices among IT experts concerning KM usage in the southern United States.

They state that the KMS usage in organizations is vital in the management of knowledge [29]. Santoro et al. in their empirical study in Italian firms concluded that KMS facilitates utilization of internal and external knowledge flows. The information technology technique used in this study is Internet of Things (IoT); this fosters knowledge flow in the organization [30]. Kumar and Srikant emphasize that the readiness of novel IT, above all the World Wide Web, has been influential in catalyzing the KM. IT, if well-resourced and applied, offers a wide-ranging knowledge base that is promptly accessed, collaborative, and of immediate worth to the SE organizations [31, 32]. Table 4 presents the summary for role of IT as an enhancer between KM and SPI.

Table 4 Role of IT as an enhancer between KM and SPI summary

Reference	Focus of the work	Findings	Conclusion
Mehta et al. [27]	Empirically investigate the effect of information technology on knowledge exchange and combination	Knowledge exchange is facilitated by IT use as compared to knowledge combination	KM phases are facilitated by IT
Moreno and Cavazotte [28]	To design a model for KM facilitated by IT	The model presented in the study involves a systematic empirical investigation and incorporates critical factors backing the success of KM initiatives and efficient usage of KMS	
Merlo [29]	Investigating KM practices among IT professionals	The effective use of KMS in organizations is decisive in the management of knowledge	
Santoro et al. [30]	An empirical study to investigate KMS facilitate exploitation of internal and external knowledge flows	IT is a necessity for increasing innovativeness in the organization and KMS as an enabler for KM	
Kumar and Srikant [31]	The study presents the role of IT as a catalyst for KM	IT offers the channel of IT in KM in organization for evolving business competitive advantage	

4 Conclusion

Complex software projects require innovative solutions to match the requirements of an increasingly unsettled business environment. The comprehensive work in this paper has created future research themes mentioned in this study for IT and KM association that influences software process improvement. The theme of KM for SPI and enhancement of software quality has received scant attention in the literature. Particularly, it intends to present research gaps that can help to develop a tentative model to describe the roles of KM and IT in the perspective of development process and product. The study can be enhanced by conceptualizing and empirically validating the research model. This comprises testing for the direct relationship between KM and SPI, KM and software product, and KM and IT. Our study contributes to academia and practice in the following ways. (i) It articulates the research gaps to investigate KM in SPI and software products and (ii) the role of IT in promoting KM. The articulation of these areas builds a spur for accelerating and directing research in these areas that will assist to establish an acknowledgment of the KM discipline within SE research in the future.

References

1. Kruchten P (2002) The nature of software what's so special about software engineering?
2. Basili V, Lindvall M, Costa P (2001) Implementing the experience factory concepts as a set of experience bases. In: Proceedings of the thirteenth international conference on software engineering & knowledge engineering (SEKE 2001), pp 102–109
3. Davenport TH, Lawrence P (1998) Working knowledge: how organizations manage what they know. Harvard Business School Press, no. January 1998, Harvard Business School Press, pp 1–15
4. Rod Iguez D, Gar EI, Anchez S, Rodríguez-Solano Nuzzi C (2010) Defining software process model constraints with rules using owl and swrl, vol 20, no 4, pp 1–16. https://doi.org/10.1142/S0218194010004876
5. Lee JC, Shiue YC, Chen CY (2016) Examining the impacts of organizational culture and top management support of knowledge sharing on the success of software process improvement. Comput Human Behav 54:462–474. https://doi.org/10.1016/j.chb.2015.08.030
6. Rus I, Lindvall M, Sinha S (2002) Knowledge management in software engineering. IEEE Softw 19(3):26–38. https://doi.org/10.1109/MS.2002.1003450
7. Basili VR et al (2001) An experience management system for a software engineering research organization. In: Software engineering workshop 2001. Proceedings. 26th Annual NASA Goddard, no. KMWorld, pp 29–35. https://doi.org/10.1109/SEW.2001.992652
8. Guzley R (1992) Cultures and organizations: software of the mind: Geert Hofstede London: McGraw-Hill, 1991, 279 pp, $39.95 (cloth). Int J Intercult Relat 16(4):455–460. https://doi.org/10.1016/0147-1767(92)90033-Q
9. Perry DE, Staudenmeyer NA, Votta LG (1994) People, organizations, and process improvement. IEEE Softw 11(4):36–45. https://doi.org/10.1109/52.300082
10. Herbsleb JD, Moitra D (2001) Global software development. IEEE Softw 18(2):16–20. https://doi.org/10.1109/52.914732
11. Abbas N, Anderson J, Weyns D (2011) Knowledge evolution in autonomic software product lines. In: Software product lines—15th international conference, SPLC, no. February 2015, pp 1–9

12. Khosravi A, Hussin ARC, Nilashi M (2018) Toward software quality enhancement by customer knowledge management in software companies. Telemat Inform 35(1):18–37. https://doi.org/10.1016/j.tele.2017.09.007
13. Serna M, Bachiller S, Serna A (2017) Knowledge meaning and management in requirements engineering. Int J Inf Manage 37(3):155–161. https://doi.org/10.1016/j.ijinfomgt.2017.01.005
14. Diaz M, Sligo J (1997) How software process improvement helped motorola. IEEE Softw 14(5):75–80. https://doi.org/10.1109/52.605934
15. Humphrey WS, Snyder TR, Willis RR (1991) Software process improvement at Hughes Aircraft. IEEE Softw 8(4):11–23. https://doi.org/10.1109/9781118156667.ch2
16. Mcgarry F, Page G, Basili V (1994) Software process improvement in the NASA software engineering laboratory
17. Sharma N, Singh K, Goyal DP (2011) An empirical investigation of critical success factors for KM Implementation in software organizations through factor analysis. Int J Recent Trends Eng Technol 6(1):67–70
18. Mitchell SM, Seaman CB (2016) Could removal of project-level knowledge flow obstacles contribute to software process improvement? A study of software engineer perceptions. Inf Softw Technol 72:151–170. https://doi.org/10.1016/j.infsof.2015.12.007
19. Schneider L, Hajji K, Schirbaum A, Basten D (2013) Knowledge creation in requirements engineering—a systematic literature review. In: 11th international conference on Wirtschaftsinformatik, March, pp 1829–1843
20. Nonaka I (1994) A dynamic theory of organizational knowledge creation. Organ Sci 5(1):14–37. https://doi.org/10.1287/orsc.5.1.14
21. De Souza ÉF, Falbo RDA, Vijaykumar NL (2015) Knowledge management initiatives in software testing: a mapping study. Inf Softw Technol 57(1):378–391. https://doi.org/10.1016/j.infsof.2014.05.016
22. Itkonen J, Lassenius C (2013) The role of the tester's knowledge in exploratory software testing. IEEE Trans Softw Eng 39(5):707–724. https://doi.org/10.1109/TSE.2012.55
23. Serna M, Serna A (2014) Ontology for knowledge management in software maintenance. Int J Inf Manage 34(5):704–710. https://doi.org/10.1016/j.ijinfomgt.2014.06.005
24. Carreteiro P, de Vasconcelos JB, Alexandre B, Rocha Á (2016) A knowledge management approach for software engineering projects development. Adv Intell Syst Comput 445, February, pp 59–68 [Online]. Available: http://www.scopus.com/inward/record.url?eid=2-s2.0-84961625750&partnerID=tZOtx3y1
25. Khalil C, Khalil S (2020) Exploring knowledge management in agile software development organizations exploring knowledge management in agile software development organizations. Int Entrep Manag J June. https://doi.org/10.1007/s11365-019-00582-9
26. Ferreira É, Souza D, De Almeida R, Vijaykumar NL (2015) Knowledge management initiatives in software testing: a mapping study. Inf Softw Technol 57:378–391. https://doi.org/10.1016/j.infsof.2014.05.016
27. Mehta N, Hall D, Byrd T (2014) Information technology and knowledge in software development teams: the role of project uncertainty. Inf Manag 51(4):417–429. https://doi.org/10.1016/j.im.2014.02.007
28. Moreno V, Cavazotte F (2015) Using information systems to leverage knowledge management processes: the role of work context, job characteristics and task-technology fit. Procedia Comput Sci 55:360–369. https://doi.org/10.1016/j.procs.2015.07.066
29. Merlo TR (2016) Factors influencing knowledge management use in technology enterprises in southern United States. Procedia Comput Sci 99:15–35. https://doi.org/10.1016/j.procs.2016.09.098
30. Santoro G, Vrontis D, Thrassou A, Dezi L (2018) The internet of things: building a knowledge management system for open innovation and knowledge management capacity. Technol Forecast Soc Change 136:347–354. https://doi.org/10.1016/j.techfore.2017.02.034

31. Kumar TV, Srikanth A (2021) Information technology—a catalyst for knowledge management in organization. ASIAN J Multidimens Res 10(5):680–686
32. Tomar A et al (2020) Machine learning, advances in computing, renewable energy and communication, vol 768, 659 p. Springer, Berlin, LNEE. https://doi.org/10.1007/978-981-16-2354-7 (ISBN 978-981-16-2354-7)

An Area and Delay-Efficient Approximate Hybrid Adder for VLSI Circuit Designs

B. Kalivaraprasad, M. V. D. Prasad, Ch. Raju Kumar Singh, S. Ravichand, and A. Pramod Kumar

Abstract Over the last few decades, a wide range of solutions for the problems had been obtained using approximate computing in prior work; the RCPA was used, and it was used in this work as well. In RCPA, signal with input carry is more significant than the output carry. In the presence of delay changes, this type of carry propagation provides more stability. Three implementations such as various amounts of delay, power energy, and accuracy are used. In our paper, the hybrid adder is implemented with configurable levels of accuracy. The parallel prefix adder is designed and applied in proposed hybrid n-bit adder to minimize the critical path time. This proposed adder is compared to other defined adders in the literature as well as other state-of-the-art known adders. The adder is implemented in this paper has more speed. The synthesis and simulation are done in Xilinx ISE 14.7 version tool.

Keywords RCPA · Energy efficient · Accuracy · DSP · Adders

1 Introduction

Adder blocks are capable part components in DSP systems arithmetic units; they consume a lot of energy with power by frequently producing hotspots on the di [1, 2]. These facts prompted the use of approximation computing [3, 4] to realiz this component. Prior research on approximate adders focused on error weight ar probability reductions [1, 2, 5]. The principal objective in the design is to mir mize the power consumption with good rapid speed. For precise processing units,

B. Kalivaraprasad (✉) · M. V. D. Prasad · Ch. Raju Kumar Singh
Department of ECE, Koneru Lakshmaiah Education Foundation, Guntur, Andhra Pradesh, Ind
e-mail: baditakali@gmail.com

S. Ravichand
Department of ECE Nalla Narashima Reddy Education Society's Group of Institutions, Hyderabad, India

A. Pramod Kumar
Department of ECE Vardhaman College of Engineering, Hyderabad, India

A. Tomar et al. (eds.), *Proceedings of 3rd International Conference on Machine Learning, Advances in Computing, Renewable Energy and Communication*, Lecture Notes in Electrical Engineering 915, https://doi.org/10.1007/978-981-19-2828-4_3

improvement in speed usually comes at the expense of increased power consumption. One strategy for increasing both power and speed is to compromise precision in computing. In situations, where some inaccuracies are acceptable, an approximate computing approach can be used. The major design goals for these adders are to reduce the summation error probability as well as the power and delay [6–10]. The RCPA (Approximation reverse carry propagate adder) is a digital signal processing approximate adder that performs addition operations. An increase in speed for precise processing units usually comes at the expense of greater power consumption. Compromising computer precision is one way to increase both power and speed. This method is successful. In applications where some inaccuracies are acceptable, approximate computing, such as RCPA, can be used. The major objective o be done is to minimize the latency and error rate of traditional adders while also educing the power consumption of the supplied adder using RCPA. The motivating eason is to implement a large number of complex components on a chip. The goal f this work is formulated to design a well-defined adder with compact delay, area, pplied in applications that uses images.

The paper is organised as follows: Sect. 2 is related to works regarding the approxiate adder, and various n-bit hybrid approximate adders over a decade are illustrated. Sect. 3, the RCPA and various approximate adder cell that are used in existing ign that will briefly tell about the modified n-bit hybrid adder and also about allel prefix adder. The results and discussion are done in Sect. 4, and finally, we conclude the paper and provide a future scope in Sect. 5.

Related Work

section discusses approximate computing approaches that have been used in nber of applications. The adder's least significant (LS) and most significant parts were independent of one another under this approach, and the MS part's nput was zero. In [11], a segmented approximate adder with a carry speculation re was proposed as a way to improve accuracy. Furthermore, an error reduction s employed to lower the adder's error.

vel Proposed Method

ated a new reverse carry propagate adder of three different designs in the pproach. They have also created a hybrid approximate adder to lower the in real-world applications. The hybrid adder is made up of two adders; one carry propagate adder (RCPA), and other is traditional ripple carry adder ork implements an n-bit-modified hybrid approximate adder. To address flaw, the ripple carry adder is replaced with the Han Carlson adder. The pproximate hybrid adder has less critical path delay (Fig. 1).

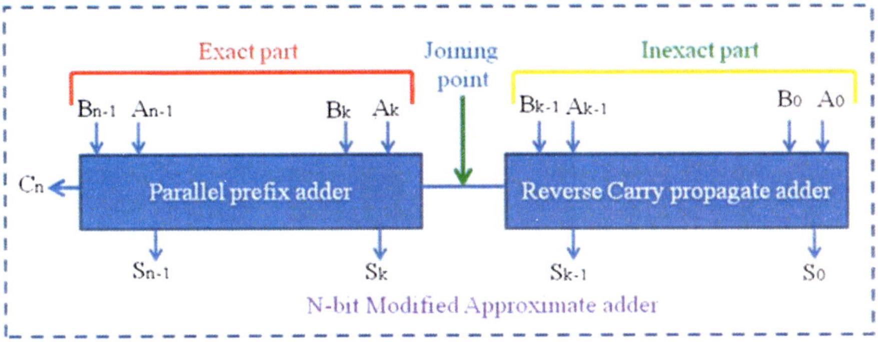

Fig. 1 n-bit modified hybrid approximate adder

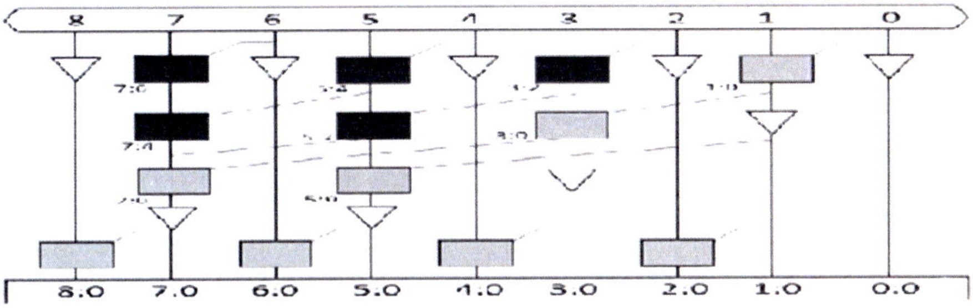

Fig. 2 Han Carlson adder with carry generation stage

Instead of employing the ripple carry adder, the Han Carlson adder is used in the modified design. The Han Carlson adder significantly reduces latency, with increased operating speed and accuracy [9]. The addition action in a parallel prefix adder is accomplished in three stages: preprocessing, carry generation, and post-processing. The propagate and generate values are implemented in the same way as the carry look ahead adder in the preprocessing step. For each input, these propagate and generate values are calculated and sent to the next stage (Figs. 2 and 3).

$$p_i = A_i \oplus B_i \tag{3}$$

$$G_i = A_i B_i \tag{4}$$

In carry generation stage, the stages are same for all parallel prefix adders. For each parallel prefix adder, the carry generating step has a different structure. The stage is implemented by using black cells and gray cells to obtain carry. The carry generation stage of Han Carlson adder is shown in Fig. 5.

In this process, final sum of the adder is calculated. XOR (Exclusive OR operation) is performed between the propagate values and previous stage carry values. The logical expression is shown below [10–14].

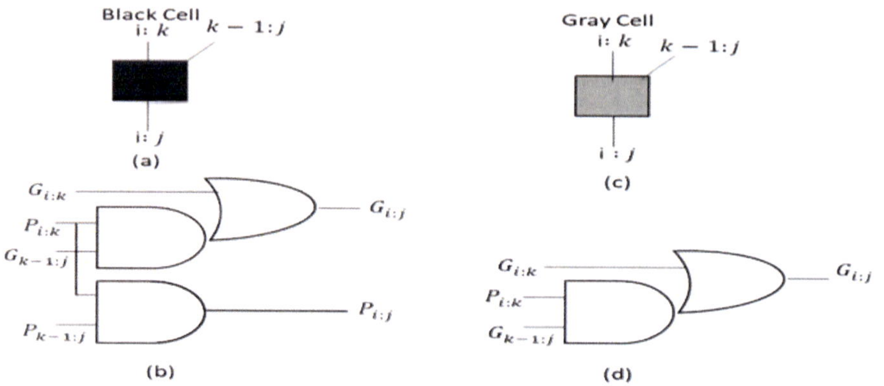

Fig. 3 Black cells **a**, **b** and gray cells **c**, **d** logic implementation

$$S_i = P_i \oplus C_{i-1} \tag{5}$$

4 Results and Discussion

This part explains about the simulation results.

The diagram shows the RTL schematic view of 16-bit RCPA Design-1. It is having two inputs, namely a and b of 16-bit each and two outputs sum(s) and carry. In this, carry output is taken as last bit of sum (S16). Figure 4 is the schematic view of the

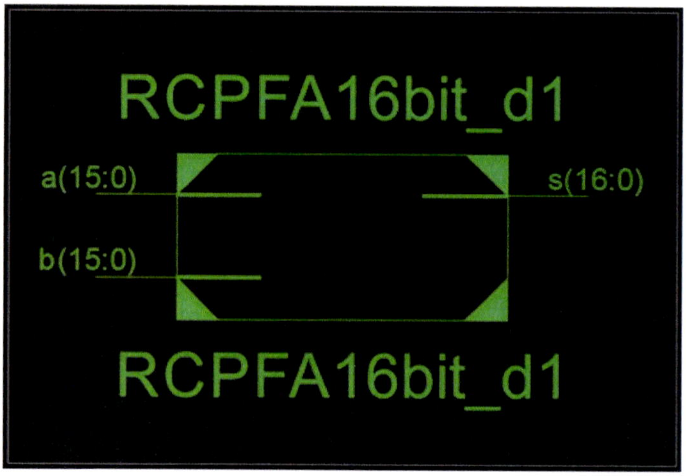

Fig. 4 RTL schematic view of RCPA design 1 with 16 bit

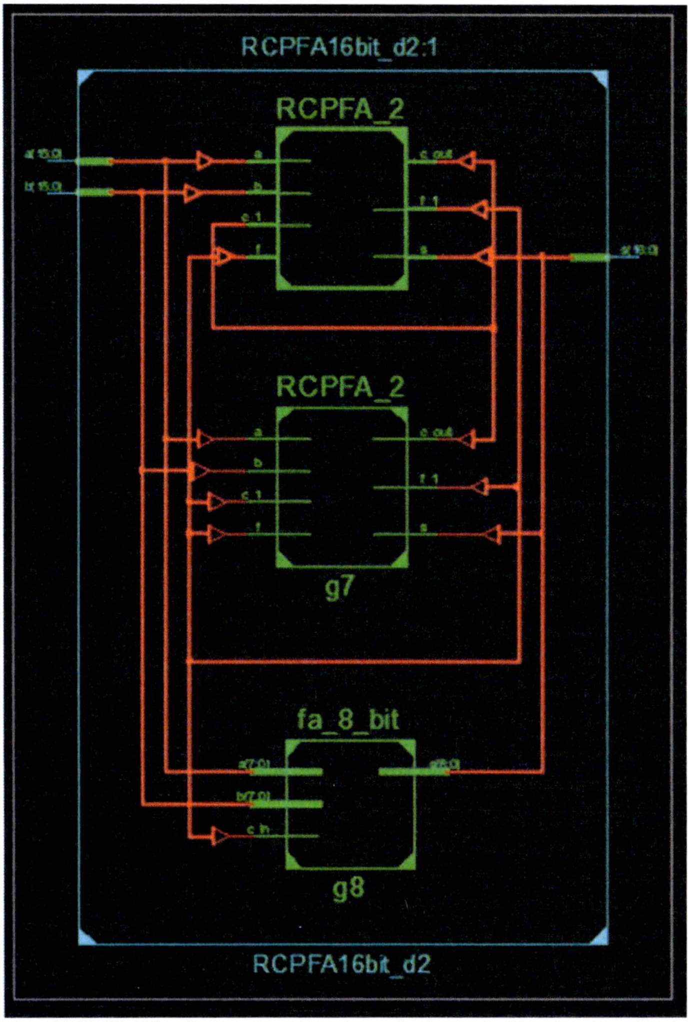

Fig. 5 RTL schematic of proposed design 2

16-bit ripple carry parallel adder design which is having two inputs and one output having 16 bit, respectively.

The proposed design of RTL schematic is shown in Fig. 5. The results of the multiplier are shown in Figs. 6, 7, and 8.

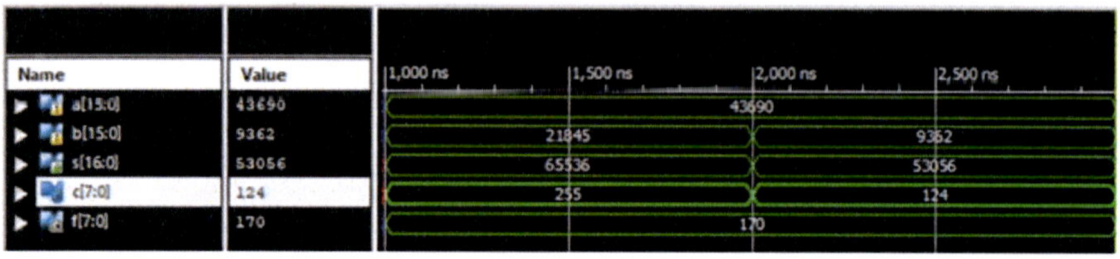

Fig. 6 Simulation output of 16-bit RCPA d1 for a given inputs

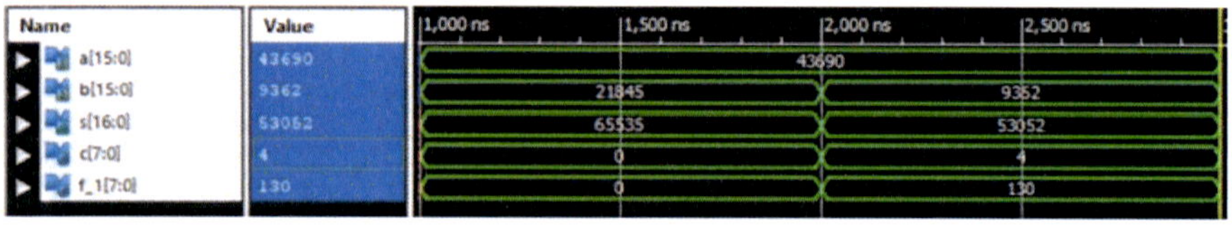

Fig. 7 Simulation output of 16-bit RCPA d2 for a given inputs

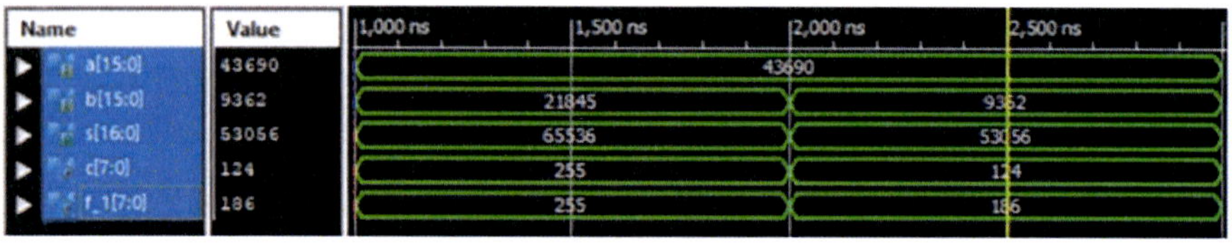

Fig. 8 Simulation output of 16-bit RCPA d3 for a given inputs

5 Conclusion and Future Work

The n-bit hybrid adder is implemented with approximation in the lower side and correct addition in the higher side The parallel prefix adder is used to improve the performance of the n-bit hybrid adder in terms of area and delay instead of Han Carlson adder. In future work, the proposed modified hybrid adder can be used for real-time applications in order to observe the performance. The synthesis and simulation are done in Xilinx ISE 14.7 version tool.

References

1. Rodrigues G, Lima Kastensmidt F, Bosio A (2020) Survey on approximate computing and its intrinsic fault tolerance. Electronics 9(4):557
2. Pejović V (2019) Towards approximate mobile computing. GetMobile Mobile Comput Commun 22(4):9–12
3. San AM, Yakunin AN (2018) Reducing the hardware complexity of a parallel prefix adder. 978-1-5386-4340-2. IEEE

4. Zeydel B, Baran D, Oklobdzija V (2010) Energy-efficient design methodologies: high-performance VLSI adders. IEEE J Solid-State Circuits 45(6):1220–1233
5. Seo H, Yang YS, Kim Y (2020) Design and analysis of an approximate adder with hybrid error reduction. Electronics 9(3):471
6. Reddy KM, Vasantha MH, Kumar YN, Dwivedi D (2019) Design and analysis of multiplier using approximate 4–2 compressor. AEU-Int J Electron Commun 107:89–97
7. Gauhar S, Sharif A, Alam N (2020) Comparison of parallel prefix adders based on FPGA & ASIC implementations. In: 2020 IEEE students conference on engineering & systems (SCES), pp 1–6. IEEE
8. Seok H, Seo H, Lee J, Kim Y (2021) COREA: delay-and energy-efficient approximate adder using effective carry speculation. Electronics 10(18):2234
9. Kishore KH, Prasad BKV, Teja YMS, Akhila D, Sai KN, Kumar PS (2018) Design and comparative analysis of inexact speculative adder and multiplier. Int J Eng Technol (UAE). ISSN No. 413-426
10. Prasad BKV, Priya RS (2016) Implementation and reconfiguration of basic digital modulation design models. J Theor Appl Inf Technol 90(1):208
11. Ghabraei S, Rezaalipour M, Dehyadegari M, Nazm Bojnordi M (2020) AxCEM: designing approximate comparator-enabled multipliers. J Low Power Electron Appl 10(1):9
12. Prasad BKV, Mazumdar D, Narendra G, Chaitanya G, Kanta TM (2018) Implementation of dual 16 QAM modulators combined with 2× 2 STBC block. In: Innovations in electronics and communication engineering. Springer, pp 445–453
13. Kalivaraprasad B, Prasad MVD, Babu KR, Shameem S, Mohan S, Vani V (2021) Comparative analysis of watermarking methods on CFRP sample thermal images. In: Computer communication, networking and IoT. Springer, pp 455–462
14. Tomar A et al (2020) Machine learning, advances in computing, renewable energy and communication. Springer, Berlin, LNEE vol. 768, 659 p. https://doi.org/10.1007/978-981-16-2354-7 (ISBN 978-981-16-2354-7)

Improvement in the Security of IoT Network Using IP Binding Technique

Shilpa B. Sarvaiya and D. N. Satange

Abstract Internet of things is the fastest going technique. In today's world, it has totally changed the way of interaction of the peoples with electronics devices and equipment. It establishes a network communication between the devices. So, the security implications matter in IoT networks. The nodes in a network need to be connected in a secured way. Node communication in IoT network sometime gets hacked, and wrong operation performs if it happened then there will be a chance to go for heavy loss. So that security in protocol of IoT needs to be improved and implemented so that the proposed methodology will help to cross check the sender and receiver node which keep the security up and maintain the variation in data handling. The proposed technique also help to improve the security of IoT network using IP binding technique.

Keywords Attacks · Encryption · Decryption · Internet of things (IoT) · IP address · IP binding · Network · Node data · Security

1 Introduction

In 2005, the world came across new emerging technology Internet of things (IoT). The IoT secure protocol system gives new dreams to the IT industries. Operating the devices from anywhere through mobile phone or laptop is a magical things. IoT brings hardware data from one place to another place as well as gives away to control any electronics devices. This domain helps to automize productions of medical devices and many more industries. In IoT usage amount of data will be collected which is too much crusial because a single change in single bit false the functioning of the big machines. Which cause very big damage to the industries.

S. B. Sarvaiya (✉)
Vidyabharati Mahavidlayala, Amravati, India
e-mail: Sarvaiya.shilpa@gmail.com

D. N. Satange
Director Students Development Sant Gadge Baba, Amravati, University, Amravati, India

A. Tomar et al. (eds.), *Proceedings of 3rd International Conference on Machine Learning, Advances in Computing, Renewable Energy and Communication*, Lecture Notes in Electrical Engineering 915, https://doi.org/10.1007/978-981-19-2828-4_4

So that dealing with the data and maintaining its security is important to the future IoT industries by considering this as a problem statement. I proposed a design of secure protocol system for IoT networks. In 2050, every electronics devices get to be controlled and can we operate through the mobile phones. So, for every second data will be stored and retrieved. IoT architecture mostly consists of sensors which are connected to IoT board so, this particular IoT board is called as node [1].

The numbers of node will be connected into single network which will be called as IoT networks. Whole data will be passed through this IoT networks so, may their will be a chance to the inturder to hack this networks which may cause a big loss. So, by implimenting Internet of things secure protocol system, it is possible to avoid this loss. Node communication in IoT network sometime gets hacked and performs wrong operation; if it happened, then there will be a chance to go for heavy loss. So that security in protocol of IoT needs to be improved and implemented so that the proposed methodology will help to cross-check the sender and receiver node which keep the security up and maintain the variation in data handling [2].

1.1 IoT Attack Surfaces

IoT attacks play an important role in data transmission; every IP address represents a device so if you want to check trustworthiness of device we can check it by IP. 70% of IoT devices have vulnerabilities hackers, and/or some other agencies find out vulnerability in such an IoT network and try to hack the data signals. In order to do that the security of IoT network compromises. So that the surface attacks will damage or more to the IoT networks as per the security concern [3].

1.2 Need of Security for IoT Devices

As above mentioned, the surface attacks will harm the IoT networks in which data can be altered by the user which may change the exact working of IoT devices. So that security in IoT devices needs to be overviewed in medical industry or automation sector security plays an important role in data. In such a way that there will be highly need of security in IoT networks [4].

- IoT devices increase day by day so maybe there are the changes to hack the device.
- Alterations in network data over the IoT network will cost big damage.
- Security gets compromise sometimes in IoT networks.
- Peer-to-peer communication and compromises sometime in some systems.

The recent evolution in current trends is Internet of Things which is introduced in the year 1998 by Kevin Ashton. Now days almost all machines requires internet services. In recent technology interaction between intelligent devices increases. In Internet of Things getting the data from devices operated on different devices which

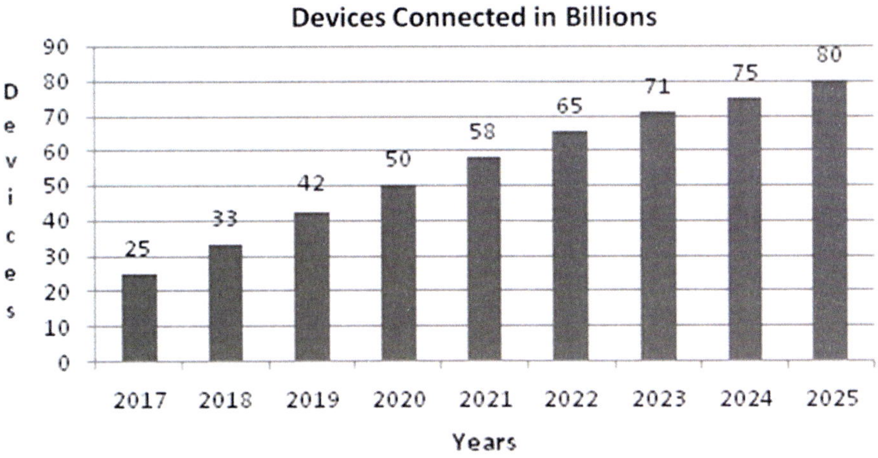

Fig. 1 Interconnected IoT devices from 2017 to 2025

are going to be operated on different platform. Figure 1 represented the number of connected devices is increasing explosively per person [5].

2 Related Work

P1: "Guo et al. [6]" Here, an author describes three new methods to look out IoT devices on the Internet: Server IP addresses in traffic, server names in name system (DNS) queries, and manufacturer information in transport layer security (TLS) certificates.

Limitations: The two main limitations of this paper for IP-based detection are: (I) Authors directly apply DNS-based detection to old network traces because server names are stable while server IP can change. (II) Authors learn new device server names during DNS-based detection by examining unknown server names DNS queried by detected IoT devices and learning those appear as if device servers.

P2: "Chen et al. [7]" Here, authors report the first systematic study on device binding mechanisms of IoT, an attempt to understand the protection implications. Also show that the threat to IoT device binding is realistic and high.

Limitations: The foremost limitations of this paper are: (I) only consider Man-In-The-Middle attacks during local binding. However, there may well be vulnerable designs within the registration process that is binding with the Cloud, which is often related to the local binding. (II) Additionally, the attacker must know which device the user has, and when the device is prepared up, the attack can only be successfully launched when the user is configuring his/her devices.

P3: "Rajashree et al. [8]" Here, authors present a scheme for internet protocol (IP) address assignment to smart IoT devices which communicate using TCP/IP and validation of source IP address within the received.

IP packets, from the IoT devices.

Limitations: Limitations of this method are: (I) it assumes that the media access control (MAC) address which is used as device identifier could also be duplicated by unauthorized users. (I) only consider IP spoofing attacks. This can be often required to prevent an unauthorized user from using IP address as source address and flooding packets to the gateway, thereby using the bandwidth allocated to authorize users.

P4: "Choi et al. [9]" Here, authors give a scheme to confirm IP continuity for near field communication (NFC)-based IoT networking. However, the low-power technology could bring potential.

Problems to device functionality.

Limitations: Limitations of this paper are: (I) NFC, which is from radio frequency identification technology, supports peer-to-peer communication on the link layer. Link addresses of NFC devices are not physically fixed values, so this gives negative influences to IP networking, especially, connection continuity. (II) Therefore, performance evaluation during this paper has only targeted to figure out whether IP continuity is guaranteed or not with Ping6 tests and also shows different results on real trip time (RTT).

P5: "Cheng et al. [10]" Here, authors utilize three-dimensional locations coordinates to assign each node a singular spatial IPv6 address supported grouping methods and scan-line scheme. Besides, assignment success rate (ASR) is utilized during this paper to determine the probability that assigns unique IP address to nodes successfully using multi-projection IP address assignment (MPIPA) scheme.

Limitations: Limitations of this paper are: (I) This scheme going to be applied in larger-scale IP-based WSNs for smart gird, especially high building scenarios. And (II) This scheme will not be targeted at developing IPv6 address auto configuration for mobile nodes.

3 Problem Statement

To improve this Sender-Receiver verification and encryption schemes in which the data is get transfer to the user with highly encryption and key transfer technique so that the data in network is remain secured. This system will also check for the sender and receiver terminal, if the sender is right and receiver is also right the cross verification of sender and receiver done as well. If the sender is right and receiver is also right, the cross verification of sender and receiver is done as well. So that the proposed methodology plays an important role in maintaining data security over the network.

4 Proposed Work and Methodology

IoT plays an important role in next generation technology; the IoT devices get controlled on to the data sent by the users; all things will be there in IoT in some next years. So, data security plays an important role in IoT platform. The signals sent in the form of data to the IoT device in which data gets transferred from one end to another. In the network if data or signal is get altered by intruder then it will highly impact on IoT devices. It will cause a big damage to the IoT devices; so to overcome this, some work is proposed in which the encryption–decryption techniques are used to make the data secured.

4.1 Architecture of Proposed Work

As per the proposed work, the data security needs to be implemented over the data network in which authentication of the both ends needs to be verified (Fig. 2).

In this, the proposed work is getting implemented with above basic architecture; the data transmission is done over the network in IoT devices which send data over the network in which the security clause is getting to be implemented over the data signal that is encryption in which the clause is getting to be implemented over the data single. In that data is encrypted with private key and sent over the network. Same key will be received by receiver node with the implementation of decryption; so, the authentication is done at the sender and receiver node with verification of sender and receiver.

In second security module, we perform the IP binding model in which the sender and receiver IP address bind together and send over the network in which the IP addresses cross verified in which verification of the sender and receiver is done; if the sender and receiver match the status, then data will be delivered successfully.

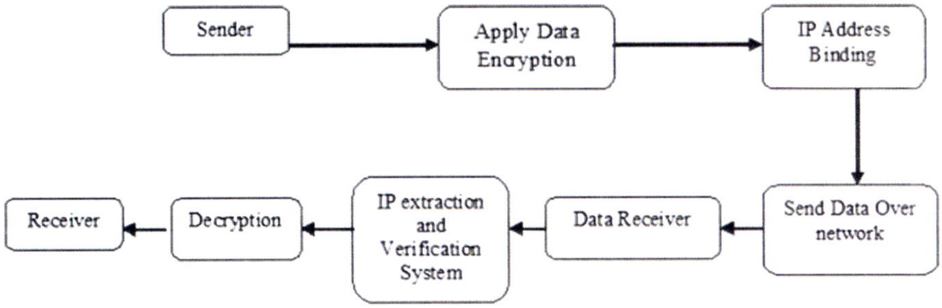

Fig. 2 Basic architecture of proposed work

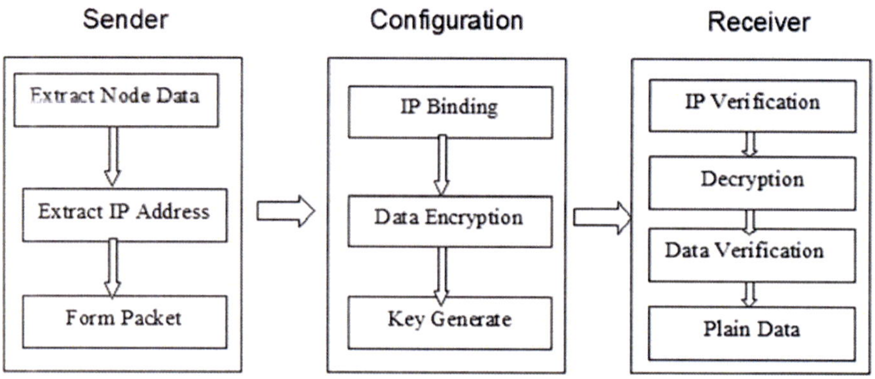

Fig. 3 Basic framework of improvement in the security of IoT network

4.2 *Methodology*

As per the proposed work, the data security is getting implemented over in which the authentication is done with both user verification. In this module the binding of IP address getting workout in which the IP of sender and receiver is get bind with packet of message and send over the network. In which the IP address also get encrypted if any message is get altered then bind IP of the sender is also get updated in which the receiver side will able to identify the exact sender IP address in which the sender can be easily identify by pattern matching implementation over the IP of sender and receiver. If the message get altered then receiver comes to know that the data signal get alter by intruder and will not perform the operation send the notification to admin IP binding sender IP and Receiver IP cross verification plays role of maintain security over the data signal transmission so that wrong or alter message will not get delivered to the IoT boards which only restrict to the right sender and receiver. In second security module, we perform the IP binding model in which the sender and receiver IP address bind together and send over the network in which the IP addresses cross verified. In which verification of the sender and receiver done if the sender and receiver match the status then data will delivered successfully (Fig. 3).

4.2.1 Encryption Module

In this, the keys get generated and perform encryption over the message whose message gets spread over the network which is in the form of encrypted message if that is retrieved by intruder then unable to decrypted as there is an exchange keys that will perform in different way. So that the messages over the network deliver at right place and decode by right node. Due to such mechanism, it will help to improve the data signal security over the network (Fig. 4).

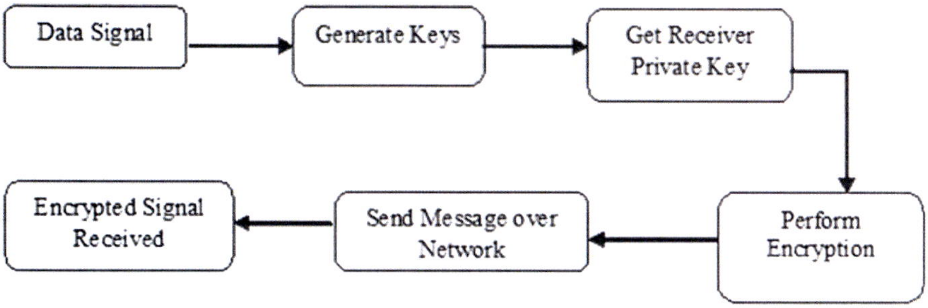

Fig. 4 Encryption module

4.2.2 IP Binding

In which there is receiver node and sender node which are connected in IoT network. In which sender node will receive the data from IoT board and perform IP binding, then encryption over the received data and form a signal packet of that which is latterly shared to the cloud via IoT network. At the other end signal is going to be received which latterly proceed with IP verification and decryption, which delivered the actual data to the receiver node according to the decision are taken by the receiver. In this the Keys are get generated and perform encryption over the message which message get spread over the network which is in the form of encrypted message if that is get retrieved by intruder then unable to decrypted as there is an exchange keys will perform in different way. So that the messages over the network deliver at right place and decode by right node. Due to such mechanism it will help to improve the data signal security over the network. So that the messages over the network deliver at right place and decode by right node. Due to such mechanism it will help to improve the data signal security over the network. If the message gets altered, then receiver comes to know that the data signal gets altered by intruder and will not perform the operation and send the notification to admin (Fig. 5).

In IP binding sender IP and receiver IP, cross verification plays role of maintaining security over the data signal transmission so that wrong or alter message will not get delivered to the IoT boards which only restrict to the right sender and receiver.

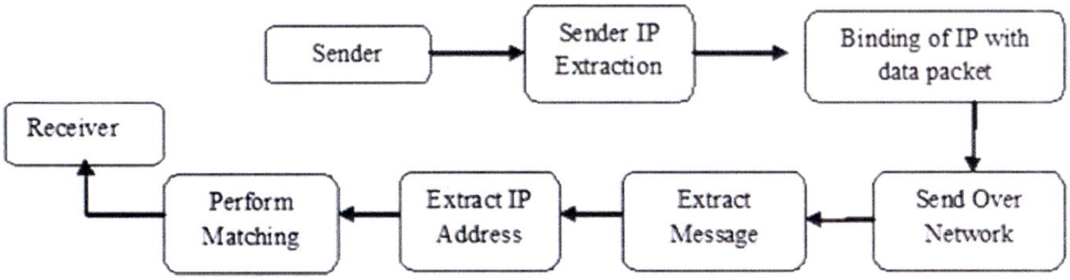

Fig. 5 Architecture of IP binding technique

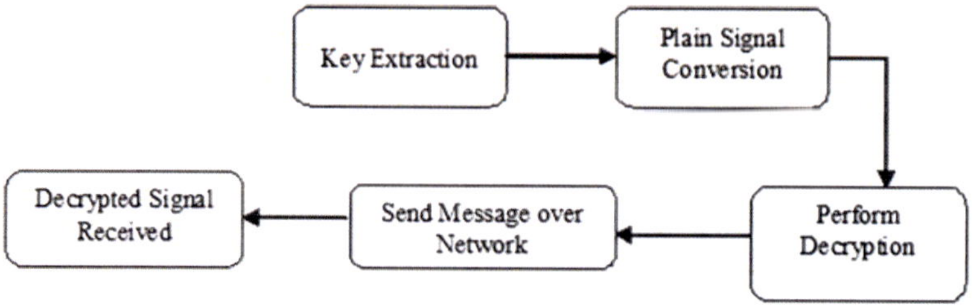

Fig. 6 Decryption module

4.2.3 Decryption Module

In this decryption module the keys are getting generated and perform encryption over the message. Which message get spread over the network. So that the messages over the network deliver at right place and decode by right node. Due to such mechanism, it will help to improve the data signal security over the network (Fig. 6).

5 Result and Discussion

In development cases of the IoT, security plays an important role in data sharing with node which are connected in IoT network, so the aim of the proposed work is to go for the approach which makes the interaction among the various nodes get bind with the highly and tightly pack with security goals for the IoT infrastructure[11]. As per the proposed methodology, evolutionary system is considered as an throughput in which the system will provide the high security over the IoT network in which the communication improves to the next level [12]. The intention of the proposed model is to build a secure protocol system for the IoT network to produce trustworthy security mechanisms for the IoT security layers. The aim of the proposed work is to help to analyze the security requirements of secure protocol system for IoT, which are authentication and access control continuously, generate and share data. Authorization, confidentiality, availability, integrity, privacy, authentication, access control and non-repudiation are very much important considerations in order to ensure the security of communication in said type of data sharing atmosphere [13]. In this situation, ad-hoc nature of networks and lack of computing resources requires major changes in existing techniques [14–18].

6 Conclusion and Future Scope

The system can be able to tackle the various attacks in IoT and make the IoT network to strong so that the proposed mechanism will help in future to integrate with the various types of projects with IoT so that the leakage found in the services can be identified and rectified as per the concerning security algorithms. The purpose system can be integrated to the real-time IoT node in which the data security can be maintained in connecting IP binding capable devices; so they can communicate with each other in a proper secured manner. The proposed methodology has limitation in only node failure cases, but the occurrence of this is to very less. The proposed system is used to prevent or decrease attacks, threats and problems as much as possible.

To improve this, sender–receiver verification and encryption scheme gets proposed in which the data gets transferred to the user with highly encryption and key transfer technique so that the data in network remains secured; in this, the system will also check for the sender and receiver terminal. If the sender is right and receiver is also right, the cross verification of sender and receiver is done as well. So that the proposed methodology plays an important role in maintaining data security over the IoT network.

References

1. Tanooj Kumar M et al (2019) A hybrid approach for enhancing security in internet of things. In: International conference on intelligent sustainable systems (ICISS 2019), ISBN:978-1-5386-7799-5. IEEE. https://doi.org/10.1109/ISS1.2019.8907974
2. Witti M, Konstantas D (2018) Internet of things and security-privacy concerns: a systematic mapping study. Int J Netw Secur Appl (IJNSA) 10(6):25–33. https://doi.org/10.1109/ISS1.2019.8907974
3. Ali H, Alauddin A (2019) Security issues in internet of things: a survey. In: 2019 IEEE international conference on innovation and intelligence for informatics, computing, and technologies (3ICT), vol 5, pp 01–06, ISSN: 7281-3012. https://doi.org/10.1109/3ICT.2019.8910320
4. Vignesh R, Samydurai A (2017) Security on internet of things with challenges and countermeasures. 2017 Int J Eng Devel Res (IJEDR) 5(1):417–423. ISSN:2321-9939. https://doi.org/10.5829/id0si.aejsr.2017.128.134
5. Karie NM, Sahri NM, Dowland PH (2020) IoT threat detection advances, challenges and future directions. In: 2020 IEEE workshop on emerging technologies for security in IoT (ETSecIoT), vol 9, 22–29. ISSN:7281-8019, November 10. https://doi.org/10.1109/ETSecIoT50046.2020.00009
6. Hang G, John H (2020) Detecting IoT devices in the internet. IEEE/ACM Trans Netw ISSN: 1063-6692, 1–14, September 11.https://doi.org/10.1109/TNET.2020.3009425
7. Chen J, Sun M, Zhang K (2019) Security analysis of device binding for IP-based IoT devices. In: The third workshop on security, privacy and trust in the internet of things, ISSN: 5386-9151, vol 9. IEEE, pp 900–905. https://doi.org/10.1109/PERCOMW.2019.8730580
8. Rajashree S, Soman KS, Gajkumar SP (2018) Security with IP address assignment and spoofing for smart IOT devices, ISSN: 5386-5314, vol 2, 1914–1918. IEEE. https://doi.org/10.1109/Cybermatics_2018.2018.00066
9. Choi Y et al (2017) Scheme to guarantee IP continuity for NFC-based IoT networking. In: ICACT 2017, ISBN: 978-89-968650-9-4, 695–698, February 19–22.https://doi.org/10.23919/ICACT.2017.7890182
10. Cheng C-Y, Chuang C-C, Chang RI (2016) Three-dimensional location-based IPv6 addressing for wireless sensor networks in smart grid. In: International conference on advanced information networking and applications, ISSN: 1550–4456, pp 824–831. IEEE.https://doi.org/10.1109/AINA.2021.42

11. Cao Z, Yang S (2019) A security communication device based on narrowband internet of things. In: 2019 IEEE 5th international conference on computer and communications, vol 7, pp 2141–2145. ISSN:7281-4743. https://doi.org/10.1109/ICCC47050.2019.9064188
12. Ahmed I, Saleel AP, Beheshti B, Khan ZA, Ahmad I (2017) Security in the internet of thing. In: The fourth HCT information technology trends (ITT 2017), Dubai, vol 4, pp 84–90. ISSN:5386-3330, October 25–26. https://doi.org/10.1109/CTIT.2017.8259572
13. Ekkachan R, Phithak T (2019) Security implementation for authentication in IoT environments. In: 2019 IEEE 4th international conference on computer and communication systems, vol 7, pp 678–681. ISSN: 7281-1322. https://doi.org/10.1109/CCOMS.2019.8821686
14. Obaidat M et al (2019) Security architecture framework for internet of things (IoT), IEEE Xplore, vol 5, pp 0154–0157, ISSN: 7281-3885. https://doi.org/10.1109/UEMCON47517.2019.8993096
15. Zhang J, Jin H, Gong L, Cao J, Gu Z (2019) Overview of IoT security architecture. In: 2019 IEEE 4th international conference on data science in cyberspace (DSC), pp 338–345. ISSN: 7281-4528. https://doi.org/10.1109/DSC.2019.00059
16. El-hajj M, Fadlallah A et al (2019) A survey of internet of things (IoT) authentication schemes, article sensors 2019, pp 1–43. https://doi.org/10.3390/S19051141
17. Vyash A, Satheesh A Comprehensive study on security issues in internet of things:a survey paper. In: Proceedings of international conference on sciences, engineering and technology, April 2–3, 2018
18. Iqbal A et al (2020) Renewable power for sustainable growth, vol 723. Springer Nature, Berlin, LNEE, 805 p. https://doi.org/10.1007/978-981-33-4080-0 (ISBN 978-981-33-4082-4)

Healthcare System with IoT Wireless Sensor Network

Raveena Yadav and Vinod Kumar

Abstract As the Internet of Things has given us a lot of smart and intelligent devices. The fields in which smart devices are available are almost in each field. The main application that has been discussed in this chapter is the healthcare system. The healthcare system carries two important functions, one is to diagnose, and the second function is giving a better treatment. Lack of time to the people are suffering from many health illnesses. They are not able to reach the hospital and meet the doctor. Health care system needs to find the symptoms before reaching the disease to a higher level and giving proper alertness to the patient. So, with the advancement of technology, there is a need to take steps forward in the healthcare system. The objective of this chapter is to give a review of the smart healthcare system with researchers' work in this field. Another objective is to provide the role of sensors in this system. The motivation for writing this chapter is to discuss and gathering knowledge about the Internet of things in the healthcare system. As health is an important and topmost thing about life, without better health, nothing is easy to achieve in the aspect of any of the sectors. This chapter is the integration of the introduction of the Internet of things, the healthcare care system with the internet of things, and the conclusion with future scope.

Keywords Internet of things · Health system · Wireless sensor network

1 Introduction

Internet of things has become an important part of our daily life. This helps most of the time such as setting a ringtone of our alarm by sensing the weather condition, morning coffee with a smart coffee maker, smart light, smartwatches, smart television, smart

R. Yadav (✉) · V. Kumar
Delhi Technological University, New Delhi, Delhi, India
e-mail: raveenakyb@gmail.com

V. Kumar
e-mail: vinod_k@dtu.ac.in

A. Tomar et al. (eds.), *Proceedings of 3rd International Conference on Machine Learning, Advances in Computing, Renewable Energy and Communication*, Lecture Notes in Electrical Engineering 915, https://doi.org/10.1007/978-981-19-2828-4_5

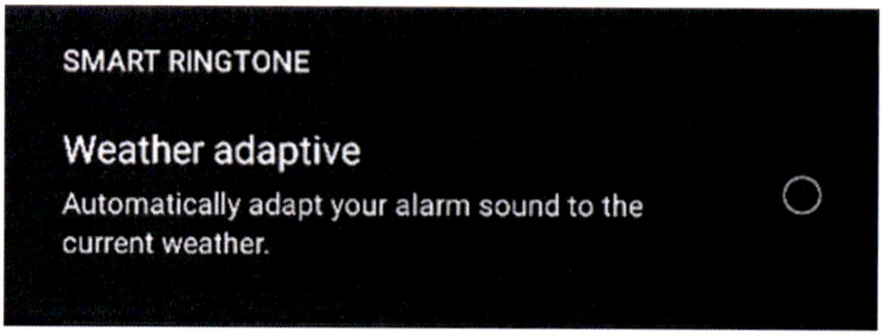

Fig. 1 Smart ringtone (realme smartphone)

lock, smart ringtone (Fig. 1), and many others devices are under the network of IoT. IoT makes a connection and helps in doing communicating among the devices.

The term Internet of things is made up of the integration of different components such as the two main components that are the Internet and other one is things. After the differentiation of these components, various components come out in the picture. As Internet is made up and follows different protocols for making a reliable connection among things. It also helps in sending information among devices of the network. Another component is the thing, things of this network have two main components that are radio frequency identification (RFID) and sensor. RFID is used for giving a unique identification number to things of this massive network. The sensor helps in gathering knowledge about its surrounding. Based on these components, different layered architectures are given by researchers as shown in Fig. 2.

Internet of things application is moving in distinct directions. It is one of the directions in the healthcare system. The motivation of moving toward this system is rising the cost of medical facilities, lacking time to people, and lack of medical resources at some places. Internet of things helps in providing a cost-effective healthcare system, and it can also handle the patients remotely. It helps all people that have lack time. In Ahmadi et al. (2019), the healthcare system is further divided into two systems. One is an e-health system, and another is a mobile health system (m-health). For the working of this system, different sensors are attached to the patient's body and nearby his body.

In this chapter, we have discussed the related work in smart healthcare systems and wireless sensor networks in health systems. This wireless sensor network helps in making this system a smart and intelligent system. This system has many different types of low-powered sensors and gathers information from its nearby area. These sensors do not make any obstruct the human daily schedule and activities. Sensors work in this system about gathering any psychological change in the human body.

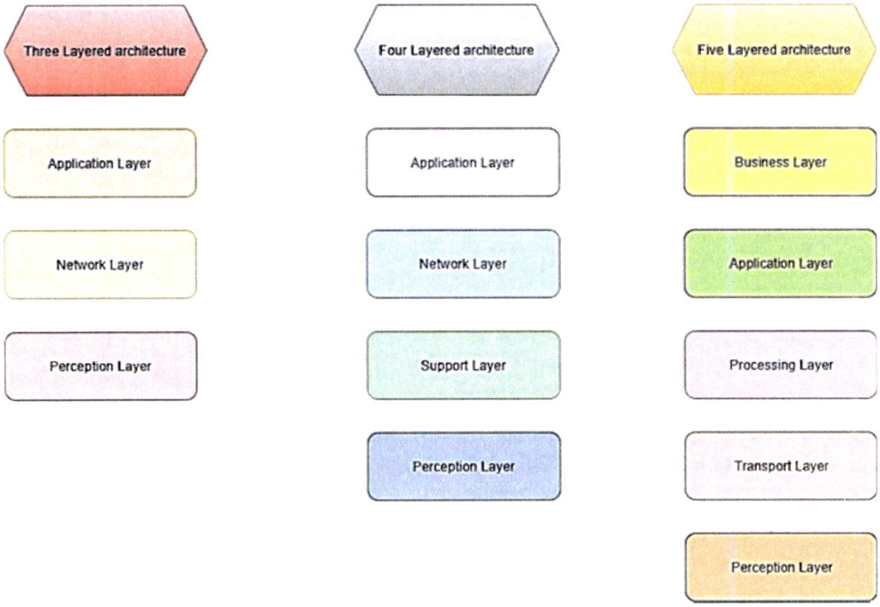

Fig. 2 Different layered architecture of IoT

2 Related Work

Budida et al. (2017) proposed a system with two functional blocks of working. The first block gathers all information from the body of the patient and the surroundings of the patient's body. The second block stores and processes the data. It also gives a final analytical view of processed data. Ahmadi et al. (2019) provided a comprehensive healthcare system. In this, they have mentioned the home healthcare system, which is useful for aged people. This system provides video conferencing, with many sensors. GPS is also used in this system, so that if any aged people lost their address, then it helps in tracking. Next is the mobile healthcare system, which contains applications in mobile devices that sensors many things such as the temperature of the body, the oxygen level in the blood, blood pressure, and also helps in tracking diet and many other features. In Shaikh et al. (2019), a smart healthcare system helps in connecting patients to doctors remotely. They have mentioned the RFID-based monitoring system that helps in tracking the objects. In (2019), Dang et al. handle the large amount of data produced in health care. This task is very complex to handle, and it also includes some confidential information; handling this sort of data is a very crucial task. With the help of cloud computing and fog computing, it can be manageable in some places. Park et al. (2017) discussed the importance of the healthcare system for old age people. They suffer from many health issues because old people are not able to go to the hospital and not able to handle the medical costs. With consideration of these issues, the mobile healthcare system and NFC are effective for taking care of old people. Ismail et al. (2020) proposed a model for disabled people by using Raspberry Pi. In this model, different sensors are used on different items and connected to one device such as a smartphone. The smartphone can be easily controlled by a disabled person with the help of speech recognition. The person can send the command via his phone and control all home devices. Jangra et al. (2018) proposed a method for sensors network of Internet of things. In this network, the sensor checks its neighboring edges for congestion. If the sensor finds it suitable for sending information to the next node, then it will first compress the data. After compressing the data, the data will be sent to the next node. Gardasevic et al. (2020) gave light to the challenges in the wireless sensor network of the healthcare system. This wireless network carries sensitive data from one sensor node to another. This makes them vulnerable to attack and theft of confidential data. The main challenge is the security of wireless sensor networks. Alamelu et al. (2017) proposed a paradigm for health care that sensors take data from the patient's body. After this, with the help of the network layer, it will transmit to the cloud server. It helps in taking the decision-making process and stores the patient's data in the record. The data can be recorded on the basis of requirement. For this, a model is proposed by Kesbi et al. (2019); four sensors are used along with a sink node and separated by room. Hence, this model is beneficial for calculating all values according to requirement. Patients also require a remote healthcare system; Rida et al. (2021) focused on this issue along with the proper management of bandwidth. According to proposed model, sensors shared the bandwidth which is bandwidth efficient model.

Due to Covid-19 pandemic we have lost many people. This disease increases the temperature of human body and causes repiratory diseases. Researchers have proposed works related to WSN related to COVID-19. Ali et al. (2020), Anjali et al. (2021) taken different parameters to their proposed model such as heartbeat and temperature of body and SpO2 level. Thus, from all these parameters, proposed model helps in giving an appropriate suggestion. Suganyedvi et al. (2021) have focused on the diet of a person as it plays an important role in living a healthy life. If it is not good, then it can give birth to disease. A smart table was proposed by author, which helps in calculating the weight and analysis of nutrients present in food via camera. After this computation, data will be sent to the cloud. Ali et al. (2021) have introduced a smart system for the dysphonia. This disease is related to the voice. Identification of this disease is done with the help of classification methid.

Hence, with all these work of researchers, we are able to find that WSN of IoTplays a crucial role in health system. After the COVID pandemic, the growth of smart health system has been increased.

3 WSN in Smart Healthcare System

The medical field is one of the fields in which a little error from the human side can lead to the death of a person. In this, doctors and nurses have complete knowledge about the disease and up-to-date knowledge. Doctors and nurses have to be attentive while communicating, giving perceptions, giving advice, and any time when they are in contact with the patient. A smart system is somewhere having positive points over the above-mentioned system [1–21]. A smart system contains devices and machines with some computation (programming). This helps in giving an error-free result and can alert the patient before the illness. This system contains many numbers sensors and makes a network from the sensors.

Wireless sensor network (WSN) helps in reducing the overall cost of the system as sensors are of low cost and help in capturing real-time data. This network helps in gathering information in many fields such as in military operation, tracking of person or animal, saving the power of electricity, and many others. A WSN smart healthcare system can be architected on three layers, and the bottom has contact with the surroundings and body. The middle layer consists of gateway that helps in connecting the sensor with the end user. The topmost layer has an application layer that helps in providing various services to the user. Smart healthcare system requires real-time data for proving itself a better system than the conventional system. Different sensors of this system use different efficient routing algorithms for sending messages to their sink node. These efficient routing algorithms can be categorized into different types such as based on energy-efficient, data security, data compression, data quality, and many other routing protocols that are available for transmission of information to sink nodes. The sink node is also known as the base station of the wireless network. It is the head of the sensor network, in which sensors send all their data to their base station. This wireless system of health care helps in getting data anytime, anyplace, and by anyone. This makes them vulnerable to attack. Because of this, many researchers have focused on security issues of the wireless system. This gives a research gap for energy-efficient routing for a wireless system, which has the same importance as security. As sensors are low-powered and resource-constrained. In the healthcare system, a patient's body is surrounded by distinct types of sensors such as airflow senor. This sensor is used for checking the breath of a person. Sensor new technology is coming up with a skin-based sensor that is something like a second layer on the skin. The fit band which is very trending these days contains many sensors. It helps in checking our daily activity like ideal state and alert for doing exercise and checking heart rate time to time. At the time of the COVID-19 pandemic, companies are adding one more feature in smartwatches, and the fitness band is checking the oxygen level. This COVID-19 pandemic gives more attention to the smart Internet-based healthcare system. A posture belt with sensors helps in detecting the posture of the person. The posture of a person can lead to many back pain diseases, so it helps in giving an alert about correcting the posture. Smart wearable device on hand and gesturing of hand helps in controlling the wheelchair of disabled person. Ultrasonic sensors and infrared sensors help in detecting any obstacle in front of the chair. All this data sent to a cloud server, and it helps in the decision-making process and stores the data into the patient's database. Cloud server also helps in removing redundant data and maintains security regarding the sensitive data. After this, information is sent to the smartphone, from which end user is able to see the location of the patient and information sent by sensors.

The wireless sensor network of the health system can be divided into four parts. The first part is about the indoor care of the patient. In this system, two types of sensor networks are used, sensors used on the body of the patient, and another one is sensors used in the home. These two networks make a network of sensors. The second part is about outdoor care; in this, if the patient is found outside, then it will send an alert to his caretaker and relative. The third part is about the decision-making process and storing and analyzing the record of a patient's illness. The fourth part is about giving information to the digital device from which the doctor, nurse, and relative can see the patient and communicate with him. The sensors used in these parts can be wearable sensors, the sensors in a form in which a person can wear that device and get the alertness from the device. Radar-based sensors, these sensors are non-contact form sensors. These sensors need not be worn by the person. There is only need to be in the range of the sensor. Radar sensors produce radiations through which they can identify the change in its surrounding.

Because of COVID-19, people are moving toward an online system whether it is about office work or anything else. The healthcare system is also one of them, and people are looking for it. People got scared to go to the hospital so there is the requirement of a smart health system that can be managed remotely. From this, patient can contact remotely to doctor, and the doctor can give a prescription. Advice and care can be given remotely. Maintaining glucose levels in the blood is very important in COVID-19. As changing in the balance of glucose level makes the body vulnerable to the attack of coronavirus. A diabetic person needs to adopt telemedicine in which a person can be remotely cared for by the doctor.

We have put an effort to propose a model (Fig. 3) for the patients who are not able to walk and need a wheel chair. In this model, we have used different types of sensors such as airflow sensor, which helps in checking the rate of breathe of patient. A fit band is attached to the wrist to patient which helps in telling the activity of patient. A posture sensor belt is used, which helps in correcting and reminding the posture of patient. Sensors on leg, infrared sensor, help in telling the movement of legs. An ultrasonic sensor is used on the wheel chair, which helps in checking if any

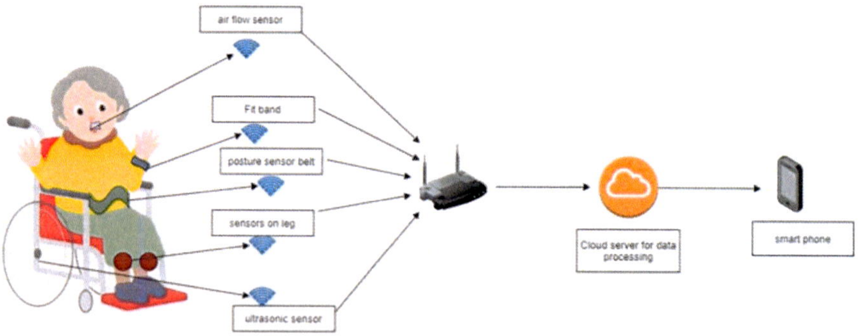

Fig. 3 Smart health system

obstrucle comes in way. All information related to the patient will send to cloud and cloud computing helps in removing the redundant data, after this processing, data send to the end user and patient's family members.

4 Conclusion

In this chapter, we have discussed the importance of health system management by using IoT. That consists of different sensors and helps in gathering information about the patient and his/her surrounding. Many researchers' work is also mentioned in this chapter. It helps in getting an idea about the smart health system and the need for improvement of the health system. As COVID-19, the health system has seen many crises, so its smart health system helps in the improvement of this system.

References

1. Budida DAM, Mangrulkar R (2017) Design and implementation of smart healthcare system using IoT. In: 2017 international conference on innovations in information, embedded and communication systems (ICIIECS). IEEE
2. Ahmadi H et al (2019) The application of internet of things in healthcare: a systematic literature review and classification. Univ Access Inf Soc 18(4):837–869
3. Shaikh Y, Parvati V, Biradar SR (2018) Survey of smart healthcare systems using internet of things (IoT). In: 2018 international conference on communication, computing and internet of things (IC3IoT). IEEE
4. Dang LM et al A survey on internet of things and cloud computing for healthcare. Electronics 8(7):768
5. Ismail A, Abdlerazek S, El-Henawy IM (2020) Development of smart healthcare system based on speech recognition using support vector machine and dynamic time warping. Sustainability 12(6):2403
6. Jangra P, Gupta M (2018) A design of real-time multilayered smart healthcare monitoring framework using IoT. In: 2018 international conference on intelligent and advanced system (ICIAS). IEEE
7. Gardašević G et al (2020) Emerging wireless sensor networks and internet of things technologies—foundations of smart healthcare. Sensors 20(13):3619
8. Alamelu JV, Mythili (2021) A prediction of speed for smart insulin pump utilizing adaptive neuro-fuzzy inference system and ANN. In: Proceedings of international conference on communication, circuits, and systems. Springer
9. Bhushan S et al (2016) A new approach towards IoT by using health care-IoT and food distribution IoT. In: 2016 2nd international conference on advances in computing, communication, & automation (ICACCA)(Fall). IEEE
10. Bhushan S, Singh AK, Vij S (2019) Comparative study and analysis of wireless mesh networks on AODV and DSR. In: 2019 4th international conference on internet of things: smart innovation and usages (IoT-SIU). IEEE
11. Bhushan S et al (2021) FAJIT: a fuzzy-based data aggregation technique for energy efficiency in wireless sensor network. Complex Intell Syst 7(2):997–1007
12. Sharma V, Kumar S, Bhushan S (2017) An overview of data redundancy reduction schemes in WSNs. In: 2017 3rd international conference on advances in computing, communication & automation (ICACCA)(Fall). IEEE

13. Anjali K et al (2021) IoT based smart healthcare system to detect and alert covid symptom. In: 2021 6th international conference on communication and electronics systems (ICCES). IEEE
14. Abbasi-Kesbi R, Asadi Z, Nikfarjam A (2020) Developing a wireless sensor network based on a proposed algorithm for healthcare purposes. Biomed Eng Lett 10(1):163–170
15. Rida JFA (2021) Development of a remote health care wireless sensor network based on wireless spread spectrum communication networks. Mat Today Proc
16. Ali S et al (2020) A review of the role of smart wireless medical sensor network in COVID-19. J Ind Integr Manage 5(04):413–425
17. Suganyadevi S, Shamia D, Balasamy K (2021) An IoT-based diet monitoring healthcare system for women. Smart Healthc Syst Des Sec Privacy Aspects 167–202
18. Ali Z, Imran M, Shoaib M (2021) An IoT-based smart healthcare system to detect dysphonia. Neural Comput Appl 1–11
19. Yadav R (2021) ADAS authentic data allowed security in internet of things. Turk J Comput Math Educ (TURCOMAT) 12(13):1761–1765
20. Jafar A et al (2021) AI and machine learning paradigms for health monitoring system: intelligent data analytics, vol 86, 513 p. Springer Nature, Berlin, SBD. https://doi.org/10.1007/978-981-33-4412-9. ISBN 978-981-33-4412-9
21. Tomar A et al (2020) Machine learning, advances in computing, renewable energy and communication, vol 768, 659 p. Springer Nature, Berlin, LNEE. https://doi.org/10.1007/978-981-16-2354-7 (ISBN 978-981-16-2354-7)

Comparative Analysis of Face Mask Detection Models

Arun Chauhan, Lokesh Khurana, Avinashwar, Prabhishek Singh, Manoj Diwakar, and Amrendra Tripathi

Abstract With an ongoing episode of Covid, the world health security and precaution need reformation and a new approach to be dealt with. The health concerns of the individual is a topic of utmost importance for every nation fighting the pandemic. With limited healthcare staff and the large public to look after, the assistance of Computer vision and AI is needed. Social distancing is a very effective way of containing the spread of a pandemic. Social distancing becomes difficult when dealing with a number of subjects like at gateways of offices, Airports, and many other sectors that have significant footfall in a day. In this paper we have tried to compare the different models for the recognition of mask on the face, for doing so we have used Real world masked face dataset (RMFD) (Iqbal et al, Renewable power for sustainable growth, Springer Nature, Berlin, LNEE, 2020) and Kaggle (Tomar et al, Machine learning, advances in computing, renewable energy and communication, vol 768. Springer Nature, Berlin, LNEE, 2020) dataset. At first we gather the images where face have actual mask on it and also augmented the image with editing the image of unmasked face with mask so that model can learn very details of the image and result will come more accurate and clean.

Keywords Face masks · MobileNetV2 · OpenCV · CNN · Facial recognition

A. Chauhan (✉) · L. Khurana · Avinashwar · P. Singh
Department of CSE, AmitySchoolofEngineering andTechnology, Amity University, Noida, Uttar Pradesh, India
e-mail: arunchauhan414@gmail.com

M. Diwakar
Department of CSE, Graphic Era Deemed to Be University, Dehradun, India

A. Tripathi
Department of CSE, UPES, Dehradun, India

A. Tomar et al. (eds.), *Proceedings of 3rd International Conference on Machine Learning, Advances in Computing, Renewable Energy and Communication*, Lecture Notes in Electrical Engineering 915, https://doi.org/10.1007/978-981-19-2828-4_6

1 Introduction

To maintain the rules of social distancing computer vision and AI can be used at these hotspots, by checking the number of people wearing masks and people maintaining a certain distance from each other, which otherwise is a difficult job for civil guards and officials. With the settling of this virus, it would be very difficult without computers and AI to monitor the aftermath.

This paper comparatively analyze different models for the detection based on a number of conditions and factors on the basis of which comparisons have been drawn and concluded the best model and approach for the same. We have extended the use of this approach to real-time video monitoring too. Face mask detection is broadly a two part process, first is the detection of human faces in then extensive streams of rich digital images and videos and then classifying the masks present on the images/streams after their successful detection, unlike other traditional approach this one also focuses more on how properly the mask is being used. We have also touched upon the shorthand problems of detecting face masks.

It makes face mask detection easier and usable by non-technical background users as well. Multiple datasets are used in this paper. This approach can be used in real-time face mask detection, it can also classify the accuracy of the mask on the face too.

2 Methodology

For the comparative analysis of Face mask detection models, we have selected three major models that are used widely with many different types of optimizations that are used for different forms of detections and classification problems. These selected models are the first approaches when dealing with detection problems. Performance and computational abilities under different parameterized conditions of these selected models are tested and investigated although they have shown beyond doubt, very effective results. With the help of cascades in OpenCV we can carry out the capturing of faces from streams of different input data. OpenCV haarcascades are trained with the help of several negative and positive images for easy and accurate detection of facial features and faces. Haarcascades have simpler architectures and works fine with multiple scales.

2.1 Models

2.1.1 CNN

The convolution layer is the building block for CNN. Features are extracted from the input datasets. Starts with an input image [1]. The structure of CNN's looks like brain neurons. Works by allocating weights or labels to segments of the input visual image applies many different filters to it to create a feature map. Applies a ReLU function to increase non-linearity. Applies a pooling layer to each feature map. Flattens the pooled images into one long vector. Connect Artificial network is inputted with vectors.

2.1.2 LSTM

In all problems dealing with the prediction of sequence, Long short-term memory networks, a.k.a LSTMs have sure shot proved to be the most effective deal [2]. The LSTM model can choose data to be kept on the basis of relevance and which information should be binned. It is an upgraded model of RNNs that handles issues like an explosion and gradient vanishing.

2.1.3 MobileNetV2

MobileNetV2 is an upgraded and reformed version of MobileNetV1 and exponentially improves the classification of objects, visual recognition, detection and segmentation to another level. MobileNets are forms of neural networks supported by Google made for systems with lesser computational supports [3]. They strive to provide the best accuracy while justifying the memory and GPU support thus making them the fattest networks in this family. MobileNets does so by cutting down dramatically the number of learnable parameters, which also allows them to be trained easily and effectively within time constraints [4]. MobileNetV2 is based on transfer learning which is easy to implement with tensor. Transfer learning allows the MobileNetV2 which has been pre-trained to be used for detecting the features. MobileNetV2 is a light weight model with faster capabilities.

2.2 Creating Database

2.2.1 RMFD Dataset

RMFD [5] is the biggest dataset created in this field of research. This particular dataset consists of masked faces images and unmasked face images (5000 masked,

90,000 unmasked). To have a balanced dataset we had chosen an equal number of masked and unmasked images (5000 each).

2.2.2 Kaggle Dataset

This dataset contains 853 images. The images are diverse and include subjects in different poses and different conditions. These images also contain multiple subjects together along with noisy objects too. This dataset is divided into 3 major classes [6]. The images in this dataset are cropped to regions of facial features showing clear faces. Both the datasets are treated to augmentation for a standardized dataset. Specifically for the true testing of the model, a new database was created with the help of OpenCV module through which 327 images were captured both with and without masks.

3 Analysis Approach

These major issues were dealt with different approaches like selecting the dataset which stimulated similar conditions and a CNN layer that could work on the input image sizes for the model to work on [7]. Since our data will be fetched from secondary devices like cameras and raspberry Pi. We need an easy to use interface that could easily work with our model. For the same, We will use OpenCV to feed the live data work with our optimized model. With testing the models on live streams, real world proposition could be drawn easily as a number of noisy elements and conditions were introduced such as lighting conditions, facial movements, shared frames and movements [8–14].

4 Comparative Observations

4.1 Advantages and Disadvantages

See Table 1.

4.2 Experimental Results

See Table 2.

Table 1 Model advantages and disadvantages

Model	Advantages	Disadvantages
CNN	1. Flexibility allows easy use for Classification problems 2. Works fine with non-linear data	1. Segmentation computation is large 2. GPU support required
LSTM	1. Can predict given lags in time of a sequence 2. Is more insensitive to gap lengths	1. Large training time before being implemented in real life 2. LSTMs are common to the issue of overfitting
MobileNet V 2	1. Lightweight model 2. Non-linearities are removed in layers	1. Relu6 is unbounded and non-differentiable at zero 2. Can produce dead neurons in layers

Table 2 Experimental result

Model	Accuracy (%)
CNN	96.07
LSTM	97.11
MobileNetV2	97.23

This Histogram plots the accuracy of three models on the same datasets. MobileNetV2 marked the maximum accuracy percentage with LSTM close to the value achieved.

CNN performed unexpectedly well with changes in poses and conditions like lighting. With multiple variations of images passed as input, The unique model of MobilNetV2 is much faster compared to the CNN version (Figs. 1 and 2).

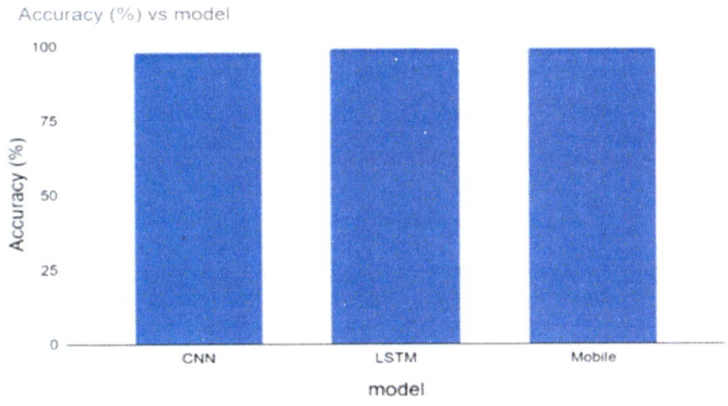

Fig. 1 Histogram of accuracy

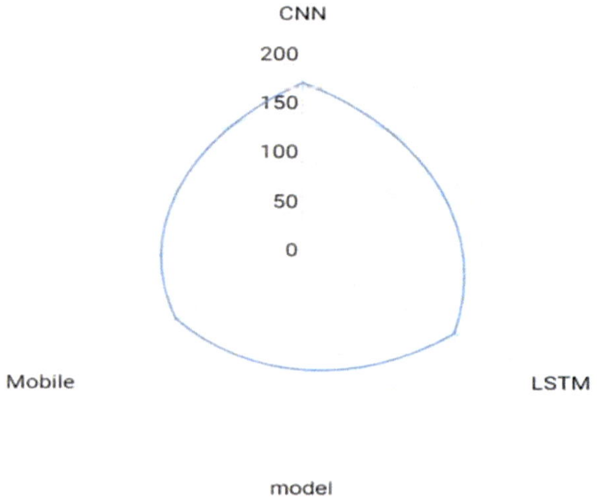

Fig. 2 Radar graph for model losses

5 Conclusion

The three chosen models have been compared on a number of factors, In the tests on similar datasets, it was observed MobileNetV2 showed a higher percentage of accuracy in the detection of face masks even with diverse subjects and provided conditions. However, all of these models can also be optimized by the addition of new classifier layer and fine-tuning them to increase the computational scope and making them more effective in the field of detection and classification.

References

1. O'Shea KT (2015) An introduction to convolutional neural networks
2. Sherstinsky A (2018) Fundamentals of recurrent neural network and long short-term memory (LSTM) network
3. Sandler M, Howard A, Zhu M, Zhmoginov A, Chen L-C (2018) MobileNetV2: inverted residuals and linear bottlenecks
4. Tsang S (2019) Review: MobileNetV2—light weight model (Image Classification)
5. Wang Z, Wang G, Huang B, Xiong Z, Hong Q, Wu H, Yi P, Jiang K, Wang N, Pei Y, Chen H, Miao Y, Huang Z, Liang J (2020) Masked face recognition dataset and application
6. Larxel (2020), https://makeml.app/datasets/mask
7. A 3D convolutional neural network method for land cover classification using LiDAR and multi-temporal Landsat imagery—scientific figure on ResearchGate. Available from: https://www.researchgate.net/figure/Flow-chart-of-3D-CNN-based-multi-stage-classification_fig3_327535503
8. Adrian Rosebrock (2018), https://www.pyimagesearch.com/2018/09/24/opencv-face-recognition/
9. Takahashi R, Matsubara T, Uehara K (2017) Multi-stage convolutional neural networks for robustness to scale transformation

10. Datahacker, https://datahacker.rs/001-how-to-read-a-video-and-access-a-webcam-with-opencv-in-python/
11. https://github.com/X-zhangyang/Real-World-Masked-Face-Dataset
12. https://www.kaggle.com/andrewmvd/face-mask-detection/metadata
13. Iqbal A et al (2020) Renewable power for sustainable growth, vol 723. Springer Nature, Berlin, LNEE, 805 p. https://doi.org/10.1007/978-981-33-4080-0 (ISBN 978-981-33-4082-4)
14. Tomar A et al (2020) Machine learning, advances in computing, renewable energy and communication, vol 768, 659 p. Springer Nature, Berlin, LNEE. https://doi.org/10.1007/978-981-16-2354-7 (ISBN 978-981-16-2354-7)

Analytical Approach Toward Face Mask Detection Using Artificial Intelligence Techniques

Avinashwar and Sanjay Kumar Dubey

Abstract By continuous hike of the deadly COVID-19 pandemic, the lifestyle of an individual has switched and changed all over the globe. Every individual has found it necessary to use a face mask in these situations. Identifying individual is wearing a face mask is very challenging due to wave of the deadly COVID-19 pandemic. The author proposed an approach in this study review work that would limit the evolution of the COVID-19 virus by personal identification who is not covering up any face mask. Many pieces of research have showed that wearing a mask reduces the possible chance of viral transmission of this life-threatening coronavirus and provides a sense of protection. The research during this zone has hiked over the past years. A typical review of the literature is studied to evaluate whether or not human beings are wearing masks, and based on these reviews, a modified analysis is done to detect which approach is feasible. This review included various search methodologies, too many research papers were recognized out of which seventeen are relevant papers. This paper will assess the research progresses related to the facial masks of an individual. It also helps the author to review out the ongoing and the forthcoming scenario of this research which have been working on facial mask detection using artificial intelligence.

Keywords Artificial intelligence · Face mask · Facial recognition · Computer vision · Neural network

1 Introduction

These days, the number of cases due to COVID-19 is hiking rapidly. Wearing a mask is mandatory with following and maintaining physical distancing. It is becoming

Avinashwar (✉) · S. K. Dubey
Amity University Uttar Pradesh, Sec-125, Noida, India
e-mail: avinashwarsharma@gmail.com

S. K. Dubey
e-mail: skdubey1@amity.edu

A. Tomar et al. (eds.), *Proceedings of 3rd International Conference on Machine Learning, Advances in Computing, Renewable Energy and Communication*, Lecture Notes in Electrical Engineering 915, https://doi.org/10.1007/978-981-19-2828-4_7

difficult for officials and guards to get control over the situation as a whole. With the deep-rooted virus, the difficulty increases, even more, to detect whether human being has put on a facial mask or no facial face. Every day great number of people have been detected for this deadly COVID-19 virus. As per the world health organization high-body temperature, tickle in throat, loss of taste, and odor are the sign of deadly virus. To avoid the extend of this virus, wearing a facial mask is necessary. To cope up with this unforeseen situation and to stop this global pandemic, artificial intelligence is advancing and contributing a lot through its various prominent branches. The main idea of artificial intelligence revolves around the nature of humans and animals that how they respond to a certain stimulus along with their decision-making abilities [1–6]. Out of those branches of AI, the most trending and useful one is named "computer vision". Vision of computer deals with the science of making machines (like computers, mobile phones, or any other suchlike devices) visually active or enabled. So, these machines can mark as identify the sphere of real-world events and tracking their current activities just like a human eye do and that can also be called a biological camera. When we provide input to our machine and in return, it detects and visualizes any images, then the concept of computer vision starts activating. Computer vision trains the model to interpret real-world phenomena on a visual basis. Its datasets consist of images that are contributed, and in interest, the output is produced according to the nature of the replication. This review paper has been categorized as follows; the paper includes a background of research in Section I and a literature review in Sect. 2. Section 3 includes research methodology which discusses the papers used to bring out the beneficial data for review. Section 4 comprises of analysis which includes the requirement of relating dataset of images for COVID-19 precautions and the importance of our review. Section 5 includes the result, followed by a conclusion in Sect. 6. After this paper, we will come to know about the best possible model in image classification.

2 Literature Review

This dreadful COVID-19 contagion is increasing all over the world perpetually. The consequence of this deadly virus is there on almost all individuals. Healthcare management is suffering a lot in this situation of crisis. The system contains a facial mask identification model where an algo like deep learning was used to imply the presence of a mask on a person's face. This model must be trained using labeled data, with the pictures constituting face images with and without masks [7]. For face mask detection, the author used a machine learning system. With the framework's training, validation, and impact evaluation, the model accurately anticipates the majority of individuals who use face masks [8]. The facial mask detecting model named SSDMNV 2 contains the training, as well as the evolution of the dataset of images, provide, which split into categories of individuals wearing masks or not and it uses an approach like OpenCV neural networks to generate fruitful results [9]. The author used the Retina facial mask model, which identifies face masks and is widely regarded as one of the first specialized facial mask detectors. This network design employs various mapping features and then use a network known as the feature pyramid to fuse the high-level information. Retina facial mask is a precise and systematic facial mask detector. It is a phase detector with a network-like feature pyramid and a distinct context module that focuses on recognizing face masks 4G-mask would be a detection and segmentation utility of facial into one frame whose aim is to obtain finer Information of face [10]. Extraction of features is done using ResNet-101 and generation of ROIs uses RPN. GIoU is a bounding box loss function to enhance detection accuracy [11]. Convolutional neural network is a model which uses TensorFlow with libraries like Keras and OpenCV to identify if an individual wearing a facial mask to protect itself [12]. The author developed a framework for facial masked detection that uses computer vision and strategies such as deep learning to diagnose. This model was used in conjunction with DL and ML approaches such as OpenCV, TensorFlow, and Keras. The reviewer used deep learning for feature extraction and combined it with three machine learning methods [13]. An explanation segmentation model for facial detection uses image classification in which each pixel is classified as face or non-facial, i.e., it builds a binary classifier and then identifies that divided region. A fully convolutional network is used to guide the semantic segmentation of the human face [14]. We have implemented an M-CNN model to detect. We have edified the model in such a way that it will work in varying dimensions and light conditions very effectively. This model will be used further by any regulating official to control the spreading of COVID [15]. The author developed an integrated computer vision and DL, which unifies models like deep learning and computations like ML using techniques like OpenCV, TensorFlow, and Keras.

It extracts features using deep transfer learning and then blends them with three machines learning traditional algorithms [16]. Adaptive boosting is a ML algorithm. It is a trans algorithm that works in tandem with several other algorithms to improve their performance [17]. SRCNet approach for distinguishing facial masks in conjunction with SR network and classification network. The recommended approach included four major steps: picture pre-processing, facial detection, and identification of mask-wearing situations [18]. Framing of this paper is done using MobileNetV2 and a technique like knowledge engineering which is implemented alongside the PyTorch and OpenCV of Python. We get a good output model that detects individuals with or without masks [19]. This proposed model has been efficient out in more than one stage, the 1st one identifies the masks that are covering a huge portion of the face, and the 2nd identifies the face that is not there in the training dataset. It applies an algorithm named GAN and uses celeb for training the data and achieved high accuracy [20]. The proposed facial detection is named LLE-CNN, which executes six state-of-the-art faces in detecting masked faces [21]. A special cascade which works on more than two layers of convolution neural network for the identifying facial mask. It uses a self-invented dataset because the already present face dataset is not sufficient to gage algorithms more categorically [22]. The author uses an algorithm like machine learning to detect facial masks and deep models of machine learning to extract features [23].

3 Review Methodology

This method is linked to a literature review that focuses on numerous research topics in order to discover, assess, choose, and include all of the high-quality research that is required for the study. Its goal is to offer a fair assessment of the review issue through the use of a reliable, rigorous, and auditable process. A lot of the previous papers on these related topics are reviewed at different stages. The review's prerequisites are specified, questions are identified, and the review procedure is explained during the preparation stage. In conducting the review stage, the preliminary studies are selected, the quality estimation used to include studies is defined. Papers before 2019 were not considered due to not much work was there in this area (Table 1).

4 Research Questions

4.1 Review Methodology

The goal of our analysis is to detect the concern and provocation faced while detecting facially masked during this deadly coronavirus using artificial intelligence. During our study, we came across various questions, and some of the questions we came across while reviewing many papers.

RQ1. Is it difficult to ascertain whether or not human being is wearing a face mask using machine learning?

RQ2. Which model is best for detecting an individual facial mask and give more fruitful results?

These review questions allow to identify and detect whether a human being is wearing a facial mask or not using ML techniques as the existing review is more conservative in the full systematic review as it did not integrate various other references such as working paper and Ph.D. thesis (Fig. 1).

4.2 Identification and Selection of Primary Studies

The primary review of this fact finding was picked for research from Springer, IEEE digital libraries, and several conferences. Because this is a fairly new and continuing issue, we have limited our study to the papers that are available online. In the very first stage, various keywords like artificial intelligence, computer vision, face mask, facial recognition, neural network, and issues to identify whether an individual is covering his face with mask or not which were used to search in the libraries. This study covers the review of research papers and conferences like Springer, Institute of Electrical and Electronics Engineering—IEEE, etc.

Table 1 Summary of literature review

Yr	Ref	Technique/too l	Model	Dataset	Issues	Challenges	Analysis
2020	[7]	Deep learning	Retina face mask	Transfer learning		The absence of large datasets of masked faces	In this study, DL method is utilized to detect facial masks using the retina face mask model, which when combined with the cross-class object deletion technique, yields a 98.7% accuracy rate
2020	[8]	Image classification	Machine learning	Kaggle dataset			In this publication, an algorithm like machine learning is trained through the classification of images method is used to edify, validate, and test. It shows an accuracy of 98.65%

(continued)

Table 1 (continued)

Yr	Ref	Technique/too l	Model	Dataset	Issues	Challenges	Analysis
2020	[9]	Deep neural network	SSDMNV2	Prajna Bhandari		This dataset was compiled from a variety of other sources, and the images were carefully cleaned to improve the accuracy of the results	In this model ,.named SSDMNV2, technique of OpenV neural network was used to generate model. Deep learning method is used to generate fruitily results. Issue of copious inaccurate conjecture has been successfully detached from the model
2020	[23]	Cross-class object removal algorithm	Retina face mask	(MAFA)			In this paper, Retina face mask consists of resents as backbone and the neck is FPN and context attention as head shows an accuracy of 94.5%

(continued)

Table 1 (continued)

Yr	Ref	Technique/too l	Model	Dataset	Issues	Challenges	Analysis
2020	[11]	Neural network	G-masked	[FDDB] and choke point datasets		Speed not soon good of the proposed method	Method G-masked was proposed in this research for face recognition and segmentation to determine whether or not an individual is wearing a face mask. It received a score of 95.97 percent
2020	[12]	Machine learning	Deep learning	OpenCV	High accuracy		In this publication, we use OpenCV, TensorFlow, keras PyTorch, and CNN to detect whether individual was wearing face masks or not. It shows accu racy of 97.86% on training set and 99.22% on test set
2020	[14]	Binomial cross entropy	Semantic segmentation	Multi-human parsing	Erroneous predict-ions		In this paper, binary face classifier can detect face in any orientation. accuracy of 93.884%

(continued)

Table 1 (continued)

Yr	Ref	Technique/too l	Model	Dataset	Issues	Challenges	Analysis
2020	[10]	1. convolution layer 2. max pooling 3. flatten 4. dropout 5. fully connected layer	M-CNN	OpenCV		The non-appearance of facial cues from the masked regions	The paper we proposed CNN architecture called m-CNN which work in varying dimensions and light conditions very effectively. It shows 91.5% accuracy
2020	[11]	Machine learning	1.Computer vision 2 Deep learning	OpenCV			In this paper, author used OpenCV, TensorFlow, keras, PyTorch, and CNN to identify an individual wearing a face mask or not shows 99.65% accuracy
2020	[12]		AdaBoost	OpenCV			In this paper adaptive boosting is used as a ma- chine learning algo with many other algo to improve their performance

(continued)

Table 1 (continued)

Yr	Ref	Technique/too l	Model	Dataset	Issues	Challenges	Analysis
2020	[13]	1. Image pre-processing 2. facial detection 3. cropping 4. image super-resolution 5. face- ask-wearing condition identification	(SRCNet)	The medical public dataset masks	We used dataset for face mask-wearing condition identification is relatively small	It does not comply with the standard video frame rate of 24 frames per second (fps)	In this paper, recognition method was put forward, which combines an SR with a classification network (SRCNet) for face classification of images. accuracy of 98.70%
2020	[14]	PyTorch OpenCV of Python	Deep learning	OpenCV	Tuning the hyperparamet er		The author of this research utilized OpenCV, PyTorch CNN to determine whether or not persons were wearing face masks
2020	[15]	1. Binary segmentation 2. RCNN	GAN	1.CelebA 2 WIDER FACE	Dataset with face mask and un face-mask is not available	It fails to remove the mask object completely	Image inspiring through GAN-based image-to-image translation approach to give possible results
2020	[16]	1. Boostin g 2 deform able part model (DPM) 3. convolutional neural network (CNN)				Incomplete and inaccurate features	In this paper ,author introduce dataset MAFA and LLE-CNN for face mask detection
2020	[17]		Gage algorithm		Self-created mask data-base		In this paper, author works on three layer of CNN to detect face mask

(continued)

Table 1 (continued)

Yr	Ref	Technique/too l	Model	Dataset	Issues	Challenges	Analysis
2021	[18]	Deep	Machine learning	Celeb a WIDER FACE dataset		Improper classification of images frames	In this paper, author uses ML to detect facial mask and deep machine learning for fracture extraction

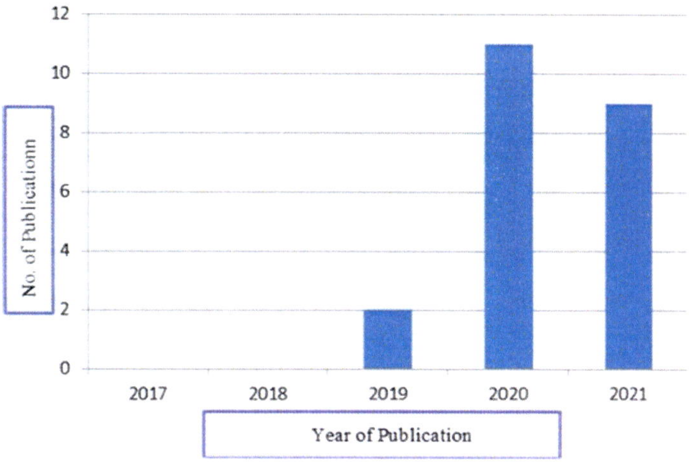

Fig. 1 Distribution of the selected papers according to their year of publish

4.3 Data Inclusion Criteria

After finalizing phase's review of these articles spanning the years 20019–2021, we compiled notes (word file) and preserved documentation of all the relevant keywords that we gathered during the research. To determine what to include and exclude from this study, we managed notes regarding each paper. These records incorporate the heading and the abstract that was obtained by reading many papers. In the next phase, we put in only those title and abstract which were linked to our study.

5 Overall Evaluation

The primary perspective of the review paper is to analyze the significance of detecting facial masks using artificial intelligence and review-related papers in this area. This part generates an overall assessment of the frame research question (Fig. 2).

RQ1. Is it difficult to identify whether a person is wearing a facial mask using machine learning?

After undergoing many research papers published in general, conference and digital libraries of IEEE explorer. It is notice that there is a quite variation in research in this area. From 2010 to 2015, the related work is not much but from 2019 to 2021, the work in facial mask using machine learning has increased as it is used to identify the facial mask of an individual and give more fruitful results.

RQ2.Which model is best for detecting an individual facial mask and give more fruitful results?

While studying many pieces of research during these last years, it is observed that there is a need for facial mask detection using artificial intelligence implementation

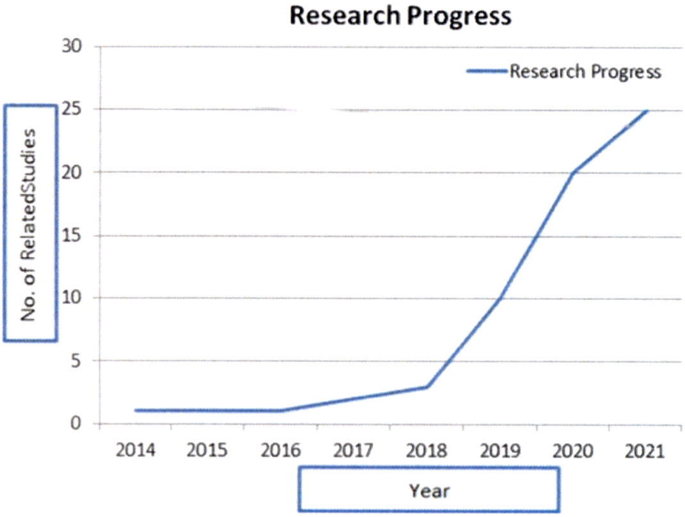

Fig. 2 Research progress (2014–21) for face mask detection

methods to stop the spread of this deadly coronavirus among the people as well as we need to maintain social distancing from each other.

6 Conclusion and Future Scope

Various methods and approaches to identify face mask detection and recognition were reviewed in this study. An analytical evaluation is performed for this reason, and reviewed research questions are included in this assessment. The selection of initial research learning and survey is also identified. The explanation and importance of the review methodology are also introduced in the paper. In this present paper, we try to detect out the respond to framed research questions. Combined various detection and identification techniques in artificial intelligence. Haar like features are used to recognize an object as its name is like Haar wavelets while it was first used in real-time facial detection as it has good calculating speed. Ada boost is less adaptive to overfitting problems than learning algorithms as it is sensitive to noisy data. The face mask detection process is very challenging. The deep learning-based method has high-recognition performance MobileNetV2 performed admirably in terms of extracting features and detecting an item. MobileNetv2 offers a highly effective mobile model that will serve as a foundation for much facial recognition technologies. This study effort, to the best of researcher knowledge, teaches us about the difficulty of face mask recognition and several techniques during the fatal COVID-19 outbreak. This is worth in saying that current research is not up to mark for COVID-19 pandemic period because many people are aware and take care of their health and wear masks to protect themselves against this virus. This method can be used in public areas with immerse system like at airport, stations, offices, school, and public places to make sure safety rules been followed.

References

1. Gupta GK, Sharma DK (2021) Depression detection on social media with the aid of machine learning platform: a comprehensive survey. In: 2021 8th international conference on computing for sustainable global development (INDIACom). IEEE, pp 658–662
2. Pradhan R, Maheshwari D, Aggarwal M, Chaturvedi A, Sharma DK (2020). IPL: from lens of data science. In: International conference on internet of things and connected technologies. Springer, Cham, pp 366–378
3. Garg S, Sharma DK (2020) Phony news detection using machine learning and deep-learning techniques. In: 2020 9th international conference system modeling and advancement in research trends (SMART). IEEE, pp 27–32
4. Agarwal Y, Katarya R, Sharma DK (2019) Deep learning for opinion mining: a systematic survey. In: 2019 4th international conference on information systems and computer networks (ISCON). IEEE, pp 782–788
5. Iqbal A et al (2020) Renewable power for sustainable growth, vol 723, 805 p. Springer Nature, Berlin, LNEE. https://doi.org/10.1007/978-981-33-4080-0. (ISBN 978-981-33-4082-4)
6. Tomar A et al (2020) Machine learning, advances in computing, renewable energy and communication, vol 768, 659 p. Springer Nature, Berlin, LNEE. https://doi.org/10.1007/978-981-16-2354-7 (ISBN 978-981-16-2354-7)
7. Rahman MM, Manic MMH, Islam MM, Mahmud S, Kim JH (2020) An automated system to limit COVID-19 using facial mask detection in smart city network. In: 2020 IEEE international IOT, electronics and mechatronics conference (IEMTRONICS). IEEE, pp 1–5
8. Nagrath P, Jain R, Madan A, Arora R, Kataria P, Hemanth J (2021) SSDMNV2: a real time DNN-based face mask detection system using single shot multi-box detector and MobileNetV2. Sustain Cities Society 66:102692
9. Sanjaya SA, Rakhmawan SA (2020) Face mask detection using mobileNetV2 in the era of COVID-19 pandemic. In: 2020 international conference on data analytics for business and industry: way towards a sustainable economy (ICDABI). IEEE, pp. 1–5
10. Suresh K, Palangappa MB, Bhuvan S (2021) Face mask detection by using optimistic convolutional neural network. In: 2021 6th international conference on inventive computation technologies (ICICT). IEEE, pp 1084–1089
11. Lin K, Zhao H, Lv J, Li C, Liu X, Chen R, Zhao R (2020) Face detection and segmentation based on improved mask R-CNN. Discret Dyn Nat Soc 2020. https://doi.org/10.1155/2020/9242917
12. MKM (2021) COVID-19: face mask detection using tensorflow- and openCV. In: Medium. https://towardsdatascience.com/covid-19-face-mask-detection-using-tensorflow-and-opencv-702dd833515b. Accessed 29 Apr 2021
13. Dey SK, Howlader A, Deb C (2021) MobileNet mask: a multi-phase face mask detection model to prevent person-to-person transmission of SARS-CoV-2. In: Proceedings of international conference on trends in computational and cognitive engineering. Springer, pp 603–613
14. Meenpal T, Balakrishnan A, Verma A (2010) Facial mask detection using semantic segmentation. In: 2019 4th international conference on computing, communications and security (ICCCS). IEEE, pp 1–5
15. Rao T, Dev S, Dileep P, Ram M (2021) A novel approach to detect face mask to control covid using deep learning. In: Ejmcm.com. https://ejmcm.com/article_2807_b1e004fc8cf0f8080144eb4707a0b85a.pdf

16. Bhadani A, Sinha A (2020) A facemask detector using machine learning and processing technique, p 13
17. Kalas MS (2014) Real time face detection and tracking using opencv. Int J Soft Comput Artif Intel 2(1):41–44
18. Qin B, Li D (2020) Identifying facemask-wearing condition using image super-resolution with classification network to prevent COVID-19. Sensors 20:5236. https://doi.org/10.3390/s20185236
19. Basha C, Pravallika B, Shankar E (2021) An efficient face mask detector with pytorch and deep learning. EAI Endorsed Trans Perv Health Technol 7:167843. https://doi.org/10.4108/eai.8-1-2021.167843
20. Loey M, Manogaran G, Taha MHN, Khalifa NEM (2021) A hybrid deep transfer learning model with machine learning methods for face mask detection in the era of the COVID-19 pandemic. Measurement 167:108288
21. Ge S, Li J, Ye Q, Luo Z (2017) Detecting masked faces in the wild with lle-cnns. In: Proceedings of the IEEE conference on computer vision and pattern recognition, pp 2682–2690
22. Bu W, Xiao J, Zhou C, Yang M, Peng C (2017) A cascade frame- work for masked face detection. In: 2017 IEEE international conference on cybernetics and intelligent systems (CIS) and IEEE conference on robotics, automation and mechatronics (RAM). IEEE, pp 458–462
23. Faizah A, Saputro PH, Firdaus AJ, Dzakiyullah RNR (2021) Implementation of the convolutional neural network method to detect the use of masks. Int J Inform Inf Syst 4(1):30–37

Kubernetes and Docker the Star Duo of Container Culture

Priyansh Pathak and Prabhishek Singh

Abstract From the dawn of the computer age we are trying to improve our technologies as much as we can and this search for improvement lead us to a new domain called cloud computing Under this domain, we found Virtualization. Virtualization has taken the industry by storm it has led the industry on a way to reduce the cost of app deployment and to use the resources very fruit fully. After the era of virtualization, we have the entered the containerization it nearly works on the same concept but it is very light weight than a virtual machine because containers shares a single host and hence it loses 99% of its weight and has only the most important dependencies thus getting a bundle of most important things and hence will help us to run many application on a single server. This chapter focuses on two major Stars of virtualization world Kubernetes and Docker. The chapter gives an insight in the beautiful of world of Containers, Kubernetes and Docker it is an exciting journey through the topic of container orchestration and creation.

Keywords Docker · Kubernetes · Containers · Virtual machine · Kubernetes architecture · Docker architecture · Docker swarm

1 Introduction

Docker and Kubernetes have revolutionized the cloud computing world they have shown us a much cleaner and efficient way of developing and managing apps. I was introduced to this amazing world of container creation and organization through google cloud platform this peaked my interest that how we are able to use these containers to deploy any application we like how we are using server resources so efficiently this efficiency made me fall in love with this concept. They have solved the major problem in the industry by providing properties like self-healing, scalability, and managing containers which have been a difficult task for companies for decades. Like many companies, shift their focus on virtualization and containerizing their

P. Pathak (✉) · P. Singh
Amity School of Engineering & Technology, Amity University Uttar Pradesh, Noida, India
e-mail: priyanshpro7072@gmail.com

A. Tomar et al. (eds.), *Proceedings of 3rd International Conference on Machine Learning, Advances in Computing, Renewable Energy and Communication*, Lecture Notes in Electrical Engineering 915, https://doi.org/10.1007/978-981-19-2828-4_8

data. Kubernetes and Docker being open source has given the companies a very versatile software to work with. This chapter is going to explore the mesmerizing world of containerization [1–23].

The chapter mainly focuses on Docker and Kubernetes Ecosystem. Docker and Kubernetes are selected for 3 reasons, firstly, the software's being open sourced they are accessible to every person from big industry leaders to a single developer who wants his application to be deployed in a cloud environment. Secondly, Docker and Kubernetes have successfully established themselves in the market as the leaders in container orchestration and creation they are also backed up by some big organizations hence perfect tools to tell the story of containerization. Third, Kubernetes and Docker containers are now synonymous with any cloud computing job. According to a survey conducted by "*Indeed*", the job shares for Kubernetes and Docker raised from 53 to 85% from October 2015 to October 2019 and there were not many experienced people to fill the position.

The jobs for containerization are growing very quickly the demand is much greater than the supply of the skilled personnel. So, to give an overview about how containerization came into play this chapter is written. Containerization is the current A-lister of the Cloud Computing World and this is going to be same for a long time. There are definitely some loop holes that should be patched for more smooth working of the concept and most of are patched and the rest problems are getting worked on but concept in its self is amazing to share the resources among the container thus making it light weight. The concept itself is brilliant but the tools used for making this concept work are much more impressive like Kubernetes and Docker. These tools make it easy to work on containers manage them and cerate them and keep them health.

Kubernetes being open source brings a lot to the table such as performance monitoring, dev tools, CI/CD workflows. This gives it an edge over other orchestration software and thus making it much more useable. Being open source also help Kubernetes to the get any recent issues resolved as there are many developers that work on to resolve the issues. Docker is perfectly tuned with container concept makes it most popular container development software its ability to pack, ship, and run the application in any environment. This give the ability of portability to the application which is very necessary now a days as there are many Operating Systems and environment available in the market this give the developer a peace of mind and ease to deploy its application any condition and environment. These two tools have helped industry to run many applications on a single server and provided the freedom to take it anywhere.

Chapter 1 referred is "*Building Modern Clouds: Using Docker, Kubernetes and Google cloud Platform*" by Jay Shah and Dushyant Dubaria The chapter gave a overview about the architecture of Docker and Kubernetes but was not able to discuss a theory of clustering and in depth analysis of services and also how it was different from its predecessor virtual machine which has been discussed these topics are discussed in this chapter. The chapter contains 4 section the first section contains Docker, its Architecture, its Networking, with facts about how Containers are different from their predecessors Virtual Machines. The second Section contains

Kubernetes, its Architecture, Cluster and major differences between the two major cluster architectures, Services and deployment are also discussed. The third section contains a qualitative difference between two orchestration system, Docker Swarm and Kubernetes. The fourth section has the concluding paragraph of this chapter.

2 Docker

2.1 Docker Containerizations

Docker is a containerization software. Containerization is a way to incorporate different types of software on a single machine. It gives the user taste of different types of software without using a different machine. These softwares are kept in an isolated environment called container. Containers plays a very vital role in the Docker. Containers are isolated virtual environments there can be many containers on your hardware. The container can be made without Docker, but this piece of software makes it easier safer, and efficient to run containers. Docker mainly uses images in its container. Containers can be created, started, stopped, moved, and deleted using Docker Command Line Interface [CLI] or Application programming interfaces [APIs].

2.2 Docker Architecture

Docker consists of a Docker client which tells Docker daemon to do all the tasks such as creating containers, running them, and efficiently distributing them. This type of architecture is named client–server architecture. The components that make Docker Architecture are:

(1) Docker Client:

This is the user Access for the Docker. It can connect with a daemon on the same host or any other daemon on any other host through remote access. It gives the user command line interface (CLI) through which the user can run, stop, and create a build. Some of the basic commands are [4] -docker pull, -docker push, -docker run, -docker build, -docker builder, -docker tag, -docker exec, -docker image, -docker config.

(2) Docker Host:

The Host provides an executable environment to execute and run applications. The Host comprises images, containers, networks, daemon, and storage. Docker daemon is where all the CLI or REST API commands are executed. The Docker daemon pulls and builds containers image. Once the image is pulled it makes a container as directed by the client it builds a working model for the container by utilizing a set of instructions called a build file.

(3) Docker Registries:

The service that accommodates from where you can download, and store images is called Docker registries. This is the place where Docker repositories are that host one or more Images for Docker. There are many Docker public registries some of them are Docker Hub, Azure Container Registry, Google Container Registry,Amazon EC2 Container Registry.

(4) Docker objects:

There are various objects that comes under Docker host and these are used to construct and deploy an application. Some these objects are given below:

(i) Images:

Image is an object that is used containers. It contains metadata that defines how container is needing to be made and what all resources is needed by the container. These images are the core part of Docker Architecture because they are the one that build, customizes, and ship applications.

(ii) Containers:

These are isolated environment that mostly defined by the images used and any additional configuration added by the user. They are made possible by using virtualization and isolating properties of Linux kernel. They have access only with directories defined in the image unless other wise directed by the user (Fig. 1).

2.3 Docker Networking

Networking in Docker is provided by network driver there are two main drivers bridge and overlay. A user can also write its own network driver plugin. An administrator can define as many users define drivers. Docker also creates a bridge interface called docker0 that is in charge for carrying traffic from and to the container. The docker0 interface acts as a router for the container. The same interface can be used for traffic routing as well as internal communication. Docker provides three modes of networking: none, bridged and macvlan, overlay, host.

2.4 Container Versus Virtual Machine

Virtual Machines are the predecessor of containers hence a qualitative analysis has been done in the Table 1.

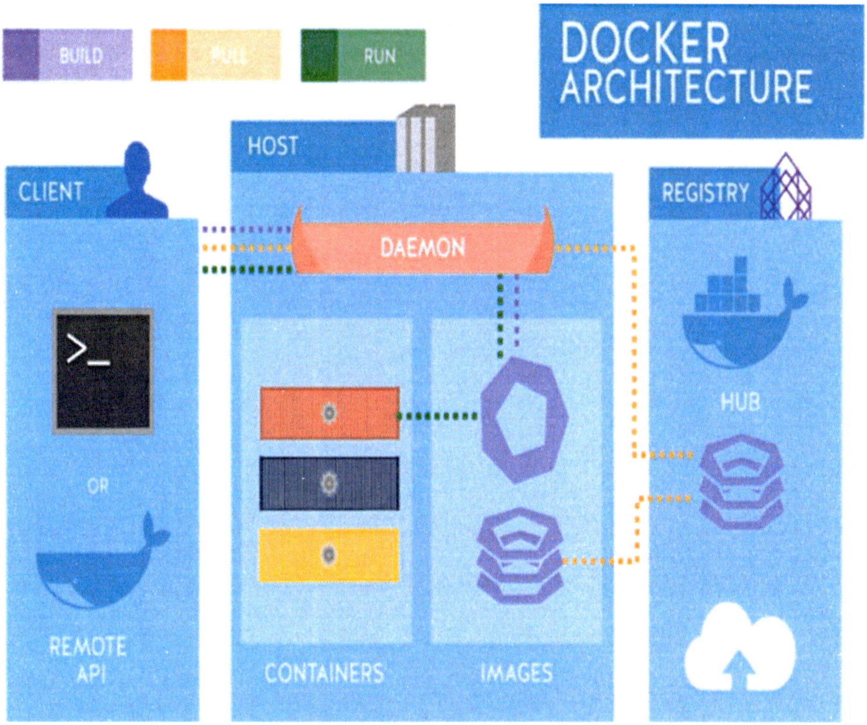

Fig. 1 Docker architecture visualization the three components of docker architecture (*Source* Docker Alternative by Aqua Security [https://www.aquasec.com/cloud-native-academy/docker-container/docker-alternatives/])

3 Kubernetes

Kubernetes is an open source container orchestration software when being an open source software its still rapidly growing. When the era of container deployment began industry felt a need of a better container management system hence Kubernetes was born (k8s). The need of Kubernetes was to get a system that is resilient while running distributed system thus making a fail-safe environment for the application running on the container Kubernetes also helps scaling the application thus using only the resources required by the application at any given time. Kubernetes and containers lie between the two platforms of Platform as a Service (PaaS) and Infrastructure as a Service (IaaS) thus giving us the best of both the platforms and giving us a new platform called Containers as a Service (CaaS).

Table 1 Comparison between containers and virtual machine

Differences	Containers	Virtual machines
Portability	As container consists only the dependencies of the system, they are highly portable	The size of the virtual machine is in gigabytes hence they are not as portable as a container
Resource usage	The code package in the container shares the resources including the OS hence managing the resources well	Resources allocated by the VM are always more than they need hence they waste and once the resources are allocated by the VM they cannot be used by any other resource
Maintenance and deployment	Containers are easier and faster to deploy and maintain as they are lightweight and work on a single OS	VMs having multiple OS And having a hefty Size is much harder to deploy and maintain
Security	Containers share the same OS hence having any fault can affect the whole system	VM is much more isolated as they have their single OS hence if one has faulted the others will not be affected

Note While container gives the user the ability to have better portability, resource usage, maintenance and deployment, Virtual machine gets an edge over container in security

3.1 *Kubernetes Architecture*

Kubernetes follows the same architecture as Docker called client-sever architecture. On the top the k8's architecture we have kubectl is Command line interface the comes the master node we can have a single master node of multiple depends on the need then comes the worker nodes (Fig. 2).

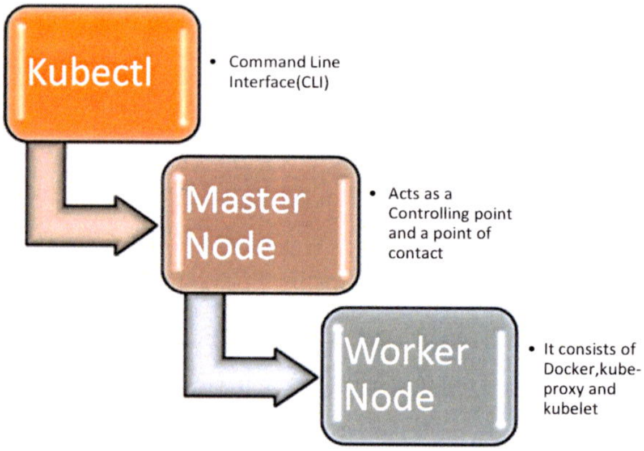

Fig. 2 Kubernetes architecture flow diagram showing the main components of kubernetes architecture (*Source* Author. Copyright Priyansh Pathak 7, August 2021)

1. kubectl: Its is the command line interface of the Kubernetes architecture it talks to kube-apiserver in master node. Some commands are kubectl get pods, kubectl get deployment, kubectl run, kubectl scale
2. Master Node: As the controlling node of the Kubernetes it has many necessary components and these components are listed below.

    (i) kube-apiserver: This the center of the master node architecture it communicates with kubectl (CLI) and the etcd cluster (make sure that data is stored in the etcd cluster or not) it gets the REST request for modification of the pods, services (running, stop, deleting, etc.). It has connection to every component in the Kubernetes architecture.
    (ii) etcd cluster: Kubernetes needed a space where they can store important data like pod name, state, number so the etcd cluster came in play. The only way to access this server is through kube-apiserver thus making it secure.
    (iii) kube-control-manager: This a manager provided in kubernetes to manage and regulate the state of a cluster. If any change in service configuration is done it spots it and start working on it so to get the changes incorporated.
    (iv) kube-scheduler: This is the decision maker ok the master node it tells where the pods will be on various nodes thus getting the best out of the available resources.

(3) Worker Node: This node is commanded by the master node and has the actual application that is accessed by the Internet its main components are

    (i) pods: It is a set of containers which together makes an application. It has application container, storage resources and a configuration how to run the containers.
    (ii) kube-proxy: It is the proxy sever that runs on each worker node this service exposes the node to internet.it does the work of request forwarding to the correct pod that has to be accessed in the cluster.
    (iii) kubelet: This main service makes sure that the pods are healthy. It makes sure that they are running in the desired state as directed by the kube-apiserver. This also tells the master node the health of the pods and where are they running.

3.2 Clustering via Kubernetes

Clustering is the heart and soul of the Kubernetes software. It makes the tiring task of managing nodes (which has pods that run containers) very easy and fault free. Depending on the application there can be several nodes that has to be run as single containerized application to give a seamless experience to the user. The max number of nodes allowed on Kubernetes environment is 5000. Cluster uses kubeconfig file for configuration it can be YAML or JSON file.

Table 2 Comparison between single and multi node architecture

Differences	Single node architecture (master node, compute)	Multi node architecture (master node, compute, etc)
Deployment	This the simplest architecture hence it is the easiest to deploy	This architecture has many master nodes hence a bit hard to deploy
Service resources	This architecture having a single node hence has less service resources	Having more nodes gives an edge to this architecture because it will have high service resources
Uses	Mainly used for testing purposes and non-production uses	This Architecture can be used in production app
Reliable	Having one master node is very unreliable as if the node fails you cannot access the worker-nodes	Here the nodes are always in the odd number to maintain the node majority hence if one more will fail user will still have $(n/2) - 1$ node left

Note While Single node is easiest to deploy and manage, Multi node gives us more reliable application

3.3 Cluster Architecture

There are mainly 2 types of Architecture as enumerated and a qualitative difference is done between them Table 2.

3.4 Deployment in Kubernetes

It is a resource object in Kubernetes that gives a declarative update to application. It gives the user the power to the user to define the application lifecycle (number of pods, image to be used) in a template form. The deployment makes sure that the no. of pods commissioned in the cluster keep running. The deployment gives us the freedom to scale the deployment, to pause and continues it, get back to previous version of failure. Thus, making these processes automated and easier for the client to manage (Fig. 3).

3.5 Services in Kubernetes

Services generally gives access to the set of pods when network access is requested through the service. It is a rest object hence its controlled through an API server there are many types of services offered in Kubernetes some of those are enumerated below:

Fig. 3 Yaml file for deployment service. The template is used to deploy nginx webserver named as priyansh-deployment (*Source* Author. Copyright Priyansh Pathak August 7, 2021)

```
  apiVersion: apps/v1
kind: Deployment
metadata:
  name: priyansh-deployment
  labels:
    app: nginx
spec:
  replicas: 3
  selector:
    matchLabels:
      app: nginx
  template:
    metadata:
      labels:
        app: nginx
    spec:
      containers:
      - name: nginx
        image: nginx:1.14.2
        ports:
        - containerPort: 80
```

(1) Cluster IP: This an internal service which provides virtual IP address to pods in a cluster so that they can communicate between them. This not an external resource hence the requests should be made from inside the cluster (Fig. 4).

Fig. 4 Yaml file for cluster IP service. The image describes a cluster IP service template name priyansh (*Source* Author Copyright Priyansh Pathak August 7, 2021)

```
apiVersion:     v1
   kind: Service
    metadata:
     name: priyansh
      spec:
       type: ClusterIP
 ports:
 - targetPort: 80
 port: 80
```

```yaml
apiVersion: v1
kind: Service
metadata:
  name: pathak
  labels:
    name: pathak
spec:
  type: NodePort
  ports:
    - port: 3036
      nodePort: 30036
      name: http
  selector:
    name: pathak
```

Fig. 5 Yaml file template for node port service. The above image a node port service named as pathak (*Source* Author. Copyright Priyansh Pathak August 7, 2021)

(2) Node Port: This Service is present on each node of the cluster and Kubernetes routes the traffic through the static port. Thus, handling the request that originate outside the cluster (Fig. 5).

(3) Load balancing: Load balancing is the technique to methodically and efficiently distribute the traffic on an application across multiple servers or in the case of Kubernetes nodes insuring that the application runs smoothly on the user's device. In Kubernetes this service must be out sourced through cloud providers such as (GCP, Azure, AWS). These cloud Service providers will make a load balancer that will be tied to Kubernetes service (Fig. 6).

4 Kubernetes Versus Docker Swarm

See Table 3.

5 Conclusion

Docker being a container creation software using containerization, it creates Linux containers, but containers are less secure then its predecessor Virtual Machine

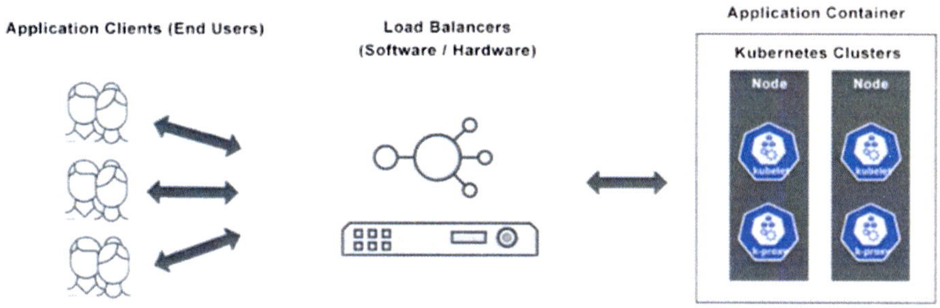

Fig. 6 Load balancing diagram. The diagram shows the location of the load balancer and its connection to the nodes in a live network (*Source* Kubernetes Load Balancer by AVI Networks [https://avinetworks.com/glossary/kubernetes-load-balancer/])

Table 3 Comparison between kubernetes and docker swarm

Features	Docker swarm	Kubernetes
Cluster deployment and management	The cluster deployment and management is fast and easy, but cluster is weak	The cluster management is tedious and slower, has a very strong cluster
GUI	Does not offer a GUI	Offers its own kubernetes dashboard GUI
Auto-scaling	Cannot do auto-scaling	Can do auto-scaling
Load balancing	Does auto load balance no need of manual intervention	Manual intervention is needed to for load balancing between pods and container
Scalability	Can do faster scalability than Kubernetes	Highly scalable

Note Docker Swarm has faster management, deployment and does auto balancing but Kubernetes has a GUI and can do auto-scaling

because it shares their host OS hence if there is a breach whole system is compromised. Kubernetes being the most famous container orchestration software, because of its open source code thus providing compatibility with many applications, it is cost and time effective with high scalability. While Kubernetes and Docker have separate purpose but when they are combined, they are much more efficient. Containerization and Virtualization is not the end of this evolution journey and we might get to see concepts like Unikernel getting into mainstream for cloud computing.

References

1. Bernstein D (2014) Containers and cloud: from lxc to docker to kubernetes. IEEE Cloud Comput 1(3):81–84
2. Vohra D (2016) Kubernetes microservices with Docker. Apress

3. Altaf U, Jayaputera G, Li J, Marques D, Meggyesy D, Sarwar S, Pash K (2018) Auto-scaling a defence application across the cloud using docker and kubernetes. In: 2018 IEEE/ACM international conference on utility and cloud computing companion (UCC Companion). IEEE, pp 327–334
4. Docker (2021). Retrieved 7 Aug 2021, from https://docs.docker.com/engine/reference/commandline/docker/
5. Mao Y, Fu Y, Gu S, Vhaduri S, Cheng L, Liu Q (2020) Resource management schemes for cloud-native platforms with computing containers of docker and kubernetes. arXiv:2010.10350.6; Marathe N, Gandhi A, Shah JM (2019) Docker swarm and kubernetes in cloud computing environment. In: 2019 3rd international conference on trends in electronics and informatics (ICOEI). IEEE, pp 179–184
6. Al-Dhuraibi Y, Paraiso F, Djarallah N, Merle P (2017) Autonomic vertical elasticity of docker containers with elasticdocker. In: 2017 IEEE 10th international conference on cloud computing (CLOUD). IEEE, pp 472–479
7. Vayghan LA, Saied MA, Toeroe M, Khendek F (2019) Kubernetes as an availability manager for microservice applications. arXiv:1901.04946
8. Overview (2021) Retrieved 7 Aug 2021, from https://kubernetes.io/docs/concepts/overview/
9. Pratama IPAE (2021) The implementation of container as a service (CaaS) cloud using openSUSE kubic. Global J Eng Technol Adv 6(1):001–009
10. Kubernetes Cluster Architecture and Considerations—Trident documentation (2021) Retrieved 7 Aug 2021, from https://netapp-trident.readthedocs.io/en/stable-v19.01/dag/kubernetes/kubernetes_cluster_architecture_considerations.html
11. Shah J, Dubaria D (2019) Building modern clouds: using docker, kubernetes & Google cloud platform. In: 2019 IEEE 9th annual computing and communication workshop and conference (CCWC). IEEE, pp 0184–0189
12. Kang B, Jeong J, Choo H (2021) Docker swarm and kubernetes containers for smart home gateway. IT Professional 23(4):75–80
13. Brewer E (2018) Kubernetes and the new cloud. In: Proceedings of the 2018 international conference on management of data, pp 1–1
14. What is a Kubernetes cluster? (2021) Retrieved 7 Aug 2021 from https://www.redhat.com/en/topics/containers/what-is-a-kubernetes-cluster
15. Deployments (2021) Retrieved 7 Aug 2021 from https://kubernetes.io/docs/concepts/workloads/controllers/deployment/
16. Modak A, Chaudhary SD, Paygude PS, Ldate SR (2018) Techniques to secure data on cloud: docker swarm or kubernetes? In: 2018 second international conference on inventive communication and computational technologies (ICICCT). IEEE, pp 7–12
17. Marathe N, Gandhi A, Shah JM (2019) Docker swarm and kubernetes in cloud computing environment. In: 2019 3rd international conference on trends in electronics and informatics (ICOEI). IEEE, pp 179–184
18. Rosen C (2021) Docker swarm versus kubernetes: a comparison. Retrieved 7 Aug 2021 from https://www.ibm.com/cloud/blog/docker-swarm-vs-kubernetes-a-comparison
19. Sawaya S (2020) If you've got it, flaunt it—kubernetes experience, that is. Retrieved 7 Aug 2021 from https://www.sdxcentral.com/articles/news/if-youve-got-it-flaunt-it-kubernetes-experience-that-is/2020/01/
20. Park S, Hwang K, Kim H, Shim B (2017) Advanced multimedia and ubiquitous engineering. Lecture notes in electrical engineering 448. FutureTech, pp 269, 276
21. Angrisani L, Arteaga M, Chakraborty S, Chen J, Chen TK, Dillmann R, Wu H et al Lecture notes in electrical engineering
22. Tomar A et al (2020) Machine learning, advances in computing, renewable energy and communication, vol 768, 659 p. Springer Nature, Berlin, LNEE. https://doi.org/10.1007/978-981-16-2354-7 (ISBN 978-981-16-2354-7)
23. Iqbal A et al (2020) Renewable power for sustainable growth, vol 723, 805 p. Springer Nature, Berlin, LNEE. https://doi.org/10.1007/978-981-33-4080-0. (ISBN 978-981-33-4082-4)

A Robust Gender Identification System for Speaker Recognition Using Linear Discriminant Analysis Stepwise Dimension Reduction

Atul Sharma and Sunil Kumar Singla

Abstract Gender identification is one of the important aspects to reduce computational time in speaker related applications because prior information of gender limits the search space. In this chapter, a robust gender identification system has been presented using a novel dimension reduction algorithm called Linear Discriminant Analysis Stepwise Dimension Reduction (LDASDR) which is based on a well-established Feature Selection algorithm called Linear Discriminant Analysis Stepwise Feature Selection (LDASFS). Three different feature sets namely, acoustic, cepstral and their combination have been separately used for the Neural Network classifier's input. For carrying out the task of gender identification, benchmark database named English Language Speech Database for Speaker Recognition (ELSDSR) has been used. The experiments have been performed both for the small and large databases. The results obtained with large training samples are better as compared with the results achieved with small number of samples. However, on applying the proposed algorithm on small sample size the classification accuracy of the system improves by 12% (approx.).

Keywords Acoustic feature · Cepstral feature · Dimensionality reduction · Gender identification · Speaker recognition · Sample size

1 Introduction

The speech signal contains the linguistic content as well as the other important information's of a speaker like gender, age and emotional state [1–26]. These informations are useful in many applications as shown in Fig. 1. Although, identification of gender is a different task, but when used before speaker recognition [3], more accurate speaker recognition models can be enabled. Moreover, in recognizing a speaker, identification of gender at first can improve the computation time as the

A. Sharma (✉) · S. K. Singla
Department of Electrical and Instrumentation Engineering, Thapar Institute of Engineering & Technology, Patiala, Punjab, India
e-mail: satul83@gmail.com

A. Tomar et al. (eds.), *Proceedings of 3rd International Conference on Machine Learning, Advances in Computing, Renewable Energy and Communication*, Lecture Notes in Electrical Engineering 915, https://doi.org/10.1007/978-981-19-2828-4_9

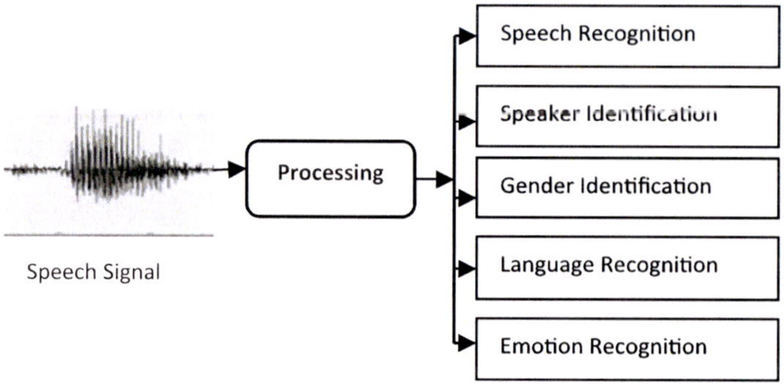

Fig. 1 Possible interpretations from speech signal

search space is limited to one gender only [4, 5]. In addition to these benefits, gender identification helps sorting telephone calls for gender sensitive surveys, providing telephone counseling in customer relationship management and making advertising gender-targeted [6]. Recently a priori gender identification system which improves the voice pathology detector accuracy has been reported [7]. Due to the above stated benefits, gender identification must be performed before the speaker authentication/recognition.

For identification of the gender, given a speech file, meaningful features are to be extracted that possess large interspeaker and small intraspeaker variations. Therefore, it is very critical to select effective features [8]. Moreover, inexact estimation of parameters while adding new features leads to increase in classification error [9]. Further, size of data for training and testing also affects the performance of gender identification system [10]. Gender identification prior to speaker recognition is a less explored field. Harb and Chen [4] identified gender for multimedia applications by applying Neural Network using features like pitch for clean speech and obtained 90% accuracy for 1 second speech files and this accuracy increased for 5 second speech files to 98.5%. Zimeng [11] assessed the gender with 40 speakers in a multi-language environment. An accuracy of 92.75% was achieved with Naive Bayes Classifier. Hu et al. [12] proposed a two stage classifier for identifying gender using features like Mel Frequency Cepstral Coefficients (MFCC) and pitch. MFCC along with Gaussian Mixture Models (GMM) incurs high complexity and computational overhead. Authors have reported accuracy in classification of 98.65%. When only pitch thresholding was used, they achieved accuracy of 96.85%. Parris and Carey [5] developed a gender identification system based on cepstral and pitch features using Hidden Markov Models (HMMs). They used Linear Discriminant Analysis (LDA) to normalize the model and reduce bias toward a particular gender. The information provided by the two features were combined on classifier level and as the LDA dimension was reduced the error rate increased. Bakir [13] trained HMM, Dynamic Time Warping (DTW) and Artificial Neural Network (ANN) with MFCC features from German speech data comprising 50 males and females each. An accuracy of 85.6 (females) and 80.3 (males) using ANN was achieved. Ahmad et. al. [14] used MFCC features along with five classifiers k-Nearest Neighbor (KNN), Multi-Layer Perceptron (MLP), Support Vector Machine (SVM), Naïve Bayes and Random Forest for gender recognition tasks. With the Korean speech data, effect of size of training data was explored. The performance degraded for small size of data as recognition rate with 20% of total data was around 65%. Gradual increase in performance of classifiers was observed with increase in length of speech. Performance of SVM classifier was the best among all the classifiers achieving an accuracy of 90%. However, they did not provide any method to counter the performance degradation in case of small training data sample size. Yusnita et al. [15] extracted LPC features from speech of 93 speakers and used ANN to recognize the gender, achieving a recognition rate of 93.3%.

Despite the potential applications for gender identification systems there are many conditions that have to be dealt with while designing such systems with high accuracy. These factors include but are not limited to sample size, choice of features, computational complexities, ease of identification and language independence. So, a system which identifies gender must furnish satisfactory and acceptable performance in all these scenarios along with language and text independency. In this proposed work, acoustic feature, cepstral feature and their combination has been compared for classification accuracy with small and large data set. Moreover, the dimensionality reduction algorithm LDASDR has been applied on the small and large data set to analyze the accuracy. Section 2 of this chapter presents the LDASDR algorithm while Sect. 3 presents proposed work. In Sect. 4, results have been presented with conclusion given in Sect. 5.

2 Linear Discriminant Analysis Stepwise Dimension Reduction (LDASDR)

LDASDR algorithm is used to reduce the dimension of feature set, such that maximum variability between classes and minimum variability among classes is maintained as it is done in LDASFS algorithm for feature selection [9]. Since the gender identification problem can be considered as a two-class classification, it can be formulated as a first-order linear multiple regression problem. The various steps of the algorithm are:

Step 1: *T*o each *k*-dimensional feature vector X_i, assign an output $c(i)$ like

$$c(i) = \begin{cases} c_1, ifi \in class1 \\ c_2, ifi \in class2 \end{cases} \tag{1}$$

where class 1 = male, class 2 = female and k denote the dimension of feature.

Step 2: Compute Discriminant score for a feature vector X_i using regression equation given in (2) which is the predicted value of $c(i)$

$$h^k(X_i) = b^T X_i + b_0 \tag{2}$$

where $b^T = [b_1, b_2, \ldots, b_k]$ and b_0 denotes the regression coefficients.

***Step 3**:* Calculate the Wilks' Lambda coefficient (λ_k) for each value of k (dimension of feature) using the Eq. (3).

$$\lambda_k = \frac{\sum_{i \in class1} \left(h^{(k)}(X_i) - m_1^{(k)}\right)^2 + \sum_{i \in class2} \left(h^{(k)}(X_i) - m_2^{(k)}\right)^2}{\sum_{i=1}^{N} \left(h^{(k)}(X_i) - m^{(k)}\right)^2} \tag{3}$$

where m_1^k, m_2^k denotes the mean of the class 1 and class 2s Discriminant scores, respectively, $m^{(k)}$ denotes the Discriminant scores mean of class 1 and 2 together and N denotes the total samples.

Smaller Wilks' lambda means relatively large separation between classes and hence better classification.

Step 4: In stepwise dimension reduction, first of all, a F_{in} value is calculated for all the dimensions of a feature as given in Eq. 4. After that, that dimension is identified for which F_{in} is maximum, which enables in making the decision about the usefulness of that particular dimension for classification.

$$F_{in} = (N - k - 2)\left(\frac{\lambda_k}{\lambda_{k+1}} - 1\right) \tag{4}$$

In Eq. 4, λ_k is Wilks' lambda prior to considering the feature dimension, and λ_{k+1} is Wilks' lambda after considering the feature dimension.

Step 5: The maximum F_{in} calculated in step 4 is compared with a threshold value and if greater, then the particular dimension is considered otherwise discarded. In this work, threshold is taken to be 1 and when the same feature repeats itself over iterations then the algorithm stops.

The flowchart for LDASDR algorithm implemented in this work is shown in Fig. 2.

3 Proposed Work

To design a gender identification system first of all, different features are extracted for effective identification and then a way to combine them has been proposed to further enhance the accuracy. In the second phase, the effect of available data size on identification accuracy has been analyzed and a way to counter the problem of small data size has been proposed. The flowchart for the proposed system is given in Fig. 3.

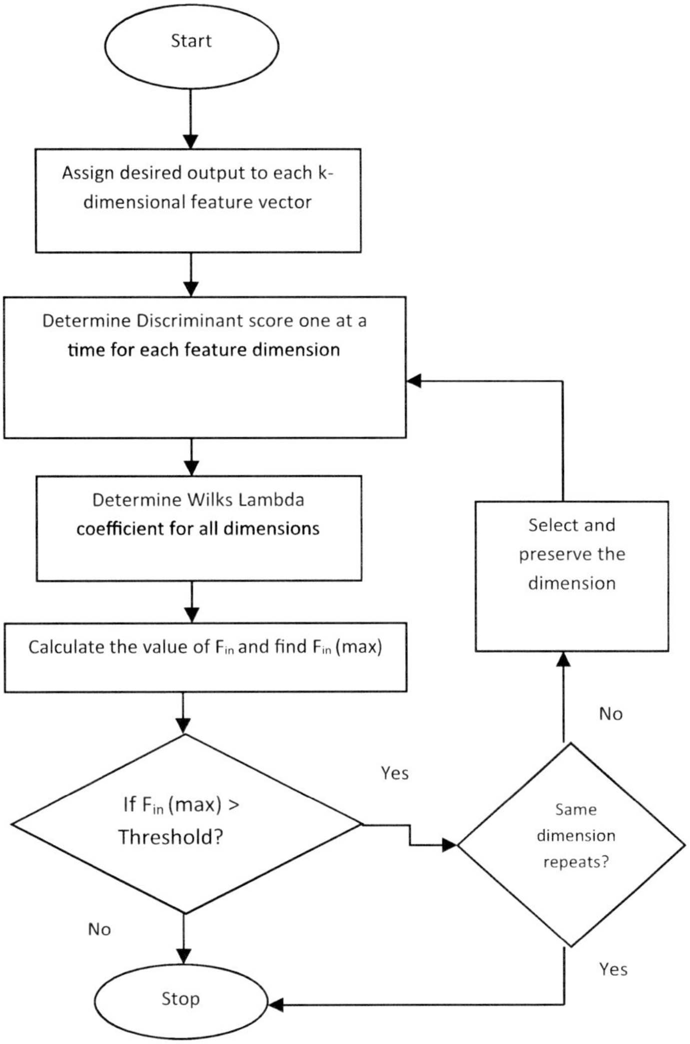

Fig. 2 Flowchart for LDASDR algorithm

As shown in Fig. 3, first of all a speech signal has been taken as an input. Then in the preprocessing step, since the silence portion in speech signal contains no information and hence it has been removed. Then two types of features namely, Acoustic (i.e., pitch) and Cepstral (i.e., MFCC) have been extracted for all the speech files in the database. Pitch or the fundamental frequency is decided by the rate at which vocal folds oscillate which further depends on vocal fold's parameters like muscle tension, mass and air pressure behind glottis in lungs and trachea. Pitch has been demonstrated to be the best discriminator among voices of both genders. The average pitch for males and females is typically in 100–155 Hz and 176–244 Hz range, respectively.

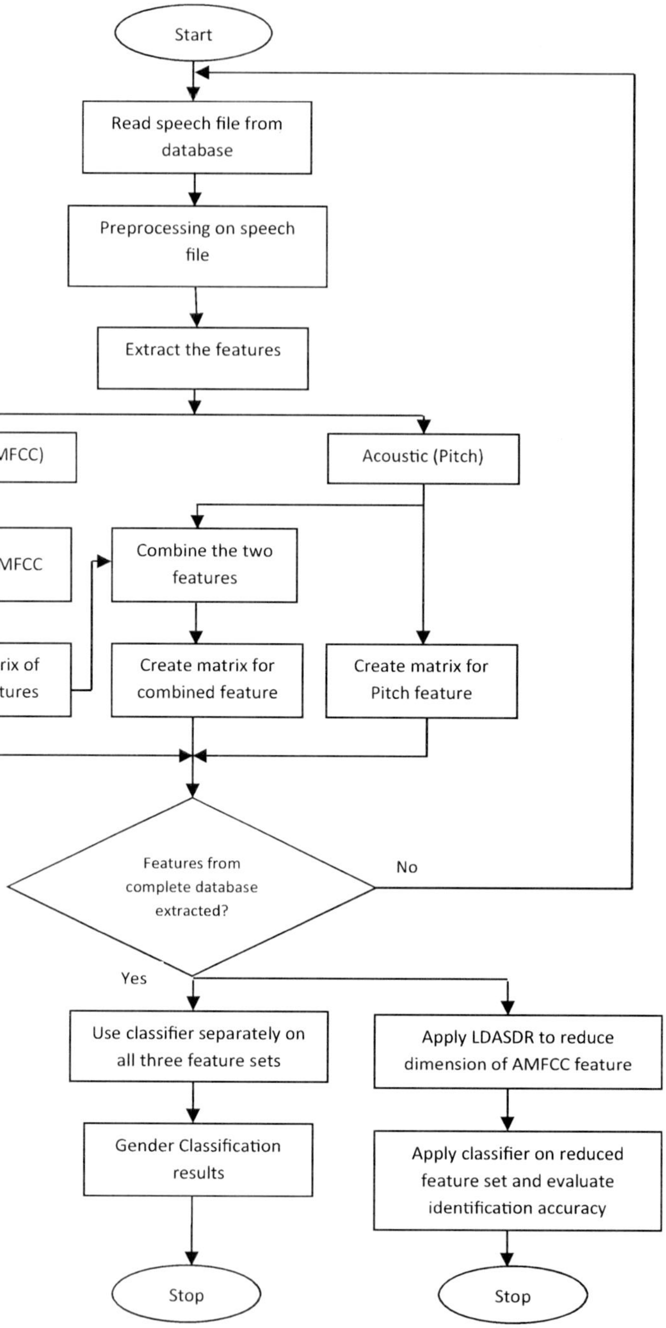

Fig. 3 Flowchart for the proposed method

Since males have massive vocal folds and longer vocal tract, the result is lower pitch for males than females [16]. As the speech signal is complex and non-stationary in nature, it is more viable to represent it in form of cepstral coefficients [17]. From speech signal MFCC features [18] have been extracted because it leads to a compact, stable and effective representation since they exploit human auditory model. These are state-of-the-art features since their introduction by Davis and Mermelstein [19]. MFCCs represent the mathematical model for shape of vocal tract. The mathematical relation between the actual frequency (Hz) and perceptual frequency (mel) is almost linear below 1000 Hz while it gets logarithmic above 1000 Hz. The relation governing physical frequency and mel frequency is given in (5)

$$\text{mel} = 2595 \log_{10}\left(1 + \frac{\text{f}}{700}\right) \tag{5}$$

In this work, MFCCs have been computed as the feature vector of each frame. As different speech files are of different lengths which leads to different number of frames, the averaged MFCCs over all frames in a speech file are used as features and called as Average Mel Frequency cepstral Coefficients (AMFCC) [20]. Therefore, the number of feature values is fixed regardless of the length of the acoustic file. After the extraction of acoustic (Pitch) and cepstral features (AMFCC) they have been combined to form a combination feature set (Pitch + AMFCC). Now, three different feature sets namely acoustic, cepstral and their combination have been separately used for the Neural Network classifier's input with 70% data for training and the rest 30% each for testing. Finally these trained classifiers have been used to classify the gender from the test samples. Since we have extracted pitch which is 1-dimensional feature and AMFCC which is 20-dimensional feature, further the LDASDR algorithm has been applied to the AMFCC feature to reduce its dimensionality and its effect on the gender identification accuracy has been analyzed for small and large databases.

3.1 Database

For carrying out the task of gender identification, benchmark database named ELSDSR has been used [21]. ELSDSR comprises speech files from 10 females and 12 males aged between 24 and 63 years [22].

In this work small sample comprises one speech file from 22 speakers and sample size is increased to 4 speech files from each 22 speakers leading to large sample size of 88.

4 Results and Discussion

Different feature sets namely, pitch, AMFCC and the combined feature set extracted from speech signals have been used for the experimentation purposes. Neural pattern recognition tool available in the MATLAB version13 has been used to perform the classification task. The experimentation has been performed on two sample sizes, small [S] and large [L], to check the effect of sample size on the recognition accuracy.

The results in Table 1 and Fig. 4 shows that both acoustic and cepstral features afford good recognition rate when sample size is large. However, a combination of the two different features lead to the better results on large sample size. But when the sample size is small then pitch feature gives better results than cepstral features.

Further, in second experimentation stepwise dimension reduction algorithm (LDASDR) has been applied to 20-dimensional AMFCC feature set obtained earlier and finally a 4-dimensional reduced feature set has been obtained. This reduced feature set has been used train the classifier and the gender identification results for small and large sample size databases have been given in Table 2.

Table 1 Recognition rate of three feature sets on different sample sizes

	AMFCC		Pitch		AMFCC + Pitch	
	S	L	S	L	S	L
Identification rate (%)	86.9	99.4	97.1	98.4	88.3	100

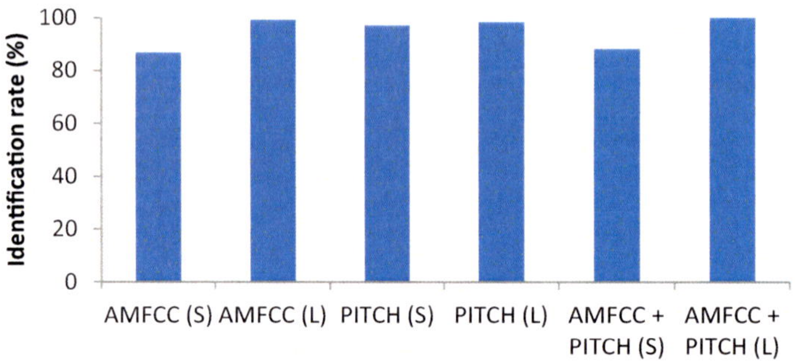

Fig. 4 Identification rate with different features using different sample sizes

Table 2 Recognition rate with and without LDASDR on different sample size

	Sample size (Small)		Sample size (Large)	
	Without dimension reduction	With dimension reduction	Without dimension reduction	With dimension reduction
Identification rate (%)	86.9	99.4	99.4	98.4

Results in Table 2 and Fig. 5 shows that when the sample size is small, then effective dimensionality reduction increases the identification rate by almost 12% (approx.) and accuracy achieved is similar to that of obtained with larger sample size. Although when sample size is large, then reducing dimensionality reduces the identification rate marginally.

Although, it is quite impractical to make comparisons among gender identification works due to variability in features, classifiers and databases used, but we have made an effort in Table 3 which shows some recent gender identification results. This table shows the contribution of this work in terms of improvement in recognition rate with small sample size as the results obtained with small sample size in earlier work [14] was approximately 65% while with proposed feature dimensionality reduction the recognition rate increased to 99.4%. Results with large sample size are also very encouraging when we combined the two features leading to a new feature set (Pitch + AMFCC).

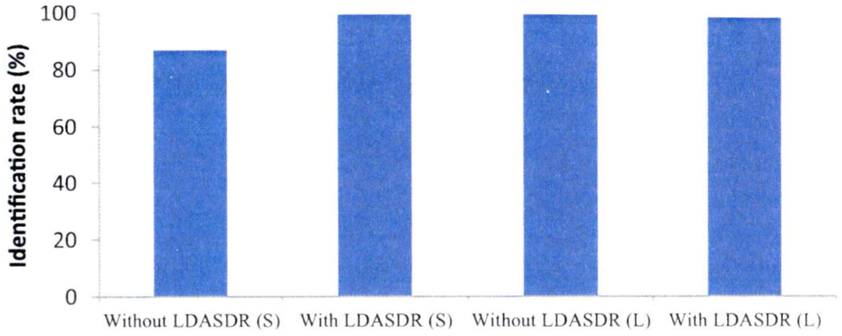

Fig. 5 Identification rate with and without LDASDR on different sample sizes

Table 3 Comparision of different gender recognition works

Year	Classifier	Features	Database size	Accuracy (%)
2005 [4]	Neural network	Pitch + Acoustic	Large	98.5
2016 [23]	4 different classifers	f0, MFCC and f0 + MFCC	Large	95.2
2017 [11]	Naive Bayes	15 features	Large	92.75
2018 [24]	Neural network	Odor	Small	66.6
Proposed work	Neural network	Pitch, AMFCC, Pitch + AMFCC	Both (large and small)	Large 100 small (with LDASDR) 99.4

5 Conclusion

In this work, a gender identification system using acoustic and cepstral features have been discussed. The experimental results reveals that a classification rate of 99.4% and 98.4% has been achieved in case of cepstral and acoustic features, respectively for large database size. However, when both features have been combined for large sample size, accuracy of 100% has been achieved. However, with small sample size the identification rate comes out to be 86.9%, 97.1% and 88.3% for cepstral, acoustic and combined feature set, respectively. With the application of LDASDR on AMFCC features for small database identification rate increases to 99.4%. Thus, it can be concluded that when available sample size for training is small, then using LDASDR is beneficial in terms of identification rate.

References

1. Yücesoy E (2020) Speaker age and gender classification using GMM supervector and NAP channel compensation method. J Ambient Intel Humaniz Comput. https://doi.org/10.1007/s12652-020-02045-4
2. Vaijayanthi S, Arunnehru J (2021) Synthesis approach for emotion recognition from cepstral and pitch coefficients using machine learning. In: Bindhu V, Tavares JMRS, Boulogeorgos AAA, Vuppalapati C (eds) International conference on communication, computing and electronics systems. Lecture notes in electrical engineering, vol 733. Springer. https://doi.org/10.1007/978-981-33-4909-4_39
3. Shumskaya O (2021) Method of real-time speaker identifying by voice. In: Radionov AA, Gasiyarov VR (eds) Advances in automation II. RusAutoConf 2020. Lecture notes in electrical engineering, vol 729. Springer, Cham. https://doi.org/10.1007/978-3-030-71119-1_37
4. Harb H, Chen L (2005) Voice-based gender identification in multimedia applicatons. J Intell Inf Syst 24(2/3):179–198. https://doi.org/10.1007/s10844-005-0322-8
5. Parris E, Carey M (1996) Language independent gender identification. In: IEEE international conference on acoustics, speech and signal processing, pp 685–688. https://doi.org/10.1109/ICASSP.1996.543213
6. Lin F, Wu Y, Zhuang Y, Long X, Xu W (2016) Human gender classification: a review. Int J Biom 275–300. https://doi.org/10.1504/IJBM.2016.082604
7. Moro-Velazquez L, Godino-Llorente JI, Castellanos-Dominguez CG, Gomez-Garcia JA (2016) An insight to the automatic categorization of speakers according to sex and its application to the detection of voice pathologies: a comparative study. Revista Facultad de Ingeniería, Universidad de Antioquia 79:50–62
8. Alkhawaldeh RS (2019) DGR: gender recognition of human speech using one-dimensional conventional neural network.Hindawi Scientific Programming, Article ID 7213717. https://doi.org/10.1155/2019/7213717
9. Sahiner B, Chan HP, Petrick N, Wagner RF, Hadjiiski L (2000) Feature selection and classifier performance in computer-aided diagnosis: the effect of finite sample size. Med Phys 27(7):1509–1522. https://doi.org/10.1118/1.599017
10. Raudys SJ, Jain AK (1991) Small smaple size effects in statistical pattern recognition: recommendations for practitioners. IEEE Trans Pattern Anal Mach Intell 13(3):252–264. https://doi.org/10.1109/34.75512
11. Zimeng H (2017) Speaker gender recognition system, University of Oulu, Department of Communications Engineering. Master's Degree Programme in Wireless Communications Engineering. Master's Thesis

12. Hu Y, Wu D, Nucci A (2012) Pitch-based gender identification with two-stage classification. Secur Commun Netw 5(2):211–225. https://doi.org/10.1002/sec.308
13. Bakir C (2016) Automatic speaker gender identification for the German language. Balkan J Elect Comput Eng 4(2):79–83. https://doi.org/10.17694/bajece.43067
14. Ahmad J, Fiaz M, Kwon S, Sodanil M, Vo B, Baik SW (2015) Gender identification using MFCC for telephone applications—a comparative study. Int J Comput Sci Electron Eng 3(5):351–355
15. Yusnita MA, Hafiz AM, Fadzilah MN, Zulhanip AZ, Idris M (2017) Automatic gender recognition using linear prediction coefficients and artificial neural network on speech signal. In: 7th IEEE international conference on control system, computing and engineering (ICCSCE), pp 372–377. https://doi.org/10.1109/ICCSCE.2017.8284437
16. Gelfer MP, Mikos VA (2004) The relative contributions of speaking fundamental frequency and formant frequencies to gender identification based on isolated vowels. J Voice 19(4):544–554. https://doi.org/10.1016/j.jvoice.2004.10.006
17. Quatieri TF (2002) Discrete time speech signal processing: principles and practice. Pearson Education, India
18. Gowdy JN, Tufekci Z (2000) Mel-scaled discrete wavelet coefficients for speech recognition. In: IEEE international conference on acoustics, speech, and signal processing, pp 1351–1354. https://doi.org/10.1109/ICASSP.2000.861829
19. Davis SB, Mermelstein P (1980) Comparision of parametric representations for monosyllabic word recognition in continuously spoken sentences. IEEE Trans Acous 28(4):357–366. https://doi.org/10.1109/tassp.1980.1163420
20. Lee CH, Lee YK, Huang RZ (2006) Automatic recognition of bird songs using cepstral coefficients. J Inf Tech Appl 1(1):17–23
21. Feng L, Hansen LK (2005) A new database for speaker recognition, Informatics and Mathematical Modelling, Technical university of Denmark [Online]. http://www.imm.dtu.dk/
22. Costantini G, Todisco M, Perfetti R, Paoloni A, Saggio G (2012) Single-sided objective speech intelligibility assessment based on Sparse signal representation. In: IEEE international workshop on machine learning for signal processing, pp 1–6. https://doi.org/10.1109/MLSP.2012.6349776
23. Mishra T, Bangalore S, Levitan SI (2016) Automatic identification of gender from speech. In: Proceeding of speech prosody, pp 84–88. https://doi.org/10.21437/SpeechProsody.2016-18
24. Sabri AQ, Alfred R (2018) Evaluation of artificial neural network in classifying human gender based on odour. In: Alfred R, Iida H, Ag. Ibrahim A, Lim Y (eds) Computational science and technology. ICCST 2017. Lecture notes in electrical engineering, vol 488. Springer. https://doi.org/10.1007/978-981-10-8276-4_31
25. Iqbal A et al (2020) Renewable power for sustainable growth, vol 723, 805 p. Springer Nature, Berlin, LNEE. https://doi.org/10.1007/978-981-33-4080-0 (ISBN 978-981-33-4082-4)
26. Tomar A et al (2020) Machine learning, advances in computing, renewable energy and communication, vol 768, 659 p. Springer Nature, Berlin, LNEE. https://doi.org/10.1007/978-981-16-2354-7 (ISBN 978-981-16-2354-7)

The Cloud-Integrated Hospital Allotment System for Dynamic Patients Integrated with IoT

Rekha S. Dange and Bharati B. Sayankar

Abstract In today's twenty-first century IoT has played a significant role in bringing automation into reality. Health monitoring systems portable or fixed can have tremendous application for injured patients if integrated with IoT, health management systems, and hospital management systems. We are proposing a review work of cloud-integrated automatic hospital and patient monitoring system to make life easy for doctors and patients for live tracking of all the individual prospects of critical patient health data. We are also proposing the IoT-enabled ambulance to auto-track the nearest hospital for critical patient treatment using machine learning algorithms. Overall, the research aims to amalgamate health monitoring systems, Hospital management systems, Ambulance tracking systems to one unified system integrated with artificial Intelligent. The technology platform powered by systematic flow diagram of server, cloud and software integrations.

Keywords Hospital management system · Ambulance tracking system · Health monitoring system · AI

1 Introduction

Before electronic health records, community and township health centres provided a variety of services [1]. Amongst the services offered were child, maternal, and geriatric care, chronic illness care, severe mental patient treatment, and health education. Public health information management systems (PHIMS) are designed to manage and organise the vast amount of data generated by public health agencies, hospitals, clinics, and other organisations. In addition, information technology may help to improve public health services. Ensuring that residents' electronic health records

R. S. Dange (✉)
Electronics and Telecommunication, G. H. Raisoni University, Amravati, India
e-mail: rekhadange73@gmail.com

B. B. Sayankar
Electronics and Telecommunication, G. H. Raisoni University, Saikheda, India

A. Tomar et al. (eds.), *Proceedings of 3rd International Conference on Machine Learning, Advances in Computing, Renewable Energy and Communication*, Lecture Notes in Electrical Engineering 915, https://doi.org/10.1007/978-981-19-2828-4_10

and other follow-up data are properly maintained and used would improve primary medical institutions' public health services and allow the health administrative department to more easily perform service statistics and analysis [2]. The healthcare system is under pressure to enhance efficiency, safety, and economic sustainability whilst maintaining or increasing affordability. Inadequate public health monitoring systems, poor screening and triage processes, and an increase in medical load all contributed to the need for improved public health crisis management. It is possible to actively learn and analyse data using artificial intelligence (AI) [3]. Artificial intelligence (AI) is changing the delivery of health care services and has shown great promise in the medical and health-care industries.

However, a substantial proportion of infected individuals have a respiratory condition that requires hospitalisation and may progress to severe sickness with hypoxic respiratory failure, necessitating long-term ventilator support [4]. As a consequence of the increased frequency of public health disasters, the usage of technology, especially artificial intelligence, has increased (AI). The requirement for this conversation subject is an understanding of new service encounters when AI is deployed in front-line services. Medical data are often represented, mined, and reasoned using AI-based clinical decision-making and action management systems. These findings show how AI may help with medical diagnosis, prevention, and therapy [5].

Data science (DS) technologies are data-driven and cover the data life cycle. Many of these phases have fewer, higher-level divisions [6]. Data science include data collection and processing, data integration, storage, and management, data analysis, and data visualisation and interpretation. This comprises data creation, collecting, processing, integration, storage, and management, data analysis (including predictive modelling), and results communication (which includes results dissemination). A focus on mental health and behavioural transformation was highlighted throughout the workshop. It was meant to establish research goals for this vital and rapidly growing area of digital health [7]. Panels on mobile sensing, user experience design, statistics and machine learning, and privacy and security were held. This study will uncover optimal practises and unresolved research issues that will progress science.

The use of Artificial Intelligence (AI)-based technologies to medical imaging is now garnering a lot of interest [8]. Whilst development is rapid and exciting, a number of steps must be accomplished before a new AI solution can be utilised in routine medical practise, especially in radiology, nuclear medicine, and radiation. The idea that new technology should aim to be as safe as the technology it is replacing. True, risk management and quality assurance will remain quiet. The IDSA and SHEA have revised its CDI Clinical Practise Guidelines for adults, recommending against repeat testing during the same diarrhoea episode [9]. At this article, we report on our CDI experience in a 280-bed community teaching hospital where the elderly makes up the majority of in-patient admissions. The aim is to enhance current healthcare services via sensor networks, medical equipment, wireless connectivity, middleware software, and end software applications. This is because indoor and outdoor health monitoring systems may help detect illness, provide emergency assistance, and save medical costs. Sensors must collect regular physiological data from individuals and provide it to medical experts through a communication system in order for health monitoring systems to work. These systems must provide monthly statistics as well as emergency reports in event of a catastrophe or emergency. When critically ill, most people die in an ambulance [10]. They need urgent help to live. In certain cases, medical staff may save their lives. As a consequence, an ambulance may only offer first aid, and any operations that are required will be performed in medical facilities. Whilst patients are still in the ambulance, this device also offers remote health monitoring and spoken advice. An ambulance is usually called to take patients to the closest medical institution (a hospital). When it comes to a patient's health and the need to transfer them to a medical institution, several difficulties may emerge. These issues are divided into two categories:

(1) evaluating patients' conditions and responding quickly, and
(2) finding the most convenient and appropriate medical facility as soon as possible.

Also known as distributed real-time systems, these two issues are linked. Technology-based solutions for ambulance management are also available. This system is based on GIS, GPS, and GSM [11]. Assisting in an emergency situation, ambulances transport injured people to the closest hospitals. In such cases, the reaction speed of our real-time technology is essential, since it affects our reputation. A faster shortest path approach may be achieved by using efficient data structures to build the road network graph.

This will decrease emergency response times and increase efficiency. A GIS, GPS, and GSM (Mouly and Pautet1992; Rahnema 1993) are used to implement it [12]. Patients are transferred to hospitals and health units whilst being treated in ambulances. The route to the most suitable hospital, an ambulance is called to the site of an emergency. The fastest routes are the most efficient. This process may take a long time in densely populated areas like Mumbai. This technique may be substantially improved in terms of processing speed by using recent data structure breakthroughs. It is difficult to decrease response time whilst increasing capabilities without incorporating the new shortest path method into a GIS.

Hospital-acquired pressure injuries (PIs) are characterised as unfavourable hospital events that cause significant pain, reduce quality of life, and may be related to an increased risk of death [13]. Rehospitalisations and healthcare costs rise as a result of these injuries, which are generally avoidable. As a consequence, hospitals employ PIs, who may face budgetary constraints or reimbursement changes. With the greater goal of treating individuals who have had many concussions. Data on PI incidence has been researched all around the globe and is being used to push hospitals to implement PI preventive initiatives. Ambulance service in the modern day is a multi-functional, dynamic system with a complicated maintenance mechanism. In the ambulance service, the significance of research decision support systems comes from the need to enhance decision-making in order to improve patient care quality. Given the importance of human life and health, we may emphasise the increasing need for "reliability" in every choice as one of the main aspects of the ambulance service's functioning. As a consequence, employing a Decision Support System (DSS) to automate the processing of massive quantities of data is a smart idea. A high degree of dynamic service work is another essential aspect of an ambulance service [14]. Because healthcare professionals usually have limited time to make a choice, there is a lot of movement. As a consequence, by automating situation analysis, the DSS may be used to reduce decision-making time in complicated situations. Transportation has changed dramatically as a consequence of technological advancements in information technology and the automotive industry.

The emergence of autonomous vehicles (AV) in particular has shown to be a game-changing alternative capable of revolutionising the slow transportation system. An AV may travel without the need for a driver by using Laser Illuminated Detection and Ranging (LIDAR), Radio Detection and Ranging (RADAR), cameras, sensors, and other technologies to monitor its surroundings. Autonomous cars will, according to some researchers, become a widespread method of transportation, surpassing current vehicles in terms of safety, efficacy, comfort, traffic density, and speed [15]. Traditional ambulances may be replaced by autonomous ambulances as a consequence of this fast development. Because both medical care professionals work together on a patient instead of one of them needing to drive, it implies more effective medical support. Patient registration, doctor management, pharmacy management, and other systems are all part of the hospital administration system which are well management by the cloud platform [16]. In the medical sector, AI, ML, and NLP have been used to create automated software that has a direct application in the entire medical management system [17]. The hospital's digitisation has contributed significantly to income creation and cost reduction [18]. Medical Al can improve the efficiency, autonomy, and functionality of clinic management system software. Custom machine learning is increasingly being used to perform tasks that were previously handled by humans. Appointment and task scheduling may be changed automatically as needed, with assignments and timetables altered on the fly to suit changing conditions and notifications issued to the appropriate physicians and other employees to match the redirected processes [19]. Clinic management system software also makes it easier for doctors to keep electronic medical records (EMR) and for patients to access them, allowing for easier access and retention of health and treatment information for faster decision-making based on a better understanding of the patient's unique health profile. Medical artificial intelligence may assist hospital management systems in the same way that clinic management systems do. Hospitals have a number of unique problems that may provide additional possibilities for improved functioning via artificial intelligence programming. Medical Al may assist with in-patient and out-patient scheduling, with choices on patient rotation based on a number of variables like as prognosis, treatment response, previous health history, payment/insurance considerations, available personnel and quarters, IPD and OPD analysis, Health care analytics, hereditary data analysis, and more [20, 21].

Ambulance systems with GPS-enabled sensors have also been shown to utilise AI and machine learning to improve hospital search efficiency and save patient lives [22, 23]. One of the most serious problems that a patient faces is travel time delays. The delay in journey time may be caused by traffic circumstances such as traffic jams, road problems, or the hospital's or patient's location. This delay may put important patients in jeopardy and, in rare cases, result in death. This issue was critically solved and was optimise using big data analytics [24–27].

2 Methodology

Methodology for the ambulance geotracking and booking the hospital beds for critically ill patients is shown in the Fig. 1. Figure 1 shows the overall flowchart with ambulance and HMS system together. The steps are given below:

Step 1: The methodology's first step is gathering the patient's data and locating an ambulance service for the patient. Patients' information is stored in the patient's business or home's database. They may be collected by utilising the Id that has been given to them. If the patient is unable to call the ambulance, someone else may do so using the patient's identification number.

Step 2: Once the ambulance arrives and the patient is transferred, the search for a nearby hospital for ICU beds or other services that the patient requires, whilst keeping in mind his or her condition, and identifying the location of the closest hospital for patient admission begins.

Step 3: This step entails searching the hospital's database for available ICU beds and other services.

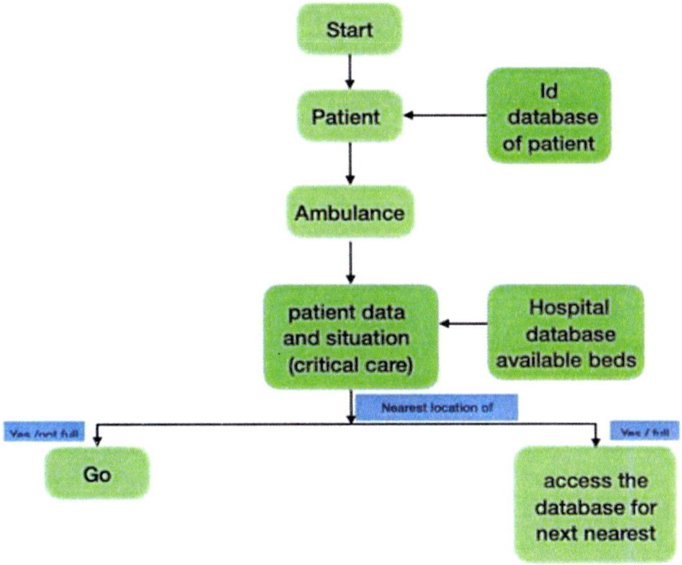

Fig. 1 Flowchart for ambulance and HMS integrated system

If beds are available, the ambulance will only go to the destination; otherwise, it will go for the next closest hospital.

Step 4: In the meanwhile, the ambulance is searching for open hospitals and will connect with a doctor to check the patient's health and make the appropriate hospital preparations.

2.1 Software Requirement Specification (SRS)

- HTML \s
- CSS \s
- JAVASCRIPT \s
- PHP \s
- PHPMySQLServer (XAMPP).

The technologies are all open source, and the technical skills required are manageable. The ease of utilising these technologies and product development time restrictions are linked [16]. The website will be hosted in a free web hosting space at first, but it will be transferred to a premium web hosting space with sufficient capacity later. The bandwidth needed is low since this software does not contain any multimedia components.

2.2 Operational Feasibility

The following resources are needed for this research work:

- Programming Device (laptop and desktop)
- Web Hosting (Freely available)
- Programming tools (free open source server)
- Individual programming, design, and planning.

Appropriate planning and execution by capable personnel under proper supervision are required. We are working on the project on the Windows 10 platform, which may be used to create and test applications. Because Windows versions are often used in workplaces, any problems that arise or modifications that are required will be easy to consult. The software will be made more user-friendly and simple to administer from the back end so that users may feel comfortable using it. Because all of the aforementioned criteria have been met, we can say that our research work is operationally viable.

2.3 The Server

XAMPP (Extremely Accelerated Multi-Platform (APACHE SERVER) which is also integrated with IoT environment of vehicle. It is a free and open source cross-platform web server solution stack bundle created by apache friends, mostly consisting of the apache web server. The scripts are written in php, html, and CSS, amongst other languages.

2.4 In the Back End

- MYSQL

Interfaces with the System.

The programme will feature a menu-based interface that is user-friendly. The screens below will be given.

i. Display Screen:
ii. Login or Sign Up Screen
iii. Menu and Reservations.

3 Particular Requirements

This section provides the software requirements at a degree of depth that allows designers and testers to build and test the system.

1. Requirements for the External Interface
2. Graphical User Interfaces.

The following screens are define as follows:

Opening Page: The menu will be shown on the left side of the screen to browse the website's pages. The user will be given an AMBULANCE INFORMATION page in the centre of the page to go through the available AMBULANCES IN THE DESIRED LOCATION based on the criteria supplied by the user.

Login Screen: After the display screen, this will be the second screen to appear. Users will be able to access various displays depending on where they wish to go. This screen's many fields will be filled in.

i. CLIENT IDENTIFICATION:
ii. PASSWORD:

4 Hospital and Location Ambulance Information (Source, Destination, Timeline)

This page will enable the user to choose which data points the passenger wishes to travel from. The following are some of the fields that will be accessible on this screen:

i. Date:
ii. DEPARTURE.
iii. ARRIVAL.
iv. AMBULANCE TYPE.

4.1 Page for Registration: (User Information)

The SIGN UP Page: This page is given to get basic information about the user and to enable the user to securely access and book AMBULANCE at any time and from any location without having to worry about payment security or other data required at the time of booking (Fig. 2). It may be booked securely at any time and from any location.

Contact Page: This page includes the ADMINS' MAIL ID and PHONE NUMBERS so that customers may contact them in an emergency or report an issue for future improvements.

A data flow diagram (DFD) visually illustrates the movement of data through an information system. It shows how data is input into and output from the system, as well as the data's sources, destinations, and storage locations. The whole system is shown in a single bubble, with input and output data indicated by incoming and exiting arrows. Login, Ambulance information, Booking information, Status, and

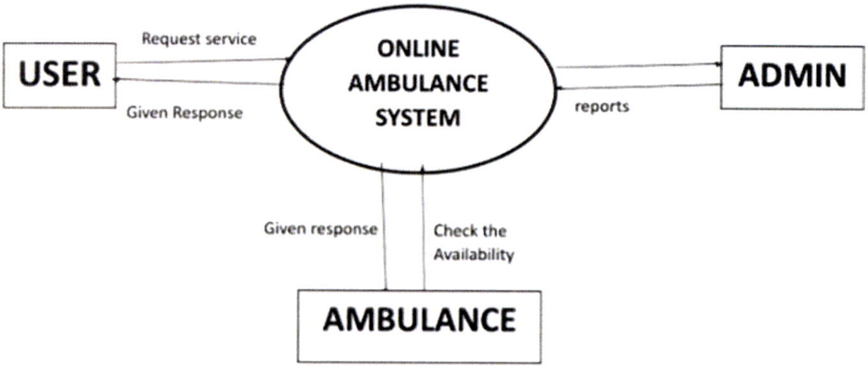

Fig. 2 Context level diagram for USER, ADMIN, and ambulance communication

Payment system are the inputs and outputs of the process known as "online Ambulance Booking System". The user interacts directly with the input/output parameters in this instance, and the user gets a single answer in terms. We'll go through a high-level overview of the system, including the databases from which the process will collect data and utilise it to execute various tasks. Figure 3 shows a level.

1 data flow diagram that explains the entire flow system.

The proposed system in Fig. 1 is the overall flow of the data with authentication and authorisation characteristics which defines the integration of cyber physical system into our proposed architecture. The add new vehicles system is integrated with database and admin authorise the vehicle adding system and our software algorithms authenticate with in the database (Fig. 3). Once the vehicle is added the driver can get authenticated with particular vehicles and each driver can get link with multiple vehicles if needed and decided by the software environment. Every driver and vehicle system represents the ambulance system which is further monitored and authorised by centrally enabled admin which interact with the users. Each user in

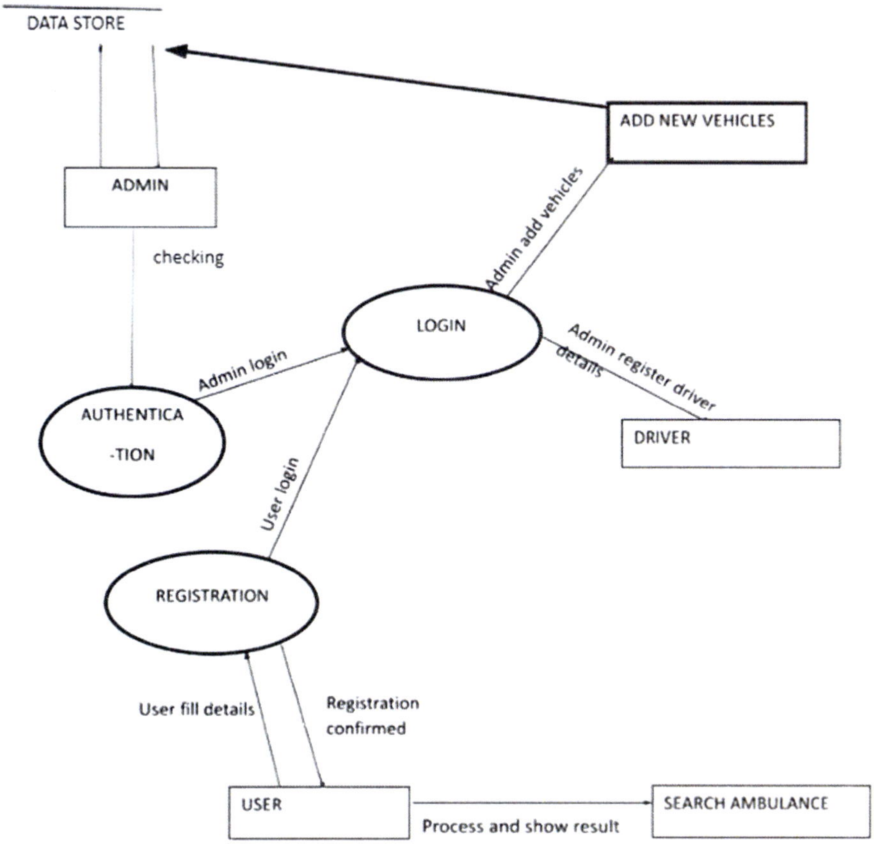

Fig. 3 Dataflow diagram for the entire proposed system

the ecosystem of our environment is stabilised used high end encrypted password protect system which is authenticated by vehicle, ambulance, admin database, and then given required authorisation. Registration of user are also taken place based on OTP authentications. This proposed architecture is for patient which is acting as user and IoT dataset are getting fetched by IoT-enabled vehicles with its GPS characteristics. This GPS dataset is again collaborated with the hospital location to find out the optimised path for the vehicle to reach in the shortest period of time (Fig. 2).

5 Conclusion

In the real-world problems related to accidents, and the management facilities of those patients. Those patients are classified in two types, conscious and unconscious. At the sight of accident our system is able to provide the patient ID to the database management of the nearby hospital if the patient is conscious. If the Patient is unconscious, it is declared as a critical patient and sends the patient data to the emergency/critical patient of the system database. It uses the real-time condition of the patient and sends the updates to the nearby hospitals. As the accidents occurs, most nearby ambulance gets deployed to the sight of incident. After getting the ID of the patient and ambulate the patient our product also finds the best shortest path to the nearby hospital. As that particular nearby hospital receives the real-time data of the patient, the hospital arranges all the requirements in prior. If the hospital updates its entries that the hospital beds are full, then our product will quickly track the location on next nearby hospital with its shortest path. At the time of real-life situations if there are more than one cases to ambulate the patients then our system checks and predicts which patient is most critical and relocates the map to that specific critical patient.

6 Future Scope

The proposed system is comprised three main system ambulance tracking system, Health monitoring system, and Hospital management system. This system when integrated with cloud technology and blockchain token system can increase the high end data analytics ability and security, respectively. The machine learning algorithms with big data analytics can bring the artificial intelligent ability to our proposed system.

References

1. Wu F, Narimatsu H, Li X, Nakamura S, Sho R, Zhao G et al (2017) Non-communicable diseases control in China and Japan. Global Health 13
2. Prinja S, Nimesh R, Gupta A, Bahuguna P, Thakur JS, Gupta M et al (2016) Impact assessment and cost- effectiveness of m-health application used by community health workers for maternal, newborn and child health care services in rural Uttar Pradesh, India: a study protocol. Glob Health Action 9(1):31473
3. Cortis LJ, Ward PR, Mckinnon RA, Koczwara B (2017) Integrated care in cancer: what is it, how is it used and where are the gaps? A textual narrative literature synthesis. Eur J Cancer Care 26(4):e12689
4. Lavis JN, Oxman AD, Lewin S, Fretheim A (2009) SUPPORT tools for evidence-informed health policymaking (STP) 3: setting priorities for supporting evidence-informed policy-making. Health Res Policy Syst. 7(Suppl. 1):S3
5. Gottret P, Schieber G (2006) Health financing revisited: a practitioner's guide. World Bank, Washington, DC
6. Erasmus E, Orgill M, Schneider H, Gilson L Mapping the existing body of health policy implementation research in lower income settings: what is covered and what are the gaps? HealthPolicyPlan
7. Aasa U, Brulin C, Ängqvist KA, Barnekow-Bergqvist M Work-related psychological factors, worry about work conditions and health complaints among female and male ambulance personnel. Scand J Caring Sci
8. Axelsson A, Jimenez Herrera M, Bång A (2016) How the context of ambulancecare influences learning to become a specialist ambulance nurse a Swedish perspective. Nurse Educ Today 37:8–14. https://doi.org/10.1016/j.nedt.2015.10.029
9. SOSFS: Socialstyrelsensföreskrifteromambulanssjukvård (Regulations from the National Board of Health and Welfare about the Ambulance service). Population Sub-site, 2013. Web Document: http://www.socialstyrelsen.se/sosfs/2009-10/. Retrieved on March 2015). Ornge [Online, https://www.ornge.ca/home. Accessed 10 November 2019
10. Pond GT, McQuat G (2018) Optimizing fleet staging of air ambulances in the province of ontario, international conference on theory and practice of natural computing, pp 215–224. https://doi.org/10.1007/978-3-030-04070-3_17
11. Carnes TA, Henderson SG, Shmoys DB, Ahghari M, MacDonald RD (2013) Mathematical programming guides air-ambulance routing at ornge. Interfaces 43(3):232–239
12. Ibanez B, James S, Agewall S, Antunes M, Bucciarelli-Ducci C, Bueno H et al (2018) 2017 ESC guidelines for the management of acute myocardial infarction in patients presenting with ST-segment elevation: the task force for the management of acute myocardial infarction in patients presenting with ST-segment elevation of the European Society of Cardiology (ESC). Eur Heart J 39(2):119–177
13. Lozano R, Naghavi M, Foreman K et al (1859) Global and regional mortality from 235 causes of death for 20 age groups in 1990 and 2010: a systematic analysis for the global burden of disease study 2010. Lancet 2012(380):2095–2128
14. Mbewu A (2009) The burden of cardiovascular disease in sub-Saharan Africa. SA Heart 6(1)
15. Spamast-Malita I (2018) Hospital management system. https://doi.org/10.13140/RG.2.2.16459.28966
16. Koyuncu B, Koyuncu H (2015) Intelligent hospital management system (IHMS). International conference on computational intelligence and communication networks (CICN) 2015:1602–1604. https://doi.org/10.1109/CICN.2015.305
17. Weber C, Reimann P (2020) MMP—a platform to manage machine learning models in industry 4.0 environments, pp 91–94. https://doi.org/10.1109/EDOCW49879.2020.00025
18. Ponmalar A, Maansi S, Mahalakshmi S, Shalini M, Madhavan R (2021) Mobile application for hospital management system, pp 1434–1437. https://doi.org/10.1109/ICICCS51141.2021.9432286

19. Tosun P, Sezgin S, Uray N (2021) Consumer complaining behavior in hospitality management. J Hospital Market Manage 1–18. https://doi.org/10.1080/19368623.2021.1941474
20. Bishnoi L, Narayan Singh S (2018) Artificial intelligence techniques used in medical sciences: a review. In: 2018 8th international conference on cloud computing, data science & engineering (Confluence), pp 1–8. https://doi.org/10.1109/CONFLUENCE.2018.8442729
21. Kaya N, Gemlik N (2021) A qualitative research on hospital managers 'perspectives on the digitalization of hospitals. J Acad Perspect Social Stud. https://doi.org/10.35344/japss.903276
22. Akca T, Sahingoz O, Koçyiğit E, Tozal M (2020) Intelligent ambulance management system in smart cities, pp 1–7. https://doi.org/10.1109/ICEE49691.2020.9249959
23. Godwin J, Santhosh Krishna BV, Rajeshwari R, Sushmitha P, Yamini M (2021) IoT based intelligent ambulance monitoring and traffic control system. https://doi.org/10.1007/978-3-030-57835-0_20
24. Wani M, Khan S, Alam M (2020) IoT based traffic management system for ambulances
25. Udawant O, Thombare N, Chauhan D, Hadke A, Waghole D (2017) Smart ambulance system using IoT, pp 171–176. https://doi.org/10.1109/BID.2017.8336593
26. Iqbal A et al (2020) Renewable power for sustainable growth, vol 723, 805 p. Springer Nature, Berlin, LNEE. https://doi.org/10.1007/978-981-33-4080-0 (ISBN 978-981-33-4082-4)
27. Tomar A et al (2020) Machine Learning, advances in computing, renewable energy and communication, vol 768, 659 p. Springer Nature, Berlin, LNEE. https://doi.org/10.1007/978-981-16-2354-7 (ISBN 978-981-16-2354-7)

Parallel Agglomerative Hierarchical Clustering Algorithm Implementation with Hadoop MapReduce

C. Maithri and H. Chandramouli

Abstract Current expectations on popular classification methods in today's data mining software are increasing due to the large volumes of generated data from online interactive systems. Hierarchical agglomerative clustering (HAC) algorithm is a popular algorithm that is adopted for clustering both similarity and dissimilarity datasets. Solutions of data analytics with parallel and distributed architectures are promising models for storing, retrieving and analyzing large datasets. Hadoop MapReduce framework is a distributed and parallel platform to handle large datasets. Classification algorithms redefined to use features of Hadoop MapReduce method to provide an ability to solve the problem of clustering on large volume of data. In this paper, we are proposing an agglomerative hierarchical clustering with parallel implementation, using MapReduce approach for large dataset. It divides dataset into data blocks; each data block is classified to groups on different distributed data node of Hadoop MapReduce with agglomerative hierarchical clustering method. The implementation is fully MapReduce programming model, which will have nodes that pull task assignment and global status for determining their individual node status. The implementation defines an execution environment that facilitates algorithm execution over different datasets. The significance of this approach provide the proposed clustering model over the distributed environment with an increase in efficiency for large data analysis.

Keywords Agglomerative hierarchical clustering · Big data mining · Distributed architecture · Hadoop MapReduce · Machine learning · Parallel algorithm

---

C. Maithri (✉)
Department of CSE, Kalpataru Institute of Technology, Tiptur, India
e-mail: maithri.c.prashanth@gmail.com

H. Chandramouli
Department of CSE, East Point College of Engineering and Technology, Bangalore, India

A. Tomar et al. (eds.), *Proceedings of 3rd International Conference on Machine Learning, Advances in Computing, Renewable Energy and Communication*, Lecture Notes in Electrical Engineering 915, https://doi.org/10.1007/978-981-19-2828-4_11

1 Introduction

Data clustering in machine learning is a first primary task performed before any analysis or synthesis of data. Fast and efficient clustering algorithms are next technological trend that will utilize robust processing systems [1, 2] to generate knowledge from large volume of unresponsive spatial data. The clustering algorithm basic operation is to generate C clusters from N dataset with subjective conditions that elements in each cluster are same and are distinct from each other out of clusters. Clustering is a mathematical model to discover data patterns in datasets, where the objects in cluster have higher degree of similarity. Clustering process is not an automatic process, but it is an iterative procedure for finding multi-objective similarity data and putting them into one group. There are many algorithmic approaches defined and implemented to perform clustering process, which parallelized to optimize computing cost. The primary algorithmic classification defined are partition method, density method, grid method, distribution-based method and connectivity-based method.

The hierarchical clustering is one popular mathematical approach that considers interconnection of dataset. The principle rule in hierarchical clustering is that the core object-related more to its nearby objects than to far away objects. The connecting rule defined based on distance or on certain similarity field values that are present in objects considered. Dendrogram represents the outcome of hierarchical clustering, which explains hierarchy of cluster formulation on given dataset [3].

There are two active implementations for hierarchical clustering method: (1) agglomerative clustering and (2) divisive clustering. The agglomerative clustering is a bottom-up method, which starts with formulating groups from each dataset considering the distance matrix and constructs a singleton cluster for all the data. The divisive clustering is a top-down method, which first considers all dataset as one singleton cluster and starts dividing to small clusters until all data objects grouped individually. Divisive clustering algorithm is advantageous only for small computations, but it is a drawback for large volume data. The agglomerative clustering has complexity of $O(n^3)$, and divisive clustering has complexity of $O(2^{n-1})$ which are both iterative approach and make slower during processing of large datasets.

Agglomerative process starts with finding the minimal distance between each cluster and performs a merge of least distance clusters together. This step is repeated till all clusters are merged into one cluster. The output of this process represented with a dendrogram that shows the merging of clusters. After formulating of one complete cluster, there will be a split process, which shows group of clusters.

The basic motivation for this work is to efficiently utilize the parallel approaches available to perform the clustering of large volume data on Hadoop. In this paper, we are proposing a parallel agglomerative clustering method implementation using the Hadoop eMapReduce framework that has multi-process design in computing the distance matrix for clusters of data. The experimental outcome of proposed algorithm implementation compared with an iterative approach of agglomerative clustering. The comparison shows an overall improvement of computation cost for a selected dataset.

2 Related Works

Most of hierarchical clustering algorithms work with using minimum spanning tree (MST) [4] and got optimized using that data structure. All these implementation differs with point of similarity calculation on existing clusters. Well-known implementation of hierarchical clustering algorithms is BIRCH [5], CURE [6], ROCK [7], CHAMELEON [8], etc. In all these algorithmic implementation, the only difference is in setting of different parameters to identify an improvement for quality of resultant clusters and reduce outliers. The work in this paper highlights only on hierarchical method to model dataset to execute in parallel way.

The popular hierarchical method implemented in many of data analysis and processing architecture is agglomerative hierarchical clustering (AHC) [9]. This method defines each data object as cluster and then repeatedly merges the neighbor nearest clusters to form a cluster tree called dendrogram. The dendrogram built after the execution of AHC from bottom leaf cluster to the root cluster by performing merge operation. Finally, the dendrogram shows that data objects are in leaf position that will be presented by user.

There are many variants of AHC implementation based on distance calculation of clusters. Whether the algorithm uses minimum or maximum distance of objects to form dendrogram that resulting in complete AHC. There are some known strategies defined as unweighted pair group method using arithmetic averages (UPGMA) and weighted pair group method using arithmetic averages (WPGMA). The details can be referred in legendre et. al. [10].

To parallelize hierarchical clustering, a MapReduce strategy conveniently discussed in the work of parallel random partition based hierarchical clustering (PARABLE) [11]. This approach is a two-step procedure. The first step performs sequential hierarchical clustering execution on distributed nodes. The second step will combine the obtained results from first step into clusters as defined using dendrogram alignment technique.

There are methods of parallel algorithm [12, 13] for hierarchical clustering in which sequential hierarchical clustering algorithm runs on parallel threads that uses data subsets to form intermediate level dendrogram result. These intermediate-level dendrogram results are merged to form final dendrogram result. The process of merging is recursive, which starts from leaves toward root with comparing nodes using similarity value. The dendrogram merging technique fails to address the fact of adding new branches into dendrogram and the effect of height at the point of merging. Many parallel hierarchical clustering algorithms are defined such as single linkage clustering algorithm (SLINK) [14], hierarchical DBScan (HDBSCAN) [15], which describe both parallel and distributed methods. The clustering through minimum spanning tree (MST) in parallel (CLUMP) [16] shows the parallel building of minimum spanning tree using distributed nodes on a set of overlapping datasets. On each node process there will be different subset of data there is a no duplication of data subset. It uses linear representation that is a list of data elements in sequential order.

The scalable algorithm proposed is shared memory single linkage (SHRINK) [17] which parallelizes the single linkage hierarchical clustering. In this method, the original data divides into sub-datasets with overlapping condition and computes the hierarchy of each subset using single linkage algorithm, which results in dendrogram. The subset dendrogram is combined for the overall dendrogram.

Another algorithm that uses single linkage [18] is distributed using single linkage algorithm with MapReduce (DiSC) [18] and is also similar to SHRINK and CLUMP. DiSC algorithm is specifically implemented using MapReduce model that divides dataset into overlapping sub-datasets. Each subset processed separately over distributed node to have intermediate results. The results combined to form overall dataset. The DiSC algorithm has two rounds of MapReduce jobs in that the first round is prim's mapper. This builds intermediate MST for giving data subset. The second round is a Kruskal reducer program, which defines a K-way merger to combine intermediate results to final output of cluster. With this discussion, it is evident that MapReduce framework will facilitate data clustering in distributed computing environment. However, this approach of MapReduce implementation may reduce the computational effort of hierarchical clustering calculations. Another reason is that many intermediate copies of data transmitted to data nodes increase the total amount of transmitted data to much higher than the normal size of dataset.

3 Preliminary Concepts

Data mining methods such as classification, clustering, regression, association and pattern matching applied for finding patterns on dataset. Data mining will be an automated method for finding the precious information and extract that information from large data repositories. This may involve in analyzing data in large volume to find some valuable information. The outcome of data mining will be clusters, graphs, time series values and equation solutions. The data mining uses two techniques, namely supervised and unsupervised learning [19]. The supervised learning technique builds model before performing actual analysis on data. In this case, the algorithms start building training data from the dataset and previous trained data to estimate patterns. Some of supervised learning methods are classification, neural networks and association algorithms [20–22]. The unsupervised learning technique applies directly on the dataset to classify them and generate the knowledge, which results with models. The unsupervised learning algorithms are clustering of data, the mining of data for knowledge discovery in data. The algorithms that deal with knowledge discovery in data (KDD) process large amount of dataset to extract knowledge out of it. The clustering process will group data objects to create a tree structure in which accessing of data object will be simple. This process is a continuous process that is defined with grouping of data points with the condition that any two clusters are distinct. The hierarchical clustering will combine data points to clusters to form a big cluster that will be in hierarchy. This hierarchical structure is called as a dendrogram. In

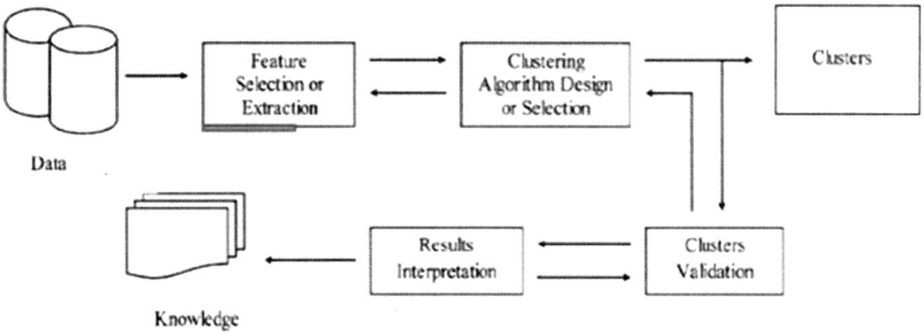

Fig. 1 Stages of hierarchical clustering

dendrogram, the root represents one cluster that contains all data points. Figure 1 shows the stages of hierarchical clustering approach.

(A) ***Hierarchical clustering algorithms***

The hierarchical clustering algorithm defined in two approaches:

(1) Agglomerative hierarchical clustering: It is a bottom-up approach to form the clusters, in which each data in dataset is defined as data object cluster at initial step and it continues to merge each cluster to form single cluster using similarity condition. The outcome is a tree dendrogram. The next step is to split the single dendrogram to form small pieces of clusters using the predefined number of clusters. Thus, with this approach, it determines exact number of clusters.

(2) Divisive: It is a top-down approach to form the clusters. Here, all dataset formed together as the single cluster. The single cluster divides into two or more clusters based on the similarity metric. This process continues until the formulation of required number of cluster. One drawback is that once the split of cluster performed there is no undo process to reformulate the original cluster. Some advantage is that it has small computation cost compared to agglomerative clustering. To improve the quality, we may use the other multi-phase clustering method along with hierarchical clustering.

(B) ***Hadoop system architecture***

Hadoop is distributed system available in open source which is useful in processing large data over a network of computers. It provides options for storing large volume of data for executing processing algorithm in distributed manner. The strength of Hadoop architecture is its file system called Hadoop distributed file system (HDFS) that is also a distributed file system and MapReduce framework used for writing applications which uses distributed processing environment in scalable manner. Hadoop has hardware failure mechanism which handles any hardware failure at node level and this failure will not effect on the application level.

The major Hadoop architectural features are

(1) Scalability: Running single program on thousands of networked machines. New machines added to Hadoop system without affecting the current execution environment.
(2) Fault tolerance: Hadoop key feature is fault tolerance, and data sent to individual node replicated in another node. On event of failure on one node data restores with using replicated copy of that data.
(3) Fast: The distributed file system of Hadoop provides a better data mapping so that the data processing algorithms run faster over large volume of data.
(4) Simplicity: It has simple APIs to deal with huge data in terms of petabytes.

(C) ***Agglomerative hierarchical clustering***

The AHC method starts with defining each data objects as clusters and merges up to form the hierarchy. With each step execution, two clusters will merge to form on cluster based on distance matrix. This merging process will be performed until the complete hierarchy of cluster is constructed. The algorithm for AHC is as follows.

Algorithm 1: AHC

1. Calculate similarity value using distance matrix of clusters. This process builds a similarity index, S_{ij}, for two clusters i and j.
2. Merge two clusters, which have closest similarity value.
3. Update similarity matrix value on merging to show out pairwise similarity value for new cluster form on step 2.
4. Repeat step 2 and 3 until the formulation of single cluster.

The similarity of two clusters may be the minimum distance value or maximum distance value. The two data clusters are called "nearest neighbor" if their distance is minimum and is computed using Eq. 1. The two data clusters are called "furthest neighbor" if their distance is maximum and computed using Eq. 2.

$$d_{\min}\left(c_i, c_j\right) = \min_{A \in c_i, B \in c_j} |A - B| \tag{1}$$

$$d_{\max}\left(c_i, c_j\right) = \max_{A \in c_i, B \in c_j} |A - B| \tag{2}$$

4 Parallel Agglomerative Hierarchical Clustering Algorithm

Figure 2 shows the MapReduce model used in PAHC algorithm, where the input data objects represented and defined to map task. The combined result is generated by the

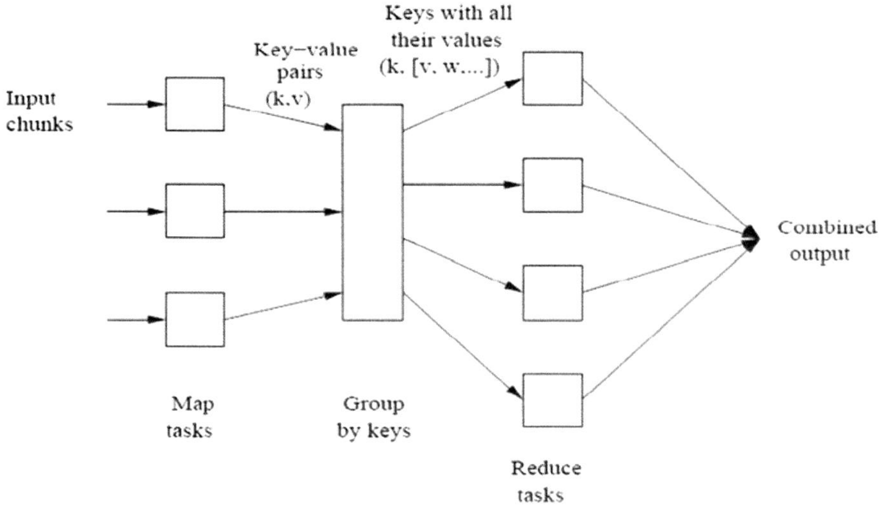

Fig. 2 MapReduce model for PAHC algorithm

reduce task. Suppose the PAHC consists of N computers, in which one is master node M, and others are data nodes or slave nodes S, then S_i ($i = 1, \ldots, N - 1$). The initial classification process starts with P initial clusters. The details of implementation of PAHC described in Algorithm 2.

Algorithm 2: PAHC

1. Initialization: This uses the data objects as clusters called P $\{p_i | i = 1..n$, where n is number of data objects$\}$. Result of initial classification is to build global mapping of the clusters to the data nodes S_i and store this mapping structure in master node M.
2. Similarity calculation process: Each data node performs the similarity calculation for assigned clusters. Suppose n (n > N) the number of cluster in this iteration, M distributes the tasks to each S_i with the cluster ID pair: assigned to S_1, assigned to S_2 and assigned to D_{N-1}.
3. Each data node performs their tasks in parallel as defined below:

   a. The vectors V_i and V_j are extracted from the mapping structure for i and j clusters from the global vector.
   b. Processes the similarity calculation using distance vector calculation, for example, cosine distance calculation method.
   c. Store back the resulted similarity value into the global mapping vector and save it in the temporary file.
   d. Waits for next cluster pair.

4. Each data node processes the similarity value in reverse sorting.

5. Each node sends result list to the master node, where the reducer thread merges the overall result and generates a sorted list.
6. Check result of this iteration: The key of first entry of global sorting result contains most similar cluster id of the iteration. When the value is less than the initial threshold t, then go to step 8, else go to step 7.
7. Merge the two clusters and adjust the similarity vector of the new cluster. Renew the global mapping vector mapping structure, and go to step 2.
8. Stop PHC process.

5 Experiments

For an experimental setup as shown in Fig. 3, we created check implementation for parallel hierarchical algorithm using the Hadoop distributed file system. As known the worst time for traditional agglomerative hierarchical algorithm is $O(n^3)$. The new parallel agglomerative hierarchical algorithm (PAHC) will have the worst time complexity $O(n^2 \log n)$. The experimental results of PAHC compared with existing hierarchical clustering. The first implementation has resulted to be $O(n^2)$ and the second which contains the grid formulation also has the runtime equivalent to $O(n^2)$ which is slightly efficient than original algorithm.

The process of priority queue algorithm first constructs a priority queue with the value of each cluster. The new cluster formed updates the queue by inserting the new

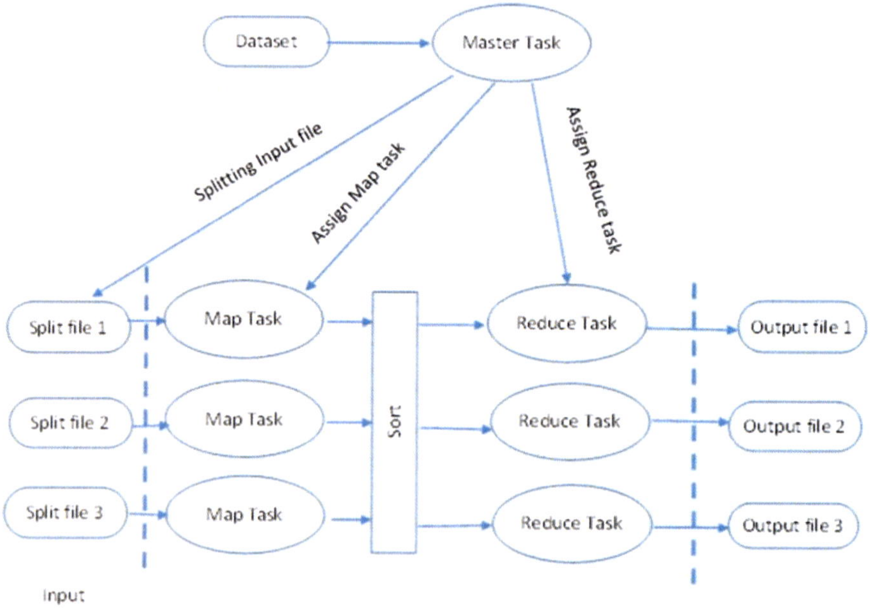

Fig. 3 Experimental setup for executing the P-AHC algorithm

cluster into it and updates all other values on queue. To insert and delete in priority queue it requires O(logn) time. To update queue entry requires two delete and one insert operation thus, resulting in $O(n^2 logn)$ time for n queues entries. Thus, total time is $O(n^2 logn)$ time compared t $O(n^3)$.

The implementation of PAHC is done on Hadoop 3.1.0 with one master node and 11 slave nodes. The interface for executing the algorithm prepared using the Java swing application with Java 11 programming and NetBeans 12 as IDE. The master node configured on server system, HP ProLiant G580 server with four processors of Xeon 2.93 GHz processor having four cores and total of 16 cores processor architecture. The server system has 128 GB RAM and 480 GB storage. Each of the slave node is configured on system, Intel core i3 CPU 2.53 GHz, 4 GB RAM and 500 GB Storage. All the nodes are running on Ubuntu 18.04 operating system. Figure 3 shows the interface of P-AHC algorithm over the Hadoop.

Datasets

The raw data is provided to parallel hierarchical agglomerative clustering algorithm to create n clusters over number of nodes. The table shows datasets generated for running of algorithm. Figure 4 shows the overall speedup for different dataset considered for

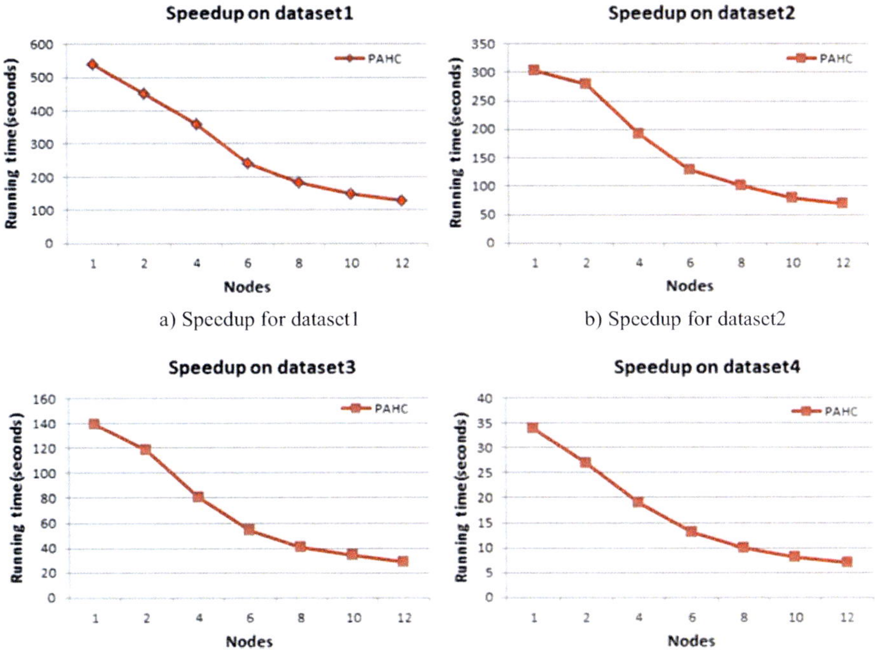

a) Speedup for dataset1 b) Speedup for dataset2

c) Speedup for dataset3 d) Speedup for dataset4

Fig. 4 Speedup in running time (seconds) for different datasets

Table1 Dataset for experiment of PAHC

Dataset				
	Dataset 1	Dataset 2	Dataset 3	Dataset 4
Nodes	20,000	15,000	10,000	5000
1	540	303	139	34
2	450	278	118	27
4	360	191	80	19
6	242	129	54	13
8	184	100	41	10
10	148	79	34	8
12	128	68	29	7

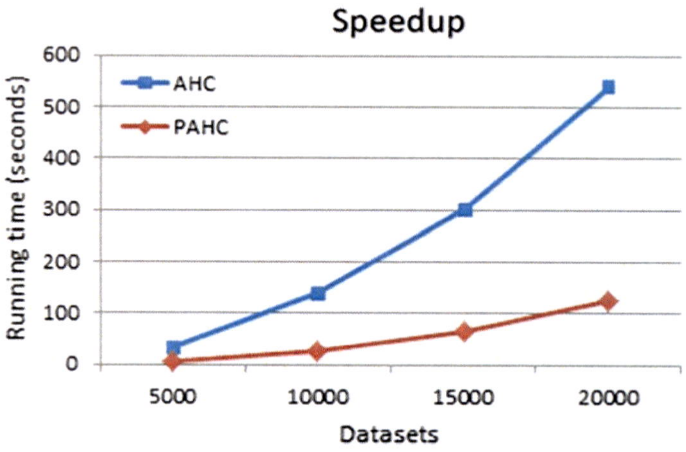

Fig. 5 Speedup comparison of AHC and PAHC

execution as in Table 1. Figure 5 provides the comparison of existing AHC with PAHC speedup.

6 Conclusion

The MapReduce implementation for parallel hierarchical agglomerative clustering algorithm proposed and executed over Hadoop distributed system shows the optimized results over a defined dataset, and the parallel computation with reduced inter-process communication shows the reduction of computing cost on considered datasets. In addition, we showed that the proposed method parallel hierarchical agglomerative clustering algorithm (PAHC) has better efficiency compared with existing algorithm and facilitates for large dataset processing. Further, the proposed algorithm can be improved for effectiveness on processing multiple documents using various memory-bounded experiments.

References

1. Pang N, Zhang J, Zhang C, Qin X (2018) Parallel hierarchical subspace clustering of categorical data. IEEE Trans Comput 68(4):542–555
2. Raghunandan GH, Rani AS, Nanditha SY, Swathi G (2017) Hierarchical agglomerative clustering based routing algorithm for overall efficiency of wireless sensor network. In: 2017 International conference on intelligent computing, instrumentation and control technologies (ICICICT). IEEE, pp 1290–1293
3. Satria BD, Barakbah AR, Sudarsono A (2021) Implementation parallel computation for automatic clustering. Jurnal Mantik 5(2):994–1005
4. Patibandla RL, Veeranjaneyulu N (2018) Survey on clustering algorithms for unstructured data. In: Intelligent engineering informatics. Springer, Singapore, pp 421–429
5. Zhang T, Ramakrishnan R, Livny M (1996) BIRCH: an efficient data clustering method for very large databases. SIGMOD Rec 25(2):103–114. https://doi.org/10.1145/235968.233324
6. Guha S, Rastogi R, Shim K (1998) CURE: an efficient clustering algorithm for large databases. In Proceedings of the 1998 ACM SIGMOD international conference on management of data (SIGMOD '98). Association for Computing Machinery, New York, NY, USA, pp 73–84. https://doi.org/10.1145/276304.276312
7. Guha S, Rastogi R, Shim K (1999) ROCK: a robust clustering algorithm for categorical attributes. In: Proceedings 15th international conference on data engineering (Cat. No.99CB36337), pp 512–521. https://doi.org/10.1109/ICDE.1999.754967
8. Karypis G, Han E-H, Kumar V (1999) Chameleon: hierarchical clustering using dynamic modeling. Computer 32(8):68–75. https://doi.org/10.1109/2.781637
9. Megarchioti S, Mamalis B (2018) The BigKClustering approach for document clustering using Hadoop MapReduce. In: Proceedings of the 22nd pan-hellenic conference on informatics, pp 261–266
10. Legendre P, Legendre L (2012) Cluster analysis. In: Legendre P, Legendre L (eds) vol 24. Elsevier, ISBN 9780444538680, pp 337e424. Developments in environmental modeling, ISSN 0167-8892. https://doi.org/10.1016/B978-0-444-53868-0.50008-3
11. Wang S, Dutta H (2011) PARABLE: A PArallel RAndom-partition Based HierarchicaL ClustEring algorithm for the MapReduce framework
12. dos Santos JA, Syed TI, Naldi MC, Campello RJ, Sander J (2021) Hierarchical density-based clustering using MapReduce. IEEE Trans Big Data 7(1): 102–114. https://doi.org/10.1109/TBDATA.2019.2907624
13. Senthilarasi S, Kamalakkannan S (2021) Unsupervised deep learning on spatial-temporal traffic data using agglomerative clustering. In: Inventive systems and control. Springer, Singapore, pp 757–776
14. Goyal P et al (2016) A fast, scalable SLINK algorithm for commodity cluster computing exploiting spatial locality. In: 2016 IEEE 18th international conference on high performance computing and communications; IEEE 14th international conference on smart city; IEEE 2nd international conference on data science and systems (HPCC/SmartCity/DSS), pp 268–275. https://doi.org/10.1109/HPCC-SmartCity-DSS.2016.0047
15. Malzer C, Baum M (2020)A hybrid approach to hierarchical density-based cluster selection. In: 2020 IEEE international conference on multisensor fusion and integration for intelligent systems (MFI), pp 223–228. https://doi.org/10.1109/MFI49285.2020.9235263

16. Hendrix W, Ali Patwary MM, Agrawal A, Liao W-K, Choudhary A (2012) Parallel hierarchical clustering on shared memory platforms. In: 2012 19th International conference on high performance computing, pp 1–9. https://doi.org/10.1109/HiPC.2012.6507511.
17. Hendrix W, Palsetia D, Patwary MMA, Agrawal A, Liao W-K, Choudhary A (2013) A scalable algorithm for single-linkage hierarchical clustering on distributed-memory architectures. In: 2013 IEEE symposium on large-scale data analysis and visualization (LDAV), pp 7–13. https://doi.org/10.1109/LDAV.2013.6675153
18. Jin C, Patwary MM, Agrawal A, Hendrix W, Liao WK, Choudhary A (2013) DiSC: a distributed single-linkage hierarchical clustering algorithm using MapReduce. In: International workshop on data intensive computing in the clouds (DataCloud)
19. Hung PD, Lien NTT, Ngoc ND (2019) Customer segmentation using hierarchical agglomerative clustering. In: Proceedings of the 2019 2nd international conference on information science and systems, pp 33–37
20. Lydia EL, Moses GJ, Varadarajan V, Nonyelu F, Maseleno A, Perumal E, Shankar K (2020) Clustering and indexing of multiple documents using feature extraction through apache Hadoop on big data. Malaysian J Comput Sci 108–123
21. Iqbal A et al (2020) Renewable power for sustainable growth. In: LNEE, vol 723. Springer Nature, Berlin, 805 p. https://doi.org/10.1007/978-981-33-4080-0. ISBN 978-981-33-4082-4
22. Tomar A et al (2020) Machine learning, advances in computing, renewable energy and communication. In: LNEE, vol 768. Springer Nature, Berlin, 659 p. https://doi.org/10.1007/978-981-16-2354-7. ISBN 978-981-16-2354-7

Design Flaws and Cryptanalysis of a Lightweight Mutual Authentication Protocol for V2V Communication in Internet of Vehicles

Vinod Kumar, Srinivas Jangirala, Hari Pratap, and Adesh Kumari

Abstract The fast development of information technology and communication includes Internet of Things (IoT), cloud computing, Internet vehicle blockchain technology, V2G technology, and artificial intelligence. The Internet of Vehicles (IoV) is one of the application of IoT. Recently, Vasudev et al. have published a paper [1] in "IEEE Transactions on Vehicular Technology" with DOI: 10.1109/TVT.2020.2986585 in IoV environment [1]. In this paper, we reviewed Vasudev et al.'s protocol and find some of the design flaws and security features like forward secrecy, vehicle server impersonation attack, insider attack, and others weaknesses. Further, we provide some valuable suggestions for future research for authentication and key agreement protocols in V2V communication.

Keywords Cryptanalysis · Design flaws · V2V communication · Internet of Vehicles

1 Introduction

Recently, the paper is published by Vasudev et al. [1]. They claim that the suggested approach is more efficient than similar protocols like Mun et al. [2], Zhao et al. [3], Mohit et al. [4], Ying et al. [5] and Chen et al. [6] in terms of computing and

V. Kumar (✉)
Department of Mathematics, PGDAV College, University of Delhi, New Delhi 110065, India
e-mail: vinod.iitkgp13@gmail.com; vinod@pgdav.du.ac.in

S. Jangirala
Jindal Global Business School, O. P. Jindal Global University, New Delhi, Haryana 131001, India
e-mail: sjangirala@jgu.edu.in

H. Pratap
Department of Mathematics, P.G.D.A.V. College Evening, University of Delhi, New Delhi 110065, India

A. Kumari (✉)
Department of Mathematics, Dyal Singh College, University of Delhi, New Delhi 110003, India
e-mail: adeshbhucker@gmail.com

A. Tomar et al. (eds.), *Proceedings of 3rd International Conference on Machine Learning, Advances in Computing, Renewable Energy and Communication*, Lecture Notes in Electrical Engineering 915, https://doi.org/10.1007/978-981-19-2828-4_12

Table 1 Notations used [1]

Symbol	Description
V_s	Vehicle user
TA	Trusted authority
RA	Registration authority
VS	Vehicle Server
$\mathcal{A}$	Attacker/Adversary
UID_i	The ith host's identification number
PW_i	UID_i's password
$h(.)$	One-way cryptographic hash
CID	The TA has a unique identifier
SID_k	The kth VS identification number
K_s	VS and TA shared a secret key
S_k	Session key
$i \stackrel{?}{=} j$	Whether i equals j
$\oplus$	Bitwise XOR operation
$\|$	Concatenation operation

communication expenses. They have also shows that Mohit et al. [4] and Li et al. [7] fails against man-in-the-middle attack, He et al. [8] fails against untraceable attack. They also untraceable attacks, stolen smart card attacks, anonymity attacks, and man-in-the-middle attacks all fail against Shi and Gong [9].

Now a days, the protocols [10–12] provided different approach in terms of security and privacy for vehicular cloud computing. Further, we have reviewed Vasudev et al. [1] and found that it fails in vehicle server impersonation attack, insider attack, and forward secrecy. We also discuss design flaws and other weakness of the same.

1.1 Motivation and Contribution

The main purpose of this comment article is to point out the key security issues in Vasudev et al.'s protocol [1]. Thus, we discuss how Vasudev et al.'s protocol is subject to vehicle server impersonation, insider attacks, forward secrecy, design errors, and other vulnerabilities. Further, we suggest the improvement for the future research in V2V communication network.

1.2 The Paper's Road Map

The following is a map of the rest of the paper. Section 2, review of Vasudev et al. protocol [1]. Section 3, statement of the problem. Section 4, suggested improvement. At last, we discuss future directions and conclusion of this work. We will start with Table 1, which contains all of the notations used in this paper.

2 Vasudev et al. Protocol [1] Evaluation Review of Vasudev

We will go over Vasudev et al. protocol's in this session. It includes "Registration phase" and "Login, authentication and communication phase" as under.

2.1 Registration Phase

V_s uses the *RA* registration form, as shown below.:

Step 1. To register with *RA*, V_s inputs UID_i, PW_i, generates random number RU_i, computes $HUID_i = h(UID_i \| RU_i)$, $HPW_i = h(PW_i \| RU_i)$ and sends $\{HUID_i, HPW_i\}$ to *RA* via secure channel.

Step 2. On receiving $\{HUID_i, HPW_i\}$, *RA* computes $A_1 = h(HUID_i \| K_s)$, where K_s is the secret shared key between *TA* and *VS*. Further, *RA* computes $B_1 = A_1 \oplus h(HUID_i \| HPW_i)$, stores $\{A_1, B1\}$ in smartcard *SC* and sends $\{A_1, B_1\}$ to V_s via secure channel.

Step 3. On receiving $\{A_1, B_1\}$, V_s computes $Z_i = UID_i \oplus RU_i$ and stores Z_i into *SC*.

The process registration of V_s is shown in Table 2.

2.2 Login, Authentication and Communication Phase

During this session, V_s and *VS* authenticate to each other using *TA* and keep the session key as shown below:

Step 1. V_s login with ID_i, PW_i and verifies $A_1 \stackrel{?}{=} B_1 \oplus h(HUID_i \| HPW_i)$. Then, V_s generates time-stamp T_u, random nonce N_u, computes $Msg_1 = h(A_1 \| T_u \| HPW_i \| N_u)$, $Y_1 = h(B_1 \| HPW_i)$, $X_1 = N_u \oplus Y_1$ and sends $\{Msg_1, X_1, T_u, SID\}$ to *TA* via insecure channel.

Step 2. On receiving $\{Msg_1, X_1, T_u, SID\}$, *TA* computes $Y_1^* = h(B_1 \| HPW_i)$, $N_u^* = X_1 \oplus Y_1^*$ and verifies $Msg_1 \stackrel{?}{=} h(A_1 \| T_u \| HPW_i \| N_u)$. Then, *TA* computes

Table 2 V_s registration phase of Vasudev et al. [1]

Vehicle (Vs)	Registration authority (RA)
Inputs UID_1,PW_1	
Generates random number RU_1	
Computes $HUID_1 = h(UID_1 \| RU_1)$	
Computes $HPW_1 = h(PW_1 \| RU_1)$	Computes $A_1 = h(HUID_1 \| K_a)$
Sends $\{HUID_1, HPW_1\}$	Where K_a is the secret shared key between TA and VS
$\cdots\cdots\cdots\cdots \rightarrow$	Computes $B_1 = A_1$ h $\bigoplus h(HUID_1 \| HPW_1)$
(via secure channel)	Stores $\{A_1, B_1\}$ in smartcard SC
	Sends $\{A_1, B_1\}$
	$\leftarrow \cdots\cdots\cdots\cdots$
Computes $Z_1 = (UID_1 \bigoplus RU_1)$ Store $\{A_1, B_1, Z_1\}$ into SC	(via secure channel)

$HCID = h(HUID_i \| CID \| SID)$, generates time-stamp T_c, computes $Msg_2 = h(HCID \| K_s \| T_c \| N_u)$, $X_2 = N_u \oplus h(K_s)$ and sends $\{Msg_2, X_2, T_c, HCID\}$ to VS via insecure channel.

Step 3. On receiving $\{Msg_2, X_2, T_c, HCID\}$, VS computes $N_u^* = X_2 \oplus h(K_s)$ and checks $Msg_2 \stackrel{?}{=} h(HCID \| K_u \| T_c \| N_u^*)$. Then, selects N_s, computes session key $S_k = h(HCID \| N_s \| N_u)$, generates time-stamp T_s, computes $X_3 = h(N_u \| N_s \| T_s \| K_s)$, $Msg_3 = N_s \oplus N_u$ and sends $\{Msg_3, X_3, T_s\}$ to TA via open channel.

Step 4. On receiving $\{Msg_3, X_3, T_s\}$, TA computes $N_s^* = Msg_3 \oplus N_u$ and verifies $X_3 \stackrel{?}{=} h(N_u \| N_s \| T_s \| K_s)$. Then, computes $w = N_s \oplus HPW_i, X_4 = h(N_u \| N_s \| HPW_i)$ and forwards $\{X_4, w\}$ to V_s via insecure channel.

Step 5. On receiving $\{X_4, w\}$, V_s computes $N_s^* = w \oplus HPW_i$ and verifies $H_4 \stackrel{?}{=} h(N_u \| N_s \| HPW_i)$. After that V_s computes $S_k = h(HCID \| N_u \| N_s)$.

This phase's process is depicted in Table 3.

3 Statement of the Problem

In this section, we discuss design flaws, cryptanalysis, and other weaknesses of Vasudev et al. [1] protocol. The details of these problems are as below.

Table 3 Login, authentication, and communication phase of Vasudev et al. [1]

Vehicle V_s	Trusted Authority TA	Vehicle Server VS
Input ID_t and PW_t		
Verifies $A_1 \stackrel{?}{=} B_1 \bigoplus h(\{HUID_t, HPW_t\})$		
Generates time-stamp T_u		
Generates random nonce N_u		
Computes $Msg_1 = h(A_1 \| T_u \| HPW_1 \| N_u)$		
Computes $Y_1 = h(B_1 \| HPW_1)$	Computes $Y_1^s = h(B_1 \| HPW_1)$	
Computes $X_1 = N_u \bigoplus Y_t$	Computes $N_a^u = X_1 \bigoplus Y_1^a$	
Sends $\{Msg_1, X_1, T_u, SID\}$	Verifies $Msg_1 \stackrel{?}{=} h(A_1 \| T_u \| HPW_t \| N_u)$	
$\cdots\cdots\cdots\cdots \rightarrow$	Computes HCID =	
	$h(HUID_t \| CID \| SID)$	
(via insecure channel)	Generates time-stamp T_e	
	Computes $Msg_2 h(HCID \| K_s \| K_e \| N_u)$	Computes $N_*^u = X_2 \bigoplus h(K_s)$
	Computes $X_2 = N_u \bigoplus h(K_s)$	Verifies $Msg_2 \stackrel{?}{=} h(HCID \| K_c \| T_c \| N_u^s)$
	Sends $\{Msg_s, X_2, T_c, HCID\}$	Generates random nonce N_s
	$\cdots\cdots\cdots\cdots \rightarrow$	Computes HCID =
	$h(HUID_t \| CID \| SID)$	
	(via insecure channel)	Generates time-stamp T_c
	Computes $X_3 = h(N_s \| N_u \| N_s \| T_s \| K_s)$	
		Computes $Msg_3 = N_s \bigoplus N_u)$
	Computes $N_s^a = Msg_3 \bigoplus N_u)$	Sends $\{Msg_3, X_3, T_u\}$
	Verifies $X_3 \stackrel{?}{=} h(N_u \| N_s \| T_s \| K_s)$	$\leftarrow \cdots\cdots\cdots\cdots$
	Computes $w = N_s \bigoplus HPW_t$	(via insecure channel)
	Computes $X_4 = h(N_u \| N_s \| HPWt$	
$Computes N_*^s = w \bigoplus HPW_t$	$\leftarrow \cdots\cdots\cdots\cdots$	
Verifies $H_4 \stackrel{?}{=} h(N_u \| N_s \| HPW_i)$	(via insecure channel)	
Computes session Key S_k	=	
$h(HCID) \| N_u \| N_s$		

3.1 Design Flaws of Vasudev et al. Protocol

Vasudev et al. [1] may have the following design flaws:

DF1. The title of this paper is not suitable because authors have discussed authentication, session key agreement, and other values between V_s and *VS* via *TA*. This is not a V2V communication. This is a V2VS communication.

DF2. There is no information how *RA* send *Vs*'s registration information to *TA*.

DF3. There is no information how to *TA* share key K_s with *VS*.

DF4. If K_s is secret for *RA* or *TA*, then some parameters are not possible like $N_u^* = X_2 \oplus h(K_s)$, $h(HCID\|K_u\|T_c\|N_u^*)$, and $X_3 = h(N_u\|N_s\|T_s\|K_s)$.

DF5. Vasudev et al. protocol did not show *VS* registration phase.

DF6. In V_s registration phase, *RA* sends smartcard *SC* to V_s which contain $\{A_1, B_1\}$. On receiving it, V_s computes $Z_i = UID_i \oplus RU_i$ and includes Z_i in *SC*. In login, authentication, and communication phase, authors did not use Z_i. If *SC* is stolen or breached, then stolen verifier attack possible in this framework.

DF7. In the presented protocol, some of the parameters used in multiple time like N_u, N_u^*, N_s and N_u^*.

3.2 Cryptanalysis of the Vasudev et al. Protocol

The protocol proposed by Vasudev et al.'s fails against the following attacks:

Forward Secrecy between V_s **and** *VS*: In step 3. of login, authentication, and communication phase of Vasudev et al. protocol, *VS* computes session key $S_k = h(HCID\|N_s\|N_u)$. Further, in step 5. of login, authentication, and communication phase of Vasudev et al. protocol, V_s computes session key $S_k = h(HCID\|N_u\|N_s)$. Thus, $h(HCID\|N_u\|N_s) \neq h(HCID\|N_s\|N_u)$ by the property of way hash function. Hence, session is not possible in [1].

Vehicle Server Impersonation Attack Assume that *VS* and V_s compute same session key $S_k = h(HCID\|N_u\|N_s)$. Further, $\mathcal{A}$ collects $K_{\mathcal{A}} = K_s$ by guessing attack. Thus, $\mathcal{A}$ impersonate to *VS* as following:

IMA1. Assume that $\mathcal{A}$ collects message $\{Msg_2, X_2, T_c, HCID\}$ in insecure channel which is sent by *TA* is step 2.

IMA2. On receiving $\{Msg_2, X_2, T_c, HCID\}$, $\mathcal{A}$ computes $N_u^{\mathcal{A}} = X_2 \oplus h(K_{\mathcal{A}})$ and verifies $Msg_2 \stackrel{?}{=} h(HCID\|K_{\mathcal{A}}\|T_c\|N_u^{\mathcal{A}})$. Then, $\mathcal{A}$ selects value $N_s^{\mathcal{A}}$, computes session key $S_{\mathcal{A}} = h(HCID\|N_s^{\mathcal{A}}\|N_u^{\mathcal{A}})$, generates time-stamp $T_s^{\mathcal{A}}$, computes $X_3^{\mathcal{A}} = h(N_u^{\mathcal{A}}\|N_s^{\mathcal{A}}\|T_s^{\mathcal{A}}\|K_{\mathcal{A}})$, $Msg_3^{\mathcal{A}} = N_s^{\mathcal{A}} \oplus N_u^{\mathcal{A}}$ and sends $\{Msg_3^{\mathcal{A}}, X_3^{\mathcal{A}}, T_s^{\mathcal{A}}\}$ to *TA* via insecure channel.

IMA3. On receiving $\{Msg_3^{\mathcal{A}}, X_3^{\mathcal{A}}, T_s^{\mathcal{A}}\}$, *TA* computes $N_s^{\mathcal{A}*} = Msg_3^{\mathcal{A}} \oplus N_u^*$ and verifies $X_3^{\mathcal{A}} \stackrel{?}{=} h(N_u^*\|N_s^{\mathcal{A}*}\|T_s^{\mathcal{A}}\|K_s)$. Then, *TA* computes $w = N_s^{\mathcal{A}*} \oplus HPW_i$, $X_4 = h(N_u^*\|N_s^{\mathcal{A}*}\|HPW_i)$ and sends $\{X_4, w\}$ to V_s via insecure channel.

IMA4. On receiving $\{X_4, w\}$, V_s computes $N_s^{\mathcal{A}**} = w \oplus HPW_i$ and verifies $H_4 \stackrel{?}{=} h(N_u \| N_s^{\mathcal{A}**} \| HPW_i)$. After that V_s computes session key $S_k = h(HCID \| N_s^{\mathcal{A}**} \| N_u)$.

Thus, $S_k = h(HCID \| N_s^{\mathcal{A}**} \| N_u) = h(HCID \| N_s^{\mathcal{A}} \| N_u^{\mathcal{A}}) = S_{\mathcal{A}}$ because$N_s^{\mathcal{A}**} = N_s^{\mathcal{A}}$ and $N_u = N_u^{\mathcal{A}}$. Hence, vehicle server impersonation attack is available in Vasudev et al. protocol.

Insider Attack There are two different insider attack possible in Vasudev et al.'s protocol which are discuss as below:

Insider Attack I:

IAI1. In step 2. of login, authentication, and communication phase of Vasudev et al. protocol, *TA* sends $\{Msg_2, X_2, T_c, HCID\}$ to *VS* via insecure channel.

IAI2. On receiving $\{Msg_2, X_2, T_c, HCID\}$, *VS* computes $N_u^* = X_2 \oplus h(K_s)$, verifies $Msg_2 \stackrel{?}{=} h(HCID \| K_u \| T_c \| N_u^*)$, selects value N_s, computes session key $S_k = h(HCID \| N_s \| N_u)$, generates time-stamp T_s, computes $X_3 = h(N_u \| N_s \| T_s \| K_s)$, $Msg_3 = N_s \oplus N_u$ and sends $\{Msg_3, X_3, T_s\}$ to *TA* via insecure channel.

IAI3. On receiving $\{Msg_3, X_3, T_s\}$, *TA* computes $N_s^* = Msg_3 \oplus N_u^*$ and computes session key $S_{TA} = h(HCID \| N_s^* \| N_u^*)$.

Thus, $S_{TA} = h(HCID \| N_s^* \| N_u^*) = h(HCID \| N_s \| N_u) = S_k$, because $N_s^* = N_s$ and $N_u^* = N_u$. Further, *TA* will collect the information from *VS* and he/she can be misuse it.

Insider Attack II

IAII1. From insider attack I, *TA* computes session key $S_{TA} = h(HCID \| N_s^* \| N_u^*)$.

IAII2. In step 4. of login, authentication and communication phase of Vasudev et al. protocol, *TA* computes $w = N_s^* \oplus HPW_i$, s $X_4 = h(N_u^* \| N_s^* \| HPW_i)$ and sends $\{X_4, w\}$ to V_s via insecure channel.

IAII3. On receiving $\{X_4, w\}$, V_s computes $N_s^{**} = w \oplus HPW_i$, verifies $H_4 \stackrel{?}{=} h(N_u \| N_s^{**} \| HPW_i)$. Then, V_s computes session key $S_k = h(HCID \| N_u \| N_s^{**})$.

Thus, $S_{TA} = h(HCID \| N_s^* \| N_u^*) = h(HCID \| N_u \| N_s^{**}) = S_k$, because $N_s^{**} = N_s^* = N_s$ and $N_u^* = N_u$. Further, *TA* will collect the information from V_s and he/she can be misuse it.

Hence, insider attack is possible in Vasudev et al.'s protocol.

3.3 Other Weakness of Vasudev et al. Protocol

There are some other weakness of Vasudev et al. protocol [1] as below:

OW1. In Vasudev et al.'s paper, the authors have compared computation and communication cost among protocols like Mun et al. [2], Zhao et al. [3], Mohit et al. [4], Ying et al. [5] and Chen et al. [6].They also compared the security qualities and features of the protocols Mohit et al. [4], Li et al. [7], He et al. [8] and Shi and Gong [9]. This is major weakness of this protocol.

OW2. In Vasudev et al.'s paper, they did not discuss password/identity change phase.

OW3. In Vasudev et al.'s paper, they used smartcard-based approach, but they did not go into detail about the revocation and re-registration phases.

4 Suggested Improvement

In this session, we discuss following suggestions required in Vasudev et al.'s protocol which will be helpful in future research:

SI1. The researchers should discuss *VS*'s registration phase with proper information of secret shared key between *TA* and *VS*.

SI2. The researchers should use biometric and fuzzy extractor-based approach to avoid guessing attack.

SI3. The researchers should discuss revocation phase and re-registration phase.

SI4. The researchers should use the clock-synchronization approach.

5 Conclusion

We have discovered that Vasudev et al. framework [1] is assailable to vehicle server impersonation, insider attack, and forward secrecy between vehicle and vehicle server in this paper. Further, we have also talked about how it has many design flaws and other flaws. As a result, it is clear that Vasudev et al. scheme is unsuitable for a V2V communication network. Moreover, we have also suggested probable improvements for future research in the same environment.

References

1. Vasudev H, Deshpande V, Das D, Das SK (2020) A lightweight mutual authentication protocol for v2v communication in internet of vehicles. IEEE Trans Veh Technol. https://doi.org/10.1109/TVT.2020.2986585

2. Mun H, Han K, Lee YS, Yeun CY, Choi HH (2012) Enhanced secure anonymous authentication scheme for roaming service in global mobility networks. Math Comput Model 55(1–2):214–222
3. Zhao D, Peng H, Li L, Yang Y (2014) A secure and effective anonymous authentication scheme for roaming service in global mobility networks. Wireless Pers Commun 78(1):247–269
4. Mohit P, Amin R, Biswas G (2017) Design of authentication protocol for wireless sensor network-based smart vehicular system. Veh Commun 9:64–71
5. Ying B, Nayak A (2017) Anonymous and lightweight authentication for secure vehicular networks. IEEE Trans Veh Technol 66(12):10626–10636
6. Chen C-M, Xiang B, Liu Y, Wang K-H (2019) A secure authentication protocol for internet of vehicles. Ieee Access 7:12047–12057
7. Li X, Niu J, Bhuiyan MZA, Wu F, Karuppiah M, Kumari S (2017) A robust ecc-based provable secure authentication protocol with privacy preserving for industrial internet of things. IEEE Trans Ind Inf 14(8):3599–3609
8. He D, Kumar N, Chen J, Lee C-C, Chilamkurti N, Yeo S-S (2015) Robust anonymous authentication protocol for health-care applications using wireless medical sensor networks. Multimedia Syst 21(1):49–60
9. Shi W, Gong P (2013) A new user authentication protocol for wireless sensor networks using elliptic curves cryptography. Int J Distrib Sens Networks 9(4):730831
10. Kumar V, Ahmad M, Mishra D, Kumari S, Khan MK (2020) Rseap: Rfid based secure and efficient authentication protocol for vehicular cloud computing. Veh Commun 22:100213
11. Kumar V, Ahmad M, Kumari A, Kumari S, Khan MK (2021) Sebap: a secure and efficient biometric-assisted authentication protocol using ecc for vehicular cloud computing. Int J Commun Syst 34(2):e4103
12. Mishra D, Kumar V, Dharminder D, Rana S (2020) Sfvcc: chaotic map-based security framework for vehicular cloud computing. IET Intell Transp Syst 14(4):241–249

Authentication and Key Update in Satellite Communication

Mukesh Kumar Mishra and Priyanka Ahlawat

Abstract Satellite communication is very popular to provide the services like audio communication, video communication, positioning of satellite, message communication, etc. Satellite mobile communication is started as first climatic development as GEO type. In satellite communication, some attacks occur due to delay in communication and connection of link intermittently. In the proposed work, we have studied the paper that reduces the effect of attack and so key algorithm has been used. Authentication of satellite system is done to restrict the message from unauthorized entity. Communication delay is also reduced to improve the performance of the communication of the satellite. In this paper, we have also reviewed different schemes and mentioned that the authentication of satellite is done and key generation for the satellite and updating of the key and privacy and integrity is maintained. Authentication of the satellite is very important to be very sure that the coming message or information is from trusted source. The satellites that send the information are authentic, and we can rely on that information. Authentication of the message is also necessary so that it maintains the integrity of the message. Authenticity of the message means that message is accurate and came from the trusted source. Key update is required for satellite communication. When new LEO satellite enters into the network, then it is necessary to update the key for those LEO satellite which is compromised by adversary, so that it can be protected from adversary.

Keywords Authentication · Security · Geostationary earth orbit · Lower earth orbit · Key update · Logical key hierarchy · Group Diffie–Hellman

1 Introduction

Terrestrial and satellite mobile system have developed independently and evolved, as well as competed consequently. In 1990s, when terrestrial mobile system entered

M. K. Mishra · P. Ahlawat (✉)
Computer Engineering Department, National Institute of Technology, Kurukshetra, Haryana, India
e-mail: priyankaahlawat@nitkkr.ac.in

A. Tomar et al. (eds.), *Proceedings of 3rd International Conference on Machine Learning, Advances in Computing, Renewable Energy and Communication*, Lecture Notes in Electrical Engineering 915, https://doi.org/10.1007/978-981-19-2828-4_13

into second generation (2G) [1], satellite mobile system emerged as a first climactic development including Immasat as geostationary earth orbit (GEO) type and Iridium as low earth orbit (LEO) type. Number of users gets increased due to terrestrial GSM system, satellite mobile systems encountered difficulties in commercialization of satellite communication system faced many difficulties and it results that some companies going to be bankrupt or reorganization [2]. In third generation (3G) and fourth generation (4G), both terrestrial mobile system and satellite mobile system proposed their diversified standards in the international telecommunication union (ITU). In ITU-R WP5D, TD-SCDMA and CDMA2000 were granted as terrestrial 3G standards, and LTE was granted as terrestrial 4G standard in 2012 [3].

From 2010s, satellite mobile system gained a second climatic development using huge LEO constellations. Both terrestrial and satellite mobile systems tried to give their best services and tried to eliminate their drawbacks. Terrestrial mobile system. Terrestrial mobile system has big data rate and low latency, but our systems covered only 20% of land area which is 6% of the entire earth surface. But satellite mobile systems have high coverage and more sustainability in case of disasters mainly in earthquake. Till date each mobile system completes the requirements of long distance transmission. So, in fifth generation, all academician and industrialist want to integrate these, which can provide high quality of service anytime and anywhere.

Communication technology has been increased with the high increment of communication, and then, satellite communication became more prevalent. Satellite communication provides audio and video communication, global positioning and message communication, and weather monitoring for different purpose such as government and military works [4].

In satellite communication, the security attacks are replay attack and man-in-the-middle attack. To protect the communication from different attacks, it is essential to make authentication of each new satellite that enters into the network. Exchange of key with the satellite attached to new satellite based on base station should be done so that authentication can be done.

Our contribution in this paper is to study the different research papers and collect the ideas by different researchers associated in different research paper. For an example, authentication of satellite is necessary for protection from different attacks so that message can be safeguarded and integrity of the message is maintained. Key is updated for LEO satellites whenever it enters into the LSN by GEO. GCC also tries to update the key, but LEO is out of range from GCC so with the help of GSN, GCC updates the key whenever LEO is compromised by adversary. Authentication of satellite and key update is delayed, so different algorithm is used to reduce the delay. Also, less resources are used so that authentication is easy.

The remaining paper is organized in such a manner that in Sect. 2, we present the related works about the satellite mutual authentication and key update. In Sect. 3. We have discussed the system modelling in satellite communication. In Sect. 6, protection from different satellite is mentioned. In Sect. 7, we brief about the security concern for satellite communication. In Sect. 8, we proposed a comparative analysis table. In Sect. 9, it is discussed about the evaluation of the satellite performance. In Sect. 11, conclusion of the paper is written. In last section, we have mentioned the references from where we have collected the ideas.

2 Motivation

In past GEO satellites were used for satellite communication (SatComs) as they avoid the fast movement between transceiver of satellite and terminals because it allows for wide cover by using a single satellite. Multibeam satellite systems are mainly created to allow to reuse the more frequency and more throughput broadband rates across the area covered, like terrestrial cellular counterparts. In this regard, number of new constellation types has been developed, which motivated by advance communication technology at less cost. Due to this, it is very needed to develop the low earth orbit (LEO) constellation that will provide high throughput with low latency. Large LEO plans include more number of satellites. Besides this our focus is on medium earth orbit (MEO) where some satellites are placed in circular orbit with the equator. Finally, it is given rise to hybrid constellation that associates the different orbits. For an example, the combination of GEO and MEO connectivity, where terminals can handover between the two orbits [5].

3 Related Works

In this review paper, it is discussed about the satellite communication that several researchers have proposed their research idea about communication of satellite, authentication of new satellite [6] that coming into the network and exchanging of key, update of key and security issues [7] during the communication. There are a number of authentication and key exchange protocol are available for authenticating the new satellite and exchanging the key by those satellite among other satellites in wireless communication.

Mutual authentication between GEO and LEO satellite and GCC is done. Here, GEO satellites work as bridge. In this communication, symmetric key cryptography protocol is used for encryption purpose [8]. In this communication GCC as base station, it sends the message to GEO, and then, GEO sends the signal to LEO. Here, GEO is not single but in group so forms a GEO satellite network (GSN) and in similar way group of LEO forms LEO satellite network (LSN). In GSN, each GEO communicates with each other and exchanges the key when new satellite enters into this network. In the same way, LEO also communicates with each other and exchanges their key when new satellite enters into the network.

When first GEO satellite is launched, then no satellite network is formed, and in this scenario, mutual authentication is done between GEO/LEO and GCC. Similarly when second satellite launched then authentication is done in similar way. It is also very important to secure the communication channel between neighbouring satellites so that GEO/LEO satellites cannot directly communicate with GCC and authentication of GEO/LEO satellite is possible only when there is secure communication channel between neighbouring satellites.

After authentication of the satellite, key update is also a very important for GEO and LEO satellite. GEO satellites control LEO satellites when adversary compromise the LEO satellites. Key updation is done in two phase. First phase is authentication of satellites, and then, the second phase is update the key for compromised LEO satellites which is compromised by adversary. When GSN and LSN is formed, then key is updated by GSN because key updation by GCC is out of range, which use GEO satellite as bridge.

For authentication purpose, different security protocols are used such as public key cryptography and secret key cryptography [9] in combination which is not good due to high computation cost. Secret key cryptography is inefficient and insecure due to vulnerability. For new session key generation Diffie–hellman key exchange is used which also has some issues of impersonation attacks, and also, data is not kept confidential.

Space information network (SIN) is a networking concept of the satellite to provide the global availability of the services to all the users. Different authentication and key update mechanism are used for this SIN, but most of them is not used in current time for the perfect forward secrecy (PFS) due to either high computational cost or several messages, needs for the exchange of session keys, otherwise not secure and vulnerable to different attacks.

4 Proposed Steps of the Algorithm

Two phases for this algorithms are as follows. One is mutual authentication and key generation, and another one is sharing of new key for the exchange of encrypted data.

Step 1. Alice wants to communicate with Bob so Alice sends his signed certificate and plain text encrypted with Bob's public key to Bob.

Step 2. Bob decrypts the cypher text using his own private key to get the original message plain text by verifying the Alice's certificate using Alice's public key.

Step 3. Bob now sends the message encrypting the message with Alice's public key and certificate signed by his own private key. If Bob cannot be able to verify the Alice's message, then Bob cannot send the message to Alice.

Step 4. After verification of the message from Alice, Bob stores it for the next time use.

Step 5. If Alice sends the message again, then it must be different from the first message; otherwise due to same message, Bob will discard it to avoid replay attack.

Step 6. When Alice gets the message from Bob, the Alice decrypts the message using his own private key and certificate by Bob's public key. If it does so then both parties are now able to exchange key. Above are called mutual authentication and key exchange protocol.

Step 7. To maintain the integrity of the message, hash is used. If generated hash is not same as the received hash, then Bob discards the message as the message is forged message. New session key is shared, and Alice and Bob decrypt the encrypted message.

5 System Modelling in Satellite Communication

GEO satellite: It is geostationary earth orbit satellite which is used as bridge between GCC and LEO satellite. It is associated with three parameters such as number of GEO satellite, security and control information of the GEO satellites.

LEO satellites: It is lower earth orbit satellite. It is same as GEO satellite associated with three parameters such as number of LEO satellites, security, and control information of LEO satellites.

GCC: It works as base station. It monitors the progress of satellite. It is associated with identity, control, and security of satellites. GCC is used to assign the key, authenticate Geo and Leo satellite, and also manage them.

GSN: It is GEO satellite network. It is group of GEO satellite. In GSN GEO-GEO communication, channel is formed.

LSN: It is LEO satellite network. It is group of LEO satellite. In LSN LEO-LEO communication, channel is formed.

Communication: Here, GEO-GEO communication, LEO-LEO communication, GEO-LEO communication, GCC-GEO communication is done.

Authentication: Authentication of satellite such as GEO and LEO is done, and also, message authentication is done to maintain the integrity of the message.

6 Protection from Different Attacks

This section presents different attacks in a network. These are summarized as follows:

(i) **Man-in-the-middle attacks**: A and B are two entities sharing their keys and message with each other. C as a intruder may modify or completely change the message between A and B. In such a case A and B gets the modified or changed message and it is done by C intentionally. It is called man-in-the-middle attack [10].

For protection from this attack, algorithm is used using public key cryptography that if C comes between A and B, C neither read the message nor modify the message. Now if suppose intruder C blocks message from A, sends message with its own certificate to B instead of A's certificate, then message is not verified at B's end. Similarly, C blocks the message coming from B, sends message with its own certificate, then again, it is not verified at A's end. So at any case C is not able to intrude the message either from A or B, and thus, we can protect the message from man-in-the-middle attack using public key cryptography.

(ii) **Replay Attack**: It is the attack when A's message is stored for future to see by replaying the message. For replay attack, if C stores the A's message for future to see by replaying it, the intruder C will not get benefit because B will discard the message by understanding it a replay message. C is not benefitted by replaying the message because cypher text, session key, and counter in combination generate the hash so by replaying the message will not provide any advantage to intruder C.

(iii) **Satellite-to-Satellite Attack**: This is cyber-attack launched by other satellite. This attack is against satellite in orbit. This attacks targets the sensors of the satellite and actuators that facilitate satellite mission capabilities and can result in cyber-physical consequences. In satellite-to-satellite attack, attacker has to destroy victim satellite by attacking on sensors and actuator [11]. The defense against this attack is to protect the satellite by combining the leveraging capabilities of robust ground station and actuator of the satellite.

(iv) **Spoofing**: It is the attack to capture the stream, alter the message, and retransmit the altered message to recipient to mislead them. It attacks on the communication segment via spoofing appear as an authorized user. Recipient does not know about this spoofing and attacker acts as trusted user. Once established as a trusted user, false instruction is sent to satellite, and the mission of the satellite is vanished [12].The protection from this attack is monitoring the network for typical activities, packet filtering to find out inconsistencies of IP address and using a network attack blocker.

(v) **Electronic Attack**: The attack which occurs due to use of electromagnetic energy and directed energy to control the electromagnetic spectrum or to attack an adversary. US space system could be neutralized by jamming and/or spoofing [13].

Jamming: It is to emit noise-like signals in such a way that prevent or mask for getting the desired signals. All military and commercial satellite systems are using the uplink and downlink jamming signals. Uplink jammer is more powerful based on ground as ground base emitter as associated with the link being jammed. Downlink jammer is less powerful so to attack on downlink jammer is more easy and reliable. Attacking with downlink jammer disrupts the mission of the satellite.

7 Security Concerns for Satellite Communication

Broadcasting of messages for satellite becomes efficient when sent to different satellites but it makes it vulnerable to eavesdropping comparison to any other services on the internet. Security for communication of satellite is necessary, and group communication security is mandatory for satellite communication. So there are different security process to secure the satellite communication.

(i) Confidentiality: It is the process through which information is kept secure. Information is not disclosed to any unauthorized entity. Privacy is mandatory part so we make the data confidential using confidentiality. In satellite communication services like pay by systems and security of sensitive distributed applications use confidentiality. Unauthorized entity is not able to access the information. When message is broadcasted, then confidentiality plays a vital role to secure the message [14]. In traffic flow analysis of the traffic is not possible due to confidentiality, it protects the traffic flow from attacker to analyse the traffic because it kept away the attacker to analyse the traffic.

(ii) Authentication of satellite: It is the main concern of this paper to authenticate the satellite when it is coming to the network. Entity authentication is to identify the authentication of the entity to recognize them that it is real. We study here three types of authentication of the satellite. It is further classified as peer to peer, authentication of origin of data source, and group message authentication. Peer-to-peer authentication: This type of authentication is done when data is transferred. Here, it is authenticated that connection is not replayed. Authentication of origin of data source: Here, origin of data source is authenticated. It should restrict the duplication and/or modification of data. For this purpose, digital signature is used. Group message authentication: The multicast message received from a group of members is authenticated using this group message authentication. The mechanism of group key communication provides the authentication of group. Each member of the group shares one key. In satellite communication, public key infrastructure (PKI) is used where key is signed by the private key of server.

(iii) Non-repudiation of message: It is process of security which defines that neither of the entity can deny about the sent and received message by them. Suppose entity A sends message Msg to entity B and entity B received the message Msg sent by A. Then, A cannot deny about the sent message, and also, B cannot deny about the received message. For satellite communication, it is vital because GEO satellite sends the message to LEO, and then, both the parties should accept about the send/receive of the message.

(iv) Integrity of the message: It is the process of security where accuracy of the message is maintained. Here, it is guaranteed that no modification of packet is done. So, for satellite communication when message is sent by one satellite to other and other satellite receives the same message as sent by the sender satellite, then it is said that integrity of the message is maintained.

(v) Key management: It is the essential part of the satellite communication. Key is important element of any cryptographic algorithms. All above-mentioned security mechanism is with the use of key. Without key, security of the communication is not appropriate, because intruder can attack easily. So to manage the key is very important. So, it is the process to establish and maintain the key. One process is to provide the group key through central server to all legitimate recipient. It is a challenging task when number of satellite is joining and leaving the group. The mechanism of key distribution is used to update the key during flow of communication. Group Diffie–Hellman (GDH) group key management is famous where there is no central server.

(vi) Multicast security: The security of group key communication is more complex than the security of two party communication. The main focus is on key updates and cost for key. So for this two approaches are used such as logical key hierarchy (LKH) and group Diffie–Hellman (GDH). LKH is used to update the key, and GDH is used to reduce the number of message with or without broadcast message to reduce key cost.

8 Comparative Analysis Table Based on Comparative Analysis

GEO: It is geostationary earth orbit satellite used in satellite communication. It is associated with three variables such that number of GEO satellites, security variables, and control information for the GEO satellites. **LEO**: It is low earth orbit satellites used in satellite communication. It is also represented as three variables like GEO such that number of LEO satellites, security variables, and control information for the LEO satellites. **GCC**: It is ground controller centre that is used to authenticate the satellite. It has three modules such that identity management module, security module, and control module. It provides pre-shared key. It authenticates and manages the LEO and GEO satellites. For communication of the satellite, LEO and GEO are the good fit to describe the communication of the satellite. GEO satellite restricts fast movement between the terminal and satellite transceiver, and it covers wide range with a single satellite. LEO satellites in present time are of more interests due to providing high throughput with very less latency (Fig. 1).

LEO satellite is between the GCC and GEO satellite. GEO satellite is much far from the earth so there is an issue of signal delay. Therefore, for the satellite communication, LEO satellite communication gained more interest to be used by

Fig. 1 Architecture for LEO satellite communication

the user for better approach. Due to LEO satellite, there is less delay and very less signal attenuation. Privacy of the users information: The identity and the locations of the users and associated information must be kept confidential so that it can avoid the attackers activity. Mutual authentication: Here, mutual authentication of the users and NCC is required so many different protocols are used to make the authentication of the users and network controller centre. Low computation cost: Authentication has low cost of computation so it makes satellite communication very fast. Perfect forward and backward secrecy: The session key which is generated previously or in future is unknown to adversary, which maintains the forward and backward secrecy (Tables 1, 2 and 3).

Table 1 Different attacks and their solutions by researchers

Attacks	Principles	Solutions
Password intrusion	Hackers hack into the system due to weak password of the system user or administrator	To protect the system from this attack, users need to have a strong password of at least eight character and need to change the password after some interval of time
Vulnerability intrusion	To perform bad activities with the system, attacker uses data of harmful database security vulnerabilities	We can protect the system from this attack by finding out the loopholes of the system and remove that loopholes instead of ignoring them by seeing it as a simple vulnerability
SQL injection	Malicious SQL code is injected by the attacker to deceive the application server to execute	It is to be restricted to input data anywhere on the site in SQL to get our details from the SQL server
Zero day ransomware	Zero day attack reaches the ICS before updating of antivirus signature or firewall, and then, it can compromise the sites	To protect from such attack, antivirus software or firewall must be installed and updated or we should not open vulnerable sites
Stuxnet	This attack used to sabotage the Iranian Nuclear Program in 2010. Stuxnet contained new forms of exploit that never seen before. It first compromises less defended system and gains information about the high defended system, heavily protected and well-designed site. Then, the adversary prepares a system, autonomous malware to exploit the protected site	There is no such good solution to defeat this attack. It can attack on the system as zero day attack. Then, it is good to have a more defended system, and we should not ignore any small loopholes of the system. If we find any loopholes, then well practice to remove those vulnerability

Table 2 Comparative analysis of Internet protocol security (IPSec) versus secure socket layer (SSL) security protocol

Parameters	SSL	IPSec
Security	Unicast	Unicast
Implementation	End-to-End	End-to-end
Implementation layer	Between application and transport layer	Network layer
overhead	Less overhead than IPSec	More overhead (about 34 bytes)
Group communication	No	No
Performance with PEP	Yes	Sever degradation

Table 3 Objective, merits, and demerits of LEO, GEO, and MEO

Satellites	Objective	Merit	Demerits
LEO	It is popular with thousands of satellites operation. Its objective is to serve communication such as mass consumer and enterprise	It supports high frequency trading, virtual gaming, and high performed computing application It is smaller, lower power satellite, and cheaper than GEO	Complex racking and ground network. Complete constellation must be in place before service starts Unproven business model, risky technology
GEO	Hundreds of satellites in orbit. Its objective is to high throughput satellite to built for data	High throughput technology which enables basic broadband applications Few satellites needed for large geographical areas	Signal power losses need more satellites and antennas
MEO	It has been used for GPS and navigation applications. Its objective is to perform in remote area where laying fibre is not viable such as cruise, and maritime area.	It is for low latency compared to terrestrial network. Simple equatorial covers 96% of global population	Dual tracking antennas required for maintaining connectivity continuously

9 Evaluation of the Satellite Performance

The performance of the satellite is evaluated based on the key exchange and key update of the satellite. For key update, LKH protocol [15]is used which is useful for satellite communication to update the key because for group key there is a central server that updates the key. And cost reduction is needed for that GDH is used that reduces the message [16]. Performance of the algorithms is affected by the number of domains. Due to mentioned reason, effect of number of domains on the key updates is also investigated.

Evaluation Parameters

Communication of the satellite among other satellites. The different parameters for the communication of the satellite are mentioned below;

(i) Authentication should be lightweighted: Number of messages is reduced, and simple communication is done to be lightweighted [17].
(ii) Security: After authentication, communication between the satellite is more secure due to using the cryptographic and perfect forward secrecy (PFS) algorithm.
(iii) Speed: Number of cryptographic algorithm use should be less to be more speedy.
(iv) Bandwidth: Minimum number of messages should be exchanged and less data to be transferred to have of more bandwidth. For more bandwidth 128 bits of data [18–21] which is equal to the key length are to be sent instead of cypher text and hash function.

10 Future Challenges

In satellite systems, the method is used for fleet management, system design and certification, and these are actually based on heuristic approach. Physical testing and statistical distribution of physical properties are not suited the future satellites that demand for lighter mass and handling the higher load for long durations. To overcome these issues, digital twin is expected to play a vital role [19]. Another future application for digital twin is to enable space-based monitoring and communication services. The two metrics cost and time to address space-based services can be drastically reduced by using software-based components in satellites which are configured from the earth [20]. Also very crucial task of digital twin is to maintain the privacy of the each entity and prevent the misuse of information [21] without the permission of the concerned entity.

11 Conclusions

Authentication of the satellite and generation of key has been taken at utmost priority. So for the security purpose public key cryptography is used to provide more security because using this algorithm the generated key is not guessed easily by intruder. Here, man-in-the middle attack using public key cryptography is good. Key is updated when new satellite enters into the network. For updating the key, LKH algorithm is used. Different security parameters have been used for the security concern such as confidentiality, integrity, and non-repudiation. For better performance to make more speedy and more prone for security, GDH algorithm is used.

References

1. Yin Y, Chen L, Lu Y et al (2019) QoS prediction for service recommendation with deep feature learning in edge computing environment. Mob Networks Appl 1–11
2. Gao H, Huang W, Yang X et al (2018) Towards service selection for workflow reconfiguration: an interface based computing. Future Gen ComputSyst (FGCS). 87:298–311
3. Gao H, Chu D, Duan Y et al (2017) The probabilistic model checking based service selection method forbusiness process modeling. J Software EngKnowl Eng. 27(6):897–923
4. Gao H, Xu Y, Yin Y et al (2019) Context-aware QoS prediction with neural collaborative filtering for internet-of-things services. IEEE IoT J. https://doi.org/10.1109/JIOT.2019
5. Ashjaee J, Rapoport LB, Kinkulkin D et al (2015) Satellite differential positioning receiver using multiple base-rover antennas. U.S. Patent No. 9,035,826
6. Dunigan T, Caw C (1998) Group key management
7. Atkinson R (1995) Security architecture for the internet protocol. RFC 1825
8. Aziz A, Patterson M (1995) Design and implementation of SKIP. In: INET'95
9. Cao C (1997) Group key management. Technical report, University of Tennessee
10. Benmalek M, Challal Y, Derhab A, Bouabdallah A (2018) VerSAMI: versatile and scalable key management for smart grid AMI systems
11. Rabieh K, Mahmoud M, Akkaya K, Tonyali S (2017) Scalable certificate revocation schemes for smart grid AMI networks using bloom filters. IEEE Trans Depend Secure Comput 14(4):420–432
12. Mahmoud M, Misicc J, Akkaya K, Shen X (2015) Investigating public-key certificate revocation in smart grid. IEEE Internet Things J 2(6):490–503
13. Ghosal A, Conti M (2019) Key management systems for smart grid advanced metering infrastructure: a survey. IEEE Commun Surv Tutor 21(3)
14. Kodheli O, Lagunas E, Maturo N, Sharma SK, Shankar B, Montoya JF, Duncan JC, Spano D, Chatzinotas S, Kisseleff S, Querol J (2020) Satellite communications in the new space era: a survey and future challenges
15. Arslan MG, Alagöz F (2006) Security issues and performance study of key management techniques over satellite links
16. Chen S, Sun S, Kang S (2022) System integration of terrestrial mobile communication and satellite communication—the trends, challenges and key technologies in B5G and 6G
17. Fourati F, Alouini M-S (2021) Artificial intelligence for satellite communication: a review
18. Ginter A (2018) VP industrial security, Waterfall Security Solutions. THE TOP 20 CYBER ATTACKS, Version 1.1
19. Altaf I, Saleem MA, Mahmood K, Kumari S, Chaudhary P, Chen CM (2020) A lightweight key agreement and authentication scheme for satellite-communication systems
20. Tomar A et al (2020) Machine learning, advances in computing, renewable energy and communication. In: LNEE, vol 768. Springer Nature, Berlin, 659 p. https://doi.org/10.1007/978-981-16-2354-7. ISBN 978-981-16-2354-7
21. Iqbal A et al (2020) Renewable power for sustainable growth. In: LNEE, vol 723. Springer Nature, Berlin, 805 p. https://doi.org/10.1007/978-981-33-4080-0. ISBN 978-981-33-4082-4
22. Gao H, Duan Y, Miao H et al (2017) An approach to data consistency checking for the dynamic replacement of service process. IEEE Access 5(1):11700–11711

An Empirical Study on Factors Affecting Selection of Software Development Life Cycle Models

Mitali Chugh

Abstract With the rapid rise of the need to improve software development, it is significant to know which factors can affect the selection of software development life cycle. A decent number of models and the unending discussion between the promoters of traditional, agile, and hybrid models choose an accurate one uneasy and complex. This study intends to provide a view in choosing development models in diverse project situations. The present survey-based study was carried out in software engineering organizations in India. Initially, the study introduces the prevalent software development life cycle models followed by the empirical results of the survey that have significant takeaways for the practitioners in software organizations. The findings of the study show that agile methodology is most prevalent for software development and user requirements understanding is the most significant factor in the selection of development model followed by project type.

Keywords Software development life cycle model · Agile methodology · Traditional methodology · Software development

1 Introduction

The software systems developed over the years have become an integral part of the organizations in diverse sectors such as healthcare, defense, banking, and transportation to compete in the dynamic competitive market environments [1–30]. To develop quality software that satisfies client requirements due to constraints of budget and schedule is challenging for software developers and thus creates a need for a systematic approach to developing software. In the earlier days, the programmers followed the build and fix approach to develop software. When following this approach, the developers were unaware that exactly how the system would look, however, due to the small size and simple software systems, the development was feasible. However,

M. Chugh (✉)
UPES, Bidholi, Dehradun, Uttarakhand 248001, India
e-mail: mchugh@upes.ac.in

A. Tomar et al. (eds.), *Proceedings of 3rd International Conference on Machine Learning, Advances in Computing, Renewable Energy and Communication*, Lecture Notes in Electrical Engineering 915, https://doi.org/10.1007/978-981-19-2828-4_14

due to the requirement of large and complex system development nowadays, development teams and stakeholders must work together to develop reliable and high-quality software to attain a competitive edge in unpredictable market conditions. Without the appropriate and systematic plan for software development, the developers might fail to recall ideas that are required for the development of high-quality software as other features need to be selected and ordered. To accomplish the systematic development for high-quality software, software development life cycle (SDLC) models are followed. The development of software that is conducted adhering to SDLC saves time facilitates management and implementation of plans along with proper documentation. In the absence of a suitable SDLC model, there is a possibility of project failure, missing deadlines, and budgets [1, 2].

The research on SDLC evolution has shown a remarkable rise, and agile methodologies are prevalent in the industry [3–5]. Many researchers and practitioners have presented a comparative account of different life cycle models to understand the suitability of the model for a project [6–9], while others have worked on the association between SDLC models and project categories [10, 11]. However, there is a dearth of studies regarding the significant factors affecting the selection of SDLC models in software industries of India. The paper is organized as Sect. 2 literature review on SDLC models employed in software development. Section 3 is the research methodology trailed by results. Section 4 is the discussion, and Sect. 5 concludes the work mentioning none of the life cycle models commonly fit all development conditions. The classical waterfall model and other iterative models such as spiral model count on static and changing requirements of clients, respectively. In addition, the agile methodologies are customer-centric and focus on the development of quality software products [9].

2 Software Life Cycle Models—An Overview

Software engineering best practices facilitate software development and ensure the development of efficient and high-quality software products. Barry Boehm has proposed seven software engineering principles for software engineering that ensure successful software development projects [12]. To develop and deploy a software product, a series of phases are followed collectively termed as software development life cycle (SDLC). At present, systematic methodologies to designing software have progressed, and notations for design have increased. The phases of SDLC comprise planning, analysis, design, development, testing, implementation, and maintenance. For each project, documentation is to capture best practices in the SE organizations for avoiding reiterating of faults in future projects. In addition, it facilitates coordination among team members [10, 13]. Diverse SDLC models exist, and SE organizations adapt their models. SE organizations can opt for the traditional, agile, and hybrid methodology for developing software.

Traditional software development methodologies are centered on predetermined phases/stages of the SDLC. The development flow in traditional models is unidirectional and for each phase, there is a particular deliverable and detailed documentation. The traditional approach includes the models such as code and fix, waterfall model, incremental build, prototype, and spiral. At the inception of traditional development methodologies, incorporated simply writing the code followed by fixing the bug referred to as code and fix model [14]. The classical and iterative waterfall models are the flow models that are the backbone of various SDLC models that were introduced in the 1970s. The classical model is rigid and does not incorporate methods for error detection and rectification that makes it unfit for real-world scenarios [15]. To understand the client requirements, prototype model is employed. The approach is to build a miniature of the system before or during the requirement phase [16]. Following the advancement of the traditional development models, the rapid application development (RAD) model was introduced by IBM in the 1980s. RAD is a variant of the waterfall model that uses a component-based approach and lays emphasis on a short development cycle [17, 18]. Next, in 1988, Boehm introduced the spiral model that had the major focus on risk analysis, covering different types of risks such as unavailability of the necessary hardware, change in requirements, cost overruns, and many more [19].

In 2001, agile methodology was introduced by Kent Beck along with 16 other software developers, and the inception of agile software development has been marked by the introduction of extreme programming methods [20]. The agile approach focuses on client satisfaction and prompt iterative software delivery. Extreme programming [21], crystal, adaptive software development (ASD), and scrum [22], are agile methodologies [4, 23]. The industry-based survey "state of agile" mentions scrum as the most prevalent agile method [24] followed by Kanban, and the least used agile method is XP. To exploit the features of both approaches of development, the software teams are using the hybrid approach.

3 Research Methodology

The objectives of the present study are:

1. To study the software development process in the SE organizations of India.
2. To recognize the factors that affect the selection of the SDLC model in Indian SE organizations.
3. To identify the significance of factors that affect the selection of the SDLC model.

To accomplish the stated objectives, a survey-based approach is undertaken in Indian SE organizations that form the universe for the study. However, it is not feasible to include the whole universe in the study, so a sampling technique is used. The sample for the study is Indian SE organizations from NCR, the self-administered questionnaire and interview technique are used to collect the information regarding

Table 1 Questionnaire

Factors for the study	
F1	Project type
F2	Project duration
F3	Project complexity
F4	Project size
F5	Risk level and its kind
F6	User requirement understanding level
F7	Interpreting application area level
F8	Involvement of client
F7	Developer's experience
F8	Size of development team
F9	Tools availability
F10	Product version
F11	Man–machine interaction
F12	Required reliability level
F13	Risk type and level
F14	Required reliability level

the selection of the SDLC model. The sample data collection was from Feb 2021 to may 2021. The survey-based methodology is employed for the present work, as it is economical and reliable. The survey results are analyzed to derive the descriptive and explanatory conclusion. In addition, interviews were carried out to gain insights into the association between factors affecting the selection of SDLC models. The three sections in the questionnaire consisted of personnel and SE organization information along with information about the software process. Following the recommendations of Straub [25], the questionnaire is designed by carrying out an exhaustive literature review for the incorporation of the prospective items in the survey [26, 27]. The questionnaire followed a five-point Likert-type scale extending from "not very significant" to "critical."

An email survey facilitates the gathering of results speedily, ensures the anonymity of the respondents, and is inexpensive [28], thus, it was the preferred method for data collection in this study. In this study, 200 questionnaires in total were emailed to software firms. The rate of response from the SE organizations was 50% (i.e., 100) which is above the minimum standard of 40%, as suggested for academic studies by Baruch [6]. The core area of SE organizations contributing in the survey included product development, consultancy, and both (Table 1).

4 Findings of Empirical Investigation

The software organizations participating in the survey dealt in the service areas of the Internet, data mining, training and education, software consultancy and services, software vendor, data warehousing, telecommunication, ERP, advanced databases, and engineering design services. About 65% of SE organizations had various service areas, and the remaining 35% had merely one service area.

Figure 1 shows the distribution of employee strength in SE organizations participating in the survey where only 5% had employees less than 100 and 50% of SE organizations had more than 1000 employees. Hence, in the survey, a major contribution is from the large organizations.

Figures 2 and 3 show the distribution of the employee designation and experience in the SE organizations of the survey. All the respondents have been members of the team in multiple projects of diverse business domains and incorporated the state of art technologies in the software industry. In addition, the participants have functioned

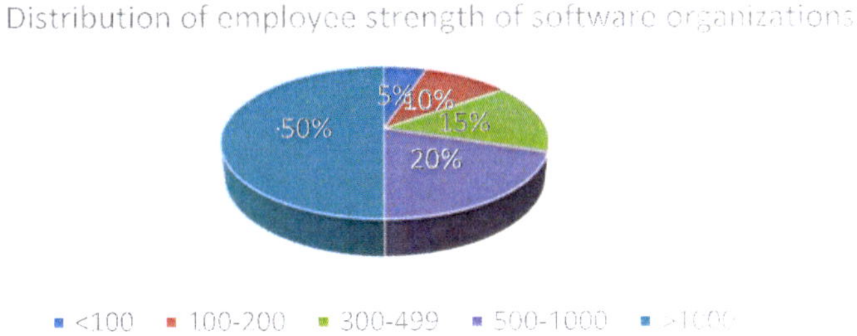

Fig. 1 Distribution of employee strength

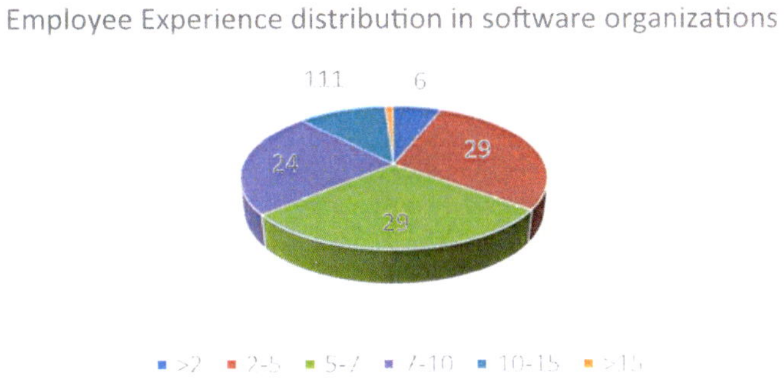

Fig. 2 Distribution of employee experience

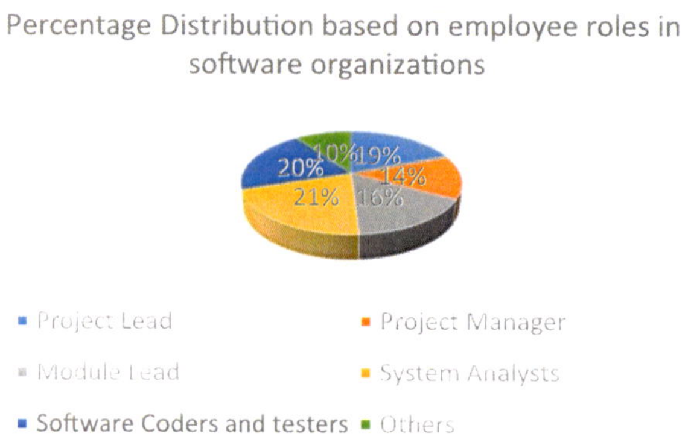

Fig. 3 Employee roles in software organizations

on various project types, and the distribution of employees involved in each project type is presented in Fig. 4.

Figure 5 shows the SDLC models used in software projects and agile methodologies are most popular with 40% of projects being developed using agile methodologies. Hybrid approach is the next methodology, which is being used in 30% of projects. The remaining 30% of projects have used traditional methodology. In addition, software professionals are at ease when they are working with the agile methodology as it emphasizes developing working products that clients can get their hands on fast, versus spending time on writing specifications beforehand. In the context of entry-exit standards, the waterfall model has a rating of 5 which represents it to be most rigid when compared to other models having a rating of 3.

The survey questionnaire also included the key attributes (14) derived from literature [27] and significant in the software development. The contributors prioritized these attributes on a scale of 1–14. '1' represented the most significant attribute and

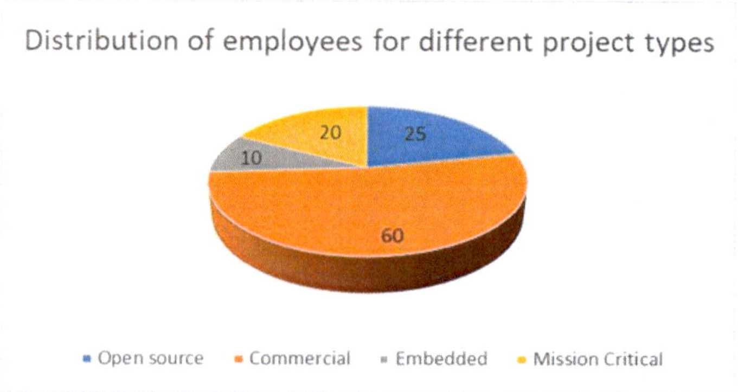

Fig. 4 Distribution of employees in different project types

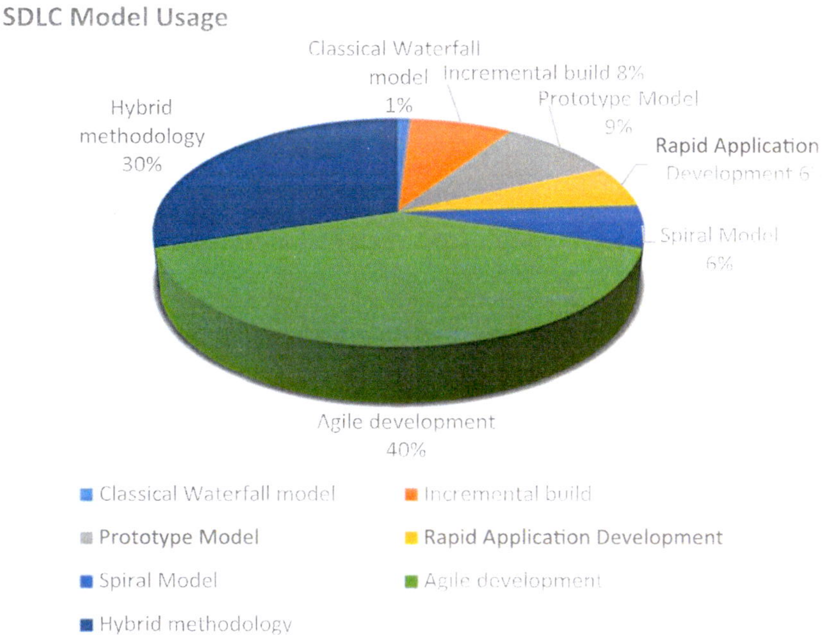

Fig. 5 SDLC models distribution in software projects

'14' represented the least significant attribute, the average scores of which are as under:

- Functionality-1.3
- Efficiency-2.5
- Timeliness, Cost-2.1
- Reliability, Correctness,-1.5
- Integrity-2.4
- Consistency-1.9
- Usability-3.1
- Portability-5.9
- Maintainability-3.3
- Reusability, complexity-3.9.

Next, the participants rated attributes on a Likert scale of 1–5 (1 = no influence and 5 = essential).

To calculate the mean for individual factors, sum of scores was divided by the number of observations per factor. The standard deviation and mean scores of factors that affected the selection of SDLC models are presented in Table 2. In addition, Fig. 6 represents mean scores for each factor where the highest score is for F6, and

Table 2 Critical factors for the selection of SDLC

Factor 1	Project type
Factor 4	Project size
Factor 6	User requirement understanding level
Factor 8	Involvement of client

Fig. 6 Scores for factors affecting selection of SDLC models

the scores are greater than and equal to 4 for F1, F4, F6, and F8. Hence, F6 is the most significant factor for the selection of the SDLC model (Table 3).

5 Conclusion

Selection process parameters are important in the selection of the SDLC models. The present study is a survey regarding the selection of SDLC models and assists to comprehend factors that are essential for the selection of SDLC models for software organizations. The respondents in the survey have rich experience in working with different project types, viz. commercial, open-source, etc., in different roles. The findings reveal that agile methodologies are most prevalent followed by the hybrid methodology in software development in India. Our research aims to facilitate the decision process regarding the selection of SDLC models by software development teams. The factors as project type, project size, involvement of client, and understanding of user requirement level are found to be significant when selecting the SDLC model. In this study, a self-administered email questionnaire is used due to constraints of time and budget. In addition, the survey is carried out only in the Indian software industry, so the results may not be generalized. To enhance the validity of

Table 3 Factor scores for individual factors of survey

Factor	Mean	Standard deviation
F1	4.556	0.7252
F2	3.78424	0.91062
F3	3.67641	0.82126
F4	4.16183	0.59364
F5	3.9	0.89431
F6	4.62288	0.72226
F7	3.66346	0.96151
F8	4.04771	0.72644
F9	3.94706	0.98523
F10	2.99881	0.77077
F11	1.44604	0.77851
F12	2.66246	0.88424
F13	2.22666	0.86667
F14	3.99334	0.69198

the outcomes in future studies, larger sample size is required and size, type, and demographic factors can be included.

References

1. Kay R (2002) QuickStudy: system development life cycle [Online]. Available: http://www.computerworld.com/printthis/2002/0,4814,71151,00.html%0A
2. Ghezzi C, Jazayeri M, Mandrioli D (2002) Fundamentals of software engineering, 2nd edn. Prentice-Hall, India
3. Al-Zewairi M, Biltawi M, Etaiwi W, Shaout A (2017) Agile software development methodologies: survey of surveys. J Comput Commun 05(05):74–97. https://doi.org/10.4236/jcc.2017.55007
4. Cockburn A, Highsmith J (2001) Agile software development, the people factor. IEEE Comput 34:131–133
5. Prabaharan S, Bhuvaneswari T, Prabaharan S (2013) A survey on software development life cycle models. Int J Comput Sci Mob Comput 2(5):262–267. [Online]. Available www.ijcsmc.com
6. Farias K, Oliveira TCDE, Scholl M (2019) Comparison of software design models: an extended. ACM Comput Surv 52(3):1–41
7. Nugroho S, Hakim L (2017) Comparative analysis of software development methods between parallel, V-shaped and iterative. Int J Comput Appl 169(11):7–11
8. Jhanjhi NZ, Naqvi M, Humayun M (2019) Analysis of software development methodologies. Int J Comput Digit Syst 8(5):446–460
9. Akinsola JE, Ogunbanwo AS, Okesola OJ, Odun-Ayo IJ, Ayegbusi FD, Adebiyi A (2020) Comparative analysis of software development life cycle models (SDLC). In: Computer science on-line conference, pp 310–322

10. Archibald RD, Voropaev VI (2003) Commonalities and differences in project management around the world: a survey of project categories and life cycle models. In: Proceeding of the 17th IPMA world congress on project management, pp 1–10
11. Desaulniers H, Douglas A, Robert J (2001) Matching software development life cycles to the project environment
12. Boehm BW (1983) Seven basic principles of software engineering. J Syst Softw 3(1):3–24. https://doi.org/10.1016/0164-1212(83)90003-1
13. Archibald DR (2004) Life cycle models for high technology projects. In: 4th International project management seminar-PMISP, pp 1–10
14. Connell MC, Carta JJ, Baer DM (1993) Programming generalization of in-class transition skills: teaching preschoolers with developmental delays to self-assess and recruit contingent teacher praise. J Appl Behav Anal 26(3):345–352. https://doi.org/10.1901/jaba.1993.26-345
15. Royce WW (1987) Managing the development of large software systems. In: 9th international conference on software engineering, pp 328–338. https://doi.org/10.7551/mitpress/12274.003.0035
16. Budde R, Kautz K, Kuhlenkamp K, Züllighoven H (1992) What is prototyping? Inf Technol People 6(2):89–95
17. Martin J (1991) Rapid application development. Macmillan Publishing Co., Inc.
18. Butler J (1994) Rapid application development in action, managing system development. Appl Comput Res 6(8)
19. Boehm B (2001) The spiral model as a tool for evolutionary acquisition. J Def Softw Eng, pp 3–11
20. Beck K (1999) Embracing change with extreme programming. Computer (Long Beach Calif) 32(10):70–77
21. Beck K (2000) Extreme programming explained: embrace change, 2nd edn. Addison-Wesley Professional, Reading
22. Schwaber K, Beedle M (2002) Agile software development with Scrum. Prentice Hall, Upper Saddle River
23. Boehm B (2002) Get ready for agile methods, with care. IEEE Comput 35:64–69
24. Incorporated Das (2020) VersionOne 14th annual state of agile report. [Online]. Available https://stateofagile.com/#ufh-i-615706098-14th-annual-state-of-agile-report/7027494
25. Straub DW (1989) Validating instruments in MIS research. MIS Q 13(2):147. https://doi.org/10.2307/248922
26. Pressman RS (2010) Software engineering practitioner's approach
27. Jalote P (2012) An integrated approach to software engineering. Springer Science and Business media
28. Sarantakos S (1998) Social research, 2nd edn. Macmillan, London
29. Tomar A et al (2020) Machine learning, advances in computing, renewable energy and communication. In: LNEE, vol 768. Springer Nature, Berlin, 659 p. https://doi.org/10.1007/978-981-16-2354-7. ISBN 978-981-16-2354-7
30. Iqbal A et al (2020) Renewable power for sustainable growth. In: LNEE, vol 723. Springer Nature, Berlin, 805 p. https://doi.org/10.1007/978-981-33-4080-0. ISBN 978-981-33-4082-4

Intelligent Approaches for Disease Diagnosis, Prevention, and Treatment

Sirineni Harshitha, Pokala Pranay Kumar, Veeramalla Rohith Goud, and Pinisetti Swami Sairam

Abstract The period of iterative insight approach is changing. With assistance of machine learning, deep learning, computer vision, one can recognize infections at premature phase and help individuals locate the early treatment which may set aside much effort for customary strategies. This chapter focuses on the methods toward early recognition, anticipation, and therapies of genuine illnesses like bosom malignancy, Parkinson's, and diabetes. The assessment of malignancy by exploring histopathological pictures personates a genuine part in the patient's turn of events, and deep learning strategies are utilized to get a bunch of boundaries from pictures used to construct convolutional networks. The models from the pretrained networks are utilized for the multi-class classification by utilizing transfer learning methods. All outcomes are utilized to feature that transfer learning gives the best assessment of breast cancer pictures. Parkinson's sickness can require around 5–6 years for location however using deep learning in medical imaging which can early recognize the ailment and start the treatment.

Keywords Machine learning · Artificial intelligence · Health care · Predictive analytics · Parkinson's disease · Breast cancer

1 Introduction

Medical or the healthcare sector is a field that plays a crucial role in life. For a decade, there are many developments taking place in the field of medicine and treatment of very complicated surgeries. Many technologies are being implemented in medicine. Robots are also used in performing complex surgeries and assist doctors while performing very complex surgeries to improve the precision, flexibility, and loss of consciousness than it is in conventional methods. Robotic surgery has been rapidly adopted by hospitals in the United States and Europe for use in the treatment

S. Harshitha · P. P. Kumar · V. R. Goud · P. S. Sairam (✉)
School of Business, Woxsen University, Kamkole, Sangareddy District, Hyderabad, Telangana 502345, India
e-mail: swami.sairam@gmail.com

A. Tomar et al. (eds.), *Proceedings of 3rd International Conference on Machine Learning, Advances in Computing, Renewable Energy and Communication*, Lecture Notes in Electrical Engineering 915, https://doi.org/10.1007/978-981-19-2828-4_15

of a wide range of conditions. Robotic knee replacement technology is one of the most successful which was being used and implemented in many multi-speciality hospitals in knee replacement surgery. Artificial intelligence and robotics combined are mostly used in the field of medicine and the healthcare sector. Robotic surgery has many benefits such as more precision surgery, significantly less pain, less risk of infection and blood loss, better clinical outcomes, early discharge from hospital, reduce recovery time or recovery period, reduce the surgery cost, and enhances visual field for surgeon, and access to hard-to-reach places for the surgeon while performing surgery [12].

During these past few years, artificial intelligence is playing a key role in making the prediction and helping the doctors to identify the problem and find a cure to it very quickly [1–23]. As we all know that there are multiple definitions of artificial intelligence. One such definition that suits best for our chapter is artificial intelligence is defined as a computational program or the logics that are used to replicate the human behaviors or human brains by performing various stages of training and testing process by using previously available data. Artificial intelligence in the medical field is used to replace the daily routine works done by humans which are time taking.

Image analysis is a current trend that is playing a key role in the medicine or healthcare sector in disease detection using the image reports such as MRI scans. Image analysis done by the human providers is a very time-taking process which can be done by the machine learning algorithms very quickly. There are many machine learning algorithms such as neural networks which analyze the images and give instant output. The accuracy of the given output depends on the training and testing dataset and how accurate the datasets which we provide. Comparatively, we can get better accuracy in a short period using machine learning algorithms and artificial intelligence.

2 Applications of ML Algorithms

Use of machine learning and artificial intelligences for patients monitoring, assisting surgeons, and hospital managers' in their day to day activities are becoming much easier. Still, there is an urge of artificial intelligence (AI) leading to dramatic changes in all facets of health care, from concept to implementation. It is commonly anticipated that rather than totally substituting the job of healthcare experts, AI technologies would encourage and strengthen human labor. AI can aid healthcare workers with duties such as data analysis, clinical reporting, and illness detection programs, as well as technological assistance in areas like as image processing, medical file system automation, and, most importantly, monitoring medical files depending on their phases. The major two diseases where the need of excessive research toward the implementation and early detection of breast cancer and Parkinson disease are explained in detail in Sects. 2.1 and 2.2.

2.1 Detection of Breast Cancer

Breast cancer is a serious health problem which doctors in the world are concerned about [14]. This type of cancer starts as a group of tumor cells in the breast cells, which can then spread to other areas of the body or invade surrounding tissues. This malignancy begins in the cells, which seem to be the basic building blocks of tissues. Breast as well as other parts of the body contains tissues [9]. Different cancer diseases impact millions of people around the world, according to estimates[18]. According to the National Breast Cancer Foundation, it is estimated in 2020, 48,530 new cases will be identified in the U.S. Not only women, but men will also be affected by breast cancer [7]. But, most women are affected by this type of cancer. The lobules (milk glands) or the ducts that attach the lobules to the nipple which is where most breast cancers start. Breast cancer commonly has no manifestations when the tumor is little and most effortlessly treated, which is the reason screening is significant for early location. The most widely recognized actual sign is an effortless knot. Breast pain or weight, industrious changes such as skin expansion, thickening, or redness, and areola changes such as unrestricted release, texture, or removal are some of the most unusual signs and symptoms [8].

Amrane et al. experimented using machine learning; classified the person suffering from breast cancer using the University of California provided dataset which contains 11 columns. In this, they removed the ID which is not a useful feature in classification. The dataset contains 683 patient's data which further divided in to benign and malignant. In this paper, the authors used Naïve Bayes and KNN algorithms which gave acceptable accuracies. Naïve Bayesian classifier got 96.19%, and KNN got 97.51% accuracy which describes according to this data, KNN gave the highest accuracy. This concluded that KNN gave accurate results for the data. The KNN algorithm structure flows where it calculates Euclidean distance after training the dataset. This Euclidean distance finds the distance between the data points and the testing instances to predict outcome [4].

Abdel-Zaher & Eldeib used deep belief networks which come under unsupervised learning which learns features on its own using the Wisconsin breast cancer dataset. The results showed as DBN gave high accuracy with 99.68%. The authors used Matlaba 2014 version where they divided training and testing as 80–20%. While training, the training data are divided into train and validation with a 70:30 ratio [1]. The author Quinlan obtained 94.74% accuracy where he used a K-fold algorithm with the C4.5 decision tree method. He defined tenfold cross-validation which gave high classification accuracy.

Hamilton et al. mentioned their usage of algorithms and resulted in 96% accuracy where they used the RIAC approach of classifying breast cancer [13]. Gunes and Polat examined using a support vector machine where they analyzed their algorithm performance using metrics like specificity, accuracy, and confusion matrix where they got 98.53% accuracy (Polat and Güneş 2007). Nauck and his colleague Kruse showed a different approach of using fuzzy techniques to classify breast cancer which gave 95.06% of accuracy [19]. Ubeyli and Derya utilized many resources to compare several algorithms. They used a support vector machine, multi-layer perceptron neural network, recurrent neural network, probabilistic neural network, a combined neural network where the analysis results showed SVM got the highest accuracy with 99.54, and second highest is RNN with 98.61%. These authors utilized Matlab as their analytics platform using the Wisconsin breast cancer dataset.

These experiments are conducted on numeric data which is stored in CSV or excel format. Now, using the latest-gen technology, many researchers started doing experiments on images. Habibzadeh et al. experimented on histopathology images using deep learning techniques. The authors used Resnet V1 50 and Resnet V1 152 pretrained neural network which gave 96.4% and 94.6% accuracies [17]. PranayKumar et al. discussed experimental results where they experimented on BreakHis dataset which gave overall accuracy as 90%.

Diagnosis of Breast Cancer

There are different ways to diagnose breast cancer, of which most suitable method is through mammogram, which can help decide whether these manifestations are demonstrative of the presence of malignant growth. When contrasted with screening mammograms, symptomatic mammograms give a more itemized X-ray of the breast-utilizing strategies. They are likewise utilized in unique conditions, for example, for patients with breast inserts. The power of a mammogram to diagnose breast cancer can be affected by the scale of the tumor, the thickness of the breast tissue, and the radiologist overseeing and analyzing the mammogram. Mammography is less likely to detect bosom tumors in women under the age of fifty than in women over the age of fifty. This may be that younger women have denser bosom tissue, which appears white on a mammogram. On a mammogram, a tumor looks white, rendering it impossible to identify [5]. The other way to detect breast cancer is an MRI scan, which utilizes radio waves and solid magnets to make itemized photos of within the breast.

The advancements in technology lead to an early detection and classify breast cancer using histopathology images or MRI scans. Using deep learning techniques, scientists and researchers can identify the cancer using images. This advance computing helps the medical field reduce the cost and early detection helps to diagnose the disease as early as possible.

2.2 *Detection of Parkinson Disease*

Parkinson is a brain disorder disease that is caused by progressive nervous system imbalance leads to symptoms like stiffness, difficulty in walking, improper posture, slowing of movement, balance, and coordination, and majorly hand tremors or hand shakings. It is seen that not everyone with the disease will have the same symptoms, and if so, they do, they don't need to experience it with the same order or at the same intensity as other person does. There are five stages of progression in PD which are typical patterns as illustrated in Fig. 1.

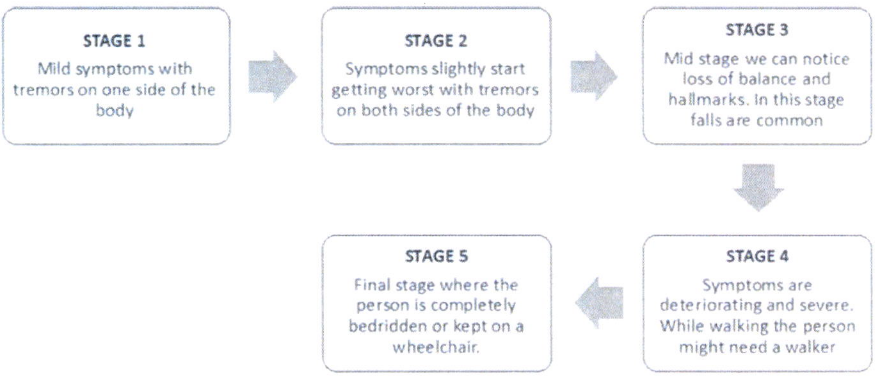

Fig. 1 Stages of Parkinson's disease

Stage One: In this initial stage, the patient will see mild symptoms which do not show any impact on the daily activities. The patient might notice tremors and other movement (improper posture, dizziness) symptoms occurring on one side of the body. Changes in walking, facial expressions, and posture might occur.

Stage Two: In this stage, the symptoms slightly start getting worst. Tremor, other movements, and rigidity are seen on both sides of the body. Walking problems and improper posture might become more severe. The person affected by it can still do his works alone, but the daily task might become difficult and lengthy.

Stage Three: Since this is the mid-stage, we can notice the loss of balance and hallmarks which means tremor (shaking) in hands, legs, arms, and face; joint stiffness; bradykinesia (slow movement), and difficulty in balance. In this stage, falls are very common.

Stage Four: In this stage, the symptoms are deteriorating and severe. Here, the person can stand without a person's help, but while walking, they might need support or a walker. In this stage, the person might need help with his daily activities, and the person is dependent on others.

Stage Five: This is the final stage very the risk is reached to greater heights, and the patient must get operated compulsory. In this stage, the joint stiffness in the legs might make it the person impossible to walk.

Early prediction of Parkinson disease

Early detection is critical for halting its progression and allowing patients to receive disease-modifying treatment. To this end, the premotor stage of the disease should be closely watched. Computer technology, especially artificial intelligence, and machine learning have a wide range of use in the diagnosis, control, and treatment of disease type is through neurologic movement disorders. Because of the high health implications heterogeneity and similarities in clinical manifestations of various neurological disorders in their early stages, these tasks are not simple. This topic aims to provide a detailed, strong review of artificial intelligence applications of kinematic analysis of movement disorders, especially Parkinson's disease, using machine learning algorithms (PD). Mainly, there are two different ways of predicting the diseases which are explained in detail below.

Early prediction of Parkinson disease through MRI Scans

The research's dataset came from the Parkinson's Progression Markers Initiative (PPMI) database; since PPMI is a multi-center study, the imaging scans acquired in the study include both temporal and spatial variations [10]. In this study, the MRI scans collected from the PPMI site are considered as the source image to do the prediction, and the data that are collected from other sources like Montreal Neurological Institute (MNI) and Individual Brain Atlases using Statistical Parametric Mapping (IBASPM) are considered as the target image [10]. The reason for selecting the PPMI dataset, PPMI is a ground-breaking project that is working with collaborators across the globe to build a strong accessible data collection and bio-sample repository to accelerate scientific discoveries and innovative therapies. PPMI's aim seems to be to characterize biochemical and therapeutic shifts throughout the illness continuum, with a focus on indicators among in persons, to detect disease, and treatment spots as quickly as feasible. The main aim of this method focuses on the detection and classification of Parkinson's disease as a healthy person or infected person with the help of 3D convolutional neural networks (CNNs).

A 3D-MRI analysis can be conducted for the study to detect Parkinson's disease with the help of using a 3D convolutional neural network. Researchers used 3D-MRI scans of the full brains to decipher complex abnormal state in all the brain's subcortical systems to diagnose Parkinson's disease. A particular performance metrics were considered in the estimation of the CNN model, and a prior test or hypothesis was designed to test the values of the performance metrics. Following the preparation of the 3D convolution layer, it was discovered that the model worked better than anticipated, strongly associating with the studies before hypothesis and displaying some impressive outcomes. The outcomes of the model are very motivating, but there is also a lot of unexplored ground in terms of designing creative frameworks that can be used to diagnose Parkinson's disease utilizing 3D CNN. Moreover, in the present study, the performance is done on the whole brain, but in future, we can just take a part or side of a brain and make the predictions. We can develop different kinds of algorithms for faster prediction and detection of the disease (Fig. 2).

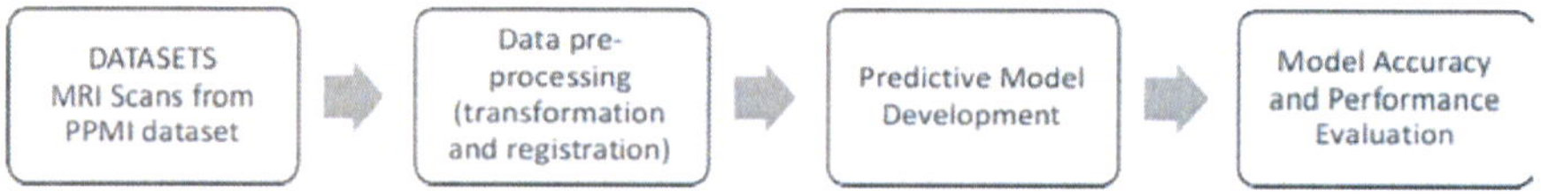

Fig. 2 Model development process

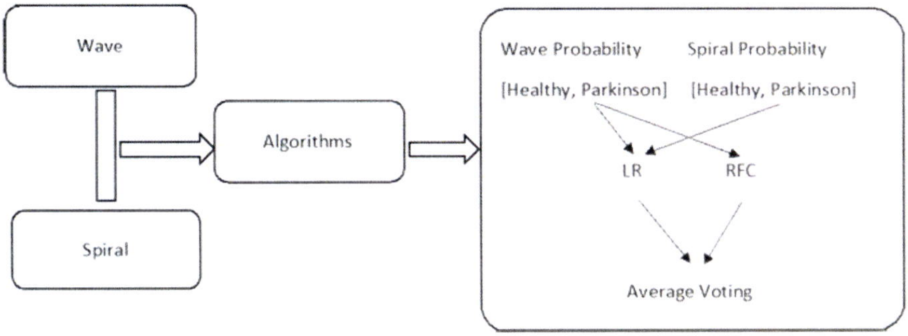

Fig. 3 Algorithm development process flow

Early prediction of Parkinson disease through spiral and wave drawings

This is another method for the prediction and early detection of Parkinson's disease. The advancement of clinical decision support necessitates the identification and recognition of the appropriate biomarkers concerning specific health problems. It has been identified that handwriting deficiency is directly related to the intensity of the people affected with Parkinson's disease (PD) in patients. In the proposed method, spiral drawings and the wave from the patients suffering from Parkinson's and healthy people were collected. Besides, in people with Parkinson's disease, the pressure and intensity applied for drawing or writing on a paper is much slower. As a result, correctly recognizing such biomarkers from the beginning of the illness can help to make more definitive clinical diagnosis. Hence, use of deep learning algorithms for analyzing the spiral and wave patterns of drawings of the people suffering from the disease can help in early identification. Figure 3 implicates the process of the algorithm development with use of wave and spiral drawings.

For better accuracy in the detection/prediction, we can use machine learning algorithms/deep learning algorithms as discussed in Table1.

3 Future Scope and Conclusion

Advancement in the healthcare sector solved many health issues. Further, the advancements must save human life by providing advanced medical care. For research, the major challenging situation is the availability of data. Since there were no technologies or resources available before, health professionals couldn't gather and interpret vast volumes of data for accurate forecasts and treatment. The main advancements are used to early prediction which is used to make next-gene medicines. Further, technology gives an overall idea to use these pretrained models to integrate with mobile applications that are used to build machine learning mobile applications. The next-gen Medicare technology helps doctors in decision-making. Machine learning advancements are used further to predict the disease,

Table 1 Algorithms and prediction accuracies

Algorithm	Dataset	Dataset attributes	Approximate accuracy rate	References
XGBoost	PPMI (MRI Scans)	MRI scan images of suffering and not suffering persons. Yes or no in both the data parts is considered	96%	[2]
Logistic regression	Spiral and wave	Yes and no of healthy person and Yes and no of unhealthy person	79%	[15]
Random forest	Voice and speech recognition	Total 26 attributes, features 1–5: Jitter features 6–11: Shimmer features 12–14: AC, NTH, HTN, features 15–19: Pitches features 20–23: Pluses, features 24–26: Fraction of locally unvoiced frames, number of voice breaks, degree of voice breaks	> 90%	[15]
Forward back-propagation artificial neural networks (FBANNs)	PPMI (MRI Scans)	MRI scan images of suffering and not suffering persons. Yes or no in both the data parts is considered	97%	[15]

and fine-tuning the algorithms gives accurate results which help in diagnosing the diseases.

References

1. Abdel-Zaher AM, Eldeib AM (2016) Breast cancer classification using deep belief networks. Expert Syst Appl 46:139–144. https://doi.org/10.1016/j.eswa.2015.10.015
2. Abós A et al (2017) Discriminating cognitive status in Parkinson's disease through functional connectomics and machine learning. Sci Rep 7(1):1–13. https://doi.org/10.1038/srep45347
3. Ahamed F, Farid F (2019) Applying internet of things and machine-learning for personalized healthcare: Issues and challenges. In: Proceedings—international conference on machine

learning and data engineering, iCMLDE 2018. Institute of Electrical and Electronics Engineers Inc., pp 22–29. https://doi.org/10.1109/iCMLDE.2018.00014

4. Amrane M et al (2018) Breast cancer classification using machine learning. In: 2018 Electric electronics, computer science, biomedical engineering' meeting, EBBT 2018. Institute of Electrical and Electronics Engineers Inc., pp 1–4. https://doi.org/10.1109/EBBT.2018.8391453
5. Barlow WE (2002) Performance of diagnostic mammography for women with signs or symptoms of breast cancer. CancerSpectrum Knowl Environ 94(15):1151–1159. https://doi.org/10.1093/jnci/94.15.1151
6. Bohr A, Memarzadeh K (2020) The rise of artificial intelligence in healthcare applications. In: Artificial intelligence in healthcare. Elsevier, pp 25–60. https://doi.org/10.1016/b978-0-12-818438-7.00002-2
7. Breast Cancer Facts—National Breast Cancer Foundation (2020) Available at https://www.nationalbreastcancer.org/breast-cancer-facts
8. Breast cancer occurrence 3 breast cancer risk factors 12 what is the american cancer society doing about breast cancer? 26 Sources of Statistics 30 References 32 (no date)
9. Carlson RW et al (2009) Breast cancer: clinical practice guidelines in oncology™. JNCCN J Natl Compr Cancer Network 7(2):122–192. https://doi.org/10.6004/jnccn.2009.0012
10. Chakraborty S, Aich S, Kim HC (2020) Detection of Parkinson's disease from 3T T1 weighted MRI scans using 3D convolutional neural network. Diagnostics 10(6). https://doi.org/10.3390/diagnostics10060402
11. Das R (2010) A comparison of multiple classification methods for diagnosis of Parkinson disease. Expert Syst Appl 37(2):1568–1572. https://doi.org/10.1016/j.eswa.2009.06.040
12. Diana M, Marescaux J (2015) Robotic surgery. Br J Surg 102(2):e15–e28. https://doi.org/10.1002/bjs.9711
13. Hamilton HJ, Cercone N, Shan N (1996) RIAC: a rule induction algorithm based on approximate classification. Department of Computer Science University of Regina, Saskatchewan, CANADA
14. Harris JR et al (1992) Breast cancer. N Engl J Med 327(5):319–328. https://doi.org/10.1056/NEJM199207303270505
15. Islam MS, Parvez I, Deng H, Goswami P (2014) Performance comparison of heterogeneous classifiers for detection of Parkinson's disease using voice disorder (Dysphonia). In: 3rd international conference on informatics, electronics & vision 2014. IEEE. https://doi.org/10.1109/iciev.2014.6850849
16. Marr B (2018) How is AI used in healthcare—5 powerful real-world examples that show the latest advances. Forbes. Available at https://www.forbes.com/sites/bernardmarr/2018/07/27/how-is-ai-used-in-healthcare-5-powerful-real-world-examples-that-show-the-latest-advances/?sh=234c9e655dfb
17. Motlagh MH et al (2018) Breast cancer histopathological image classification: a deep learning approach. bioRxiv. bioRxiv, p 242818. https://doi.org/10.1101/242818
18. Nahid AA, Mehrabi MA, Kong Y (2018) Histopathological breast cancer image classification by deep neural network techniques guided by local clustering. BioMed Res Int. https://doi.org/10.1155/2018/2362108
19. Nauck D, Kruse R (1999) Obtaining interpretable fuzzy classification rules from medical data. Artif Intell Med 16(2):149–169. https://doi.org/10.1016/S0933-3657(98)00070-0
20. Noorbakhsh-Sabet N et al (2019) Artificial intelligence transforms the future of health care. Am J Med, pp 795–801. Elsevier Inc. https://doi.org/10.1016/j.amjmed.2019.01.017
21. Oh SL et al (2020) A deep learning approach for Parkinson's disease diagnosis from EEG signals. Neural Comput Appl 32(15):10927–10933. https://doi.org/10.1007/s00521-018-3689-5
22. Tomar A et al (2020) Machine learning, advances in computing, renewable energy and communication. In: LNEE, vol 768. Springer Nature, Berlin, 659 p. https://doi.org/10.1007/978-981-16-2354-7. ISBN 978-981-16-2354-7

23. Iqbal A et al (2020) Renewable power for sustainable growth. In: LNEE, vol 723. Springer Nature, Berlin, 805 p. https://doi.org/10.1007/978-981-33-4080-0. ISBN 978-981-33-4082-4

Evaluation of Trust Establishment Mechanisms in Wireless Networks: A Statistical Perspective

Ashutosh Kumar Choudhary and Surendra Rahamatkar

Abstract Establishing trust in wireless networks requires the network designer to incorporate different levels of security protocols in the network. These security protocols enable certain checks in the network which if found to be followed properly increase the trust level of nodes. A very simplistic trust establishment example is a network where in all nodes have IP (internet protocol) addresses starting with prime numbers. In such a network, if a node with a non-prime IP join in, then the other network nodes can identify this new (or discovered) node as an attacker and block communications with it. But for large-scale networks, complex trust establishment algorithms are needed, which can provide multi-level checks and thereby enhance the overall security of the network. In order to design a trust-based network, it is necessary for the network designers to evaluate and select the best protocol(s) suited for their network. To perform this task, a large amount of design time is dedicated to protocol evaluation, which delays network deployments. Thus, this text provides a statistical comparison of different trust establishment protocols applied to wireless networks and assists network designers to select the best suited protocol for their application. Moreover, this text also recommends certain upgradations to the existing protocols in order to further improve their security levels.

Keywords Trust · Security · Checks · Validation · Attacks · Blockchain · Machine learning

A. K. Choudhary (✉)
Department of Computer Science & Engineering, Amity School of Engineering & Technology, Amity University Chhattisgarh, Raipur, India
e-mail: ashutoshchoudhary23@gmail.com

S. Rahamatkar
Amity School of Engineering & Technology, Amity University Chhattisgarh, Raipur, India
e-mail: srahatmatkar@rpr.amity.edu

A. Tomar et al. (eds.), *Proceedings of 3rd International Conference on Machine Learning, Advances in Computing, Renewable Energy and Communication*, Lecture Notes in Electrical Engineering 915, https://doi.org/10.1007/978-981-19-2828-4_16

1 Introduction

Trust establishment in any wireless network requires the network designers to study, analyse and implement node-level, router-level and network-level checks in the network. These checks require a large amount of data analysis to be done before actual application to the networks. An example of this data analysis can be observed from Fig. 1, wherein the case of a mobile ad hoc network (MANET) [1, 2]-based trust establishment scenario is taken. In this scenario, the network requires a large amount of information from the nodes and the routers (this information is termed as first-hand and second-hand information). Using this information, the network evaluates different trust-based parameters which include, eigen-values, direct trust

Fig. 1 Example trust establishment scenario

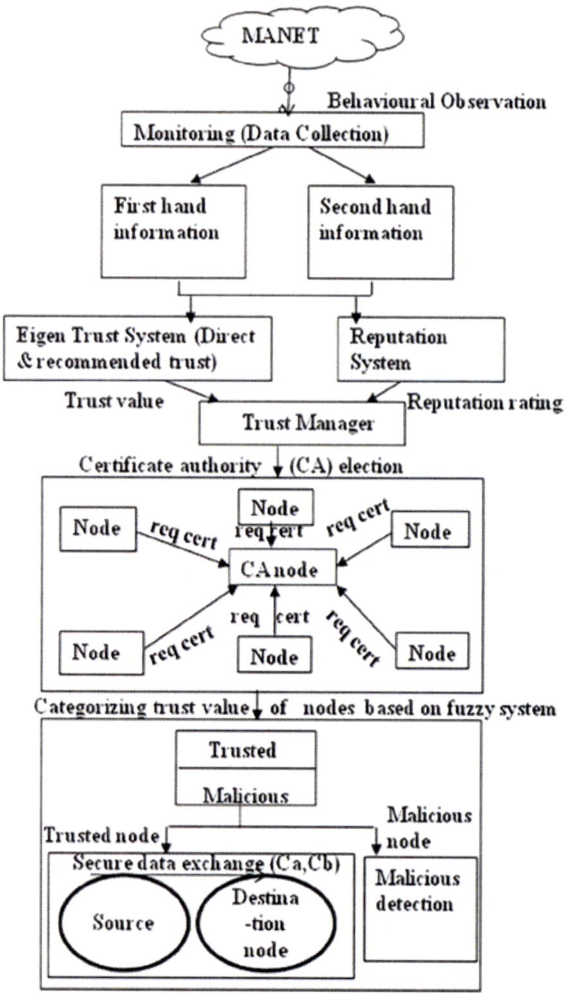

values, indirect trust values, reputation values, etc. All these values are given to a trust manager, which evaluates these values at node level, router level and network level. These evaluated values are given to central processing layer, which is depicted by certificate authority (CA) in Fig. 1. The CA requests data from each of the nodes and validates this data for integrity checks. Once the nodes are authorized by this CA, then a certificate (or a token) is issued at node level. Using the contents of this certificate, nodes can communicate with each other with high level of security. The issuing of a trust certificate involves a lot of complex classification and prediction operations. These operations are taken care by a high efficiency classifier.

Once the certificates are distributed, then the prediction engine also evaluates if the given node is malicious or not. In order to make this comparison effective, prediction engines are needed. Establishing trust in wireless networks requires the network designer to incorporate different levels of security protocols in the network. These security protocols enable certain checks in the network which if found to be followed properly increase the trust level of nodes. A very simplistic trust establishment example is a network where in all nodes have IP (internet protocol) addresses starting with prime numbers. In such a network, if a node with a non-prime IP join in, then the other network nodes can identify this new (or discovered) node as an attacker and block communications with it. But for large-scale networks, complex trust establishment algorithms are needed, which can provide multi-level checks and thereby enhance the overall security of the network. In order to design a trust-based network, it is necessary for the network designers to evaluate and select the best protocol(s) suited for their network. To perform this task, a large amount of design time is dedicated to protocol evaluation, which delays network deployments. This requires a large amount of delay, computational cost and deployment cost in the network, which can be reduced if these models are statistically reviewed and evaluated on a common platform. Thus, the main motivation of this study is to perform this computation and provides researchers with clear conceptual and statistical analysis about the reviewed models. Based on this motivation, a study of these certificate issuing protocols along with their respective predictive and classification engines is given in the next section. This is followed by a statistical performance comparison between these algorithms to conclude the best available trust establishment techniques in a given network. Finally, this text concludes with some interesting observations about the reviewed protocols and recommends different protocols to improve the same.

2 Literature Review

Due to the limitation in communication range for wireless networks, establishment of long-ranged trust protocols is very difficult. Thus, researchers in [3] introduce a notion of belief-based protocol which can be used for any wireless network. This protocol uses both network and node parameters like number of frames received, number of data packets forwarded, number of control packets forwarded, number of data packets received, number of streams established, etc. Using these parameters, the trust value is quantified, and each node is given a particular trust level. This trust level classifies the nodes into either acknowledgement based, packet precision based, gratuitous route reply based, blacklist based, beacon based, destination unreachable, salvaging and authentication objects. Each of these categories has its own trust quantifications and based on this quantification, the final trust level is evaluated. Due to such a complex trust evaluation scenario, the overall packet loss percentage increases as the number of malicious nodes is increased, which also increases the packet and byte overheads. But the average latency and throughput are improved as the chances of attacks are reduced in the network. A Low Energy Adaptive Clustering Hierarchy (LEACH) based trust establishment protocol for wireless multimedia networks is proposed in [4], wherein the responsibility of trust establishment is given to the cluster heads. This reduces computational overheads on the member nodes, thereby improving the network lifetime. Trust evaluations are kept simple in this work, and only successful and failed communications are taken for final trust value computation. The trust evaluation formula can be observed from Eq. 1, wherein the duration of communication along with packet delivery and packet loss values is considered for finding trust. These values decide the quality with which a node is able to communicate with the neighbouring nodes, which allows the cluster head to identify nodes that are creating large packet drops and thereby classify them as non-trustworthy nodes.

$$T_r = F\left(\frac{\Delta t, S}{D}\right) \tag{1}$$

where T_r is the trust level, F is the trust function, S is the number of successful packets, D is the number of dropped packets and Δt is the time duration for which this trust value is being evaluated. Due to this simple evaluation by the cluster head, the packet loss is reduced along with the energy consumption and the end-to-end communication delay. But this protocol requires further evaluation and comparison with other trust establishment algorithms. A highly complex privacy aware trust management algorithm which considers socio-psychological measures is described in [3]. Here, Elliptic-Curve Cryptography Diffie–Hellman (ECC-DH) algorithm is used for encryption and hashing, which forms the initial privacy aware data dissemination layer. This is followed by the consideration of social factors like node's ability, benevolence and consistency (ABC) in order to evaluate its trust levels. The block diagram of this system can be observed from Fig. 2, wherein information gathering, information sharing, privacy integration and trust mapping are done with the help of the ABC parameters. Due to socio-psychological features, the algorithm is able to improve the malware detection rate to nearly 100% even for higher number of malicious neighbouring nodes, and this improves the overall QoS for the network. This algorithm must be further evaluated using the attack-based features discussed in [5], wherein many trust-based network-protocols are discussed in detail. Researchers have listed that bad-mouthing attack, ballot stuffing attack, black hole attack, collusion attack, conflicting behaviour attack, data forgery attack, denial of service attack,

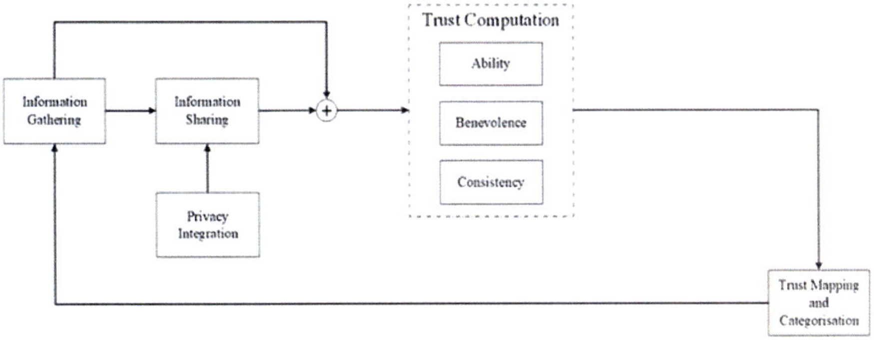

Fig. 2 Socio-psychological trust establishment architecture [3]

garnished attack, node replication attack, ON–OFF attack, reputation time-varying attack, selective forwarding attack, selfish attack, sinkhole attack, sleeper attack and Sybil attack majorly affect wireless networks.

They also analyse that though schemes like a beta distribution-based Trust and Reputation Evaluation System, Distributed Reputation based Beacon Trust System, Hybrid Trust Management Scheme, etc. improve the trust levels of the system but energy efficiency, risk-evaluations and joint optimization mechanisms with information forwarding and trust management are the main issues with these existing schemes and must be solved by researchers. In order to verify, trust-level algorithms are devised that utilize request and reply packet analysis. One such algorithm is described in [6], wherein both direct and indirect trust levels are verified with the help of immediate neighbours via checking their request and response packets. An example of such checking can be observed from Fig. 3, wherein a direct trust chain

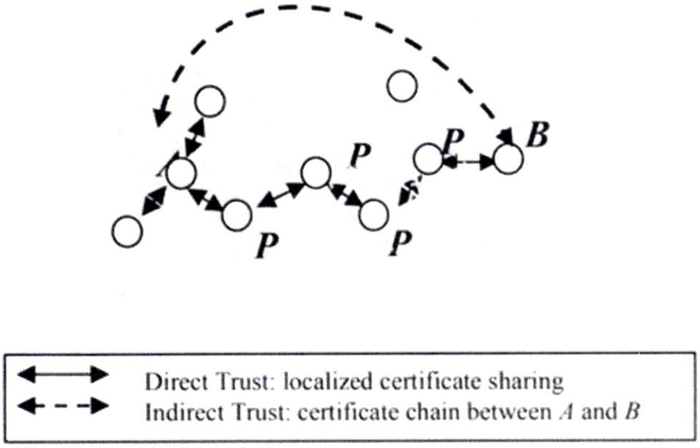

Fig. 3 Direct and indirect trust verification [6]

(route) is established between nodes A and B, while an indirect trust request/response mechanism is established between the same set of nodes.

Due to this 2-level trust establishment and verification process, the packet delivery ratio improves and the average delay of communication reduces. But this increases the routing load and the packet overhead on the routing protocol, which can be reduced with the help of light weight route establishment techniques. Moreover, this verification can be simplified with the help of higher-order cryptography and hashing protocols. For instance, the work in [7] proposes a trust management scheme based on hybrid cryptography to secure wireless networks. This technique symmetric hash message authentication code (SHMAC) for hashing the data between nodes, and asymmetric identity-based digital signature (AIBDS) in order to establish addressing in the network. Due to the hybrid combination of these protocols, the final trust values are improved. In order to compute the trust value, a cluster head (or agent trusted authority) is selected, while for verification of trust levels, a routing node (roadside unit) is used. The proposed scheme works in three phases, which can be observed from Fig. 4, wherein phase 1 is used for providing addresses to the nodes, phase 2 is used for verification of these addresses and finally phase 3 is used for removing any malicious nodes and establishing communication routes in the network.

The system is applied to vehicular ad hoc networks (VANETs) but can be extended to any kind of wireless networks with proper selection of address assignment and trust verification nodes. The proposed system can reduce the communication delay and the overheads during communication via the hybrid trust establishment methodology. While this protocol uses network-level trust establishment, the work in [8] uses node-level trust establishment to obtain a similar security and quality of service (QoS)

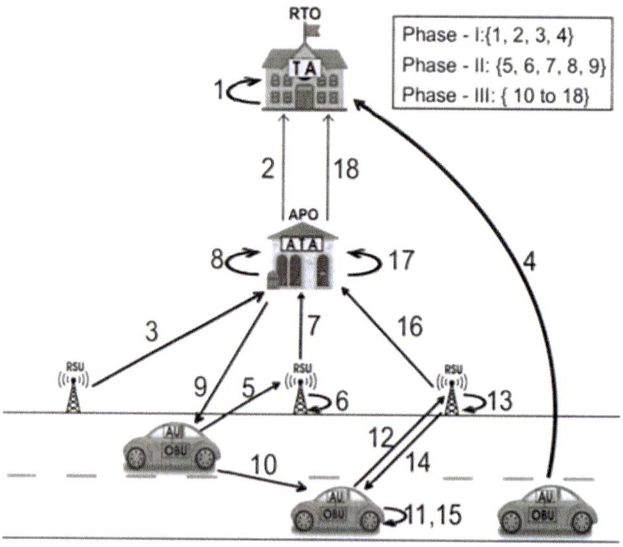

Fig. 4 Hybrid trust management system [7]

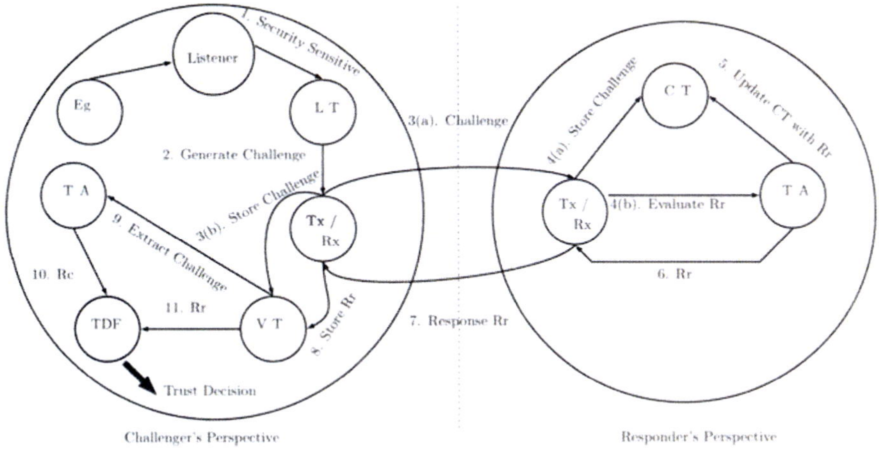

Fig. 5 Challenge-response mechanism for node trust-level evaluation [8]

performance. Using this research, the designers can create a challenge-response model for node-level trust evaluation. This when combined with peer-to-peer properties like self-scrutiny and self-attestation at node level, further enhances the node's peer-to-peer trust levels. The self-attestation and self-scrutiny techniques can be observed from Fig. 5, wherein the node internally generates challenges and then either solves them internally or gives them to other nodes for solving. Based on the responses of this solution, the verifying node either passes or fails the solver nodes.

If verifier node passes the solver node, then a trust level is established between them and the network in general. Due to these techniques, the overall delay and energy consumption of trust establishment and verification process are reduced when compared with state-of-the-art algorithms. But this performance reduces as the number of nodes increase due to a large number of self-attestation and self-scrutiny operations occurring in the network per unit time, which requires the nodes to consume higher power and increase the delay of communication. In order to reduce this effect, the work in [9] can be used, wherein a network with large number of nodes is clustered into smaller segments or clusters. Each of the node-level trust evaluation computations are applied at intra-cluster level, which reduces the number of computations needed for trust establishment. Moreover, the network-level trust evaluation techniques are applied at inter-cluster level (between cluster heads), thereby providing a 2-level light weight trust computation check. This improves the energy efficiency and delay performance when compared with [8], and thus can be used for real-time wireless applications. An extension of the work done in [8] can be observed in [10], wherein a self-certification scheme is developed in order to improve the trust level of nodes in a small-scale network. The performance of this network is similar to [8] and can be improved with the help of clustering techniques as mentioned in [9]. The process of cluster head selection for trust establishment can be further improved using semi-Markov models are proposed in [11]. The work proposes an availability

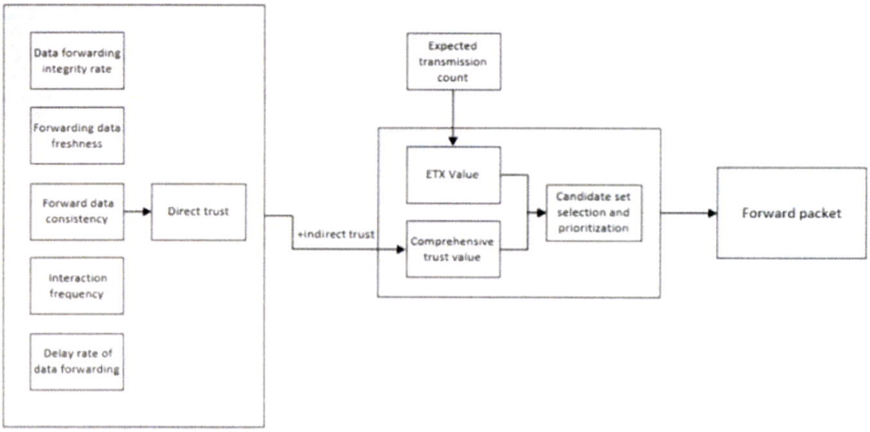

Fig. 6 Node-level trust evaluation [11, 12]

predictive trust factor-based semi-Markov mechanism (APTFSMM) to select cluster heads that have enhanced lifetime in the network. This is done by selecting a cluster head that requires lower energy while communication with other nodes and can compute complex sequence of trust values. Due to this selection, energy consumption of the network reduces, thereby increasing the number of alive nodes per unit time in the network. An extension to this work can be observed in [12], wherein trust calculations are done at node level. These calculations combine expected transmission count (ETC) with node trust value in order to evaluate the final trust level of any node. In order to evaluate the node-level trust values, the architecture proposed in Fig. 6 is used, wherein data forwarding rates, data consistency, interaction frequency and delay rate of forwarding are considered to form direct trust, while indirect trust is evaluated using neighbouring node data.

Due to this node-level trust evaluation, the trust-based routing performance of the network improves in terms of security and QoS levels. This performance includes a reduction in end-to-end delay and increases in packet delivery ratio and an improvement in overall network throughput. While improving the trust level of the network, there is an increase in the security performance of the network. This improvement can be utilized to deploy effective intrusion detection systems. Such a layer trust-based intrusion detection system (LB-IDS) is proposed in [13], wherein trust levels are evaluated at physical layer, medium access control (MAC) layer and network layer. Each of these trust levels are given to a summation unit, where the final trust computation is done. The flow of this process can be observed from Fig. 6, wherein parameters like energy, number of packets, back-off time, number of hops, etc. are considered while evaluating trust at different layers of the network.

The protocol embeds direct and indirect trust calculations to the AODV protocol in order to make it trust-aware. The proposed BT-AODV possesses all the advantages of AODV like reduced delay, increased throughput, reduced power consumption and increased packet delivery ratio and adds trust-based routing capabilities to it. Due to this, the overall security of the AODV protocol improves, and it is able to defeat denial of service, masquerading and Sybil attacks which are most prevalent in wireless networks. A survey of these attacks can be observed in [14], wherein different vulnerabilities that are inherent to wireless networks have been discussed. Moreover, the work also proposes a light weight, low-memory overhead and low-energy trust management protocol named GATE is proposed. But the protocol is effective only for static networks, where the node configurations do not change very often. Due to this limitation, the protocol has limited application capabilities for real-time wireless networks. This capability can be improved with the help of the work in [15], wherein a clustering approach is used for trust establishment in short-ranged networks. The cluster head selection is done using firefly algorithm followed by a hybrid encryption–decryption technique for secure and effective communication. The firefly algorithm selects a cluster head using minimum delay, maximum throughput and minimum transmission energy parameters. This reduces energy consumption in the network, thereby improving network lifetime and increasing the network throughput. While traditional techniques are able to improve the trust level in nodes, but none of the previously discussed protocols are 100% effective in terms of security and QoS parameters. In order to reduce this drawback, network designers use peer-to-peer distributed ledgers named as blockchains for QoS aware trust improvement in the network. An example of such a protocol is discussed in [16], wherein blockchain is used for Internet of Multimedia Things (IoMT) network under 5G network conditions. It uses similar trust management entities like certificate authority, trust calculations, trust evaluation, etc. but all these operations are performed using a decentralized approach. An example of such a decentralized mechanism can be observed from Fig. 7, wherein different certificate authorities are deployed for different kinds of applications, and all these authorities are combined using a certification blockchain network.

The performance evaluation of this protocol is still under process, and it is recommended that researchers perform this analysis before deploying this network for their network use case. Another research that removes the probability of distributed DoS in the network is proposed in [17], wherein a blockchain signalling system (BloSS) is proposed. It uses a combination of blockchain and software-defined networks (SDNs) to establish high-level trust in the network. The architecture for the blockchain network can be observed from Fig. 8, wherein the targets node 'T' and mitigator node 'M' are communicating with the help of a signalling scheme controlled by the software-defined network.

Due to use of BloSS scheme, the delay and energy requirement of the network reduces. Further due to the immutable nature of the blockchain, the network possesses a 100% detection rate for DDoS attacks. In order to further scale the applicability to other attacks, the work in [1] can be considered. Using this work, the network designers can deploy smart contracts with physically unclonable hashing functions (PuF) and dynamic proof of work (dPoW)-based consensus algorithms. This physical un-clonability and dPoW algorithms enable the network to be applied for a wide variety of network attacks with reduced delay and improved throughput performance. But as the length of this blockchain increases, there is an increase in complexity and

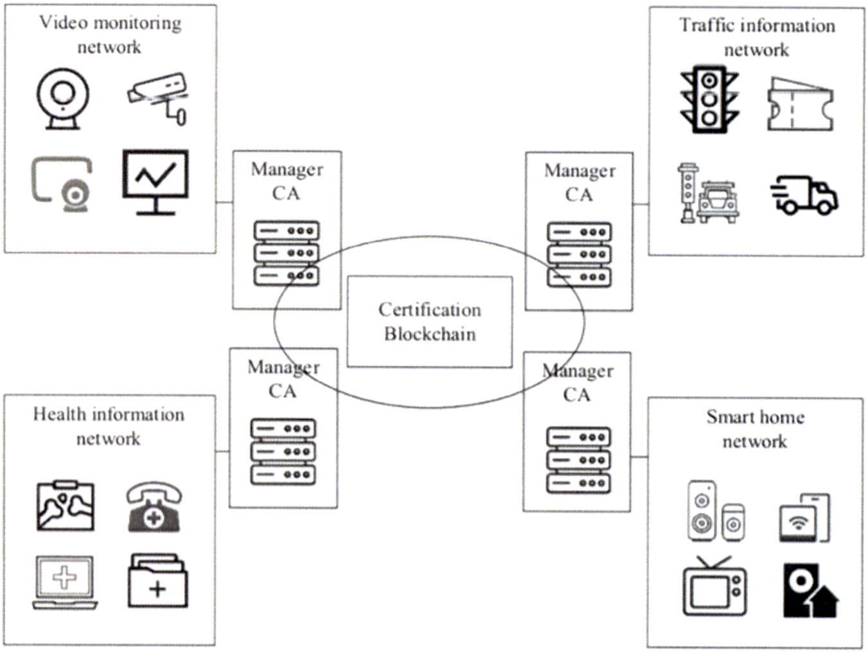

Fig. 7 Blockchain-based trust establishment [16]

delay of block creation, which reduces the network lifetime. This lifetime can be improved with the help of offloading these computations on the edge nodes. Such an edge-node computation offloading algorithm is discussed in [2]. The proposed algorithm is able to reduce the energy consumption of the nodes for computing block hashes but increased the communication delay in the network. This increase in delay reduces the throughput but increases the packet delivery ratio of the network. The effects of this drawback can be reduced by using a distributed computational mechanism as discussed in [18–21]. Using this mechanism, the heavy computations are divided into sequential parts, and each part is executed on a different distributed computational unit. Due to the peer-to-peer nature of this protocol, the overall delay is reduced, and throughput of the network is optimized. The work in [18] showcases this distributed trust evaluation mechanism between the Chinese and the European trade unions, and a high level of security with good QoS performance is observed. Another blockchain powered similar to [17] is proposed in [19], wherein the DDoS attack is removed using IP filtering. The performance of this approach is similar to [17], but is computationally light weight as compared to the work in [17]. The prediction performance for the work in [19] can be extended to larger attack scenarios using better prediction techniques. Such a support vector regressive linear programme boost technique for defeating multiple attacks using blockchain is proposed in [20]. This uses combined knowledge from a set of weak learners in order to improve the overall system knowledge about attackers and non-attacker trust levels. The learning

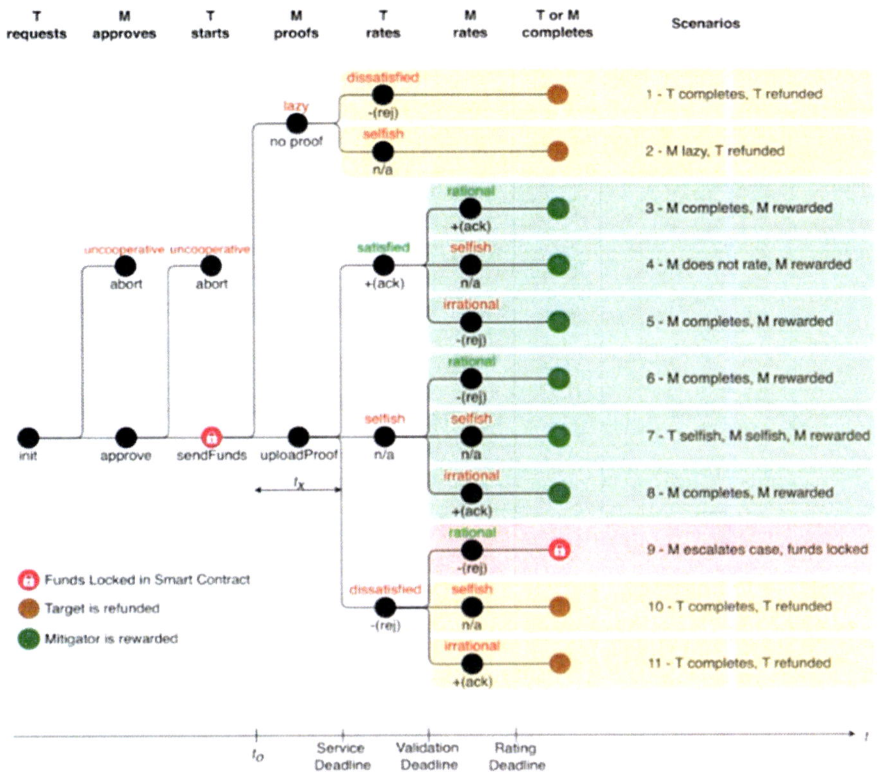

Fig. 8 The BloSS scheme [17]

process for this system can be observed from Fig. 9, wherein MANET application is considered for trust establishment and trust evaluation. While establishing high level of security in the network this also improves the packet delivery ratio, reduces the delay and increases the throughput of the network due to the distributed learning approach. An application of such a distributed approach when applied to E-commerce enterprises is mentioned in [21–23], wherein parameters like ease of use, behavioural analysis, usefulness of information and user intentions are considered while evaluating trust in the network. This trust evaluation is combined with trust perception in order to incorporate distributed trust levels in the network. Due to this, 3rd party vendors and providers are able to communicate in the network with high level of trust and security.

But this requires a large number of computations in the network, thereby limiting its fairness to access network parameters. An algorithm that ensures fair access of resources to all nodes is proposed in [21], wherein access patterns like grant, delegate, revoke and get access are discussed. Here, the network nodes follow the given steps for establishing access-level trust in the network,

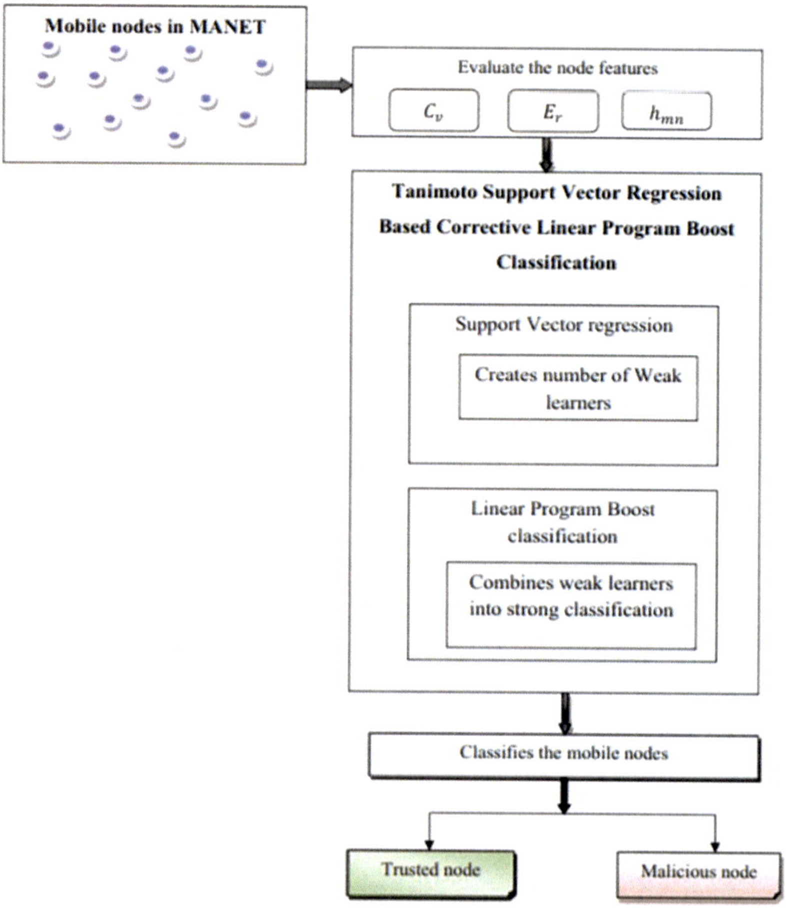

Fig. 9 Distributed learning approach for trust management [20]

- Initially permissions are set and requested by the nodes
- Using these permissions, the granted nodes are empowered to use certain services
- The nodes consider the different QoS constraints in the network while accessing it
- Once these constraints are met, then use of devices is permitted by the network controller to the nodes
- Finally, hold operations are performed in order to improve the overall network security.

3 Statistical Analysis

In order to perform statistical analysis of the reviewed protocols, the following parameters are considered,

- Energy consumption, which is a measure of the amount of energy needed by the network to perform trust establishment-based communication
- Delay needed for performing these communications
- Throughput performance obtained during these communications
- Packet delivery ratio of the network post deployment of these techniques.

As all the reviewed algorithms have been tested on different network scenarios, and each of these network deployments have different node configurations, thus a fuzzy statistical evaluation of these values is done. The fuzzy levels of very low (VL), low (L), moderate (M), high (H) and very high (VH) are considered while comparison of these protocols.

The following Table 1 evaluates the energy consumption of the protocols and suggests the applications where it can be applied,

A similar comparison of other QoS and security parameters can be observed from Table 2, wherein it can be observed that blockchain-based networks are the most effective in terms of security and QoS performance for wireless networks. Here, Thr. indicates throughput, PDR indicates packet delivery ratio and Acc. indicates the accuracy of attack prediction in the network. A high value of accuracy indicates high level of trust establishment in the network.

From the comparative analysis of QoS parameters, it can be observed that the semi-Markov model outperforms other methods, while its performance can be improved using fog computing (edge-based computing) models. Moreover, the security performance is the most optimum with blockchain computing models. The performance of the blockchain-based trust models can further improve with the help of edge-based computing models.

4 Conclusion and Future Work

From the extensive review about trust establishment protocols, it can be concluded that the overall energy consumption of node-level trust mechanisms and the semi-Markov models is the most optimum, which makes them applicable for both low-power and high-power applications. Moreover, this performance can be extended for other QoS parameters like end-to-end delay, throughput and packet delivery ratio. With the help of blockchain-based methods the overall security of the trust establishment networks improves drastically, which is a result of the immutable nature of these algorithms. The trust computation performance of these networks can be further improved with the help of edge-based or fog-based computing combined with peer-to-peer distributed protocols. In future, researchers can try to combine the

Table 1 Energy comparison of different trust establishment approaches

Method	Energy	Applicability
General trust model [3]	M	Any wireless network
Cluster-based with LEACH [4]	VL	Low powered large-scale networks
PSTRM [3]	M	Battery powered networks
Certificate distribution [6]	H	Large node networks
Hybrid cryptography [7]	VH	Vehicular networks and other high-power (and highly secure) networks
Node-level trust [8]	L	Battery powered networks
Clustering approach with comprehensive trust establishment [9]	M	Sensor networks
Self-certified networks [10]	L	Peer-to-peer networks
Semi-Markov mechanism [11]	L	Large-scale cluster-based networks
Opportunistic Routing [12]	M	Sensor networks
LB-IDS [13]	H	High-powered sensor networks
BT-AODV [24]	M	General wireless networks
GATE [15]	VL	Static networks
BAN-trust [16]	L	Low ranged wireless networks
Distributed CA [17]	M	Large organizations with wireless capabilities
BloSS [17]	H	Large-scale IoT
PUF with dPoW [1]	VH	Vehicular and other high-powered networks
Falcon [2]	M	Generic wireless networks with central processing units
TSVRCLPBC [20]	H	MANETs and other moderate powered networks
Fair Access [22]	M	Resource usage intensive networks
Extended control model [23]	H	Large-scale networks with high usage capacity
IETF for LLNs [29]	L	Low power and lossy networks

blockchain protocol with fog computing models along with peer-to-peer processing models in order to further improve trust levels in the network. Moreover, it is recommended to use machine learning models for further enhancing the QoS and security performance of the network.

Table 2 Comparison of QoS and security parameters

Method	Delay	Thr	PDR (%)	Acc. (%)
General trust model [3]	M	M	95	85
Cluster-based with LEACH [4]	M	H	97	89
PSTRM [3]	M	M	97	91
Certificate distribution [6]	H	M	98	90
Hybrid cryptography [7]	VH	M	96	95
Node-level trust [8]	M	H	99	91
Clustering approach with comprehensive trust establishment [9]	M	H	99	93
Self-certified networks [10]	L	H	97	90
Semi-Markov mechanism [11]	L	H	99	95
Opportunistic routing [12]	M	H	98	94
LB-IDS [13]	H	H	98	95
BT-AODV [24]	M	H	99	93
GATE [15]	L	VH	99	85
BAN-trust [16]	M	H	97	94
Distributed CA [17]	H	M	97	99
BloSS [17]	M	M	97	99
PUF with dPoW [1]	H	M	99	100

References

1. Bhushan S, Singh AK, Vij S (2019) Comparative Study and Analysis of Wireless Mesh Networks on AODV and DSR. In: 2019 4th International conference on internet of things: smart innovation and usages (IoT-SIU), pp 1–6. https://doi.org/10.1109/IoT-SIU.2019.8777466
2. Chauhan S, Tyagi SB (2014) Performance evaluation of reactive routing protocols in VANET. Int J Innov Adv Comput Sci 3(9):189–193
3. Pirzada A, Mcdonald C (2006) Trust establishment in pure ad-hoc networks. Wireless Pers Commun 37(1–2):139–168. https://doi.org/10.1007/s11277-006-1574-5
4. Ramesh S, Yaashuwanth C (2019) Enhanced approach using trust based decision making for secured wireless streaming video sensor networks. Multimedia Tools Appl 79(15–16):10157–10176. https://doi.org/10.1007/s11042-019-7585-5
5. Fang W, Zhang W, Chen W, Pan T, Ni Y, Yang Y (2020) Trust-based attack and defense in wireless sensor networks: a survey. Wireless Commun Mob Comput 2020:1–20. https://doi.org/10.1155/2020/2643546
6. Trust management scheme based on hybrid cryptography for secure communications in VANETs. IEEE J Mag (2021). Retrieved 19 Jan 2021, from https://ieeexplore.ieee.org/document/9040657
7. Node-level trust evaluation in wireless sensor networks. IEEE J Mag (2021). Retrieved 19 Jan 2021, from https://ieeexplore.ieee.org/document/8620267/
8. A novel and comprehensive trust estimation clustering based approach for large scale wireless sensor networks (2021). Retrieved 19 January 2021, from https://ieeexplore.ieee.org/document/8705286

9. Amuthan A, Arulmurugan A (2019) An availability predictive trust factor-based semi-Markov mechanism for effective cluster head selection in wireless sensor networks. Int J Commun Syst 33(6):e4298. https://doi.org/10.1002/dac.4298
10. Ghugar U, Pradhan J, Bhoi S, Sahoo R (2019) LB-IDS: securing wireless sensor network using protocol layer trust-based intrusion detection system. J Comput Netw Commun 2019:1–13. https://doi.org/10.1155/2019/2054298
11. Wang T, Hu K, Yang X, Zhang G, Wang Y (2018) A trust enhancement scheme for cluster-based wireless sensor networks. J Supercomput 75(5):2761–2788. https://doi.org/10.1007/s11227-018-2693-y
12. Vamsi P, Kant K (2016) Trust aware cooperative routing method for WANETs. Secur Commun Netw 9(18):6189–6201. https://doi.org/10.1002/sec.1765
13. Sahoo R, Ray S, Sarkar S, Bhoi S (2018) Guard against trust management vulnerabilities in wireless sensor network. Arab J Sci Eng 43(12):7229–7251. https://doi.org/10.1007/s13369-017-3052-7
14. Rodrigues B, Scheid E, Killer C, Franco M, Stiller B (2020) Blockchain signaling system (BloSS): cooperative signaling of distributed denial-of-service attacks. J Netw Syst Manage 28(4):953–989. https://doi.org/10.1007/s10922-020-09559-4
15. Zhang X, Wu W, Yang S, Wang X (2020) Falcon: a blockchain-based edge service migration framework in MEC. Mob Inf Syst 2020:1–17
16. Qian J, Wu W, Yu Q, Ruiz-Garcia L, Xiang Y, Jiang L, Shi Y, Duan Y, Yang P (2020) Filling the trust gap of food safety in food trade between the EU and China: an interconnected conceptual traceability framework based on blockchain. Food Energy Secur 9(4)
17. Singh R, Tanwar S, Sharma T (2019) Utilization of blockchain for mitigating the distributed denial of service attacks. Secur Privacy 3(3)
18. Bhushan S, Bohara B, Kumar P, Sharma V (2016) A new approach towards IoT by using health care-IoT and food distribution IoT. In: 2016 2nd International conference on advances in computing, communication, & automation (ICACCA) (Fall), pp 1–7. https://doi.org/10.1109/ICACCAF.2016.7748955
19. Diwakar M, Singh P, Kumar P, Tiwari K, Bhushan S (2022) A critical review on secure authentication in wireless network. In: Tomar A, Malik H, Kumar P, Iqbal A (eds) Machine learning, advances in computing, renewable energy and communication. Lecture Notes in Electrical Engineering, vol 768. Springer, Singapore. https://doi.org/10.1007/978-981-16-2354-7_55
20. Bhushan S, Kumar M, Kumar P et al (2021) FAJIT: a fuzzy-based data aggregation technique for energy efficiency in wireless sensor network. Complex Intell Syst 7:997–1007. https://doi.org/10.1007/s40747-020-00258-w
21. Sharma V, Kumar S, Bhushan S (2017) An overview of data redundancy reduction schemes in WSNs. In: 2017 3rd International conference on advances in computing, communication & automation (ICACCA) (Fall), pp 1–8. https://doi.org/10.1109/ICACCAF.2017.8344670
22. Tomar A et al (2020) Machine learning, advances in computing, renewable energy and communication. In: LNEE, vol 768. Springer Nature, Berlin, 659 p. https://doi.org/10.1007/978-981-16-2354-7. ISBN 978-981-16-2354-7
23. Iqbal A et al (2020) Renewable power for sustainable growth. In: LNEE, vol 723. Springer Nature, Berlin, 805 p. https://doi.org/10.1007/978-981-33-4080-0. ISBN 978-981-33-4082-4
24. Ren Y, Zhu F, Zhu K, Sharma P, Wang J (2020) Blockchain-based trust establishment mechanism in the internet of multimedia things. Multimedia Tools Appl. https://doi.org/10.1007/s11042-020-09578-y

Comparative Investigation and Determination of Partial Discharge Source Using Self-organizing Map and K-nearest Neighbor Methods of Artificial Neural Network

Kajol Chaudhari, Priyanka Kothoke, Akhilesh Kumar Mishra, and Yogesh Chaudhari

Abstract Partial discharge (PD) patterns are a valuable diagnostic instrument for high voltage (HV) insulation arrangements. Experts' human can detect potential insulation flaws in various PD data representations. Phase-resolved PD (PRPD) One of the most often utilized ways to express information is through patterns. When it comes to HV equipment reliability, there must be a connection between the statistical features of PDs to the defect's characteristics in order to figure out what kind of defect it is. For example, a void, surface, or Corona partial discharge can be seen on the graphical user interface (GUI) of the model, which was constructed in Python in Google Colaboratory by K-nearest neighbor (KNN) method, and later that output is compared with the characteristics of heat map in self-organizing map (SOM).

Keywords Partial discharge · Self-organizing map · K-nearest neighbor · Google Colab · Heat map

1 Introduction

Partial discharge is a limited discharge that only partially crosses the insulation between conductors and which can or cannot occur next to a conducting object," according to the IEC60270 standard. Electrical stress concentrations in the insulation or on the insulation surface are the most common cause of partial discharges. Pulses with a duration of less than 1 ns are the most common kind of such discharges. Pulse-less discharges in gaseous dielectrics, on the other hand, can arise in more continuous forms. The emission of sound, light, heat, and chemical reactions is frequently seen during partial discharges [1]. Partial releases can cause both physical and organic deterioration of insulation materials, and if not detected for an extended period of

K. Chaudhari (✉) · A. K. Mishra · Y. Chaudhari
Shri J. J. T. University, Jhunjhunu, Rajasthan, India
e-mail: kajolchaudhari20@gmail.com

P. Kothoke · A. K. Mishra · Y. Chaudhari
MGM College of Engineering and Technology, Navi Mumbai, Maharashtra, India

A. Tomar et al. (eds.), *Proceedings of 3rd International Conference on Machine Learning, Advances in Computing, Renewable Energy and Communication*, Lecture Notes in Electrical Engineering 915, https://doi.org/10.1007/978-981-19-2828-4_17

time, they can cause electrical breakdown of HV equipment. Internal discharge, surface discharge, and Corona are all examples of partial discharges. When a charge builds up inside a conductor or cavity within a solid or liquid insulation, this is known as an internal discharge. An insulation material's surface can experience a discharge known as a surface discharge. Corona is a partial discharge in gases surrounding conductors that are not protected by solid or liquid insulation. According to [2], "Corona (in air) occurs when the voltage gradient of an electrical conductor exceeds some threshold value, causing an ionization of the air around it.". Batteries that have been partially discharged undergo a stochastic physical process [3]. Charge displacement, acoustic and electromagnetic waves, and light are all accompanying phenomena that can be used to locate the source of the discharge. A variety of metrics are utilized in the detection of Parkinson's disease because of this. This Ph.D. thesis focuses on the electrical measurement of PD, which originated in high-voltage AC systems based on charge displacement. The device under test is given a calibration pulse, and the measurement procedure and system calibration were carried out in accordance with the IEC60270 standard. Before delving into the specifics, consider the following definitions.

- ***Charge apparent q***: "According to this definition, "appeared charge" refers to that amount of charge that, when applied between two points of an object's terminals in a test circuit, gives the same reading on a measuring instrument as a pulse of direct current (PD). Pico-coulombs are commonly used to express the apparent charge (pC). Charge locally at the discharge location is not equivalent to the apparent charge because it cannot be measured directly" [4].
- ***Discharge inception voltage***: "Voltage at which a PD pulse quantity becomes equal or exceeds a predefined low value is known as "inception voltage Ui" in practice [5].
- ***Discharge extinction voltage***: "Extinguishing voltage At what voltage does the magnitude of a selected PD pulse quantity become equal to, or lower than, the specified low value?" [6–8].

  Python is used to build the model in Google Colaboratory; statistical parameters are entered to determine the type of partial discharge on the GUI (e.g., Corona discharge versus void discharge versus surface discharge, etc.); the output is compared to the heat map's features in SOM.

  Literature Survey

  The discharge type is currently identified using signal processing and image processing instruments such as the wavelet transform, image processing software, and distance classifiers, among others. These techniques have the following drawbacks:

- Signal processing tools that take frequency into consideration rather than time are used to extract the PD patterns. Because insulation flaws aren't something that can wait, researchers are working to quantify PD patterns and types instantly. There have been numerous applications of the wavelet transform to time–frequency analysis. Continuous wavelet transforms can be used to extract time-varying pattern information by assessing the similarity between the signal and a collection of quickly decaying and oscillating functions known as wavelets. However, this procedure requires a lot more time and effort to determine the type of discharge. Consequently, in an effort to save time, we try to detect PD using both the support vector machine and the neural network methods.

- Modern image processing techniques may be distinguished using three-dimensional -q-n patterns, which are viewed as images in image processing tool. This image will not generate reliable results about PD pattern clusters or insulation faults when used as an input to distance classifiers. Distance classifiers classify patterns according to the smallest minimum distance. However, they cannot accurately compare freshly formed patterns to already existing ones [8].

2 Self-organizing Map (SOM)

In a typical neural system, the SOM maps a multidimensional space onto a two-dimensional space by fulfilling the first request. This neural system does not require any outside help. It imitates the human brain's ability to sort out component guides by simulating this in the computer. There are two layers to this neurological system, as depicted in Fig. 1, one for information and the other for yield, both of which have neurons in two-dimensional lattices.

Using SOM's number crunching, unpredictable measurement input vectors are transformed into a few scattered measurements. In the information space, weight vectors would isolate one sort of example depiction from the others and structure it with aggressive learning that is consistent. As a result, SOM defines how neurons in the neuron layer that are physically close to one another react to comparison information vectors when they see groups of similar information vectors. Instead of focusing on the two-dimensional goal space, group mapping shows analogous information vectors in the source space that are authentically nonlinear in nature [10]. SOM, in conjunction with the closeness of the information, may result in an order outcome [11]. As well as validating PD designs for turbo generators and gas-protected switchgear [6], it evaluated the voltage soundness of the power framework [12].

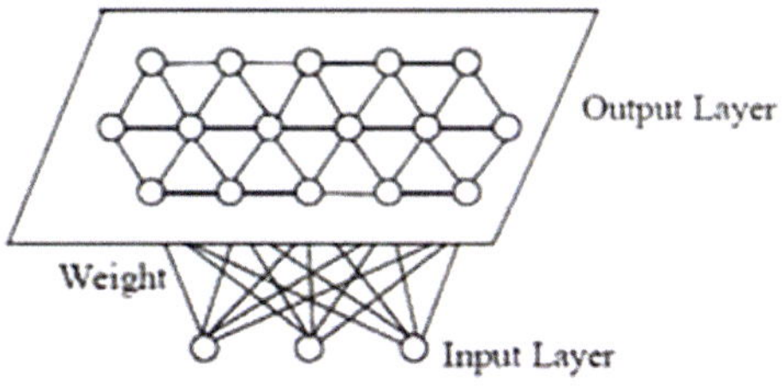

Fig. 1 System structure of SOM [9]

3 K-nearest Neighbor (KNN)

To classify an unknown sample, a KNN classifier considers all data samples to be equal in importance. When the data classes overlap, classification becomes more difficult as a result. There are no different weights for each sample in KNN. Unknown samples cannot be classified with the help of the kNN algorithm because it is incapable of doing so. Because of this, fuzzy set theory has been used to improve the KNN algorithm and eliminate its flaws. There are several ways in which it outperforms traditional KNN, including by producing membership values in addition to the label, as well as confidence measures for new samples. Especially if the new data sample does not fit into any of the available classes, this benefit could be highlighted. It is important to note that fuzzy KNN is used to determine how much a new sample belongs to each class. With this algorithm, all samples in data classes are assigned a class membership (weights) based on a variety of techniques.

4 Experimental Setup

Google Colab in Python is used in this study to simplify things because it's free and open source, and it may be customized to fit the project's needs in the future. These two approaches, the SOM method of ANN and the KNN method, both take as input the five parameters: Discharge mean, standard deviation, variance, skewness, and kurtosis.

An initial model for these technologies is created in the Google Colab laboratory. When creating a model, statistical parameters are obtained and used for both training and testing, as described above. The final processed file is then formed by integrating all six sets of known and unknown data. The Google Colab model is then placed in a Python server and used to create a GUI. The GUI will display the kind of discharge for the KNN method.

A heat map is used to organize the final processed file using self-organizing technique libraries. With this method, the structure (heat map) will be organized in such a way that one neuron will be the winner. The winner is the neuron that indicates that its major characteristics match those of its neighbors (adjacent neurons) when compared to other neurons. SOMs are used in ANN. Target 0 is assigned to Corona, whereas surface target 1 and void target 2 are both handed to Corona. The target is nothing more than a label, and the criteria for labeling are c for Corona, s for surface, and v for void, in that order, with target 0 being the lowest and target 1 being the highest. The winning neuron on the heat map will display three different colors for three different discharges, as seen below. It is important to remember that the color that will be present in greater amounts symbolizes the discharge, or anticipated output. If many discharges occur in insulation from a single source, this approach can also determine the exact percentages of discharges.

4.1 SOM Method of ANN

A map of itself (SOM) or a self-ordering feature map (SOFM) is a kind of artificial nerve network (ANN), skilled in using unattended learning to harvest a low-dimensional, (typically double-dimensional), discrete illustration, known as a map, of the input range of training samples.

4.2 Optimization

According to cross-validation, it is the most reliable way to estimate how well a model performs on different test sets (generalization). If you want to select the best model from a family of models that are parameterized by a grid of parameters, you can use the grid search technique (Table 1).

In this method, SOM libraries are read and the final csv (comma-separated value) file is called. Winner neuron will be the leader neuron to confirm the matching characteristics with neighbors. Target 0, target 1, and target 2 will be given different colors and will be allotted in alphabetical order to Corona, surface, and void, respectively. It will plot the figure depending upon the variations and show number of winner neurons in Fig. 2. Some colors in this figure are dark and some are faint depending upon the level of matching characteristics. Strong matching of characteristics will be indicated by dark color.

SOM will generate the activation function. Final processed file will be called to check for the output of partial discharge using SOM as it will display the colors of the winning neuron. Any color which is present in large quantity will confirm the type of discharge allotted with that color (Fig. 3).

Table 1 Choice of hyperparameters used for SOM method of ANN:

Sr. no.	1	2	3
Plan type	Winner neurons	Activation frequencies	Class Pie
Authentic diagram	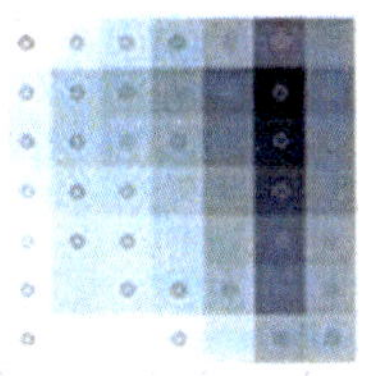	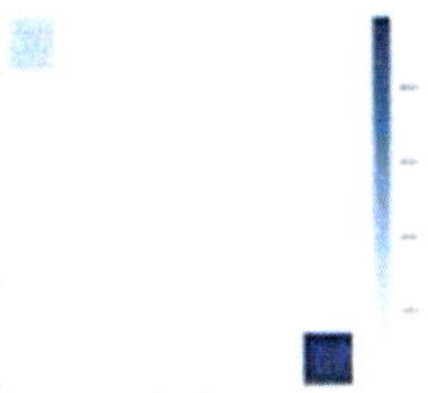	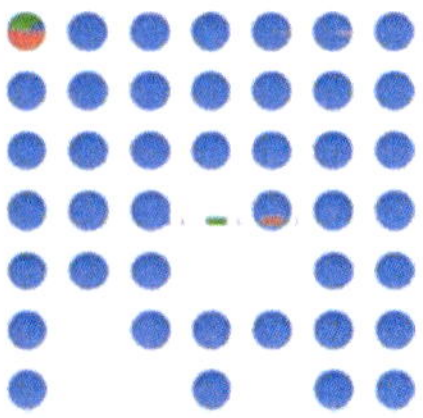

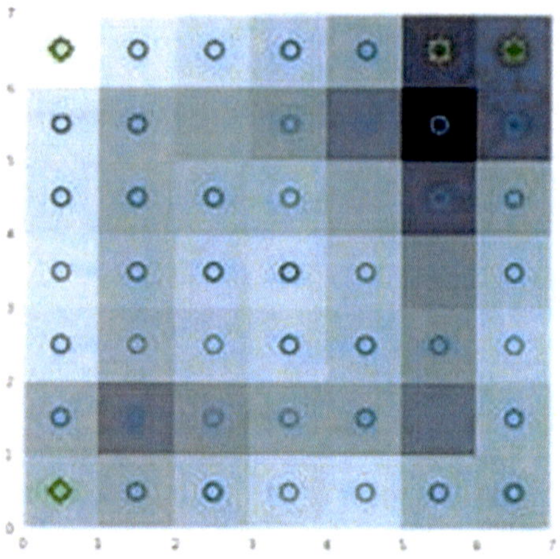

Fig. 2 Characteristics of winner neurons called as heat map

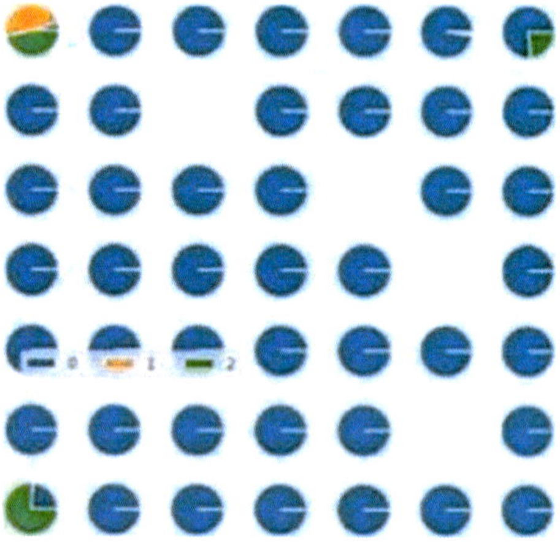

Fig. 3 Characteristics showing final output of SOM

4.3 KNN Method of ANN

Input settings for the aforesaid method, viz. KNN will reveal the outcome (kind of discharge) individually by entering (statistical parameters). Input values will be entered using a single icon in the window called the input data label, while the prediction label will display the output. Figure 4 shows the five parameters that will be used in the input data label. Figure 5 shows the expected output from the machine learning model, and Figure 5 shows the prediction label.

A Corona discharge is produced by picking any value from a set of five input factors and using that value to define the type of discharge. This is shown in Table 2. Surface and void discharges can also have their result displayed in table style.

The values for these statistical parameters are obtained using following relationships

Fig. 4 Entering the input values on GUI

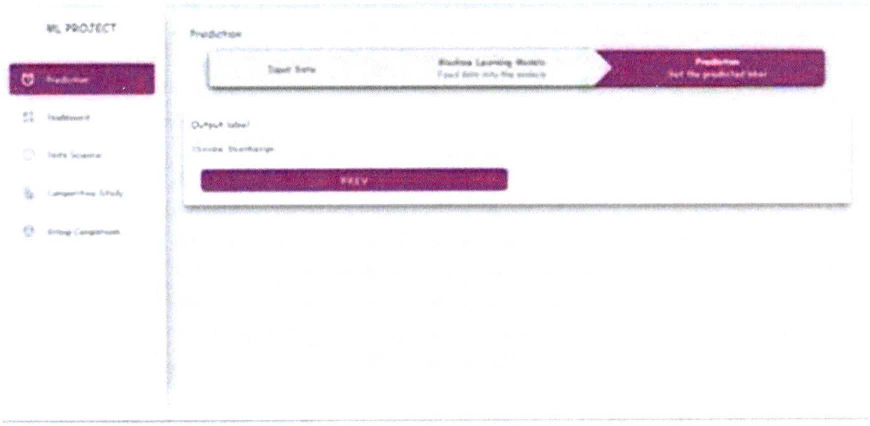

Fig. 5 Output by KNN method

Table 2 Table showing five entered values by user, i.e. input values and GUI showing output, i.e. type of discharge using KNN method

Sr. No.	Input limits standards	Output on GUI of group evaluation by KNN Method
1	Mean = 0.5316	Corona discharge
2	Standard deviation = 0.032169	Corona discharge
3	Skewness = 0.861068	Corona discharge
4	Kurtosis = 2.89909	Corona discharge
5	Variance = 0.001048	Corona discharge

Table 3 Comparative table of parameters between KNN and SOM methods of ANN

S. No.	Attribute	Col: mean	Col: standard deviation	Col: skewness	Col: kurtosis	Col: variance
1	Count	7119.000	7129.000	7119.000	7129.000	7.13E + 03
2	Mean	5201.310	3069.452	0.975	3.476	6.59E + 07
3	Std	12,843.394	7515.014	0.877	3.994	3.06E + 08
4	Min	0.000	0.000	-1.155	0.000	0.00E + 00
5	25%	1.929	1.706	0.459	1.585	2.91E + 00
6	50%	135.903	109.113	0.787	2.276	1.19E + 04
7	75%	1487.186	845.449	1.346	3.839	7.15E + 05
8	Max	105,051.589	136,527.796	8.929	81.822	1.86E + 10

$$\text{Mean Value: } (\mu) = \frac{\sum_{i=1}^{N} (x_i f(x_i))}{\sum_{i=1}^{N} f(x_i)} \tag{1}$$

$$\text{Variance: } (\sigma^2) = \frac{\sum_{i=1}^{N} (x_i - \mu)^2 f(x_i)}{\sum_{i=1}^{N} f(x_i)} \tag{2}$$

$$\text{Skewness } (S_k) = \frac{\sum_{i=1}^{N} (x_i - \mu)^3 f(x_i)}{\sigma^3 \sum_{i=1}^{N} f(x_i)} \tag{3}$$

$$\text{Kurtosis: } (K_u) = \frac{\sum_{i=1}^{N} (x_i - \mu)^4 f(x_i)}{\sigma^4 \sum_{i=1}^{N} f(x_i)} - 3 \tag{4}$$

$$\text{Standard Deviation} = \sqrt{\text{Variance}} \tag{5}$$

where x = number of pulses n, $f(x)$ = PD charge magnitude q, μ = average mean value of PD charge magnitude q, and σ = variance of PD charge magnitude q.

The above results are verified by using SOM method of ANN method as explained in Sect. 4.2 which discharge is present in what quantity (Table 3).

5 Discussion and Conclusion

The output from KNN method is Data1 and Data2 as surface discharge and Data3 as void discharge. In Fig. 3, 0 stands for Corona which is in blue color, 1 for surface in orange color, and 2 for void in green color. In Fig. 3, the neuron in first row and first column is leading one, and it shows orange and green color in more percentage indicating that surface and void discharge are more as compared to corona discharge. Thus, the winning neuron confirms the percentages of different discharges present as 52% surface type, 45% void type, and 3% Corona type.

Table 4 Output by KNN method

Sr. No.	Unknown data	Partial discharge source
1	Value of data 1	Exterior
2	Value of data 2	Exterior
3	Value of data 3	Void

The obtained percentage results of SOM method can be compared with that obtained from KNN method in Table 4. Data 1 and Data 2 show surface discharge in Table 4 similar to that of 52% from SOM method. Data 3 is showing void discharge in Table 4 similar to that of 45% from SOM method, also in Table 4, corona discharge is not predicted, so it is showing in only 3% in SOM method that can be considered as negligible. By using both the methods, the results are verified and validated.

References

1. Yazici B (2004) Statistical pattern analysis of partial discharge measurements for quality assessment of insulation systems in high-voltage electrical machinery. IEEE Trans Ind Appl 40:1579–1594
2. Hirata A, Nakata S, Kawasaki Z-I (2006) Toward automatic classification of partial discharge sources with neural networks. IEEE Trans Power Deliv 21(1):526–527. https://doi.org/10.1109/TPWRD.2005.848439
3. Hyde JR, Kemp IJ (1994) Partial discharge pattern recognition and associated technologies. In: IEE colloquium on monitoring technologies for plant insulation, pp 8/1–8/5
4. Park SH, Lee KW, Lim KJ, Kang SH (2003)Classification of external and internal PD signals generated in molded transformer by neural networks. In: Proceedings of the 7th international conference on properties and applications of dielectric materials (Cat. No.03CH37417), vol 1, pp 463–466. https://doi.org/10.1109/ICPADM.2003.1218451
5. Ping S, Dake X, Guoli W, Yanming L (2002)Application of neural network with genetic algorithm to UHF PD pattern recognition in transformers. In: Annual report conference on electrical insulation and dielectric phenomena, pp 732–735. https://doi.org/10.1109/CEIDP.2002.1048900
6. Gulski E, Kreuger FH (1992) Computer-aided recognition of discharge sources. IEEE Trans Electr Insul 27(1):82–92. https://doi.org/10.1109/14.123443
7. Tomar A et al (2020) Machine learning, advances in computing, renewable energy and communication. In: LNEE, vol 768. Springer Nature, Berlin, 659 p. https://doi.org/10.1007/978-981-16-2354-7. ISBN 978-981-16-2354-7
8. Iqbal A et al (2020) Renewable power for sustainable growth. In: LNEE, vol 723. Springer Nature, Berlin, 805 p. https://doi.org/10.1007/978-981-33-4080-0. ISBN 978-981-33-4082-4
9. Ghosh S, Kishore NK (2002) Modelling of partial discharge inception and extinction voltages of sheet samples of solid insulating materials using an artificial network. Proc IEEE 149(2)
10. Peng X et al (2019) A convolutional neural network-based deep learning methodology for recognition of partial discharge patterns from high-voltage cables. IEEE Trans Power Delivery 34(4):1460–1469. https://doi.org/10.1109/TPWRD.2019.2906086

11. Catterson VM, Sheng B (2015) Deep neural networks for understanding and diagnosing partial discharge data. In: 2015 IEEE electrical insulation conference (EIC), pp 218–221. https://doi.org/10.1109/ICACACT.2014.7223616
12. Salama MMA, Bartnikas R (2002) Determination of neural-network topology for partial discharge pulse pattern recognition. IEEE Trans Neural Networks 13(2):446–456. https://doi.org/10.1109/72.991430

Review of Sentiment Analysis on COVID-19 and Lockdown Twitter Data: Novel Techniques

Sudeep Kisan Hase and Rashmi Soni

Abstract Novel coronavirus (COVID-19) saw a daily major rise after August 2021. Health ministry of India shows more than 1000 deaths in 24 h. Reviewing this pandemic situation, there are again possibilities of lockdown in India. Daily updates of vaccinations, government permissions are changing day by day. Live updates and its insights should be reached to the common man. People are creating rumors about COVID-19. Social media like Twitter can ease the lifestyle in pandemic situations. The survey of COVID-19 sentiment analysis of Twitter data included in this paper. Novel techniques like BERT model, TextBlob, and VADER libraries are also discussed.

Keywords Sentiment analysis · BERT · ASUM · NLP

1 Introduction

A new strain of coronavirus was identified in Wuhan, China, and it is referred as 2019-nCoV, and it is from the family of SARS. Transmission of this virus is going through droplets of infected person or by touching the contaminated surfaces.

Up to mid-August 2021 the coronavirus cases are 212,598,874. Total deaths are 4,444,709, and recovered patients are 190,221,796. USA is on top having cases 38,545,144, and India is second having cases 32,449,306. In first wave of coronavirus, all people even doctors are unaware about diagnosis and treatment. Vaccines are not manufactured and distributed on time. In second wave, young people mostly affected, and death cases are more as compared to the first wave. Now, there is possibility of third wave as cases are increasing after August 2021. National Institute of Disaster Management (NIDM) India warned PMO that there will be outbreak of COVID-19

S. K. Hase (✉) · R. Soni
Department of CSE, Oriental University, Indore, India
e-mail: hase.sudeep@gmail.com

R. Soni
e-mail: drrashmicse@gmail.com

A. Tomar et al. (eds.), *Proceedings of 3rd International Conference on Machine Learning, Advances in Computing, Renewable Energy and Communication*, Lecture Notes in Electrical Engineering 915, https://doi.org/10.1007/978-981-19-2828-4_19

cases, and children will affect more in third wave. In August 2021, there are 3.28 active cases of coronavirus found in India. In third wave, people who already affected with disease are getting caught by this virus second time though they have taken two doses of vaccines.

In first and second wave, the symptoms are cold, cough, flu, pneumonia, breathing disease, no taste and smell. With fever severe diarrhea, vomiting or abdominal pain are the new symptoms in third wave.

Communication for Development (C4D) program has been launched by UNISEF in India. Health, Nutrition, Wash, Child Protection, Education all these plans were implemented by UNISEF during COVID-19. We have to overcome this coronavirus by implementing strategies. Rate of active coronavirus cases has been reduced by enforcing lockdown in India. Though the percentage is less still active cases reduced so as the deaths. Awareness about this novel coronavirus has been created by social media. People are posting text regarding coronavirus cases, infection, spreading, symptoms, deaths, etc.

Social networking sites are very important in today's era. Not only in India but in whole world social networking sites have peak visits. Billions of megabytes are used to store the data. If we compare site visits before 2020 and after 2020, there is huge difference. Among social networking sites, Twitter is very famous nowadays. Though Twitter produces huge data but analysis of these data is also important. India is affected by first corona wave in 2020 and affected more badly in second wave. Government officials, actors, socialists, and politicians are using Twitter as official handle. Opinion of politicians, socialists, medical officers, and activist is very important. Analysis of these handles in terms of opinions is called sentiment analysis.

In India, most of the businessman, politician, film fraternity, government, and health organizations are using social media. Some people use print media, electronic media to express their views. After twenty-first century, most of the people were attracted toward social media. The social media apps like Facebook, WhatsApp, LinkedIn, Instagram, WeChat, Snapchat, Skype, Viber, Pinterest, Telegram, etc., are more popular today. Limited authenticity is the drawback of some social media application. Through these applications, views and personal information are shared through text, audio, video, and emoticons. After posting such thoughts, people give different opinions, and these opinions are important to analyze sentiment of different people, government, and organization. Some of the apps like YouTube and Tiktok changed the way of interaction and communication. All the aspects of individual's life such as entrepreneurship, personal issues, organization, online education, medical, politics, and advertisement. Thus, all over the world big organizations and in India MSME Organizations are concentrating on social media movements to analyze live orientation, future business by evaluating system contents on social networking sites like Twitter, Facebook, and WhatsApp.

However, it becomes difficult to predict user sentiment because several factors affect some content of social app dataset. This paper elaborates work on Twitter sentiment evaluation and gives a live orientation detection process over the tweets and retweets regarding COVID-19.

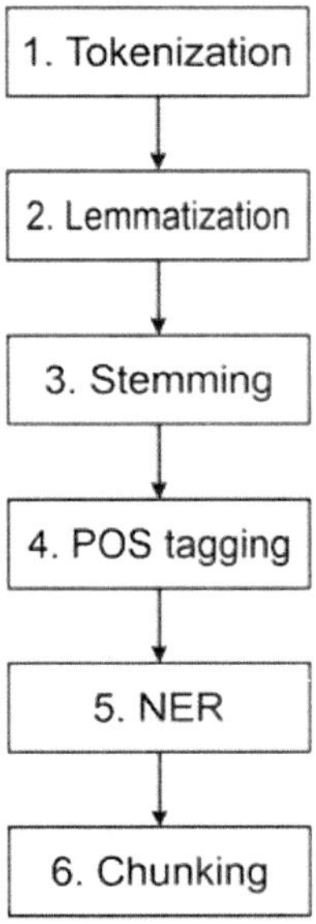

Fig. 1 Natural language processing steps

Figure 1 first step is tokenization in which the sentences are split into words so that interpretation of the word can be done easily. It means inflectional forms of the words are removed and it returns dictionary word. E.g. Protecting to protect. It is rule-based approach, so accuracy is less in stemming. E.g. output of the word "Studies" will be "Studi". Lemmatization knows the context of the word. It is dictionary-based approach and accuracy of lemmatization is more as compared to the stemming. E.g. the output of word "Studies" in Lemmatization will be "Study".

Part of speech (POS) is tagging process where sentence is divided into list of words, and according to the noun, adjective, verb tagging is done. e.g. ("Krishna", "NN") (noun). Named entity recognition (NER) is the next step which is primarily used in information extraction. Terms are used for the words which represent specific entities related with real-world objects, e.g., (Krishna, "Person"), (India, "GPE") where GPE represents country, state, or city.

Sentiment analysis is the part of NLP. With this analysis, human emotions can be mined. As per Fig. 1, this mined data will help government officials to control the coronavirus situation. Natural language processing classifiers and machine learning classifiers can perform sentiment analysis. Different APIs can be used to fetch this Twitter data. Survey of corona sentiment analysis and its results have been discussed in following section.

2 Related Work

Lockdown, vaccine, symptoms, and current cases in particular areas are hot topics nowadays. Frequently used Twitter data can be a great source to analyze opinions which will be helpful for the administration. The Thesaurus (semantic orientation of words) approach, ML findings, and combination of these both approaches are used for sentiment categorization. Data preprocessing should be done on the raw data so that algorithm efficiency will be maintained. This can be done with natural language processing.

Wu et al. [1] analyzed the flight booking data in China inter-country. He has counted the number of positive cases traveled throughout the country with flights. Data acquired by this method is used to stop traveling of such person and how many person should be quarantined.

Medford et al. [2] evaluated keywords and its data like infections, preventions, COVID-19, COVID positive, etc. Topic modeling has been applied in this study. They have carried out above 50 k users post having 1 lakh tweets. The result shows 50% people are in fear and above 30% people are in surprise. Political post, financial condition, and racial post are also fetched by Medford and team.

Li et al. [3] access Weibo post with the help OER based on machine learning algorithms. They have evaluated emotions in the form of positive and negative. T-test is performed on the post and evaluated that people are in stress and in negativity.

Pandey et al. [4] use input of post in the form of Hindi and mined the authenticated information with the help of Cohen's Kappa which depending in probability of random agreement.

Kayes et al. [5] collected more than 1lakh tweets in Australia and extracted more than 3000 tweets containing lockdown and social distancing. Some of the tweets they have used for training and validation purpose and some of the tweets for testing purpose. Test data gives 83% accuracy, and training data gives 80% accuracy on these keywords. They have carried out the result that people of Australia are supporting lockdown. They are supporting and positive about social distancing.

Dubey [6] extracted opinions of USA and Indian tweets. He has performed lexicon-based analysis to find out positivity and negativity of President Donald Trump and Prime Minister Narendra Modi. He carried out that above 60% tweets are positive about Narendra Modi while 40% tweets are positive about Donald Trump.

Barkur and Vibha [7] used R programming to find out sentiments of different people who have posted keywords with hastag IndiaLockdown and IndiaFight-sCorona. R studio gives word count which shows fear, anger, and sadness of COVID-19. Kolya [8] data visualization approach to determine insights about pandemics and use artificial intelligence applications.

A. Abd-Alrazaq collects over 2 lakh tweet and categorized it into four different topics like virus, origin, countries, economy, etc. They have observed that racism and economy are the two tweets sections which show negativity.

Naiknaware [9] extracts tweets from APIs then preprocessed it. After preprocessing, they stored data in csv file format. Depending on the polarity, they have applied scores and analyzed polarity. Prediction is based on positive, negative, and neutral tweets.

Wu [10] used GARCH modeling and applied this text to support vector machine. With lexicon approach, accuracy is up to 75%.

Ding [8] twitter comments are collected using SentiSW. They have categorized these tweets into positive, negative, and neutral. Tenfold cross-validation technique is used to carry out precision, recall, and accuracy values.

Pota et al. [11] used neural network-based approach to carry out polarity on tweets. SemEval-2017 data set is used for training and testing. They have used convolutional neural network technique and lexicon-based technique. As compared to lexicon-based, CNN gives good performance in classification.

Das and Kolya [12] gathered data set with the help of NodeXL soft and Twitter API. They performed polarity on the GST data in India. They also performed Naïve Bayes classifier. Most of the Twitter user are negative about GST India.

3 Approaches

Gupta et al. [13] and his team use new methodology in which Tweepy API, TextBlob library, and VADER tool are used. Data preprocessing, CountVectorizer, and supervised machine learning classifiers are used to perform the polarity.

As per Fig. 2, particular data is found out from social media with the help of _Tweepy API of Python. To find out polarity of the tweet data, they have used TextBlob library and VADER tool in Python shown in Fig. 3.

The result from both the TextBlob and VADER tool is intersected to check polarities. Most of the tweets are positive, average polarity is for neutral, and remaining are less for negative polarity.

Tokenization, punctuation stop words removal, lemmatization, and stemming are done in preprocessing [14, 15]. NLTK is used to clean the data for further processing.

The libraries like Pandas, sklearn, and matplotib are also used in this experiment.

To convert text data into features, they have used CountVectorier feature to calculate frequency of the word. Eight machine learning classifiers are used by these authors for training and testing [14].

As soon as the processing has been done, this extracted data is passed through different machine learning classifiers. Perceptron, Passive_Aggressive_Classifier, Ridge_Classifier, Ada_Boost_Classifier, Linear_Support_Vector_Machine, Bernoulli_Navie_Bayes, and Multinominal_Naive_Bayes classifiers are used by [13].

For this experiment, they have taken real-time Twitter data from April 2020 named "Indialockdown" using tweepy API.

As per shown in Fig. 4, results are drawn by Gupta et al. [13] with the help of recall, precision, accuracy, F1-score with unigrams, bigrams, and trigrams. With the help of these evaluators, tweets regarding lockdown were analyzed and it was found that approximately 48% of tweeters are in favor of lockdown, 29% are in opposition,

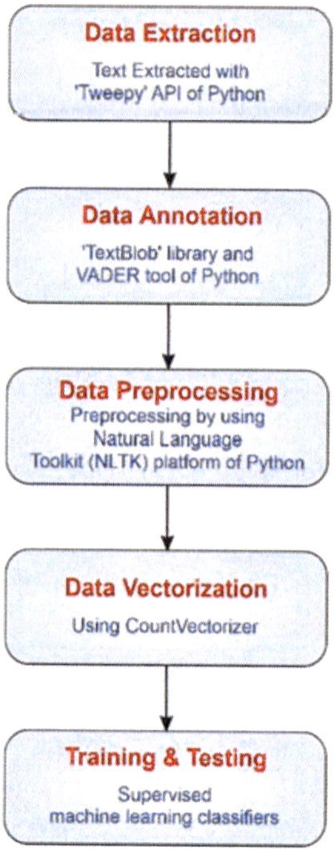

Fig. 2 Methodology used for sentiment analysis

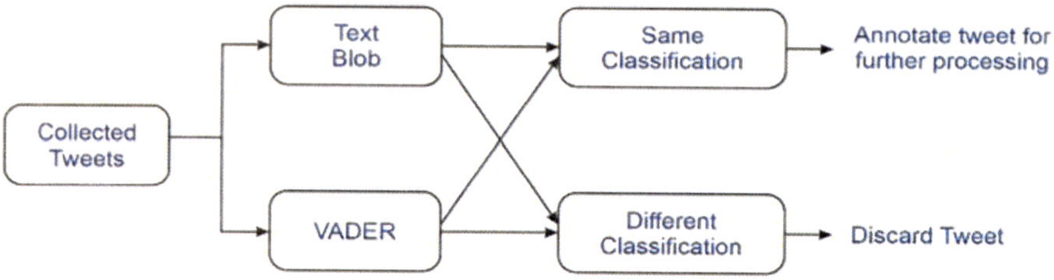

Fig. 3 Data annotation process

and 21% are neutral. For Linear_support_Vector_classifier with unigram accuracy is 84.4% which is highest amongst all eight classifiers. For perceptron classifiers with bigram accuracy is 82% which is highest amongst all eight classifiers.

Perceptron classifiers also give more accuracy with trigrams. Multinominal Navie Bayes (67.5%) and Bernoulli_Naive_Bayes (55.2%, 52%) have less accuracy with unigrams, bigrams, and trigrams, respectively.

Vu Le Anh et al. [16] and team explore streaming aspect sentiment analysis. They have used aspect and sentiment unification model where documents are split into sentences, sentences are splits into words, words are splits into aspects, and then sentiments are evaluated. In this model, they have used Frank–Wolfe algorithm,

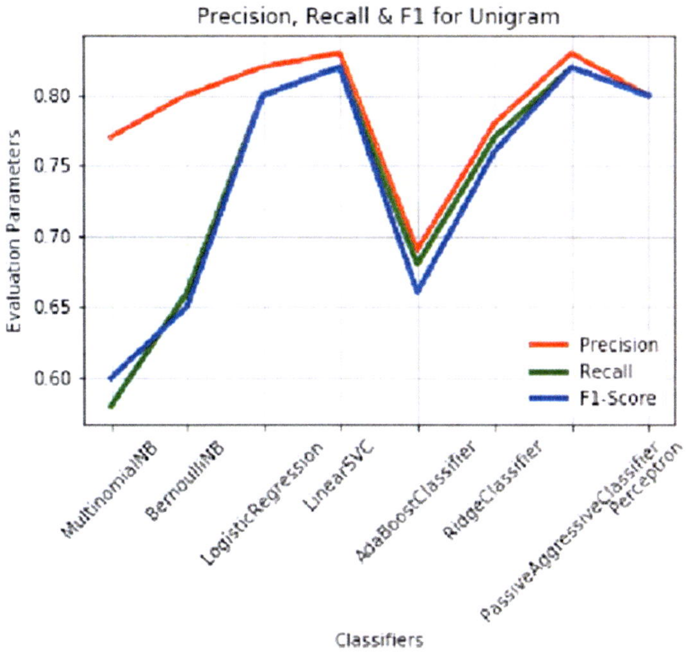

Fig. 4 Extracted result for lockdown data

MAP estimate, and streaming ASUM algorithms. This method gives scalable and faster results.

Singh et al. [17] and his team classified the human emotions with the help of BERT language modeling. They have collected the data and divided it into training and testing set with the help of tran-test split. Then, training set converted into torch tensor. BERT model is trained using parameters. This model gives more accuracy with the help of BERT model.

As per Fig. 5, this system has been developed by Hase [18] in which we send request to fetch the Twitter data through REST API. This API connects Twitter with OAuth, and fetched data has been fetched in JSON format. This API accessed with the help of Java library called Twitter4j.

Our Java application and Twitter API connection are done with the help of Ttwitter4j. The algorithm for preprocessing is given below.

for (int i = 0; i < tweets.size(); i++) {Status t = (Status) tweets.get(i); user = t.getUser().getScreenName(); msg = t.getText(); UserMentionEntity[] ue = t.getUserMentionEntities(); for (int j = 0; j < ue.length; j++) {msg = msg.replace("@" + ue[j].getText(), " "}HashtagEntity[] h = t.getHashtagEntities(); MediaEntity [] m = t.getMediaEntities(); URLEntity[] u = t.getURLEntities(); for (int j = 0; j < ue.length; j++) { msg = msg.replace("@" + ue[j].getText(), ""); } for (int j = 0; j < h.length; j++) { msg = msg.replace("#" + h[j].getText(), ""); } for (int j = 0; j < m.length; j++) { msg = msg.replace(m[j].getText(), ""); }

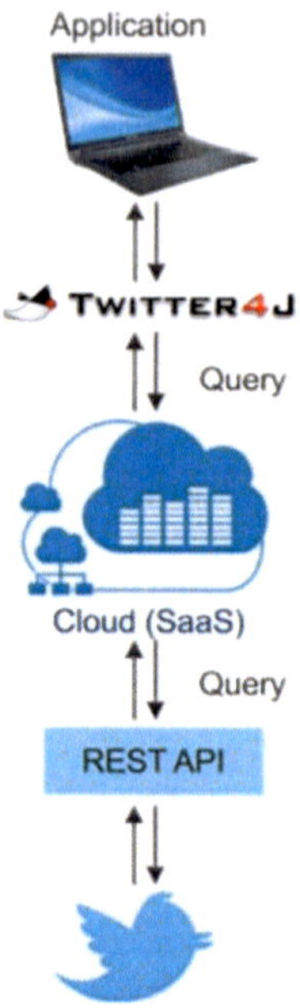

Fig. 5 Experimental setup

After preprocessing, the sentiment analysis is processed via Standford coreNLP. This Java-based library assigns integer values to the data denoting highest positivity with 5 and negativity with 0. As per keyword, we can find positivity and negativity of any keyword with help of this system. Preprocessing, analysis, and data are stored in SaaS application [18].

4 Conclusion

Social media users are increasing day by day. People want to know honest opinions and want to share true data about pandemic and COVID-19. Twitter is the best and trusted social media application. Health emergencies, positive cases, vaccines, and vaccine advantages are now shared within the people. Most of the people supporting government decisions and using this system mental stability also maintained during the pandemic.

Sentiments and polarity of the users can be acquired with the help of sentiment analysis. In this paper, we have discussed different techniques to extract the data from Twitter. Various preprocessing techniques are also discussed in this paper. The classifiers, machine learning techniques, and algorithms are also discussed.

For validation unigrams, bigrams, trigrams techniques discussed, and result of precision, recall, and F1-score with these techniques can give better results for polarity.

In future, we can extract real-time data from the Twitter and with the help of Python and R, we will see insights of the data.

References

1. Wu JT, Leung K, Leung GM (2020) Nowcasting and forecasting the potential domestic and international spread of the 2019-nCoV outbreak originating in Wuhan, China: a modeling study. Obstetrical Gynecol Surv 75(7):399–400
2. Medford RJ, Saleh SN, Sumarsono A, Perl TM, Lehmann CU (2020) An 'infodemic': leveraging high-volume Twitter data to understand public sentiment for the COVID-19 outbreak. medRxiv. https://doi.org/10.1101/2020.04.03.20052936
3. Li S, Wang Y, Xue J, Zhao N, Zhu T (2020) The impact of COVID-19 epidemic declaration on psychological consequences: a study on active Weibo users. Int J Environ Res Public Health 17(6):2032
4. Pandey et al (2020) A machine learning application for raising WASH awareness in the times of COVID-19 pandemic. arXiv:2003.07074. [Online]. Available: http://arxiv.org/abs/2003.07074
5. Kayes AS, Islam MS, Watters PA, Ng A, Kayesh H (2020) Automated measurement of attitudes towards social distancing using social media: a COVID-19 case study. Tech Rep
6. Dubey AD (2009) Decoding the Twitter sentiments towards the leadership in the times of COVID-19: a case of USA and India. SSRN Electron J
7. Barkur G, Vibha GBK (2020) Sentiment analysis of nationwide lockdown due to COVID 19 outbreak: evidence from India. Asian J Psychiatry 51(Art. no. 102089)
8. Das S, Kolya AK (2017) Sense GST: text mining & sentiment analysis of GST tweets by naive Bayes algorithm. In: Proceedings of 3rd international conference on research in computational intelligence and communication networks (ICRCICN), pp 239–244
9. Ding J, Sun H, Wang X, Liu X (2018) Entity-level sentiment analysis of issue comments. In: Proceedings of 3rd international workshop emotion awareness software engineering, pp 7–13
10. Pota M, Esposito M, Palomino MA, Masala GL (2018)A subword- based deep learning approach for sentiment analysis of political tweets. In: Proceedings of 32nd international conference on advanced information networking and applications workshops (WAINA), pp 651–656
11. Hase SK, Hase AK, Aher PS (2016) Collective intelligence and sentimental analysis of twitter data by using StandfordNLP libraries with Software as a Service (SaaS). In: 2016 IEEE international conference on computational intelligence and computing research (ICCIC). https://doi.org/10.1109/ICCIC.2016.7919697, Electronic ISBN: 978-1-5090-0612-0, Electronic ISSN: 2473-943X. https://ieeexplore.ieee.org/abstract/document/7919697
12. Tomar A et al (2020) Machine learning, advances in computing, renewable energy and communication. In: LNEE, vol 768. Springer Nature, Berlin, 659 p. https://doi.org/10.1007/978-981-16-2354-7. ISBN 978-981-16-2354-7
13. Gupta P, Kumar S, Suman RR, Kumar V (2021) Sentiment analysis of lockdown in India during COVID-19: a case study on Twitter. IEEE Trans Comput Soc Syst 8(4). https://doi.org/10.1109/TCSS.2020.3042446. Electronic ISSN 2329-924X
14. Nemes L, Kiss A, Social media sentiment analysis based on COVID-19. J Inf Telecommun 5(1):1–15. https://doi.org/10.1080/24751839.2020.1790793
15. Vijay T, Chawla A, Dhanka B, Karmakar P (2020) Sentiment analysis on COVID-19 Twitter Data. In: 5 IEEE international conference on recent advances and innovations in engineering—ICRAIE 2020.https://doi.org/10.1109/ICRAIE51050.2020.9358301. Electronic ISBN:978-1-7281-8867-6

16. Le Anh V, Van CP, Cao CV, Van LN, Than K (2016) Streaming aspect-sentiment analysis. In: The 2016 IEEE RIVF international conference on computing & communication technologies, research, innovation and vision for future. https://doi.org/10.1109/RIVF.2016.7800291, Electronic ISBN: 978-1-5090-4134-3
17. Singh M, Jakhar AK, Pandey S (2021) Sentiment analysis on the impact of coronavirus in social life using the BERT model, 33. Springer. https://doi.org/10.1007/s13278-021-00737-z
18. Iqbal A et al (2020) Renewable power for sustainable growth. In: LNEE, vol 723. Springer Nature, Berlin, 805 p. https://doi.org/10.1007/978-981-33-4080-0. ISBN 978-981-33-4082-4

A Comparison of Hourly Solar Energy Generation Forecasting Using RNN and LSTM Network

Neeraj, Pankaj Gupta, and Anuradha Tomar

Abstract Nowadays, various advanced techniques, like big data and deep learning, have been used in energy management systems for improving energy efficiency. Various research has been done on solar power generation prediction using multiple machine learning approaches. However, it is not easy to make exact predictions due to the alternative nature of solar energy. Therefore, this work aims to accurately implement and compare the neural network models to predict solar power generation. The models in this study are implemented using Python, TensorFlow, and Keras. The performance of the models is evaluated using 500 KWp grid-connected plant data. This work forecasted a solar power generation for 72 h using a time series dataset. Approximately, three years of the dataset are collected from January 2019 to October 2021 with hourly resolution from Airport Depot, Delhi, India. This paper compared the performance of the recurrent neural network (RNN) and long short-term memory (LSTM) in the form of accuracy. The three-year energy generation dataset is split into train and test data. The train data uses three years of energy generation data from January 2019 to September 2021, while the test data are used for October 2021. As a result, the determinant of coefficient (R^2) is a statistical error matrix representing the predicted values' accuracy. The results show that the LSTM model predicts solar power generation with high accuracy ($R^2 = 0.92$).

Keywords Solar power · Energy generation · Recurrent neural network · Long short-term memory

Neeraj (✉) · P. Gupta
Department of Electronics and Communication Engineering, Indira Gandhi Delhi Technical University for Women, New Delhi, Delhi, India
e-mail: neeraj42819@gmail.com

A. Tomar
Department of Instrumentation and Control Engineering, Netaji Subhas University of Technology, New Delhi, Delhi, India

A. Tomar et al. (eds.), *Proceedings of 3rd International Conference on Machine Learning, Advances in Computing, Renewable Energy and Communication*, Lecture Notes in Electrical Engineering 915, https://doi.org/10.1007/978-981-19-2828-4_20

1 Introduction

Renewable energy generation has risen exponentially in the last few decades. This growth has been significantly troubling the power providers [1]. So, energy forecasting models using a data-driven approach play an essential role in enhancing the plant's power efficiency through various parameters such as energy management, operations, and control approaches. Energy generation is an aspect that performs a significantly in the prediction [2]. Due to that reason, forecasting requirements for renewable energy generation will provide a better view to the power provider [3].

For residential, industrial, and commercial applications, solar energy is now one of the most important forms of energy production. Due to weather conditions, solar energy nature is alternating, which is the main challenge for energy generation. Therefore, electricity production is being affected due to climate change. In actual solar power plants, a decline of more than 25% can be seen in solar power generation. Generally, this fact can limit the grid-connected photovoltaic system. Therefore, accurate power generation forecasting plays an essential role in energy management in microgrid systems [4, 5].

In the literature, generation forecasting mainly focuses on long-term forecasting for a period of one year. Still, the one-year period is very long, so the predicted results are not reliable. Many machine learning models: support vector methods, numerical weather prediction(NWP), autoregressive integrated moving average (ARIMA), fuzzy logic, and combined neural networks have been proposed, but their work is limited for prediction; they cannot predict for short periods.

In the literature of solar energy forecasting, various approaches have been observed. These approaches can be divided into four parts: (a) statistical methods using data-driven approaches depends on historical time series data, (b) machine learning and deep learning approaches using neural networks, (c) physical algorithms using satellite images, (d) hybrid or combined models [6]. In [7], a gradient boosted regression method has been applied for hourly ahead photovoltaic power generation forecasting. In comparison with autoregressive models, this model is more accurate. In [8], a hybrid model is used for short-term photovoltaic energy generation estimation. The autoregressive model, a neural network, and the discrete wavelet transform are all merged in this model. It is found that the hybrid model produces more precise results than the single model. In [9], an ensemble technique with neural networks is proposed for solar energy prediction. In [10], a support vector regression method is combined with a seasonal decomposition method for energy forecasting. This model represents a good prediction with high accuracy compared to regression, neural networks, ARIMA, and SARMIA.

With the giant dataset, most of the generalized forecasting techniques for photovoltaic energy prediction are not suited for accurate forecasting. Nowadays, deep learning techniques have been used for forecasting applications [11]. The deep learning methods are becoming popular for representing the time series data in terms of accurate prediction. Deep learning is a branch of machine learning and artificial intelligence that use more traditional methods for extracting representations from data [12]. It is also a machine learning approach that uses neural network design to learn characteristics and tasks directly from data. The data are processed through numerous hidden layers in these neural networks. Various deep learning approaches have been used in the past few years, including backpropagation, gradient descent, and neural networks [2].

RNN is a neural network that operates with sequential input. It provides accurate prediction with good efficiency in the various domains using time series data [13]. LSTM is a form of neural network that stands for long short-term memory. It replaces every hidden layer with an LSTM cell [14]. In this paper, RNN and LSTM model is applied on time series sequential data for solar energy generation forecasting. The Part 2 introduces the basic idea of LSTM and RNN models. Then, Part 3 represents the forecasted results using the LSTM and RNN model. Lastly, the conclusion is presented in Part 4.

2 Methodology

This paper uses two data-driven approaches for energy forecasting: recurrent neural networks (RNNs) and LSTM networks. These methods belong to AI-based machine learning methods. The detailed information of both the models is provided in the following.

2.1 *Recurrent Neural Network (RNN)*

RNNs employ a feedback loop to feed the results from the previous time step in the sequence as part of the current time step's input. This is the same as with regular feed-forward neural networks. A single neural network with the same weights processes with time step in the sequence. Neural network parameters are common for each time step in the sequence. It is trained by inputting data and then optimizing the time sequence parameters so that the output value of the loss function is minimized for each time step in the sequence [15]. The RNN can be configured in various ways, depending on the application. Features can be extracted from an input sentence then used to produce time series data as the output. As such, RNNs can be applied to many different applications.

$$S_t = F_w(S_{t-1}, X_t) \tag{1}$$

This equation explains what RNNs are and how they work. X_t denotes the input at time step t, S_t denotes the state at time step t, and F_w is the recursive function.

$$S_t = \tanh(W_s S_{t-1} + W_x X_t) \tag{2}$$

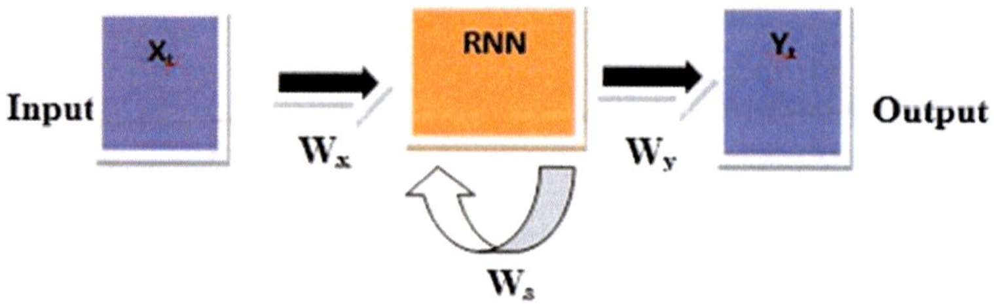

Fig. 1 Simple RNN

A tanh function is a recursive function. W_x multiplies with the input state, while W_s multiplies with the prior state. Then, it passes through a tanh activation to get the new state. The weights are W_x and W_s. The new state S_t is multiplied with W_y to produce the output vector. It can be seen in Fig. 1, the input and output state are calculated using the previous and new state [16].

2.2 *Long Short-term Memory (LSTM)*

Long-range dependencies are difficult to capture with typical RNNs. In other words, while working with a vast dataset and multiple RNNs layers, the risk of vanishing gradient problems is observed. LSTM stands for a long short-term memory, and every hidden layer is replaced with an LSTM cell, and the cell state is added to every cell. The LSTM algorithm was designed to address the problem of vanishing gradients. Each LSTM cell has a cell state vector apart from the hidden state vector. Input, forget, and output are the three gates on each unit. They all have a sigmoid activation, but their smooth curves are in the 0 to 1 range. Thus, the model is still differentiable. $\overline{C}$ modifies the cell state, apart from these gates. It has the tanh activation, centered with a zero range. LSTM model helps to solve the vanishing gradients problems [17].

$$i^t = \sigma(W^i[h^{t-1}, x^t] + b^i) \tag{3}$$

$$f^t = \sigma(W^f[h^{t-1}, x^t] + b^f) \tag{4}$$

$$o^t = \sigma(W^o[h^{t-1}, x^t] + b^o) \tag{5}$$

$$\overline{C}^t = \tanh(W^c[h^{t-1}, x^t] + b^c) \tag{6}$$

$$C^t = f^t C^{t-1} + i^t \overline{C}^t \tag{7}$$

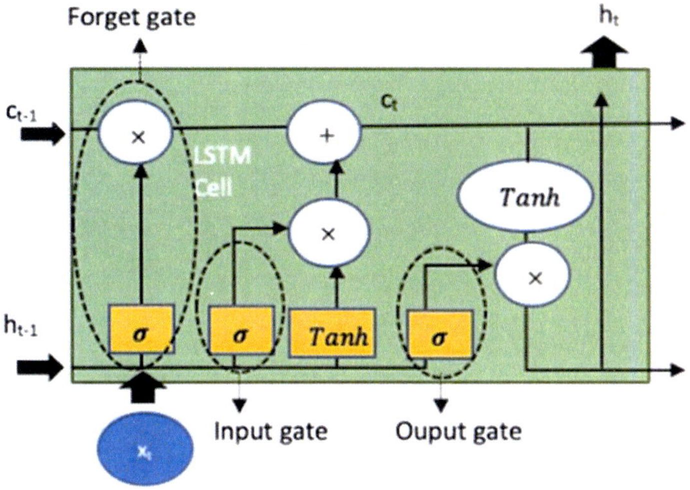

Fig. 2 LSTM cell

$$h^t = \tanh(C^t) * o^t \tag{8}$$

i^t, f^t, and o^t are the input, forget, and output gates of the LSTM cell. W represents the recurring construction between the previously hidden and the existing layers. The hidden layers are connected to the input through the weight matrix. The cell state $\overline{C}$ is calculated and depending on the current and the previous input. C stands for the unit's internal memory. Figure 2 shows the equations that describe the behavior of all gates in the LSTM cell. As inputs, each gate accepts the hidden state and the current input x. The vectors are concatenated, and a sigmoid is applied. $\overline{C}$ is a new potential value for the cell's state. The input gate controls the memory cell's updating. As a result, it's applied to the $\overline{C}$ vector, which is the only one that can change the state of the cell. The forget gate determines how much of the previous state should be remembered. To get the hidden vector, this state is applied to the output gate [18].

3 Results and Discussion

3.1 Structure of the Model

This paper aims to forecast solar energy generation using deep learning approaches. The main steps of these approaches are described in Fig. 3.

These approaches contain the five main steps:

- Collect the data from the solar power plant.

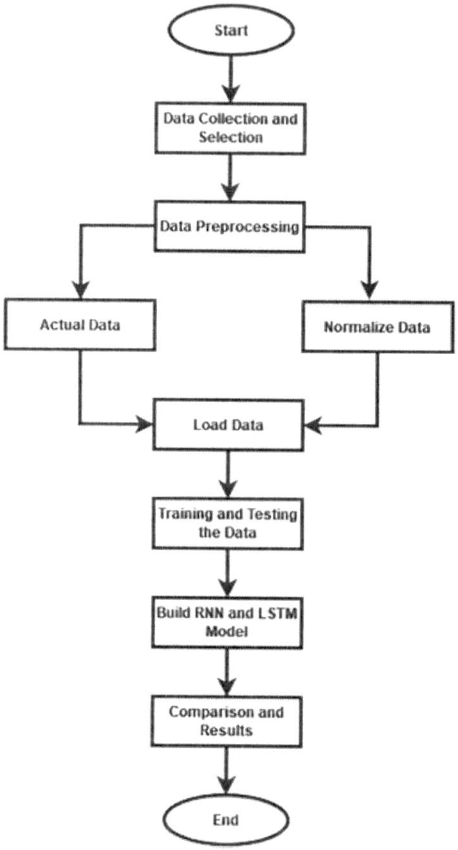

Fig. 3 Block diagram of forcasted model

- Preprocessing the data: Cleaning and checking the missing data.
- Data visualization: Normalize the actual data
- Training and testing
- Build RNN and LSTM model.

The accuracy of the models is measured using a variety of characteristics. Finally, a graph is used to compare the results of the LSTM and RNN models.

3.2 *Data Collection*

The work emphasizes the prediction of solar energy generation. The actual solar generation dataset is taken from the Airport Depot, New Delhi, India and was collected from January 2019 to October 2021 with hourly intervals. The dataset is used to predict the energy generation from the solar plant.

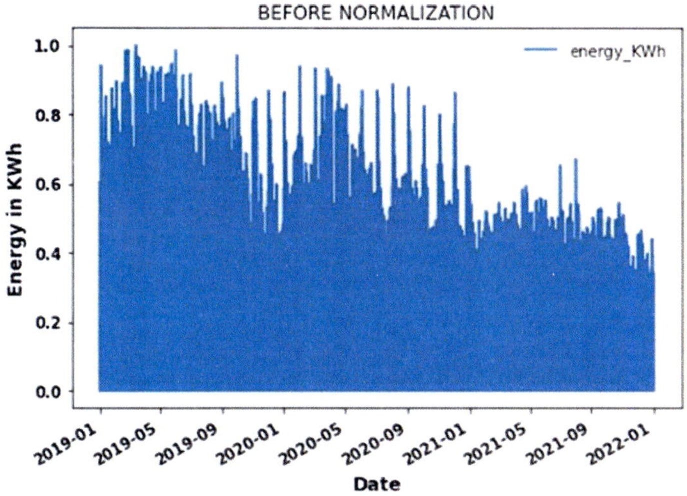

Fig. 4 Solar energy generation per hour from January 2019 to October 2021—before normalization

3.3 Data Visualization

3.3.1 Before Normalization

This work is based on data gathered with hourly resolution from a 500 KWp solar power plant in Delhi, India, from January 2019 to October 2021. The hourly distribution of the solar energy generation dataset is depicted in Fig. 4. At mid-day, the solar energy reaches the highest, and after that it is reduced to zero overnight. The generated solar energy is relatively high from April to August. Also, it is observed that the solar power production is relatively low in January, February, March, November, and December.

3.3.2 After Normalization

In this work, to improve efficiency, the dataset is normalized to a range between 0 and 1, represented in Fig. 5. MinMaxScaler is used to normalize the dataset, then used by the transform() method for training and testing. The new data are prepared and will be used to make a prediction. The formula is used for normalization the date is

$$\begin{aligned}&\text{NormalizedData}\\&=(\text{Data}-\text{Minimum}(\text{Data}))/(\text{Maximum}(\text{Data})-\text{Minimum}(\text{Data}))\end{aligned}$$

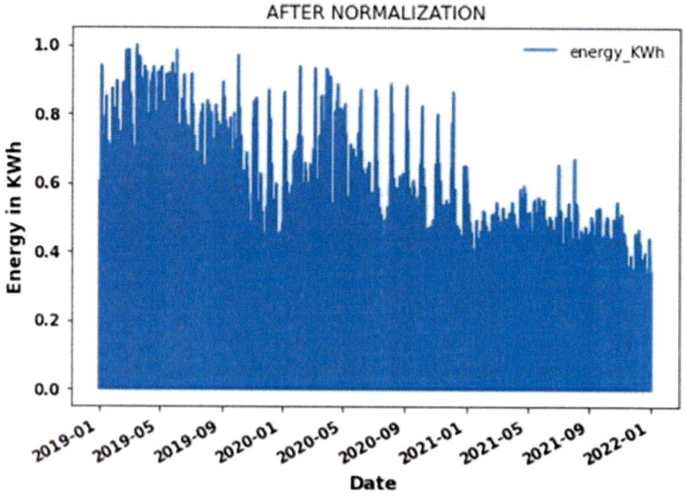

Fig. 5 Solar energy generation per hour from January 2019 to October 2021—after normalization

3.4 *Test Results Using Normalized Data*

Figures 6 and 7 represent the test results using a normalized dataset for RNN and LSTM models. The number of hidden layers for RNN and LSTM models is determined through trial and error in this study. The real dataset is separated into training and testing sets to construct the RNN and LSTM models. From the entire dataset, 95% behaves as the training dataset, and 5% is treated as the test data. The two-layered neural network is fed to the training set with one node as input and output layer. The loss function is MSE, and epochs are 50 with 150 batches of data are stored in history. In this work, RNN and LSTM models have used two hidden layers. Each input has 30 characters and 20 units per RNN cell, so there are 2101 parameters to train. Table 1 represents the parameters used by the RNN and LSTM models.

The predicted results for short-term solar energy generation are represented in Fig. 6. Once the RNN model is built based on the training data, it predicts the energy output. After that, check the forecasting ability of the model. Figure 7 represents the predicted energy generation's compared graph with actual data on a time horizon using an LSTM model. The value of R^2 is observed as 0.80 for RNN and 0.92 for LSTM, which are close to 1. Thus, the LSTM model delivers the greatest accuracy. Figure 8 shows the compared predictions that RNN and an LSTM model make. Because the recurrent RNN model layer incorporates feedback loops, LSTM stores the memory for a short time. However, the RNN model will be challenging to train and solve problems requiring long-term temporal dependence learning. It happens because the loss function gradient decreases rapidly with time which is known as

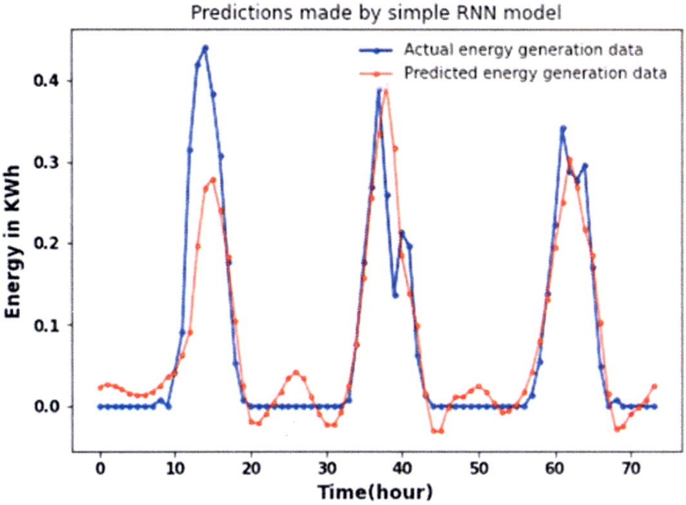

Fig. 6 Plot of actual and predicted energy generation using RNN model

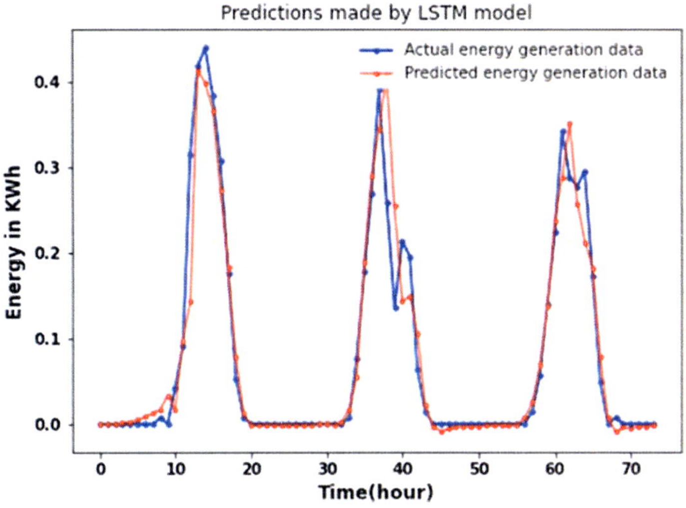

Fig. 7 Plot of actual and predicted energy generation using LSTM model

vanish gradient problem. LSTM is a kind of RNN model, but it uses some special units additionally. LSTM model included a memory cell that can store information for a long time and used a group of gates to control the information. The long-term dependency problem is resolved by increasing the repeated layers in the LSTM model.

Table 1 RNN and LSTM parameters

Model	Parameter	Value
RNN, LSTM	Number of hidden layers	3
	Loss function	MSE
	Number of units	20
	Optimizer	Adam
	Epochs	50
	Batch size	100
	Sequence length	30
	Activation	Tanh

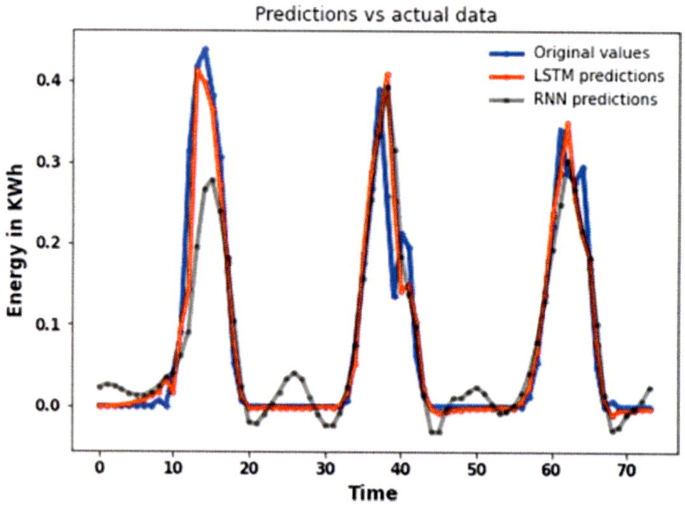

Fig. 8 Plotting of compared predictions by RNN and LSTM model in a single graph

4 Conclusion

In this paper, the short-term solar energy generation can be accurately estimated using RNN and LSTM models. RNN and LSTM models were trained to predict solar energy generation in Airport Depot, Delhi, India. The three-year historical datasets have been collected from January 2019 to October 2021. The prediction process involved five stages: data collection, data preprocessing, data normalization, building RNN, and LSTM model and comparing the model based on accuracy. Python is used to implement the forecasted model. The number of hidden layers for both the models is two. After testing or training, the results obtained from the normalized data show that the LSTM model's performance behaves better than that of the RNN model. LSTM model provides the best performance in terms of determinant of coefficient (R^2=92%). Therefore, the LSTM model yielded the best short-term solar power generation prediction results. To further improve the prediction efficiency in future, we will use weather and generation data together and create a combined model by merging the LSTM model with another model.

References

1. Sweeney C, Bessa RJ, Browell J, Pinson P (2020) The future of forecasting for renewable energy. Wiley Interdisc Rev Energy Environ 9(2):e365
2. Ugurlu U, Oksuz I, Tas O (2018) Electricity price forecasting using recurrent neural networks. Energies 11(5)
3. Ahmed A, Khalid M (2019) A review on the selected applications of forecasting models in renewable power systems. Renew Sustain Energy Rev 100:9–21
4. Antonanzas J, Osorio N, Escobar R, Urraca R, de Pison FM, Antonanzas-Torres F (2016) Review of photovoltaic power forecasting. Solar Energy 136:78–111
5. Lotufo ADP (2020) Solar photovoltaic power forecasting. J Electr Comput Eng 020:8819925
6. Sobri S, Koohi-Kamali S, Rahim NA (2018) Solar photovoltaic generation forecasting methods: a review. Energy Convers Manage 156:459–497
7. Persson C, Bacher P, Shiga T, Madsen H (2017) Multi-site solar power forecasting using gradient boosted regression trees. Solar Energy 150:423–436
8. Kushwaha V, Pindoriya NM (2019) A sarima-rvfl hybrid model assisted by wavelet decomposition for very short-term solar pv power generation forecast. Renew Energy 140:124–139
9. Raza MQ, Mithulananthan N, Summerfield A (2018) Solar output power forecast using an ensemble framework with neural predictors and Bayesian adaptive combination. Solar Energy 166:226–241
10. Lin K-P, Pai P-F (2016) Solar power output forecasting using evolutionary seasonal decomposition least-square support vector regression. J Cleaner Prod 134:456–462
11. Chen Z, Chen Y, Wu L, Cheng S, Lin P, You L (2019) Accurate modeling of photovoltaic modules using a 1-d deep residual network based on i–v characteristics. Energy Convers Manage 186:168–187
12. Zheng J, Xu C, Zhang Z, Li X (2017) Electric load forecasting in smart grids using long-short-term-memory based recurrent neural network. In: 2017 51st Annual conference on information sciences and systems (CISS), pp 1–6
13. Ma J, Ma X (2018) A review of forecasting algorithms and energy management strategies for microgrids. Syst Sci Control Eng 6(1):237–248
14. Sun Y, Venugopal V, Brandt AR (2018) Convolutional neural network for short-term solar panel output prediction. IEEE, pp 2357–2361
15. Fan C, Wang J, Gang W, Li S (2019) Assessment of deep recurrent neural network-based strategies for short-term building energy predictions. Appl Energy 236:700–710
16. Mishra S, Palanisamy P (2018) Multi-time-horizon solar forecasting using recurrent neural network. In: 2018 IEEE energy conversion congress and exposition (ECCE). IEEE, pp 18–24
17. Xiaoqiao H, Zhang C, Li Q, Yonghang T, Gao B, Shi J (2020) A comparison of hour-ahead solar irradiance forecasting models based on lstm network. Math Probl Eng 2020:1–15
18. Srivastava S, Lessmann S (2018) A comparative study of lstm neural networks in forecasting day-ahead global horizontal irradiance with satellite data. Solar Energy 162:232–247

Supervised Machine Learning Approach for Crack Detection in Digital Images

Deepti Vadicherla and Vijay Gadicha

Abstract Cracks are very common in concrete wall, railway sleeper, pavement construction, old paintings and underwater walls of dam. Crack detection is a challenging task in terms of accuracy. Manual inspection of cracks has many qualitative drawbacks. Therefore, automation in the process of crack detection is essential. Machine learning has become fast growing area of research that covers many fields including image classification. This paper reviews few supervised machine learning-based crack detection methods, namely random forest, support vector machine and Naive Bayesian. These methods have been investigated for solving many challenging problems in classification. Literature in the paper presents machine learning techniques to detect cracks on the surface of images automatically. In this paper, some of the recent research articles on crack detection have been reviewed and also checked their feasibility toward the crack detection task.

Keywords Crack detection · Supervised machine learning · Classification of images · Support vector machine

1 Introduction

Crack detection is the process of detecting and positioning of cracks on different surfaces like concrete, metals, ceramics, canvas and plastic. Detection of cracks is a crucial task in many fields such as in reinforced concrete structures where two materials act together in resisting forces are usually designed to allow cracking under service loads which leads to importance of structural integrity. Crack detection helps in monitoring structural condition and ensures structural safety by preventing damage and future failure. Metal surface crack testing helps in industries like pressure vessels and pipes. It also helps in the process of restoration of ancient paintings and sculptures. Traditionally, human inspection is used for crack evaluation which has many disadvantages like more time consumption, poor repeatability and reproducibility,

D. Vadicherla (✉) · V. Gadicha
Department of Computer Science and Engineering, GHRU, Amravati, Maharashtra, India
e-mail: deepti.vadicherla@gmail.com

A. Tomar et al. (eds.), *Proceedings of 3rd International Conference on Machine Learning, Advances in Computing, Renewable Energy and Communication*, Lecture Notes in Electrical Engineering 915, https://doi.org/10.1007/978-981-19-2828-4_21

lack of expertise, in some cases safety issues and use of large amount of labor. Automation in this field has been applied to overcome the disadvantages of manual crack detection process. There are many methodologies which are being used to automate the process of crack detection like image processing and machine learning [1]. Machine learning has shown the good potential in the digital image-related research besides many other fields. With the help of machine learning concepts, it is possible not only to automate the process of crack detection but also to improve its result. Supervised machine learning-based crack detection methods [2], namely random forest [3], support vector machine [4] and Naive Bayesian [5], can be used in image classification for detecting cracks in the predefined feature extracted images.

Crack detection is the process of finding the line like structure in the image which was not part of the original image. Like many other fields, machine learning plays an evolutionary part in the process of detecting cracks too. In case of crack detection, the difference between image processing and machine learning technique is that the earlier one can do analysis of exterior defects while the later one can do analysis of detailed defect features. This is because of the ability of machine learning in learning from the same type of defect features, setting rules and predicting the similar type of defects in new data [6].

The paper is structured as below. Section 2 focuses on the motivation and problems present the existing system. In Sect. 3, the process of detecting cracks with the help of machine learning is discussed, with a brief overview of unsupervised machine learning and supervised machine learning. In Sect. 4 conclusion is mentioned. This paper provides overview of the supervised machine learning algorithms for detecting cracks in the digital images. Discussions about the existing problems are done followed by conclusion.

2 Motivation

With a few measures and precision, the new idea and innovation changes. Cracks can be different in shapes, sizes and colors. Detecting different cracks automatically on different surfaces is a challenging task. Table 1 shows the list of artifacts. Limitations of existing methods in the process of crack detection are mentioned in Table 2. The motivation to achieve accuracy and speed remains the vital parameters in the process of crack detection. Significance of crack detection is stated in many industries like nuclear power plant where its components need to be checked on a regular basis, which is time intensive if it is performed manually. In the realm of national infrastructure maintenance and restoration, automatic pavement crack detection is gaining popularity [7]. Surface defect identification is a critical component of industrial quality assurance [8]. Automation in crack detection provides help in restoration of ancient paintings and art analysis [9–16].

Table 1 List of artifacts

Author	Artifacts
Chen [5]	Noisy pattern, stitching image
Xu [8]	Blur
Yang [10]	Reflection
Dong [11]	Hole
Ding [12]	Difference in textures
Tangsakul [13]	Haze
Nguyen [14]	Inhomogeneity, contrast between cracks and background

Table 2 List of limitations

Sr. No.	Technique	Limitations
1	Crack forest	Crack dimensions are not measured Works for only still images
2	SVM	Not all cracks with different geometries can be detected
3	Naïve Bayesian	Needs large training dataset (more than 1 lakh) for better accuracy Detects cracks but does not quantify attributes

3 Detecting Cracks Using Machine Learning

The scientific field where a computer program is said to learn from the different past experiences of some type of feature categories, setting some measures, predicting the case with same type of feature category and at the same time improving its ability to do so is nothing but machine learning [17, 18]. In this particular case, it is the method which learns from the past experiences of detecting cracks in digital images of different surfaces. The method improves its performance by detecting cracks more accurately. The ability to improve itself in the process of prediction is the key point behind the usefulness and uniqueness of machine learning. Unsupervised machine learning and supervised machine learning can be used in the process of detecting cracks [6].

3.1 Unsupervised Learning Methods

Unsupervised learning method is a type of algorithm that learns patterns from untagged data. Training samples in this type of machine learning method do not have data labels, which reduces the human influence of the subjective perception on the result. Another disadvantage of unsupervised learning method is that it gives less accuracy in results because the precise information regarding input data is not known [6].

3.2 Supervised Learning Methods

Supervised learning method needs the training data with labels. Naïve Bayesian, random forest, support vector machine, regression and artificial neural network are some of the widely used supervised learning algorithm. In this paper, three classification-based methods, namely random forest, Naive Bayesian and support vector machine, are reviewed [6].

The process of applying supervised machine learning here and now is explained in Fig. 1. Classification is the standard formulation of these types of methods. The learning process includes the extracting information and forms a function from the prelabeled categorized data. In this process, classification model predicts the class-wise category of give example.

Random Forest. Random forest is used for both classification as well as regression. It is an example of supervised machine learning algorithm. Random sample is selected in this method from a given set of data. From every sample in the dataset, it creates

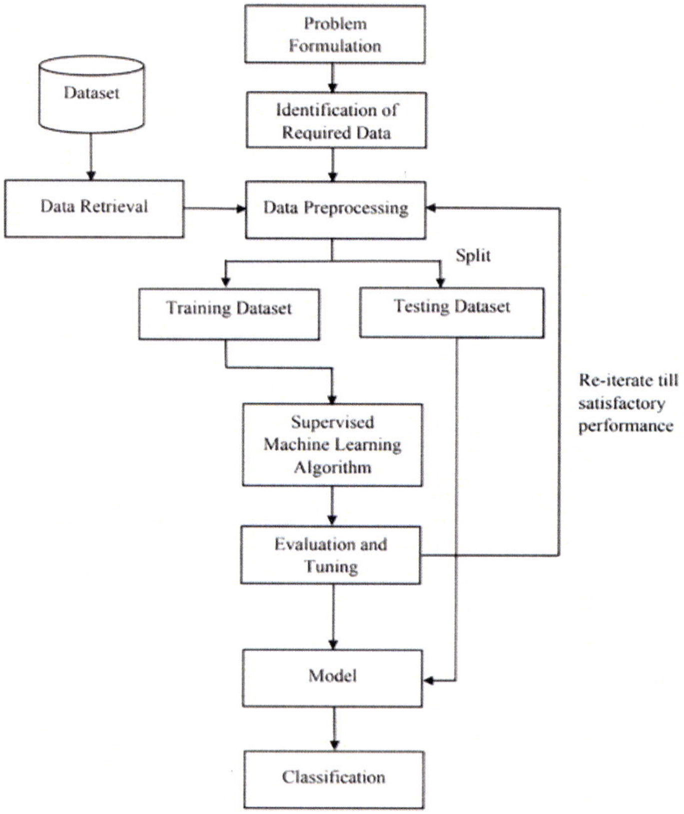

Fig. 1 Flowchart of supervised machine learning method

a decision tree which helps to predict the result and deciding the best result among them with the help of voting. Random forest method reduces the over-fitting by taking average of result [19].

Using random structured forest, crack detection can be done accurately. In this process, features are extracted by using integral images resulting in clear recognition of cracks. Random forest is introduced at this stage for more accurate crack detection as shown in Fig. 2. It consists of individual decision tree for each data sample with the help of feature randomness which is then used to predict uncorrelated forest of trees resulting in more accuracy [3, 20].

Support Vector Machine. Support vector machine (SVM) is one of the most powerful, task-oriented and flexible supervised machine learning algorithms. This method is different in many ways from other supervised machine learning algorithms. It became popular among researchers because it can process number of variables with finite number of categories and the continuous variables which have any value within a particular range [19]. In SVM, each data sample with multiple categories is plotted

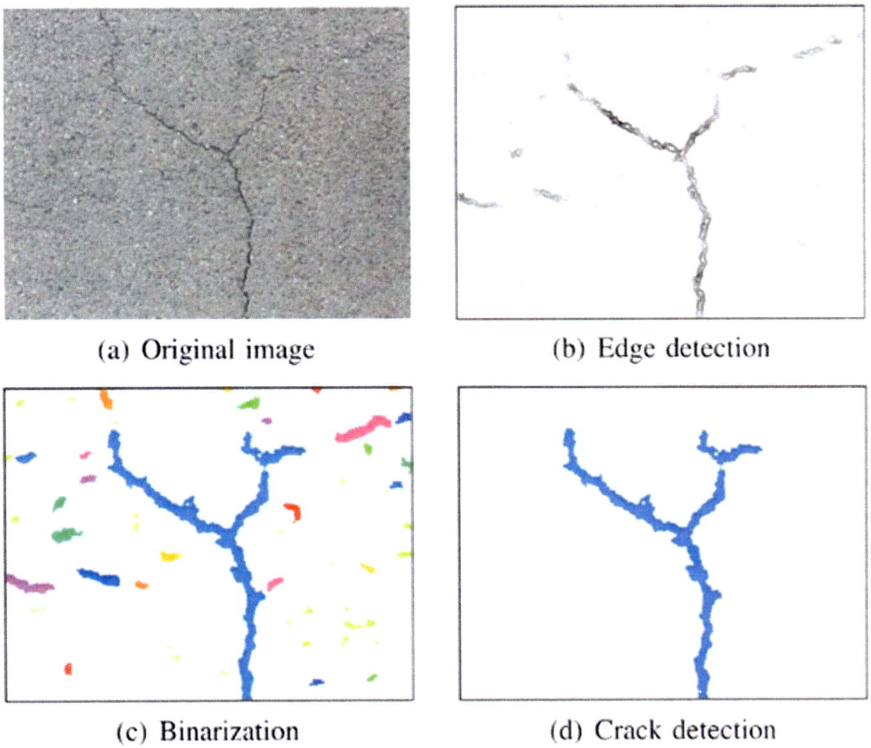

Fig. 2 **a** Original image of pavement surface. **b** Result of random forest edge detection. **c** Result of eroding and dilating. **d** Classification result. Darker-colored pixels in (**c**, **d**) indicate the higher possibility of being a crack. (*Source* ARCD 2552248, p. 3)

on the multidimensional space near hyperplane. The aim is to choose a hyperplane with largest margin between hyperplane and support vector.

SVM as a classifier uses the data of digital images of cracks after the feature extraction as shown in Fig. 3a. Extracted features from the crack images are used in the training process. In Fig. 3b, it is shown that how the pre-trained SVM classifier helps in the process of crack detection in new sample images [4]. A multi-level SVM model can also be used to improve the classification results [21].

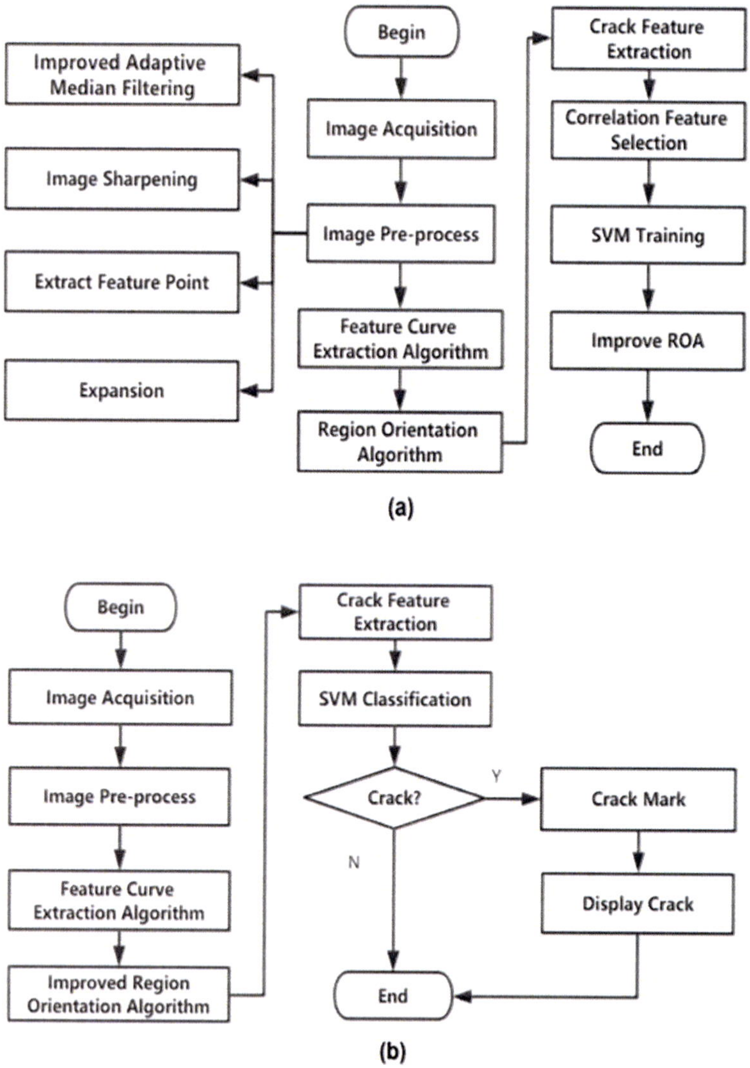

Fig. 3 **a** Flowchart of feature extraction followed by training using SVM. **b** Steps of detecting cracks using SVM. (*Source* MDMC 2870435, p. 53149)

Naïve Bayesian. Conditional independence regarding features present in a particular category is the additional quality present in Naïve Bayesian classification if it is compared with the Bayes' theorem. As the name 'Naïve' suggests, the classifier accesses the feature samples which are conditionally independent. But the problem with this method is that, even if the features are dependent in a class, it considers it as independent feature. Naïve Bayesian classifier uses probability for prediction [19]. It calculates posterior probability of class L by using the formula as below,

$$P(L|\text{features}) = \frac{P(L)(\text{features}|L)}{P(\text{features})}$$

where

$(L|\text{features})$ indicate the posterior probability of class L when *features* are happening.
(L) is the prior probability of L.
$(\text{features}|L)$ indicate the posterior probability of *features* when L is happening.
(*features*) is the prior probability of *features*.

To determine whether a small part of image consists crack structure or damage free image, Naïve Bayesian classifier is useful when there is a case of varying data samples with variety in the quantity of cracks in the images [5]. It is considered in the image that an image consists of n cracks and $P\left(C_{crk}\middle|S_1^c, \ldots, S_n^c\right)$ is posterior probability of images with damage and $P\left(C_{ncrk}\middle|S_1^c, \ldots, S_n^c\right)$ is posterior probability of images without damage. If the condition mentioned in the below formula is true, patch of the image is considered as damaged,

$$\frac{P\left(C_{crk}\middle|S_1^c, \ldots, S_n^c\right)}{P\left(C_{ncrk}\middle|S_1^c, \ldots, S_n^c\right)} < \theta$$

4 Conclusion

Automation in the process of crack detection in digital images of different surfaces is reviewed. Current research in this area focuses on detecting cracks with different approaches such as image processing method. Another approach is machine learning. It's classifiers learn from the past experiences of detecting cracks and can improve its performance. With its basic categories, classification methods are divided into supervised and unsupervised learning methods. This paper briefly discusses the three methods, namely random forest, support vector machine and Naive Bayesian which are classification methods of supervised machine learning algorithm. Further improvement can be done in terms of accuracy, and scope of problem can be extended as detecting the cracks in real-time environment.

References

1. Mohan A, Poobal S (2018) Crack detection using image processing: a critical review and analysis. Alexandria Eng J 57(2):787–798. https://doi.org/10.1016/j.aej.2017.01.020. Elsevier Publication
2. Ali L, et al (2021) Performance evaluation of different algorithms for crack detection in concrete structures. In: IEEE international conference on computation, automation and knowledge management, pp 53–58. https://doi.org/10.3390/s21051688
3. Shi Y et al (2016) Automatic road crack detection using random structured forests. IEEE Trans Intel Transp Syst 17(12):3434–3445. https://doi.org/10.1109/TITS.2016.2552248
4. Haodi H et al (2018) Method for detecting micron cracks on a magnetic rotor surface based on a support vector machine. IEEE Access J 20:1–13. https://doi.org/10.1109/ACCESS.2018.2870435
5. Chen F-C et al (2018) NB-CNN: deep learning-based crack detection using convolutional neural network and Naıve Bayes data fusion. IEEE Trans Ind Electron 65(5):4392–4400. https://doi.org/10.1109/TIE.2017.2764844
6. Cao W et al (2020) Review of pavement defect detection methods. IEEE Access J 8:14531–14544. https://doi.org/10.1109/ACCESS.2020.2966881
7. Peng C, Yang M et al (2020) A triple-thresholds pavement crack detection method leveraging random structured forest. J Constr Build Mater, Elsevier Publ 263:1–10. https://doi.org/10.1016/j.conbuildmat.2020.120080
8. Liang X, Shuai LV et al (2020) A weakly supervised surface defect detection based on convolutional neural network. IEEE Access J Art 8:42285–42296. https://doi.org/10.1109/ACCESS.2020.2977821
9. Sizyakin R, Cornelis B et al (2020) Crack detection in paintings using convolutional neural networks. IEEE Access J Art 8:74535–74552. https://doi.org/10.1109/ACCESS.2020.2988856
10. Yang J, Wang W et al (2019) Infrared thermal imaging-based crack detection using deep learning. IEEE Access J Art 7:2169–3536. https://doi.org/10.1109/ACCESS.2019.2958264
11. Dong X et al (2019) Learning-based texture synthesis and automatic inpainting using support vector machines. IEEE Trans Ind Electron 66(6):4777–4787. https://doi.org/10.1109/TIE.2018.2866043
12. Ding D et al (2019) Image in painting using nonlocal texture matching and nonlinear filtering. IEEE Trans Image Proc 28(4):1705–1719. https://doi.org/10.1109/TIP.2018.2880681
13. Tangsakul S et al (2020) Single image haze removal using deep cellular automata learning. IEEE Access J Art 8:103181–103199. https://doi.org/10.1109/ACCESS.2020.2999076
14. Nguyen HN, Kam TY, Cheng PY (2018) Automatic crack detection from 2D images using a crack measure-based B-spline level set model. ACM J Multidimension Syst Signal Process 29:213–244. https://doi.org/10.1007/s11045-016-0461-9
15. Tomar A, et al (2020) Machine learning, advances in computing, renewable energy and communication, vol 768. Springer Nature, Berlin, LNEE, p 659. https://doi.org/10.1007/978-981-16-2354-7. (ISBN 978-981-16-2354-7)
16. Iqbal A, et al (2020) Renewable power for sustainable growth, vol 723. Springer Nature, Berlin, LNEE, p 805. https://doi.org/10.1007/978-981-33-4080-0. (ISBN 978-981-33-4082-4)
17. Mitchell T (2013) Machine learning. In: Liu CL (ed). McGraw-Hill Education, Chennai, pp 2–3
18. Yokoyama S, Matsumoto T (2017) Development of an automatic detector of cracks in concrete using machine learning. Elsevier Publ, Sci Dir J Sustain Civil Eng Struct Constr Mater, J 171:1250–1255. https://doi.org/10.1016/j.proeng.2017.01.418
19. https://www.tutorialspoint.com/machine_learning_with_python/classification_introduction.htm
20. Wang S et al (2018) Panoramic crack detection for steel beam based on structured random forests. IEEE Access J 6:16432–16444. https://doi.org/10.1109/ACCESS.2018.2812141
21. Pasadas DJ, et al (2020) Detection and classification of defects using ECT and multi-level SVM model. IEEE Sens J 20(5):2329–2338. https://doi.org/10.1109/TIM.2019.2893009

Efficient Attack Model-Based Resilient Key Predistribution for WSN

Priyanka Ahlawat

Abstract A sensor network is prime arrangement of different miniature sensors placed in hostile environments. Security of these sensors becomes extremely important for applications like border security. The paper investigates the dilemma of physical node capture from an adversarial vision. In this paper, we present improved attack model for calculating the route node participation to find the most vulnerable nodes in a network. The vulnerability of the network is characterized by a route node participation which considers the number of times its key appears as a link key in the routes/paths found between source and sink. The vulnerable nodes are assigned lesser keys as compared to other nodes. The node rank is calculated based on link constancy ratio. The proposed attack model is analyzed with other attacking strategies based on the attacking rounds and path compromise ratio. It is shown that proposed model is more robust, and if an attacker attacks, then it would require greater number of attacks for entire network to be disrupted. We also apply hash-based mechanism to further secure the link keys having higher key utilization. It is based on node rank. The results show that the proposed algorithm is enhanced than other schemes, namely EG, Sarmad scheme and Walid scheme in terms of fraction of additional compromised links.

1 Introduction

Wireless sensor network (WSN) is a set a great numeral of minute multifunction and spatially distributed sensors communicate wirelessly over very short distances. A great diversity of the applications in the areas is related with household monitoring and ranges up to critical military applications. It is very challenging to provide security in transmitting critical information in a WSN. In such scenario, we have to focus on the security issues as well. Taking into account minimization of the transmission cost and security aspect of entire network poses a great deal of challenge in WSNs [1].

P. Ahlawat (✉)
National Institute of Technology, Kurukshetra, India
e-mail: priyankaahlawat@nitkkr.ac.in

A. Tomar et al. (eds.), *Proceedings of 3rd International Conference on Machine Learning, Advances in Computing, Renewable Energy and Communication*, Lecture Notes in Electrical Engineering 915, https://doi.org/10.1007/978-981-19-2828-4_22

Our work focuses on the optimization problem of minimizing the cost of transmission of routes/paths from the multiple nodes acting as source to a sink node under the restriction that susceptibility of the entire network is within given level [2–6]. We have designed the proposed scheme by considering the adversary viewpoint while defining vulnerabilities. Such networks are used in many applications like industry automation, tracking patients, medical exploration, environment monitoring, etc. Following reasons post a big challenging to provide security to these networks. Sensors are greatly resource inhibited which makes public-key ciphers infeasible. An adversary can without difficulty eavesdrops the communication. WSNs can be broadly classified into homogenous and heterogeneous sensor networks. Sensor nodes are of same type in terms of energy, memory and communication capabilities in homogenous. Heterogeneous networks consist of mixture sensor nodes with varying capabilities. However, homogeneous networks show good results in many application scenarios, but it is found that incorporating a mixture of sensor nodes can significantly improve the security and lifetime of the network. Due to applicability of WSN in wide range of applications, these are subjected to number of threats [7–10]. The adversarial attacks on the network can be passive or it can be active. These attacks destroy the confidentiality of the network. WSN is deployed in hostile environments, thus subjected to various risks [1, 2]. Capture of node is one of the foremost problems in such network [3, 4]. This work aims to formulate the attacking actions of the adversary to build an vulnerability model. The attacker tends to confine least amount number of nodes and aims to demolish the absolute the network in minimum possible time. A lot of papers are found in literature where authors formalized the activities of attack. Most primitive scheme was in [4] that was probabilistic where two sensor nodes can communicate only if they contribute to frequent key. Key server distributes the keys from a great pool to sensor nodes. The nodes broadcast their key identifiers to find common key. If they are unable to find any common key, they go for path key establishment. An attack model is given by Lin et al. in [5] in which dominating set are used as vulnerable points of the network that are probable to be attacked. A minimum key set is obtained by the defender in [6] to establish an attack model. It was given by Wu et al. Its focus is to enhance the competence of attacker. A deployment-based attack model to increase the security of the network is given in [7]. A path vulnerability matrix is given in [8] to increase the usefulness of the security mechanisms of the WSN. It is done by incorporating various vulnerable points in a network. A lot of sensors are randomly placed in a field, and hence, they necessitate security for transferring network data through the communication network. It becomes very essential to be concerned about the security of the data traffic and should be implemented with least storage overhead. It can be done by selecting paths which minimize the transmission cost. The schemes are based on the link joint by numerous paths that frequently endure from a greater danger of being compromised by attackers [11, 12]. A key as a link used in larger number of routing paths will definitely have a greater impact on the network if gets compromised. It makes the network designer to prevent such attack by reducing a particular link's or node's vulnerability, i.e., the chance of being attacked. Thus, we have used the adversary model based on the link key connected to the contribution in position of routing paths.

The main aspire of paper is to present an improved algorithm making the network designer to further endure a better number of adversarial attacks. Such attacks can be categorized as physical in which nodes are captured. It is done through compromising a link where the keys used in communication link have been eavesdropped by attacker during eavesdropping.

The paper is planned as: Sect. 2 details the accessible works that are carried out in this domain. Section 3 presents the models and preliminaries used in this scheme. Section 4 provides the detailing of the proposed efficient key management in great detail. Section 5 gives a comparative study of proposed scheme with further schemes. Finally, we conclude our paper in Sect. 6.

2 Existing Schemes

WSNs is large collection of tiny multifunction and spatially distributed sensor that generally transmit and communicate wirelessly over short distances. These sensors incorporate different property for to sense the environment. These sensors process data and communicate among other sensors. WSNs provide a convenient, cost-efficient way to integrate information with other networks and network components. However, use of wireless technology also creates many novel pressure such as eavesdropping. There are different attacks possible in WSN, namely node capture, denial of service (DOS), attack on routing paths, traffic analysis, replication and replay attacks. Several researchers have presented papers on how to quantify the vulnerability of the network. To protect the data between the sensor nodes, keys are need that can encrypt and decrypt the ongoing communication between sensors. To provide security to WSN, numerous KMS are proposed. One of the most primitive KMS is given by Eschenauer and Gligor [4]. A huge pool of keys is generated by a centralized entity. A subset of keys are given to every sensor nodes earlier than deployment in to sensor nodes. The nodes broadcast their key identifiers. The node having common key identifier set up a pair-wise key with that sensor nodes. In basic scheme, a subset of keys are randomly picked by the key distribution server and assign to the sensor nodes. Wu et al. [5] quantified the vulnerability of a WSN based on the degree of an edge or a node that is being shared among multiple routing paths. WSN is modeled as an undirected graph. Ahlawat et al. in [7, 8] present a graph-based attack model to further enhance the attacker capability to attack the network. Zhang et al. in [9] computed routing paths based on Dijkstra algorithm in graph-based network model. Chan et al. in [10] presented a q-composite scheme. It uses more than one key to establish the link key. It is resistance to small-scale attacks. Another enhancement to this scheme is given by Chan et al. in which the nodes share at least q keys to establish a secure link. This scheme performs well in small-scale attacks but fails in large-scale attack. One way hashing is used on the pre-distributed keys to increase security. These schemes induce some computation overhead. Du et al. [13] presented a key distribution method for heterogeneous networks. Qiu et al. in [14] presented a hybrid scheme to prolong the life time of sensor nodes. A cryptographic one way hash function-based scheme is proposed by Shan et al. [11] where hashing is applied on the key predistribution to increase the resilience of the scheme. A hash chain and deployment-based scheme is given in [15] by Lin et al. where long chains are assigned to nodes after their deployment. Nodes establish the pair-wise key from these chains. After that the chains are deleted from memory of nodes. A asymmetric key predistribution is presented by Khan et al. [16]. A hash chain approach based on node identifier is given by Bechkit et al. [12] in which keys of the nodes are hashed based on the node identifier to increase the resistance of the scheme. Zhang et al. [17] presented an improved key predistribution in which different security levels are defined for keys. Higher security level keys will disclose less information in uncompromised nodes that lower security level. The scheme has good resilience against capture of node capture. A method for key organization is given by Mirvaziri et al. in [18] for hierarchical sensor networks. Symmetric key cryptography is used to secure the data traffic between the nodes. It is shown that this method has reduced the memory consumption and energy consumption. It has maintained a favorable level of security also. Xiong et al. [19] presented a novel key predistribution based on hash chaining. Several parameters are proposed such as hash chain length and number of chains to increase the safety of the scheme. It is shown that small size key ring is used to establish the connection.

We observed that existing key distribution schemes are unable to meet all requirements related to security, storage, computation and communication of WSNs [12, 20–25]. Thus, there is still a requirement of a new key management scheme that should be competent to present an exchange between the dissimilar necessities in the network.

3 Models and Preliminaries

Models and possible vulnerabilities in a network have been discussed in the section. The models for proposed key management scheme are as under:

(i) Model of network: The network is a static graph. Vertices are the nodes. The edges set is cryptographic links between two nodes. During predistribution phase, key server allocate keys which are further hashed on link constancy to each node. In this phase, the nodes transmit their node identifiers to find the common key. There are quite a lot of routes possible from source to node to destination node. Sink collects and source senses the various physical parameters.

(ii) **Adversary model**: Adversary is intelligent one that has absolute facts about topology of network and routing. It also has information of keying information through eavesdropping during shared key discovery phase. The adversary is required to extract the cryptographic keys in a polynomial time during a node capture.

(iii) **Key predistribution**: During predistribution, keys are assigned to every node of the network from a large generated offline key pool. Data exchange between the nodes is possible only when two communicating nodes have at least one familiar key identifier among these nodes. In order to protect the network, it is enviable to have less number of common keys among the sensor nodes. The node capture impact gets increased when the adversary extracts the cryptographic keys of the network. This further increases the strength of the network.

(iv) **Ratio of link constancy** computes the occurrence of inclusion of a exacting link key in all the probable paths. If the link key with higher link key utilization value is captured by attacker, obviously larger number of routing paths becomes vulnerable. The attacker being intelligent calculates the rank, then it attacks the node with maximum link. It results in greater number of destroyed paths in a minimum time and with least resources.

4 The Proposed Key Management Scheme

The proposed efficient key management scheme is presented in this section. In the proposed scheme, network is a static undirected graph where nodes cannot move after being deployed in network. In the graph, vertex is the sensors. Edges are the communication cryptographic links between two sensor nodes. We use random key predistribution for the proposed scheme. Key distribution server allocates the keys in an offline manner from pool of cryptographic keys. During the sharing of key identifiers, two neighboring sensor nodes found whether they have some common key or not. We have considered following metrics to calculate the link constancy ratio for each link key used in routing paths:

(i) **Node route participation**: It is ratio of the incidence of a node in all possible routing paths from a source to sink node. It is calculated as the ratio of routing paths in which node appears to all the possible routing paths in a network. All possible routes from source to destinations are calculated. After that the occurrence of incidence of every node is designed. Finally, the rank of every node of the network is assigned according to this occurrence number. Higher value of this metric indicates a node more vulnerable to adversary attacks.

(ii) **Vulnerable node set**: It is set of vulnerable nodes that have more possibility of being attacked by adversary. The vulnerability set of nodes is generated based on the maximum and minimum value of node route participation for each node. We calculate the threshold value by taking average of these two values. We consider the nodes to be included in vulnerable set only if their route participation value is $> =$ the defined threshold value by network designer.

The proposed scheme is divided into different phases. In the first phase, attack model is constructed using node route participation. Once it calculated, we identify the nodes belonging to vulnerable set. In proposed scheme, we apply unbalanced key predistribution. So, during key predistribution phase, keys belonging to vulnerable set nodes if used as link key may pose a greater risk to safety of that link. It thus poses a threat to complete route that uses that link. Thus, to improve the resilience of the network, we apply hashing to such nodes based on link constancy ratio. The proposed scheme utilizes unbalanced key predistribution with hashing. It is done to increase the resistance against impact of node capture. Lesser number of keys are given to the vulnerable nodes. It is done to further enhance the safety of sensor network. If rank based on the link constancy ratio becomes more than its average. It is used to check whether it is incorporated in vulnerable set else not. The next step is to generate a set link keys of nodes in the vulnerable set. Link constancy ratio is the frequency of their occurrence in vulnerable node set. During key predistribution phase, keys are randomly distributed by key distribution server and hashed according to link constancy ratio. In this case, we take maximum value of link constancy ratio as the rank of the node.

5 Comparative Analysis

Comparative analysis of the proposed algorithm based on number of attacking rounds, path compromise ratio, resilience against node capture and storage overhead is presented in this section.

(i) **Attacking rounds** is defined as the quantity of nodes compromised by the attacker to completely demolish the routing paths between sources and sink in a network. In each attacking round, only one node is considered. We have implemented the algorithms in C++. We will find all routing paths in the network. After that, we try to find the nodes that have maximum participation in these paths. Set of vulnerable nodes are computed based on the node route participation. Finally, we calculate link constancy ratio for each link key. The keys are hashed according to the link constancy ratio. In proposed scheme, various aspects such as node route participation and link constancy ratio are used to improve the system potential to put off those attacks. Path compromise ratio is used for comparative analysis between basic scheme and improved. The most imperative intend of the attacker is to be cooperation the network in least amount probable time and with smallest amount resources. We observe that in improved scheme, attacker captures the least number of nodes and in a lesser time.

(ii) **Ratio of Path compromise**: It is stated as the proportion of routing paths destroyed to all available paths in a network. The scheme that destroy larger number of paths with small number of nodes will obviously have good path compromise ratio.

Figure 1 depicts the path compromise ratio for different attacks based on attacking rounds. In each attacking round, one node is taken by the attacker. In proposed scheme, the node with higher destructiveness value, i.e., destroy maximum paths

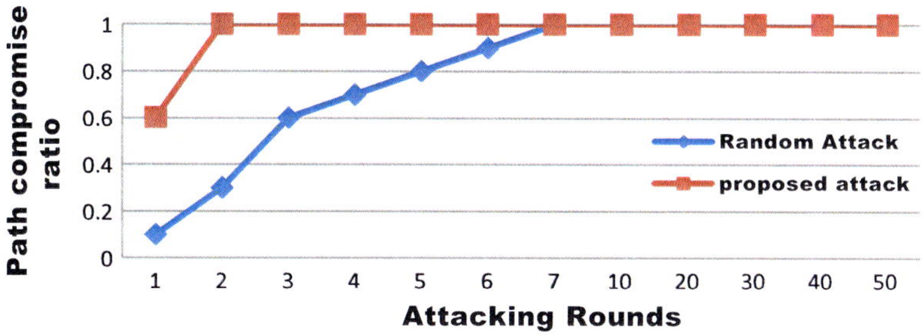

Fig. 1 Path compromise ration versus attacking rounds

is selected in every round. It makes the path compromise ratio equals to one with minimum number of nodes. In random attack, no strategy is used which results in higher values of attacking rounds.

(1) *Storage overhead*

It is defined as the memory required for storing security credentials. Complex applications need more credentials and thus usually have high storage overhead. The nodes store the value of hash function along with the keys and their identifiers. As the proposed scheme assigns different key ring sizes to the nodes belonging to vulnerable set and vice versa. It results in reduced overhead as compared to basic EG scheme We have considered only the overhead incurred by predistribution of keying information for analysis.

(2) *Anti-capture analysis of the proposed scheme*

Due to the heavy deployment and resource constrained sensors, the sensors are usually not tamper resistant making capturing of node as one of the important attack in WSN. The key predistribution from a single pool also induces the vulnerabilities in the security structure of the network. The additional compromised links are a fraction of external uncompromised links that get compromised. Let x be the number of nodes possessed by the adversary [9, 14]. We deal the compromise of the vulnerable set nodes. It is fraction of uncompromised cryptographic keys when x number of sensors possessed by attacker during a node capture. The probability that a node is captured is T and is given as $\frac{m}{P}$. The portion of uncompromised will then result 1-T when x sensors are captured by attacker $(1-T)^x$. Thus, likelihood that a cryptographic key is identified to attacker is (1-$(1-T)^x$). Captured key is hashed AC mod R times. Hash function is applied r times where $0 \leq \text{r} \leq \text{R}$ with probability 1/R. When a key is hashed, then probability that given key is revealed to attacker becomes $R-1/R$. Thus, probability that key is revealed to attacker is $\sum_{r=0}^{R-1} \frac{1}{R} \cdot \frac{R-r}{R} = \frac{R+1}{2R}$. Each sensor node has m keys, and thus, it is modified as link key constancy $= 1-(1-\frac{(R+1).m}{2R.P})^x$, where link key constancy is the probability of additional compromised links due to hashed EG scheme. m is the key ring.

In Fig. 2, it is shown that the proposed scheme has least number of fraction additional compromised links. It is compared with various existing schemes when

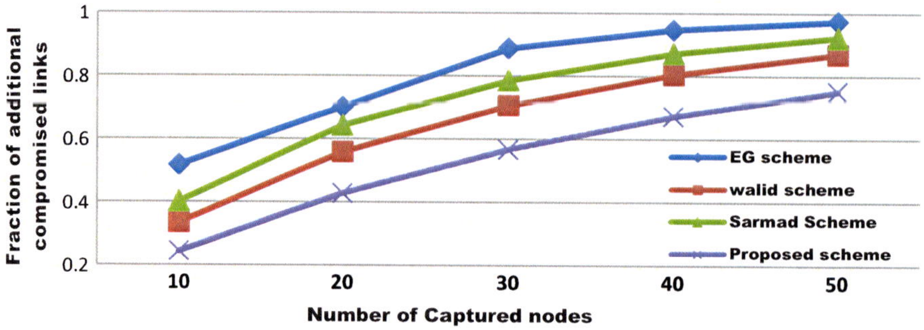

Fig. 2 Experiment result for the fraction of additional compromised links of different key predistribution schemes, where $m = 50$, $P = 1000$, $R = 10$, $n = 100$, for EG and Walid $m = 70$

compromised/captured nodes are kept same. We have only considered vulnerable nodes. This is owing to the hashed distribution that mitigates the probability of key compromise. Larger nodes will reveal larger fraction of keys to an adversary, thus lowering the resilience of the scheme. In EG scheme, no hashing is applied and thus has highest fraction of additional links for captured nodes. In Sarmad scheme, small key ring is used and thus has lower fraction of compromised links than EG scheme. In Walid scheme, hash function is applied based on node identifier and thus lower value than EG scheme and Sarmad scheme. In proposed scheme, hash function is applied on vulnerable nodes, and thus, it is more efficient as hash function is not applied on all nodes. It is better than EG and Sarmad scheme because it has lower key ring size as well hashed keys. It is lower than Walid scheme because of smaller key ring size. In Walid scheme, nodes are hashed without considering their route participation ration. Whereas in proposed scheme, this factor is considered which makes it more resistant.

For Fig. 3, the following values are assumed: $x = 1500$, $m = \{2, 5, 10\}$, $P = 1000$. When the key ring stored in the node gets increased, the probability of key compromise also increases. Larger key ring reveals more information. It results in revelation of larger portion of the key pool.

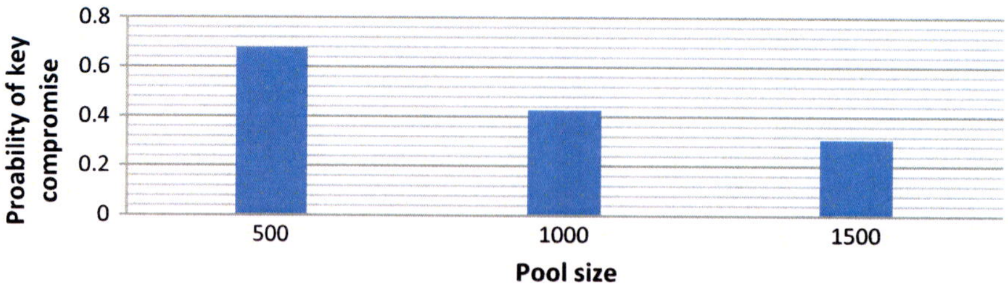

Fig. 3 Consequence of the variation in pool size on the probability of key compromise of the proposed scheme

6 Conclusion

The capturing of sensor node is one of the well-known attacks in wireless sensor network (WSN) security. This paper focuses on mitigating the effect of node capture on the key management scheme. An proficient attack representation is given based on route node participation and link constancy ratio. To counteract such attack, an hash-based mechanism is done on the key predistribution phase. The results also confirm that the projected scheme gives good security against capturing of node capture. It has least number of additional captured links. It is shown that it has mitigated the effect of ratio of path compromise. To incorporate this model on mobile nodes is our future work.

References

1. Zhang J, Varadharajan V (2010) Wireless sensor network key management survey and taxonomy. J Netw Comput Appl 33(2):63–75
2. Akyildiz IF, Su W, Sankarasubramaniam Y, Cayirci E (2002) Wireless sensor networks: a survey. Comput Netw 38(4):393–422
3. Xiao Y, Rayi VK, Sun B, Du X, Hu F, Galloway M (2007) A survey of key management schemes in wireless sensor networks. Comput Commun 30(11–12):2314–2341
4. Eschenauer L, Gligor VD (Nov 2002) A key-management scheme for distributed sensor networks. In: Proceedings of the 9th ACM conference on computer and communications security. pp 41–47
5. Lin C, Wu G, Xia F, Yao L (2013) Enhancing efficiency of node compromise attacks in vehicular ad-hoc networks using connected dominating set. Mob Netw Appl 18(6):908–922
6. Wu G, Chen X, Obaidat MS, Fellow of IEEE, Lin C (2013) A high efficient node capture attack algorithm in wireless sensor network based on route minimum key set. Secur Commun Netw 6(2):230–238
7. Ahlawat P, Dave M (2018) Deployment based attack resistant key distribution with non overlapping key pools in WSN. Wireless Pers Commun 99(4):1541–1568
8. Ahlawat P, Dave M (2017) A hybrid approach for path vulnerability matrix on random key predistribution for wireless sensor networks. Wireless Pers Commun 94(4):3327–3353
9. Zhang L, Yang W, Rao Q, Nai W, Dong D (2013) An energy saving routing algorithm based on Dijkstra in wireless sensor networks. J Inf Comput Sci 10(7):2087–2096
10. Chan H, Perrig A, Song D (May 2003) Random key predistribution schemes for sensor networks. In: 2003 Symposium on security and privacy, 2003. IEEE, pp 197–213
11. Shan TH, Liu CM (Dec 2008) Enhancing the key pre-distribution scheme on wireless sensor networks. In: 2008 IEEE Asia-Pacific services computing conference. IEEE, pp 1127–1131
12. Bechkit W, Challal Y, Bouabdallah A (2013) A new class of Hash-Chain based key pre-distribution schemes for WSN. Comput Commun 36(3):243–255
13. Du X, Xiao Y, Guizani M, Chen HH (2007) An effective key management scheme for heterogeneous sensor networks. Ad Hoc Netw 5(1):24–34
14. Qiu B, Chen X, Wu Q (2016) A key design to prolong lifetime of wireless sensor network. Chaos, Solitons Fractals 89:491–496
15. You L, Yuan Y, Yu W, Wang Q (2015) A key distribution scheme for WSN based on hash chains and deployment knowledge. Int J Distrib Sens Netw 11(7):640792
16. Sarmad UK, Lavagno L, Pastrone C (Oct 2010) A key management scheme supporting node mobility in heterogeneous sensor networks. In: 2010 6th international conference on emerging technologies (ICET). IEEE, pp 364–369

17. Zhang J, Li H, Li J (2020) An improved key pre-distribution scheme based on the security level classification of keys for wireless sensor networks. Int J Inf Comput Secur 12(1):40–52
18. Yu W (March 2010) A pairwise key management scheme based on hash function for wireless sensor networks. In: 2010 second international workshop on education technology and computer science, vol 2. IEEE, pp 198–201
19. Xiong P, Su Q (2021) Key distribution strategy of wireless sensor network based on multi-hash chain. J Web Eng 713–742
20. Bhushan S, Kumar M, Kumar P, Stephan T, Shankar A, Liu P (2021) FAJIT: a fuzzy-based data aggregation technique for energy efficiency in wireless sensor network. Complex Intell Syst 7(2):997–1007
21. Singh AK, Alshehri M, Bhushan S, Kumar M, Alfarraj O, Pardarshani KR (2021) Secure and energy efficient data transmission model for WSN. Intell Autom Soft Comput 27(3):761–769
22. Bindra HS, Maakar SK, Sangal AL (2010) Performance evaluation of two reactive routing protocols of MANET using group mobility model. Int J Comput Sci 7(3):38–43
23. Bhushan S, Singh AK, Vij S (April 2019) Comparative study and analysis of wireless mesh networks on AODV and DSR. In: 2019 4th international conference on internet of things: smart innovation and usages (IoT-SIU). IEEE, pp 1–6
24. Iqbal A et al (2021) Intelligent data-analytics for condition monitoring: smart grid applications. Elsevier, 268 pp. https://www.sciencedirect.com/book/9780323855105/intelligent-data-analytics-for-condition-monitoring. (ISBN: 978-0-323-85511-2)
25. Tomar A et al (2020) Machine learning, advances in computing, renewable energy and communication, vol 768. Springer Nature, Berlin, LNEE, 659 pp. Doi: https://doi.org/10.1007/978-981-16-2354-7. (ISBN 978-981-16-2354-7)

Liver Cancer Detection Using Hybrid Approach-Based Convolutional Neural Network (HABCNN)

Shashi Bhushan

Abstract Liver cancer is one of the serious disease these days. Every year almost 800,000 people get detect of liver cancer and 700,000 died out of that. But it is very difficult to detect it at early stage. To detect this manually is a tedious task and cannot be perform easily as there was a serious of tests and all through which patient has to go through and still did not have any surety of detection. So CT images after applying watershed segmentation can be used to get this liver cancer detected by assessing tumor load, clinical treatment response by using the concept of deep learning. In this paper, a Hybrid Approach-Based Convolutional Neural Network (HABCNN) is proposed for liver tumor segmentation which has been mathematically modeled so that issue of cancer detection get resolved. The kidney and spleen is segmented initially. After cancer tissue segmentation, GLCM was used to extract features from tumor parts. Eventually, Hybrid Approach-Based Convolutional Neural Network (HABCNN) was used to identify the hepatocellular and metastatic carcinoma for hepatic cancer. A better 99.44% identification precision came from the suggested classifier. The modern segmentation approach should insure that the precise boundary structure of the cancer area is known so that important information can be better diagnosed. A comparison based on specificity, sensitivity, accuracy and DSC has been done with some existing techniques like Multilayer Perception [14] and C4.5 [15]. Without human intervention, the classification method allows successful cancer damage defection. So this can be used by medical practitioner for early detection of liver cancer.

Keywords Liver cancer detection · Deep learning · Hybrid approach-based convolutional neural network

S. Bhushan (✉)
School of Computer Science, University of Petroleum and Energy Studies, Dehradun, India
e-mail: tyagi_shashi@yahoo.in

A. Tomar et al. (eds.), *Proceedings of 3rd International Conference on Machine Learning, Advances in Computing, Renewable Energy and Communication*, Lecture Notes in Electrical Engineering 915, https://doi.org/10.1007/978-981-19-2828-4_23

1 Introduction

Since liver cancer signs appear in late stages, it cannot be identified early such that they are challenging to detect, which causes a high mortality rate among all other cancers. The most significant cause of death among humans is liver cancer. Therefore, the detection of liver cancer is required in the early diagnosis and this early detection offers the most possibility to treat cancer patients safely and effectively. However, it is the most complicated approach to boost the likelihood for the individual to live. Segmentation and classification of liver tumors from computed tomography images are essential to the advancement of early diagnosis and treatment of liver cancer. Nevertheless, the varying morphologies, boundaries, different densities, and sizes of the lesions make this job complicated. In the second stage of the work, to design a new method for distinguishing and classifying tumors from computed tomography images using a Hybrid Approach-Based Convolutional Neural Network (HABCNN) is employed [1–11]. The findings are correlated with scientific specialists' results. The ultrasound image of liver is shown in Fig. 1. The world's most common cause of mortality is liver disease [1]. The early diagnosis of the occurrence of liver cancer via optimum therapies is vital to enhance survival chances. At present, biopsy for cancer is regarded as the golden standard, depending on the tumor's location, although it is uncomfortable, invasive, and not always a viable alternative. Noninvasive treatment of liver disorders can be performed using techniques of diagnostic imaging. Computed Tomography (CT) is the best useful imaging device for oncological and follow-up treatments. In the United States, in 2007, more than seventy million CT scans take effect in the recognition, diagnosis, and observing of liver lesions. Images are collected from a contrast agent before and during intravenous infusion. The

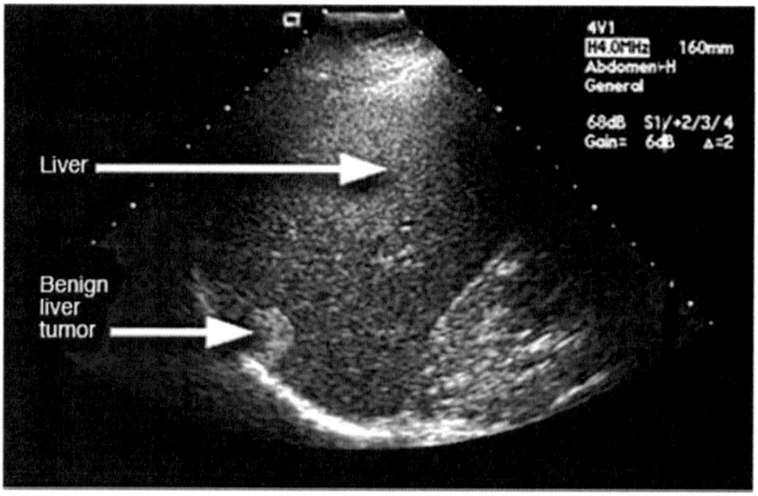

Fig. 1 Ultrasound image of liver cancer

liver is usually detected and diagnosed by radiologists [2, 3]. Lesions are based on various densities presented at specific scan time and optimal identification in the portal stage in Hounsfield Units (H.U.). A variety of cancer lesions are identified in the liver including both benign and malignant. The primary source of mortality from cancer is metastasis. Metastatic diseases usually arise from prime locations in the intestine, breast, heart, pancreas, and stomach affect the liver. In the number, size and general appearance of metastatic liver cancer lesions, there is a considerable variability, rendering computerized metastatic liver injury diagnosis a challenging one. Radiological procedure is available to evaluate the hepatitis manually [4]. This process takes time for a 3-D CT scan of several hundred slices and several lesions to be looked for by the radiologist. This is not a definitive examination. In the first time, it is especially important to identify so-called "too small to distinguish" cancer lesions. For radiology to perceive this as a demanding job it takes additional time and concentration. Over the modern years, the introduction of these methods into clinical procedures would lead to even greater information and practical evaluation and care, as well as more over depth research strategies. The development of predictive tools for computed tomography (CT) clarifications has both a significant clinical and economic value. Throughout the area of medical science and analysis photos such as MRI, X-ray, C.T, quickly evolved deep learning algorithms, mainly CNNs and FCN's, and DBN's. Comprehensive learning strategies are used to identify, classify and diagnose information for images, the segment organs and lesions. The identification, diagnosis, marking, and evaluation of CT images and X-rays have been the most challenging for medical image processing. For example, large data sets are used in the application of liver cancer research to train the model that combines deep learning methods, including CNN, DBNs for image recognition, and NLP [5, 6].

2 Medical Image Processing

Even though human experts are available to diagnose and predict the kind of diseases and cancer cells, an opinion based on machine intelligence aids the decision. In this connection, there is the applicability of various devices employed for diagnosing and computing the pathologies of human diseases (Fig. 2).

3 Proposed Methodology

In Fig. 3, a detailed description about the proposed methodology is given. First dataset in the form of images has to be taken as an input, preprocessing is done in step 2, in step 3 segmentation is done through watershed segmentation, in step 4 Features get extracted from segmented region then a training and testing set has to be generated to get the model trained and tested. After performing training and testing, in next step images will be get processed, classified and segmented using proposed

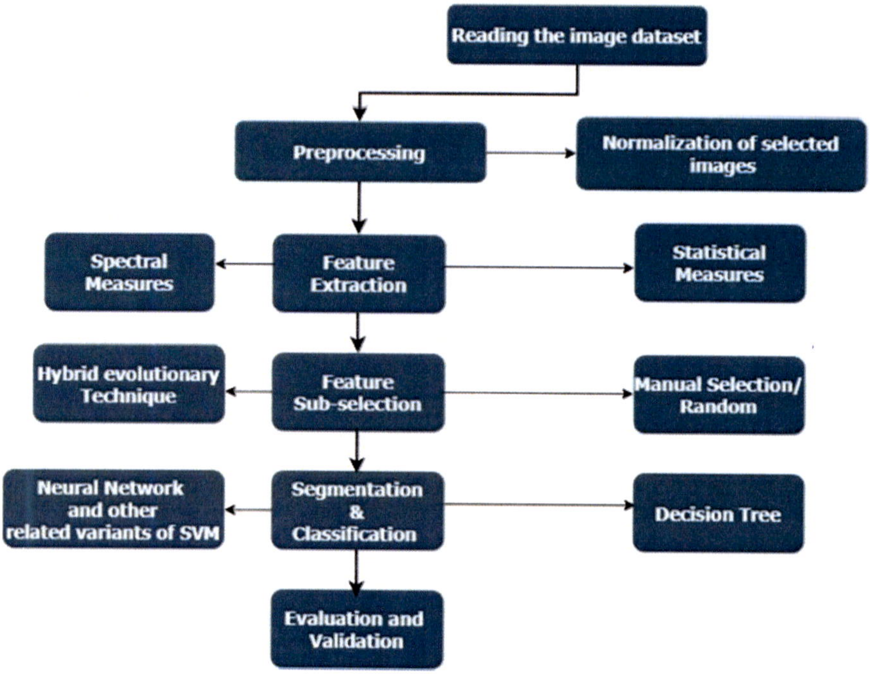

Fig. 2 Medical image processing

Hybrid Approach-Based Convolutional Neural Network (HABCNN). And at last performance analysis is done based on sensitivity, specificity, accuracy, and DSC.

Step 1: Dataset of Liver CT Images

CT scan is a noninvasive diagnostic image method in which it is the mixture of X-ray and CT to produce the horizontal or axial images of the liver. Hence, image processing-based research work facilitates liver cancer treatment. In computer tomography, X-ray beam passes in a circle movement around the body. By this, it offers the different view of similar organ. CT scan is done with or without "Contrast" (type of material that taken by mouth and/or injected into an intravenous (IV) line that cause liver organ or tissues under supervision to become clearer).

Step 2: Preprocessing

The first step of the diagnosis of liver carcinoma is preprocessing. To ensure the durability and usability of a database, preprocessing is important. For this, any step seems to be essential to the workflow of image processing. The process carries out preprocessing of unnecessary error identification using filters and histogram equalizing techniques. Here, with CT image the noises can be removed in this step. The Adaptive Median Filter (AMF) is often used as a non-linear optical filtering system to remove noise from an image or signal. A noise reduction is a typical

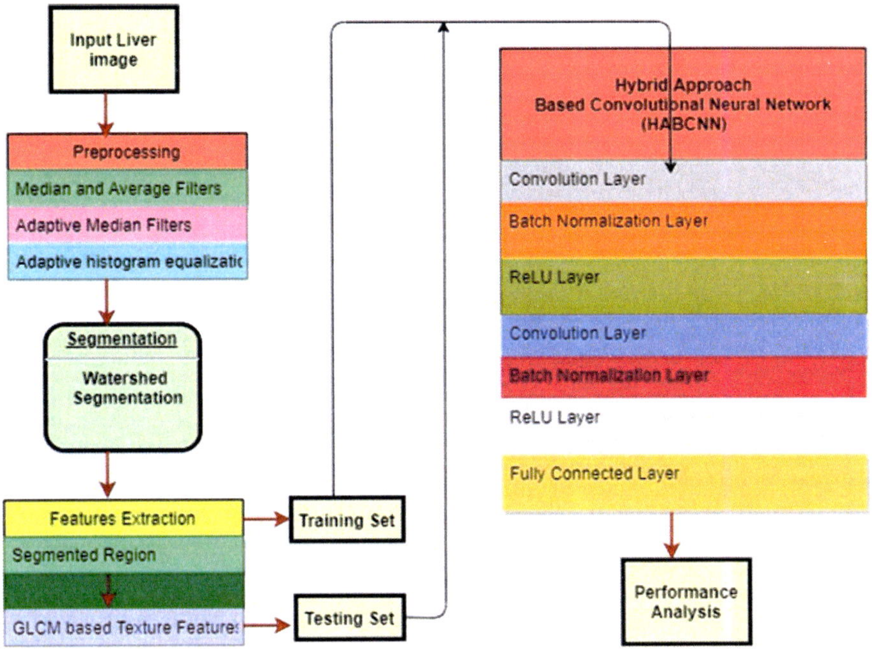

Fig. 3 Proposed methodology

preprocessing method for efficiency enhancement. Preprocessing is carried out to enhance the contrast of the image in the CT image. Typically, histogram equalization is achieved to increase image consistency. Histogram Equalization is a technique used to enhance the contrast of pictures. It is also used to improve the sensitivity values effectively, i.e., the image intensity range is broadened. In order to boost relations between regions, it allows less local contrast. Therefore, after implementation of the histogram equalization, the average contrast of the images is improved. Consider h denote the normalized histogram of an image. Hence,

$$h^x = (\text{Total number of picture element in} \times \text{intensity/total no. of pixel}) \quad (1)$$

where $x = 0, 1, \ldots, x - 1$.

The histogram equalized image can be defined as

$$H_{i,j} = \text{base}((x - 1)\sum_{x=0}^{b_{i,j}} p^x \quad (2)$$

While the result indicates that the equalization process used is exactly flat histograms, it can soften them and improve them.

Step 3: Segmentation

Segmenting both the liver and tumor regions is an important and demanding task in the recent days. For this purpose, various segmentation techniques are developed, which segment the image with the help of modalities. There are different types of techniques for segmentation which include manual segmentation, semi-automatic segmentation, and fully-automatic segmentation. In manual segmentation, the medical experts segment the image slice by slice, and the liver frontiers are recognized contrarily by a dissimilar radiologist or the same radiologist at a different time [7]. This type of segmentation explains the structure and observation of the medical images. In this category, the difficulties are mainly related to reducing image quality and increasing artifacts. The major drawbacks of manual segmentation are large amount of image slices, intensive time, and lack of results. Further, creation of distinct dataset is a highly complex process. Due to these issues, the semi-automatic segmentation is developed, which interactively identifies the seed points for extracting the boundary of liver from the selected regions. However, it has some important disadvantages such as more amount of time, inefficient segmentation results, and it needs a user interaction to segment the liver and exactly locate the tumor. In order to reduce the workload of medical experts, and avoid inaccurate measurements, the fully-automatic segmentation is developed to automatically. Segmenting the liver and tumor from the medical images without user interaction helps the radiologists to assist diagnosis. It works based on prior knowledge of the image that includes the shape and localization of liver. When compared to the semi-automatic segmentation, it has the advantages of reduced time consumption, user interaction and full automation. Thus, fully-automatic segmentation is highly preferred for liver and tumor identification during diagnosis. The class of image is distinct, and it never remains the same because the input image can vary from a scanner to scanner. It leads to the problem of automatic recognition, and so the segmentation depends on the acquired input image. Still, the extraction of liver and tumor from the abdominal images is a challenging and demanding task. Based on the contrast enhanced CT scans, the diagnosis is provided during the clinical practice for liver tumor. The contrast media reaches the liver via the hepatic artery or the portal vein if the enhancement of two vascular trees is possible. The evolution of liver tissue region is the discriminating factor of liver tumor. The visual analysis of CT scans provided by the radiologists is not enough for an accurate tumor detection and diagnosis because only the small part of information stored in the images is identified using CT scans. Here, the segmentation was done utilizing the watershed algorithm. The separation of an image into several parts is image segmentation. Segmentation of the pixel range is meant to split the CT image into segments. Frame segmentation is used to identify points and borders, image curves. The segmentation comprises of a series of parts describing the whole entity or an image contours array. The image characteristics can easily be predicted by using this process. In this study, Gray scales are used for the segmentation method. The variation between large pixels and small pixels around the object's boundaries are evaluated.

$$W^{\text{segment}} = \sum_{\{i,j\}\varepsilon Q_2}^{n} K_2(Z_i, Z_j).w.\log_{ci} + \gamma \int c_i dx \quad (3)$$

Then, the features can be selected with the GLCM after the segmentation stage. The Grey Level Co-occurrence Matrix method (GLCM) is a means of obtaining second-order statistical texture properties. The technique has been used in many applications and the presence of three or more pixels is seen in the third and higher-order textures. The GLCM is an arithmetical capacity that can ordinarily dispense with the curios productively. The exactness of the picture may likewise be kept clear. For the review cycle, the picture might be extricated. GLCM might decide the recurrence of the pixels in a predefined accuracy. The single pixel is to be addressed here and another pixel is to be known as the Ø course l and the adjoining worth detachment of m. commonly, m obtains a single worth, and Ø can benefit directionally. Then, the got directional worth can dispense with the qualities of the photos used for the division cycle. The GLCM cycle may be set as continues in condition 3:

$$P(m, s) = G(m, s, 0, \emptyset) \sum_{m=1}^{H} \sum_{s=1}^{H} G(m, s, 0, \emptyset) \quad (4)$$

where G is the recurrence vector, m, s, o is the recurrence of the specific part will by and large have the pixel upsides of l and m, P addresses the 88 provisions of a picture, (m, s) was the part of the m and l, Ø addresses the standardized steady. By carrying out the GLCM, the various traits can be acquired.

Step 4: Classification using Hybrid Approach-Based Convolutional Neural Network (HABCNN)

After extracting the features, the classification process is applied in the last stage for classifying the normal tissues and liver lesions. The main intention of this stage is to apply a machine learning model for an accurate disease classification during diagnosis. Based on the feature vectors, the classifier can accurately classify the liver and tumor regions from the given image. The advantage of classification is that it separates the objects into diverse classes and identifies the healthy and the disease-affected tissues. The stage of cancer then be calculated after isolation of its characteristics whether it is mild, moderate or otherwise extreme [8]. Hybrid Approach-Based Convolutional Neural Network (HABCNN), one of the well establishing algorithms, was included in this classification. The goal is categorized as probable. It is an algorithm for pre-trained convolution. In this case, HABCNN enables measurement of discrepançies between a single variable dependency and one or more different variables. The HABCNN anticipates the possibilities and has a strategy. Statistical models occur. During this process, HABCNN reads the image first and redraws it, then measures the class likelihood for the grading process. The neural network of convolution is one of the deep neural networks of learning. In the image recognition and classification process, HABCNN represents a major breakthrough. The most widely used smash the visual meaning is the function of optics in the interpretation and characterization of images.

A HABCNN has been masterminded as the layers as ReLU layers, Convolutional layers, pooling layers, and A Fully associated layer. While contrasting other picture characterization calculations, CNNs have exceptionally less pre-preparing steps. This CNN must be utilized in different fields for a considerable length of time. Convolution the critical job of this convolution cycle is to focus the data on picture features. The underlying cycle in CNN is the convolutionary layer. During this stage the layer usefulness is recognized and the element map is made from the information picture. ReLU layer the subsequent stage is the straightforwardly rebuilt unit layer. In this review, the institution technique has been presented on the element maps to expand the organization's nonlinearity. Here, the unwanted qualities successfully can be erased.

Pooling layer: The most common way of pooling can continuously diminish the size of the info. The progression of pooling can bring down over fitting. It will quickly call attention to the vital boundaries by raising the quantity of boundaries required.

Leveling layer: It is a significant simple move whereby the surveyed work guide ought to be straightened into the numbers consecutive segment. Completely associated layer commonly the traits which can be combined with the qualities. The grouping technique should be possible with the high percentile in precision. The mistake will predominantly be estimated and recorded.

Softmax: Softmax is likewise utilized in neural frameworks to plan the un-standard organization movement over expected execution gatherings to a probabilities appropriation. The Softmax has been applied for some issues in different review fields. The likelihood of the decimal will infer 1.0. Taking the connected Softmax varieties: Full Softmax is the Softmax, which can register probability for each possible class. Softmax registers probability for every one of the positive names; but only for a self-assertive illustration of negative names. This CNN empowers the inconsistency between a solitary variable and at least one unique factors to be estimated. For CNN, chances and a capacity are determined [9]. This is the gathered administering. In this strategy, CNN would first be able to peruse and rearrange the picture and afterward utilize its class probability to gauge the characterized methodology.

Process starts from Neuron activation by

$$a_j^1 = \sigma \sum K.W_{j,k}^1.a_K^1 + b_j^i \tag{5}$$

Equation (5) in vectored form as

$$a^l = \sigma\left(W^l a^{l-1} + b^l\right) \tag{6}$$

Training set can be merged in this as

$$C = \frac{1}{2}\left\|Y - a^l\right\|^2 = \frac{1}{2}\sum_j \left(y_j - a_j^l\right)^2 \tag{7}$$

HABCNN can be classified as

$$F = a_j^1 b_j^1 - \left(a_j^1 b_j^{12}\right) \tag{8}$$

where m is empirical constant, F is the feature.

4 Results

The data sets of the liver cancer CT images were obtained from LiTS (Figs. 4 and 5).

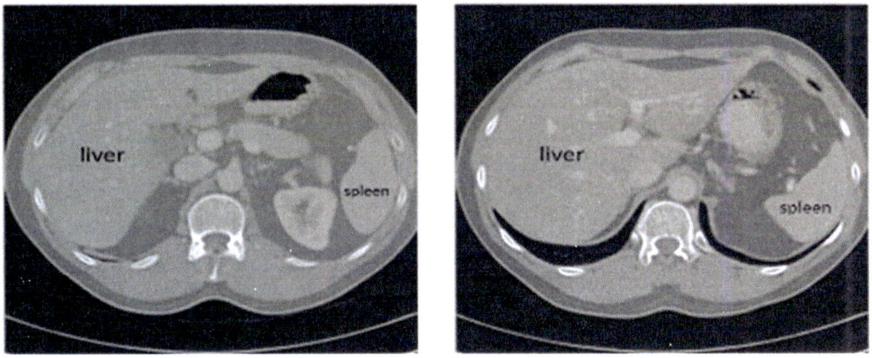

Fig. 4 Liver CT image

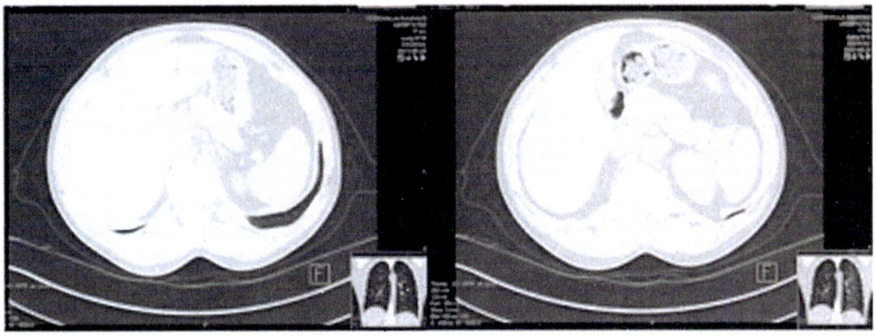

Fig. 5 Preprocessing

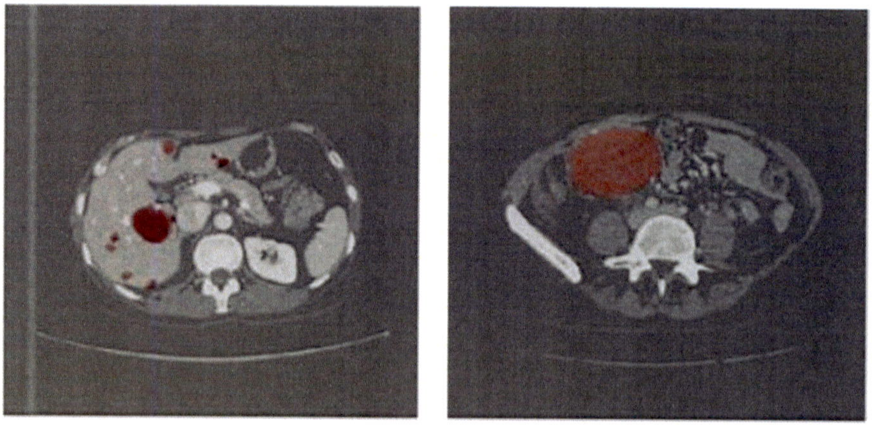

Fig. 6 Tumor detection using hybrid approach-based convolutional neural network

The process of preprocessing and segmentation can be done using histogram equalization and Hybrid Approach-Based Convolutional Neural Network will give the peak up results as shown in Fig. 6. The proposed classification methodology is simulated for the performance assessment. This analyzes was carried out using the LiTS data collection for liver cancer. The neural network training test ratio set constitutes 70% of the total data collection and the test set contains 30% of the overall results.

Accuracy: It is a measure of arithmetical predisposition; the accuracy is the proportion of true results (both true positive and true negative) in the total data.

$$\text{Accuracy}(A) = (TP + TN)/(TP + TN + FP + FN)$$

Sensitivity: Sensitivity, which is also called the true optimistic rate, the recall, or possibility of detection in some fields, measures the proportion of actual positives that are correctly identified.

$$\text{Sensitivity} = TP/(TP + FN)$$

Specificity: Specificity, which is also called the true positive rate, measures the amount of the actual negatives that are correctly identified.

$$\text{Specificity} = TP/(TP + FP)$$

Dice Similarity Coefficient: The Sørensen Dice Similarity coefficient is a mathematical method that calculates the correlation of two datasets.

$$DSC = 2TP/(2TP + FP + FN)$$

Table 1 Comparison table: proposed versus existed

Performance metrics	Multilayer perceptron [14]	C4.5 [15]	HABCNN (Proposed)
Specificity	88.02	96.02	97.1
Sensitivity	90.38	93.78	95.99
Accuracy	89.20	95.01	99.44
DSC	89.04	94.40	98.1

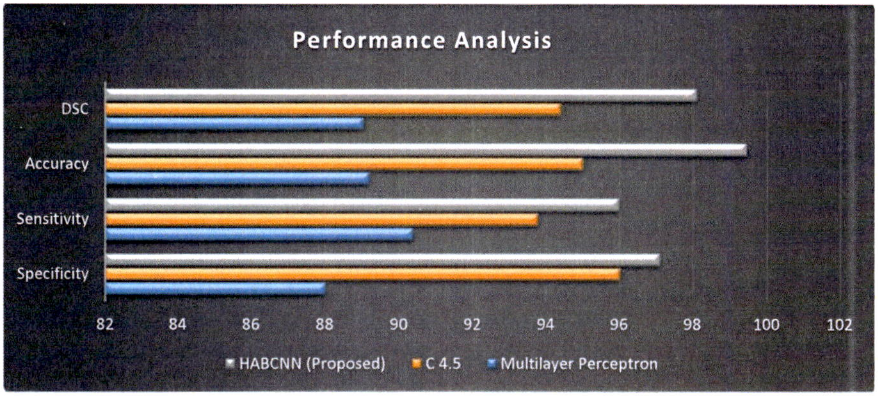

Fig. 7 Performance analysis

The comparison table of the proposed vs the existing detection mechanisms is shown in Table 1. From the Fig. 7. It can be inferred that the diagnosis and classification of hepatocellular carcinoma and liver cancer form metastasis can be effectively accomplished by using improved residual Google net CNN classifier. The proposed classifier HABCNN perform better than existing methodologies in all aspects.

5 Conclusion

Therefore the CT image segmentation is a critical operation during the second step of the study to determine the picture representing some portion of the renumber, spleen, etc. Contour and clustering methods have been used extensively for the segmentation of the medical image. This chapter therefore concentrates on the watershed segmentation of liver cancer. The kidney and spleen is segmented initially. After cancer tissue segmentation, GLCM was used to extract features from tumor parts. Eventually, Hybrid Approach-Based Convolutional Neural Network (HABCNN) was used to identify the hepatocellular and metastatic carcinoma for hepatic cancer. A better 99.44% identification precision came from the suggested classifier. The modern segmentation approach should insure that the precise boundary structure of the cancer area is known so that important information can be better diagnosed. Without human intervention, the classification method allows successful cancer damage defection. This technique may be particularly valuable for the early detection of cancer by physicians and nurses in patients.

References

1. Guo L-H, Wang D, Qian Y-Y, Zheng X, Zhao C-K, Li X-L, Bo X-W, Yue W-W, Zhang Q, Shi J, Xu H-X (2018) A two-stage multi-view learning framework based computer-aided diagnosis of liver tumors with contrast enhanced ultrasound images. Clin Hemorheology Microcirc 69(3):343–354
2. Vorontsov E, Cerny M, Régnier P, Di Jorio L, Pal CJ, Lapointe R, Vandenbroucke-Menu F, Turcotte S, Kadoury S, Tang A (Mar 2019) Deep learning for automated segmentation of liver lesions at CT in patients with colorectal cancer liver metastases. Radiol Artif Intell 1(2):180014
3. Deshwal V, Sharma M (2019) Breast cancer detection using SVM classifier with grid search techniques. Int J Comput Appl 178:18–23
4. Lakhani P, Sundaram B (2017) Deep learning at chest radiography: automated classification of pulmonary tuberculosis by using convolutional neural networks. Radiology 284(2):574–582
5. Yan K, Wang X, Lu L, Summers RM (2018) DeepLesion: automated mining of large-scale lesion annotations and universal lesion detection with deep learning. J Med Imag 5(3):036501
6. Bayrak EA, Kirci P, Ensari T (2019) Comparison of machine learning methods for breast cancer diagnosis. In: 2019 Scientific meeting on electrical-electronics and Biomedical Engineering and Computer Science (EBBT) 1–3
7. Kaarthik K, Sridevi A, Vivek C (2017) Image processing based intelligent parking system. IEEE International conference on electrical, instrumentation and communication engineering
8. Mohapatra P, Panda B, Swain S (2019) Enhancing histopathological breast cancer image classification using deep learning. Int J Innovative Technol Exploring Eng 8:2024–2032
9. Shamy S, Dheeba J (2019) A research on detection and classification of breast cancer using k-means GMM & CNN algorithms. Int J Eng Adv Technol 8:501–505
10. Iqbal A, et al (2021) Intelligent data-analytics for condition monitoring: smart grid applications. Elsevier, p 268. https://www.sciencedirect.com/book/9780323855105/intelligent-data-analytics-for-condition-monitoring. (ISBN: 978-0-323-85511-2)
11. Tomar A, et al (2020) Machine learning, advances in computing, renewable energy and communication, vol 768. Springer Nature, Berlin, LNEE, p 659. https://doi.org/10.1007/978-981-16-2354-7. (ISBN 978-981-16-2354-7)

The Use of LSTM Models for Water Demand Forecasting and Analysis

Shashi Bhushan

Abstract Demand assessment in a water appropriation network gives significant information to checking and controlling frameworks. In light of budgetary and actual limitations, there is a need to appraise water demand from a set number of sensor estimations. The demand assessment issue is underdetermined due to the restricted sensor information and the understood connections between nodal demands and tension heads. Climate change and global warming is one of the major issue which effecting this water demand/supply issue. In this paper, a real world problem of water demand and supply has been taken into consideration. Guaranteeing ongoing water the board and streamlining becomes obligatory for settling the limitations of water supply/request and to consent to biodiversity necessities. To address this issue, Long Short-Term Memory network (LSTM) forecast is used to check about consumption and absolute requirement. Calculation of accuracy and percentage error is done by using LSTM and same has been compared with other technique called ARIMA. Accuracy of around 60% has been calculated by using LSTM as compared to 49% by using ARIMA. Similarly Mean absolute percentage error has been calculated by using LSTM as 39.41% and by ARMA it is 59.60%. By this daily-based demand of water is forecasted by using deep learning-based forecast analysis methods.

Keywords Long short-term memory (LSTM) · Water distribution · Deep learning

1 Introduction

Now a days the technologies like IoT, Artificial Intelligence, Deep Learning [1], cloud computing and fog computing are very useful to address any kind of real world problem. Out of them one real world problem of water demand and supplying which arises because of global warming and climate change can be addressed and get a solution by using above mentioned techniques. While thinking about the Indian conditions, there could be no appropriate water management system (WMS) other

S. Bhushan (✉)
School of Computer Science, University of Petroleum and Energy Studies, Dehradun, India
e-mail: tyagi_shashi@yahoo.in

A. Tomar et al. (eds.), *Proceedings of 3rd International Conference on Machine Learning, Advances in Computing, Renewable Energy and Communication*, Lecture Notes in Electrical Engineering 915, https://doi.org/10.1007/978-981-19-2828-4_24

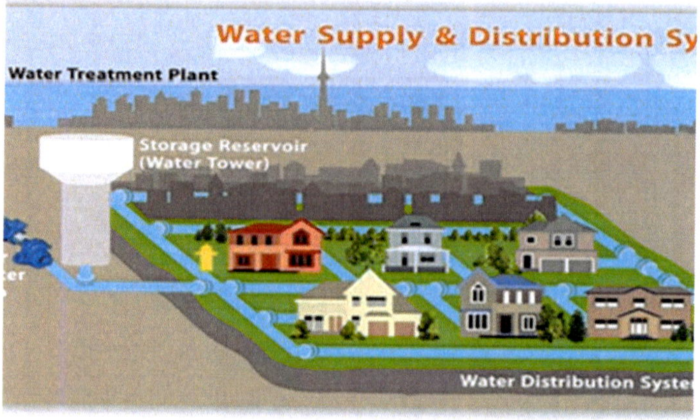

Fig. 1 Water distribution system [5]

than some metering arrangements that are continued in metro urban areas. The water management system will be the main need of great importance arrangement instead of searching for the elective wellsprings of water [2]. The utilization of water shifts generally between different sorts of customers for example, private, business, modern and furthermore the utilization fluctuates as indicated by the seasons. It is exceptionally expected to set up the WMS for effective dissemination of assets [3, 4]. For a proficient Water Management System, there is need for appropriate water dissemination organization (WDN) that comprise capacity supplies, principle siphoning station, sponsor siphoning station, fire hydrants and administration lines. The WDN can be comprehensively grouped under two classifications which are as per the following:

a. Under-Ground WDN
b. Over-Ground WDN.

A basic water dispersion system is represented in the Fig. 1 which reproduces how water is shipped from its source and how it is provided to the purchasers. There are different parts like valves, intersections, joints and lines utilized in the WDN development.

2 Deep Learning

Machine Learning is a part of AI which facilitates the system with the self-learning module without the intervention of programmers. It is helpful in performing data prediction. The internal process architecture is illustrated in the Fig. 2. Figure 2 Machine Learning Process Architecture Machine Learning is classified under three major categories based on the method of learning activity they employ.

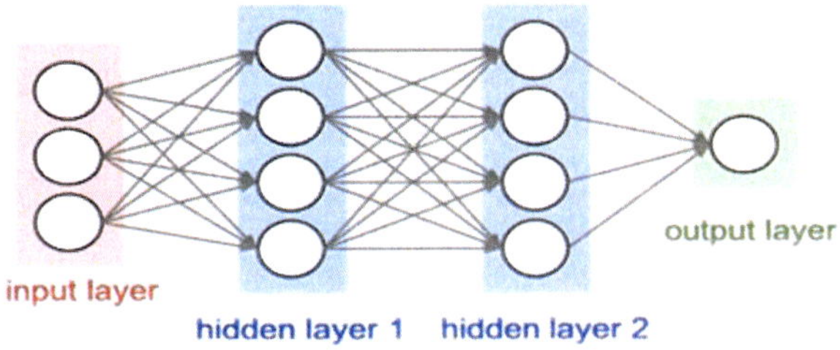

Fig. 2 [3]: Neural network model

a. Reinforcement Learning—This learning is executed on the merits of results or rewards they obtain.
b. Supervised Learning—This learning is done on the basis of training they have undergone on the basis of training dataset.
c. Unsupervised Learning—The patterns are being discovered on the basis of unlabeled data. The following are the problems that are being addressed by Machine Learning.
i. Classification
ii. Regression
iii. Clustering
iv. Anomaly Detection.

The algorithms such as Linear SVM, Naïve Bayes, K-Nearest Neighbors and Decision Tree are used to perform various analytics operation using ML [6, 7]. Deep Learning is the most popular technique to be used to connect various interconnected layers in neural network. In data learning field, Machine learning is very effective. First neural network has to be trained by using large dataset and complex algorithms. The algorithms such as Linear SVM, Naïve Bayes, K-Nearest Neighbors and Decision Tree are used to perform various analytics operation using ML. Now a day's people are moving toward deep learning because of presence of different interconnected layers in neural network.

Neural network models are forecasting methods by which forecasting has to be done. These are totally based on the working of human brain. Figure 2 consist of three basic parts: Input, one or more than one hidden layer and an output layer. Input layer is used to represent the input fields, hidden layers are used for computation, and output layers represents the target fields. But using different algorithms, neural network can find hidden patterns and correlation in processed/raw data, perform clustering and classify the data as well [8]. The DNN is built with Layers that follows certain hierarchy. This phenomenon of extraction of features from the input is utilized to make futuristic predictions.

Figure 3 illustrates the training process of deep learning. Here the difference in the desired outcome and obtained error value is accounted which improves the prediction accuracy of the system. Convolution Neural Network (CNN) helps to perform visual representation of complex data using huge volume of available data. It also helps in self-learning of multiple transformation structure of different layers of the input. In

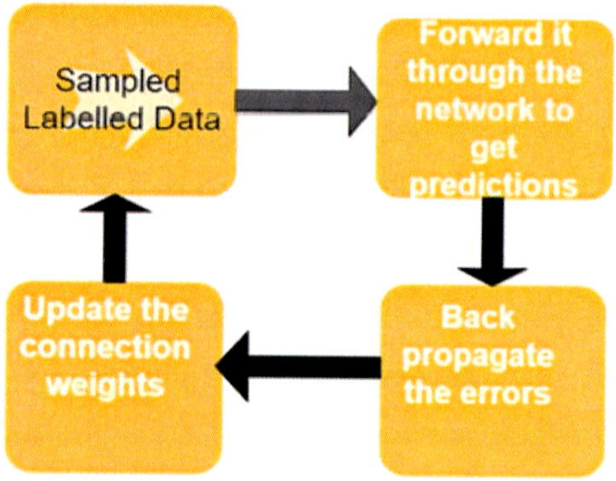

Fig. 3 Deep learning training process

CNN the input is abstracted into huge number of 3-Dimensional layer and provides the output in the same format with volume of numbers [9, 10].

3 Long Short-Term Memory Networks (LSTM)

Prerequisite of LSTM is recurrent neural network (RNN). RNN performs mapping function between input sequence and output sequence by learning the essence of algorithms. It performs the operations similar to that of the general purpose computers. It is handful in performing the time-series analysis, image recognition, pattern analysis, stock market prediction and various logical sequential of various logical sequential operations (Fig. 4).where h_t and x_t are output and input for processing. h_{t-1} and h_{t+1} are the previous and next output whereas x_{t-1} and x_{t+1} are the previous and next inputs. In this RNN, three states are there as input, hidden and output.tan(h) Is the geometrical operation. h_t (Hidden layer vector) can be calculated as:

Fig. 4 Processing states: RNN

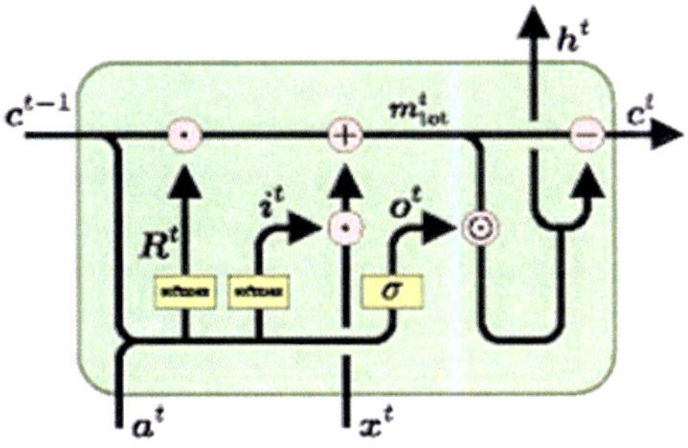

Fig. 5 LSTM

$$h(t) = \sigma_h(W_h X_t + U_h h_{t-1} + B_h) \quad (1)$$

where is the activation function (hidden layer) based on W, U and B as metrics and vectors? One to one RNN operation is performed between the fixed size *i/p* and fixed size *o/p*. It is otherwise called as Vanilla processing mode which is capable enough to perform without RNN. Classification of Image is the typical application performed by one to one RNN. In one to many RNN, the *o/p* is obtained in sequential form and *i/p* is of fixed size. Here captioning for images is the example for the one to many application whereas the pictures are captioned with text based on the features extracted. This is more applied in performing the textual analysis of sentiments. Whereas from the given sequence of the sentiments, the resulting expression is determined as happy or sad and +ve (positive) or −ve (negative). In many to many, the inputs are obtained in a sequence and the output is also delivered in sequence.

LSTM is an extraordinary kind of RNN which learns the drawn out conditions at a quick speed. It has a special element consolidated in it viz. monitoring data throughout a more drawn out span of time. LSTM follows an exceptionally unique approach to communicate between the four distinct layers present in the design. Figure 5 shows the various layers in LSTM.

4 Step by Step Procedure of LSTM

Step 1: To know about how much previous data can be stored. The first thing done by LSTM is to remove the irrelevant data from input cell and also it has to be done in a specific time stamp and by using sigmoid function σ. This will decide that which data has to be forwarded in the next step to get processed. A function forget gate $f(t)$ is also used to know about the data which will not be forwarded and is calculates as Eq. (2):

$$f(t) = \sigma\left(W_f.\left[h_{t-1}, X_t\right] + B_f\right) \quad (2)$$

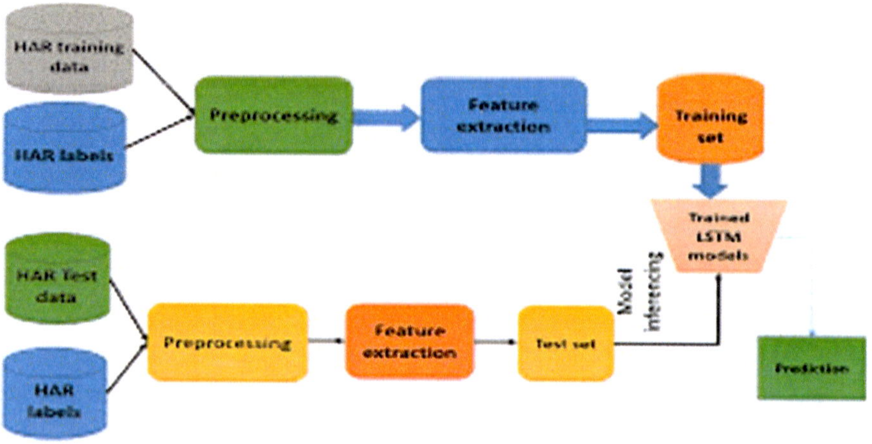

Fig. 6 LSTM: training model

Step 2: How to update cell values. In previous step, the information about the memory storage is clear. Now a comparison has to be done between current input and previous output $(X_t V_s h_{t-1})$. This will help LSTM to know about how much past data has to be used along with the current input. Equation (3) is used for this calculation as:

$$i(t) = \sigma\left(W_i.\left[h_{t-1}, X_t\right] + B_i\right) \tag{3}$$

where $i(t) inputgate$ will decide about the data to be forwarded.

Step 3: Select about the data to be processed from this stage to output stage. In this, output gate $(o(t))$ is decided. This decision has to be taken by sigmoid layer and can be calculated by Eq. (4) as (Fig. 6):

$$o(t) = \sigma\left(W_o.\left[h_{t-1}, X_t\right] + B_o\right) \tag{4}$$

5 Result Analysis

In view of the LSTM by applying on water request prediction by using the dataset of water utilization of Austin city, which will momentarily talk about forecast investigation bring about terms of accuracy of forecast and level of blunder in prediction. In view of this forecast examination, an IoT-based appropriation network configuration is talked about in the accompanying segment. The arithmetical articulation which assumes an essential part in deciding adequacy of the prediction model as far as productivity is portrayed underneath.

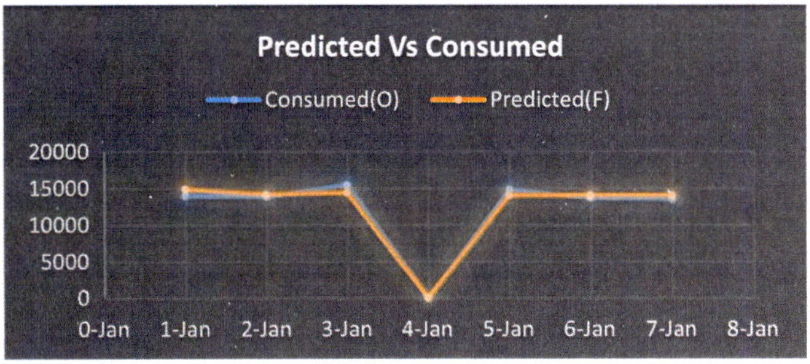

Fig. 7 LSTM: predicted (forecast) versus consumed

$$e_t = O_t - F_t \tag{5}$$

where e_t is calculated error, actual water consumption is calculated by O_t and predicted demand by F_t. % of error (PE) can be calculated as Eq. (6):

$$PE = \frac{O_t - F_t}{O_t} * 100 \tag{6}$$

The forecast as shown in Fig. 7 is made by applying respective models which reflects the drop in accuracy of the model.

$$\text{Mean Error}(ME) = \frac{1}{n}\sum_{i=1}^{n} e_t \tag{7}$$

$$\text{Mean Absolute Error}(MAE) = \frac{1}{n}\sum_{i=1}^{n} |e_t| \tag{8}$$

Table 1 showcases how water demand forecast for the quarter is performed using LSTM. Similarly, Table 2 describes the calculated Mean Absolute Percentage Error (MAPE) and Accuracy of LSTM implementation and the same is graphically illustrated in Fig. 8.

6 Comparison of Statistical and Deep Neural Analysis

The efficiency of the statistical analysis method between (ARIMA) [7] and LSLR [11] method is discussed and it is evidently proven that the accuracy of ARIMA is found to be better than the other forecast method. Here the accuracy of the statistical method ARIMA is compared with the Recurrent Neural Network-based method called LSTM [12]. Table 3 shows the forecast results which are discussed (Fig. 9).

Table 1 Forecast analysis using LSTM

Date	Consumed (O)	Predicted (F)	Error	Mod (error)	% error	Actual error (O-F)	SQ. error
1-Feb	13,953.40	14,850.30	0.05	0.06	6.46	– 890.16	810,280.80
2-Feb	13,861.10	14,160.71	0.02	0.02	2.23	– 305.50	94,570.50
3-Feb	15,432.60	14,399.65	– 0.06	0.96	93.25	1035.64	1,078,775.50
–	–	–	–	–	–	–	
28 Apr	14,766.20	14,034.50	– 0.05	0.98	95.10	730.20	536,210.30
29 Apr	13,680.10	14,034.50	0.03	0.03	2.66	– 360.50	130,180.80
30 Apr	13,550.20	14,034.50	0.04	0.04	4.40	– 595.40	352,130.10
				Total	**39.41%**		

Table 2 Mean absolute % error and accuracy of LSTM

Technique	LSTM (Quarter) (%)
Mean absolute % error (MAPE)	39.41
Accuracy	59.60

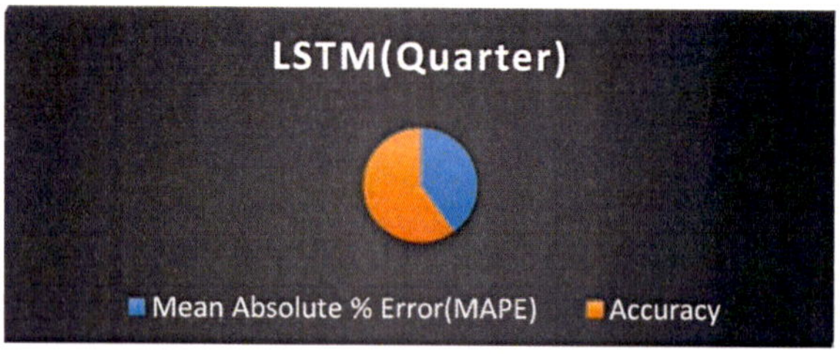

Fig. 8 MAPE versus accuracy using LSTM

Table 3 LSTM versus ARIMA accuracy and error

	LSTM	ARIMA
MAPE	39.41%	52.45%
MSE	3.6×10^5	9.7×10^6
ACCURACY	59.60%	47.55%

LSTM received an accuracy of 60% which is better than the forecast analysis by ARIMA which is having an accuracy of 40%.

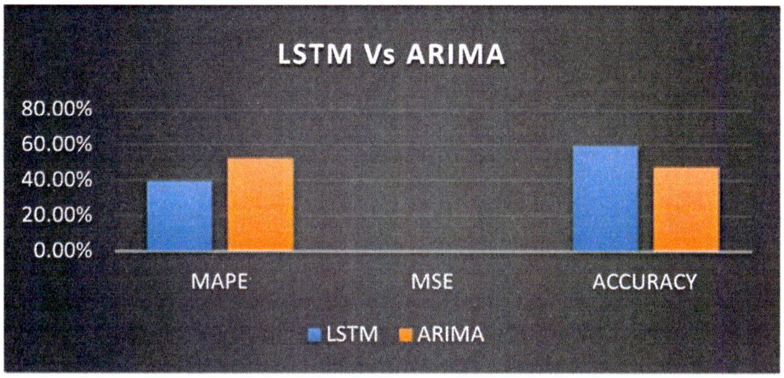

Fig. 9 LSTM versus ARIMA (MAPE, MSE and ACCURACY)

7 Conclusion and Future Work

By utilizing Deep Learning, governments can conjecture the organizations' and clients' requirements, robotize protection activities, and tailor their administrations and items dependent on quantitative and subjective measures. So LSTM model is giving better result as compare to ARISE as far as accuracy and Mean Absolute Percentage Error is concern is calculation the view of water demand and supply. The accuracy of the LSTM obtained is nearly 60 percentage whereas ARIMA is only 49 percentage. The Mean Absolute Percentage Error of the LSTM obtained is nearly 40 percentage only, whereas ARIMA is 60 percentage. For future work an IoT fog based methods smart water meters has been installed in every users house, communication has been done by using 6lopan and radio frequency so that water consumption data can be calculated and after that data has been used for analysis.

References

1. Adamowski J, Fung Chan H, Prasher SO, Ozga-Zielinski B, Sliusarieva A (2012) Comparison of multiple linear and nonlinear regression, autoregressive integrated moving average, artificial neural network, and wavelet artificial neural network methods for urban water demand forecasting in Montreal, Canada. Water Resour Res 48(1)
2. Alvisi S, Franchini M, Marinelli A (2007) A short-term, pattern-based model for water-demand forecasting. J Hydroinf 9(1):39–50
3. Babel MS, Shinde VR (2011) Identifying prominent explanatory variables for water demand prediction using artificial neural networks: a case study of Bangkok. Water Resour Manage 25(6):1653–1676
4. UNESCO (2015) The United Nations world water development report 2015: water for a sustainable world; United Nations world water assessment programme—WWAP; 9231000713. UNESCO: Paris, France
5. Pulido-Calvo I, Montesinos P, Roldán J, Ruiz-Navarro F (2007) Linear regressions and neural approaches to water demand forecasting in irrigation districts with telemetry systems. Biosyst Eng 97:283–293

6. UNESCO (2016) The United Nations world water development report 2016: water and jobs—facts and figures; United Nations world water assessment programme—WWAP; 9231002015. UNESCO: Paris, France
7. Ho SL, Xie M (1998) The use of ARIMA models for reliability forecasting and analysis. Comput Ind Eng 35(1–2):213–216. ISSN 0360-8352
8. Kumura T, Suzuki N, Takahashi M, Tominaga S, Morioka S, Ivan S (2015) Smart water management technology with intelligent sensing and ICT for the integrated water systems. NEC Tech J 9:103–106
9. Iqbal A, et al (2021) Intelligent data-analytics for condition monitoring: smart grid applications. Elsevier, p 268. https://www.sciencedirect.com/book/9780323855105/intelligent-data-analytics-for-condition-monitoring. (ISBN: 978-0-323-85511-2)
10. Tomar A, et al (2020) Machine learning, advances in computing, renewable energy and communication, volume 768. Springer Nature, Berlin, LNEE, p 659. https://doi.org/10.1007/978-981-16-2354-7. (ISBN 978-981-16-2354-7)
11. Kaur A, Kaur K, Chopra D (2015) Entropy based bug prediction using neural network based regression. International conference on computing, communication and automation, pp 168–174. https://doi.org/10.1109/CCAA.2015.7148399
12. Nasser AA, Rashad MZ, Hussein SE (2020) A two-layer water demand prediction system in urban areas based on micro-services and LSTM neural networks. IEEE Access 8:147647–147661. https://doi.org/10.1109/ACCESS.2020.3015655

Machine Learning-Based Analytical and Predictive Study on Formula 1 and Its Safety

S. Dhanvanth, Rohith Rajesh, S. S. Samyukth, and G. Jeyakumar

Abstract *Formula 1*, in short *F1*, is an international racing sport. It is of the highest class among the category. It is a premier sporting event right from the time it started in 1950. It consists of a series of races, known as Grand Prix, taking place across the world on a wide range of circuits. There is enormous amount of data being captured, analyzed, and used to design, build, and drive the *F1* cars. The primary goal of this paper is to increase the safety in *F1* races by predicting if a race can result in accidents based on the conditions and factors under which the race will be held. This paper also proposes to use different racing datasets from various sources in order to analyze, visualize, and make fruitful predictions along with some kind of analysis using different types of plots, based on the requirements. The data are visualized using different plots after doing required preprocessing. A few conclusions were made based on the patterns observed through comparing different plots. This paper also presents the comparative performance analysis of cars over the years. The performance measurement considered in this study (based on the race results) is the time taken for laps/races. Apart from this, a comparative analysis of two drivers on different race tracks was depicted using plots. Finally, the paper presents a machine learning model to predict the winner of the next race based on the previous results including important factors like the weather conditions.

Keywords Data visualization · Data analysis · Regression model · Machine learning · Formula 1

1 Introduction

Formula1 being a sport where the cars are racing at extremely high speeds, it is a risky sport unless proper precautions are taken. The *FIA* has come up with stringent rules to ensure the safety of drivers. In spite of this, there have been many dark hours

S. Dhanvanth · R. Rajesh · S. S. Samyukth · G. Jeyakumar (✉)
Department of Computer Science and Engineering, Amrita School of Engineering, Amrita Vishwa Vidyapeetham, Coimbatore, India
e-mail: g_jeyakumar@cb.amrita.edu

A. Tomar et al. (eds.), *Proceedings of 3rd International Conference on Machine Learning, Advances in Computing, Renewable Energy and Communication*, Lecture Notes in Electrical Engineering 915, https://doi.org/10.1007/978-981-19-2828-4_25

in this sport. Many of these accidents have resulted in fatalities making this a matter of deep concern. This paper attempts to make the sport safer by predicting in advance the possibility of accidents in an upcoming race by considering factors like the driver, his grid position, the weather, and the past records of accidents at that circuit. This prediction could indeed result in avoiding fatalities and accidents in races to a great deal [1–22].

Data visualization is a vast field dealing with the graphic representation of data. It is a very useful way to depict and communicate especially when the data are entirely numeric. It provides an accessible way to note different trends, outliers, and patterns in data. Translating information into a visual context, such as a map or graph makes it easier to get much better insights. Effective visualization helps users analyze and reason about data with evidence. Effective visualization implies making the right choice from the huge pool of choices like scatterplots, line plot, pie charts, histograms, and a lot more.

Machine learning provides the luxury of training models where the models learn by experience. It can be used in different applications where algorithms are used to train models. This learning capability of models is used in the work presented in this paper. In this work, a model is trained to predict the race winners after training the model using the previous results. This paper is an attempt to demonstrate the various visualization techniques by analyzing the *Formula 1* datasets using a machine learning model. The proposed model is used to predict the winner of the next *Formula 1* race. The sources of the datasets used in this study and the conclusive plots are presented in this paper.

Some insights and rules in Formula 1 are 'for a year, in a particular circuit, the race held is called a Grand prix', 'there are many teams as per the Federation Internationale de l'Automobile (FIA) rules', 'one team can send only 2 drivers for a race', 'the teams are called constructors', and 'each driver belongs to a team (constructor)'. The *F1 Grids* determine the start position of a racer, and it is based on their performance in the qualifying round.

Rest of the paper is organized as follows—Sect. 2 narrates few important related works, Sect. 3 discusses the details of the proposed work, Sect. 4 explains the results and discussion of the machine learning models, and the Sect. 5 concludes the paper.

2 Related Works

Formula 1 undoubtedly strives on the results from various kinds of data. The impact of analyzing data is huge in this sport. There are numerous data scientists working for each team in Formula 1. Few interesting related works are highlighted in this section.

There have been many attempts to visualize the data. The work presented in [1] shows a year wise overview in a calendar representation and the lap time overview of drivers using visualization tools. [2] is another work where data are collected from scratch, then analyzed using a few plots and finally a prediction of winning races using a regression model is done. In [3], the *Formula 1* data have been interpreted in the best way possible which could help to arrive at conclusions. Tools like *Tableau* have been used in [3] to explore the data in the form of plots. The importance of word cloud is well written in [4]. The correct use of the type of regression is vital in getting better predictions. Hence, choosing the right learning model is important. The articles [5, 6] have sighted the use cases where the logistics regression should be used.

Before giving the data to any model, the data have to be preprocessed in such a way that it helps the model to make near-accurate predictions. [7] explains, in detail, the library called pandas which is very vital in the various stages of data preprocessing. [8, 9] explain how the data visualization plays a major role not just in understanding the data but also for presenting data in an attractive way.

There are also many interesting articles in the literature presenting the aerodynamic design of race cars [10] and the mechanical and technical aspects for *Formula 1* technology [11].

Using machine learning approaches for predicting various factors in the real-world problems around us is a common trend of research nowadays. In [12], the basic machine learning models have been used in the ideal way possible to get good predictions of placement possibility of students. A supervised machine learning algorithm-based system to process and classify the images of the license plates is presented in [13]. The authors of [14] have done bitcoin price prediction using time series and roll over. The comparison of random forest regression and logistic regression and the ideal use cases for the techniques is presented in [15]. The idea of using neural networks to make strategical changes in races is well presented in [16]. An efficient way of analyzing the presence of gender bias in text data systems in terms of the occupations, using a pre-trained model, is proposed by authors of [17]. Ref. [18] presents a model to translate the sign language words to English word, using deep learning techniques. In [19], the paper presents an approach where artificial neural networks are used to predict race results.

Considering the study presented above, this paper presents an experimental work which focuses to use these resources to assimilate the necessary information and use the impactful parameters in the analysis and design of a machine learning model.

3 The Proposed Work

The work presented in this paper aims at predicting accidents in an upcoming race in order to enhance safety of driver, carrying out visualization to bring out meaningful observations, and then using the best suitable machine learning model to predict the winner of *Formula 1* race.

The primary goal of increasing safety in the sport by predicting the accidents in an upcoming race is carried out by preprocessing the data to extract the correlated attributes and then using the new processed dataset with an appropriate learning model in order to predict the accidents. Three models using random forest, *KNN*, and logistic regression are proposed to be tested and the one better with the better performance has to be chosen.

The following are the observations that are intended to be found out by the visualization techniques.

1. How the speeds of the cars have changed over the years? Has the speed of cars increased continually due to the tweaks in the car by the constructors?
2. Is the grid position a major deciding factor in the race?
3. A comparison between two rivals of this era.

This work also proposes a machine learning model to predict the winner of a race based on the correlated factors. To reach the goal of using visual techniques to analyze the data, the data preparation and exploration were done first.

3.1 Data Preparation and Exploration

The datasets for this study are taken from following sources.

- A group of datasets from Kaggle [20] giving details of Formula 1 race from 1950 to 2017. This dataset gives the information about constructors, race drivers, lap times, status, pit stops, and race results.
- In order to get more recent data, few datasets were formed after scraping details from Websites which also gives additional information like the weather condition during the race.

After preparing these datasets, the next task of preprocessing is done as and when required based on the requirements. For the purpose of prediction of accidents, the preprocessing steps before training the model involved:

- Scaling the data using min–max scaler in order to restrict each feature to a specific range.
- *SMOTE*, which is an oversampling technique, is used in order to generate synthetic values for minority class, thus making the dataset balanced.

On observing the final preprocessed data frame used for the prediction of accidents, it was observed that the data are imbalanced as the number of records available for accident data is much smaller than the non-accident data. This imbalance can lead to a biased model which will result in many false negatives (predicting that no accidents have occurred in a race where accidents actually occurred).

To handle such an imbalance, synthetic minority oversampling technique (SMOTE) is used to oversample the minority class.

3.2 Treating Missing Value

The missing values in the dataset were minimal, and in order to avoid discrepancies by using methods to fill missing values, the missing values are dropped.

4 Results and Discussion

The experimental results are presented in three phases. The Phase I focuses on the proposed model to predict accidents in an upcoming race. The Phase II focuses on the analyzing the observations made through a set of data visualization plots. The Phase III is discussing the proposed machine learning model to predict winner of a race.

4.1 Phase I—Predicting Accidents in an Upcoming Race

In this phase, the preprocessed datasets were used to train three different models in order to predict accidents in an upcoming race. The three different machine learning algorithms used for training were logistic regression, random forest, and *KNN*. Logistic regression is a parametric model used for linear solutions. *KNN* is a non-parametric model which supports non-linear solutions although it takes more time than logistic regression. Random forest follows the principle of ensemble learning wherein it functions by constructing multiple decision trees to carry out predictions.

The features that were chosen for training the models were decided based on correlation analysis. The features chosen were the driver *ID*, the grid position of the driver, the weather condition where the race is going to be held and the past record of accidents on that particular circuit for a particular constructor.

The results upon training models using the mentioned features are compared using accuracy as the measure. Accuracy is a suitable measure here as the data have been made balanced in the preprocessing phase. The logistic regression-based model gave an accuracy of only 44.74% while the models based on random forest and *KNN* gave accuracy of 72.20% and 85.93%, respectively. The poor accuracy of logistic regression is because it does not work well for non-linear predictions. *KNN* is working with the nearest neighbors, and hence, it tends to give better results in some particular cases like in our model, thereby giving a higher accuracy.

This model with an accuracy of 85.93% can indeed help in prediction of accidents and thereby influence the safety factor of the sport. The *FIA* could in fact use these predictions to make key decisions like rescheduling a race which could in turn avoid fatal accidents.

4.2 *Phase II—'Data Visualization and Analysis'*

This section discusses the findings in the order of the observations.

Observation 1: 'There is no constant increase in speed of cars over the years'

A part of the plot used to bring out this observation is shown in Fig. 1. 27 such subplots were generated. This plot consists of subplots for each race track, initially the result and the race datasets were merged. Then, the required preprocessing was done in order to obtain uniformity in the lap times for plotting the values. Then, the fastest lap time for each track was taken out year wise and stored separately. This was then plotted using subplots where each subplot is for a different Grand Prix, and the x and y axes denote the year and fastest lap time, respectively.

On observing the line plotted in each subplot, the speed of the cars has reduced during the middle years showing the variations in the line plot. This variation occurred as a result of the decision by the officials to enforce safety in the sport and thereby bring in more stringent rules which have to be enforced by the constructors.

Observation 2—'The grid position plays a deterministic role in the final race position'

This observation is made with the help of a scatterplot (shown in Fig. 2) using the *seaborn* library. *Seaborn* is a library in *Python* for making statistical graphs to understand and explore data.

Upon using the alpha parameter in the scatter plot, the points where maximum occurrences happen are darker. In the plot, shown in Fig. 2, the darker points are clearly visualized around $x = y$. This indicates that the grid position and the final position are approximately the same in majority of the cases. This makes perfect sense as the person who has an advantage in the grid position has less traffic ahead of him and hence, he has high chances of finishing near the same position from where he started the race.

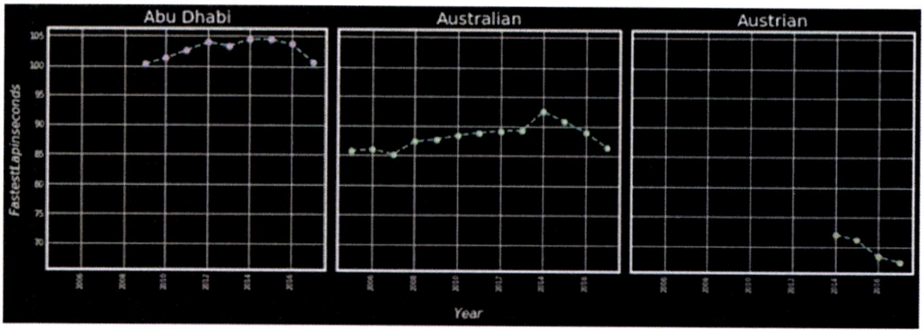

Fig. 1 Year wise plots of fastest lap times in different tracks

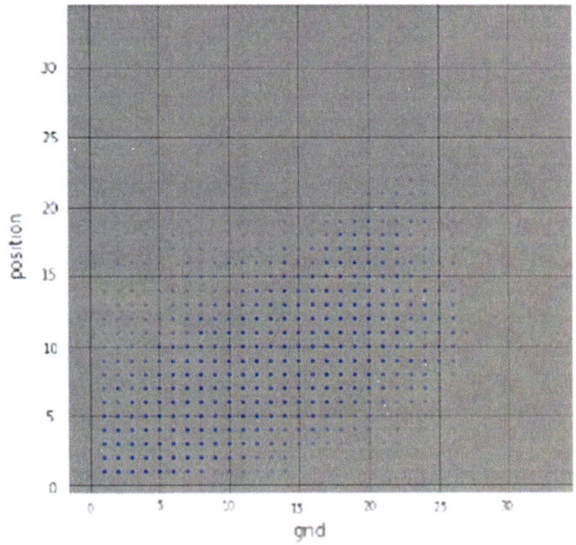

Fig. 2 Scatterplot for the grid versus position using alpha parameter to emphasize the majority of the values

Observation 3—Comparison

This is a comparative plot between two greats of the current era *Vettel* and *Hamilton*. The neck-to-neck competition of these two drivers is shown using two-line plots (one for each driver) on the same plot having year in the x axis and fastest lap time in the y axis. Upon observing the plots shown in Fig. 3, there is a clear change in performance of the drivers based on tracks. Such plots were generated for 26 different circuits. The performance depends upon factors that differ in a track like the ground force. Thus, a driver may have an advantage in a particular track over the other driver.

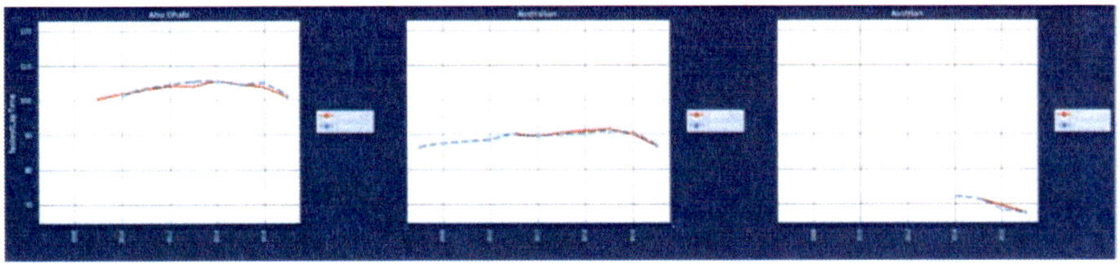

Fig. 3 Comparative performance of the two drivers in different tracks over the years

4.3 Phase III—Predicting the Winner of the Race

In Phase III, a machine learning model was designed to predict the winner of a race. This prediction was done using logistic regression. Regression is to determine the strength and character of relationship between a dependent variable and one more independent variable. There are different types of regression like linear regression, polynomial regression, logistic regression, and many more. Logistic regression is used when the dependent variable is of binary nature. This implies that the dependent variable falls into either of two specific categories. Hence, logistic regression was used in this experiment to determine who is going to win the next race. To perform logistic regression, the required datasets were collected from reliable sources.

If the driver and constructor are included in the set of independent variables, the model tends to be biased toward *Hamilton* and *Mercedes* as they have been very successful for the past 7 years. Hence, instead of taking these two variables, the current form of the driver and the current performance of his car were measured by taking into account the number of wins in the last 5 races and the number of podiums in the last 5 races.

Then after checking the correlation, only the attributes with higher correlation were chosen as dependent variables for the model. The attributes chosen for dependent variables were '*grid position*', '*number of wins in last 5 races*', '*number of podiums in last 5 races*', '*circuit*', and '*weather*'. The '*weather*' was classified as *wet* and *dry* as these are two extremes and tend to affect the result. '*Circuit*' is a categorical variable, and hence, it has to be encoded before using it in the regression model. The '*circuit*' was encoded using one hot encoding, and then, the model was trained.

After training the model, the correctness of the model was checked using the races in the year 2020. Upon observing the results of the prediction, it was found that the correct winner in a race was predicted in 11 out of the 17 races, held in 2020. This gives an accuracy of 64.7%. When the actual races were checked, it was found that the model still managed to predict winner. This prediction was successful if unforeseen circumstances like accidents and penalties were not considered.

5 Conclusion

In this paper, the prediction of accidents in an upcoming race, use of data visualization using plots to draw meaningful inferences and using logistic regression to predict the winner of a race has been presented. This study was performed using multiple datasets with the Formula 1 data from 1950 to 2020. The comparative study of models to predict the accidents was helpful in finding an ideal model to predict if accidents could occur in a race and thereby the model will positively influence the sport by enhancing the safety aspects. Using line plots, scatter plots, word cloud, and subplots, the data were interpreted. The meaningful conclusions obtained were

(i) there is no constant increase in speed of cars over the years. (ii) The grid position plays a deterministic role in the final race position. (iii) A comparative plot between two greats of the current era—Vettel and Hamilton shows that racers have advantages based on tracks. The machine learning model with a 64.7% accuracy predicted who wins the race based on the past performances. The accuracy is reduced to 64.7% as the model cannot take into account the unforeseen circumstances.

For future work, we plan to take more aspects into consideration like adding more features into the dataset which will lead to better accuracy in the model. The future work also has scope in predicting additional variables related to the sport like how often a driver crashes or get a penalty.

References

1. Tobias Lampprecht, David Salb, Marek Mauser, Huub van de Wetering, Michael Burch Tobias Lampprecht, David Salb, Marek Mauser, Huub van de Wetering, Michael Burch and Uwe Kloos, "Visual analysis of formula one races", in the proceeding of 23rd International Conference Information Visualisation, (2019).
2. Veronica Nigro, "Formula 1 race predictor", towards datascience, weblink: https://towardsdatascience.com/formula-1-race-predictor-5d4bfae887da, (2020).
3. Chinmay Wyawahare, "Formula 1 grand prix analysis", towards datascience, weblink: https://towardsdatascience.com/formula-1-grand-prix-analysis-d05d73b1e79c, (2020).
4. Depaolo C, Wilkinson K (2014) Get your head into the clouds: using word clouds for analyzing qualitative assessment data. TechTrends 58(3):38–44
5. Joanne Peng, Kuk Lida Lee and Gary M. Ingersoll, "An introduction to logistic regression analysis and reporting", The Journal of Educational Research, Vol. 96., No. 1., pp. 3–14., (2002).
6. Saishruthi Swaminathan, "Logistic regression — detailed overview", towards datascience, weblink: https://towardsdatascience.com/logistic-regression-detailed-overview-46c4da4303bc, (2018).
7. Wes Mckinney, "pandas: a foundational python library for data analysis and statistics", weblink: https://www.researchgate.net/publication/265194455_pandas_a_Foundational_Python_Library_for_Data_Analysis_and_Statistics/citations, (2011).
8. Rahul Raoniar, "Generate publication-ready plots using seaborn library", towards datascience, weblink: https://towardsdatascience.com/generate-publication-ready-plots-using-seaborn-library-part-1-f4c9a6d0489c, (2020).
9. Badreesh Shetty,"Data Visualization using Matplotlib", towards datascience, weblink: https://towardsdatascience.com/data-visualization-using-matplotlib-16f1aae5ce70, (2018).
10. Agathangelou B and Gascoyne M, "Aerodynamic design considerations of a Formula 1 racing car", SAE Technical Paper 980399, (1998).
11. Wright P, Matthews T (2001) Formula 1 Technology. Premier Series Books, SAE International
12. Tadi Aravind, Bhimavarapu Sasidhar Reddy, Sai Avinash and Jeyakumar G, "A comparative study on machine learning algorithms for predicting the placement information of under graduate students", In proceedings of Third International conference on I-SMAC (IoT in Social, Mobile, Analytics and Cloud) (I-SMAC), (2020).
13. Anjali Suresan, Divyaa Mahalakshmi G, Meenakshi Venkatraman, Shruthi Suresh and Supriya, "Comparison of machine learning algorithms for smart license number plate detection system", Image Processing and Capsule Networks, ICIPCN 2020 - Advances in Intelligent Systems and Computing, Vol 1200., (2021).

14. Dhanya N M, "An empirical evaluation of bitcoin price prediction using time series analysis and roll over", Inventive Communication and Computational Technologies, Lecture Notes in Networks and Systems, Vol 145., (2021).
15. Couronné, Raphael, Philipp Probst, and Anne-Laure Boulesteix. "Random forest versus logistic regression: a large-scale benchmark experiment." BMC bioinformatics 19, no. 1 (2018).
16. Heilmeier, Alexander, André Thomaser, Michael Graf, and Johannes Betz. "Virtual Strategy Engineer: Using Artificial Neural Networks for Making Race Strategy Decisions in Circuit Motorsport." Applied Sciences 10, no. 21 (2020).
17. Pranav Rai and Kumar P N, "An analytical approach for recognizing the occupational gender bias in the data systems", Image Processing and Capsule Networks, ICIPCN 2020 - Advances in Intelligent Systems and Computing, Vol.1200., (2021).
18. Kartik P.V.S.M.S, Sumanth K.B.V.N.S, Ram V.N.V.S and Prakash P, "sign language to text conversion using deep learning", Inventive Communication and Computational Technologies - Lecture Notes in Networks and Systems, Vol. 145., (2021).
19. Stoppels, Eloy. "Predicting race results using artificial neural networks." Master's thesis, University of Twente, (2017).
20. Chris G, "Formula 1 Race Data", kaggle, weblink: https://www.kaggle.com/cjgdev/formula-1-race-data-1950, (2017).
21. A.Iqbal et al., Intelligent Data-Analytics for Condition Monitoring: Smart Grid Applications, (Elsevier, 2021), 268 p. https://www.sciencedirect.com/book/9780323855105/intelligent-data-analytics-for-condition-monitoring. (ISBN: 978–0–323–85511–2)
22. Anuradha Tomar et al., Machine Learning, Advances in Computing, Renewable Energy and Communication, (Springer Nature, Berlin, LNEE volume 768, 2020), 659 p. Doi: https://doi.org/10.1007/978-981-16-2354-7. (ISBN 978–981–16–2354–7)

Missing Person Identification System Using Deep Learning Algorithm (CNN) and Machine Learning Classifiers

M. Chenna Keshava, C. Srilakshmi Prasanna, V. Kavitha, K. Mahima, V. Padmavathi, and P. Srinivasulu

Abstract Missing our loved ones is the most painful situation. India, as the world's second most populated country, reports a countless number of missing person cases and many of those remain untraced. With the use of deep learning methods, a novel strategy is employed to locate individuals who have gone missing. People can upload images of shady people along with the necessary information to a common webpage. The submitted image is immediately verified to the registered images of missing people. The given input person image goes through a categorization process, and the image with the perfect match is chosen from the missing person database. Deep learning type is trained to accurately recognize the missing person photo in order to execute the selection task. Face recognition is performed using the convolutional NN and a robust deep learning method. Face descriptors are derived from photos using VGG-face deep architecture. The retrieved descriptors are matches with nearest missing person testing photos and classified using the KNN and SVM classifiers. The convolutional algorithm, unlike other deep learning methods, is utilized as a high-level feature extractor. The obtained outcomes are invariant to noise, light, contrast, occultation, pose, and the person's age because the best-performing algorithms were chosen. As a consequence, the results are obtained with a high degree of precision.

Keywords CNN · Deep learning · VGG-face · Face recognition · KNN · SVM

M. C. Keshava (✉) · C. S. Prasanna · V. Kavitha · K. Mahima · V. Padmavathi · P. Srinivasulu
Department of Computer Science and Engineering, JNTUA College of Engineering, Pulivendula, Andhra Pradesh 516390, India
e-mail: keshava1047@gmail.com

C. S. Prasanna
e-mail: srilakshmi1023@gmail.com

V. Kavitha
e-mail: kavithareddy.velagalapalli@gmail.com

K. Mahima
e-mail: mahimakonda99@gmail.com

V. Padmavathi
e-mail: padmavathividem8@gmail.com

A. Tomar et al. (eds.), *Proceedings of 3rd International Conference on Machine Learning, Advances in Computing, Renewable Energy and Communication*, Lecture Notes in Electrical Engineering 915, https://doi.org/10.1007/978-981-19-2828-4_26

1 Introduction

As a densely populated country, India records a significant number of missing person instances, many of which go unsolved. This paper describes a new model for detecting missing people that uses deep learning algorithms and machine learning classifiers [1–14].

The application will perform an automatic search of the entered photo among the missing person photographs. This aids policemen in tracking the missing individual in any region. The major goal is to compare and analyze the KNN and SVM classifiers. Our dataset has been used to train and test both classifiers. The accuracy of the training and testing processes, as well as the overall time spent on each, are calculated. The acquired values are compared, and the classifier with the largest accuracy and the shortest processing time is chosen. Finally, the user interface is linked to the chosen algorithm.

The visual geometry group (VGG) Facial models are a collection of face recognition algorithms developed by members of the University of Oxford's visual geometry group (VGG) and tested on common computer vision datasets. VGG Face and VGGFace2 are the two most popular VGG models for face recognition.

The K-nearest neighbor method is one of the most basic machine learning algorithms. It uses the supervised machine learning approach. The KNN technique considers that the new case/data and current cases are linked, and it assigns the new case to the category that is closest to the existing categories. As illustrated in Fig. 1, the KNN technique retains all available information and classifies new data points based on their similarity to current data. This means that the KNN technique can swiftly sort data into a well-defined category.

SVM is a supervised machine learning method that can be wield to solve classification and regression problems. However, it is mostly wield to solve classification task. In the SVM method, each data item is represented as a point in n-dimensional

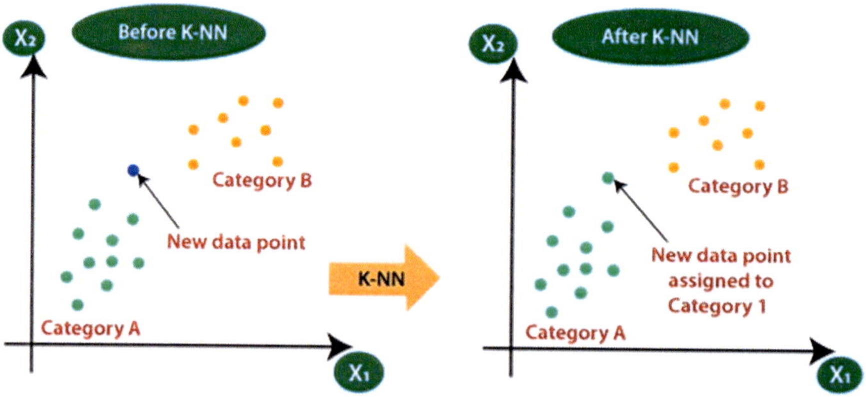

Fig. 1 Working of KNN algorithm

Fig. 2 Working of SVM algorithm

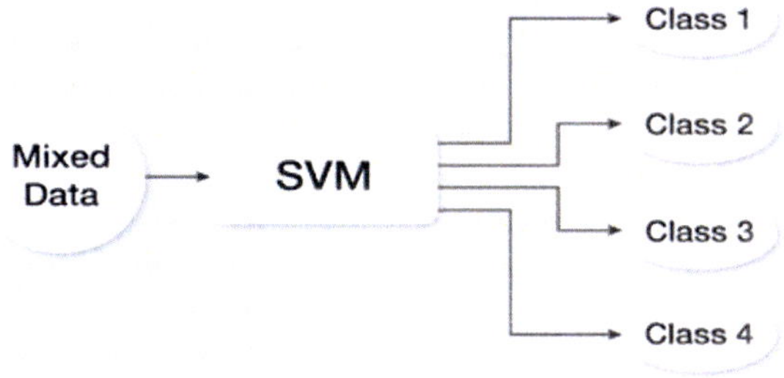

space, with the feature vector indicating the value of a specific position. The classification is then completed by identifying the hyperplane that clearly distinguishes the two classes, as shown in Fig. 2.

2 Literature Survey

The authors described principal component analysis (PCA) as an eigenface method for reducing the dimensionality of the original data space in their study [1]. PCA solves the recognition problem in a lower-dimensional representation space than picture space.

Linear discriminant method is an appearance-hinged technique for dimensionality reduction that has shown excellent results in face recognition, according to study [2]. The main disadvantage of using LDA is that it may run into the difficulty of a small sample size.

The scale-invariant feature transform (SIFT) recovers the key points (location and descriptors) for all database photos, according to the author of this work [3]. Then, given a changed image, SIFT extracts the image's main points and compares them to the dataset's points. To provide a feature description, the SIFT algorithm extracts the interesting key points from an image.

Speeded-up robust features (SURF) detectors are used to discover the interest spots in an image, and SURF descriptors are used to extract the feature vectors at each interest point, according to the authors of paper [4]. To lower the time cost of feature matching and computation, SURF uses 64 dimensions to encode the feature vectors at each interest point. To spin a sphere, the SURF approach is not stable.

The major goal of the research [5] is to look at using protection pursuit (PP) to reduce the complexity of a big number of users' support vector machine (SVM) classifier. The proposed method will aid in the accurate establishment of biometric identification systems for massive datasets. Preliminary results on the YALE face database indicated that the proposed strategy is effective in increasing user recognition rates while reducing SVM complexity.

The author demonstrated the consistent benefit of replacing the SoftMax layer with a linear support vector machine in article [6]. Deep learning using neural networks has claimed state-of-the-art outcomes in a wide variety of tasks.

Face recognition was the most extensively used for image analysis, according to the authors of research [7]. Its appeal stems from the fact that it has a wide range of business and law enforcement uses, as well as the availability of cutting-edge methodologies.

The authors of paper [8] offer a new database dubbed "ImageNet," a large-scale ontology of images built on the WordNet structure's backbone. ImageNet seeks to fill the bulk of WordNet's 80,000 synsets with 500–1000 clean, full-resolution images on average. The digital revolution has resulted in a massive data explosion. According to the most recent estimates, Flickr has over 3 billion photos, YouTube has a similar amount of video clips, and the Google Image Search database has an even higher number of images.

The paper [9] provides a method for identifying missing children that blends deep learning-based face feature extraction with KNN-based matching. Unlike most deep learning algorithms, their technique simply uses a convolutional network as a high-level feature extractor, while the trained KNN classifier does the child recognition.

The research study [10] focused on face detection system as well as two application fields: querying human beings and searching similar backdrops. The rest area application can be further divided into two scenarios: Querying human identification in general, such as photographs of people on the beach, and identifying specific individuals, such as photographs of Hilary Clinton giving a speech.

An innovative strategy to exact recognition and tracking of human faces in videos is presented in the publication [11]. The goal is to spread the information of a collection of seed faces that have been discovered offline with high certainty in order to recover faces that have been missed or detected with low confidence. In particular, their method generates a person-specific skin color model from the color of the seed face, which is then used to perform particle filtering for sequential face tracking. Then, to improve the overall smoothness of the localization findings, a backward propagation strategy is designed.

The paper [12] addresses a simple yet effective method for face recognition that makes robust use of histogram of oriented gradients (HOG) features. The following are the three primary contributions of this work: To begin, we propose extracting HOG descriptors from a regular grid to compensate for mistakes in face feature detection caused by occlusions, posture, and lighting variations. Second, by combining HOG descriptors at multiple sizes, significant structures for face recognition can be captured. Third, we determine that dimensionality reduction is required to eliminate noise and make the classification process less prone to overfitting.

3 Problem Statement

A notion is suggested for creating a virtual environment in which recent pictures of people submitted by family/relatives at the time of reporting missing cases are maintained in a repository. The public is encouraged to take images of people in suspicious situations and post them to the platform. This photo will be automatically searched among the images of the missing persons. This aids police officers in locating the individual in any location. When a person is located, the photograph taken at the moment is compared to the images posted by the police at the time of the missing person's abduction. Occasionally, the person has been missing for an extended period of time. This age gap is shown in the images as the shape of the face and the texture of the skin alter with age. It is necessary to use an age-insensitive feature discriminator. Position, orientation, lighting, occlusions, background noise, and other elements may all have an impact on a person's face appearance. The photograph shot by the public may not be of high quality, since some of them may have been taken from a remote location without the person's knowledge. To overcome these challenges, a performance comparison of SVM and KNN based on time and accuracy was conducted, and the most efficient method was employed to match the photos.

4 WorkFlow

The steps are followed to accomplish desired results are:

1. Training the dataset with suitable images.
2. Implementing the both SVM and KNN classifiers on the trained dataset.
3. Results comparison.

Almost 1000 images are collected from Internet and from our co-students and created a file folder for each person with the name of that person. Each folder contains at least two images of single person with at most age gap of 10 years. The path of these directories is given as training directory in the training function. Images which contain noises like age, brightness, position of person, etc., to make sure that algorithm is working fine with any type of images.

The performance of both SVM and KNN is tested by measuring parameters like accuracy and total time taken to execute the entire program. The resulted values are tabulated and graphs are drawn from those values. Using those graphs, the most efficient algorithm is obtained and used in the project to match the person images.

5 Results Analysis

Time and Accuracy at different number of images

The time and accuracy of SVM and KNN at 50 images are calculated. The time taken by the SVM is greater than the KNN. In Fig. 3, the time and accuracy of SVM and KNN at 100 images are shown. The time taken by the SVM is greater by 150 s than the KNN.

The time and accuracy of SVM and KNN at 150 images are evaluated. The time taken by the SVM is greater than the KNN. In Fig. 4, the time and accuracy of SVM and KNN at 200 images are shown. The KNN takes less time and both the SVM and KNN given approximately same rate of accuracy.

The time and accuracy of SVM and KNN at 250 images are calculated. The time taken by the SVM is greater than the KNN. Figure 5 illustrates the time and accuracy of SVM and KNN at 300 images are shown. The time taken by the SVM is much greater than the KNN to process 300 images.

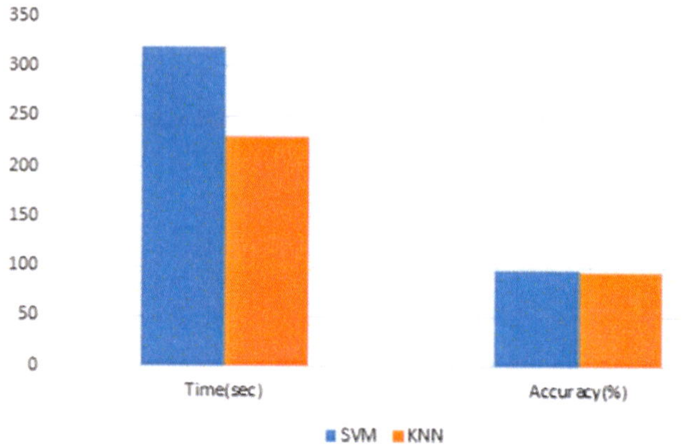

Fig. 3 Time and accuracy for 100 images

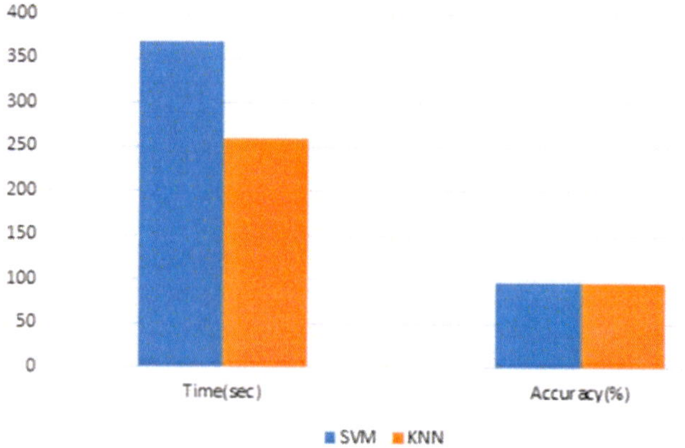

Fig. 4 Time and accuracy for 200 images

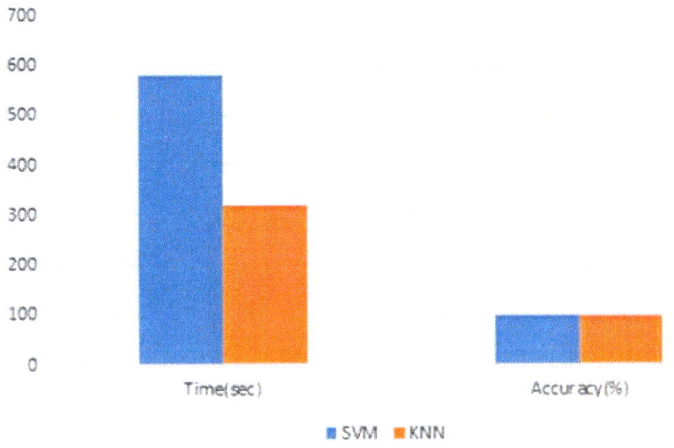

Fig. 5 Time and accuracy for 300 images

The time and accuracy of SVM and KNN at 350 images are evaluated. The time taken by the SVM is greater than the KNN. Fig. 6 shows the time and accuracy of SVM and KNN at 400 images. The time taken by the SVM is greater than the KNN.

The time and accuracy of SVM and KNN at 450 images are evaluated and compared. The time taken by KNN is lesser than the SVM. In Fig. 7, the time and accuracy of SVM and KNN at 500 images are shown. The time taken by the SVM is little greater than the KNN.

Table 1 shows accuracy rate in percentage of SVM and KNN at different number of images. The SVM accuracy rate is decreasing with increasing the number of images, whereas KNN accuracy rate is increasing with increasing number of images. The KNN algorithm achieves highest accuracy with large number of images when compared to SVM.

In Fig. 8, the accuracy of SVM and KNN at different number of images is shown. At first, i.e. from 50 to 150 images, SVM obtained a little high rate of accuracy. Then

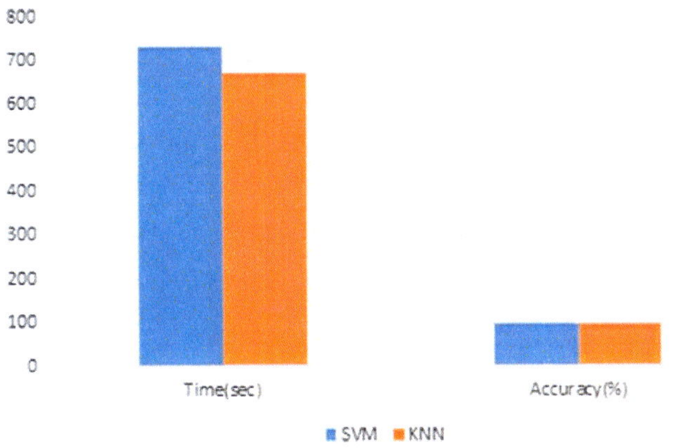

Fig. 6 Time and accuracy for 400 images

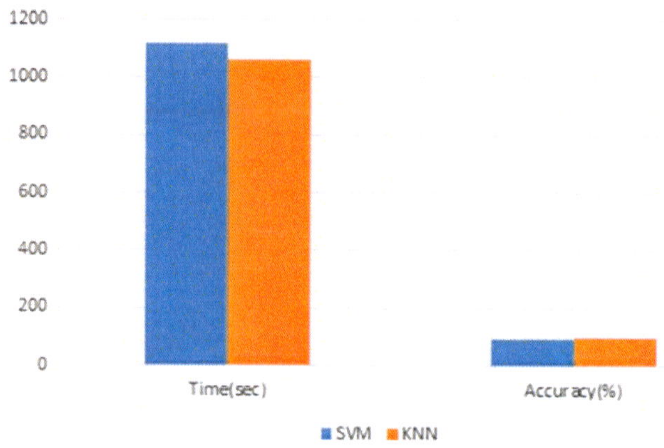

Fig. 7 Time and accuracy for 500 images

Table 1 Accuracy (in %) of SVM and KNN at different number of images

No. of images	Accuracy rate of KNN (in %)	Accuracy rate of SVM (in %)
50	90.0	95.0
100	95.0	97.0
150	96.6	98.3
200	97.5	98.7
250	97.7	97.8
300	97.8	97.7
350	97.7	97.2
400	97.6	96.2
450	97.0	96.2
500	97.0	96.8

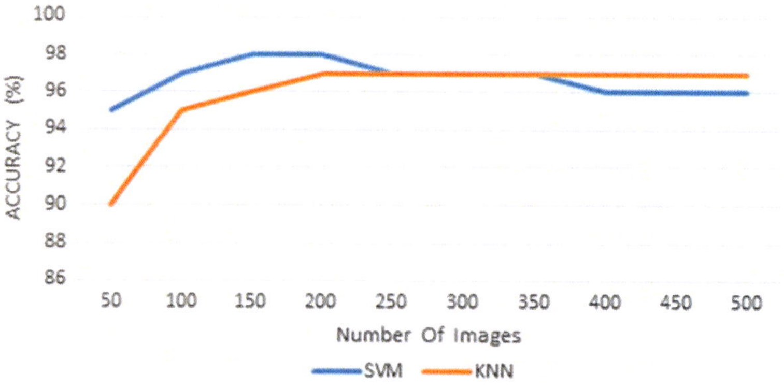

Fig. 8 Number of images versus accuracy for KNN and SVM

the number of images increased from 200 to 250 images both SVM and KNN have given same accuracy rate. But with increasing number of images, i.e. from 300 to 500 images, KNN showed a greater rate of accuracy.

Table 2 shows total time taken in seconds by SVM and KNN to give results at different number of images. The SVM takes more time to process than KNN. With increasing number of images, time taken by both SVM and KNN increases, but KNN takes less time when compared to SVM.

Fig. 9, the time taken by SVM and KNN at different number of images is shown. The time taken by the SVM is greater than the KNN.

Table 2 Time taken (in sec) by SVM and KNN

No. of images	KNN (Sec)	SVM (Sec)
50	169.88	288.35
100	232.00	314.37
150	240.30	345.88
200	260.65	367.06
250	280.85	397.17
300	320.70	582.37
350	500.48	684.72
400	670.92	732.44
450	860.92	932.45
500	1060.92	1120.26

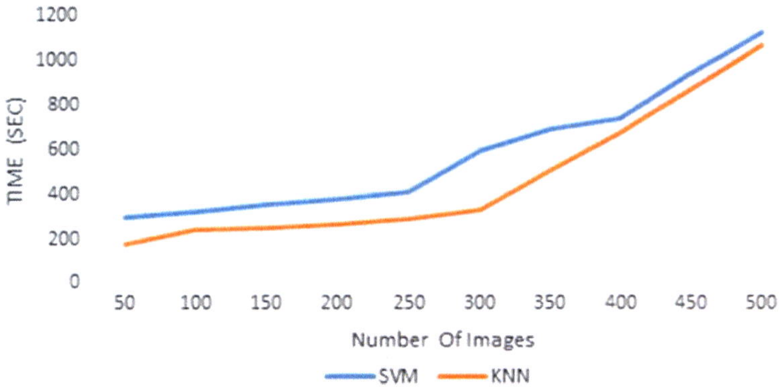

Fig. 9 Number of images versus time taken of KNN and SVM

6 Conclusion

A missing person identification framework is implemented, which combines an effective CNN-hinged deep learning strategy for feature extraction with KNN and SVM classifiers for categorical classification. The two classifiers are compared using different parameters that have been learned using human face feature representations. When compared to SVM, the KNN achieves higher accuracy in less time. As a result, KNN is integrated into the user interface in order to match missing person photos. The developed face recognition technology could be used to identify missing people in a convenient way.

References

1. Gumus E, Kilic N, Sertbas A, Osman NU (2010) Face recognition using PCA (principal component analysis). Expert Syst Appl 37:6404–6408
2. Bansal A, et al (2012) Face recognition using PCA and LDA algorithm. In: 2012 second international conference on advanced computing and communication technologies, vol 5. pp 251–254
3. Luo U, Ma Y, Takikawa E, Lao S, Kawade M, Lu B-L. Person specific SIFT (Scale invariant feature transform). J Appl Res Technol Versión Online 13. ISSN 2448-6736 versión impresa ISSN 1665-6423
4. Du G, Su F, Cai A, Face recognition using SURF features. Proceedings of SPIE—the international society for optical engineering, vol 4, pp 1–9. https://doi.org/10.1117/12.832636
5. Ghouzali S, Larabi S, Mislov A, Wilson C, Chen L (2020) Face identification based bio-inspired algorithms. Int Arab J Inf Technol 17(1):1–5
6. Tang Y (Feb 2015) Deep learning using linear support vector machines. International conference on machine learning 2013: challenges in representation learning workshop. Atlanta, Georgia, USA arXiv:1306.0239-v4-21, pp 6–9
7. Mishra1 S, Dubey A (Jan 2015) Face recognition approaches: a survey. Int J Comput Bus Res (IJCBR). 6(1):1–5. ISSN (Online): 2229-6166
8. Deng J, Dong W, Socher R, Li L, Li K, Fei-Fei L (2009) ImageNet: a large-scale hierarchical image database. 2009 IEEE conference on computer vision and pattern recognition, pp 248–255. https://doi.org/10.1109/CVPR.2009.5206848
9. Suman (Dec 2020) Nearest neighbor for missing child identification. Int J Eng Res Manage (IJERM) 07(12):49–58. ISSN: 2349-2058
10. Srihari R, Zhang Z, Rao A. Image background search: combining object detection techniques with content-based image retrieval (CBIR) systems 1:1–5. https://doi.org/10.1109/IVL.1999.781131
11. Tao J, Tan Y-P (2005) Accurate face localization in videos using effective information propagation. IEEE international conference on image processing 2005, pp 2-72. https://doi.org/10.1109/ICIP.2005.1530506
12. Déniz-Suárez O, et al (2011) Face recognition using histograms of oriented gradients. Pattern Recognit Lett 32:1598–1603-v2, 1–8
13. Iqbal A, et al (2021) Intelligent data-analytics for condition monitoring: smart grid applications. Elsevier, p 268. https://www.sciencedirect.com/book/9780323855105/intelligent-data-analytics-for-condition-monitoring. ISBN: 978-0-323-85511-2
14. Tomar A, et al (2020) Machine learning, advances in computing, renewable energy and communication, vol 768. Springer Nature, Berlin, LNEE, p 659. https://doi.org/10.1007/978-981-16-2354-7. ISBN 978-981-16-2354-7

A Genetic Algorithm Framework to Solve Two-Dimensional Maze Problem

K. Harshak Krishnaa, Kaluri Vijay Jonathan, Konjeti Harsha Saketh, Aravind Tadi, and G. Jeyakumar

Abstract *Genetic algorithm* (*GA*), which is an instance under the repository of *evolutionary algorithm* (*EA*), follows the problem-solving approach inspired from biological activities like mutation, crossover and selection. As an attempt of showcasing the performance of *GA* on simple real-world optimization problems, this paper proposes to present a genetic algorithm framework to solve the two-dimensional toy maze problem. The proposed framework is named as $GA_{maze.}$ The novelty of GA_{maze} lies in the way it updates the population from generation to generation. The GA_{maze} makes use of fitness information of the candidate solutions to optimize the individuals' learning capacity and to make them reach the final goal by making them to understand how to overcome obstacles and find a path. The detailed methodology of the proposed GA_{maze} is presented in this paper. The experimental results revealed that the GA_{maze} could solve different instances of the maze problem. The detailed inferences along with the limitations are presented in this paper.

Keywords Genetic algorithm · Evolutionary algorithm · Maze solving algorithm · Population diversity · Population updating

1 Introduction

A *maze* is a complicated structured form of play, usually undertaken for entertainment or fun. The *maze* games are sometimes played purely for enjoyment, sometimes for achievement or reward as well. *Maze* games have emerged very famous, exciting stuff and fascinating area from mathematical point of view. These games can also be additionally used to instruct human brains. The evolutionary computing (*EC*) domain of computer science comes under the soft computing approaches for bringing artificial intelligence (*AI*) to computing systems. *EC* includes variety of algorithms,

K. H. Krishnaa · K. V. Jonathan · K. H. Saketh · A. Tadi · G. Jeyakumar (✉)
Department of Computer Science and Engineering, Amrita School of Engineering, Amrita Vishwa Vidyapeetham, Coimbatore, India
e-mail: g_jeyakumar@cb.amrita.edu

A. Tomar et al. (eds.), *Proceedings of 3rd International Conference on Machine Learning, Advances in Computing, Renewable Energy and Communication*, Lecture Notes in Electrical Engineering 915, https://doi.org/10.1007/978-981-19-2828-4_27

commonly known as evolutionary algorithms (*EAs*), to simulate the human intelligence on searching through simple algorithmic structures. These algorithms are used for "generate-and-test" approaches of solving optimization problems. For the given optimization problem, the *EA* generates a set of possible solutions and then follows an evolutionary search strategy. This evolutionary search strategy is inspired by the natural evolution with the operations of parent selection, mutation, crossover and survivor selection. Thus, *EAs* are bringing artificial intelligence to the computing systems for generating-testing-searching the optimum solution for the given problem.

This paper presents an attempt made by the authors, in bringing out the potential of an *EA* is solving the real-world toy game problem. The experiment presented in this paper uses *GA* for its study to solve the two-dimensional maze problem.

The remaining part of the paper includes Sect. 2 to present the related works, the Sect. 3 to present the proposed methodology, the Sect. 4 for results and discussion and finally the Sect. 5 to conclude the paper.

2 Related Works

This is a common practice among the *EC* researchers that they explore the working of their *EA* of interest on simple and common optimization problems around them, to start with their research work. Few recent as well as interesting such works are summarized in this section, first, and then, the works related to maze problem solving are described.

In video analytics, video summarization is a critical as well as important task to create a compressed sequence of important frames (key frames) of the videos. An evolutionary computing-based approach for generating the key frames of the given video is developed in [1]. In the smart building applications, identifying the location and tracking the movement of the critical resources are a hectic task. The most commonly used technology for this purpose is radio frequency identification (*RFID*). In [2], an evolutionary algorithm-based framework is demonstrated to find the optimal positions for the *RFID* readers to cover more resources with a smaller number of *RFID* readers. A comparative study on dynamic programming and *GA*-based approaches for solving the mathematical subset problem is presented in [3]. The study presented in [4] also explains the process of getting an optimized solution for the subset sum problem using *GA*. Similar to that, a *GA*-based system with different mutation and crossover operators to solve the Sudoku problem is presented in [5]. A study on differential evolution algorithm with different crossover operators to solve the benchmarking global optimization problems is described in [6]. In computer vision-based applications, identifying the position of the object of interest is done often with different approaches. The authors in [7] have presented a simple but efficient *EC*-based system for object identification. An artificial bee colony algorithm-based image contrast enhancement system is presented in [8]. Interestingly, an attempt to merge the *EC* domain with computer malware domain was made by the authors of [9], and their insights are presented in [9].

Having presented different interesting real-world optimization problems solved by different *EAs*, now the next section presents the works in the literature related to using *EAs* for solving the maze problems.

The research work explaining, in depth, generating mazes using different algorithmic approaches is presented in [10]. The algorithms considered in this study are depth-first search (*DFS*), Kruskal's algorithm and Prim's algorithm. The authors in [11] explain how A* search algorithm can be effectively used to find the shortest path from the starting point to the ending point without facing any obstacles during the process. The work demonstrating the solution for an automatically generated rectangular maze is presented in [12]. The maze is first randomly generated using the A-Mazer algorithm and effectively concluded that the modulo operator is found to be competent for generating the best solutions. In [13], two maze problem-solving approaches are presented—(1) the traditional *GA* approach and (2) the *GA* with domain-specific knowledge. An implementation of *GA* used to solve randomly generated mazes that are of varying sizes and complexity is presented in [14]. This work evaluated the performance of *GA* and compared it with several non-*AI* techniques such as depth-first search (*DFS*), breadth-first search (*BFS*), A-star algorithm ($A*$), Dijkstra algorithm (*DA*) and greedy best-first search (*GBFS*). Generation of various mazes using evolutionary computing approach is presented in [13]. This work also used *GA* to generate optimal maze structures [14].

After analyzing the applicability of *EAs* on solving the real-word problems around us and after reviewing the advantages shown in the state of the art in using *GA* for solving maze problem, this paper focuses on presenting a novel population diversity-based approach of solving the maze problem with *GA*. The proposed methodology is presented in next section.

3 Proposed Methodology

A *GA*-based algorithmic framework (denoted as GA_{maze}, henceforth) to solve the given two-dimensional maze problem is proposed, and the methodology followed in the framework is presented in this section.

The design of experiment of the GA_{maze} includes the following parameters and assumptions—population size = 1000, maximum unit steps in a move (max_unitmove) = 10, mutation factor = 0.3, maximum moves to solve the maze = 20, number of generations = 1000, diversity factor = 0.0, directions: 0-up (Yellow), 1-right (Green), 2-down (Blue), 3-left (Purple), maze size: 20 × 10 (as shown in Fig. 1), and the starting and ending points are taken as *S* and *F*, respectively.

The first step in GA_{maze} is the creation of initial population of the game. The algorithmic procedure implemented in GA_{maze} for initial population generation includes the following steps—(a) the length of each chromosome is generated in random with a minimum length of 5 and maximum length of *maximum* of (*no_of_rows*, *no_of_columns*), (b) every move (i.e., every gene) is of the format (*direction*, *unit_steps*) where both of them are taken random, (c) the direction in a particular

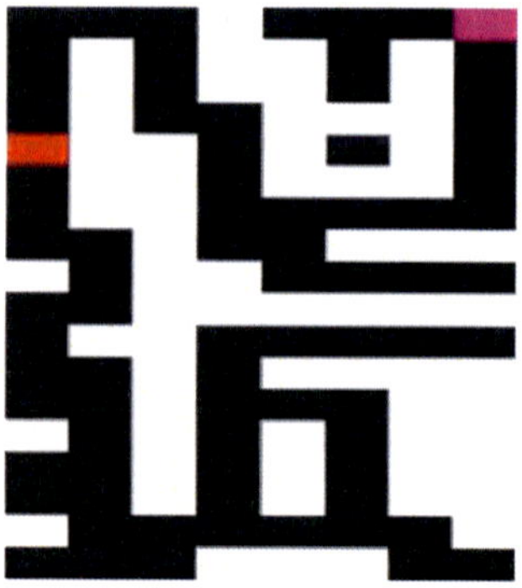

Fig. 1 Simulated maze environment for the game

gene is calculated using the function *nextmovecalculator* ()—in which the conditions used are no two opposite moves can be consecutive and the same move cannot be applied (up–down, left–right, up–up, down–down, right–right and left–left cannot be in consecutive moves), (d) Maximum unit step in a move is in the range of [1, *max_unitmove*] and (e) 1000 such chromosomes are generated to form the initial population.

After the population initialization, the next critical and important task is to find the fitness of the individual chromosomes in the population. The GA_{maze} followed the following algorithmic procedure for this purpose—(a) fitness of the chromosome is calculated as from the current position the agent moves in the direction present in the gene for particular units moves present in the gene, (b) in the process for every correct unit move, it is awarded with a fitness of *1/unit move* (i.e., at the end of each command (direction, number of unit steps) the agent gets a fit value of 1), (c) if in any situation an agent hits the wall then a penalty of -1 is awarded and the Euclidian distance from that current position to the final position is subtracted from the fitness to give final fitness. The Eq. (1) is used for this purpose

$$\text{fitness} = \text{fitness} - (\text{penalty} + \text{distance}(\text{current position, final position})) \quad (1)$$

Once the population is ready with fitness assigned for each candidate, the next process is to start the evolutionary cycle of *GA*. This cycle includes *parent_selection, crossover, mutation, survivor_selection* and *checking_for_termination.*

In $GA_{maze,}$ the parent_selection is done based on the fitness values of the candidates. The top two fittest chromosomes in the population are selected, and the crossover between them is done following the *k-point crossover* logic. In *k-point crossover*, *k*-value is defined as "The point till which the penalty of the chromosome is 0 is taken as the *k*-point of the parent chromosome".

The anomalies in this crossover are (assuming *k*-values of parent 1 and parent 2 as k_1 and k_2, respectively)—(a) If $k_1 > k_2$, then the genes from parent 1 are taken till k_1, and remaining genes are taken from parent 2. Within parent 2 only the direction is taken, and the *no_unitmoves* is taken in random (b) If $k_1 < k_2$, then the genes from parent 2 are taken till k_2, and remaining genes are taken from parent 1. Within parent1, only the direction is taken, and the *no_unitmoves* is taken in random and vice-versa, (c) If $k_1 = k_2 = 0$, A *k*-value is selected in range of *min (length of parent 1, length of parent 2)*. Till *k*, genes are taken from parent 1 in which direction is same as parent 1 and *no_unitmoves* is generated in random. Similarly, from *k-point* till the length of parent 2 genes is taken from parent 2 in which direction is same as parent 2 and *no_unitmoves* is generated in random, and (d) If $k_1 = k_2 > 0$ (i.e., the moves are good till point k_1), then genes from parent 1 are taken till k_1, and remaining genes (both direction and *no_unitmoves*) are generated in random till length of parent 2. After generating the child chromosome, a validation check is done at point *k* in chromosome so that no rules taken in the formation of initial population are violated.

After the crossover operation, the mutation is done on random basis—(a) *K*-values of both parent and child are calculated, and then a random gene is replaced with the compatible parent in the range of child *k* to length of selected parent, (b) after mutation a validation of $i = check$ is done to ensure that no rule is violated. If any rule is violated, then those two genes are combined with a *unitmove* of average of two genes, which is same as the validation check in crossover.

Next, the survivor selection takes place by following the logic of replacing the least fit chromosome in the population by the child chromosome.

Finally, to check the condition for repetition of the above evolutionary cycle of GA_{maze}, the termination condition is checked. The termination condition used here is "Repeat until find a solution".

The GA_{maze} also incorporates a diversity-based population updating mechanism. In this mechanism, the population diversity of new population is measured. The diversity measure used in GA_{maze} is the standard deviation of the fitness of the candidates in the population. If there is a drop in population diversity, then the new population is generated, and a population with a combination of 50% of current population and 50% of new population is taken to the next generation. This process was done because the experimental observations during the evolutionary cycle revealed that the diversity of the population is lost after a certain number of generations.

4 Results and Discussions

In GA_{maze}, the starting and ending points were fixed. The GA_{maze} makes the candidates or phenotypes to reach the goal by making them evolve into strong maze solvers by mutating or altering their *genotype* (properties) using *GA* steps. Initially, the number of moves to solve the maze and the initial direction in which the individual needs to start exploring are randomized. This is for the candidates to evolve over generations and in turn to solve the maze.

The proposed GA_{maze} incorporated with the population diversity-based population updating mechanism was experimented with different runs, each with different initial populations. The experimental results depicting the performance of the GA_{maze} are presented in Table 1. In Table 1, G—generation at which the solution is found and k-value—total number of commands (direction, number of unit steps) taken to solve the

Table 1 Experimental results of GA_{maze}

Run	G	K-value	Output chromosome	Maze solved
1	140	9	[(0, 4), (1, 2), (2, 3), (1, 1), (2, 4), (1, 1), (0, 1), (1, 3), (0, 6)]	Yes
2	124	7	[(0, 4), (1, 2), (2, 3), (1, 1), (2, 3), (1, 4), (0, 6)]	Yes
3	80	7	[(0, 4), (1, 2), (2, 3), (1, 1), (2, 3), (1, 4), (0, 6)]	Yes
4	66	9	[(0, 4), (1, 2), (2, 3), (1, 1), (2, 4), (1, 1), (0, 1), (1, 3), (0, 6)]	Yes
5	1000	3	[(0, 4), (1, 2), (2, 3), (3, 4)]	No
6	148	7	[(0, 4), (1, 2), (2, 3), (1, 1), (2, 3), (1, 4), (0, 6)]	Yes
7	112	7	[(0, 4), (1, 2), (2, 3), (1, 1), (2, 3), (1, 4), (0, 6)]	Yes
8	1000	5	[(0, 4), (1, 2), (2, 3), (1, 1), (2, 3), (1, 4), -0.0]	No
9	1000	5	[(0, 4), (1, 2), (2, 3), (1, 1), (2, 3), (1, 3)]	No
10	199	7	[(0, 4), (1, 2), (2, 3), (1, 1), (2, 3), (1, 4), (0, 6)]	Yes
Average	**387**	**6.6**	**Best Solution [(0, 4), (1, 2), (2, 3), (1, 1), (2, 3), (1, 4), (0, 6)]**	

maze. Based on the randomness in the initial population, the GA_{maze} showed different performances at different runs; however, it was able to solve the maze problem in 7 out of 10 runs.

As shown in Table 1, in Run 7 GA_{maze} could solve the maze in a smaller number of generations (112). Hence, the solution [(0, 4), (1, 2), (2, 3), (1, 1), (2, 3), (1, 4), (0, 6)] is regarded as the best solution out of all the runs. On average, the number of generations taken by GA_{maze} was 387, and the average K-value was 6.6. The trajectory of the fitness value and the K-value of the best chromosome across the generations for the Run 7 is shown in Figs. 2 and 3, respectively. Fig. 4 shows the output of how the maze was solved where yellow is up, green is right, blue is down, red is the starting point, and pink is the ending point.

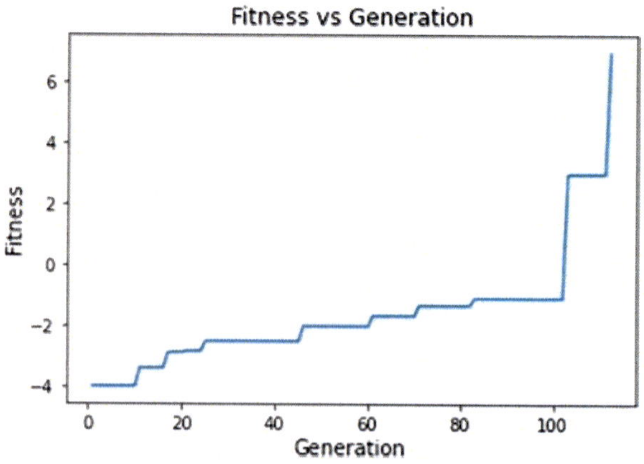

Fig. 2 Generation versus fitness value (at Run 7)

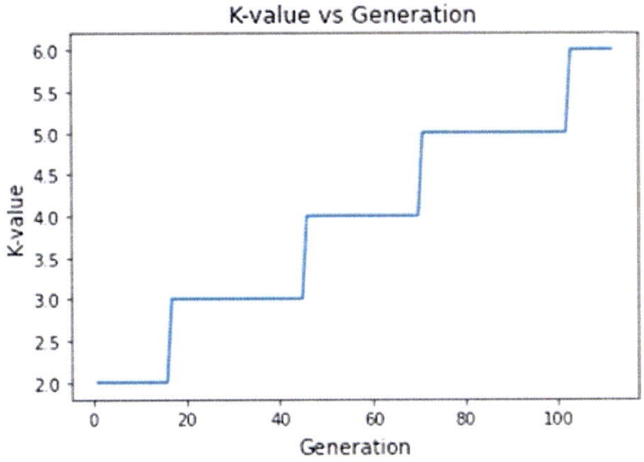

Fig. 3 Generation versus K-value (at Run 7)

Fig. 4 Visualization of maze solving by by GA_{maze}

5 Conclusions

This work is an attempt of showcasing the applicability of evolutionary algorithms in solving the real-world optimization problems around as with a simple toy two-dimensional maze game, which is known for everyone. A genetic algorithm (*GA*) with population diversity-based population updating mechanism to solve the traditional two-dimensional toy game problem was proposed. The proposed *GA*-based framework was named as GA_{maze}. The detailed methodology of GA_{maze} was presented in this paper. The working of the proposed framework is verified with different runs each with different initial populations. The experimental results attained and the inferences gained were presented in this paper.

The limitation of the GA_{maze} is—it finds difficult to solve bigger mazes; it finds difficult to solve maze when the possibilities are very less, and it loses the population diversity often. Considering the strength and weakness of the GA_{maze}, the future work planned includes trying on bigger mazes where the probability to find solution is less and also the number of solutions to solve the maze is high and optimizing the number of steps taken by the algorithm to solve the maze.

References

1. Abraham KT, Ashwin M, Sundar D, Ashoor T, Jeyakumar G (2017) An evolutionary computing approach for solving key frame extraction problem in video analytics. In: Proceedings of ICCSP-2017—international conference on communication and signal processing
2. Jeyakumar G, Nagarajan R (2018) Algorithmic approaches for solving RFID reader positioning problem with simulated and real-time experimental setups. In: Proceedings of 7th international conference on advances in computing, communications and informatics, ICACCI 201830, pp 1383–1387
3. Saketh KH, Jeyakumar G (2020) Comparison of dynamic programming and genetic algorithm approaches for solving subset sum problems. In: Smys S, Tavares J, Balas V, Iliyasu A (eds) Computational vision and bio-inspired computing. ICCVBIC 2019. Advances in intelligent systems and computing, vol 1108, pp 472–479
4. Srivatsa D, Teja TPVK, Prathyusha I, Jeyakumar G (2019) An empirical analysis of genetic algorithm with different mutation and crossover operators for solving Sudoku. In: Deka B, Maji P, Mitra S, Bhattacharyya D, Bora P, Pal S (eds) Pattern recognition and machine intelligence. PReMI 2019. Lecture notes in computer science, vol 11941
5. Saketh KH, Sumanth KBVNS, Kartik PVSMS, Aneeswar KSS, Jeyakumar G (2020) Differential evolution with different crossover operators for solving unconstrained global optimization algorithms. International conference on image processing and capsule networks, proceedings by Springer-AISC
6. Sree KV, Jeyakumar G (2020) An evolutionary computing approach to solve object identification problem for fall detection in computer vision-based video surveillance applications. Studies in computational intelligence, vol 873. Springer, pp 1–18
7. Keerthanaa K, Radhakrishnan A (2020) Performance enhancement of adaptive image contrast approach by using artificial bee colony algorithm. In: Proceedings of the 4th international conference on computing methodologies and communication, ICCMC, pp 255–260
8. Murali R, Velayutham CS (2020) A conceptual direction on automatically evolving computer malware using genetic and evolutionary algorithms. In: Proceedings of the 5th international conference on inventive computation technologies, ICICT 2020, pp 226–229
9. Shah SH, Mohite JM, Musale AG, Borade JL (2017) Survey paper on maze generation algorithms for puzzle solving games. Int J Sci Eng Res 8(2)
10. Thada V, Shrivastava U (2014) Solution of subset sum problem using genetic algorithm with rejection of infeasible offspring method. Int J Emerg Technol Comput Appl Sci pp 259–262
11. Barnouti NH, Al-Dabbagh SSM, Naser MAS (2016) Pathfinding in strategy games and maze solving using A* search algorithm. J Comput Commun 4:15–25
12. Choubey NS (2012) A-mazer with genetic algorithm. Int J Comput Appl 58(17):0975–8887
13. Iqbal A, et al (2021) Intelligent data-analytics for condition monitoring: smart grid applications. Elsevier, p 268. https://www.sciencedirect.com/book/9780323855105/intelligent-data-analytics-for-condition-monitoring. ISBN: 978-0-323-85511-2
14. Tomar A, et al (2020) Machine learning, advances in computing, renewable energy and communication, vol 768. Springer Nature, Berlin, LNEE, p 659. https://doi.org/10.1007/978-981-16-2354-7. ISBN 978-981-16-2354-7

A Novel Secure Vector Product for Protecting the Privacy of Data in Vertically Partitioned Dataset

Vibhor Sharma, Shashi Bhushan, Anuj Kumar Singh, and Pramod Kumar

Abstract The contemporary advancements in the field of data collecting, storage and networking have resulted in data being saved and kept at many locations. Data mining methods allow hidden knowledge to be extracted from massive data sources. Data on the centralized server are not always present. Data might be obtained at individual or several places. Data are divided vertically, horizontally or at both sites and disseminated between places. The sharing of data repositories existing at many stages in the mining process leads to discomfort in data privacy. Data mining protection is a technology that prevents sensitive information from being disclosed to others during a mining operation. Simple multiparty calculations and anonymization are the most common privacy protecting technology for data mining. Secure multiparty (SMC) computing allows parties to do distributed data mining activities in separate places without disclosing more personal data. The SMC protocol employs mostly encryptions that offer high security and enhance the complexity of protocols. Before it is utilized for mining, perturbation alters the original sensitive data. Distortion of the data may diminish the mining performance accuracy to decrease the usefulness. A novel vector scalar product is suggested to save execution time and to protect privacy data privacy in the vertically partitioned dataset, retaining the induction of decision tree. A downward approach to the safe scalar product is taken. Each node is a radix or fewer party groups. The binary vector is divided as well. Each group discovers a safe sum of binary vectors with the secret sharing of Shamir. The running time of the CPU was established between sequences and when running downstream.

V. Sharma (✉)
HSST, Swami Rama Himalayan University, Dehradun, India
e-mail: vibhor087@gmail.com

S. Bhushan
School of Computer Science, University of Petroleum and Enegy Studies, Dehradun, India
e-mail: tyagi_shashi@yahoo.com

A. K. Singh · P. Kumar
Krishna Engineering College, Ghaziabad, India
e-mail: anuj.singh@krishnacollege.ac.in

P. Kumar
e-mail: drpramoderp@live.com

A. Tomar et al. (eds.), *Proceedings of 3rd International Conference on Machine Learning, Advances in Computing, Renewable Energy and Communication*, Lecture Notes in Electrical Engineering 915, https://doi.org/10.1007/978-981-19-2828-4_28

The execution time was determined. The proposed bottom-up approach results in comparison with following sequential approach.

Keywords Classification · Cryptography · Decision tree · Privacy preserving data mining · Vector product · SMC

1 Introduction

The rapid development of technologies in collecting and storing data required the organizations to gather colossal data volume. There is an assessment that the volume of the information is doubled in size for every 20 months and the number of databases is also increased faster [1]. It is tremendously a challenging task to extract the required information and also the traditional techniques, and tools are failed to handle these gigantic dimensions of data. Data mining handles huge volume of data with the combination of sophisticated algorithms and available data analysis methods. In general, the transaction data is known as raw material and data mining is a crawler which screens the valuable items of information from large amounts of raw data. Data mining is the combination of several automated techniques required to extract hidden or concealed information from huge databases. The word data mining denote the nontrivial abstraction of implicit, effective, potentially helpful and eventually meaningful information from huge databases with the support of latest computing devices [2, 3]. Many popular applications in data mining from last few decades have been informed from various areas like finance, banking, marketing, medical diagnosis, telecommunications and manufacturing. Apart from the advantages of using data per se (e.g., maintaining the profiles and purchases of the customers up to date, preserving list of products, its prices and quantities, etc.), the usage of traditional mining tools [4] for mining these datasets discloses priceless knowledge which should be known earlier to the data holder. The knowledge patterns extracted from the database educate the data holders and play vital role in decision making and business planning [5] using data mining techniques, and also, it is successful. However, it is very difficult to maintain individual privacy while collection, processing and mining of data [6]. Preserving individual privacy is a significant feature of business planning with data mining. It is observed that the process of discovering valued, non-obvious information from huge databases with data mining techniques constantly is prone to vulnerable and misuse [7, 8]. Therefore, privacy and data mining might be always conflicted, but no need to intrude privacy in reality for data mining.

A protected vector product that fills in as the center activity in large numbers of the information mining calculations is one of the SMC conventions. This part centers around planning an effective, commonsense protection saving secure vector product calculation as to correspondence and calculation cost. The calculation processes the security saving decision tree enlistment over vertically parceled information effectively with secret sharing and participating site of participating sites to frame a tree.

The DM approach followed by most commercial products is decoupled architecture-based. DM works on data taken from a database and modified to an appropriate form. But the disconnected method has disadvantages. It needs a data analyst to prepare extensively to extract data and preparing it to achieve a specific fixed form needed by the tool. It offers a fixed DM paradigm that occasionally corresponds to specific requirements leading to selection of a tool without offering sufficient generality when requirements change. Rules, after extraction, are placed in a DM tool, and it is difficult to merge information in them with the data in a database.

Besides introduction section, related work is given in second section. Third section describes the problem definition. Initial-level decision tree induction algorithm is defined in section four. Further, decision tree induction over vertically partitioned data is explained using algorithm in section five. Sixth section represents proposed methodology followed by experimental results and discussion in seventh section. Conclusion and future work are mentioned in section eight.

2 Related Work

Cryptographic-based strategy is a significant method to determine the precision protection compromise. The security saving information mining result is strategy precise, and the protection of private data is saved under predefined assurance limitations. Certain cryptographic properties or highlights are utilized to execute the protection saving conventions. The property of the cryptographic technique utilized in security safeguarding information mining incorporates added substance homomorphic encryption property and commutative encryption property. The center segments that most protection saving disseminated information mining conventions depend on these properties. The most widely recognized test in security safeguarding appropriated information mining calculations is the effectiveness with enormous data sets and versatility of the calculations. The vast majority of the cryptographic calculations for security protecting information mining are created to address these difficulties. Authors in [9] introduced the paper association rules, deep learning for cryptographic calculation for ensuring security. Finding knowledge from huge quantity of information base was alluded as data mining. Moderate demotion of information was acquired by information mining. Safeguarding privacy is significant for security viewpoints. Information assortment was worked with by versatile terminals [10]. Authors in [11] developed the paper Agent Technology Role for Web Usage Mining by Homomorphic Encryption relying on Recommendation in E-trade applications. Productivity, time intricacy of framework was decreased by utilizing computerized reasoning for E-commerce applications. Ref. [12] represented the paper Protecting Privacy for information mining in Paillier cryptoprocessor. Protection awareness was improved by preserving privacy data mining calculations. Paillier cryptoprocessor embraces public key encryption conspire. Framework incorporates encryption, key generation and decryption. Ref. [13] introduced the paper Cryptography, Stenography, Data mining for Secure E-bank. This paper gives simple admittance to 24 h banking, and online fraud banking was recognized, forestalled for viable, secure e-banking. Client card details were given online site installment door. Undeniable level security was acquired by encryption calculation. Ref. [14] Represented the paper Privacy saving ID3 Decision Tree calculation by scrambled informational indexes. Precise market data for business was given ID3 algorithm. Putting away crude information in focal site to mine the information was done in ID3 choice tree calculation. Reevaluating of information stockpiling was done in ID3 calculation. Every client's dataset was scrambled. Parceling of information bases was finished by protection safeguarding ID3 calculation. All informational indexes of each client were encoded in symmetric homomorphic encryption. A cryptographically secure scalar item [15] was analyzed which utilizes private and public key for dividing information among two gatherings. The calculation utilized for experimentation utilizes added substance secret divides among two gatherings, and the creators inferred that calculation delay was the significant bottleneck of execution of any cryptographic execution which can be improved utilizing useful advancements or specific equipment. Subjectively apportioned information is a mix of on a level plane and vertically divided data. The idea of pseudo scalar product (PSP) was proposed to do the protected scalar item activity [16] with less correspondence and calculation cost and safer than existing methodologies. The projected calculation depends on double vector where a twofold vector addresses every extraordinary worth in the quality. The estimation of PSP is comparable to the cardinality of set crossing point of different double vectors, and homomorphic encryption is utilized for the protected calculation. A scalar product (SSP) convention was proposed [17] with improved proficiency utilizing homomorphic encryption. The convention accepts two vectors as information and utilizing homomorphic encryption, and the portions of each gathering are registered. A safe scalar item convention [18] was introduced within the sight of malevolent foes. The contribution to the calculation is private vectors of two gatherings, and the yields are u and v, where v is the amount of the relative multitude of estimations of one vector, and u is the amount of result of the relative multitude of private estimations of the two vectors in addition to the estimation of v. The creators utilized the confirmation of information on a discrete logarithm and the evident encryption. Secure multiparty calculation utilizes cryptographic techniques or offers to play out the information mining activity without uncovering the information. The information is circulated in various areas, and the information can be on a level plane or vertically or mixture divided [19, 20]. The works did around here generally get the correspondence part of the mining calculation like secure set association, secure set convergence, supplement, secure scalar item, absent information move and so on. The space of examination in SMC is focused on limiting the correspondence cost, calculation cost of the calculation and expanding the protection level of the information.

3 Problem Definition

Perhaps the most pervasive information mining issues found in genuine life is order. Decision tree enlistment is a broadly recognized arrangement approach for characterization. The protected vector product fills in as core in a large portion of the decision tree enlistment over the vertically apportioned dataset. To work with the protected vector product, the segment of informational index held by various participating sites is considered as a double vector. The double vectors address the segment vectors of datasets, and the vector product is found on these segment vectors from various participating sites. Despite the fact that cryptographic procedures are free from any danger, a lot estimation and correspondence cost associated with them makes it unreasonable for circumstances including an enormous number of participating sites. The issue is to process the safe vector product activity in a proficient manner to make the mining calculation effective.

4 Proposed Methodology

The vector product of binary vectors is computed by pseudo vector product. The binary vector is created with ones and zeroes, and the vector is considered as a number with base r, where $r <= 35$. The binary vector is treated as a number with base 'r'. With the help of mentioned process of sharing, total addition of all the values with base 'r' of all participating sites is calculated. The computed sum of the numbers to the base 'r' is then converted to base two by changing the bit position as '1' if the digit value is equal to the number of parties of the group else it is '0'. The sum of binary vector of each group is combined together to form the resultant binary vector. This combining takes place till the number of groups becomes one. The count of 1s in the binary vector in the resultant binary vector of the top node gives the pseudo vector product. The final resultant vector formed by combining the resultant vectors of each group is computed using a bottom-up tree. All the parties are present in the bottom layer of the tree. The parties are then formed as groups. In each group, the first participating site is assumed as the representative, and it finds the resultant value of its group and then converts it into binary vector form. The resultant binary vector is present with the representative of each group. The parties of the bottom layer which has the resultant binary vector combine to form groups for the next higher layer. The formation of higher level layers thus continues by grouping representative parties, till a single representative alone is present in the top layer. The representative of the top layer is the participating site with the class attribute. The original value is increased by adding random number by representative site of each group. The random number is removed after finding the resultant sum that is computed using Shamir's secret sharing [21] in that particular group. The random number present in the resultant vector makes the original sum hidden from the other parties.

Stage 1: Setup maintenance

Objective: Steps for setup maintenance

Begin

First Step: Calculate $m_1 = \frac{m}{d}$ and $n_1 = \frac{n}{r}$.

Second Step: Following are calculated by each participating site p_i where $i = 1$ to n

- If maximum information gain condition of a field is satisfied, then initialize a vector by specifying 1 as the position of bits else specify 0.
- Decompose the binary vector into m_1 parts where all parts contain d digits except the last part.
- The vector is represented as $(0/1)_r^m$, where r is radix.
- A random number is generated with r and concatenate with all vectors.

Third Step: Set c_count to 0 where c_count is the counting of cardinality of vector.

End

The representative alone knows the sum, but the other parties do not know the private value of any other participating site. The representative will not be able to get the original values of remaining participating sites unless $n - 1$ parties collude with each other to find the value of the other one when there are 'n' number of parties. Position of bit is replaced by 1 if a specific bit position has the value 'x' else it is 0, where X is the number of parties in a group. The resultant vector in a group now has only 1s and 0s. Count the representative parties in a particular layer.

Stage 2: Computing the final vector of each partition

Objective: Calculate the final vector for each instance of partitioning

begin

for each partition of transaction t_j where $j = 1$ to m_1

do

{

1.1: $N = n_1$

1.2: repeat

{

1.2.1: for each group of participating sites Pi where $i = 1$ to N

do

{

(continued)

(continued)

1.2.1.1: A random participating site is chosen as representative. Let total number of participating sites be '*f*'

1.2.1.2: Calculate the secure sum of P_i using Shamir's secure sum protocol

1.2.1.3: Representative site receives random numbers from all participating sites of a group

1.2.1.4: Representative site calculates the sum using mentioned Shamir's method. Sum of all random numbers received is subtracted from the calculated sum. Position of digit is set to 1 if value in a digit position is obtained is '*f*' else digit position is set to 0

}

1.2.2: Set N = total number of representative sites

} until ($N < 0$)

1.3: x = Total count of '1's in the final specified vector

1.4: c_count = c_count + x

}

End

The representatives' are considered as parties, and groups are formed as mentioned before. These constitute a particular level in the bottom-up tree.

5 Experimental Work and Discussion

The proposed scalar product algorithm was implemented in MATLAB, and the execution time was measured in dual core 2 GHz processor with 512 M memory. The execution time of an algorithm varies according to the data size which depends on the number of instances and the number of attributes. The attributes are split among the various parties in vertically partitioned dataset. The number of participants that have the vertically partitioned dataset also affects the execution time of the algorithm due to communication needed between the participants. The participants can be located anywhere around the world. Experiments were conducted by assuming the participants are in the same environment. The process cost alone is experimentally shown. The execution time was measured by varying two different parameters, namely the number of participants and the number of instances which is same in all the participants for vertically partitioned datasets. Experimental results were shown by computing execution time of the algorithm by varying the number of parties with fixed number of instances and by varying the number of instances with fixed number of parties.

Consider 'm' as the total number of instances, 'n' as the total number of participating sites, 'd' as the number of digits in maximization and the total number of partitions be$m1 = \frac{m}{d}$. To compute the scalar product, each participating site selects a random polynomial and a publically known value. Then, each participating site finds its share for each publically known value and sends the corresponding share to the respective parties. The final sum can be found from the shares of all the parties. All the groups in the same level can operate independent of each other so that they can perform the group operations in parallel with other groups. The scalar product protocol is executed $u \times m_1$ times for each attribute, where 'u' is the possible distinct values of the attribute. The number of public values is equal to the number of parties in each group. This will have the maximum value r. The vector product of each participating site requires r polynomial evaluation. The process cost of finding the scalar product for an attribute is $r \times u \times m_1$. For $n_1 = \frac{n}{r}$ participating site groups, the process within a group can be performed in parallel. If there are $\log_r n$ levels in the tree, then the cost is $\log_r n \times r \times u \times m_1$. The communication cost within a group to compute the scalar product of an attribute is $r \times u \times n_1 \times \log_r n \times m_1$. The comparison of process and communication cost is given in Table 1.

First the experiment was performed by keeping the number of instances 'm' as stable and the number of participants as changing. Figure 1 shows execution time

Table 1 Comparison of process and communication cost

Analysis parameter	Homomorphic public key encryption (Paillier)	Proposed algorithm
Communication cost	$(m * n * 2 * \log m)$ bits for encrypting $(t + 2) \log m$ bits for decryption	$r * u * n1 * m1 * \log_r n$
Processing Cost	$[(m + 1) * n * 2 * \log m]$ bits multiplication and $2(t + 1)$ exponentials, where t is secret key	$\log_r n * u * r * m1$

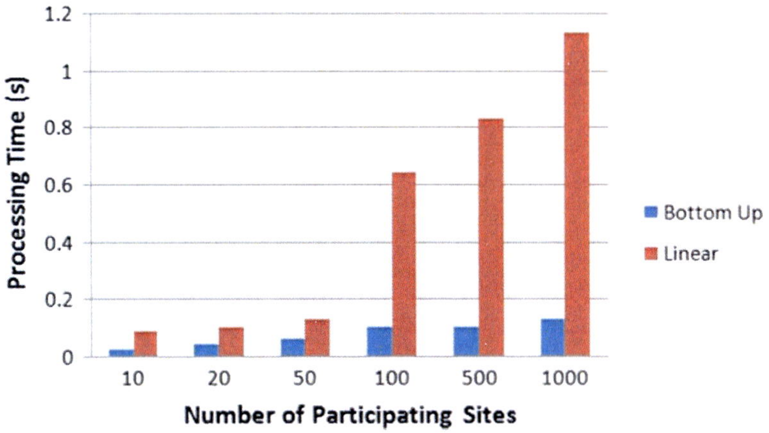

Fig. 1 Processing time variation based on increasing number of participating sites

of the algorithm when the numbers of parties 'n' varies by keeping the number of instances or the number of transactions 'm' constant. For this, the CPU dataset from UCI machine repository was considered. This has 209 instances; the ARFF format was taken for the experimental analysis. The aim was to shown the execution time by varying the participants, so one dataset was considered.

The value of m is set as 209. The number of participating sites is varied from 10 to 1000 to show the scalability of the algorithm. The parties are grouped in the order of radix 'r'. The parties in each group communicate only among themselves in each level. So in each level the group can perform process in parallel with other groups. One among the participating site will be selected for processing in the next higher level. Then in that level the parties are group in the order of 'r'. This continues till the root node with r or less elements is reached. In this experiment, the value of 'r' is fixed as ten. Increase in the number of parties will not cause bottleneck when a tree is followed. That is why this algorithm scales well with the increase in the number of parties. The result shows that the execution time of the algorithm figures out a negligible increase when the groups are executed in parallel. When the groups are executed in sequential, the execution time varies from 0 to 1.2 s. The experiment shows that the proposed algorithm scales well with the increase in the number of parties.

The experiment was then performed by keeping the number of participant as fixed and the number of instances 'm' as changing. Figure 2 shows execution time of the algorithm when the numbers of instances are varied by keeping the number of parties constant. The number of participants is fixed as 20. The value of m is varied from 1000 to 30,000. The number of instances is also grouped into 'd' which is maximum digits in an integer. In this experiment, this was fixed as ten. This too can be performed parallel among the parties.

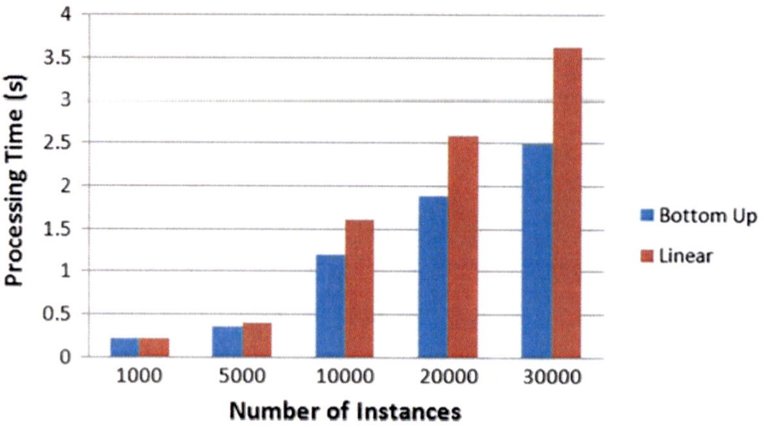

Fig. 2 Process time variation based on increasing number of instances

6 Conclusion and Future Work

The outcome of the experiment shows that the completing time of the algorithm figures out steady increase in execution time when executed in bottom-up fashion. But the completing time increase is drastic when executed sequentially. This illustrates that by grouping the participating parties and by splitting the data instances into different partitions, the completing time of the cardinality of scalar product is very much reduced. Using this secure scalar product algorithm in the decision tree algorithm gives out an efficient output in terms of execution time which is very essential for data mining algorithms operating over very large datasets. The process and communication overhead is very much reduced with the privacy preservation implementation because of the tree-based structure, alliance of the parties and partitioning of data instances. In future, another privacy protection technique may test the safe scalar product. Noise addition may be investigated using several ways in perturbation over vertically partitioned data.

References

1. Friedman A, Schuster A (25–28 July 2010) Data mining with differential privacy, KDD 10. https://doi.org/10.1145/1835804.1835868
2. Agarwal CC, Yu PS (2004) A condensation approach to privacy preserving data mining. In: Advances in database technology—EDBT. Lecture notes in computer science. Springer, Berlin, Heidelberg, 2992: 183–199
3. Chouhan A, Kumari A, Saiyad M (2019) Secure multiparty computation and privacy preserving scheme using homomorphic elliptic curve cryptography. 2019 international conference on intelligent computing and control systems (ICCS). Madurai, India, pp 776–780
4. Gascón A, et al. Privacy-preserving distributed linear regression on high-dimensional data. Proc Priv Enhancing Technol 2017(4):345–364

5. Chouhan A, Patel S, Jinwala DC (2013) Comparative analysis of elliptic curve cryptography based algorithms to implement privacy preserving clustering through secure multiparty computation. Publ J Inf Sec, Sci Res
6. Balaji M, Rao G (2013) An adaptive implementation case study of apriori algorithm for a retail scenario in a cloud environment. 13th IEEE/ACM international symposium on cluster, cloud, and grid computing. Delft, pp 625-629. https://doi.org/10.1109/CCGrid.2013.104
7. Bourse F, Sanders O, Traoré J (2020) Improved secure integer comparison via homomorphic encryption. In: Jarecki S (eds) Topics in cryptology—CT-RSA 2020. CT-RSA 2020. Lecture notes in computer science, vol 12006. Springer, Cham, pp 391–416
8. Keshavamurthy BN, Sharma M, Toshniwal D (2010) Privacy-preserving Naive Bayes classification using trusted third party and different offset computation over distributed databases. 2010 first international conference on parallel, distributed and grid computing (PDGC 2010), Solan, pp 362–365
9. Zhao J, Chen Y, Zhang W (2019) Differential privacy preservation in deep learning: challenges, opportunities and solutions. IEEE Access 7:48901–48911
10. Li L, Lu R, Choo KR, Datta A, Shao J (2016) Privacy-preserving-outsourced association rule mining on vertically partitioned databases. IEEE Trans Inf Forensics Secur 11(8):1847–1861. https://doi.org/10.1109/TIFS.2016.2561241
11. Ahila S, Shunmuganathan KL (2016) Role of agent technology in web usage mining: homomorphic encryption based recommendation for e-commerce applications. Wireless Pers Commun 87:499–512
12. Koundinya AK, Gautham SK (2021) Two-layer encryption based on paillier and elgamal cryptosystem for privacy violation. Int J Wirel Microw Technol 3:9–15. https://doi.org/10.5815/ijwmt.2021.03.02
13. Devadiga N, Kothari H, Jain H, Sankhe S (2017) E-banking security using cryptography, steganography and data mining. Int J Comput Appl 164:26–30. https://doi.org/10.5120/ijca2017913746
14. Yi-bin L, Ying-ying W, Xue-wen R (2017) Improvement of ID3 algorithm based on simplified information entropy and coordination degree. Chin Autom Congr (CAC) 2017:1526–1530. https://doi.org/10.1109/CAC.2017.8243009
15. Yang Z, Wright RN, Subramaniam H (2006) Experimental analysis of a privacy-preserving scalar product protocol. Int J Comput Syst Sci Eng 21(1):47–52
16. Shuguo HAN, Wee Keong NG (2007) Multi-party privacy-preserving decision trees for arbitrarily partitioned data. Int J Intell Control Syst 12(4):351–358
17. Goethals B, Laur S, Lipmaa H, Mielikäinen T (2004) On private scalar product computation for privacy-preserving data mining. Lect Notes Comput Sci 3506:104–120. https://doi.org/10.1007/11496618_9
18. Yang B, Yong Y, Yang C-H (2013) A secure scalar product protocol against malicious adversaries. J Comput Sci Technol 28(1):152–158
19. Iqbal A, et al (2021) Intelligent data-analytics for condition monitoring: smart grid applications. Elsevier, p 268. https://www.sciencedirect.com/book/9780323855105/intelligent-data-analytics-for-condition-monitoring. ISBN: 978-0-323-85511-2
20. Tomar A, et al (2020) Machine learning, advances in computing, renewable energy and communication, vol 768. Springer Nature, Berlin, LNEE, p 659. https://doi.org/10.1007/978-981-16-2354-7. ISBN: 978-981-16-2354-7
21. Shamir A (1979) How to share a secret. Commun ACM 22:612–613

A Bibliometric Analysis of Anomaly Detection for IoT-Enabled Smart Cities

Neeraj Chugh

Abstract Anomaly detection for IoT-enabled smart cities is a significant field in the functioning of smart cities. To better characterize and understand anomaly detection for IoT-enabled smart cities literature studies, there is a requirement for widespread bibliometric analysis in this area. The aim of this work is to characterize the research literature over the last ten years (2011 to 2020) using citation and thematic analysis which is an automated process. To achieve the said objective, a bibliometric study is conducted on Web of Science (WoS) data comprising 375 research papers. The thematic analysis depicts the most studied topic distributions. It has been observed that significant areas are security, privacy, machine learning, and fault detection, respectively, associated with current research subject. The results of the study reveal that publications per year have grown swiftly in the last five years and on average 67 papers are published per year. The largest share of papers has been contributed by Asia followed by Europe and then North America in this field. Through this paper, researchers can get an overview of the studied field and contribute to the progression of this domain.

Keywords Outlier detection · IoT · Anomaly detection · Smart cities

1 Introduction

The broad acceptance of the modern IoT model has resulted in the development of smart cities in recent years. Smart cities work to facilitate convenience and improve life quality in metropolitan cities. As these IoT devices are linked to sensors that are directly linked to large cloud servers, the network traffic of a smart city through IoT systems is rising at a high rate and leading to new cybersecurity challenges

N. Chugh (✉)
School of Computer Science, University of Petroleum and Energy Studies, Dehradun, Uttarakhand, India
e-mail: nchugh@ddn.upes.ac.in; neerajatmit@gmail.com

Uttarakhand Technical University, Sudhowala, Uttarakhand, India

A. Tomar et al. (eds.), *Proceedings of 3rd International Conference on Machine Learning, Advances in Computing, Renewable Energy and Communication*, Lecture Notes in Electrical Engineering 915, https://doi.org/10.1007/978-981-19-2828-4_29

[1]. To mitigate these cyberattacks, new techniques for detecting anomalous IoT devices need to be built by developers. Detection of anomalies is a crucial problem that has been studied in different research and application environments [2–5]. For some domains, various anomaly detection techniques have been developed specifically, while others are generally applicable. Anomaly detection for IoT-enabled smart cities is a progressively growing research theme that has many applications in the actual world. Still, various anomaly detection techniques are unsuccessful to maintain adequate precision because of the existence of different types of devices, data generated by them, and high-velocity data generated by a variety of sources [6]. Anomaly detection for IoT-enabled smart cities is a progressively growing research theme that has many applications in the actual world. Still, various anomaly detection techniques are unsuccessful to maintain adequate precision because of the existence of different types of devices, data generated by them, and high-velocity data generated by a variety of sources [6]. Anomaly detection for IoT-enabled smart cities is a progressively growing research theme that has many applications in the actual world. Still, various anomaly detection techniques are unsuccessful to maintain adequate precision because of the existence of different types of devices, data generated by them, and high-velocity data generated by a variety of sources [6]. Anomaly detection literature has shown immense development, and hence, there is a need for bibliometric studies in anomaly detection for IoT-enabled smart cities. On searching the different databases like Web of Science, Scopus, and Google Scholar, etc., none of bibliometric studies have been reported on anomaly detection for IoT-enabled smart cities. Hence, the present work is one of its kind among anomaly detection studies.

The present work comprises: Sect. 2 covers allied works in the arena to visualize the significance of the bibliometric studies in the areas of IoT, anomaly detection, and smart cities. Section 3 converses the data extraction process and research method, which has been implied to form the pool of research papers analyzed later in the study. Section 4 includes the results of this work and represents results respective to research questions designed for the work. Section 5, discussion and conclusion, concludes this study, and to close, Sect. 6 states the forthcoming work and limitations.

2 Background Work

This section presents the studies related to the bibliometric analysis in the fields of IoT anomaly detection and smart cities to explore the various perspectives in the previous bibliometric studies [1–29].

Mora et al. [1] reports on in a bibliometric study cover the first two decades of research from 1992 to 2012 on smart cities, and the analysis displays that smart-city research is disjointed and interrelation is not present in the studies. The developments in the field are categorized into two diverse tracks: (a) European university comprehensive peer-reviewed publications on smart cities and (b) American business community techno-centric publications on smart cities. As suggested, the future development of this novel and upcoming research area intimidations being challenged

[1]. Zhao et al. [2] bring up complete scientometric analysis and knowledge mapping relating to variety, time carving, and dynamics, employing VOSviewer and CiteSpace to analyze the research work in the sphere of smart cities. As a final point in the study, the research cutting edge terms are urban-development, artificial intelligence, cloud computing, based on co-occurrence analysis of keywords [2]. Another significant work carried out by Guo et al. [3] through bibliometric analysis on smart cities research presented a valued and influential orientation for scholars and practitioners in the domain of smart cities. Their work visually exhibits a broad overview of the area in reference to the regular publications production, chief area of smart cities scholars, most leading countries (organizations, sources, and authors), and motivating research directions in the smart city studies [3]. Wamba and Queiroz [4] performed bibliometric and network analysis to comprehend the developments and research directions in the smart cities research. The results reveal the top authors and articles in the field of study. In addition, the work presents research developments built on cluster classification, and finally, their study suggests academicians and managerial takeaways [4]. Janik et al. [5] in their study conducted current wide-ranging bibliometric analysis to present and evaluate the systematic background of smart and sustainable cities (SSC) literature [5]. González-Zamar et al. (2020) investigated smart cities research in relation to IoT applications. They acknowledged that the exceptional thematic areas were computer science and engineering and scientific production has improved yearly, such that the most recent triennium has added 83.23% of the publications [6]. Nascimento et al. (2020) in a bibliometric study mapped the methodical production on the Internet of things (IoT), and findings reveal top year of production, reputed venues for publications, top author based on number of publications, and top areas of publications to present a picture of IoT evolution from 2010 to 2018 [7]. Perez et al. [8] analyze top journals from 1991 to 2019 through bibliometric analysis, and the findings reveal insights about the document types, the scattering of countries/territories, the dissemination of institutions, the authors' terrestrial dissemination, the most active authors and their research interests or fields, the relationships between chief authors and pertinent publications, and the most cited articles. As per the authors, the quantitative description of smart cities literature facilitates the development of a methodology for the analysis of the maturity of a smart city [8]. Thus, from the above discussion, we can infer that the bibliometric studies are significant for presenting the trends in the spheres of the study and are performed periodically to report on the top authors, venues of publications, top organizations and countries, citation landscape, and similar related information along with the research trends in domain under analysis. Hence, the present study intends to offer a full overview of anomaly detection for IoT-enabled smart cities through bibliometric analysis of the relevant literature.

3 Research Methodology and Data Extraction

The quantitative analysis can measure the progress, variation of articles in the journals or to measure the outmodedness of literature in diverse fields. Bibliometrics is defined as "the application of mathematical and statistical methods to measure quantitative and qualitative changes in dissimilar media" [9]. The present study conducts the bibliometric analysis for research papers on anomaly detection for IoT-enabled smart cities. To accomplish the stated aim, the following research questions have been framed:

RQ 1. What is the number of studies published each year and annual citation trend in anomaly detection for IoT-enabled smart cities studies, published each year from 2011 to 2020?

RQ 2. Which continents and countries are leading in terms of contributed papers and terms of citation?

RQ 3. What papers have been highly cited in anomaly detection for IoT-enabled smart cities study?

RQ 4. What is the trending highest occurrence of keywords in the fetched data set from WoS?

RQ 5. What are the prominent thematic trends for anomaly detection for IoT-enabled smart cities literature?

The review protocol of the present bibliometric study is divided in six phases. Each phase has certain activities as (i) define research questions, (ii) design review protocol, (iii) identify the inclusion and exclusion criteria, (iv) search strategy and study selection process, (v) perform a quality assessment, and (vi) data extraction and synthesis.

Few techniques are used to obtain the publications for the successful writing of this research paper.

To choose the digital library, we explored the literature on bibliometric studies that comprise IoT, anomaly detection, and smart cities [3, 5, 8]. Clarivate Analytics Web of Science and Scopus are the predominant databases that are frequently used for examining and extracting citation data from the literature [10, 11]. On comparison of Scopus and WoS, studies remark: WOS has rigorous acquaintance since 1990 and majority of its journals are written in English [12].

The citation analysis in WoS offers graphics that are more explicable and complete when compared to Scopus's citation analysis [11, 12].

Thus, we decided to use WoS in our bibliometric study. It begins by using ("anomaly detection" or "outlier detection") and ("IoT" or "smart city") as the combination of keywords for searching research papers containing similar terms or titles. The keywords are critical because it offers context for the research focus and does not deviate from it [13]. The search fetched the data set of 375 documents from the WoS database. WoS includes different components such as IEEE explore ACM and Elsevier and was selected because it hosts almost all areas of study and includes data in categories such as books, journals, and conferences [14]. Various organizations across the globe rely on the WoS database, and hence, the standard of publications indexed by this database is widely accepted. This study will offer an enhanced understanding of the diverse directions in which research on the explored topic is conducted.

The retrieved documents along with citations of retrieved papers are included in present study. The data was downloaded from WoS as plain text and entered into Biblioshiny, science mapping tool for analysis [15] to build the knowledge base and R reads the export files.

4 Result

The findings relate to the research questions that guided the study. This section comprises the findings of the bibliometric analysis conducted on the documents retrieved from WoS.

4.1 Publications and Annual Citation Trend (RQ1)

Figure 1 shows the development trend of the research literature based on the count of papers incorporated in WoS by their year of publication. The citations function generates the frequency table of the most cited references. Only few papers were published in the year 2012 to 2013. As observed from the graph, the major growth is from the year 2015, and the annual growth rate is 53.92%. It is also important to notice that the values for the year 2020 are partial.

Figure 2 clarifies the numbers of five types of publications: journal articles, study papers, data papers, conference proceedings, and early access. The majority of the pool comprises journal articles and proceeding papers, i.e., 44% and 50%, respectively, while data papers have the lowest percentage of 0.27%

There are no papers published in the year 2011 as shown in Figs. 1, 2, and 3. As depicted in the graph of Fig. 4, there is an estimated continuous growth in the count

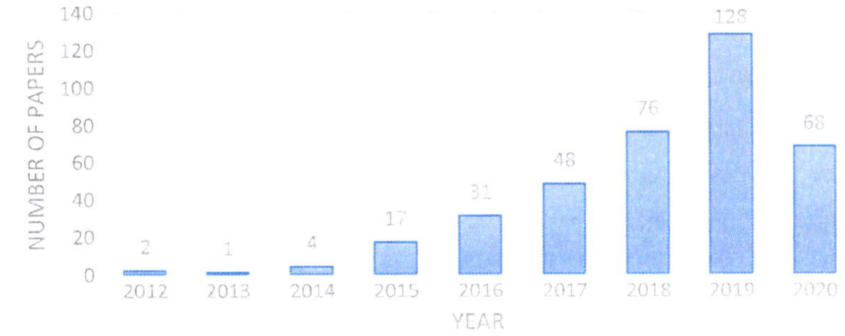

Fig. 1 Number of papers by their publication year

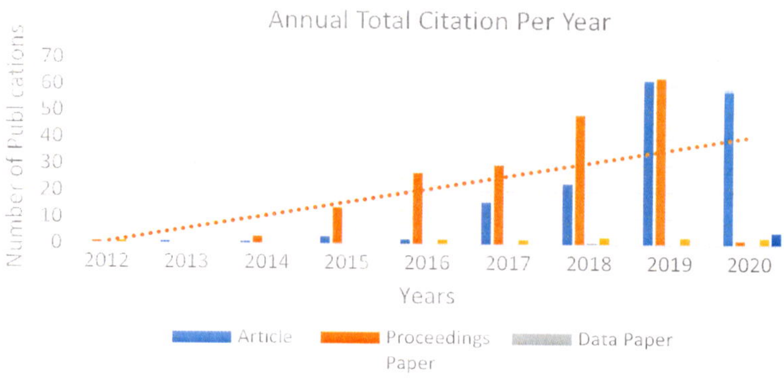

Fig. 2 Annual total citation per year

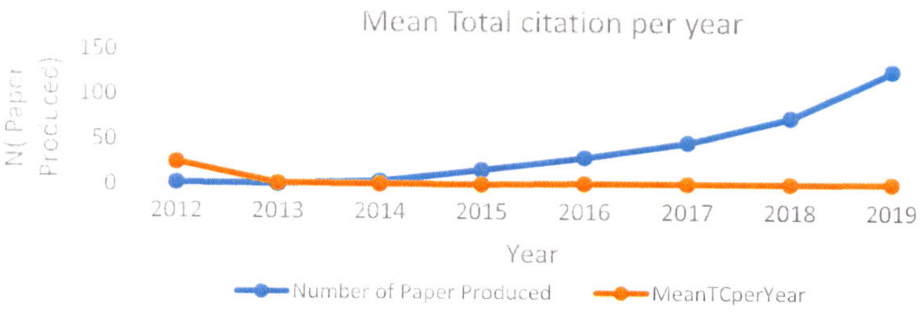

Fig. 3 Mean total citation per year

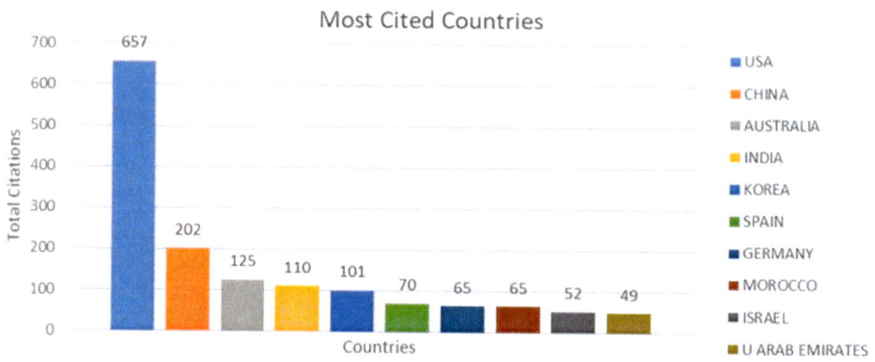

Fig. 4 Most cited countries

of research papers published per year which indicates the importance of anomaly detection for IoT-enabled smart cities among the researchers and practitioners.

A valuable metric to measure the annual impact of the author or journal is the average paper citations per year. The significant factor that affects the paper citations is the cited paper's age [16, 17]. While older papers receive more citations than younger papers. We should assume that the probability of being cited reduces dramatically with age [18], and over time, the newer studies are more cited when compared to older ones [19].

4.2 Leading Continents and Countries in Publications (RQ2)

The country rankings based on citations are shown in Fig. 5. For citations, the USA is the topmost country with 657 citations, China having second with 202, third Australia with 125, and fourth India with 110 highest citations. An analysis was carried out by evaluating the events in Asia, North America, Europe, Australia, South America, Africa, and which are the six major continents.

This analysis shows that Asia is predominant in this the research field with 44.27% of the largest number of publications, followed by Europe with 30.66%. The analysis is represented in Table 1.

Biblioshiny menu option supports to access statistics of countries centered on affiliations of authors. Hence, our pool incorporated those statistics. The top six countries (China, USA, India, Korea, Spain, and Australia) have contributed nearly

Fig. 5 Word cloud for titles

Table 1 Continents contribution to the research on anomaly detection on IoT-enabled smart cities

Continent	Publications (%)
Asia	44.27
Europe	30.66
North America	16.8
Australia	4
Africa	2.4
South America	1.87

half (~50%) of the total studies, and this shows that across countries' contribution pool is very non-normal.

4.3 Highly Cited Papers in Anomaly Detection for IoT-Enabled Smart Cities Study (RQ3)

To recognize the highly cited papers, the metric of total citations, local citations, global citations, and the average annual number of citations has been applied. Tables 2 and 3 are providing the details of the five most global cited papers and five most local cited papers, respectively.

To have an overview regarding the applied metric total citations is the total count of citations through all the papers. Total citations per year include the citation count divided by the total years included in the result set. Local citations assess the impact of the document in the considered pool of research papers. Local citations identify number of times an author or document involved in this pool have been cited by other pool authors [15]. The global citations are related to the impact of a research paper in the complete bibliographic database.

Table 2 Five most global cited papers

Paper title	DOI	TC per year	Global. citations	Year
Sensor Mania-the internet of things, wearable computing, objective metrics and the quantifies self 2.0 [20]	https://doi.org/10.3390/JSAN1030217	44.111	397	2012
Data quality in internet of things: A state-of-the-art survey [21]	https://doi.org/10.1016/J.JNCA.2016.08.002	13	65	2016
A feature-based learning system for internet of things applications [22]	https://doi.org/10.1109/JIOT.2018.2884485	30.5	61	2019
DDoS attack detection and mitigation using sdn: methods, practices, and solutions [23]	https://doi.org/10.1007/S13369-017-2414-5	12.25	49	2017
N-BaIoT-network-based detection of IoT botnet attacks using deep autoencoders [24]	https://doi.org/10.1109/MPRV.2018.03367731	15.667	47	2018

Table 3 Five most local cited papers

Paper title	DOI	Local citations	Year
N-BaIoT-network-based detection of IoT botnet attacks using deep autoencoders [24]	https://doi.org/10.1109/MPRV.2018.03367731	11	2018
A comparative study of anomaly detection techniques for smart city wireless sensor networks [25]	https://doi.org/10.3390/S16060868	10	2016
Anomaly detection and privacy preservation in cloud-centric Internet of things [26]	NA	6	2015
IoT healthcare analytics: the importance of anomaly detection [27]	https://doi.org/10.1109/AINA.2016.158	6	2016
Fog-empowered anomaly detection in IoT using hyper ellipsoidal clustering [28]	https://doi.org/10.1109/JIOT.2017.2709942	5	2017

4.4 *Trending Keywords in the Studies (RQ 4)*

An overview of commonly used keywords is presented by observing the patterns of the highest keywords occurrences in the dataset is provided in this section. This conversation is important because it supports papers to be correlated and thus identified in current and past journal. To examine and recognize research gaps and directions, WoS produces complete records. We analyze the fashions in the highest keyword occurrences in the fetched data. We find anomaly detection and IoT to be most recurrently used in the author keywords category. However, detection and anomaly are the two highest frequent words in the title category which represent detection of anomaly is trending as these are being used regularly by the authors for writing. It can be also inferred that anomaly detection has been done by using different machine learning algorithms based on classification and build or proposed different frameworks for the same.

Figure 5 representing word cloud generated based on titles, which give us insight about the data that the words anomaly detection, Internet of things, data, and IoT are the more frequent words of this research.

4.5 *Thematic Analysis (RQ 5)*

This question aims to identify, by using thematic analysis, which research themes have attracted the attention of the researchers. From this question, we will identify which topics need further attention to be explored. The function thematicEvolution () performs on the basis of clustering and analysis of co-word network. The method

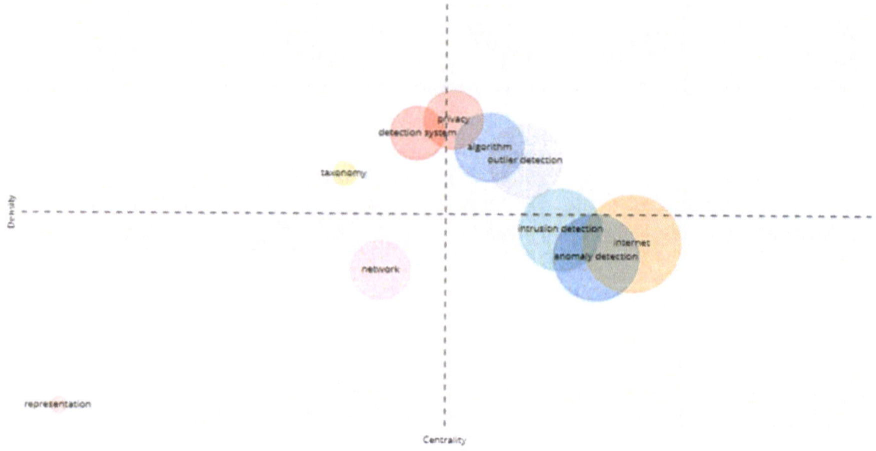

Fig. 6 Thematic map (2018–2020)

is based on Cobo et al. [29]. In Fig. 6, anomaly detection and intrusion detection are in the lower right quadrant and overlapped with the Internet which indicates these are strongly correlated and some part of these terms are in the upper right quadrant and adjacent to outlier detection which further overlapped with an algorithm to deduce that the linking between them. It can be concluded from this that the focus of research in the last few years became Internet-related things with anomaly detection.

5 Discussion and Conclusion

All the present work reveals a study of scientific publications based on the theme anomaly detection for IoT-enabled smart cities in the WoS database, in the temporal cut of 10 years (2011 to 2020). The evaluation of systematic production using bibliometric permits an assessment of the performance of scholars, organizations, continents, countries, and research documents (journal articles, conference papers, etc.) through a mapping of journal articles, technical and other systematic publications, applying quantitative and qualitative metrics.

In this research, 375 publications focusing on the theme anomaly detection for IoT-enabled smart cities were identified, considered a period that corresponds to an average of 67 papers published per year. The maximum number of papers on the subject was published in the year 2019, with 128 publications. All the publications identified in this study were analyzed based on knowledge areas, journals or venues where they were published, continents, and countries from where articles have been published, highly cited articles that included both local and global citations.

6 Future Work and Limitations

In the study, we have presented the trending and highly cited topics in anomaly detection for IoT-enabled smart cities from 2011 to 2020 and identifies the trends in this field, so future works can conduct bibliometric analysis periodically to comprehend dynamics in the studied area and present a comparative account with the result set of the current study. For future studies, this study suggests the use of other databases of research data and the use of new keywords related to the theme anomaly detection for IoT-enabled smart cities. We intend to work with the recognized research gaps in the future with the application of state-of-the-art technologies and empirical studies.

References

1. Mora L, Bolici R, Deakin M (2017) The first two decades of smart-city research: a bibliometric analysis the first two decades of smart-city research: a bibliometric. J Urban Technol 24(1):3–27
2. Zhao L, Tang Z, Zou X (2019) Mapping the knowledge domain of smart-city research: a bibliometric and scientometric analysis. Sustainability 11(6648):1–28
3. Guo Y, Huang Z, Guo J, Li H, Guo X (2019) Bibliometric analysis on smart cities research
4. Wamba SF, Queiroz M (2019) A bibliometric analysis and research agenda on smart cities a bibliometric analysis and research agenda on smart cities. Sustainability 9:0–12
5. Janik A, Ryszko A (2020) Scientific landscape of smart and sustainable cities literature: a bibliometric analysis
6. Mariana-Daniela G-Z, Abad-Segura E, Vázquez-Cano E, López-Meneses E (2020) IoT technology applications-based smart cities: research analysis. Electronics 9(1246):1–36
7. Alves D et al (2020) A bibliometric study about internet of things. Int J Adv Eng Res Sci 6495(4):213–220
8. Perez LM, Oltra-Badenes R, Gutierrez JVO, Gil-Gomez H (2020) A bibliometric diagnosis and analysis about smart cities. Sustainability 12(6357):1–43
9. Pritchard A (1969) Statistical bibliography or bibliometrics? J Documentation 25(4):348–349
10. Aghaei Chadegani A, et al (2013) A comparison between two main academic literature collections: web of science and scopus databases. Asian Soc Sci 9(5):18–26
11. Mongeon P, Paul-Hus A (2016) The journal coverage of web of science and scopus: a comparative analysis. Scientometrics 106(1):213–228
12. Fingerman S (2006) Web of science and scopus: current features and capabilities. Issues Sci Technol Librariansh 48(fall)
13. Amina S, Vera R, Dargahi T, Dehghantanha A (2019) A bibliometric analysis of botnet detection techniques. Handb Big Data IoT Secur 345–365
14. Merigó JM, Blanco-mesa F, Gil-lafuente AM, Yager RR, College I (2016) A bibliometric analysis of the first thirty years of the international journal of intelligent systems. In: IEEE symposium series on computational intelligence (SSCI), pp 1–6
15. Aria M, Cuccurullo C (2017) Bibliometrix: an R-tool for comprehensive science mapping analysis. J Informetr 11(4):959–975
16. Lachance C, Poirier S, Larivière V (2014) The kiss of death? The effect of being cited in a review on subsequent citations. J Assoc Inf Sci Technol 65(7):1501–1505
17. Peng TQ, Zhu JJH (2012) Where you publish matters most: a multilevel analysis of factors affecting citations of internet studies. J Am Soc Inf Sci Technol 63(9):1789–1803
18. Lynn FB (2014) Diffusing through disciplines: insiders, outsiders, and socially influenced citation behavior. Soc Forces 93(1):355–382

19. Sin S-CJ (2011) International coauthorship and citation impact: A bibliometric study of six LIS journals, 1980–2008. J Am Soc Inf Sci Technol 62(9):1770–1783
20. Swan M (2012) Sensor mania! the internet of things, wearable computing, objective metrics, and the quantified self 2.0. J Sens Actuator Netw 1(3):217–253
21. Karkouch A, Mousannif H, Al H, Noel T (2016) Data quality in internet of things: a state-of-the-art survey. J Netw Comput Appl 73:57–81
22. Wu D, Shi H, Wang H, Wang R, Fang H (2018) A feature-based learning system for internet of things applications. IEEE Internet Things J 6(2):1928–1937
23. Bawany ZN, Shamsi JA, Salah K (2017) DDoS attack detection and mitigation using SDN: methods, practices, and solutions. Arab J Sci Eng 42(2):425–441
24. Meidan Y, et al (2018) N-BaIoT-network-based detection of IoT botnet attacks using deep autoencoders. IEEE Pervasive Comput 9:12–22
25. Garcia-Font V, Garrigues C, Rifà-Pous H (2016) A comparative study of anomaly detection techniques for smart city wireless sensor networks. Sensors (Switzerland) 16(6)
26. Butun I, Kantarci B, Erol-kantarci M (2015) Anomaly detection and privacy preservation in cloud-centric internet of things. In: IEEE ICC 2015—workshop on security and privacy for internet of things and cyber-physical systems, pp 2610–2615
27. Ukil A, Bandyoapdhyay S, Puri C, Arpan P (2016) IoT healthcare analytics: the importance of anomaly detection. In: 2016 IEEE 30th international conference on advanced information networking and applications, pp 994–997
28. Lyu L, Member GS, Jin J, Rajasegarar S (2017) Fog-empowered anomaly detection in IoT using hyperellipsoidal clustering. IEEE Internet Things J 4(5):1174–1184
29. Cobo MJ, Herrera F (2011) An approach for detecting, quantifying, and visualizing the evolution of a research field: a practical application to the fuzzy sets theory field. J Informetr 5(1):146–166

Modelling and Performance Analysis Using Fuzzy Logic MPPT Controller for a Photovoltaic System Under Various Operating Conditions

Shafqat Nabi Mughal, Md. Sabir Hassan, Yog Raj Sood, and R. K. Jarial

Abstract One of the most widely used energy technologies is photovoltaics. However, because sun irradiation is irregular, getting the most electricity out of a photovoltaic system is difficult. Due to the nonlinear characteristics of power obtained from solar modules, it is difficult to retrieve maximum power at a single point. To obtain maximum power output at single point, there is a requirement of maximum power point tracking mechanism. The goal of this research is to create an accurate and efficient MPPT algorithm for peak power extraction based on artificial intelligence (AI). In this study, a photovoltaic system is modelled and simulated in MATLAB using perturb and observe and the proposed fuzzy logic based controller. It was observed as compared to P&O under the same operating conditions, the suggested controller is faster, more accurate with less oscillation around the maximum power, and has more peak power. The proposed approach gave results that were very close to the data sheets provided by manufacturers of solar modules.

Keywords Solar power · Photovoltaic · Renewable energy · Artificial intelligence · Fuzzy logic · DC-DC converter

1 Introduction

Fossil fuels are gradually depleting, but carbon emissions from these sources are resulting in environmental pollution and affecting climate to a large extent. The power sector is rapidly moving towards renewable energy sources, to meet the future

S. N. Mughal (✉)
Electrical Engineering Department, SoET, BGSB University Rajouri, Jammu and Kashmir, India
e-mail: snmughal@bgsbu.ac.in

S. N. Mughal · Y. R. Sood · R. K. Jarial
Electrical Engineering Department, NIT Hamirpur, Himachal Pradesh, Hamirpur, India
e-mail: jarial@nith.ac.in

Md. S. Hassan
Electrical and Electronics Engineering Department, Maulana Azad National Urdu University, Telangana, Hyderabad, India

A. Tomar et al. (eds.), *Proceedings of 3rd International Conference on Machine Learning, Advances in Computing, Renewable Energy and Communication*, Lecture Notes in Electrical Engineering 915, https://doi.org/10.1007/978-981-19-2828-4_30

energy demands as well as in reducing the carbon emissions [1–24]. Solar energy is commonly employed among various forms of renewable energy sources, due to its availability, clean nature, and dependability. One of the most extensively utilised technologies for converting solar energy to electricity is solar photovoltaic (SPV) [1]. The solar cells have nonlinear properties, and there is a single point at which maximum power is extracted under given solar irradiation and temperature [2]. This particular location is known as the maximum power point (MPP). The peak power of the SPV module is shown in Fig. 1 at various levels of irradiation. Similarly, Fig. 2 shows peak power at different temperatures.

To get the most out of an SPV system, it need to be operated at its most powerful setting [3, 4]. As a link between the SPV input and the load, a DC-DC converter creates this peak point. By adjusting the operating point of the DC-DC converter, the load impedance sensed by the source can be matched and adapted to maximise the power recovered from the PV panel [5]. This research uses MATLAB to model a PV system with a 213.15 Wp capacity and an upgraded fuzzy logic controller.

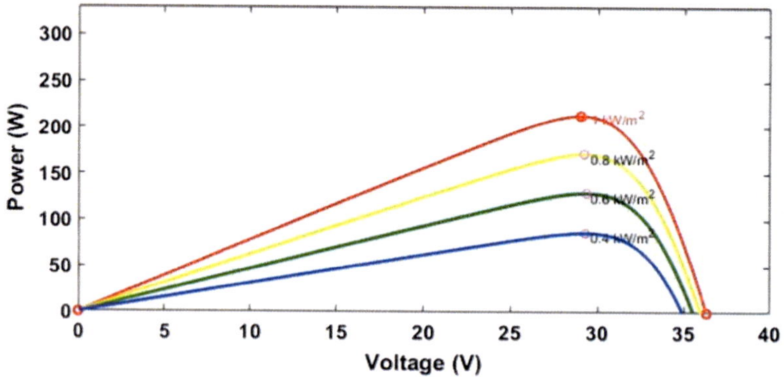

Fig. 1 PV characteristics curve (with different irradiation levels)

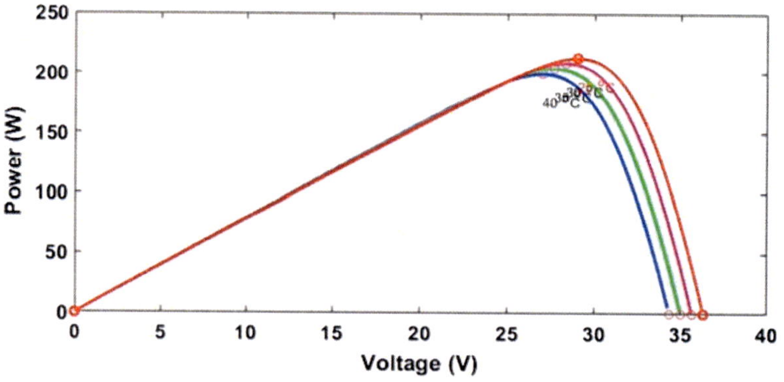

Fig. 2 PV characteristics curve (with different temperature levels)

This fuzzy logic controller is compared to the perturb and observe MPPT method (P&O). A DC-DC boost converter is being investigated in order to produce the most power possible. The simulation of a solar array, the MPPT converter, and the results comparison are all exhibited.

Many traditional algorithms are found in the literature which are used in the modelling of the MPPT systems. These methods are implemented in various models, and these are perturb and observe and modified perturb and observe. Apart from these, there are various other algorithms also, such as fractional short circuit current and fractional open circuit voltage which are extensively used as well.

More methods known as sliding control and incremental method are also documented in the literature. The most common method is P&O method. However, it suffers from problem that when solar irradiance is highly variable if faces difficulty in locating the maximum power point. Hence, in this research work, we are proposing an improved fuzzy logic controller MPPT technique to perform better and improve the system performance.

2 PV Cell Modelling and Converter

2.1 PV Cell Model

Solar PV cells are made up of semiconductor PN devices that convert light energy directly into electricity using the photovoltaic effect. The current control source is substituted with an antiparallel loss diode to produce an equivalent model of a cell, and non-ideality is demonstrated by the inclusion of parallel and series resistance, as seen in Fig. 3 [6].

The output current of a solar cell is mathematically given by following equation

$$I = I_\rho - I_o\left(\frac{q(V + R_s * I)}{e^{A.K.T} - 1}\right) - \frac{V + R_s * I}{R_\rho} \tag{1}$$

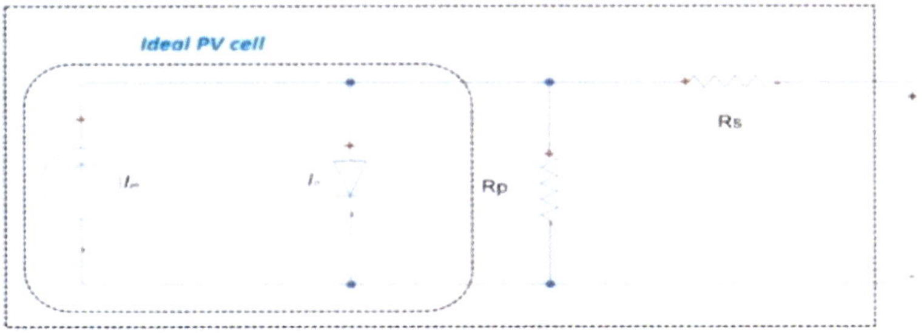

Fig. 3 Equivalent circuit of SPV cell

where

I_ρ is photon current (amperes)
I_o is saturation current (ohms)
R_s is resistance series (ohms)
R_ρ is resistance parallel (ohms)
A is the ideality factor (diode)
Q is charge (electron)
K is Boltzmann's constant
T is ambient temperature (kelvin)
V is output voltage for cell (volts)
I_I is output current for cell (amperes).

2.2 Design of DC-DC Converter

A DC-DC converter is utilised in solar photovoltaic systems for getting the power output from the solar modules. The use of MPPT techniques is utilised to adjust the converter's duty cycle for getting the maximum power. In this research, we have utilised DC-DC converter known as boost converter. Figure 4 depicts the DC-DC boost converter. It comprises of input source and output load.

The following formulas describe the relationship between the boost converter's output voltage and input voltage, as well as its output current and input current:

$$\frac{V_{\text{out}}}{V_{\text{in}}} = \frac{1}{1-d} \tag{2}$$

$$\frac{I_{\text{out}}}{I_{\text{in}}} = 1-d \tag{3}$$

As far as the selection of the inductor and capacitor value is concerned, it can be calculated by Brigitte.

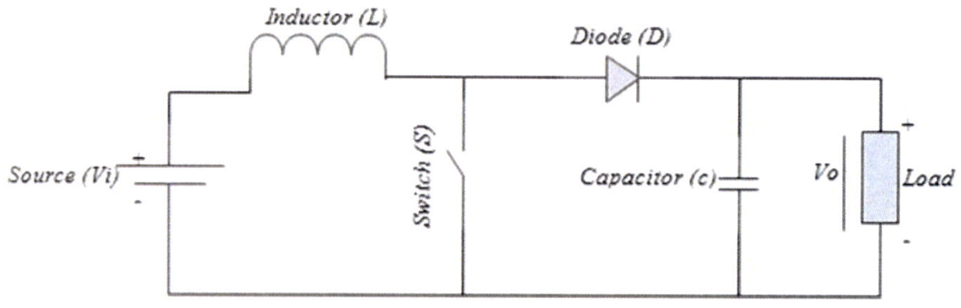

Fig. 4 Typical DC-DC converter

2.3 Inductor Value Selection

With the use of a boost converter, the value of an inductor can be calculated as follows using:

$$L = \frac{V_{in}}{f_s \Delta I_L} d \tag{4}$$

where

f_s switching frequency
ΔI_L input current ripple

The ratio of input to output current ripple factor (CRP). And for better assessment of inductor value, CRF is taken inductor 30%, i.e.

$$\frac{\Delta I_L}{I_{out}} = 0.3 \tag{5}$$

2.4 Capacitor Value Selection

With the use of a boost converter, the value of a capacitor can be estimated as follows using the following equation.

$$C = \frac{I_{out}}{f_s \Delta V_o} d \tag{6}$$

where

ΔV_o Output voltage ripple

And for better assessment of capacitor value, output voltage ripple is taken 5%, i.e.

$$\frac{\Delta V_o}{V_o} = 0.05 \tag{7}$$

Input voltage, output voltage, output current, and switching frequency are all taken into consideration when constructing the DC-DC boost converter: 25 V, 38 V, 5A, and 20 kHz, respectively. Also the value of the actual duty cycle is taken as 0.35.The components and their values for converter are given in Table 1

Table 1 Converter parameter

Name of the parameter	Value
Inductor	0.275 mH
Input capacitor	66 μF
Output capacitor	45 μF
Resistor	8 Ω

3 MPPT Mechanisms

In solar PV system, an MPPT is an essential part in which point is sustained where maximum power is available. Solar PV system has nonlinear characteristics so it has a unique point at which maximum power exist [7, 8]. But every instant peak power of solar PV system changes, due to its environmental condition and solar irradiation. So extraction of maximum power available from PV system utilises MPPT technique. Generally, MPPT technique depends on the solar PV system structure and measure the value of either voltage, current, or power [9, 10]. But in AI-based MPPT technique, there is no need of overall knowledge of the solar PV system and there is no limitation in sensing the parameters.

3.1 Perturb and Observe

It has a straightforward feedback design and only requires a little amount of hardware. The interface converter's duty cycle changes based on the solar panel's power output. When the panel's power is raised, the perturbation voltage increases in the same direction; otherwise, it increases in the other direction. The flowchart in Fig. 5 [11] makes it simple to understand. This strategy, on the other hand, has several drawbacks, which are detailed below:

- An oscillation occurs around the MPP at steady state due to which power loss occurs in the PV system.
- Rapidly change in atmospheric condition and solar irradiation it lost to track peak power.
- Peak power is far, then its tracking is slow if step cycle is small.

3.2 Fuzzy Logic Control-Based MPPT

Microprocessor-based FLC solves nonlinearity very easily and does not depend on mathematical equation and system design. It can be construct with the predict input and output. It can be operating with two or more input variable and output can be taken one or more variable.

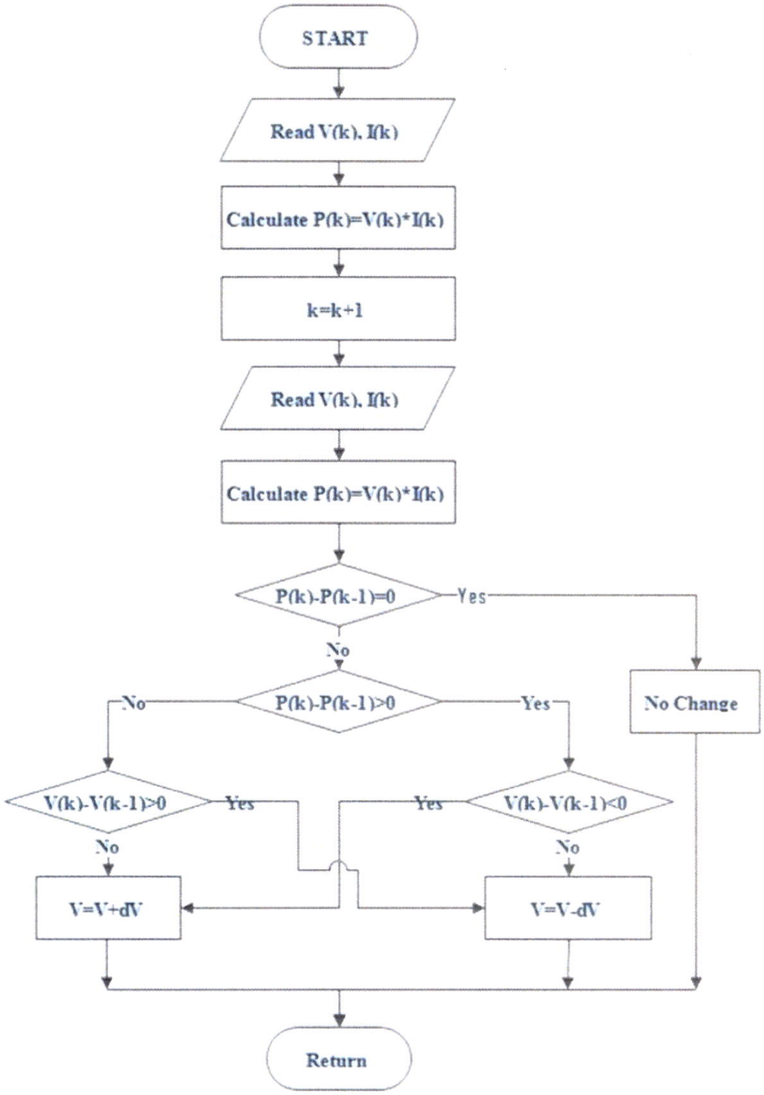

Fig. 5 Flow chart for P&O

In this work, change in power and voltage are taken input parameter. While, change in duty cycle is taken as output parameter.

$$\Delta V = V(t) - V(t - 1) \tag{8}$$

$$\Delta P = P(t) - P(t - 1) \tag{9}$$

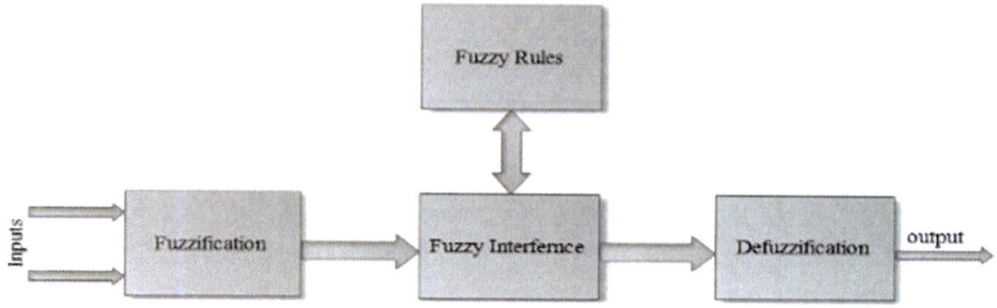

Fig. 6 Fuzzy logic controller

where P and V denote the PV array's power and voltage, respectively, and t denotes the instantaneous time.

There are three major stages of fuzzy logic controller. First stage is fuzzification in which real data is modified in to the fuzzy linguistic variable through membership function (MF). The resolution of the MF and fuzzy levels is used to achieve accuracy. Big positive (BP), medium positive (MP), small positive (SP), zero positive (ZP), zero (ZE), zero negative (ZN), small negative (SN), medium negative (MN), and big negative (BN) are the nine fuzzy levels employed in this research.

The second step is a fuzzy interface system that employs a knowledge-based reasoning mechanism. This component was created with MATLAB using the Mamdani method [12–18]. The third step is defuzzification, which is accomplished with the help of the centroid approach. When a linguistic output variable is defuzzed, it becomes a controller signal that may be used by a programme. Fuzzy inference system establishes IF THAN relationships between the input variable and the output signal [19–24]. With the fuzzy logic controller process depicted in Fig. 6, the flow chart can be seen in Fig. 7.

4 Simulation of Photovoltaic System in MATLAB

All simulation results using model in MATLAB. The simulated model parameter are shown in Table 2.

4.1 Generic Model of SPV System

The above chosen solar PV module has a maximum output power of 213.15 W under standard testing conditions (STC). However, its peak power is 189.4 W under normal operating circumstances (NOC). Table 2 lists the data and parameters for the module. The PV module model is meant to be a general-purpose tool for analysing any commercial PV module. Figure 8 illustrates a MATLAB/Simulink model of a

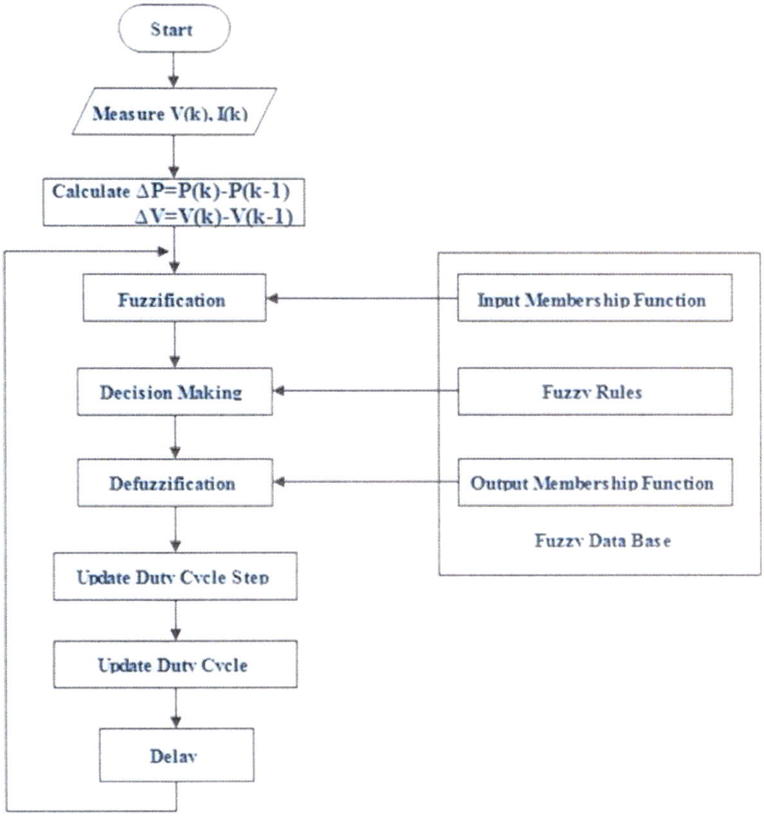

Fig. 7 Flow chart for FLC-based MPPT

Table 2 Module specification

Name of parameter	Parameter value
Max power	213.15 W
V_{oc}	36.3 V
V_{mp}	29 V
I_{sc}	7.84 A
I_{mp}	7.35 A
I_p	7.86 A
I_o	$2.926e^{-10}$ A
Diode ideality factor	0.98117
R_{shunt}	313.3991 Ω
R_{series}	0.39383 Ω

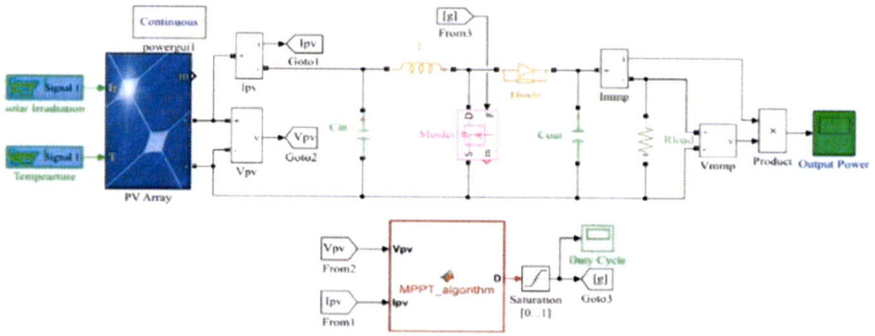

Fig. 8 Simulation model of the generalised PV system

generalised PV module based on mathematical equations. This model may be used to investigate the features of any commercial PV module by altering the variables in the subsystem dialogue box.

4.2 *Proposed Model Using FLC Method.*

Figure 9 shows a flow chart illustrating the suggested model's operation. Figure 10 shows the FLC MATLAB/Simulink model. In the generalised model, current and voltage are used as inputs for ports 1 and 2. For FLC, changes in power and voltage are used as inputs, whereas changes in duty cycle are considered outputs. This signal is sent to a pulse width modulator, which generates a duty cycle for the converter.

5 Result and Discussion

In a SPV system, electrical or physical parameters can be employed as inputs. Electrical parameters were employed as input in this model. Changes in power and voltage have been taken into consideration by FLC. Irradiation and temperature have been taken into account for the solar PV module. Any irradiation level and temperature can be used to construct the model. Irradiation and temperature were recorded over a period of 0–10 s. Temperatures are affected by the amount of irradiation. Figures 11 and 12 depict irradiation and temperature.

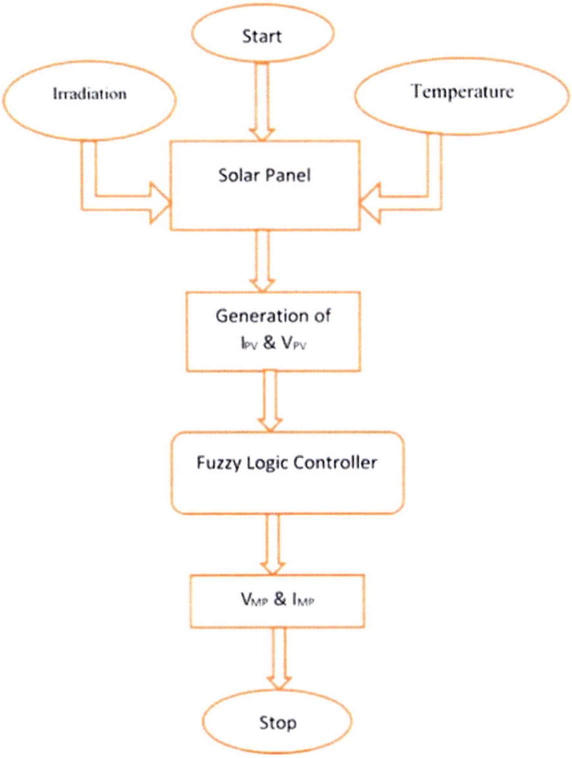

Fig. 9 Flow chart depicting the working of the proposed method

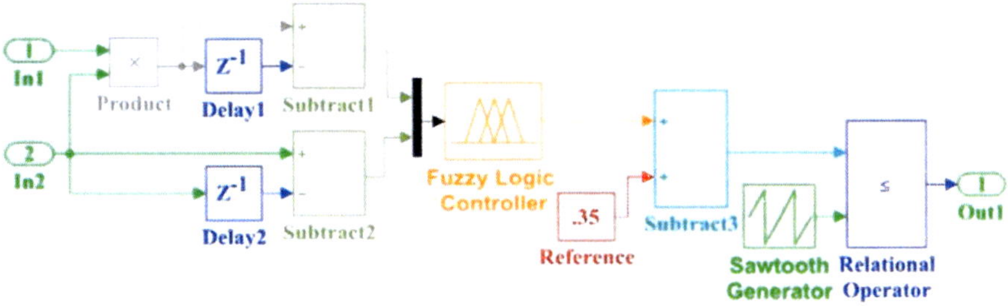

Fig. 10 Simulation model of the proposed FLC method

5.1 *Simulation Result (P&O Method)*

The results of the P&O approach are given in Fig. 13, and it shows that when the irradiation is continuous, gradual oscillation occurs at MPP. Changes in irradiation produce greater oscillation and require longer to reach MPP.

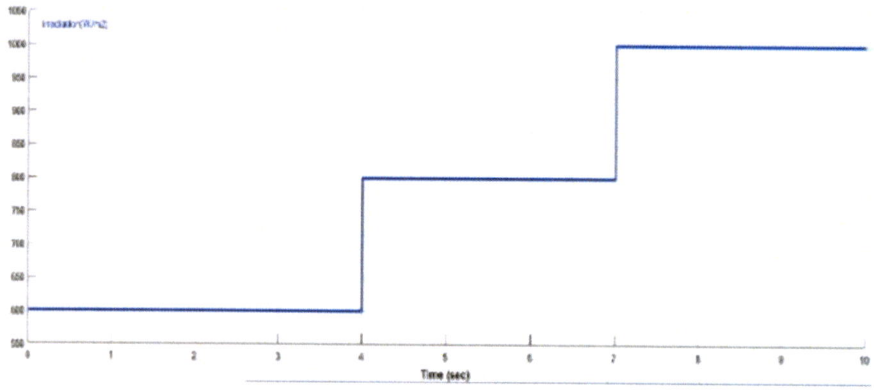

Fig. 11 Response at different irradiation levels

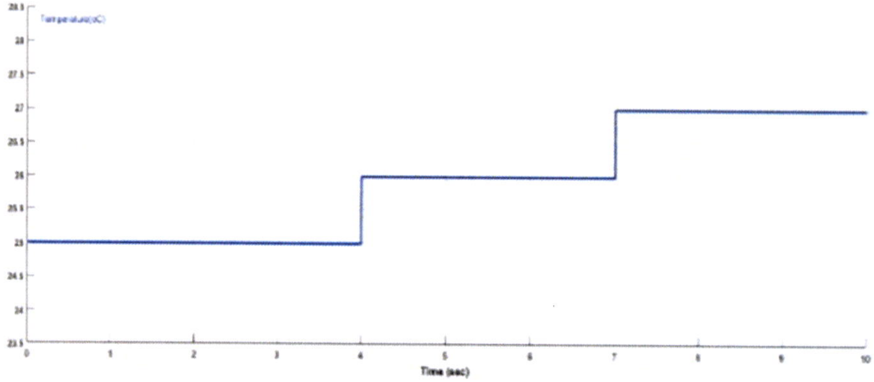

Fig. 12 Response at different temperature levels

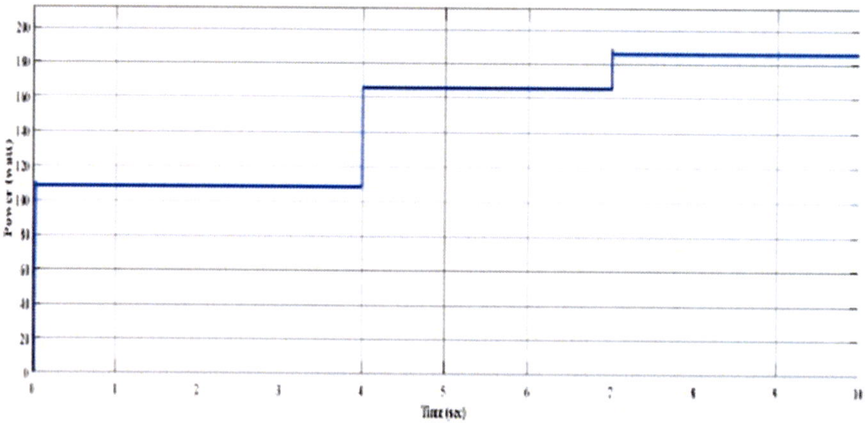

Fig. 13 Power versus time curve using P&O technique

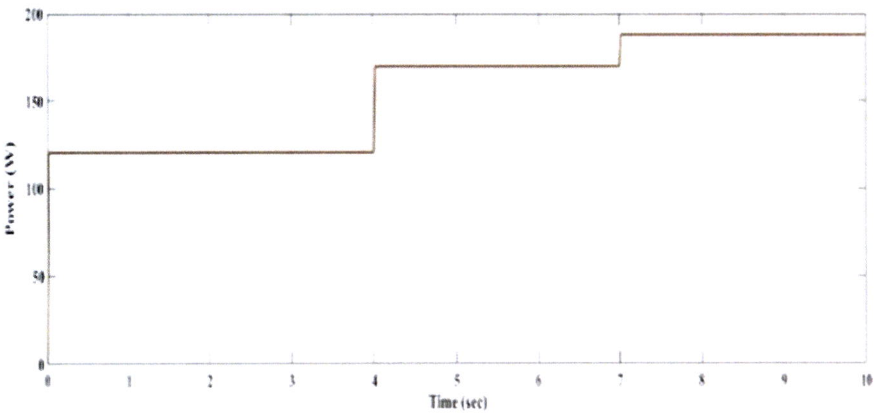

Fig. 14 Power versus time curve using the proposed FLC method

5.2 *Simulation Result (FLC Method)*

Figure 14 shows the power outcome of a PV module using the FLC technique, and it can be observed that it does not oscillate at MPP under constant irradiation, and that as the irradiation level increases, it hits peak power more rapidly and with less effort.

Peak power extract of P&O technique and proposed FLC technique result at different irradiation and temperature has depicted above, and comparisons are in Table 3.

When a conventional technique's performance is compared to an AI-based technique's performance under various radiation and temperature situations, it is evident that the proposed FLC technique is more cost effective. Compared to the standard method, the proposed strategy outperformed it in all areas: tracking speed, accuracy, implementation difficulty, dynamic responsiveness, and steady-state inaccuracy. An array of comparisons can be found in Table 4.

Table 3 Comparative analysis using P&O method w.r.t. the proposed FLC method

Irradiation and temperature	P&O technique			Proposed FLC-based technique		
	I_{max} (A)	V_{max} (V)	P_{max} (W)	I_{max} (A)	V_{max} (V)	P_{max} (W)
600 W/m^2 at 25 °C	4.661	23.30	108.6	4.74	25.4	120.5
800 W/m^{2}at 26 °C	5.760	28.80	165.9	6.00	28.3	169.8
1000 W/m^2 at 27 °C	6.099	30.49	186.0	6.52	28.8	187.9

Table 4 Parameters comparison

Name	P&O	FLC
Tracking speed	Slow	Fast
Implementation complexity	More	Less
Tracking accuracy	Less	High
Dynamic response	More oscillatory	Less oscillatory

6 Conclusion

There is evidence to suggest that the proposed method extracts more power than more traditional methods like P&O MPPT. If you compare this method to the P&O method, it is faster, more exact, and has less volatility around the maximum power point, making it more accurate. The power output from the proposed technique using a 213.15 Wp SPV system is also found close to the data sheets. These outcomes increase the system performance and efficiency. The proposed work has many future extensions as follows:

- By improving the membership function, the system efficiency can be increased.
- To extract peak power from SPV systems in accordance with the data sheet can be easily done.
- To further improve the system performance, new combinations of the proposed technique with new AI technique can be investigated.
- To make AI-based MPPT techniques more attractive among the users, new cost-effective measure need to be taken to reduce the total costs
- AI-based techniques are expert knowledge based and require professional manpower to operate solar energy systems. A graphical user interface (GUI) system might be presented.

References

1. Mughal SN, Sood YR, Jarial RK (2020) A proposal on techno-financial design aspects of photovoltaic system for the twin districts of Rajouri and Poonch (Jammu & Kashmir), EAI Endorsed Trans Energy Web 8(31):1–11. Online First. https://doi.org/10.4108/eai.13-7-2018.165521
2. Mughal SN, Sood YR, Jarial RK (2021) A neural network-based time-series model for predicting global solar radiations. IETE J Res https://doi.org/10.1080/03772063.2021.1934576
3. Hassan MS, Mughal SN, Jarial RK, Sood YR (2019) A comparative analysis of different maximum power point tracking algorithms of solar photovoltaic system, applications computing, automation and wireless systems in electrical engineering. Lecture notes in electrical engineering, vol 553. Singapore, pp 217–229
4. Mughal SN, Sood YR, Jarial RK (2019) Design and techno-financial analysis of solar photovoltaic plant for school of engineering and technology at BGSB university, Rajouri (J&K). Applications of computing, automation and wireless systems in electrical engineering. Lecture notes in electrical engineering, vol 553. Springer, Singapore, pp 231–243

5. Kumar G, Mughal S, Banoth S (Oct–Dec 2019) Design and development of an AC/DC micro-grid for isolated power system and grid connected system. Think India J 22(4):7031–7035 ISSN: 0971-1260
6. Arnett JC, Schaffer LA, Rumberg JP, Tolbert REL (1984) Design, installation and performance of the ARCO Solar one-megawatt power plant. In: 5th Photovoltaic solar energy conference, pp 314–320
7. Mughal S, Sood YR, Jarial RK (2018) A review on solar photovoltaic technology and future trends. Int J Sci Res Comput Sci Eng Inf Technol (IJSRCSEIT) 4(1)
8. Villalva MG, Gazoli JR, Ruppert Filho E (2009) Comprehensive approach to modeling and simulation of photovoltaic arrays. IEEE Trans Power Electron 24(5):1198–1208
9. Salas V, Olias E, Barrado A, Lazaro A (2006) Review of the maximum power point tracking algorithms for stand-alone photovoltaic systems. Solar Energy Mater Solar Cells 90(11):1555–1578
10. Schoeman JJ, Wyk JDV (1982) A simplified maximal power controller for terrestrial photovoltaic panel arrays. In: 1982 IEEE power electronics specialists conference. IEEE, pp 361–367
11. Koutroulis E, Kalaitzakis K, Voulgaris NC (2001) Development of a microcontroller-based, photovoltaic maximum power point tracking control system. IEEE Trans Power Electron 16(1):46–54
12. Kumar A, Sah B, Singh AR, Deng Y, He X, Kumar P, Bansal RC (2017) A review of multi criteria decision making (MCDM) towards sustainable renewable energy development. Renew Sustain Energy Rev 69:596–609
13. Wasynezuk O (1983) Dynamic behavior of a class of photovoltaic power systems. IEEE Trans Power Appar Syst 9:3031–3037
14. Hussein KH, Muta I, Hoshino T, Osakada M (1995) Maximum photovoltaic power tracking: an algorithm for rapidly changing atmospheric conditions. IEE Proc-Gener, Transm Distrib 142(1):59–64
15. Sahid MR, Yatim AHM, Taufik T (2010) A new AC-DC converter using bridgeless SEPIC. In: IECON 2010–36th annual conference on IEEE industrial electronics society. IEEE, pp 286–290
16. Iqbal A, et al (2021) Intelligent data-analytics for condition monitoring: smart grid applications. Elsevier, p 268. https://www.sciencedirect.com/book/9780323855105/intelligent-data-analytics-for-condition-monitoring. (ISBN: 978-0-323-85511-2)
17. Tomar A, et al (2020) Machine learning, advances in computing, renewable energy and communication, vol 768. Springer Nature, Berlin, LNEE, p 659. https://doi.org/10.1007/978-981-16-2354-7. (ISBN 978-981-16-2354-7)
18. Iqbal A, et al (2020) Renewable power for sustainable growth, vol 723. Springer Nature, Berlin, LNEE, p 805. https://doi.org/10.1007/978-981-33-4080-0. (ISBN 978-981-33-4082-4)
19. Ahmad MW, et al (2022) Intelligent data-analytics for power and energy systems, vol 802. Springer Nature, Berlin, LNEE, p 641. https://doi.org/10.1007/978-981-16-6081-8. (ISBN 978-981-16-6081-8)
20. Iqbal A, et al (2020) Soft computing in condition monitoring and diagnostics of electrical and mechanical systems. Springer Nature, Berlin, p 496. https://doi.org/10.1007/978-981-15-1532-3. (ISBN 978-981-15-1532-3)
21. Iqbal A, et al (2020) Meta heuristic and evolutionary computation: algorithms and applications. Springer Nature, Berlin, p 949. https://doi.org/10.1007/978-981-15-7571-6. (ISBN 978-981-15-7571-6)
22. Jafar A, et al (2021) AI and machine learning paradigms for health monitoring system: intelligent data analytics, vol 86. Springer Nature, Berlin, SBD, p 513. https://doi.org/10.1007/978-981-33-4412-9. ISBN 978-981-33-4412-9

23. Srivastava S, et al (2019) Applications of artificial intelligence techniques in engineering, SIGMA 2018, vol 1, 698. Springer Nature, AISC, p 643. https://doi.org/10.1007/978-981-13-1819-1. (ISBN 978-981-13-1818-4)
24. Srivastava S, et al (2019) Applications of artificial intelligence techniques in engineering, SIGMA 2018, vol 2, 697. Springer Nature, AISC, p 647. https://doi.org/10.1007/978-981-13-1822-1. (ISBN 978-981-13-1821-4)

An Optimum GA-Based Solution for Economic Load Dispatch for Clean Energy

Prashant, Anwar Shahzad Siddiqui, and Md. Sarwar

Abstract This article shows the optimal cost of the solar thermal generating cost. The negative environmental consequences of gaseous and particulate pollution generated by thermal plants can be minimized by properly allocating load among some of the plant's multiple producing units. The thermal generating units consist of thermal emission cost and fuel cost, while the cost of solar photovoltaic generation is a totally dependent on available solar radiation. This optimal cost of the system has to be minimized. In this paper, a GA-based multi-objective approach is suggested for solving the load dispatch problem in a cost-effective way. The existing methods involve more cost for both thermal and solar generating units. In order to overcome such issues, genetic algorithm is applied. It is observed that cost of total thermal generating units and solar generating units is found to be lesser with genetic algorithms in comparison with existing techniques like swarm optimization and evolutionary algorithm. The effectiveness of the topic is tested on IEEE 9 bus system using Simulink.

Keywords Emission cost · Fuel cost · Solar · Genetic algorithm

1 Introduction

Renewable energies have received much interest in recent times as a replacement for thermal production [1–17]. The rapidly dwindling stocks of fossil energy, the large increase in fuel costs and also the environmental impacts connected with thermal fuels all contributed towards this change in attitude [1]. The operational cost of conventional economic dispatch is minimized by properly allocating the quantity of electricity to be produced by multiple power sources. Economic dispatch (ED) is known optimization challenge that seeks to find the most cost-effective combination of energy sharing from contracted producing units. The ED issue has been solved using a variety of optimization approaches, according to the research [1]. Several

Prashant (✉) · A. S. Siddiqui · Md. Sarwar
Department of Electrical Engineering, Faculty of Engineering and Technology, Jamia Millia Islamia, New Delhi, India
e-mail: prashant.pacificcold@gmail.com

A. Tomar et al. (eds.), *Proceedings of 3rd International Conference on Machine Learning, Advances in Computing, Renewable Energy and Communication*, Lecture Notes in Electrical Engineering 915, https://doi.org/10.1007/978-981-19-2828-4_31

artificial intelligence-based optimization approaches have been presented in the field to handle the complicated ED challenge such as neural network [2], tabu search [3]and swarm optimization [4]. Thermal power plants emit pollutants that have a negative influence on the climate [5]. If the environmental impact of emissions is taken into account, the economic dispatch challenge becomes considerably more difficult. The combined emission economic dispatch issue is a multi-objective ED challenge that includes of both, i.e. cost of fuel plus emissions (CEED). PSO is used to solve the CEED issue in [6] underneath the restrictions of power balancing and electricity production capacity restrictions. By incorporating a price penalty element, the multi-objective CEED issue was transformed into a single-objective optimal function in [7]. The artificial bee colony technique was used to address the issue. The goal of economic dispatch model is to minimize two conflicting objective functions, fuel cost and emissions, while satisfying a number of equality and inequality constraints [8]. The dual-objective dispatch issue was successfully solved in [9] using a fuzzy satisfaction-maximum selection technique, but expanding the approach to incorporate more goals remains a major challenge. The researchers examined ED with significant renewable power penetration and assessed the effects of uncertainty including grid congestions on dispatch methods in [10]. Furthermore, [11] looked into the solar integrating power grid and provided an efficient PSO-based outcome to a more general CEED optimization issue with power balancing, generator limitations, renewables limits and ramp rate restrictions as constraints. Traditional search strategies demand exacting continuity; however, genetic algorithm optimization approach enables nonlinearity and discontinuities to emerge in the optimal solution. The pay-off knowledge from the objective function is used to establish optimum solution through the implementation of this method to the economic dispatch issue. In this work, GA-based multi-objective strategy has been proposed for solving load dispatch challenge in the economic manner. The suggested methodology has been tested on IEEE 9 bus system, and the results are compared with other techniques. The results analysis shows the superiority of the proposed methodology over other techniques.

2 Configuration of IEEE 9 Bus System

The six lines, nine buses and three generators make up the IEEE 9 bus system [12–14]. Figure 1 depicts the conventional architecture of the IEEE 9 bus system [15]. Three generators are connected to bus nos. 1, 2, 3, and three loads are connected to bus nos. 5, 7, 9 in a conventional IEEE 9 bus system. [15].

The internal parameters of the IEEE 9 bus system are given in [15].

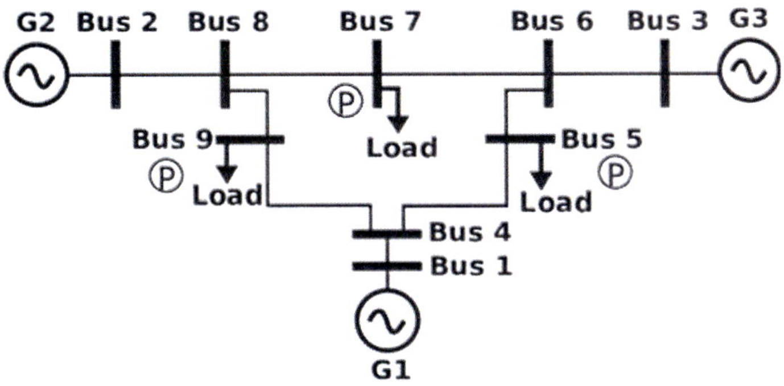

Fig. 1 Structure of IEEE 9 bus system [15]

3 Design of Optimal Economic Generation Using Genetic Algorithms (GAs)

The given system comprises conventional thermal generating units and solar units. The combined cost of the system will be sum of cost of both these generating units. It is divided into two sections as follows.

3.1 Optimal Generating Cost of Thermal Generating Units

Thermal generating units are composed of two types of cost: emission and fuel cost. The general expression of fuel cost is taken from [14] as shown in Eq. (1)

$$F(P_i) = \sum_{i=1}^{3} \left(a_i P_i^2 + b_i P_i + c_i\right) + e_i \mathrm{Sin}(d_i(P_i - P_{\max}))\$/h \quad (1)$$

where a_i, b_i, c_i, d_i and e_i are fuel coefficients of ith generating units represented in units of \$/W^{2}h, \$/W h, \$/h, 1/rad.MW and \$/h, P_i is actual power flow, and $P_{\max}$ is maximum power flow. The general expression of emission cost is also taken from [14] as shown in Eq. (2).

$$E(P_i) = \sum_{i=1}^{3} \left(h_i P_i^2 + j_i P_i + k_i + l_i e^{\delta_i P_i}\right)\$/h \quad (2)$$

where h_i, j_i, k_i, l_i, δ_i are the emission coefficients of the ith generating units represented \$/MW^{2}h, \$/MWh, \$/h,, \$/MWh.

The combined cost of the thermal generating units (OT) will be sum of Eqs. (1) and (2) as shown in Eq. (3)

$$\text{OP}(P_i) = \sum_{i=1}^{3} \left(h_i P_i^2 + j_i P_i + k_i l_1 e^{\delta_i P_i}\right) + \sum_{i=1}^{3} \left(a_i P_i^2 + b_i P_i + c_i\right) + e_i \sin(d_i (P_i - P_{\max})) \tag{3}$$

3.2 *Optimal Generating Cost of Solar Generating Units*

The general expression of solar generating units is taken from [14] as shown in Eq. (4)

$$P_{\text{gi}} = P_{\text{rated}} \frac{(1 + (T_{\text{ref}} - T_{\text{amb}})\alpha) S_j}{100} \tag{4}$$

wherein P_{rated} represents rated power, T_{ref} denotes reference temperature, T_{amb} depicts ambient temperature, α **denotes coefficient of temperature,** S_{j} represents irradiation of solar energy, and m represents solar generating units and the solar cost be Eq. (5)

$$\text{solar generation cost } = \sum_{j=1}^{m} \text{PU}_j P_{gi} \tag{5}$$

wherein PU_j denotes per unit cost of jth solar unit. The overall equation of solar cost is given in Eq. (6)

$$\text{OCS} = \sum_{j=1}^{m} \text{PU}_j P_{gj} + K \left(\sum_{j=1}^{m} P_{gj} - \sum_{j=1}^{m} \text{PU}_j P_{gj} \right) \tag{6}$$

3.3 *Optimal of Total Generating Cost by Genetic Algorithm*

The genetic algorithm mimics natural genetics' behaviour. There are three steps to it: reproduction, crossover and mutation. In the replication process, the best string is chosen from among the available strings for the next iteration's variable selection. Two strings are swapped during the crossover procedure to choose the best string.

Variables are transformed from 1 to 0 or vice versa in mutation. The inequality cost constraints are a_i, b_i, c_i, d_i, h_i, j_i, k_i, l_i, δ_i, K, α and e_i. The process of estimation of optimal cost of both generating units through genetic algorithm is illustrated through the flow chart as shown in Fig. 2.

This procedure starts with the selection of a binary string, which is considered as follows: population size: 5, length of the complete string: 5, crossover probability, $P_c = 0.9$ and mutation probability, $P_m = 0.06$. Let $x \in (a_i, b_i, c_i, d_i, h_i, j_i, k_i, l_i, \delta_i, K, \alpha, e_i)$. It can be realized that x belongs to ten variables. All these variables are substituted in Eq. (7) one by one.

$$x^r = x^{\min} + \frac{x^{\max} - x^{\min}}{2^l - 1} y^r \tag{7}$$

Fig. 2 Flow chart explaining the GA

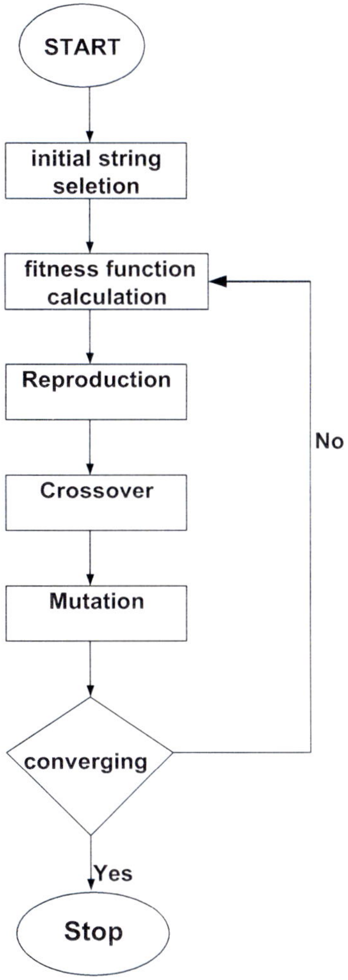

where r is iteration and y^r is binary string length.
Now use Eq. (8) for getting the new value of constraint variables

$$(x^r)_{\text{new}} = \frac{1}{\left(1 + \frac{\alpha' \varepsilon^r}{x^r}\right)} \quad (8)$$

α' is step size and which is equal to $0.5, \varepsilon^r = OC^r - OC^{ref}$.

4 Analysis of Cost Assessment

Three thermal generating units and three solar generating units are included in the system. The three thermal generating units and thirteen solar generating units will be installed in various locations throughout the IEEE 9 bus system. Estimating the true power loss of the entire system corresponding to the particular bus is used to determine where the thermal generating units should be placed on the IEEE 9 bus. The placement of a solar plant is calculated in the same way. In [15], the procedure of measuring the overall real power loss associated with a certain bus is described. The real power loss at each bus is given in Table 1, and the exact location of bus is given in Table 2 on the basis of minimum real power loss.

Table 2 shows that bus numbers 5, 6, 7 are the optimum locations for thermal producing units, while bus numbers 4, 8, 9 are the best locations for solar power plants in terms of real power loss. Once the locations of various thermal producing

Table 1 Real power loss at various locations of IEEE 9 bus

Line		Real power loss ($p.u$)
From bus	To bus	
1	4	0.45
4	5	0.36
5	6	0.35
3	6	0.31
6	7	0.32
7	8	0.48
8	2	0.45
8	9	0.42
9	4	0.37

Table 2 Positioning of thermal generating and solar power units at various buses

Type of plant	Buses Nos
Thermal generating unit placing	5,6,7
solar plant unit placing	4,8,9

Table 3 Cost coefficient values of thermal units

Unit	*ai*	*bi*	*ci*	*di*	*hi*	*ji*	*ki*	*li*	δ_i	*ei*
1	0.17	365	787	24	0.009	0.34	14.6	22	15	14
2	0.15	42.1	499	32	0.007	0.35	15.9	25	14	11
3	0.09	41.3	968	43	0.008	−10.44	41.3	19	16	12

units and solar power units have been determined, the cost of both generating units must be calculated.

4.1 *Optimum Cost of Thermal Generating Units*

The design aspects of optimal cost of thermal generating cost which consist of emission and fuel cost are already discussed in Sect. 3. From the Eq. 4, it is clear that there are ten cost coefficients and its values are given in Table 3.

Wherein the values of $a_i(\$/MW^2h)$ $b_i(\$/MWh)$, $c_i(\$/h)$ $h_i(\$/MW^2h)$, $j_i(\$/MWh)$ and $k_i(\$/h)$ are taken from [14]. The new variables are $d_i(\text{rad}/MW)$, $l_i(\$/h)$ *and* $\delta_i(1/MW)$. The minimum and maximum values of generating units in MW are 10, 10, 40, 35, 130,125 and 125, 150, 250, 210,325, 315.

4.2 *Optimum Cost of Solar Generating Units*

The design aspects of a solar power plant's optimal cost, which include emission and fuel costs, have already been examined in section III. Using Eq. 6 as a starting point, 0.5 is the value of α and 0.5 is the value of k. Table 4 lists the internal parameters of a solar power plant [14]. The parameters such as T_{amb}, irradiation and load demand are referred through [14] which are represented in Table 5.

The optimal generation cost provided by a solar power plant is calculated using the parameters given in Tables 4 and 5. After examining the available parameters and designing the system using the suggested approach, fuel cost, emission cost and overall optimal cost of the thermal generating unit are computed. In addition, the most cost-effective solar power is estimated. Table 6 shows the different cost analyses of both thermal and solar generation units using the proposed method for

Table 4 Parameters of solar units [14]

UUnit no.	Rated power (MW)	Unit rate ($/kwh)
4	20	0.22
8	25	0.23
9	25	0.23

Table 5 Irradiations, ambient temperature and load demand of solar generation units [14]

Time (am to pm)	Irradiation (W/m^2)	T_{amb} (degree Celsius)	Load demand (MW)
1	0	30	955
2	0	29	942
3	0	28	953
4	0	28	930
5	5.4	28	935
6	101	–	963
7	253	29	989
8	541	31	1023
9	530	33	1126
10	793	34	1150
11	1078	35	1201
12	1125	36	1235
13	1013	37	1190
14	848	37	1251
15	326	37	1263
16	654	38	1250
17	392	38	1220
18	215	37	1200
19	38	35	1160
20	0	34	1090
21	0	34	1020
22	0	33	980
23	0	32	975
24	0	–	960

a 24-h period. Also, the thermal cost of generating units with existing methods [11, 13, 14] is given in Table 7.

The estimated value of thermal generating cost by using genetic algorithm and its comparison with existing methods (Refs 14, 13, 11) is given in Table 7. In the similar manner, the estimated value of solar generating cost by using genetic algorithm and its comparison with existing methods (Refs 14, 13, 11) is given in Table 8. On the basis of above comparison shown from Tables 8 and 9, it is concluded that total thermal generating cost and solar generating cost are found to be lesser with genetic algorithm in comparison with existing methods.

Table 6 Costs assessment of thermal and solar generating units using genetic algorithm

Duration (am to pm)	Fuel cost($/h) × 10^4	Emission cost($/h) × 10^4	Total thermal generation cost($/h) × 10^4	Solar generation cost ($/h) × 10^4
12:00–01:00	13.4	11.1	24.5	6.02
01:00–02:00	17.7	11.9	29.6	6.23
02:00–03:00	5.8	16.9	22.7	5.75
03:00–04:00	7.7	11.2	18.9	5.99
04:00–05:00	17.7	13.7	31.4	5.66
05:00–06:00	13.5	10.2	23.7	6.15
06:00–07:00	4.8	8.2	13	6.31
07:00–08:00	3	9.2	12.2	5.91
08:00–09:00	5.5	11.2	16.7	6.03
09:00–10:00	4.7	4.4	9.1	5.95
10:00–11:00	3.5	3.8	7.3	6.32
11:00–12:00	4.5	4.5	9	6.14
12:00–13:00	3.8	4.1	7.9	5.95
13:00–14:00	3.6	4.2	7.8	6.02
14:00–15:00	4	3.8	7.8	5.75
15:00–16:00	8.4	15.7	24.1	5.83
16:00–17:00	10.3	13.7	24	5.65
17:00–18:00	20.3	14.7	35	5.95
18:00–19:00	16.1	16.7	32.8	6.13
19:00–20:00	7.4	9.9	17.3	6.24
20:00–21:00	5.6	9.5	15.1	6.32
21:00–22:00	8.1	10	18.1	5.89
22:00–23:00	7.3	9.6	16.9	5.94
23:00–24:00	6.1	11	17.1	6.05

5 Conclusion

Thermal power plants emit pollutants that have a negative influence on the environment. As a result, in addition to cost reduction, a significant effort has been made to cut emissions to a limit by integrating solar generating units. Thermal generating units have two costs: thermal emission and fuel, but solar generation has only one cost dependant on available solar irradiation. The system's ideal cost must be kept to minimum. For both thermal and solar generating units, the existing approaches are more expensive. The application of a genetic algorithm is employed to overcome such difficulties. In comparison with conventional methodologies, the cost of total thermal producing units and solar generating units is determined which comes out to be lower with genetic algorithms. The topic's effectiveness is evaluated using

Table 7 Total thermal generating cost comparison of proposed techniques with existing techniques

Duration (am to pm)	Genetic algorithm total thermal generation cost($/h) × 10^4	Ref [14] Total thermal generation cost($/h) × 10^4	Ref [13] Total thermal generation cost($/h) × 10^4	Ref [11] Total thermal generation cost($/h) × 10^4
12:00–01:00	24.5	25.3	26.7	26.2
01:00–02:00	29.6	31.4	32.8	32.3
02:00–03:00	22.7	25.5	26.9	26.4
03:00–04:00	18.9	22.7	24.1	23.6
04:00–05:00	31.4	36.2	37.6	37.1
05:00–06:00	23.7	29.5	30.9	30.4
06:00–07:00	13	19.8	21.2	20.7
07:00–08:00	12.2	20	21.4	20.9
08:00–09:00	16.7	25.5	26.9	26.4
09:00–10:00	9.1	18.9	20.3	19.8
10:00–11:00	7.3	18.1	19.5	19
11:00–12:00	9	20.8	22.2	21.7
12:00–13:00	7.9	20.7	22.1	21.6
13:00–14:00	7.8	21.6	23	22.5
14:00–15:00	7.8	22.6	24	23.5
15:00–16:00	24.1	39.9	41.3	40.8
16:00–17:00	24	40.8	42.2	41.7
17:00–18:00	35	52.8	54.2	53.7
18:00–19:00	32.8	51.6	53	52.5
19:00–20:00	17.3	37.1	38.5	38
20:00–21:00	15.1	35.9	37.3	36.8
21:00–22:00	18.1	39.9	41.3	40.8
22:00–23:00	16.9	39.7	41.1	40.6
23:00–24:00	17.1	40.9	42.3	41.8

the IEEE 9 bus system. Future research will focus on analysing the cost and emission challenges incorporating non-convex thermal generating plants, accounting for losses and implementing the suggested methodology on big electric power networks.

Table 8 Solar generation cost comparison of proposed techniques with existing techniques

Duration (am to pm)	Genetic algorithm solar generation cost($/h) × 10^4	Ref [14] total solar cost($/h) × 10^4	Ref [13] solar generation cost($/h) × 10^4	Ref [11] solar generation cost($/h) × 10^4
12:00–01:00	6.02	6.61	6.84	7.21
01:00–02:00	6.23	6.82	7.05	6.73
02:00–03:00	5.75	6.34	6.57	6.97
03:00–04:00	5.99	6.58	6.81	6.64
04:00–05:00	5.66	6.25	6.48	7.13
05:00–06:00	6.15	6.74	6.97	7.29
06:00–07:00	6.31	6.9	7.13	6.89
07:00–08:00	5.91	6.5	6.73	7.01
08:00–09:00	6.03	6.62	6.85	6.93
09:00–10:00	5.95	6.54	6.77	7.3
10:00–11:00	6.32	6.91	7.14	7.12
11:00–12:00	6.14	6.73	6.96	6.93
12:00–13:00	5.95	6.54	6.77	7
13:00–14:00	6.02	6.61	6.84	6.73
14:00–15:00	5.75	6.34	6.57	6.81
15:00–16:00	5.83	6.42	6.65	6.63
16:00–17:00	5.65	6.24	6.47	6.93
17:00–18:00	5.95	6.54	6.77	7.11
18:00–19:00	6.13	6.72	6.95	7.22
19:00–20:00	6.24	6.83	7.06	7.3
20:00–21:00	6.32	6.91	7.14	6.87
21:00–22:00	5.89	6.48	6.71	6.92
22:00–23:00	5.94	6.53	6.76	7.03
23:00–24:00	6.05	6.64	6.87	7.09

References

1. Cho HH, Strezov V (2021) Comparative analysis of the environmental impacts of Australian thermal power stations using direct emission data and GIS integrated method. Energy 231:120898. https://doi.org/10.1016/j.energy.2021.120898
2. Elsayed WT, El-Saadany EF (2015) A fully decentralized approach for solving the economic dispatch problem. IEEE Trans Power Syst 30(4):21792189. https://doi.org/10.1109/TPWRS.2014.2360369
3. Mantawy AH, Soliman SA, El-Hawary ME (2002) A new tabu search algorithm for the long-term hydro scheduling problem. In: LESCOPE'02. 2002 Large Engineering Systems Conference on Power Engineering. Conference Proceedings, pp 29–34. https://doi.org/10.1109/LESCPE.2002.1020663

4. Jeyakumar DN, Jayabarathi T, Raghunathan T (2006) Particle swarm optimization for various types of economic dispatch problems. Int J Elect Power Energy Syst 28(1):3642. https://doi.org/10.1016/j.ijepes.2005.09.004
5. Brown MT, Ulgiati S (2002) Emergy evaluations and environmental loading of electricity production systems. J. Cleaner Prod. 10(4):321334. https://doi.org/10.1016/S0959-6526(01)00043-9
6. Kumar AIS, Dhanushkodi K, Kumar JJ, Paul CKC (2003) Particle swarm optimization solution to emission and economic dispatch problem, In: Proceedings Conference on Convergent Technologies for Asia-Pacific Region (TENCON), vol. 1. pp 435439, October. doi:https://doi.org/10.1109/TENCON.2003.1273360
7. Sakthivel VP, Suman M, Sathya PD (2021) Combined economic and emission power dispatch problems through multi-objective squirrel search algorithm. Applied Soft Comp 100:106950. https://doi.org/10.1016/j.asoc.2020.106950
8. Abido MA (2003) Environmental/economic power dispatch using multiobjective evolutionary algorithms. IEEE Trans Power Syst 18(4):1529–1537. https://doi.org/10.1109/TPWRS.2003.818693
9. Huang C-M, Yang H-T, Huang C-L (1997) Bi-objective power dispatch using fuzzy satisfaction-maximizing decision approach. IEEE Trans Power Syst 12(4):1715–1721. https://doi.org/10.1109/59.627881
10. Zhang B, Rajagopal R, Tse D (2014) Network risk limiting dispatch: Optimal control and price of uncertainty. IEEE Trans Autom Control 59(9):24422456. https://doi.org/10.1109/TAC.2014.2325640
11. Maleki A, Pourfayaz F (2015) Optimal sizing of autonomous hybrid photovoltaic/wind/battery power system with LPSP technology by using evolutionary algorithms. Solar Energy 115:471–483. https://doi.org/10.1016/j.solener.2015.03.004.2015
12. Anderson PM, Fouad AA (1977) Power system control and stability. The Iowa State University Press, Ames, IA
13. Swetapadma A, Yadav A (2017) A novel decision tree regression based fault distance estimation scheme for transmission lines. IEEE Trans Power Delivery 32(1):234–245. https://doi.org/10.1109/TPWRD.2016.2598553
14. Khan NA, Awan AB, Mahmood A, Razzaq S, Zafar A, Sidhu GAS (2015) Combined emission economic dispatch of power system including solar photo voltaic generation'. Energy Convers Manag 92(1):82–91. https://doi.org/10.1016/j.enconman.2014.12.029
15. Moradi MH, Abedini M, Mahdi Hosseinian S (2016) A combination of evolutionary algorithm and game theory for optimal location and operation of DG from DG Owner Standpoints. IEEE Trans Smart Grid 7(2):608–616. https://doi.org/10.1109/TSG.2015.2422995
16. Tomar A, et al (2020) Machine learning, Advances in Computing, Renewable Energy and Communication, LNEE volume 768. Springer Nature, Berlin, 659 p. https://doi.org/10.1007/978-981-16-2354-7. ISBN 978-981-16-2354-7
17. Iqbal A, et al (2020) Renewable power for sustainable growth, LNEE volume 723. Springer Nature, Berlin, 805 p. https://doi.org/10.1007/978-981-33-4080-0. ISBN 978-981-33-4082-4

Wireless Power Transmission: A Review

Sachin Pachauri, Rajat Kumar, Abhinav Saxena, Ramashankar Yadav, Gaurav Verma, and Nirmal Kumar Agarwal

Abstract Wireless power transmission (WPT) is the vast area subject because this technology has the ability to enhance the upcoming future wireless power transmission will have great impact in the future because of its ability to conduct through air without interconnecting wires. In this paper, we will try to review the history and the technology and try to conclude its advantages and disadvantages by analyzing this technology.

Keywords WPT · MFT · Tesla · A4WP · AC

1 Introduction

In the modern world, electrical energy is the most common requirement for everyday life and we cannot imagine living without electricity [1–17]. As a result, it must be transported to the distribution lines. One of the major issues while this process is the loss of electrical power in the conductors. Since the demand is increasing losses are also increasing. As a result, it is critical to avoid losses and can be a cost-cutting option. Since the concept of wireless power transmission is not new [1]. Researchers came up with the idea, but have not been able to put it into practice yet. Wireless power transmission has been relevant to wireless battery charging to various

S. Pachauri · G. Verma
Department of Electrical Engineering, Rajkiya Engineering College, Mainpuri, Uttar Pradesh 205001, India

R. Kumar
Department of Electrical & Electronics Engineering, Vardhaman College of Engineering, Kacharam, Shamshabad, Hyderabad, Telangana 501218, India

A. Saxena · R. Yadav (✉) · N. K. Agarwal
Department of Electrical Engineering, JSS Academy of Technical Education, Noida, Uttar Pradesh 201301, India
e-mail: rsy15796@gmail.com

A. Saxena
e-mail: abhinaviitroorkee@gmail.com

A. Tomar et al. (eds.), *Proceedings of 3rd International Conference on Machine Learning, Advances in Computing, Renewable Energy and Communication*, Lecture Notes in Electrical Engineering 915, https://doi.org/10.1007/978-981-19-2828-4_32

electronic gadgets in the modern electronic technology. This technology can also be employed in situations where using conducting wires is dangerous or difficult. Many scientists attempted to conduct electricity without the need of a physical medium at first, each with their own methods and applications [2]. This article gives a brief overview of wireless networks, its benefits over wired networks, including security challenges that need to be addressed right now. The 802.11 design and numerous services given by it are next discussed, accompanied by the objective for performing the research. Following that, a statement of the problem highlighting the faults in the current IEEE 802.11i standard is offered [14]. Wireless communication network can enable users roam about and stay connected to the network without losing control of their position, resolving a variety of mobility challenges. The 802.11 architecture resembles that of an unit [15].

2 Wireless Power Transfer (WPT)

Inductive power transfer, also called wireless power transfer, can be utilized across short or significant distances without the utilization of wires. In comparison with other technologies, this one is quick and low maintenance. It enables electrical devices to charge without the use of physical wires [3].

In comparison with other wire sources, however, power loss owing to this technique is quite minimal. A wireless power system's main purpose is to allow gadgets to be charged indefinitely without the usage of physical connections. The three primary approaches for wireless power transfer are microwaves, resonance, and solar cells. Devices that transport electromagnetic energy from a source to a receiver use microwaves. Nikola Tesla developed the radio and is regarded as the "Father of Wireless" because he pioneered wireless power transfer and proved it in 1891. At the World Columbian Exposition in Chicago in 1893, he showed the illumination of vacuum lights without the utilization of links as an actual medium [4].

Tesla constructed tower for wireless power transmission of electrical power. The tower was known as warden cliff tower and is shown in Fig. 1.

Nikola Tesla was the founder of the AC electricity and continuously perform experiments on wireless power transmission. The thought came in the wake of realizing that earth is itself a conductor and can move charge through its surface. The examinations were not making power while they were simply transferring it. The ideas can be involved in the problem of energy crises. Since his experiments has potential to save the planet by creating power with its respective limitations (Fig. 2).

A wireless power transmission framework has extraordinary characteristics that nearly guarantee its significance as a part. The unique properties are:

- Between the wellspring of energy and the mark of utilization, no mass as wires is vital.
- At the speed of light, energy can be exchanged.
- The energy transfer direction can be quickly adjusted.

Fig. 1 Nicola Tesla tower

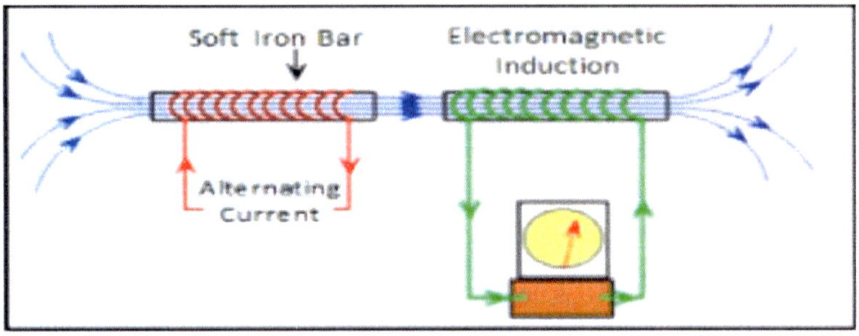

Fig. 2 Electric current induces in secondary coil

- At longer microwave wavelengths, no energy is lost in the vacuum of space, and only a small amount is lost in the Earth's atmosphere.
- Because of the operating at microwave frequencies, the mass of the power converters at the system terminals can be low.
- The difference in gravitational potential between two sites has no effect on energy transfer between them.

3 Techniques

The two techniques are discussed as:

3.1 Near-Field Technique

Near-field techniques are used to measure appliances that are close to the power source. The three types of electromagnetic radiation are discussed below. These methods can be used to solve problems caused by bad weather or security concerns [5].

- *Electromagnetic (EM) Radiation:* Emission by EM radiation is the transfer of energy from a power source's transmission antenna to a receiving antenna using radioactive EM waves. Omnidirectional and unidirectional radiations are the two sections of this two-part classification in terms of energy emission direction. Though omnidirectional radiation makes information exchange easier and more suitable, it has a severe efficiency problem in energy transfer, resulting in quick breakdown when the distance is vast [6].
- *Inductive Coupling:* The coupling of two LC circuits with the same resonance frequency is known as inductive coupling. Magnetic field induction, which is a natural aspect of current flow across wire, is used to power it. The essential and auxiliary loops in inductive coupling are two distinct coils. For the purposes of simplicity, convenience, and safety, they were all connected wirelessly. Without the need of cables, inductive coupling has long been a popular means of transporting power.
- *Magnetic Resonant Coupling*—Due to the mix of inductive coupling and reverberation, Kurs et al. fostered this innovation, which permits us to make extremely amazing contacts between two separate things. The magnetic field around the coil will also transfer energy to the electric field surrounding the capacitor. Because of alternating current in an essential curl that makes a changing attractive field that incites a voltage across the terminals of an auxiliary loop at the recipient, energy can be effectively communicated from a source curl to a collector loop with low energy misfortune. Compared to inductive coupling, this method has various advantages, including high efficiency, minimal radiation loss, and significantly larger range and directionality. The typical structure is shown in Fig. 3

3.2 Far- Field Technique[7]

Far-field approaches require line of sight and aim for great power transfer. Microwave power transmission and laser power transmission are the two kinds of power transfer [7].

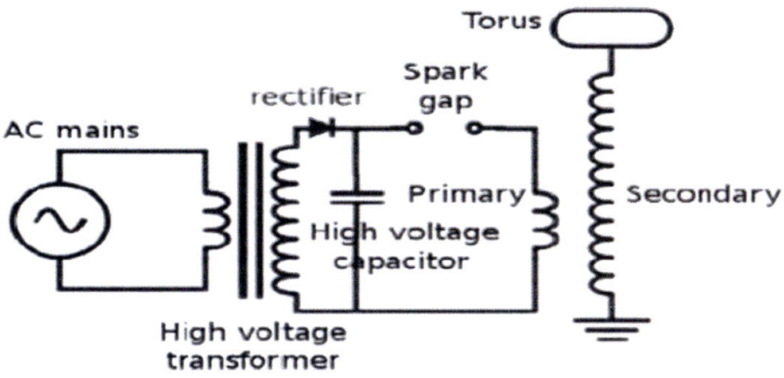

Fig. 3 Magnetic resonant coupling

Microwave Power Transmission (MPT): This system uses geosynchronous receiving and transmission satellites to allow items to obtain power from the base station through magnetrons. MPT enhances energy conversion efficiency, although focusing the beam in a tiny region can be problematic. In addition, this technology may easily travel through the atmosphere. The transformation of electrical energy into microwave energy, which is subsequently captured via a rectenna, is the first step in power transmission. As a result, AC must be converted to DC before being transformed to microwaves using a magnetron. The rectenna receives and transforms the transmitted waves back to AC.

Laser Power Transmission: This innovation is marginally unique in relation to MPT in that it utilizes a mirror to concentrate power in a restricted locale. This technology also generates high, coherent, and non-dispersed powers. Laser technology loses power as it passes through the atmosphere.

3.3 Tesla Method for Wireless Transfer [13]

This critical analysis describes contemporary magneto-inductive research activities on wireless power transfer with a transmission distance higher than the transmitter coil diameter, starting with Tesla's ideas of wireless power transfer a century ago. It condenses the operational principles of a variety of wireless power research into two categories: I maximum power transfer and (ii) greatest energy efficiency. In terms of energy efficiency and transmission distance capabilities, the distinctions and implications of these two approaches are explained. The discrepancies between the system and transmission energy efficiency are also noted.

Related topics like as human exposure and winding resistance reduction are also considered. According to the review, using the maximum energy efficiency concept in 2-coil systems is better for brief rather than semi applications, using the maximum output principle in 4-coil systems is satisfactory for maximizing transmission distance but has a low system energy efficiency (less than 50%), and using the peak power efficiency principle in relay may be a decent solution for good system energy efficiency and good transmission distance.

It is worth noting that wireless power transfer has been widely used in ac machines, which Tesla also pioneered [2]. Energy is transported from the stimulated stator windings over the air gap to the rotor cage in a cage induction machine, for example: in electric machines, the primary principle is energy transfer via coupled windings. As a result, electric circuit theory for magnetically coupled circuits may be used to quantitatively explain wireless power systems. Since the 1960s [3–7], wireless power transfer has become a hot research area for transversal energy systems for medical implants and induction heaters [8].

4 Recent Techniques

The use of resonance to improve the efficiency of wireless energy transfer in a range of applications has increased dramatically in recent years. Electronic firms are also working on the fundamental components of electronic gadgets in order to speed up the technology's adoption in niche applications. This will improve our ability to think creatively and deliver more reliable outcomes. Some of these game-changing apps have already been released, while others are still in the works. Automotive charging, for example, is a revolutionary idea that has yet to reach the market due to the requirement for uniformity in its charging infrastructure. A consortia firm, on the other hand, has already produced a game-changing update for traditional inductive charging in mobile electronics. SDOs are aiming to develop interoperability standards for highly resonant wireless power transfer in mobile devices, allowing multi-vendor goods to charge anywhere in a shared wireless environment [8]. All of this work is paving the way for a new wireless transmission technology trend that can be applied to a wide range of applications.

4.1 Qi Technology

Little inductors are being utilized in this strategy to convey power over higher frequencies while likewise taking into account a charging distance of a couple of centimeters. As a result, to avert a lack of a significant magnetic field, portable devices must be put perfectly on the dock. Qi parts can utilize various resonator exhibits to create a more prominent charging region in light of the fact that to its restricted charging region. Individual curls turned on, on one or the other hand, do not decrease the issue and surprisingly squander a ton of power. Users must line their gadgets exactly with the magnetic fields in order to establish a good enough connection as shown in Fig. 4

A limited communication protocol is included in the Qi standard to reduce the amount of power required by many coils. This technique could be utilized by the

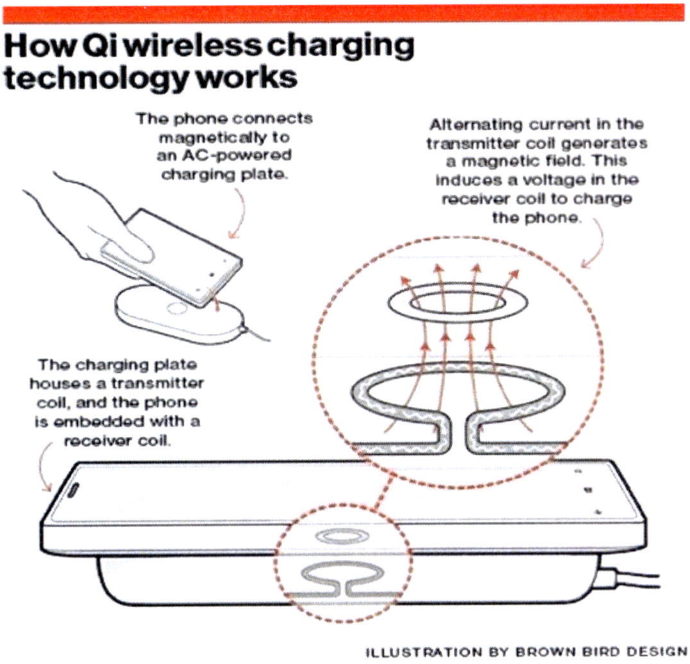

Fig. 4 How Qi technology works

getting gadget to let the charger know how frequently it need power and when it is completely energized [9].

4.2 Alliance for Wireless Power (A4WP) Technology

A4WP is a cutting-edge wireless power transfer convention that considers more proficient force conveyance to electronic gadgets. This depends on resonators that send and get reference power without the utilization of interconnecting links. This strategy permits a solitary transmitter to at the same time accuse numerous gadgets of contrasting force necessities. Since this technique utilizes a more extensive electromagnetic field rather than little inductor loops, gadgets could be charged without fixing up delightfully with the curl. Despite the fact that A4WP has yet to be commercialized, its presence enables electronic items to be charged in any configuration, including in the Z-axis. Additional feature of A4WP is that it enables chargers to be embedded in items while still allowing magnetic fields to emanate energy from the objects [10].

4.3 PMA Technology

Power Matters Alliance is another new technology (PMA). This is the organization that has been working with a gathering of examination bunch pioneers to assemble an effective power system in a worldwide, not-revenue driven area for battery-prepared devices that utilization remote charging advances PMA has developed at a quick speed since its foundation in 2012. Over 100 members from a variety of industries have only lately begun to engage in this new technological standard, including telecommunications, consumer products, automotive, retail, furniture, surfaces, and others. PMA's turn of events and achievement might be because of the association's remarkable way to deal with giving remote charging to where shoppers need it most, just as its individuals' persistent effort and commitment [11]. The overall complete analysis is shown in Table 1

5 Application of Wireless Power Transmission

Low-power devices with a rated power of less than 1 W, such as wireless sensors or other electronic mobile devices, and high-powered devices with a power range of more than 1 W in the industrial field are used in various applications. Direct wireless

Table 1 Comparison between Qi standard, PMA, and A4WP standard [12]

S. no	Recent technologies		
	Qi standard	PMA	A4WP standard
1	Induction charging through magnetism	Induction charging	Charging through magnetic resonance
2	Large magnetic field deficiency	Sufficient magnetic field	Large magnetic field
3	Charging distances few centimeters at most	Less than 10 cm	Much larger charging distances
4	Not easy to charge more than one device at a time	One device at time	Multiple devices can be charged from a single transmitter thanks to the design
5	Heat up the back of the devices	Heat up the devices	Do not heat up the devices
6	Uses small precise inductor coil	Small coil	Much larger coil
7	Need to line up perfectly within the coil	Line up with the coil	No need to line up
8	137-memberorganization with100products	More than 100-member organization	Very few companies working for the development

powering is defined as items like led lights in which the energy source is directly connected to the load and various charging techniques, such as battery or capacitor charge, are necessary. Wireless charging systems can be employed in two different ways [13–17].

5.1 Field of Electronics

Wireless charging systems are most commonly used in the sector of electronics, with wireless power sources installed behind the corkboard in electronic devices such as laptops. This device is capable of delivering more than 20 watts of power. It can also be charged up to 40 cm away from the wireless charging source. The resonators of the source and device are perpendicular to one another.

Due to the advantages of charging over distance and with spatial freedom, scientists predict that by 2020, highly resonant wireless power transfer will account for over 80% of all wireless charging systems. This technique can also be used to charge mobile gadgets like smart phones with enough computing power to be charged by a wireless charger. Wireless charging technology, similar to how other goods such as iPads or cameras may be charged at any time which is shown in Fig. 5, anyplace, even in public places (as seen in the picture), can be extremely beneficial to humans.

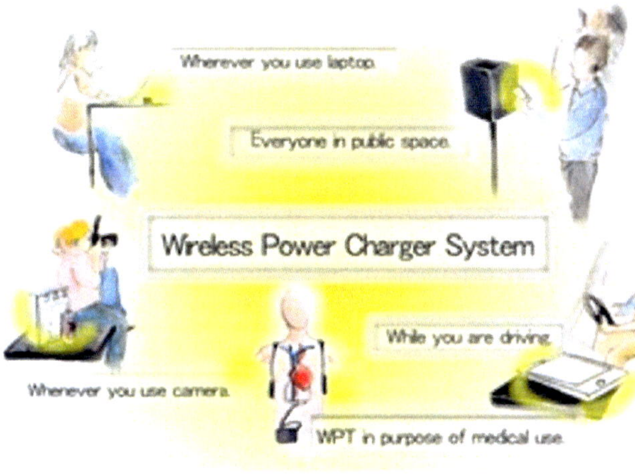

Fig. 5 Power charging system in our daily life

5.2 *Mobile Devices*

Wireless power move has helped embedded clinical gear, for example, LVAD heart help siphons, pacemakers, and implantation siphons. Utilizing this procedure, power might be productively conveyed to clinical gear immovably embedded inside the human body. It could likewise get rid of the requirement for drivelines that pass through the body and essential battery substitution systems.

5.3 *Defense Systems*

Wireless charging increases the reliability, aesthetics, and security of electronic gadgets in military systems, and designers are coming up with new concepts for future defense technologies. The remote charging capacities of the Talon tele-worked robot, for instance, permit it to be re-energized while being shipped by truck starting with one region then onto the next. One more utilization of protection frameworks is cap mounted hardware with night vision and radio gadgets that can be fueled remotely from a battery pack conveyed in the trooper's vest, eliminating the requirement for expendable batteries or a force line interfacing the cap to the vest-mounted battery pack. Several standards development groups and industrial alliances have been working on specifications and standards related to the use and monetization of wireless power in recent years.

5.4 *Solar Power Satellites (SPS)*

Because it uses massive solar arrays in geosynchronous Earth's orbit, it has the largest application of wireless power transmission. These satellites play a critical role in the transmission of microwaves as a source of power on the planet. Wireless sensors and RF power adaptive rectifying circuits are two more prominent applications of the wireless power source.

6 Aspects of Wireless Power Transmission

6.1 *Advantages*

- The wireless power transmission would eliminate the existing system of high-power cables and wires throughout the towers and stations.
- It could connect the generating stations on global scale.

- It has freedom to choose from large range of transmitters and receivers. Even the mobile transmitters and receivers can be used for WPT system.
- Power can be transferred to places to nowhere physical wires cannot reach.
- Since the loss in this system is minimum so the efficiency would be higher.
- Power failure due faults and power theft would not be possible in this method.

6.2 Disadvantage

- The initial capital cost would be very high for the implementation.
- It can have interference of microwaves with the current communication system.

6.3 Biological Impacts

People are generally afraid about microwave radiation, yet studies have shown that microwave radiation will never be higher than the dose received. It will be a tad greater than the emissions produced by cell phones. The power densities of ANSI/IEEE exposure requirements are met by cellular phones. This public display will be done in accordance with safety regulations.

7 Conclusion

The notion of WPT is examined, as well as modern innovations that have made human life more meaningful. New wireless power technology scopes are now available, and they are in fierce rivalry with one another. We will have to wait and see which of them dominates the other in the future. We will have to see who has the better applicability in the future among these. The A4WP norms, which have an enormous attractive field and charging distance, will be in front of different advancements, as per the examination Table 1. Other standards, such as Qi and PMA, have also been fast advancing. Many under-researched applications will benefit if this technology continues to improve at a quick pace.

References

1. Franceschetti G, Gervasio V (2012) Wireless power transmission. A new science is borne. IEEE, Kyoto, Japan
2. Shidujaman M, Samani H, Arif M (2014) Wireless power transmission trends. ICIEV, Dhaka, Bangladesh
3. Sample A, Smith JR, Experimental results with two wireless power transfer systems. Department of Electrical Engineering, University of Washington, WA 98195

4. Abd Aziz PD, Razak ALA, Bakar MIA, Aziz NA (2016) A study on wireless power transfer using Tesla Coil technique. Electrical Technology Section, University Kuala Lumpur—British Malaysian Institute Gombak, Malaysia
5. Carvalho NB, Georgiadis A, Costanzo A, Rogier H (2014) Wireless power transmission: R&D activities in Europe. Microwave Theory Tech 62(4), April
6. Brown WC (1996) The history of wireless power transmission. Microwave Power Transmission Systems, Perry Lane, Weston, Vol 56 No 1
7. Chen X, Shengbao YK (2017) Hybrid wireless power transfer. Instrumentation and Electrical Engineering, Jilin University, Changchun, China
8. Mou X, Sun H (2015) Wireless power transfer: Survey and roadmap. Department of Electrical and Electronics Engineering, Xiangdhan University, China, Vol 53, No 2, April
9. Shinohara N (2013) Wireless power transmission progress for electric vehicle in Japan. Research Institute for Sustainable Humanosphere, Kyoto University, Uji, Kyoto, Japan
10. Choudhary V, Singh SP, Kumar V, Prashar D (2011) Wireless power transmission: An innovative idea, Graphic Era University, UK, Vol 1, No 3
11. Khorashadi-Zadeh MP (2006) Correction of saturated current transformers secondary current using ANNs. IEEE Trans Power Delivery 21(1):73–79
12. Rim T, Mi C (2017) Wireless power transfer for electric vehicles and mobile devices. Wiley-IEEE Press
13. Hui SYR, Zhong W, Lee CK (2014) A critical review of recent progress in mid-range wireless power transfer. IEEE Trans Power Electron 29(9):4500–4511
14. Diwakar M, Singh A, Kumar P, Tiwari K, Bhushan S, Kaushik M (2021) Secure authentication in WLAN using modified four-way handshake protocol. In: Machine learning, Advances in computing, renewable energy and communication, pp 635–643, August
15. Diwakar M, Singh P, Kumar P, Tiwari K, Bhushan A (2021) A critical review on secure authentication in wireless network machine learning. In: Advances in computing, renewable energy and communication, pp 623–633, August
16. Iqbal A, et al (2021) Intelligent data-analytics for condition monitoring: smart grid applications. Elsevier, 268 p. https://www.sciencedirect.com/book/9780323855105/intelligent-data-analytics-for-condition-monitoring. ISBN: 978-0-323-85511-2
17. Tomar A, et al (2020) Machine learning, advances in computing, renewable energy and communication, vol 768. Springer Nature, Berlin, LNEE, 659 p. https://doi.org/10.1007/978-981-16-2354-7. ISBN: 978-981-16-2354-7

A Comprehensive Performance Evaluation of Novel Big Data Log Analytic Framework

Kiran Deshpande and Madhuri Rao

Abstract Every organization's top priority is to safeguard its vital assets, as it faces several external threats that might impair end-user service, resulting in financial and reputational disasters. By improving knowledge and comprehension of the information recovered from logs, log analysis may assist in uncovering a gap in protection systems, which will in turn aid in understanding the vulnerabilities and gaps in IT infrastructure security. Continuously produced, diverse in format, and massive log data must be analysed concurrently in an organization highly dependent on IT services in order to achieve real-time security event detection. To overcome these challenges, a considerable research is being conducted. Research has reached the stage where Big Data technology can be employed to overcome with the challenges in log analytics. By integrating functionalities of open source platforms and services at every step of Big Data log analytic, framework that offers the core components for analysing large scale log and network traffic data can be developed. In this research, we offer a method for conceptualizing and developing a real-time log acquisition, analysis, visualization, and correlation setup for tracking and identifying major security occurrences that might result in security vulnerabilities. With the goal of determining an in depth correlation of identified security events, we also fed stream processed logs into the batch processing framework. Finally, we compared the performance of the novel framework developed for Big Data log analytic with well-known open source and commercial log analysis platforms and found that novel framework outperformed in terms of real-time search and analytic criteria in consideration.

Keywords Big Data log analytic · Performance evaluation · Open source frameworks

K. Deshpande (✉) · M. Rao
Thadomal Shahani Engineering College, University of Mumbai, Maharashtra, India
e-mail: deshpande.82@gmail.com

A. Tomar et al. (eds.), *Proceedings of 3rd International Conference on Machine Learning, Advances in Computing, Renewable Energy and Communication*, Lecture Notes in Electrical Engineering 915, https://doi.org/10.1007/978-981-19-2828-4_33

1 Introduction

In recent years, research community is seen focussed towards development of scalable data processing systems with different characteristics and capacities as an impact of the rising need for applying analytics to real-world problems [1]. In today's Big Data revolution, organizations will benefit more from a platform that can handle both Big Data computing paradigms, i.e. batch and stream processing [2, 3]. Stream-based applications can be monitored in real time through stream processing to discover unusual and hazardous activities. Further, using offline analysis, the incident may be further categorized, correlated, and appropriate actions can be taken [4]. For designing applications that employ both real-time and offline analytics, a modelling framework that offers both batch and stream processing is essential [2]. Such a system would be able to address the demands of next-generation Big Data log analytics in a major way.

Log generation is a constant process that generates massive volumes of logs in various forms and speeds that may be analysed to acquire valuable insights. In the light of the present Big Data revolution and the relevance of log analytics, several open-sourced and proprietary systems for performing log collection, storage, discovery, evaluation, and visualization have been created. Considering the benefits of Apache Spark and ELK Stack for log processing, combining Spark and ELK Stack abilities with open source platforms and services helpful at different steps of log analytics, can prove beneficial for constructing a sustainable log monitoring and analysis framework.

A lot of research seen in recent literature is aimed at dealing with real-time log analytic challenges. Stream and batch log analysis may play a crucial role in supporting administrators in maintaining track of multiple security incidents and taking preventative steps, taking into consideration research focus on log analysis. Any organization's activities are greatly influenced by the Internet. Analysing network activity is crucial for detecting unusual and harmful occurrences that might compromise an organization's security and credibility. Log data is usually massive and ever-increasing. In order to identify in depth linkage, forensic investigation of events requires the use of both real-time and historical events. For real-time log ingestion, parsing, storing, retrieving, and searching of such log data, trending advanced analytics platforms and services might be effective.

In this research, we present a consolidated heterogeneous log analytics system that incorporates open source platforms and services. Furthermore, development of such system might serve as a reference for integrating features of stream and batch processing capabilities for scalable log analytics looking at the present need for Big Data log analytic. By combining the capabilities of all of the platforms listed, the framework seeks to deliver a system that integrates both computing paradigms while also enabling log exploration and analysis in real time in order to fulfil contemporary Big Data log analytic requirements. Because the ingested logs have already been preprocessed and indexed through the stream processing approach put in place through Kafka and the ELK Stack in this research, indexed logs could be used directly for batch processing executed through Spark without the need for any additional ETL processing, resulting in cost savings in terms of time and efforts. Furthermore, with the approach used in implementation since batch processed logs can move to stream processing framework graphical visualizations of batch processed logs are possible through visualization tool used for stream processed logs, which eliminates the need of an independent visualization tool for batch processed log data.

The rest of contents organized as follows. Section 1 depicts introduce idea and background in brief. Section 2 presents related research. The workflow of implementation architecture using FOSS tools is covered in Sect. 3. The newly proposed framework's performance is assessed in Sect. 4. The concluding portion, Sect. 5, covers the research findings.

2 Related Work

This study explores the many approaches and procedures created by the research and industrial groups to address the problems of next-generation Big Data log analytics [1–30]. In addition, we provided a review of related research on the current state of acceptance and maturity of open source platforms and services used for resolving Big Data log analytic challenges in recent years. Through this study, we also aimed to emphasize various methodologies from investigators on the integration of stream and batch analysis.

2.1 Big Data Analysis: Conceptual Understating and Recent Work in Integrating Its Computing Paradigms

Investigation into analysis methodologies and technology related to Big Data is now becoming extremely popular, keeping eye on gaining value from massive information available [1–30]. The goal of the research studies [3, 6] is to look at the concept, architecture, and exemplary Big Data computing systems. This emphasizes on the necessity for composite processing systems as well as the conceptual depiction and comparative analysis of batch and stream data processing systems. The research [2] presents a unified computing architecture for stream and batch analysis. It investigates the building of a layer between MapReduce workloads and the streaming platform in order to merge the advantages of stream processing with the coding convenience and acquaintance with MapReduce batch processing. Stream analytics is an emerging topic for researchers that focuses on significant concerns such as scaling, interoperability, high availability, immediacy, variability, and optimized task scheduling, as mentioned in the pioneering work [25]. The researchers of the study [28] highlighted issues and opportunities in presenting Big Data as well as discussed an unique strategy for charting students' learning activities on social network systems.

2.2 *Comparative Assessment of Big Data Log Analytic Frameworks*

The article [13] examines the merits and drawbacks of stream analytics frameworks for Internet of things applications. The researchers of the seminal article [7] evaluated and contrasted Spark and ELK Stack, two prominent frameworks for analysing log data. Through their work authors performed experiments on different problem solutions with different complexity to evaluate impact on non-functional features, like processing time and resource usage differ between them. The seminal paper [17] highlighted the usage of open source platforms, demonstrating their practicality by comparing log analysis performance between commercial and open source solutions. The benefits of open-sourced alternatives over propriety security log analysis solutions with high cost, sophistication, and resource needs have been highlighted by S. J. Son et al. [17]. Due to Big Data characteristics such as volume, variability, and speed, conventional detection algorithms have a difficult time in identifying fraudulent events. The application of Big Data analysis for threat intelligence has been illustrated by R. More et al. [12].

3 Survey Regarding Status of Adoption and Maturity of Open Source Platforms into Big Data Log Analytics

In [27], authors discussed emerging trends in Big Data analytics to handle Big Data accurately. The key article [18] discusses batch analysis through Spark for analysing traffic patterns and detecting risks for of Squid logs to generate web traffic statistics such as top destinations viewed and top users. B. H. Park, S. Hukerikar et al. [26] propose a log analysis solution that combines distributed non-relational database technology with Spark to provide precise findings for both end users and administrators. Parallel computing platforms like Spark and Hadoop can increase log analytics performance through using data parallelism. Therdphapiyanak et al. [19] used Hadoop to showcase scalable log analysis for effectively reporting suspicious traffic from massive logs. Authors advocated [13, 14] using ELK Stack to simply and swiftly manage Big Data technical challenges. Liu, JC et al. developed an ELK Stack-based cyberattack detection strategy related to network log visualization and analysis [8]. The purpose of this article's network log management and analytics system is to develop capabilities for classifying, evaluating, and displaying network log data to be processed. Software development teams are already using Docker Container innovation to provide software portability as well as a quicker software delivery cycle and better operational performance. L. Chen et al. developed a Docker Container log analytics system using the ELK Stack, the lightweight Filebeat for log collection, and Kafka in their study [9]. Sanjappa and Ahmed presented a platform that aggregates [16] heterogeneous log data through Logstash for discovering malicious network activity. The author demonstrated how to leverage the entire ELK ecosystem for effective log analytics and easy-to-understand insights [16, 17]. YuanTing Wang et al. reported research targeted at constructing a monitoring system that uses the ELK Stack to address problems with weak or disappearing Wi-Fi connections [22]. Because of its abilities for automatically collecting, classifying, consolidating, and displaying log data, the ELK Stack may play a vital role in constructing robust log analysis systems for heterogeneous logs. Prakash et al. demonstrated their work in which the ELK Stack is utilized for effectively geo locate website visitor activity using produced logs [20]. The use of log analysis to address IT infrastructure security challenges is widespread in the literature. In [10], B. Debnath et al. have demonstrated implementation where machine learning was utilized to detect patterns in application logs, which were then combined with real-time log processing to create advanced log analysis applications. Data mining techniques for providing a strategy that facilitates widespread usage of logs for system management with assured reliability is suggested by P. He [11]. Metha et al. devised a framework utilizing ELK Stack, Apache Spark, and Apache Hadoop to identify anomalies in network interface log records [21]. Few researchers tried integrating Elasticsearch and Spark separately in their architecture and developed a strategy for faster log analytics using ELK Stack and Apache Spark.

4 Implementation Architecture

The envisioned framework's operational architecture is depicted in Fig. 1. This section covers the intended framework's operating process is in place in depth. Flume is used for the log ingestion. Being source, it aggregates real-time logs from internal network, peripheral firewall, and application servers, and send collected logs to Kafka as a destination. Kafka creates topics on logs collected and use java objects to handle same. Kafka makes use of Zookeeper at the backend. To clone, restore, and maintain these topics, Kafka integrates Zookeeper at backend. Flume delivers log data to Kafka producer, while Kafka consumer feeds the logs to Logstash for further processing. Logstash is responsible for processing and classifying Apache Kafka's unprocessed

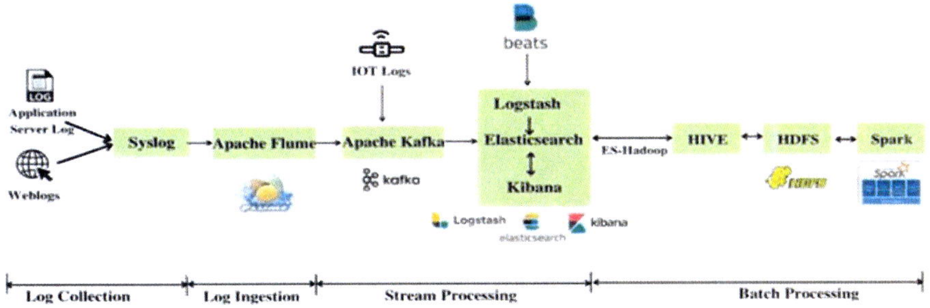

Fig. 1 Implementation architecture

logs. It parses the data by creating attribute names from topics retrieved from the Kafka consumer. When building an Elasticsearch index, these attribute names come in handy. Simultaneously, Logstash screens content using grok, a filter plugin that finds certain patterns and eliminates stuff that does not fit in that pattern. In Elasticsearch, this makes data organization and index construction easier. Index mapping created through timestamps may be employed on log data to provide visualization through Kibana. Visualizations based on Elasticsearch indices let people understand real-time log data in terms of graphical analysis. Packetbeat and Metricbeat are in charge of supervising the performance of IT assets, services, and Intranet in real time by sending metrics related to ELK Stack [5].

To transport tagged logs to HDFS, the ES-Hadoop connector is used by Elasticsearch. Hive is Hadoop's user interface. For data administration, Hive features a SQL-like interface. The MapReduce technique is used by the Hadoop architecture to perform batch analysis through HDFS. Because of the slowness of file system read–write operations, MapReduce has processing limits. Spark provides better approach to MapReduce latency during batch processing in this scenario. Spark-scala user interface is used to store the processed logs in HDFS. These batch-analysed logs would be sent to Elasticsearch through the ES-Hadoop and Hive. Finally, Elasticsearch can use Kibana to provide the data visualizations needed for comprehensive, low-latency analysis of batch-analysed historical log information.

5 Performance Evaluation of Framework

This part compares the proposed framework's performance in analysing voluminous logs created through experimental environment created, with open source log analysis platforms such as Graylog and Nagios, as well as commercial product such as Splunk. To evaluate performance, various log sizes were analysed. Keeping in mind the log volume sizes given previously and the queries specified below, the time necessary to find and analyse a certain log data pattern from labelled and categorized logs is documented. The proposed framework outperforms for queries I and II over a range of log volume sizes when compared to commercial and open source log analysis platforms. The novel framework performs remarkably well for query III and IV for log volumes of 5 Million and 10 Million. However, for larger log sizes such as 15 million and 20 million, the innovative framework performs equally well when compared to log analysis platforms under consideration. Adding dynamic horizontal scalability to novel framework core components can improve its performance even further over large log sizes.

1. Query I: Analysing Apache Logs to visualize Number of Unique Daily Hosts per hour over day.
2. Query II: Analysing Apache logs to visualize 404 Errors per Day over week period.

3. Query III: Analysing Apache Logs to visualize the Top Twenty 404 Response Code Hosts over month period.
4. Query IV: Analysing Apache Logs to visualize the Top Twenty 404 Response Code Endpoints over month period.

Figure 2 depicts a graphical analysis of the numbers reported during performance evaluation with respect to Query I. It has been noticed that novel framework outperforms conventional log analysis tools such as Graylog, Splunk, and Nagios. As far as Query I is concerned, for larger log sizes such as 15 Million and 20 Million, log search and analytic time increase in proportion to log size used for other tools, while for novel framework search and analytic time is nearly constant for log sizes 15 Million and 20 Million.

Figure 3 displays a graphical breakdown of the data reported during performance evaluation with respect to Query II. In terms of query II, novel framework surpasses competing commercial and open source tools across all log volumes. Another item to consider is the performance of Nagios in comparison with Graylog and Splunk. Nagios outperforms Graylog and Splunk for log size 20 Million.

Figure 4 is a graphical representation of the data obtained during performance evaluation with respect to Query III. In terms of query III, novel framework outperforms commercial and open source technologies for log sizes of 5 and 10 Million. However, for larger log volume sizes, novel framework performance does not match that exhibited in Query I and II. This could be because if log ingestion exceeds a certain size, Elasticsearch becomes overburdened while indexing. Elasticsearch performance may be influenced by simultaneous search queries because it serves

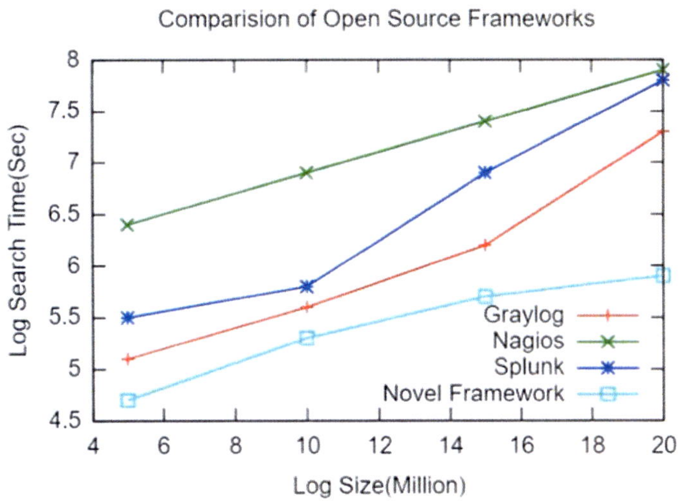

Fig. 2 Performance analysis of frameworks for query I

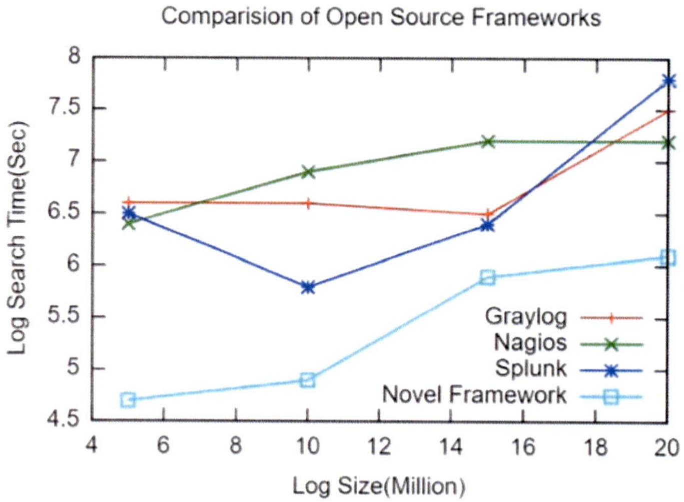

Fig. 3 Performance analysis of frameworks for query II

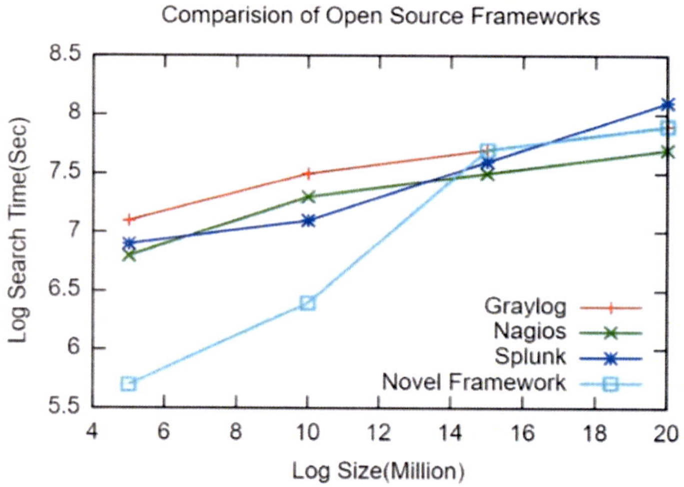

Fig. 4 Performance analysis of frameworks for query III

essential functions such as log indexing, real-time search, and analytic simultaneously in the modelled framework. Elasticsearch cluster with horizontal dynamic scalability can result in improved performance over larger log volumes.

Figure 5 depicts the data observed during performance evaluation in reference to Query IV graphically. For query IV also, more or less novel framework performed in same way as experienced in Query III when we did performance evaluation of novel framework with Graylog, Nagios, and Splunk in terms of search and analysis time

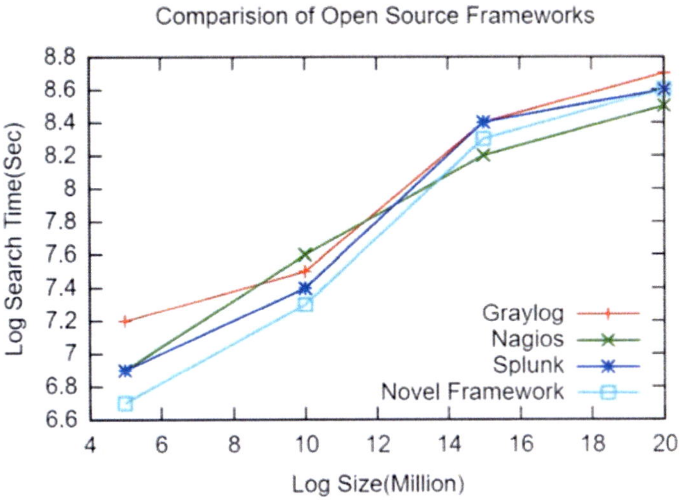

Fig. 5 Performance analysis of frameworks for query IV

required. Enhancement in performance can be observed with the same mitigation plan discussed during performance evaluation of novel framework for query III.

6 Conclusion

Because of the various format, ongoing nature, and large number of log and traffic data, we used a Big Data analytic approach for monitoring security issues at their initial occurrence in order to prevent larger future security disasters. Our approach to gather and analyse raw heterogeneous log data relating to vital IT infrastructure within the educational organization is addressed in this paper. The results achieved are encouraging since we acquired extensive understanding about different safety events, well in advance, and so could prevent severe safety catastrophes. The most serious security occurrences identified include attempting attacks on port 22 and 23 of the routers, accessing websites with 404-Page not found or 403-disallowed status code, and access from low reputation IPs which were analysed further through proposed framework. The novel framework is created and deployed in a real-world context to collect and mine actionable intelligence from massive log data, which is not achievable with human data review. The setup is particularly successful in developing centralized security monitoring and analysis system with searchable information in JSON format, dashboards for analysing various log events, and its visualization for monitoring. The configuration generates efficient and searchable indexes for rapid searches while consuming less disc space than raw logs after stream processing which can be further used for Apache Spark-based batch processing meant for forensic analysis of events identified. The configuration has improved the IT security of the organization under study, and it provides an ideal framework for debugging security concerns, performance monitoring, security analysis, and predictive analysis. We next compared the performance of the novel framework to that of commercial and open source log analysis frameworks, and discovered that the novel framework excelled or performed equally in terms of the various search and analytic criteria considered throughout the study.

References

1. Lv Z, Song H, Basanta P et al (2017) Next generation Big Data analytics: State of the art, challenges, and future research topics. IEEE Trans Industr Inf 13(4):1891–1899. https://doi.org/10.1109/TII.2017.2650204
2. Harvan M, Locher T, et al (2016) Cyclone: Unified stream and batch processing. In: 2016 45th International Conference on Parallel Processing Workshops (ICPPW), pp 220–229. https://doi.org/10.1109/ICPPW.2016.42
3. Hu H, Wen Y, Chua T-S et al (2014) Toward scalable systems for Big Data analytics: A technology tutorial. IEEE Access 2:652–687. https://doi.org/10.1109/ACCESS.2014.2332453
4. Chaudhari S, Maurya VK et al (2019) Real time logs and traffic monitoring, analysis and visualization setup for IT security enhancement. Next Generation Computing Technologies (NGCT-2019)
5. Deshpande K et al (2021) An open source framework unifying stream and batch processing. In: 3rd International Conference on Inventive Computation and Information Technologies, ICICIT-2021
6. Yu S, Sun X, et al (2021) Data processing and development of big data system: A survey. In: Advances in Artificial Intelligence and security. ICAIS 2021. Communications in Computer and Information Science, vol 1423. Springer, Cham. https://doi.org/10.1007/978-3-030-78618-2-34
7. Boros A, et al (2020) A comparative evaluation of Big Data frameworks for log processing. In: Proceedings of the 11th International Conference on Applied Informatics Eger, Hungary, January 29, 31, 2020, published at http://ceur-ws.org
8. Liu JC, Yang CT et al (2021) Cyberattack detection model using deep learning in a network log system with data visualization. J Supercomputing. https://doi.org/10.1007/s11227-021-03715-6
9. Chen L, Liu J, Xian M, Wang H (2020) Docker Container Log Collection and Analysis System Based on ELK. International Conference on Computer Information and Big Data Applications (CIBDA) 2020:317–320. https://doi.org/10.1109/CIBDA50819.2020.00078
10. Debnath B, et al (2018) LogLens: A real-time log analysis system. In: IEEE 38th International Conference on Distributed Computing Systems (ICDCS), pp 1052–1062. https://doi.org/10.1109/ICDCS.2018.00105
11. He P, Zhu J, et al (2018) Towards automated log parsing for large scale log data analysis. IEEE Trans Dependable Secure Comp 15(6):931–944, 1 Nov–Dec. https://doi.org/10.1109/TDSC.2017.2762673
12. More R, Unakal A, et al, Real time threat detection system in cloud using big data analytics. In: 2nd IEEE International Conference on Recent Trends in Electronics, Information and Communication Technology, Bangalore, pp 1262–1264
13. Nasiri H et al (2019) Evaluation of distributed stream processing frameworks for IoT applications in Smart Cities. J Big Data 6:52. https://doi.org/10.1186/s40537-019-0215-2

14. Bajer M (2017) Building an IoT data hub with Elasticsearch, Logstash and Kibana. In: 5th International Conference on Future Internet of Things and Cloud Workshops (FiCloudW), pp 63–68. https://doi.org/10.1109/FiCloudW.2017.101
15. Al- IYM et al (2017) Network security enhancement through effective log analysis using ELK. International Conference on Computing Methodologies and Communication (ICCMC) 2017:566–570. https://doi.org/10.1109/ICCMC.2017.8282530
16. Sanjappa S, Ahmed M, Analysis of logs by using Logstash. In: Satapathy S, Bhateja V, Udgata S, Pattnaik P (eds), Proceedings of the 5th International Conference on Frontiers in Intelligent Computing: Theory and Applications. Advances in Intelligent Systems and Computing, vol 516. Springer, Singapore. https://doi.org/10.1007/978-981-10-3156-4-61
17. Son SJ, et al (2017) Performance of ELK stack and commercial system in security log analysis. In: IEEE 13th Malaysia International Conference on Communications (MICC), pp 187–190. https://doi.org/10.1109/MICC.2017.8311756
18. Mishra DD, Pathan S et al (2018) Apache spark based analytics of squid proxy logs. IEEE International Conference on Advanced Networks and Telecommunications Systems (ANTS) 2018:1–6. https://doi.org/10.1109/ANTS.2018.8710044
19. Therdphapiyanak J, Piromsopa K, Applying Hadoop for log analysis toward distributed IDS. In: Proceedings of the 7th International Conference on Ubiquitous Information Management and Communication (ICUIMC '13). Association for Computing Machinery, New York, NY, USA, Article 3, 16. https://doi.org/10.1145/2448556.2448559
20. Prakash T, Kakkar M, Patel K (2016) Geo identification of web users through logs using ELK stack. In: Proceedings of the 2016 6th International Conference Cloud System and Big Data Engineering (Confluence),Noida, India, 14, 15 January, pp 606–610
21. Mehta S, Kothuri P; Garcia DL (2018) Anomaly detection for network connection logs. arXiv:1812.01941
22. Wang YT, Yang CT, Kristiani E, Chan YW (2018) The implementation of Wi-Fi log analysis system with ELK stack. In: Hung J, Yen N, Hui L (eds), Frontier Computing. FC 2018. Lecture Notes in Electrical Engineering, vol 542. Springer, Singapore. https://doi.org/10.1007/978-981-13-3648-528
23. Yang CT, Kristiani E, Wang YT, et al (2020) On construction of a network log management system using ELK Stack with Ceph. J Supercomputing 76, 6344–6360. https://doi.org/10.1007/s11227-019-02853-2
24. Kolajo T, Daramola O, Adebiyi A (2019) Big data stream analysis: a systematic literature review. J Big Data 6, 47. https://doi.org/10.1186/s40537-019-0210-7
25. Xie W, Li P, Xu H (2018) Architecture and implementation of real-time analysis system based on cold chain data. In: Barolli L, Javaid N, Ikeda M, Takizawa M (eds) Complex, intelligent, and software intensive systems. CISIS 2018. Advances in Intelligent Systems and Computing, vol 772. Springer, Cham. https://doi.org/10.1007/978-3-319-93659-8-44
26. Park BH, Hukerikar S, Adamson R, Engelmann C (2017) Big data meets HPC log analytics: Scalable approach to understanding systems at extreme scale. In: IEEE International Conference on Cluster Computing (CLUSTER), pp 758–765. https://doi.org/10.1109/CLUSTER.2017.113
27. Naga Rama Devi G, Kumar A, Mozar S (2018) Emerging trends in Big Data analytics: A study ICCCE 2018. Lecture Notes in Electrical Engineering, vol 500. Springer, Singapore. https://doi.org/10.1007/978-981-13-0212-1-57
28. Tam NT, Song I, Kim K, Joukov N (2016) Big Data visualization in information science and applications (ICISA) 2016. Lecture Notes in Electrical Engineering, vol 376. Springer, Singapore.https://doi.org/10.1007/978-981-10-0557-2-40

29. Iqbal A, et al (2021) Intelligent data-analytics for condition monitoring: Smart grid applications. Elsevier, , 268 p. https://www.sciencedirect.com/book/9780323855105/intelligent-data-analytics-for-condition-monitoring. ISBN: 978-0-323-85511-2
30. Tomar A, et al (2020) Machine learning. In: Advances in computing, renewable energy and communication. vol 768. Springer Nature, Berlin, LNEE, 659 p. https://doi.org/10.1007/978-981-16-2354-7. ISBN 978-981-16-2354-7

Polarities Inconsistency of MOOC Courses Reviews Based on Users and Sentiment Analysis Methods

Sandeep Kumar, Anuj Kumar Singh, Shashi Bhushan, and Arun Vashishtha

Abstract Sentiment classification has been perceived as an important feature in today's scenario, as it provides strategies for repeatedly assessing popular assessments to excerpt meaningful information about a product or amenity. The most important source of information about the social interactions taking place in Massive Open Online Courses (MOOCs) is forum messages. Patterns and learner behaviors can be discovered by analyzing forum messages. In our review, we contended that this number, which we alluded to as user polarity, may not precisely reflect the sentiment communicated in all of the lines that make up the assessment. We check out conclusions from six MOOC course audits and track down the sentiment analysis methods that naturally infer polarities have irregularities. Users tend to review their courses positively, although negative language and elements do arise from time to time, which these algorithms discover. To beat these issues, we present a polarity aggregation model that thinks about the two polarities while following the mathematical mean. We research its exhibition by removing components from MOOC audits and doling out added extremity to them. The benefit is that it combines the context's sentiment with the sentiment collected by a pre-trained algorithm.

Keywords MOOC · Sentiment analysis · Polarity

1 Introduction

Sentiment analysis (SA) is a space of affective research that has detonated in prevalence lately because of the ascent of web or web-based media [1–18]. This field has been assigned as a domain which concentrate on line that investigations individuals' perspectives, assessments, or thoughts regarding things, organizations, or

S. Kumar (✉) · A. K. Singh · A. Vashishtha
Krishna Engineering College, Ghaziabad, India
e-mail: sandeep.kumar@krishnacollege.ac.in

S. Bhushan
School of Computer Science, University of Petroleum & Energy Studies, Dehradun, India

A. Tomar et al. (eds.), *Proceedings of 3rd International Conference on Machine Learning, Advances in Computing, Renewable Energy and Communication*, Lecture Notes in Electrical Engineering 915, https://doi.org/10.1007/978-981-19-2828-4_34

occasions to decide their feeling, sentiment analysis methods (SAMs) are calculations created by a few groups that can consequently distinguish the hidden opinion of a distributed survey [2]. Numerous organizations are carrying out these calculations to improve decisions and gain a superior comprehension of their purchasers' conduct or thoughts regarding their business or items. Massive online open courses (MOOCs) are an example of this, where learners can access great-quality learning resources from around the world. In the meantime, online education is transitioning from a content gathering phase to a content filtering phase. As a result, the students have been eager to select appropriate courses for themselves, and they may share their views and ideas about the courses they desire in the comment section. These evaluations include essential feedback from a variety of people, perceptive what students think about the course and its primary tools is critical for improving MOOCs. "Who is teaching the course?" is the key element that students investigate, based on the comments gained from student reviews. MOOCs provide students with the opportunity to participate in social learning through course forums. Even if post-course surveys are unavailable, the needed information about attitude can be deduced by analyzing the data from the course [3]. In this paper, we had tried to analyze a solution for the existing problem and as a result, we have developed the following research questions. "Is it common for people to create statements with conflicting polarities in the same sentence?" "Does the Coursera reviews rating accurately reflect the polarity of each statement inside an opinion?" On finding the solution to existing problem, we hope to solve these questions. Each sentence's polarity can be detected by SAMs. As a result, the Coursera bubbles rating cannot be used to indicate polarity in all sentences or aspects. We scraped the Kaggle web sites of six MOOC courses for the experiment, garnering a total of 76,382 Coursera student reviews. We utilize four sentiment analysis methods and check out how their polarities connect with client evaluations. Our tests obviously uncover a low coordinating with rate with regards to perceiving great, unbiased, and negative audits, showing that there is an inactive disparity between them [5]. The conduct of the proposed extremity model is then examined, thinking about its boundaries, and its presentation is assessed utilizing a perspective-based sentiment analysis (PBSA) system. We separate angles and allot them to the model's polarities.

2 Literature Survey

Walsh [15] utilized various order calculations to examine the opinion of Stanford course surveys in his undertaking, including the MaxEnt classifier, the Naive Bayes classifier, and the Stanford classifier. It could help teachers, students, and heads in grouping course surveys and recognizing how students feel about the courses they are taking. As per the discoveries of the examinations, the unigram model in the Naive Bayes (NB) classifier performs better, but the freedom suppositions make it wrong in some cases. Wen et al. [16] proposed aggregate feeling investigation in MOOC to survey students' viewpoints with respect to the course and its devices. It might uncover a solid connection between the mind-set communicated in the course discussions and the quantity of understudies who exited. In any case, an exhaustive MOOC dictionary is needed to work on the strategy. The presentation of these methodologies was then assessed utilizing the accuracy, review, f-measure, and exactness measurements. The restrictions are as per the following: (1) inability to analyze pictures with images for investigation; (2) exactness diminishes for trigram, four-gram, and five-gram classification. Sadhasivam et al. [11] proposed a technique for preparing opinion classifiers by joining feeling information from an assortment of sources, including opinion vocabularies, feeling classifiers from various source domains, and unlabeled and named information in the objective domain. In expansion, to tackle the model, a streamlining calculation was made.

3 Results

3.1 The Sentiment Analysis Problem

Sentiment analysis is a new concept of NLP domain that focuses on analyzing people's attitudes toward a product, service, organization, issue, or person in written text. The goal is to create computer approaches for identifying feelings and extracting information to help decision-makers. The two types of machine learning algorithms widely employed in sentiment analysis are supervised and unsupervised learning approaches [13]. The dataset is labeled and trained using supervised learning techniques to produce a useful output that aids in making the best decision possible. Unsupervised learning approaches contrast from supervised learning in that they do not require any data labels, making them more challenging to implement. Sentiment classification, which divides reviews into positive and negative categories, has a variety of uses. Numerically, an assessment can be characterized as a 5-tuple [14]:

$$\left(E_i,\ A_{ij},\ S_{ijkl},\ H_k,\ T_l\right)$$

where E_i is the Ith assessment entity, A_{ij} is the jth characteristic, a property identified with the element E_i, S_{jkl} is the feeling of the assessment toward a trait A_{ij} of element E_i by the assessment holder Hk at time T_l, hk is the kth assessment holder, and tl is lth time when the assessment was radiated.

3.2 Sentiment Analysis Methods

The development of SAMs that can identify polarization in an instinctive and efficient manner has centered on polarity detection. These SAMs are designed to analyze a variety of texts, including tweets and reviews. There are various studies in the

Table 1 Four SAMs that we used

SAM	Group	Numerical values	References
Affine	MLD	{−5,−4,−3,−2,−1, 0,1,2,3,4,5}	36
Bing	MLD	{−1,0,1}	11
Core NLP	MDL	{0,1,2,3,4}	17
Meaning cloud	MML	{0,1}	41

literature that compare the performance of different SAMs across multiple texts. These techniques are categorized into three categories [10]:

1. **Method based on Lexicon Dictionary (MLD)**: These method entails establishing a sentiment lexicon, or a list of terms with a sentiment orientation. The dictionary can be built using seed words, corpus words (words relevant to a given subject), or a combination of the two. Synonyms and antonyms are frequently added to the lexicon.
2. **Method based on machine learning (MML)**: It creates statistical models using classification methods. There are two types of methods: supervised and unsupervised. The fundamental distinction is that the first group constructs the model using labeled opinions. The feature extraction to be anticipated is one of the most basic undertakings in these methodologies.
3. **Method based on deep learning (MDL)**: Deep learning has seen significant growth in recent years as a result of its excellent performance in a variety of sectors. When it comes to identifying correlations from raw data, sentiment analysis methods focused on ANN learning have demonstrated to produce excellent outcomes when compared to other methods. We recommend separating deep learning from machine learning-based methods based on the facts present to us (Table 1).

3.3 *The Data*

Our tests are based on MOOC courses reviews of six different courses: machine learning, Python, Data Science Tools, IBM data science data processing using Python and R programming, we collect the reviews about different Coursera and NPTEL-MOOC courses reviews from Kaggle. As a result, six datasets containing a total of 97,456 reviews were constructed.

As shown in Table 2, the artificial intelligence has the greatest number of comments (23.88% of the total). With an average word count of 98.2 and a sentence length of 4.75, machine learning has the longest reviews. IBM data science has the highest average rating on Coursera courses with a rating of 4.97. With a score of 4.00, the data processing using Python is the least valuable monument.

Table 2 Summary of the six datasets' text attributes

Courses	Reviews	Words	Sentences	Average words	Average sentences	Average user polarity
Machine learning	57,599	5,654,780	273,486	98.2	4.75	4.39
Python	36,708	3,450,702	168,923	94.0	4.60	4.67
Data Science Tools	43,710	3,925,195	146,975	89.8	3.36	4.19
IBM data science	54,333	5,240,704	235,734	96.5	4.34	4.97
data processing using Python	67,504	5,850,328	279,974	86.7	4.15	4.00
R programming	41,568	3,655,499	188,605	87.9	4.54	4.21

3.4 A Study of Polarity Discrepancies Between Users and SAMs

Many research projects have used students opinions about a course, content and trainee as a source of data. People evaluate user opinions to derive information about their views regarding any restaurant, hotel, or tourist destination. Notwithstanding, as far as we could possibly know, the connection between polarities of users review and the polarities of each sentence inside the assessment has never been explored. Numerous organizations that show up on the web might feel that the guest is fulfilled simply by taking a gander at the normal rating, yet by not digging further into each remark, they might be passing up important data [11]. As a result, we believe it is vital to conduct research comparing the association between sentiment analysis methods and their polarities. Finally, we believe that focusing the study on cultural monuments is noteworthy because as per literature review, we had studied and have used them as the subject of investigation. This section contains an in-depth examination of sentiment analysis by users and their polarities.

We identify that the vast majority of MOOC courses students reviews are positive which means that students are happy with the content delivered to them and it is shown in Fig. 1. Notwithstanding, this sharing is not kept up with all through SAMs. We see that Affine Fig. 2 and meaning cloud (Fig. 5) acquire a comparative extremity appropriation to the users. Notwithstanding, Affine does not identify any adverse conclusions and Meaning cloud identifies 1,985 more bad audits in IBM Data Science dataset. Bing (Fig. 3), core NLP (Fig. 4) show altogether different circulations: they recognize a lot more nonpartisan and negative audits. As a rule, when contrasting the extremity dispersions of clients and SAMS, we discover little cross-over. Clients are more sure, while SAMs are more impassive and negative. This reality shows an unmistakable befuddle in deciding an assessment's inclination, which could be

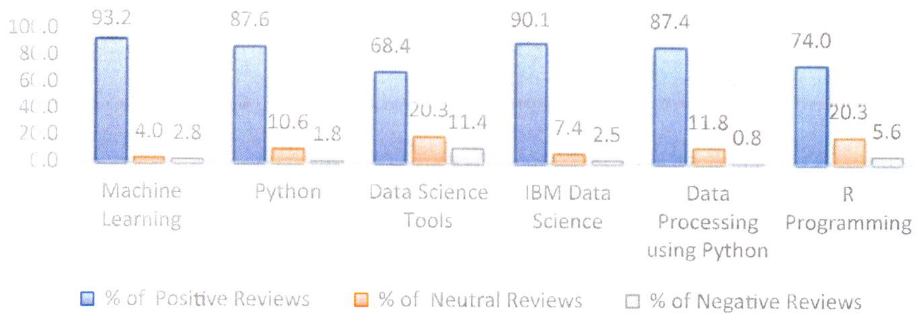

Fig. 1 Polarities inconsistency of courses reviews based on user polarity

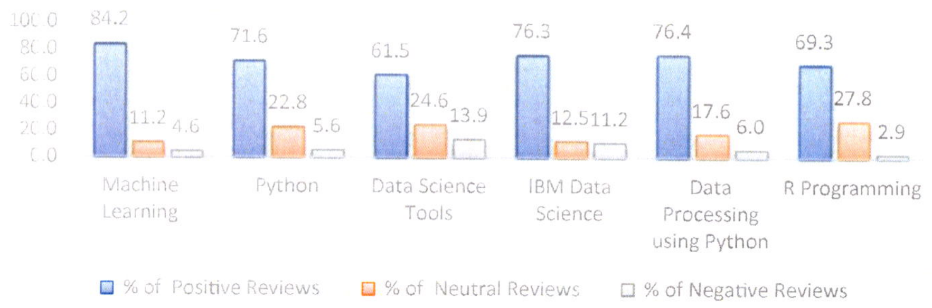

Fig. 2 Polarities inconsistency of courses reviews affine polarity

identified with the different extremity that happen in sentences. It is additionally apparent in Fig. 1, where the client positions with five air pockets (a good assessment), yet there are sentences with a negative extremity inside a similar assessment.

Figure 6 portrays the extremity coordinating with proportion among clients and SAMs: each line of the network reflects client grouped polarities, though every segment addresses SAM-ordered polarities. We show the normal rates over the six

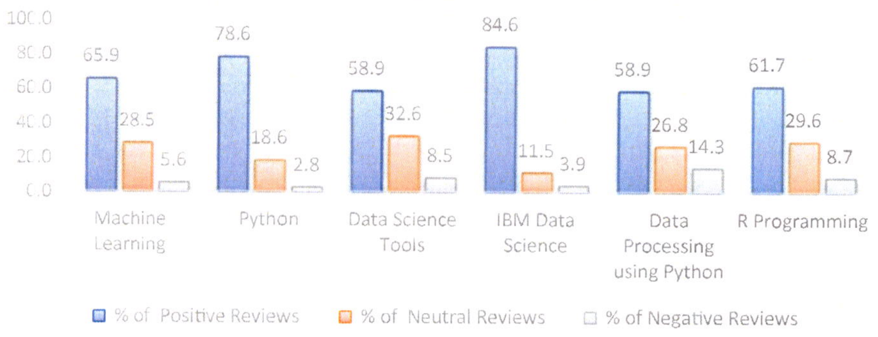

Fig. 3 Polarities inconsistency of courses reviews bing polarity

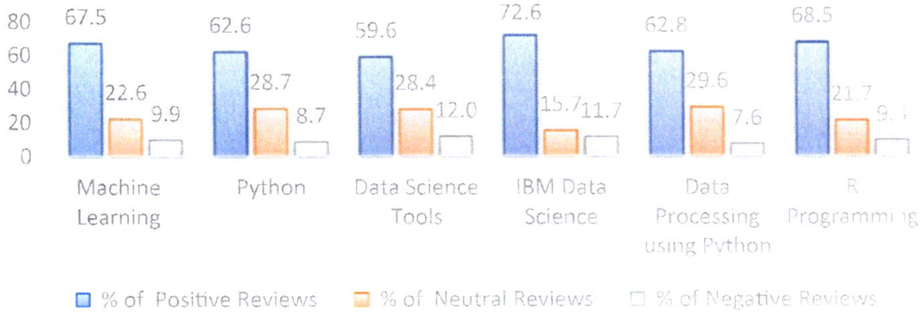

Fig. 4 Polarities inconsistency of courses reviews core NLP

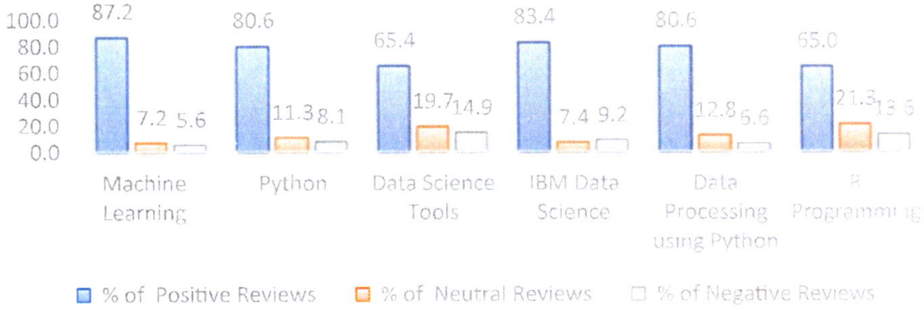

Fig. 5 Distribution of polarities of courses reviews meaning cloud

User	Positive	81.50%	14.50%	4.00%
	Neutral	66.70%	21.80%	11.50%
	Negative	42.10%	31.60%	21.30%
		Positive	Neutral	Negative
		Affine		

User	Positive	74.50%	16.50%	9.00%
	Neutral	56.20%	29.60%	14.20%
	Negative	28.10%	42.60%	29.30%
		Positive	Neutral	Negative
		Bing		

User	Positive	47.70%	32.50%	19.80%
	Neutral	21.20%	39.30%	39.50%
	Negative	11.60%	34.70%	53.70%
		Positive	Neutral	Negative
		Core NLP		

User	Positive	87.40%	9.30%	3.30%
	Neutral	69.90%	18.40%	11.70%
	Negative	53.80%	26.90%	19.30%
		Positive	Neutral	Negative
		Meaning Cloud		

Fig. 6 Level of coordinating between users (lines) and SAMs (sections) polarities

courses to advance the format (4 SAMs 6 courses = 24 networks). Since the appropriation on the four tables is so close, this is defended (the greatest standard deviation of all courses is 0.176). Since blue tones win in practically all certain positive cells, SAMs work well in distinguishing energy. In the other hand, on unbiased nonpartisan and negative cells, pale blue tones are the most predominant, showing a low relationship proportion. Core NLP qualifies the least (47.70%). This one makes a superior showing with distinguishing awful client surveys (68.10%), however, all of the others do inadequately (proportions underneath 46%). Most of them order them as certain. The extremity that shows the most unfortunate results is lack of bias. There is anything but a solitary SAM that stands apart with regards to distinguishing this halfway extremity. Information exhibits an extensive bungle among clients and SAMs extremity, as found in Fig. 6. When identifying polarities, we show that there is a low measure of matchings. We found through investigating message information that clients tended to compose negative sentences on certain audits as well as the other way around. Therefore, we ought to prompt against utilizing a client's extremity as the general opinion of their audits since we will miss out on a ton of data.

4 Conclusions

This paper resolved an issue with the MOOC courses review rating that it has never been addressed previously as per best of our knowledge. We demonstrated that the students tend to rate the quality of content of the MOOC course, yet there are sentences with the opposite polarity. As a result, those reviews cannot be applied to all sentences. We developed our theory and examined the polarity matching between polarity of users review and six student analysis methods to demonstrate this. We discovered that they have a low correlation when it comes to detecting polarities. We also spoke about how the average of matches on over 47% of people can distinguish three polarities (positive, neutral, and negative). This is because, as previously said, individuals do not utilize the same attitude in every written text; rather, people change over time, and sentiment analysis methods can identify those changes.

References

1. Abdi A, Shamsuddin SM, Aliguliyev RM (2018) QMOS: Query-based multi-documentsopinion-oriented summarization. Inform Process Manag 54(2):318–338
2. Adinolfi et al (2016) Sentiment analysis to evaluate teaching performance. Int J Knowl Soc Res (IJKSR) 7(4):86–107
3. Banic L, Mihanovic A, Brakus M (2013) Using big data and sentiment analysis in product evaluation. In: Information and Communication Technology Electronics and Microelectronics, 36th International Convention on IEEE, 1149–1154
4. Duan W, Cao Q, Yu Y, Levy S (2013) Mining online user-generated content: using sentiment analysis technique to study hotel service quality. In: System Sciences (HICSS), 2013 46th

Hawaii International Conference on IEEE, 3119–3128
5. ElSahar H, El-Beltagy SR (2015) Building large arabic multi-domain resources for sentiment analysis. In: International Conference on Intelligent Text Processing and Computational Linguistics in Springer International Publishing, 23–34
6. Gonçalves P, Dalip DH, Costa H, Gonçalves MA, Benevenuto F (2016) On the combination of "off-the-shelf" sentiment analysis methods. In: Proceedings of the 31st Annual ACM Symposium on Applied Computing, 1158–1165. https://doi.org/10.1145/2851613.2851820
7. Liyanagunawardena TR, Parslow P, Williams SA (2017) Exploring "success" in MOOCs: Participants' perspective. Massive Open Online Courses and Higher Education: Where to Next ?, 92–108
8. Niu Z, Yin Z, Kong X (2012) Sentiment classification for microblog by machine learning. In: Proceedings—4th International Conference on Computational and Information Sciences, ICCIS 2012, 286–289
9. Ren Y, Ji D (2017) Neural networks for deceptive opinion spam detection: An empirical study. Inf Sci 385:213–224
10. Ribeiro FN, Araújo M, Gonçalves P, Gonçalves MA, Benevenuto F (2016) Sentibench-a benchmark comparison of state-of-the-practice sentiment analysis methods. EPJ Data Science 5(1):1–29
11. Sadhasivam J, Kalivaradhan RB (2018) A hybrid approach for deep belief networks and whale optimization algorithm to perform sentiment analysis for MOOC courses. Int J Adv IntellParadig
12. Sadhasivam J, Kalivaradhan RB (2019) An empirical comparison of supervised learning algorithms and hybrid WDBN algorithm for MOOC courses. J Ambient Intelligence and Humanized Computing
13. Tang D, Wei F, Qin B, Dong L, Liu T, Zhou M (2014) A joint segmentation and classification framework for sentiment analysis. In: Proceedings of the 2014 conference on empirical methods in natural language processing (EMNLP), pp 477–487
14. Tripathy A, Agrawal A (2016) Rath SK Classification of sentiment reviews using n-gram machine learning approach. Expert Syst Appl 57:117–126
15. Walsh RJ (2018) Sentiment analysis of Stanford course reviews
16. Wen M, Yang D, Rosé CP (2018) Sentiment analysis in MOOC discussion forums: what does it tell us? In: Proceedings of educational data mining, (Edm), pp 1–8
17. Iqbal A, et al (2021) Intelligent data-analytics for condition monitoring: Smart grid applications. Elsevier, 268 p. https://www.sciencedirect.com/book/9780323855105/intelligent-data-analytics-for-condition-monitoring. ISBN: 978-0-323-85511-2
18. Tomar A, et al (2020) Machine learning, advances in computing, renewable energy and communication, vol 768. Springer Nature, Berlin, LNEE, 659 p. https://doi.org/10.1007/978-981-16-2354-7. ISBN 978-981-16-2354-7

Neural Augmentation Using Meta-Learning for Training of Medical Images in Deep Neural Networks

Tina, Sanjay Kumar Dubey, and Ashutosh Kumar Bhatt

Abstract Image data augmentation techniques are state-of-art method for drive of classification and segmentation of multidimensional medical imaging by method of deep neural network approaches. Neural augmentation is considered as type of data augmentation in medical image processing which is widespread and actively used method in enhancing the measures of performance of neural network. The novel approaches introduced on the grounds of deep network which expanded tremendously over the years as it improves the clinical assistance for the doctors in the treatment of patients. The major concern in healthcare domain relies on identification of tumor in the medical imaging modalities at probable early stage in humans, and deep neural network plays vigorous role in the detection of tumors using neural network with possible less layers. Neural network augmentation with the gradient of meta-learning in deep learning models will assess performance of convolutional neural network (CNN) model based on assessment parameters like accuracy and loss function along with augmented images. Deep convolutional network get to trained using meta-learning approach, and then, augmentation will be accomplished. This proposed work implements the neural augmentation with meta-learning approach in neural network subsequently training the dataset to extemporize accuracy rate in training and validation.

Keywords Neural augmentation · Meta-learning · Convolutional neural network · Medical imaging · Brain tumor MRI

Tina (✉) · S. K. Dubey
Amity University, Uttar Pradesh, Noida, India
e-mail: dudeja.tina@gmail.com

S. K. Dubey
e-mail: skdubey1@amity.edu

A. K. Bhatt
Uttarakhand Open University, Haldwani, Uttarakhand, India

A. Tomar et al. (eds.), *Proceedings of 3rd International Conference on Machine Learning, Advances in Computing, Renewable Energy and Communication*, Lecture Notes in Electrical Engineering 915, https://doi.org/10.1007/978-981-19-2828-4_35

1 Introduction

The initiation of artificial intelligence in field of healthcare informatics instigates the improvised and advanced in transformation of clinical technology. The illustrations of abnormalities in parts of organ in body are rapidly intensifying for nearly entire age groups in human beings. However, the initiation of deep learning models in field of medical domain and clinical diagnosis aids in the enhanced medical procedures and better decision-making with improvised image visualization of medical image modalities. The illustrations of abnormality in structures such as brain tumor cases which are rising at a quick rate for certain reasons, for example, age of patient, chemical exposure, radiations, etc. Researchers aim to deliver improved visualization of organ which is affected, so early diagnosis can be apprehended. Over the years, artificial intelligence made substantial improvement in the healthcare informatics which comprises preprocessing of medical computed tomography (CT) image which is accommodating in diagnosing the ailment at initial stage in easy way. Now a days, health care needs well-organized and reliable practices to analyze diseases such as cancer which is the reason of mortality worldwide for humans. The best practice for investigating and identifying tumor is magnetic resonance imaging (MRI). Advanced level grades evaluate that in most cases upper level-staged tumors are damaging as these are developed in brain at a fast rate. In the underlying stage, tumor does not spread much; however, in later stage, there are chances of spreading to parts of the brain. Early diagnosis of tumor is hence recommended as in affected portion of brain, cells are destructive; subsequently, these cells may pass into the circulation systems of brain and spread into the encircling cells [1]. Consequently, it is imperative to identify early phase of brain tumors with the most effective accuracy.

MRI imaging avails a technique for classification of tumors by means of image augmentation and models of deep neural network. Image data augmentation techniques applied on medical imaging modalities provide the diversity on the available dataset by inclusion of image cropping, flips, Gaussian filters, principal component analysis (PCA), color jittering, etc., and through these techniques, capturing of noteworthy structures of actual image dataset can be done. The prospect of enhancement in the data visualization entirely counts on the learning strategies and deep learning strategies including data augmentation diminishes overfitting models by training the dataset [2].

Image augmentation addresses the issue of overfitting in labeled dataset which provides a procedure of intensifying a dataset being derivative from information accessible in the prevailing dataset. When the implementation is done on images, then the insight for that is ubiquitous, as its application involves translation, rotation, and other type of modifications in the healthcare imaging which are predominant in the medical dataset to generalize the learning model. Analyzing medical image segmentation and augmentation plays important part suing deep neural learning models such as deep convolutional neural networks, growing convolutional neural networks [3], and support vector machines. As the mixing of varied samples is quite challenging tasks through similar feature space representations for newer sample generation [4].

This paper instigates in generating the augmented images through the process of self-learning methodology which will help to make learn normal and abnormal structures by themselves in directive to reduce complexity in training dataset of medical image besides reducing overfitting. So, the concept of neural augmentation can be utilized in aggregation with meta-learning approach in deep learning framework.

2 Related Work

Data augmentation methods alike adding noise, rotation, flip determine to degree the performance gain specified by augmentation techniques. They avail recommended data augmentation methods which will be beneficial in procedure of decision-making in clinical assistance. F. Baseslice et al. [5] proposed an approach for sparse tensor prototype which is built on weighted regularization by means of MRI through efficient computation. This projected model implements MRI image modalities for given dimensions for 20 slices. The experimentation was exhibited on four diverse contrast models through SSIM and PSNR estimation. The outcomes demonstrate that sparse tensor technique outperformed the improvised regularization for weighted pixels. M. Jaderberg et al. formulated a framework for the intent of synthetics data generation with the intent of word recognition in deep learning architecture. In this paper, NDB, i.e., natural data blending is used to prepare the synthetic data which seems to be further realistic and highly sufficient in order to replicate the real-word data through infinite amount of training data [6]. I. Goodfellow et al. introduced the generative adversarial networks (GANs) which considered as unsupervised learning practice which implements minimum and maximum strategy in the two networks, wherein the first network efforts to make images and detect the counterfeits to make another network engrosses the same method. This is done in course to generate the augmented images from datasets obtainable [7]. J. Lemley introduced the concept of smart augmentation defined by author which addresses the matter of training dataset to lower the overfitting and improvisation in regularization through deep learning approach for combining images to resolve the above stated issues that are reducing the overfitting and regularization method [8]. H.A. Khan proposed the concept for classification of CT images in direction to locate the region of interest (ROI), and then, cropping it for augmentation purpose on large training dataset, they applied edge detection technique. Also, according to the proposed methodology, afterward, training of small dataset, accuracy, and rate of accuracy can be attained which are showed after experiments [9]. The results of proposed models are compared to ResNet framework, VGG-16, and Inception model. V. Olsson proposed practices for augmentation called ClassMix in "ClassMix: Segmentation-Based Data Augmentation for Semi-Supervised Learning." In this paper, augmentations of medical image were generated using mixing of samples which are unlabeled, and in further procedures, these samples were leveraged based on prediction of network with respect to limitations of any corresponding object. The evaluation of samples was built on semantic segmentation and secondly diverse design and training managements [10].

K. Nishi proposed the strategies of augmentation in approaches of deep learning for the labeled noise. The paper "Augmentation Strategies for Learning with Noisy Labels" has advantage for higher noise ratios, and in that case, copious augmentation needs to be avoided [11]. S. Mounsaveng addresses the issue of hyper-parameters for data augmentation through the proposal of an effective system to train network that will learn to effectively distribute the conversion in directive to improvise the generalization through bi-level optimization. This technique helps to optimize the parameters of data augmentation along with validation set. Also, experiment results justify that proposed method generates the accuracy of image classification better restored from existing data augmentation [12]. A. Krishna proposed the DRL, i.e., deep reinforcement learning that fit in the neural style transfer toward creating the anatomical shapes. It will generate the higher-resolution medical images at varied quantities with smaller datasets [13].

In above mentioned review, it has been seen that labeled dataset in medical imaging is primary concern in the dispersal of neural network in healthcare informatics. Medical image augmentation generally trained the neural network after initial phase or from preprocessing in command to classify whether the brain MRI encompasses tumor or not. The regularization techniques applied are generally pretrained through the deep learning protype which will decrease the generalization error by accumulation of parameters into it. So, effective regularization practice is somewhat necessary to decrease the variance via reduction in the error for validation testing. We applied the neural augmentation can be used in unification through meta-learning approach in learning framework to provide image augmentation where the dataset will be classified by learning the new tasks from existing set of jobs. This technique attempts to generate the augmented medical images with the process of learning approach which will help to make learn images themselves in order to reduce complexity in training data and reduce overfitting.

3 Materials and Methods

3.1 *Neural Image Augmentation*

It is being suggested to train the neural model on a dataset to learn it in the most ideal manner and furthermore to maintain a strategic distance from the overfitting of issues. Fundamental deep learning models can be improvised over massive dataset by preparing a counterfeit dataset with adjustments in available dataset [1–25]. Neural image augmentation (NIA), as shown in Fig. 3, approach aids in training the dataset to anticipate the matter of overfitting main origin basis with sight of the prospect that further information can be mined or extracted through original dataset available. NIA exaggeratedly inflates the size of training dataset by using the concept of oversampling or data warping, i.e., transformation of prevailing images in the way that their labels are preserved involving augmentations (color, flips, geometric

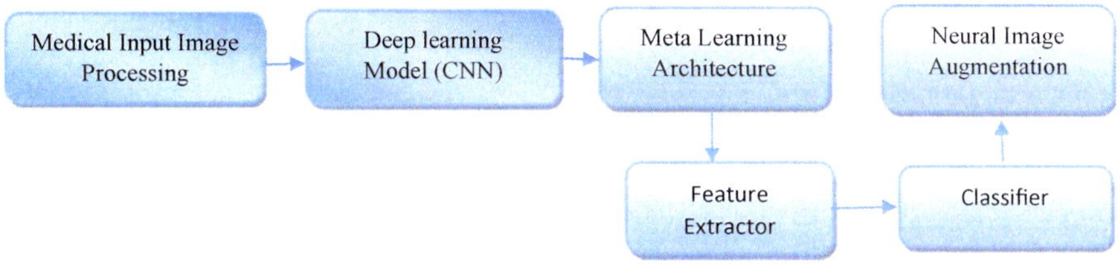

Fig. 1 Flow of neural image augmentation (NIA)

transformations, generative adversial network, and neural style transfer) as shown in Fig. 1. However, oversampling generates synthetic instances thereby adding it to training dataset including feature spacing and mixing of images. Although both the augmentations available does not produce exclusive contrast between dataset. While in training phase, the network takes at most two images as input from same set of similar class, and it get to pass through a layer, and this layer returns the single same-sized image usually called as image from augmentation or augmented image. This output augmented image along with input images again passes through a network called as classification, and its loss is entropy (cross) loss via sigmoid of class score. NIA usually improvised the accuracy in classification besides reduction in overfitting. This approach is quite cooperative in convergence of network in rapid manner.

The persistence of neural style transfer (NST) is extensively castoff in 3D image to synthesize the image to envisage the possible representations. As pragmatic, the un-balanced datasets every so often encountered in procedure of diagnostic systems which foresees the inconsistency for equally training in context to testing set. The neural style algorithm optimizes the loss of data where loss might be style or content and in command to disparity of style of testing data slice to equivalent targeted training dataset portion. This case simultaneously enforces the constraint on newly generated image in order to uphold its content. The neural style transfer algorithm inescapably modifies the testing set images from given dataset through determination of training dataset as it is being laid to get processed through same framework.

3.2 Meta-Learning

The implementation procedure of augmentation plays vital role in choosing the type of network architecture. Through this way, deep convolutional neural network learns the best utilization of feature space representations in the medical dataset with some definite class to obtain the detailed target network by merger structures of further samples. The practices of augmentation can be yet again utilized in image classification in procedure of discourses the issue of envisaging the unlabeled images in the imaging dataset.

The influencing factors in the meta-learning framework are as follows:

Meta-Train:

In meta-learning framework, the aim is to train a meta-model by viewing it solitary for fewer examples per class. Besides that, it tests in contradiction of examples from similar classes which have remained kept out afterward for original dataset. It will be verified in much similar way when obtainable with fewer training examples since novel class of set. Individually, every training instance comprises of set of test and train data points which is termed as episode. Also, it can be alternately called as support (train) and query (test) sets. The classes in training set define N-class classification or N-way task along with labeled examples. The dataset for meta-training has episodes, each having test set also training set, and also, each test dataset comprises hyper-parameters fine-tuning and generalization.

Meta-Optimizer:

Meta-optimizer is process of meta-learning framework enhances the arrangement of the neural components which are formerly linked with existing deep neural network. Through this strategy, performance can be optimized by fine-tuning the hyperparameters of neural network to improve it. At training dataset, the computational procedure for each task parameters ϕ will be done, and then, updating meta-parameters by means of the gradient descent objective will be computed (with meta-parameters). Then, repeat this iteratively using optimizer like SGD.

Meta-Loss Function:

Though loss function in meta-learning will specify how fine the model along with neural model is performing on the job but in outcome, the meta *learner* is accustomed by means of a gradient descent method for each new task in the cluster of episodes. The objective for learning meta-loss function is castoff to train a classifier through the specified parameters of function M_φ as meta-loss to optimize the trained data f_θ to generate the output of loss value $\mathscr{L}_{\text{Learned}}$. The parameter for θ to compute gradient descent in training of f_θ for the above function is as follows [15]:

$$\theta_{\text{new}} = \theta - \alpha \nabla_\theta \mathcal{L}_{\text{Learned}} \tag{1}$$

$$\mathcal{L}_{\text{Learned}} = M_\varphi(y, f_\theta(x)) \tag{2}$$

wherever, y perceives ground truth value in supervised learning. In order to learn a loss function that could be used recommend learning model to acquire parameters for meta-loss function φ through gradient descent calculation. The substantial task is to originate a training indication to acquire loss parameters φ.

4 Proposed Methodology

4.1 Convolutional Neural Network with Meta-Learning

These systems are usually getting train after being visible to a variety to tasks; besides that its capability to acquire new tasks is simultaneously tested. Meta-learning approach trains the model in order to absorb newer tasks and that too in meta-training set. It deals with few optimizations; one is learner who is learning new tasks and other meta-learner who is providing training to learner. Some of methods which are defined in meta-learning approach comprise of learning optimize, metric learn, and recurrent model. The strategy trains a model which can be recurrent model in command to take the dataset in systematic and ordered way. It will process the input which are in training dataset (Fig. 2).

This incorporates the proposed methodology by showing the conjunction of deep learning model laterally with the pretrained model before classification using meta-learning technique for self-learning training to reduce its loss and time complexity. Numerous frameworks in deep learning undergo speedy enhancement in domain of medical image processing which is performed on various platforms. Each hidden layer signifies the image filters for feature extraction, i.e., to excerpt the differential features from input image. The output layer entails of feature maps which has innumerable filtered forms which denote resolution, classification of image, segmentation, image synthesis, etc. [16–25].

In Fig. 3, flow of generation of augmented medical images using meta-learning style has been revealed where the meta-learning approach is used after image preprocessing to pretrain the model in order to quote the features after the medical images in dataset. The neural image augmentation is generally used to develop the existing model or the inference by neural model components. This will integrate the forward pass along with deep learning architectures.

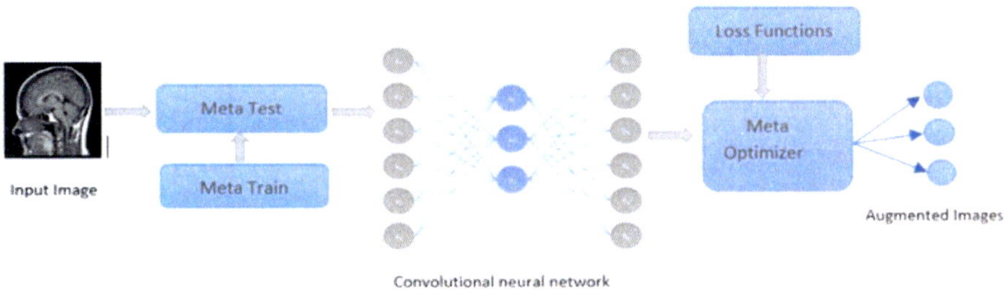

Fig. 2 Convolutional neural network model with meta-learning approach

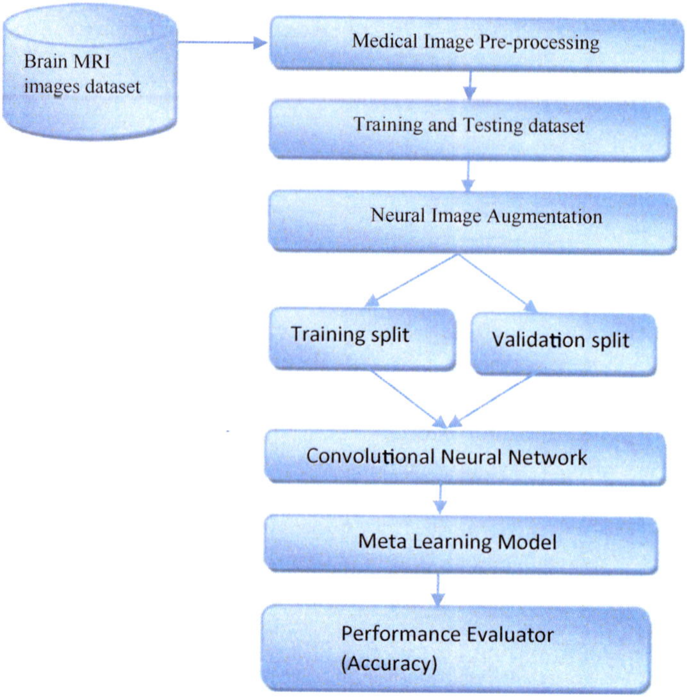

Fig. 3 Flow graph for implementation of meta-learning approach in neural augmentation

4.2 *Training Procedure for Meta-Learning*

Deep networks are generally get trained for purpose of object recognition where the visual representation of brain MRI which will make information of the inherited object explicit with the contented in hierarchy of its processing. The focus point of the representation is to compare the actual content with that of comprehensive pixel values. The information which is being inherited can visualize directly and that too layer by layer through the reconstruction method from feature maps. In higher layers, the pixel values do not constraint to the reconstruction; besides that it will capture the higher-level content and arrangement in given input image.

In Fig. 4, model of meta-learning insights the hyperparameters aimed at the training in regular mode via consideration of omega (ω) in command to specify either learning rate, also the strength ("regularization"). The parameter θ is implemented on model f_θ, and through that multi-tasks are adapted by the approach of set of meta-batches $\mathcal{T}_m$. The tasks of batch set $\mathcal{T}_m$ are sampled either randomly or through MAB, i.e., multi-armed bandit in which sampling is done through observation and for each task of $\mathcal{T}_m$, a sample from T_j. After that sample A_j^{tr}, a training set with N_{tr} volume and A_{val}, a validation set with N_{val} volume. The adaptation along with learning rate α is calculated as:

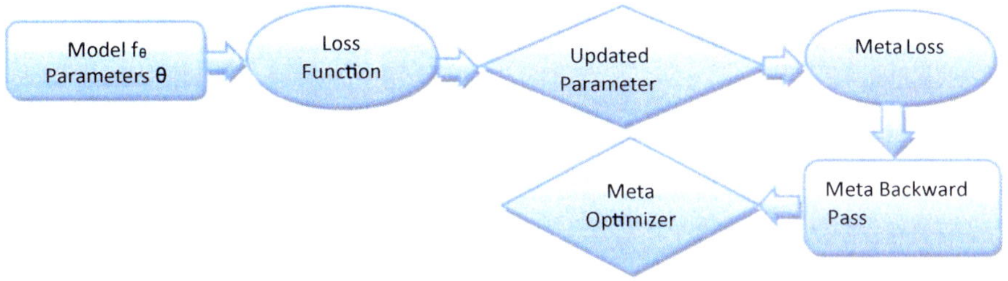

Fig. 4 Training process for meta-learning model

$$\theta' j(t) = \theta(t) - \alpha \frac{\alpha \partial (lk) f\theta(t) Dj(\mathrm{tr})}{\partial \varsigma}$$

4.3 Augmentation Using Meta-Learning Procedure

Algorithm: Augmentation Using Meta-Learning Approach on Dataset

1. Input: $\omega_0, \theta_0, w_\theta, n_w, n_\theta$
2. Input: Train the dataset trainset, validation dataset valset, test dataset testset.
3. for $i = 1$ to epoch do
4. for $t = 1$ to T dob, then generate N_w policys according to $p_{\theta t}$
5. Call Procedure: Meta-train for augment training from from A_{train} with N_w policy, respectively;
6. Obtain the validation loss $f_{\text{val}}(\text{wt})$ j on A_{val}; Restore the network parameters, ˆwt = wt; end for
7. Utilize validation loss $f_{\text{val}}(w_{\text{t}})\, j$,policys c_j $(j = 1,\ldots,N_\theta)$ to update θ_t according to Eq. 8; end for
8. end for, est the network on A_{test};
9. Return final validate set.

5 Experiments and Discussions

The dataset consists images which are anatomical in nature of around 120 patients from The cancer Genome Atlas (TCGA) and The cancer imaging Archive (TCIA). The accurate dimensions of each 3D images are 256 × 320 × 320 with fluid-attenuated inversion recovery (FLAIR) sequence. The state of dataset is heterogeneous, and foundation is also nonuniform in nature, whereas the dataset splits into non-overlapping subset each batch size. The experiment illustrates that meta-learning procedure can implement loss function to include further information accessible in

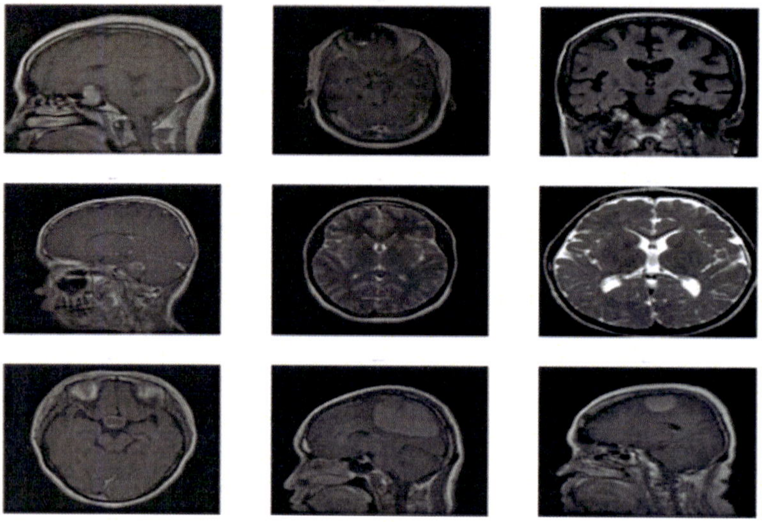

Fig. 5 Training set for applying meta-learning parameters

duration of meta-training. In Fig. 5, the training medical image dataset is attained after implementation of hyperparameters involved in meta-learning.

In above Fig. 6, gap among validation and training accuracy determines the attained gap to estimate the overfitting. The proposed model emphasizes on reduction of gap among dataset of training and validation to interpret the accuracy (Fig. 7).

6 Conclusion

The neural augmentation has the potential to improvise the performance in addition to accuracy on the available testing medical dataset. The mentioned approach will train the medical imaging data inevitably using prior learning approaches in our trained dataset in process of augmentation in deep neural network. The training loss besides validation accuracy perhaps achieves better results. In future, the application of generative adversial networks can be implemented on deep neural network utilizing a superfluous discriminator model-based model can be obtained for better results. The foremost focus is on mutual information present in a class so that a model could be constructed to provide augmentation. The proposed approach "Neural Augmentation" with inbuilt concept of meta-learning which will be helpful in automatic training of appropriate augmentations in deep neural network. Meta-learning framework in deep learning model leverages to self-learn new tasks which embraces wide spectrum for learning techniques in neural network.

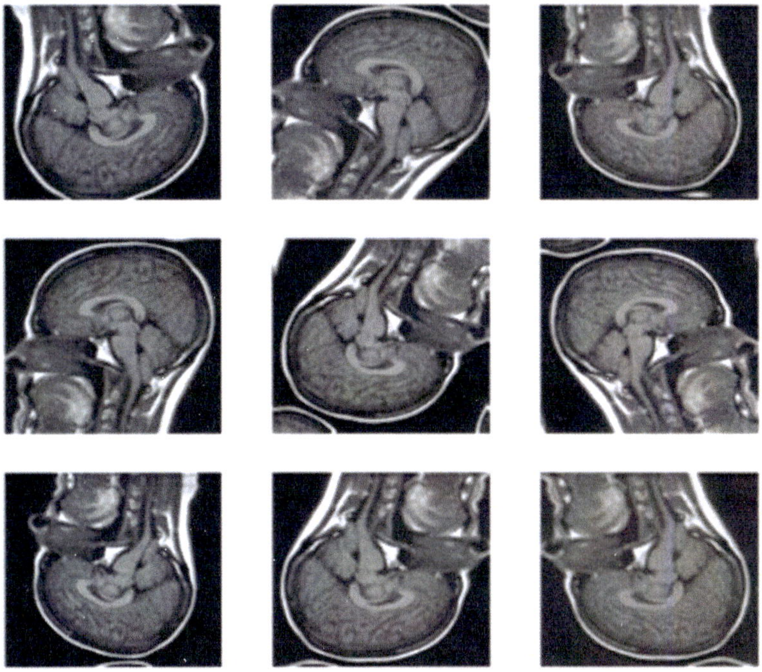

Fig. 6 Augmented set of images after implementing meta-learning approach

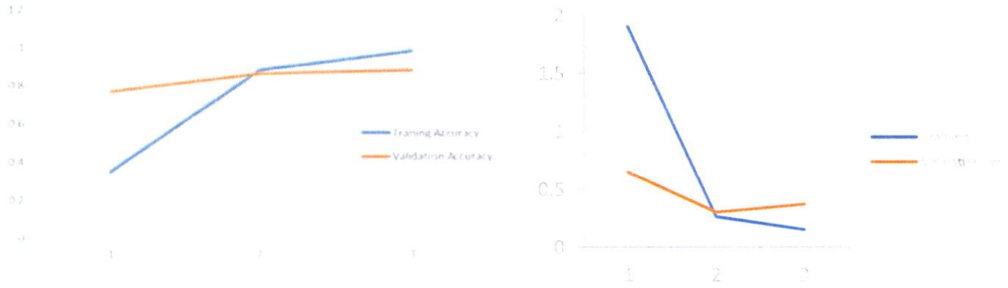

a) Training and Validation Accuracy b) Traning and ValidationLoss

Fig. 7 Accuracy and loss line graphs with respect to training and validation

References

1. Yao JC, Hassan M, Phan A, Dagohoy C, Leary C, Mares JE et al (2008) One hundred years after "carcinoid": epidemiology of and prognostic factors for neuroendocrine tumors in 35,825 cases in the United States. J Clin Oncol 26(18):3063–3072. https://doi.org/10.1200/JCO.2007.15.4377.PubMed
2. Dong C, Loy CC, Tang X (2016) Accelerating the super-resolution convolutional neural network. In: European Conference on Computer Vision

3. Mittal M, Mohan L, Kaur S, Kaur I, Verma A (2019) Deep learning based enhanced tumor segmentation approach for MR brain images. Applied Soft Computing Journal 78:346–354. https://doi.org/10.1016/j.asoc.2019.02.036
4. Long J, Shelhamer E, Darrell T (2015) Fully convolutional networks for semantic segmentation. In: Proceedings of the IEEE conference on computer vision and pattern recognition, pp 3431–3440
5. Baselice F, Ferraioli G, Pascazio V, Sorriso A (2019) 'Denoising of MR images using Kolmogorov-Smirnov distance in a non local framework.' Magn Reson Imag 57:176–193. https://doi.org/10.1016/j.mri.2018.11.022
6. Verma V, Lamb A, Kannala J, Bengio Y, Lopez-Paz D (2019) Interpolation consistency training for semi-supervised learning. In: Proceedings of the 28th International Joint Conference on Artificial Intelligence, pp 3635–3641. AAAI Press
7. Jaderberg M, Simonyan K, Vedaldi A, Zisserman A (2014) Synthetic data and artificial neural networks for natural scene text recognition [Online]. Available: http://arxiv.org/abs/1406.2227
8. Goodfellow I, et al (2014) Generative adversarial nets. Proc Adv Neural Inf Process Syst, 2672–2680
9. Lemley J, Bazrafkan S, Corcoran P (2017) Smart augmentation learning an optimal data augmentation strategy. IEEE Access 5:5858–5869. https://doi.org/10.1109/ACCESS.2017.2696121
10. Khan HA, Jue W, Mushtaq M, Mushtaq MU (2020) Brain tumor classification in MRI image using convolutional neural network, 17(10):6203 6216
11. Olsson V, Tranheden W, Pinto J, Svensson L (2021) Proceedings of the IEEE/CVF Winter Conference on Applications of Computer Vision (WACV), pp 1369–1378
12. Nishi K, Ding Y, Rich A, Höllerer T (2021) Augmentation strategies for learning with noisy labels. http://arxiv.org/abs/2103.02130
13. Mounsaveng S, Laradji I, Ben Ayed I, Vazquez D, Pedersoli M (2021) Proceedings of the IEEE/CVF Winter Conference on Applications of Computer Vision (WACV), pp. 1691–1700
14. Bhujle HV, Vadavadagi BH (2019) 'NLM based magnetic resonance image denoising—A review.' Biomed Sig Process Control 47:252–261. https://doi.org/10.1016/j.bspc.2018.08.031
15. Bechtle S, Molchanov A, Chebotar Y, Grefenstette E, Righetti L, Sukhatme G, Meier F (2021) Meta learning via learned loss, pp 4161–4168. https://doi.org/10.1109/icpr48806.2021.9412010
16. Elhoseny M, Shankar K (2019) 'Optimal bilateral filter and convolutional neural network based denoising method of medical image measurements.' Measurement 143:125–135. https://doi.org/10.1016/j.measurement.2019.04.072
17. Yun S, Han D, Chun S, Joon Oh S, Yoo Y, Choe J (2019) Cutmix: Regularization strategy to train strong classifiers with localizable features. In: 2019 IEEE/CVF International Conference on Computer Vision, ICCV 2019, Seoul, Korea (South), October 27–November 2, pp 6022–6031. IEEE
18. Venu SK, Ravula S (2021) Evaluation of deep convolutional generative adversarial networks for data augmentation of chest x-ray images. Future Int 13(1):1–13. https://doi.org/10.3390/fi13010008
19. Singh SP, Wang L, Gupta S, Goli H, Padmanabhan P, Gulyás B (n.d.) Review_3D_Deep_Learning_Medical_Images, 1–13
20. Kim M, Yan C, Yang D, Wang Q, Ma J, Wu G (2020) Deep learning in biomedical image analysis. Elsevier, In Biomedical Information Technology (Second Edi). https://doi.org/10.1016/B978-0-12-816034-3.00008-0
21. Milletari F, Navab N, Ahmadi S (2016) V-Net: Fully convolutional neural networks for volumetric medical image segmentation. In: 2016 Fourth International Conference on 3D Vision (3DV), pp 565–571
22. Guo L, Zhao L, Wu Y, Li Y, Xu G, Yan Q (2011) Tumor detection in MR images using one-class immune feature weighted SVMs. IEEE Trans Magn 47(10):3849–3852
23. Singh L, Chetty G, Sharma D (2012) A novel machine learning approach for detecting the brain abnormalities from MRI structural images. In: IAPR International Conference on Pattern Recognition in Bioinformatics, Springer, Berlin, Heidelberg, November, pp 94–105

24. Jafar A, et al (2021) AI and machine learning paradigms for health monitoring system: Intelligent data analytics, SBD, volume 86. Springer Nature, Berlin, 513 p. https://doi.org/10.1007/978-981-33-4412-9. ISBN 978-981-33-4412-9
25. Tomar A, et al (2020) Machine learning, advances in computing, renewable energy and communication, LNEE volume 768. Springer Nature, Berlin, 659 p. https://doi.org/10.1007/978-981-16-2354-7. ISBN 978-981-16-2354-7

Effects of Common-Mode Voltage in ZSI-Based Induction Motor Drive for EV Applications

Dogga Raveendhra, M. Prashanth, and K. Sudha

Abstract In this manuscript, buck-boost type single-stage inverter is used to drive induction motor is adopted, and common mode voltages are investigated. Further, the effect of the parasitic parameters of motor due to stator to frame, stator to rotor, rotor to shaft, and shaft to ground etc., on common mode voltages is investigated in three-phase impedance source inverter. Common mode voltages sluggish the overall performance in case of drives due to drawing high-frequency components of currents to meet the demands of parasitic capacitances. Equivalent circuit model by considering various practical parameters of 5HP motor such as above mentioned parasitic capacitances is developed on MATLAB/Simulink platform. In this paper, bearing voltages, shaft voltages, and common mode currents are investigated in case of impedance source inverter.

Keywords ZSI · Impedance source inverter · Single-stage inverter · Induction motor drive and common-mode voltages

1 Introduction

Now-a-days, increase in the fossil fuels consumption had led to increased pollution and environment concern. If the consumption of fossil fuels is at the same rate, then the fuels will get perish after few decades. To prevent this, electric vehicle plays a major role. Advancements in the electric vehicles are possible with the improvement

D. Raveendhra (✉) · M. Prashanth · K. Sudha
EEE Department, Gokaraju Rangaraju Institute of Technology, Hyderabad, Telangana 500090, India
e-mail: doggaravi19@gmail.com

M. Prashanth
EEE Department, NITW, Warangal, Telangana 506004, India

K. Sudha
EEE Department, SR University, Warangal, Telangana 506371, India

A. Tomar et al. (eds.), *Proceedings of 3rd International Conference on Machine Learning, Advances in Computing, Renewable Energy and Communication*, Lecture Notes in Electrical Engineering 915, https://doi.org/10.1007/978-981-19-2828-4_36

of the power electronics. These vehicles are driven only by electric motor and are fed by power electronic converters to obtain variable speeds.

So, DC motor can be used for electric propulsion because of its good torque-speed characteristics, but they are bulky, less reliable, low efficiency; frequent maintenance is required due to commutator. As there is advancement in the field of electric drives, AC motors became famous owing to less maintenance, high reliability. Although permanent magnet motors have high power density but less constant power region. Permanent magnet made the constant power region fixed but can be varied by keeping an extra field winding. Switched reluctance motor is fault tolerant, rugged, but it creates EMI noise, torque ripple, uncontrollable bus current ripples, new topology of converter to drive the motor. From these comparisons with DC motor, switched reluctance motor, permanent magnet motor, cage induction motor is best suited for electric propulsion [1]. Induction motor has good capabilities and is mostly accepted for propulsion in the electric vehicle because of its ruggedness, low cost and maintenance, reliability. It can also be operated in adverse environment. In rigors environment like tractive and industrial drives, induction motor drives are well adapted, and they are highly preferred [2, 3]. Many advancements had already been done in the induction motor drive technology compared to the commutator fewer motor drives. Induction motor drive is fed through the inverter.

Generally, there are two types of inverters; they are voltage source inverter and current source inverter. However, these are suffered from the drawbacks. The main drawbacks of voltage source inverter (VSI) are the output voltage of VSI is less than the DC link voltage; the devices in the same phase leg shouldn't be turned ON simultaneously since it leads to short circuit to avoid this dead time has to be provided. Considering dead time causes distortion in the output waveform. Misgating due to EMI noise has to be avoided; otherwise, it will cause shoot through fault. And also, in VSI, to obtain sinusoidal waveform at the output of inverter, LC filter is used. On the other hand, in current source inverter (CSI), the input source is a constant current source which is a voltage source in series with the inductor. CSI allows unidirectional flow of current and blocking voltage bi-directionally. The output voltage of the current source inverter is high; then, the input DC voltage during DC to AC conversion the output voltage is boost inverter, whereas for AC to DC conversion, the output voltage is buck converter [4]. In CSI, either the lower or upper device should be turned ON at any instant, otherwise, the devices get damage when the inductor is open-circuited. Misgating can be caused due to EMI noise which leads to the short circuit. So, certain overlap time has to be provided for commutation of the device. Combined disadvantages of both the VSI and CSI are they cannot be buck-boost converter; it can be a buck converter or boost converter; it cannot be exchanged the power circuit of the converter, i.e., power circuit of current source converter cannot be replaced with voltage source converter, and VSI and CSI are mostly effected due to EMI noise [5].

To overcome the aforementioned drawbacks, Z-source inverter (Z.S.I) is the best converter between the source and load with inbuilt buck/boost capabilities. Z.S.I has the same characteristics of AC-DC converter with low cost and decreased complexity.

Shoot through in phase leg doesn't short circuit the inverter leg and improves the reliability. Modulation index and shoot through duty ratio will simultaneously control the voltage across the motor, power regulation between the super capacitor, and battery is also possible. Distinct values of phase current make effective use of switching, and conduction losses are possible with the leg numbers during shoot through scheme [6, 7]. One of the remedies for minimizing the harmonics in the inverter output is rise the switching frequency. But, this advancement is followed by some problems; we know that among them, common mode current (CMC) is one. The CMC is related to the common mode voltage (CMV) that leads to defective operation of the current detection circuits, unwanted electromagnetic interference (EMI), and destructs the bearings in motors [8]. Bearing currents flow through the parasitic capacitances between the stator to frame, stator to rotor, etc. In many cases, the effect of CMV is neglected, but it has to be considered for practical applications to obtain the absolute results. The above effects of CMV necessitate mitigation of CMV [9, 10].

2 Proposed System

The basic structure of electric vehicle is battery, bidirectional DC-DC converter (BDC), DC-AC converter and motor. The operating voltage of the motor is high, but the voltage of the battery pack is low. The voltage has to be boosted to operate the motor. A BDC inverter is a two-stage process implemented in most of the electric vehicles to boost the voltage. Alternatively, this voltage boosting can be done with the help of Z-source topology.

2.1 Battery

Toyota Prius uses nickel cadmium battery which has good performance and occupies little space. The nominal voltage of the battery pack is 201.6 V; it has each battery of voltage 7.2 V with 28 modules connected in series. It is charged from the generator of the HEV and meets power requirements of the motor. The battery is sized to meet specified output power of 21 kW [11–13].

2.2 Power Control Unit (PCU)

The main subassemblies of the power control unit are as follows: DC-DC buck converter, air conditioning compressor inverter, bidirectional DC-DC converter, motor inverter, and generator inverter. The battery nominal voltage 201.6 V is boosted up to 500 V by the bidirectional DC-DC converter (BDC). And it can also be used to charge the battery from the generator. The power rating of the BDC is 25 kW. The

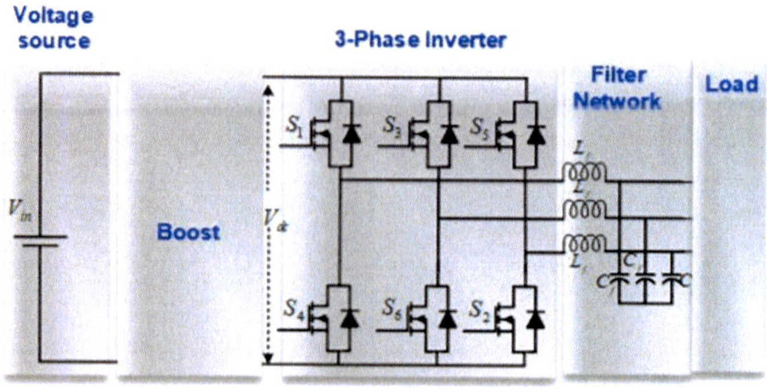

Fig. 1 Conventional two-stage power conversion

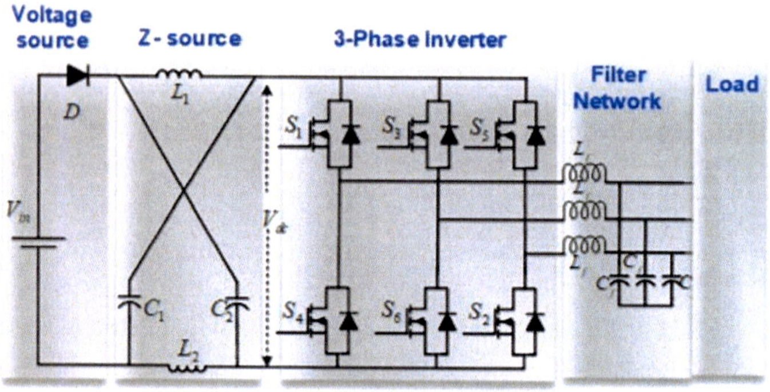

Fig. 2 Z-source inverter

output of BDC forms DC link and feeds to inverter for the DC-AC conversion to achieve 3-ph AC supply, which can be used to drive the motor. This type of scheme is called two-stage power condition unit as shown in Fig. 1. PCU also contains 12 V DC-DC buck converter which meets the requirements of the auxiliary power supply. Auxiliary power supply is used for lighting, audio, etc. During the period of acceleration or at peak period, the power requirement is 50 kW; it is not only fed by the generator (30 kW) and also fed by the battery (20 kW) via a BDC to the motor. Z-source inverter plays equivalent role of boost converter fed inverter in a single stage as shown in Fig. 2.

2.3 Z-Source Inverter (ZSI)

The ZSI overcomes the disadvantages of the conventional voltage source and current source inverter. The below Fig. 2 depicts the configuration of the ZSI. In this inverter, an impedance network is implanted in the power circuit. The impedance source is made up of capacitor (L_1, L_2) and inductor (C_1, C_2) connected in X shape [3].

Battery and X-shaped impedance network composed of two inductor and two capacitor acts as the impedance source. In the switching network, power semiconductor device and diode are connected in anti-parallel manner to form a bidirectional switch. In DC-AC, DC-DC, AC-AC, AC-DC power converters, Z-source can be implemented. The operating principle of the ZSI is explained with an example: Battery is used as the source for the ZSI in DC to AC conversion. The figure shown above is ZSI for battery-powered application. Depending on the current drawn from the stack, there will be variation in the voltage that is generated by the battery. The voltage at the input of the switching network has to be greater than the AC voltage, so impedance source is proposed for battery-powered vehicles. ZSI with battery as a source can be used directly to produce high AC voltage compared to the input battery voltage. Current flow in opposite direction to the DC source is avoided by connecting a diode in series. Irrespective of the battery voltage, the output voltage value has wide range of variation in between zero and infinity which is an advantage of ZSI. The output voltage of ZSI has a huge variation, and it is buck-boost inverter. This unique characteristic is not possible in the conventional voltage source and current source inverters. In VSI, there will be eight possible states where as in the ZSI, there will be nine operating states of switches. There will be two zero states and six active states that can be obtained in the voltage source inverter. The zero states are attained by shorting the upper or lower power devices, respectively. In ZSI, there is an additional switching state which is introduced by turning ON the upper and lower devices of the same leg, i.e., shorting across the load terminals, two phase legs, or all three-phase legs. This state is known as the shoot through state, and it is avoided in the conventional voltage source inverter because shoot through will occur. In seven different methods, shoot through can be introduced into the inverter circuit either by shorting the same leg, two phase legs, or all three-phase legs. The advantage of the shoot through state is that the output voltage of the inverter can be either bucked, or it can be boosted. Zero shoot through states is assigned evenly in each phase without changing the total time interval for zero state. This means that there is no change in the active states. The equivalent DC-link voltage of the inverter is increased due to the shoot through conditions. Equations (2) and (3) will explore the detailed relationships for boost factor and output voltages in this case. It is clear that the equivalent switching frequency observed from the Z-source network is six times the switching frequency from the main inverter, significantly lowering the desired passive components.

Boost factor:

$$B = \frac{T}{T_1 - T_o} = \frac{1}{1 - 2\frac{T_o}{T}} \tag{1}$$

Output peak phase voltage

$$\widehat{v}_{ac} = M.\frac{\widehat{v}_i}{2}$$
$$\widehat{v}_{ac} = M.B.\frac{V_o}{2} \quad (2)$$

$$\text{Capacitor voltage } V_{C1} = V_{C2} = V_C = \frac{1 - \frac{T_o}{T}}{1 - 2\frac{T_o}{T}} V_o$$

2.4 *Motor*

Torque and field can be separately controlled by implementing vector control in induction motor (IM) drive. During the constant power mode, the speed range improves. Operating the induction motor above the rated speed by weakening the flux in a constant power mode is achieved. The working of induction motor in the constant power region is limited due to its breakdown torque. Breakdown torque is achieved at critical speed. The critical speed occurs approximately two times the synchronous speed. Motor will stall when it is operated above the critical speed taking maximum current. The efficiency of IM is less than permanent magnet motor. At high speeds, efficiency declines because the rotor winding and rotor copper losses are zero. The induction motor propulsion has many disadvantages; they are poor power factor, low efficiency at high speed, less inverter usage, large power motor, high loss. These difficulties are already discussed in several papers. So, to overcome these issues, researchers suggested changes in the design of the IM for electric vehicle applications, and detailed modeling of such systems will be helpful for attaining accurate responses. Constant power region can be extended without increasing the size of the motor by applying multiphase pole induction motor drive method. In traction application, multiphase pole induction motor improves the constant power region without increasing the size of motor. Another method is modifying the sinusoidal PWM method along with shoot through duty ratio insertion in PWM technique will be employed to compare with two carriers for the generation of firing signals to ZSI. Complete block diagram of ZSI-powered induction motor drive is shown in below Fig. 3.

2.5 *Common Mode Voltage*

The CMV is the voltage difference between the neutral point of the star-connected load and the midpoint of the VSI DC bus. CMV can be written as

$$V_{\text{no}} = \frac{V_{\text{ao}} + V_{\text{bo}} + V_{\text{co}}}{3} \quad (3)$$

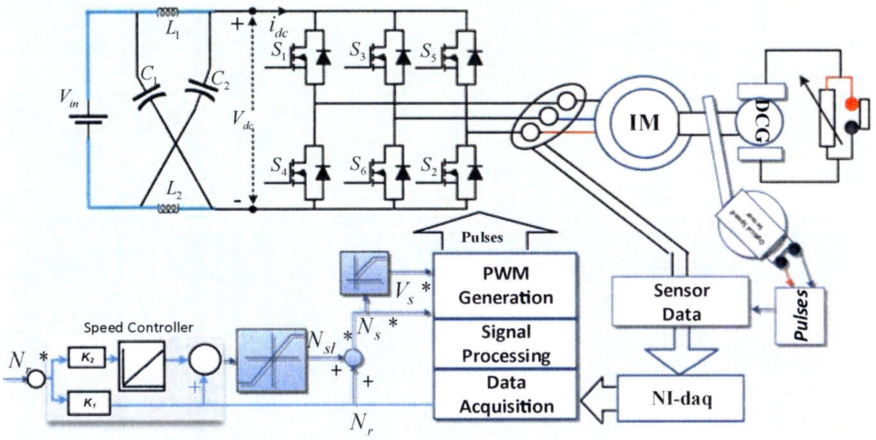

Fig. 3 Slow shows the proposed system for electric vehicle

As it is known that two-level three-phase VSI provides only discrete output voltages instead of the sinusoidal voltages. The CMV instantaneous values will be varying between $\pm V_{\text{dc}}/6$ and $\pm V_{\text{dc}}/2$ instead of zero and depends on the inverter switching state. Based on the switching states, the CMV is explained, when the switching state changes the CMV varies suddenly. Due to the sudden variations, there is problem of sudden damage in the power electronic drive applications. The motor windings, rotor frame, motor frame are close to each other; therefore, stray capacitances come into effect at high frequencies between the layers as shown in Fig. 4a. So, the capacitances can be represented as effective capacitance. The modeled circuit for CMV in the inverter-fed induction motor drive is shown below Fig. 4b. The connection of effective resistance, inductance, and stray capacitance of the motor forms

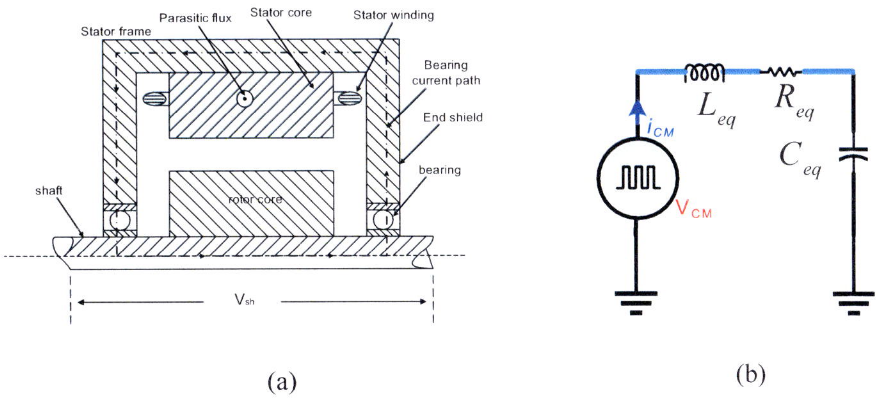

Fig. 4 Equivalent circuit of inverter with IM drive

the series RLC resonant circuit for the inverter-fed drive. The frame of the motor is grounded for the purpose of safety in practical appliances. At high frequencies, when the CMV is high with the rate of change of voltage is high causes CMC to flow through the stray capacitance and between the motor to the ground.

The motor-rated current and the maximum value of the CMC are similar in some applications. When the value of CMV/CMC is high, then the bearing fails. EMI is caused in the AC grid that supplies power to the drive and the control circuit because the CMV and CMC are at high frequencies. Due to this, undesirable tripping is caused, and electronic circuits fail. So, to avoid this, the mitigation of CMV and CMC is needed, and various methods that reduce CMV are improved [10].

Installing of the active and passive filters will rise the size and cost. Since the improved CMV techniques have some drawbacks in the practical condition and limited performance but mitigates CMV/CMC. CMV/CMC can also be minimized by optimization of the pulses pattern in the PWM strategy. Some methods are theoretically feasible, but practical systems have low performance compared to developed method. Implementing some methods to the practical systems will create issues like high THD when load current flows. When the AC motor drive is connected through a long cable to the inverter, then it produces high peak over shoots in the line to line voltage in some methods.

3 Results and Discussion

ZSI-fed IM drive to know the importance of CMV in electric vehicle application, simulation studies were performed with the help of MATLAB Simulink. Figures. 1, 2, 3, 5, and 6 shown below are results of the induction motor drive without and with considering CMV. The parasitic effects of induction motor have to be taken into account for proper operation of drive system switches, elements, and they have to be designed for peak values. When the CMV is taken into account, the line current peak value has increased, and parasitic capacitances will be formed due to which the current flows through them, they are stator to frame, stator to rotor, ground currents. The parameters are shown in the following Table 1 (Figs. 7, 8, 9, and 10).

4 Conclusion

The effect of the parasitic components of motor on CMV is successfully investigated in three-phase ZSI-fed IM drive. For the investigation of various parameters related to common mode voltage effects, simulation model has been developed by considering various practical parameters such as parasitic capacitances between various parts of induction motor such as stator to frame, stator to rotor, rotor to shaft, and shaft to ground. Stator to frame, stator to rotor, and ground currents have been presented for the successful demonstration of the effect of common mode in single-stage drive.

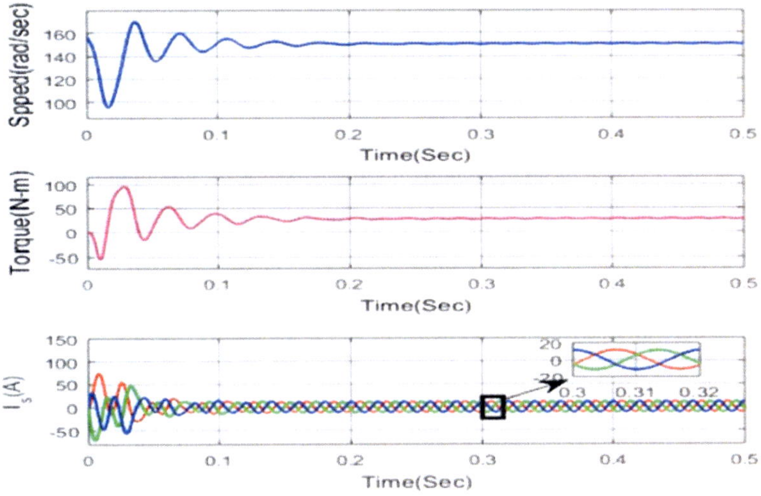

Fig. 5 Speed, torque, stator current of induction motor drive without considering the effect of CMV

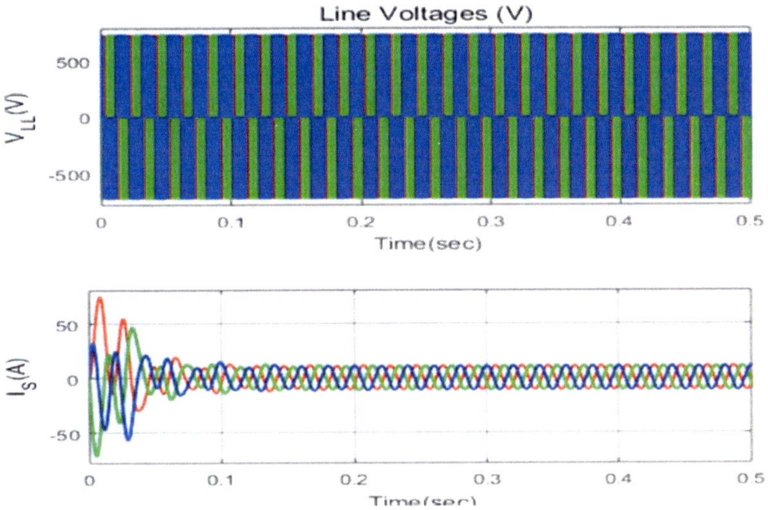

Fig. 6 Line voltage and line current of the IM drive without considering the effect of CMV

Table 1 Various parameters with CMV and without CMV in an induction motor drive

Parameters	With CMV	Without CMV
Rotor speed	154 rad/sec	150 rad/ sec
Electrical torque	20 N-m	25 N-m
Stator current	10 A	10 A
Line to line voltage	750 V	750 V
Line current	19 A	15 A
Bearing voltage	10 V	–
CMV	400 V	–
Stator to frame current	0.025 A	–
Stator to rotor current	0.025 A	–
Ground current	0.05 A	–

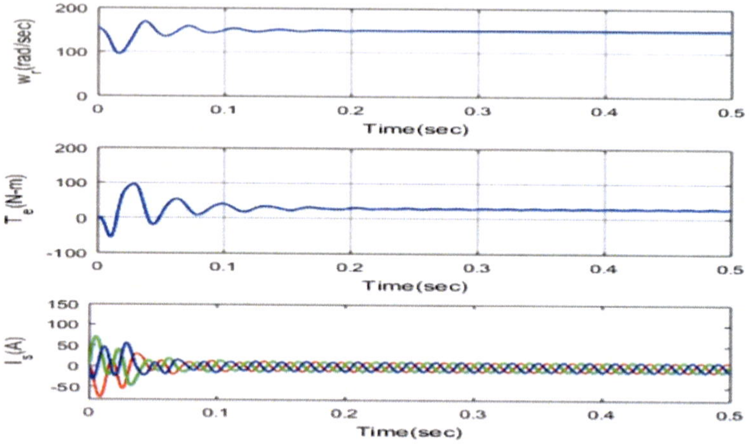

Fig. 7 Speed, torque, stator current of induction motor drive after considering the effect of CMV

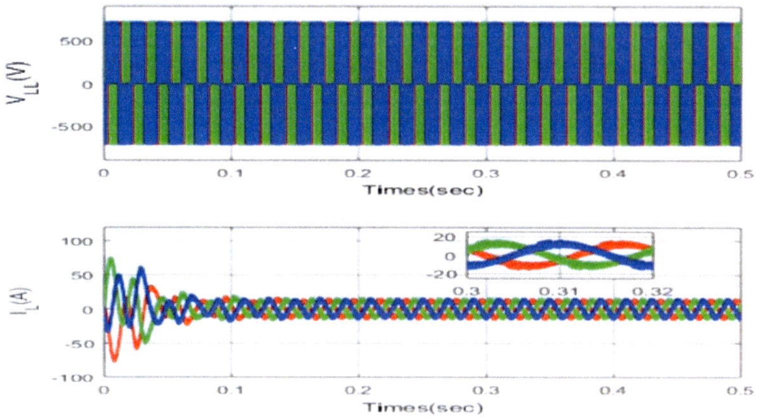

Fig. 8 Line to line voltage and line current of induction motor drive after considering the effect of CMV

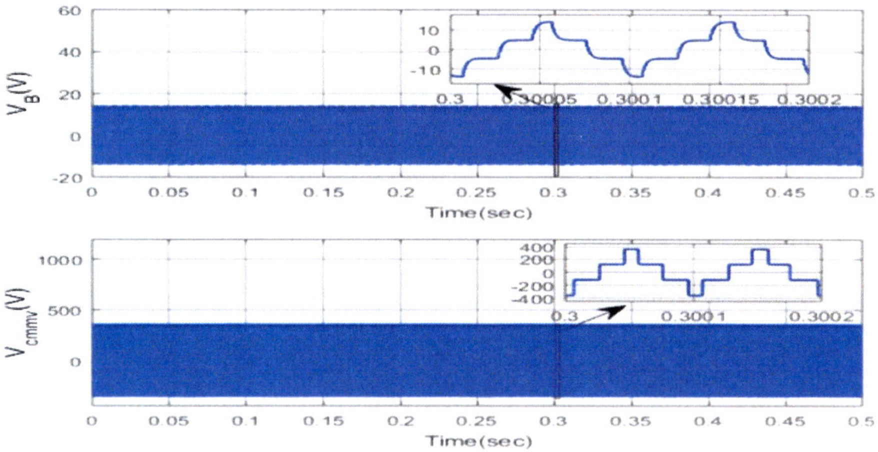

Fig. 9 Bearing voltage and CMV of induction motor drive

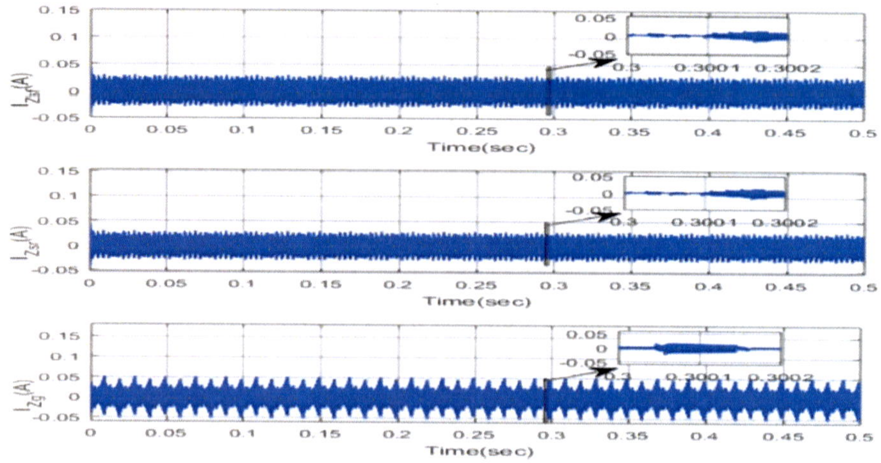

Fig. 10 Stator to frame, stator to rotor, ground currents

References

1. Zeraoulia M, Benbouzid MEH, Diallo D (2006) Electric motor drive selection issues for HEV propulsion systems: A comparative study. IEEE Trans Veh Technol 55(6):1756–1764
2. Cao J, Emadi A (2012) A new battery/ultra capacitor hybrid energy storage system for electric, hybrid, and plug-in hybrid electric vehicles. IEEE Trans Power Electron 27(1):122–132
3. Peng FZ (2003) Z-source inverter. IEEE Trans Ind Appl 39(2):504–510
4. Kumar L, Jain S (2012) A multiple input dc-dc converter for interfacing of battery/ultra-capacitor in EVs/HEVs/FCVs. In: 2012 IEEE 5th India International Conference on Power Electronics (IICPE), Delhi, pp 1–6
5. Baba M, Lascu C, Boldea I (2012) Z converter control of a V/f induction motor drive. In: 2012 13th International Conference on Optimization of Electrical and Electronic Equipment (OPTIM), Brasov, pp 529–534
6. Lai JS, Nelson DJ (2007) Energy management power converters in hybrid electric and fuel cell vehicles. Proc IEEE 95(4):766–777
7. Raveendhra D, Pathak MK (2018) Three-phase capacitor clamped boost inverter. IEEE J Emerging Selected Topics Power Elect 7(3):1999–2011
8. Lai Y-S, Chen P-S, Lee H-K, Chou J (2004) Optimal common-mode voltage reduction PWM technique for inverter control with consideration of the dead-time effects-part II: applications to IM drives with diode front end. IEEE Trans Industry Appl 40(6):1613–1620
9. Erdman JM, Kerkman RJ, Schlegel DW, Skibinski GL (1996) Effect of PWM inverters on AC motor bearing currents and shaft voltages. IEEE Trans Ind Appl 32(2):250–259
10. Ün E (2007) Common mode voltage and current reduction in voltage source inverter driven three phase AC motors. In: M.S. thesis, Middle East Tech Univ, Ankara, Turkey, November
11. Evaluation of 2004 Toyota Prius Hybrid Electric Drive System manual, U.S. Department of Energy

12. Iqbal A, et al (2021) Intelligent data-analytics for condition monitoring: Smart grid applications. Elsevier, 268 p. https://www.sciencedirect.com/book/9780323855105/intelligent-data-analytics-for-condition-monitoring. ISBN 978-0-323-85511-2
13. Tomar A, et al (2020) Machine learning, advances in computing, renewable energy and communication, LNEE volume 768. Springer Nature, Berlin, 659 p. https://doi.org/10.1007/978-981-16-2354-7 ISBN 978-981-16-2354-7

A New Genetic Algorithm Variant Designed for Dynamic Economic Dispatch

Harish Pulluri, Preeti, B.Vedik, and T. Anil Kumar

Abstract One of the key tasks of power production operations and control is dynamic economic dispatch (DED). It defines the optimum settings of generators for a given period with a projected load requirement. The aim is to run an electricity system cheaply as long as it operates within its safety limitations. Therefore, this article aims to propose a hybrid technique to solve DED. The basic genetic algorithm (GA) when used as a search level takes longer to get nearly optimal results. The proposed technique uses a three-parent crossover and diversity operator resulting in increasing the potential for both exploration and exploitation of the algorithm technique. Two test cases with quadratic cost function are employed to demonstrate the efficacy and validity of the proposed method. Experimental findings compared with many DED solution techniques, namely differential evolution (DE), hybrid DE, sequential quadratic programing, artificial bee colony, and other recently published results, and these results proved that proposed technique achieved superior solutions.

Keywords Dynamic economic dispatch · Three-parent crossover · Transmission loss · Diversity factor

1 Introduction

DED is an extension of the issue of static economic transmission (SED). SED scenario finds the cost-efficient production combination of generators to fulfil the anticipated demand for a single load at a particular time hour. Because of the high-power system

H. Pulluri · T. A. Kumar
Electrical Engineering Department, Anurag University, Hyderabad, Telangana, India
e-mail: harisheee@anurag.edu.in

Preeti (✉)
Skill Faculty of Engineering and Technology, Shri Vishwakarma Skill University, Palwal, Haryana, India
e-mail: preetieednith@gmail.com

B.Vedik
Electrical Engineering Department, SR University, Warangal, Telangana, India

A. Tomar et al. (eds.), *Proceedings of 3rd International Conference on Machine Learning, Advances in Computing, Renewable Energy and Communication*, Lecture Notes in Electrical Engineering 915, https://doi.org/10.1007/978-981-19-2828-4_37

load fluctuation, SED could not meet the operating restrictions of the generators. The primary aim of the DED is to reduce the overall cost of production while meeting the limitations of equality, inequality and dynamic restrictions. Moreover, owing to look ahead inability, the outcomes of SED will be suboptimal when evaluating a time horizon moment of a time instance [1]. The balance of load demand is the constraint on equality, and the restrictions on the forbidden area and limitations of capacity generation are the constraints of inequality. The solution of the DED issue is more complex by considering these dynamic restrictions. Much work has been expended in trying to successfully address the essential but complex DED issue, and a variety of solution approaches have been suggested. Until now, these techniques have been experimentally divided into two groups: classical and heuristic methods. Classical methods include Lagrangian method [2], quadratic programing [3] and dynamic programing [4], etc., and while they offer some benefits like great calculation efficiency and theoretically optimal [5], they have several drawbacks as well. As a substitute for traditional methods, heuristic techniques have received much attention and proven their efficacy as strong optimizers for the issue of DED in the past several decades, like evolutionary programing [6] particle swarm optimization (PSO) [7], differential evolution (DE) [8], artificial bee colony (ABC) [9], krill herd algorithm (KHA) [10] and artificial immune system (AIS) [11].

In 1960, John Holland invented the genetic algorithm (GA) [12]. To date GA has been used to resolve a number of real-world issues of optimization [13–15]. It may quickly reach the global minimum search area, and it takes more time to converge. A hybrid approach is one way of tackling this problem. Several GA variations were thus presented to avoid the disadvantage trap in local optima and reach global solution with in less time [16]. The major contributions in this paper are as follows:

(i) Consider DED problem instead of classical ED, since introduction of dynamic constraints makes the DED problem more complicated.
(ii) DED problem is solved using newly created a variant of GA with three-parent crossover. This method introduced a three-parent crossover and a typical mutation via a diversity operator, resulting in maintain efficient chromosomes.

The effectiveness of GA-TPC is shown with two distinct test systems. The remaining paper is arranged as follows, Sect. 2 provides a mathematical model of DED problem considering valve points, Sect. 3 offers about GA-TPC algorithm, Sect. 4 shows three different cases, and achieved results compare with the outcomes of the latest techniques and the final conclusion in Sect. 5.

2 Mathematical Model

DED is required to optimize the overall cost of all thermal generators exposed to different restrictions on a regular basis over a time horizon. The thermal cost characteristics, associated constraints and basic formulations are discussed more below [7].

2.1 Optimization of Total Cost (TC)

Usually, the DED problem's goal function may be approximated by a simple quadratic equation [7].

$$\min f = \sum_{t=1}^{T} \sum_{m=1}^{NG} a_m + b_m P_{Gm,t} + C_m P_{Gm,t}^2 \tag{1}$$

where f gives TC of all generators; $P_{Gm,t}$ indicates active power of mth generator at tth hour.

2.2 Optimization of TC with Valve Points (TCV)

However, the production curve for multi-valve steam units differs considerably in comparison with the quadratic function of the active power output. The inclusion of a valve point effect on the fuel cost of the producing unit provides a better representation of the cost of fuel. As the valve point is completed with spiking, the fuel price function includes more nonlinear series. A non-convex function to assess the effect of the valve points is thus employed in the study given below [7].

$$\min f = \sum_{t=1}^{T} \sum_{m=1}^{NG} a_m + b_m P_{Gm,t} + C_m P_{Gm,t}^2 + \left| d_m \times \sin(e_m (P_{Gm,t}^{\min} - P_{Gm,t})) \right| \tag{2}$$

where $a_m, b_m, c_m, d_m \& e_m$ indicate cost coefficients of mth generator.

2.3 Constraints

The limitations in the current work are briefly described below [7].

Equality constraints: It is a real power balance constraint and is given below,

$$\sum_{n=1}^{NG} P_{Gnt} = P_D(t) + P_{\text{loss}}(t) \qquad t = 1, 2, \ldots T \tag{3}$$

where P_D reports load demand, and P_{loss} indicates transmission loss and is calculated as follows,

$$\sum_{k=1}^{NG}\sum_{m=1}^{NG} P_{kt}B_{km}P_{mt} + \sum_{k=1}^{NG} B_0 P_{kt} + B_{00} \qquad t = 1, 2, ...T \tag{4}$$

where B_{km}, B_k&B_{00} are called loss coefficients.

Inequality constraints: These are expressed among their low and high limits and are given below,

$$P_{Gn}^{\min} \le P_{Gnt} \le P_{Gn}^{\max} \quad n = 1, 2, \ldots, N_G \qquad t = 1, 2, \ldots, T \tag{5}$$

3 Proposed Genetic Algorithm with Three-Parent Crossover

Different GAs for many real-world numerical problems have been presented over several decades. However, the effectiveness of the various approaches is dependent only on features of the objective function. In certain instances, GA did not perform nor was compared with other algorithms [17, 18]. Therefore, GA performance is improved by adding three-parent crossover instead of a typical two-point crossover, and diversity operator is applied instead of a fairly regular mutation [17]. The current crossover uses three parents to produce three new children, helping explore and leverage the diversity operator.

Crossover is a GA operator of great importance. It is responsible for recombination structure and GA convergence speed. The conventional GA combines the chromosomes from the two chosen parents to produce a new chromosome which inherits information regions contained in parent chromosomes. The crossover suggested in the GA-MPC is based on an idea of heuristic crossover, and here, a child (c) is created from a set of two parents (a, b), like $c = a + rand(a - b)$, where 'rand' is a random number among 0,1. The GA-MPC nevertheless uses three rather than two parents.

The procedure for the proposed algorithm is explained below.

(i) **Selection**

Selection of the parents is a simple process by which parents are chosen based on fitness of the chromosomes. The likelihood of adding additional offspring to the following generation is that solutions with high fitness ratings. A basic selection of roulette wheels rule utilized in our approach [19].

(ii) **Proposed three-parent crossover**

Crossover procedure is very important in GA. To generate new offspring, the crossover must be able to use search space information. Offspring distribution should neither be disproportionately narrow or disproportionately large compared to that of their parents. It is possible that the offspring will lose diversity and converge early if their distribution is much smaller than that of their parents. The opposite may be true if the children are dispersed extensively, in which case they may be too varied and require an excessively long time to converge to optimality. There should be a balance between exploration and exploitation in the next generation. Based on the aforementioned idea, in the proposed work, three parent crossover based on random procedure is used rather than regular two parent crossover. The procedure is given below [17].

1. Select the parent individuals by using selection process.
2. If any two individuals are similar, then one is replaced with randomly from selection pool.
3. Arrange those three individuals according to best to worst fitness value.
4. A number 'ε' is produced randomly;

    (a) New off springs are produced by using following equations

$$\begin{aligned} \mathrm{OF}_1 &= x_1 + \varepsilon(x_2 - x_3) \\ \mathrm{OF}_2 &= x_2 + \varepsilon(x_3 - x_1) \\ \mathrm{OF}_3 &= x_3 + \varepsilon(x_1 - x_2) \end{aligned} \tag{6}$$

where x_1, x_2 & x_3 are the selected parents by using selection process, and OF_1, OF_2 & OF_3 denote newly generated off springs.

(iii) **Diversity operator**

To improve the exploitation capability in the individuals, diversity operator introduced in [14] considered here.

The step-wise procedure of GA-TPC to solve ED is given below:

Step 1: Initialize GA-TPC variables, max generations ($\mathrm{G}_{\max}$).

Step 2: Each chromosome in GA-TPC is a solution to a DED issue. The kth chromosome in mth generation is expressed in below given form

$$X_k^m = \begin{bmatrix} P_{g1,1,k}^m & P_{g1,2,k}^m & \cdots & P_{g1,t,k}^g \\ P_{g2,1,k}^m & P_{g2,2,k}^m & \cdots & P_{g2,t,k}^m \\ \vdots & \vdots & \ddots & \vdots \\ P_{gNg,1,k}^m & P_{gNg,2,k}^m & \cdots & P_{gNg,t,k}^m \end{bmatrix} \begin{matrix} k = 1, 2 \ldots \mathrm{NP} \\ g = 1, 2 \ldots G_{\max} \end{matrix} \tag{7}$$

where t indicates number of intervals in the dispatch period.

Step 2: Evaluate fitness of every individual using Eq. 8.

$$|F| = f + w_P\left(|P_{G1} - P_{G1}^{\lim}|\right)^2 \tag{8}$$

where w_P indicates penalty value of slack bus real power.

Step 3: Apply the selection, proposed crossover, diversity operator, and create new generation.

Step 4: If any variable exceeds its existing limits, then it will be set to inline high or low value.

Step 5: Terminate the process, if utmost iterations are marked, and take the best result from previous iteration as best solution. Else, go to Step 2.

4 Simulation Results

Two different modules are investigated to assess the feasibility and efficacy of the GA-TPC technique suggested in the solution of the DED issue. The dispatch time is chosen as 24 h for one day. The number individuals and utmost iterations in all the cases are considered 40 and 300, respectively. The following are the two cases:

M1: a three-generator system without point loadings.

M2: a ten-generator system with valve point loadings.

4.1 M1: 3 Unit System

The proposed system consists of three generators and complete data for this system that includes cost characteristics of generators, generator limits, and load demand in each interval is referred from reference [20]. The optimal set of active powers obtained to this system with GA & GA-TPC are given in Table 1. These results are compared with CSA [20] and ISA [20], RGM [21] and ACO [21] and are given in Table 2. From this table, it is noticed that the suggested approach provides a superior way to discover solutions to such complex DED issues, with minimum, average and maximum costs. A minimum of 176,017.5363 ($/day) and a minimum of 176,059.3264 $/day achieved utilizing the formulations of proposed GA-TPC and original GA, showing the remarkable nature of the suggested method. In addition, the convergence characteristic of the method suggested is compared and shown in Fig. 1 with the original GA. This figure indicates that both convergence speed and optimum objective function of the proposed GA-TPC beats conventional GA. Here, 3-unit system size is very small, so the deviation of optimal cost from GA to GA-TPC is very small. Thus, the convergence curves are much closer to each other.

4.2 M2: 10 Unit System

GA-TPC performance is identified by considering 10 unit systems for solving DED problem with inclusion of valve points. This system is believed to have a complete date from [7]. Table 3 illustrates the findings achieved for the 10-unit system with

Table 1 Optimal solution of 3-unit system without VPL using GA & GA-TPC

	GA				GA-TPC			
t (h)	P_1 (MW)	P_2 (MW)	P_3 (MW)	Cost (S/h)	P_1 (MW)	P_2 (MW)	P_3 (MW)	Cost (S/h)
1	41.341	48.659	50	6150.808	37.909	42.5476	59.543	6148.4312
2	43.748	42.72	63.532	6377.581	40.711	45.7443	63.545	6376.6675
3	43.961	54.113	56.926	6493.501	43.022	47.1747	64.803	6491.4402
4	38.382	40	81.618	6614.415	43.423	49.8616	66.715	6606.4585
5	48.998	54.571	61.43	6723.759	43.688	52.0029	69.309	6722.0393
6	44.832	47.487	77.681	6839.822	45.969	52.9109	71.121	6837.9471
7	54.631	53.428	66.942	6955.989	48.22	53.6396	73.14	6954.2919
8	46.12	62.795	71.085	7073.065	49.569	55.1802	75.251	7071.0227
9	63.42	56.313	90.266	7781.993	59.315	63.64	87.045	7779.7419
10	67.692	68.29	94.018	8260.228	66.437	69.6138	93.95	8260.1533
11	69.608	74.686	95.705	8503.082	69.568	72.2599	98.172	8502.7299
12	79.374	72.439	98.187	8748.433	72.68	74.5422	102.78	8746.9002
13	71.719	74.971	93.309	8503.619	69.735	71.9122	98.353	8502.7284
14	66.244	65.051	88.705	8019.541	63.295	65.9794	90.725	8019.16
15	58.386	60.831	80.783	7542.076	56.408	60.8439	82.748	7541.9093
16	53.472	52.366	74.163	7071.661	50.478	55.3434	74.179	7071.0302
17	43.3	56.332	70.368	6838.516	47.312	52.192	70.496	6837.9732
18	48.842	56.649	79.508	7188.46	52.24	56.1071	76.652	7188.1803
19	45.17	59.664	95.166	7548.116	57.969	60.1938	81.837	7542.0067
20	72.003	64.392	103.6	8505.098	69.57	71.5997	98.83	8502.7353

(continued)

Table 1 (continued)

	GA				GA-TPC			
t (h)	P_1 (MW)	P_2 (MW)	P_3 (MW)	Cost (S/h)	P_1 (MW)	P_2 (MW)	P_3 (MW)	Cost (S/h)
21	67.307	65.251	92.442	8139.833	64.692	67.3155	92.992	8139.4628
22	54.213	55.534	80.253	7305.985	51.872	58.4052	79.723	7305.7212
23	48.387	49.286	62.327	6607.532	43.017	49.8323	67.151	6606.4604
24	45.796	49.204	50	6266.213	38.384	47.6841	58.932	6262.345
TC ($/h)				**176,059.3264**	**TC ($/h)**			**176,017.5363**

Table 2 Comparison of the statistical analysis for 3-unit system with the other methods

Method	Minimum cost ($/h)	Average cost ($/h)	Maximum cost ($/h)	ET (min)
RGM [21]	177,291	–	–	–
ACO [21]	176,212	–	–	–
CSA [20]	176,370	–	–	–
ISA [20]	176,320	–	–	–
GA	**176,059.3264**	**176,066.6535**	**176,095.8222**	**0.38**
GA-TPC	**176,017.5363**	**176,019.1552**	**176,028.3286**	**0.42**

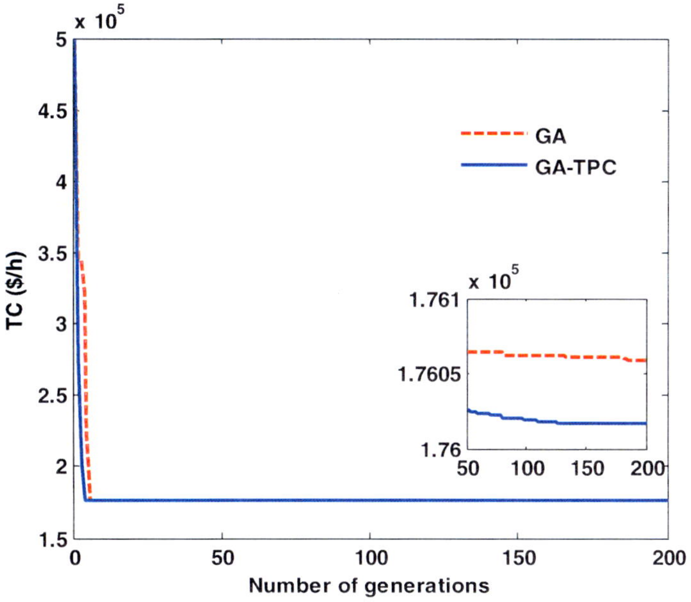

Fig. 1 Convergence curve of 3-unit system

valve point loading effect. These findings are compared to those of previously developed algorithms such as DE [8], hybrid EP-SQP [6], hybrid PSO-SQP [18], deterministically guided PSO (DGPSO) [7], hybrid DE (HDE) [7], improved DE (IDE) [7], ABC [6], modified DE (MDE) [7], AIS [12], AIS-SQP [12], chaotic DE (CDE) [7] and improved PSO (IPSO) [7]. This table shows a comprehensive comparison of solution quality, including lowest, average and maximum cost, as well as simulation time, and it is confirmed that the proposed method produces more optimum results outcomes that the methods described in the literature. Tables 4 and 5 shows the optimal set of active powers obtained to this system with GA-TPC and GA respectively. The suggested algorithm's convergence characteristic is shown in Fig. 2 and compared to the original GA. As can be seen from the graph, the suggested method beats the original GA in terms of convergence speed and optimality. The variation

Table 3 Comparison of the statistical analysis for 10-unit system with the other methods

Method	Minimum cost ($/h)	Average cost ($/h)	Maximum cost ($/h)	ET (min)
DE [8]	1,019,786	0	0	11.25
EP-SQP [6]	1,031,746	1,035,748	0	20.51
PSO-SQP [7]	1,027,334	1,028,546	1,033,986	16.37
DGPSO [7]	1,028,835	1,030,183	0	15.39
HDE [7]	1,031,077	0	0	0
IDE [7]	1,026,269	0	0	0
ABC [6]	1,021,576	1,022,686	1,024,316	2.6029
MDE [7]	1,031,612	1,033,630	0	12.5
AIS [12]	1,021,980	1,023,156	1,024,973	19.01
AIS-SQP [12]	1,029,900	0	0	0
CDE [7]	1,019,123	1,020,870	1,023,115	0.32
IPSO [7]	1,018,217	1,018,965	1,020,418	2.8
GA	**1,029,091.80**	**1,029,189.56**	**1,029,455.20**	**1.2**
GA-TPC	**1,015,473.71**	**1,015,536.60**	**1,015,823.71**	**1.1**

of TC with 20 trials is shown in Fig. 3 for 6-unit system, and it is observed that 17 trials were achieved optimal cost by the GA-TPC method over 20 trials and indicates the precision of the proposed method. Aforementioned simulation results depict that GA-TPC is successful in addressing small-scale test systems and using it to solve multi-objective DED for large and practical power systems would be an extension of the current study.

5 Conclusion

To address the dynamic economic dispatch issue of power systems with valve point loading effects, this article proposes a novel method termed genetic algorithm with three-parent crossover. Two different test scenarios are used to validate the technique. Comparing the suggested technique to other previously published approaches, including lowest, average and maximum costs as well as simulation time, provides a thorough understanding of the pros and cons of each. The findings of the study show that GA-TPC was able to find solutions that were more cost-effective. The comparison of suggested algorithm's convergence characteristics with conventional GA also confirms the speed and ability of the GA-TPC method to discover superior solutions. These facts suggest that the technique under consideration is capable of resolving DED problems.

Table 4 Optimal solution of 10-unit system using GA-TPC

t (h)	P_1(MW)	P_2(MW)	P_3(MW)	P_4(MW)	P_5(MW)	P_6(MW)	P_7(MW)	P_8(MW)	P_9(MW)	P_{10}(MW)	Cost ($/h)
1	226.62	135.00	73.00	60.00	167.33	122.45	129.59	47.00	20.00	55.00	28,359.6021
2	379.87	135.00	88.087	60.00	73.00	122.45	129.59	47.00	20.00	55.00	30,005.0057
3	455.69	222.26	73.00	60.00	73.00	122.45	129.59	47.00	20.00	55.00	33,095.292
4	303.25	309.53	286.17	60.00	73.00	122.45	129.59	47.00	20.00	55.00	36,169.1056
5	226.62	396.79	311.98	60.00	73.00	160.00	129.59	47.00	20.00	55.00	37,922.3118
6	379.87	396.79	306.73	60.00	73.00	160.00	129.59	47.00	20.00	55.00	41,117.5852
7	456.50	396.79	291.79	60.00	122.86	122.45	129.59	47.00	20.00	55.00	42,560.2457
8	379.87	396.79	305.13	60.00	222.59	160.00	129.59	47.00	20.00	55.00	44,339.2702
9	456.50	396.79	340.00	60.00	222.59	123.51	129.59	120.00	20.00	55.00	47,894.9451
10	456.50	396.79	307.81	241.25	222.59	122.45	129.59	120.00	20.00	55.00	51,346.7458
11	456.50	396.79	297.39	300.00	222.59	148.11	129.59	120.00	20.00	55.00	53,239.2323
12	456.50	460.00	298.94	300.00	222.59	160.00	129.59	85.31	52.05	55.00	55,271.1464
13	456.50	460.00	297.39	241.25	222.59	142.66	129.59	47.00	20.00	55.00	51,623.3424
14	379.87	309.53	300.40	300.00	222.59	160.00	129.59	47.00	20.00	55.00	48,140.8612
15	379.87	396.79	297.39	60.00	172.73	144.60	129.59	120.00	20.00	55.00	44,559.2939
16	303.25	309.53	284.57	60.00	222.59	122.45	129.59	47.00	20.00	55.00	39,418.2678
17	226.62	396.79	311.98	60.00	73.00	160.00	129.59	47.00	20.00	55.00	37,922.3118
18	456.50	309.53	305.06	60.00	122.86	122.45	129.59	47.00	20.00	55.00	40,981.7692
19	456.50	396.79	315.93	60.00	172.73	122.45	129.59	47.00	20.00	55.00	44,266.6374
20	456.50	460.00	317.61	241.25	222.59	122.45	129.59	47.00	20.00	55.00	51,600.466
21	456.50	396.79	340.00	60.00	222.59	126.144	129.59	85.31	52.05	55.00	47,921.2908

(continued)

Table 4 (continued)

t (h)	P_1(MW)	P_2(MW)	P_3(MW)	P_4(MW)	P_5(MW)	P_6(MW)	P_7(MW)	P_8(MW)	P_9(MW)	P_{10}(MW)	Cost ($/h)
22	456.50	135.00	304.00	60.00	222.59	160.00	129.59	85.31	20.00	55.00	41,303.7322
23	303.25	396.79	190.36	60.00	73.00	57.00	129.59	47.00	20.00	55.00	34,804.1756
24	456.50	135.00	85.46	60.00	73.00	122.45	129.59	47.00	20.00	55.00	31,611.0776
TC ($/h)											**1,015,473.713**

Table 5 Optimal solution of 10-unit system using the GA

t (h)	P_1 (MW)	P_2 (MW)	P_3 (MW)	P_4 (MW)	P_5 (MW)	P_6 (MW)	P_7 (MW)	P_8 (MW)	P_9 (MW)	P_{10} (MW)	Cost ($/h)
1	295.18	135.00	73.00	60.00	116.58	104.23	130.00	47.00	20.00	55.00	28,835.4134
2	383.81	135.00	73.00	60.00	73.00	133.18	130.00	47.00	20.00	55.00	30,081.4690
3	234.70	135.00	311.54	60.00	73.00	118.74	130.00	120.00	20.00	55.00	33,514.2946
4	444.25	135.00	300.01	60.00	73.00	141.72	130.00	47.00	20.00	55.00	36,581.9959
5	452.82	399.24	73.00	60.00	123.12	125.20	124.58	47.00	20.00	55.00	38,159.0757
6	458.88	135.00	340.00	60.73	221.38	160.00	130.00	47.00	20.00	55.00	41,522.4435
7	450.83	391.59	289.50	60.00	84.679	144.79	130.00	47.00	48.58	55.00	43,423.0248
8	381.89	394.32	312.71	60.00	184.16	160.00	123.86	84.04	20.00	55.00	44,949.1070
9	455.97	460.00	316.42	165.23	112.55	160.00	130.00	48.81	20.00	55.00	48,765.2458
10	460.24	460.00	298.52	197.25	234.30	160.00	130.00	47.00	29.67	55.00	52,222.2846
11	378.76	460.00	314.69	300.00	243.00	124.53	130.00	120.00	20.00	55.00	53,935.9627
12	469.28	460.00	340.00	300.00	243.00	121.46	130.00	77.61	23.63	55.00	55,997.8655
13	391.29	419.50	340.00	300.00	243.00	126.19	130.00	47.00	20.00	55.00	52,381.0673
14	442.68	460.00	293.19	60.00	243.00	126.34	130.00	83.61	30.16	55.00	48,784.4983
15	456.99	460.00	340.00	60.00	73.00	134.01	130.00	47.00	20.00	55.00	44,967.6019
16	374.39	310.12	340.00	60.00	73.00	109.03	128.57	83.87	20.00	55.00	40,057.3805
17	464.17	135.00	292.44	60.00	157.71	120.45	128.21	47.00	20.00	55.00	38,271.5555
18	444.25	396.70	73.00	60.00	218.73	112.94	130.00	84.33	53.02	55.00	41,875.6074
19	379.08	384.23	340.00	60.00	243.00	126.28	121.39	47.00	20.00	55.00	45,209.9895
20	457.65	395.86	340.00	171.32	218.01	143.55	130.00	80.58	80.00	55.00	52,359.3581
21	467.50	309.39	318.71	170.77	211.23	160.00	130.00	81.38	20.00	55.00	48,655.8495

(continued)

Table 5 (continued)

t (h)	P_1 (MW)	P_2 (MW)	P_3 (MW)	P_4 (MW)	P_5 (MW)	P_6 (MW)	P_7 (MW)	P_8 (MW)	P_9 (MW)	P_{10} (MW)	Cost ($/h)
22	454.19	306.56	276.10	60.00	119.13	160.00	130.00	47.00	20.00	55.00	41,456.3079
23	231.34	135.00	322.07	60.00	171.58	160.00	130.00	47.00	20.00	55.00	34,990.4801
24	226.30	135.00	157.17	60.00	235.83	114.92	130.00	49.75	20.00	55.00	32,093.9229
TC ($/h)											**1,029,091.801**

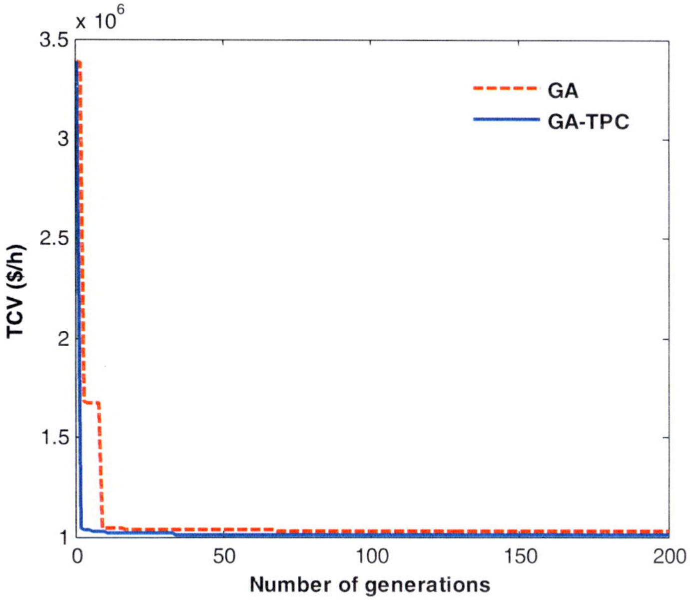

Fig. 2 Convergence characteristics of 10-unit system

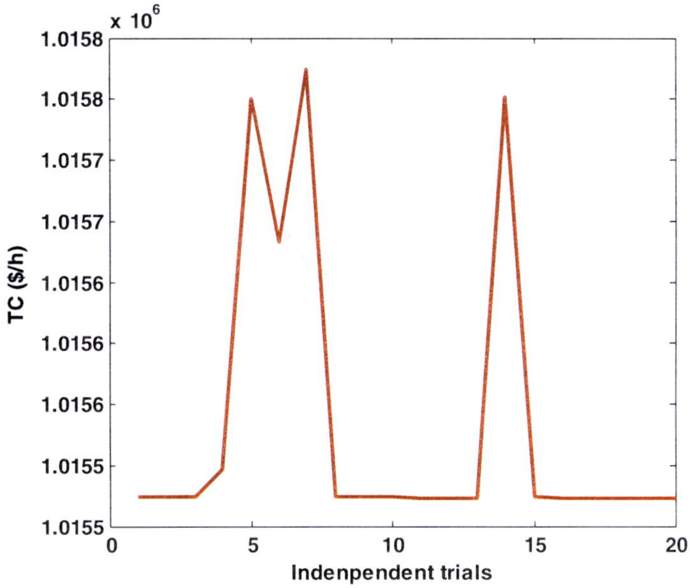

Fig. 3 Variation of TCV for 10-unit system with 20 trials

References

1. Li F, Morgan R, Williams D (1997) Hybrid genetic approaches to ramping rate constrained dynamic economic dispatch. Elect Power Syst Res 43:97–103
2. Hindi SK, Ab Ghani MR (1991) Dynamic economic dispatch for large scale power systems: a Lagrangian relaxation approach. Int J Elect Power Energy Syst 3:51–56
3. Chen CL, Wang S (1993) Branch-and-bound scheduling for thermal generating units. IEEE Trans Energy Conversi. 8(2):184–189
4. Travers D, Kaye RJ (1998) Dynamic dispatch by constructive dynamic programming. IEEE Trans Power Syst 13:72–78
5. Xia X, Elaiw AM (2010) Optimal dynamic economic dispatch of generation: a review. Electr Power Syst Res 80(8):975–986
6. Victoire T, Jeyakumar AE (2005) A modified hybrid EP–SQP approach for dynamic dispatch with valve-point effect. Int J Electr Power Energy Syst 27(8):594–601
7. Rabiee B-I, Ehsan M (2012) Time-varying acceleration coefficients IPSO for solving dynamic economic dispatch with non-smooth cost function. Energy Convers Manage 56:175–183
8. Balamurugan R, Subramanian S (2008) Differential evolution-based dynamic economic dispatch of generating units with valve-point effects. Elect Power Compon Syst 36:828–843
9. Hemamalini S, Simon S (2011) Dynamic economic dispatch using artificial bee colony algorithm for units with valve-point effect. Euro Trans Electr Power 21:70–81
10. Pulluri H, Kumar NG, Rao UM, Kumar MG (2019) Krill Herd algorithm for solution of economic dispatch with valve-point loading effect. Appl Comput Auto Wireless Syst Electr Eng. Lecture Notes in Electrical Engineering 553, 383–392
11. Hemamalini S, Simon SP (2011) Dynamic economic dispatch using artificial immune system for units with valve-point effect. Int J Elect Power Energy Syst 33:868–874
12. Pulluri H, Vyshnavi M, Shraddha P, Priya BS, Hari TS (2020) Genetic algorithm with multi-parent crossover solution for economic dispatch with valve point loading effects. Innovations in Electr Electro Eng Lecture Notes in Electrical Engineering 6, 429–438
13. Sloiman HA (2011) Modern optimization techniques with applications in electric systems. Springer Publications. https://doi.org/10.1007/978-4614-1752-1
14. Malik TN, Asar AU, Wyne MF, Akhtar S (2010) A new hybrid approach for the solution of nonconvex economic dispatch problem with valve-point effects. Int J Elctr Power Syst Res 80, 1128–1136
15. Celal Y, Serdar O (2011) A new hybrid approach for nonconvex economic dispatch problem with valve-point effect. Energy 36:5838–5845
16. ElsayedRuhul SM, SarkerDaryl A, Essam L, A comparative study of different variants of genetic algorithms for constrained optimization simulated evolution and learning, vol 6457, pp 177–186
17. Saber N, Ruhul A, Dary L (2011) GA with a new multi-parent crossover for constrained optimization, in IEEE Congress on Evolutionary Computation, pp 857–864
18. Tomar A, et al (2020) Machine learning, advances in computing, renewable energy and communication, LNEE Vol 768. Springer Nature, Berlin, p 659. https://doi.org/10.1007/978-981-16-2354-7. ISBN 978-981-16-2354-7
19. Goldberg DE (1989) Genetic algorithms in search, optimization and machine learning, reading. Addison-Wesley, Reading, MA
20. Trivedi IN, Jangir P, Bhoye M, Jangir N, An economic load dispatch and multiple environmental dispatch problem solution with microgrids using interior search algorithm Neural Comput & Appl. https://doi.org/10.1007/s00521-016-2795-5
21. Esmat A, Magdy A, ElKhattam W, ElBakly AM (2013) A novel energy management system using ant colony optimization for micro-grids. In: 2013 3rd International conference on Electric power and energy conversion systems (EPECS), Istanbul, pp 1–6

Single-Stage Power Conditioning Unit for Battery-Assisted, Solar-Powered Remote Area Power Supply

Dogga Raveendhra, K. Sudha, and M. Prashanth

Abstract This work presents a solar-powered, battery-assisted remote area power system (RAPS) that employs a single-stage power conditioning unit. This manuscript involves comprehensive mathematical modeling of power converters and their various modes of operation, as well as design guidelines. The solar-powered, battery-assisted arrangement creates a dc bus, which is then fed to an inverter for dc–AC conversion. Between the battery and the dc bus, a new dc–dc bidirectional converter is used. When compared to a conventional bidirectional dc–dc converter, this converter has twice the gain. This aids in sizing the battery with a lower dc level, which provides better safety. A 3 kW remote area power system (RAPS) created in the MATLAB Simulink environment is used to validate the proposed system. The critical performance of this converter is validated under varying environmental and load conditions. The results validated the converter's feasibility and demonstrated that it outperforms existing solutions in terms of dynamic performance under both line and load varying conditions.

Keywords PV module · Bidirectional dc–dc converter · SPV · BDC · Battery

1 Introduction

The necessity of introducing nonconventional energy sources is due to environmental pollution and saves the conventional energy sources to future generation [1]. So the new technologies were adopted with cheap, less carbon compound emissions, and

D. Raveendhra (✉)
EEE Department, Gokaraju Rangaraju Institute of Technology, Hyderabad, Telangana 500090, India
e-mail: doggaravi19@gmail.com

K. Sudha
EEE Department, SR University, Warangal, Telangana 506371, India

M. Prashanth
EEE Department, NITW, Warangal, Telangana 506004, India

A. Tomar et al. (eds.), *Proceedings of 3rd International Conference on Machine Learning, Advances in Computing, Renewable Energy and Communication*, Lecture Notes in Electrical Engineering 915, https://doi.org/10.1007/978-981-19-2828-4_38

less generation cost [2]. Among all renewable energy sources, the solar PV system is meet to regulate such problems and will expect better results, but the problem is efficiency. However, there are several maximum power point tracking (MPPT) methods. [3] proposed to excerpt maximum power from the system under various weather conditions. But in practice, there are a few MPPTs adopted by the market [4]. Because our main motto is to fulfill issues like power supply reliability and power quality. To integrate the PV system with DC bus [1], the system must meet the standards [5]. Maximum utilization of power electronic systems can result in network nonlinearity and its effects on overall system performance. A bidirectional dc–dc converters (BDC) [6–11] are required for interfacing the battery [4] in stand-alone solar photovoltaic (SPV) systems. This manuscript proposes a high-gain, high-efficiency single-stage BDC suitable for PV applications. By adopting a buck-boost, BDC developed in this study with a lithium-ion rechargeable battery to properly manage the PV system power [12–16].

Designing an inverter for an islanded (stand-alone) mode power system [6, 14] is a very complex and challenging task in commercialization because the loads are nonlinear and the power generation is also not constant due to atmospheric conditions. Because there is no grid and we are concentrating on islanded mode, maintaining constant DC link voltage is a major issue. Furthermore, nonlinear loads make energy management difficult. In this paper, a PV system is proposed that generates a UPF supply to the utility as well as nonharmonic current to the loads.

2 Proposed System

BDC employed to form a stiff dc bus with good voltage regulation capability in battery-assisted solar-powered system is shown in Fig. 1. From this figure, it can be noticed that solar modules, MPPT algorithm, battery, BDC along with dual loop control are the essential components in this system (Fig. 2).

A. **Solar PV module:** In a PV module, multiple PV cells are connected in series to increase voltage and in parallel to increase current. The industry standard for significant power generation is 36 cell modules. On the front, tempered glass (or another transparent material) is used, and on the back, a protective and water-proof substance is used. From the PV module, output is produced in variable dc form based on the temperature and illumination. Sadly, PV systems have two major disadvantages: conversion efficiency is very low (9–17%), specifically, in low-irradiation situations, and the amount of electric power generated by solar arrays varies continuously with weather situations. To enhance the conversion efficiency, below discussed, MPPT technique is employed in the control loop of dc–dc converter.

B. **MPPT Technique:** There are several MPPT techniques [2, 3] are presented to increase the efficiency of the solar module by tracking maximum output in different atmospheric conditions, here in this paper, incremental conductance method is adopted due to its merits in terms of accuracy.

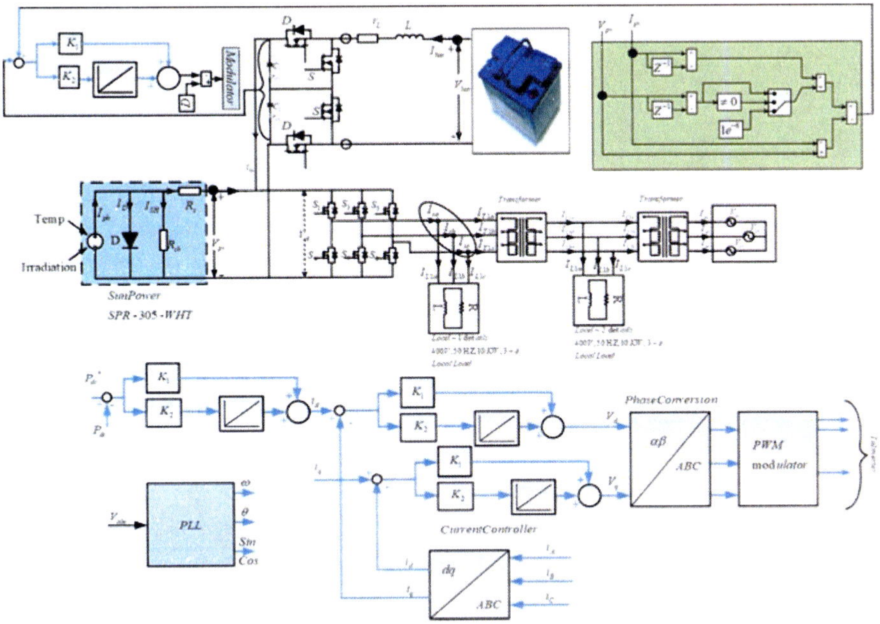

Fig. 1 Battery-assisted solar PV system

Battery: The below Fig. 3 represents nonlinear standard battery model. The selection of battery is also taken into main consideration based on its size, full capacity [12], and its type. The battery type selection is based on six factors such as how much period of time that the battery should supply power to the load continuously, battery aging factor, temperature correction factor, capacity rating factor, rated battery voltage, and depth of the battery. Over or under charging battery not efficient, it may lose efficiency and its life time. Hence, to operate the battery in proper operating region, there is need of efficient converter and control technologies which must be adopted as discussed in the following subsections.

BDC operation: The charging/discharging of the battery is communicating through BDC. The above Fig. 4 demonstrates the proposed BDC which is having broad voltage conversion range [8] when compared to conventional bidirectional buck-boost converter. While analyzing the BDC modes of operation, The following assumptions [8, 9] are taken into account such as the devices ON state resistance and capacitors equivalent series resistance are ignored, and the capacitors C_{H1}, C_{H2}, and C_L used in this converter are enough large and the voltage across them is assumed as constant, and the capacitor values C_{H1} and C_{H2} are identical; therefore, voltage across each capacitor is $\frac{V_H}{2}$.

Step-down mode of operation:

The below Fig. 4b demonstrates the proposed converter operating in buck mode. PWM techniques employed to control the devices S_1 and S_4, and the devices S_2 and

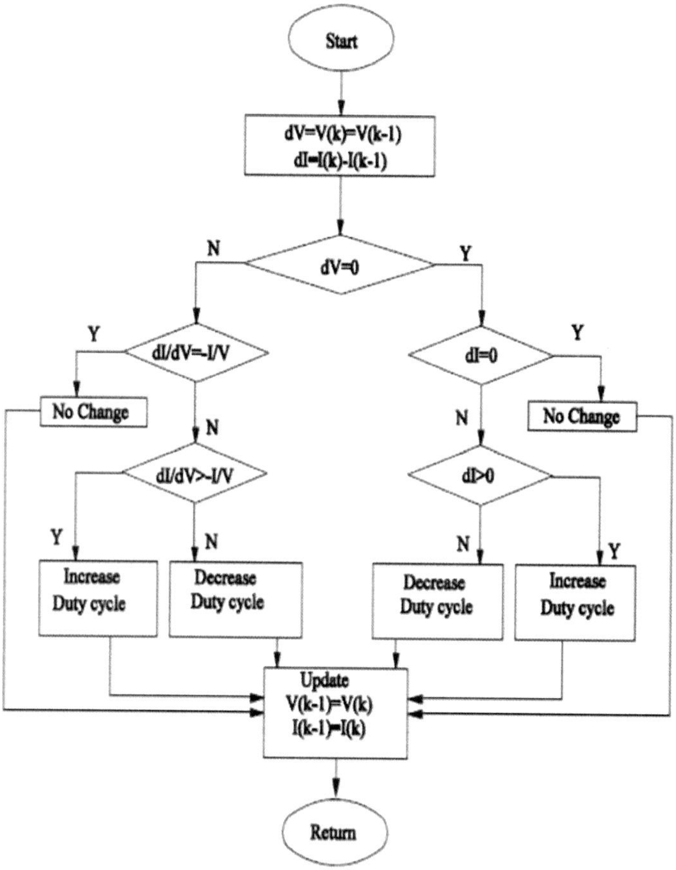

Fig. 2 Incremental conductance method

Fig. 3 Nonlinear standard battery model

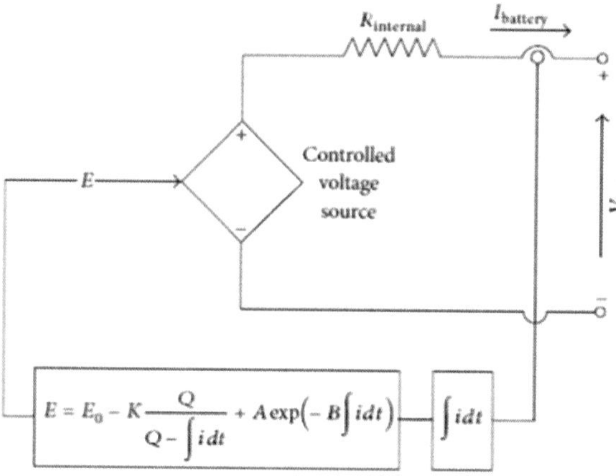

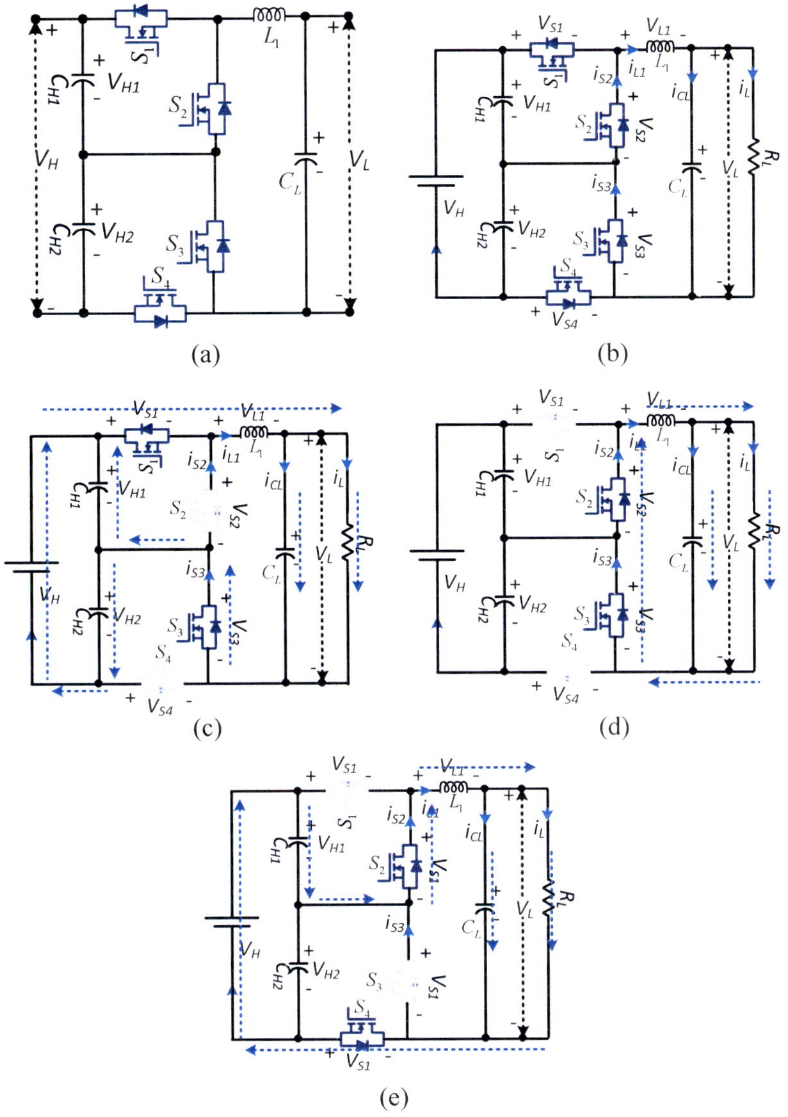

Fig. 4 **a** Proposed bidirectional buck-boost converter, **b** Equivalent circuit in buck mode **c** During interval 1, **d** During interval 2, and **e** During interval 3

S_3 are used as active rectifiers. Some characteristic graphs in continuous conduction mode (CCM) are depicted in Fig. 5.

The above step down mode can be subdivided into four different intervals as explained below.

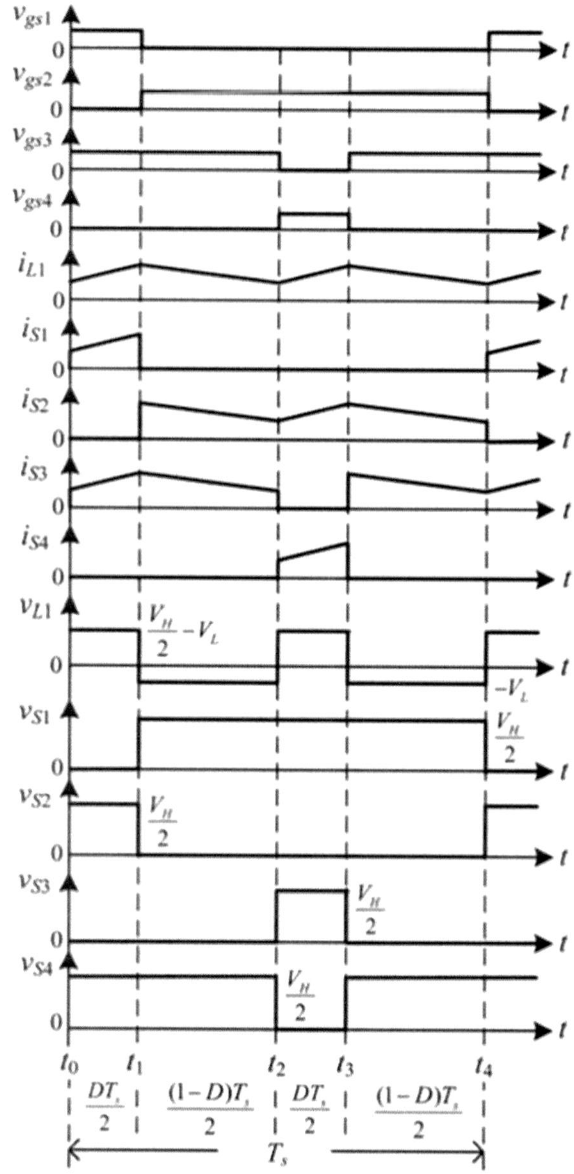

Fig. 5 Waveforms for buck mode in CCM

Interval 1 (t_0–t_1): The devices S_1 and S_3 are ON and devices S_2 and S_4 are OFF. Meanwhile, the active (synchronous) rectifier is controlled by switch S_3. The current direction is depicted in Fig. 4c. During this interval, inductor L_1 stores energy and the load feed by half of the supply voltage $\left(V_{H1} = \frac{V_H}{2}\right)$. Thus, the inductor L_1 voltage is given by

$$V_{L1} = \frac{V_H}{2} - V_L \tag{1}$$

The same inductor current is obtained as

$$i_{L1}(t) = \frac{1}{L_1}\left(\frac{V_H}{2} - V_L\right)(t - t_0) + i_{L_1}(t_0) \tag{2}$$

Interval 2 (t_1–t_2): The devices S_2 and S_3 are ON and devices S_1 and S_4 are OFF. Meantime, the devices S_2 and S_3 are employed for the active (synchronous) rectifiers. The direction of current flow is shown in Fig. 4d below.

During this interval, the inductor starts releasing energy to the capacitor C_L and load R_L. Thus, the inductor voltage is

$$V_{L1} = -V_L \tag{3}$$

The same inductor current is achieved as

$$i_{L1}(t) = \frac{-V_L}{L_1}(t - t_1) + i_{L1}(t_1) \tag{4}$$

Interval 3 (t_2–t_3): The devices S_2 and S_4 are ON and S_1 and S_3 are OFF. Meantime, the switch S_2 is utilized for the active (synchronous) rectifier. The direction of current flow is shown in Fig. 4e below.

During this interval, inductor L_1 stores energy and the load feed by half of the supply voltage $\left(V_{H2} = \frac{V_H}{2}\right)$. Thus, the inductor L_1 voltage is given by

$$V_{L1} = \frac{V_H}{2} - V_L \tag{5}$$

The same inductor current is achieved as

$$i_{L1}(t) = \frac{1}{L_1}\left(\frac{V_H}{2} - V_L\right)(t - t_2) + i_{L1}(t_2) \tag{6}$$

Interval 4 (t_3–t_4): In this interval, operation is same as interval 2. Therefore, the inductor voltage is

$$V_{L1} = -V_L \tag{7}$$

The current through the same inductor is achieved as

$$i_{L1}(t) - \frac{-V_L}{L_1}(t - t_3) + i_{L_1}(t_3) \tag{8}$$

By using volt-sec balance theory on the inductor L_1, one can obtain

$$\int_0^{DT_s/2} V_{L1}(interval1)dt + \int_0^{(1-D)T_s/2} V_{L1}(interval2)dt + \int_0^{DT_s/2} V_{L1}(interval3)dt + \int_0^{(1-D)T_s/2} V_{L1}(interval4)dt = 0 \tag{9}$$

By substituting Eqs. (1), (3), (5), and (7) into Eq. 9, the voltage gain is given by

$$\int_0^{DT_s/2} \left(\frac{V_H}{2} - V_L\right)dt + \int_0^{(1-D)T_s/2} -V_L dt + \int_0^{DT_s/2} \left(\frac{V_H}{2} - V_L\right)dt + \int_0^{(1-D)T_s/2} -V_L dt = 0 \tag{10}$$

$$2\left(\frac{V_H}{2} - V_L\right)\frac{DT_s}{2} - 2V_L\left(\frac{(1-D)T_s}{2}\right) = 0 \tag{11}$$

$$\text{Voltage gain } M_{\text{sd}} = \frac{V_L}{V_H} = \frac{D}{2} \tag{12}$$

The inductor current peak value is found to be

$$i_{L_1\text{Peak}} = \frac{DT_s}{2L_1}\left(\frac{V_H}{2} - V_L\right) \tag{13}$$

From the boundary condition mode [8] by applying amp-sec balance equation on capacitor C_L, the subsequent equation can be achieved as

$$i_{C_L} = \frac{(1/2)[(DT_s/2) + (1-D)T_s/2]i_{L1peak}2 - i_L T_s}{T_s} = 0 \tag{14}$$

Step-up mode of operation: The below Fig. 6a depicts the equivalent circuit of this converter while operating in step-up mode. PWM techniques utilized to control the devices S_2 and S_3 and the S_1 and S_4 are used for active rectifiers. Some typical waveforms in CCM are depicted in Fig. 7.

Interval 1 (t_0–t_1): The devices S_2 and S_3 are ON and S_1 and S_4 are OFF. The direction of current flow is illustrated in Fig. 6b below. During this period, the inductor L_1 stores energy and both the capacitors starts discharging through load.

The voltage across the inductor is given by

$$V_{L1} = V_L \tag{15}$$

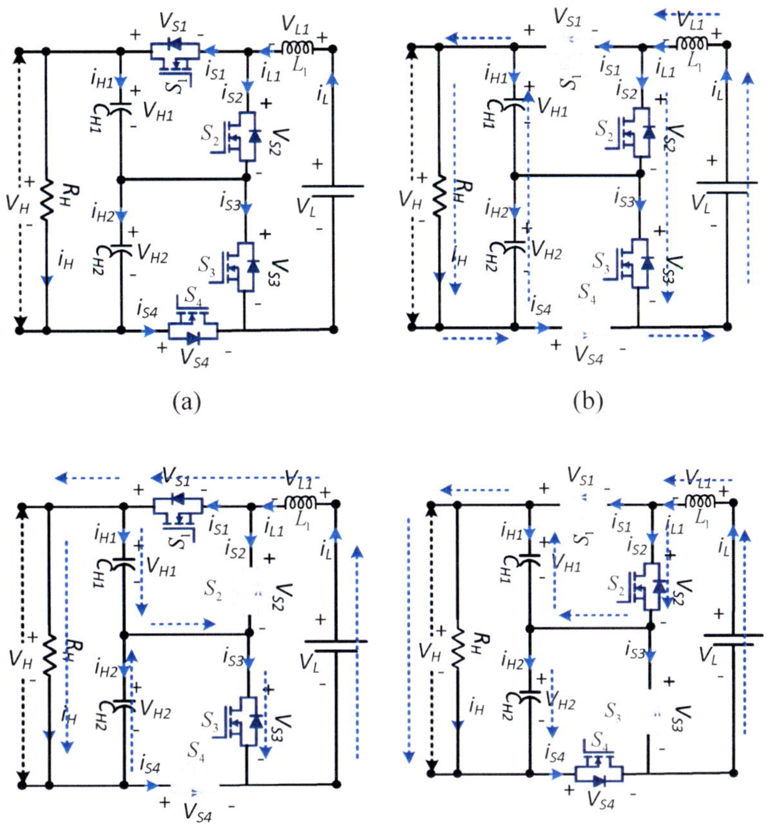

Fig. 6 **a** Equivalent circuit in step-up mode **b** During interval 1 **c** During interval 2 **d** During interval 4

The inductor current is achieved as

$$i_{L1}(t) = \frac{V_L}{L_1}(t - t_0) + i_{L1}(t_0) \tag{16}$$

Interval 2 (t_1–t_2): The devices S_1 and S_3 are ON and S_2 and S_4 are OFF. Meantime, the switch S_1 is utilized for the active (synchronous) rectifier. The direction of current flow is shown in Fig. 6c below.

During this period, the inductor starts releasing energy to capacitor C_{H1} along with supply voltage V_L and both the capacitors combined together to discharge for the load. Thus, the inductor voltage is

$$V_{L1} = V_L - \frac{V_H}{2} \tag{17}$$

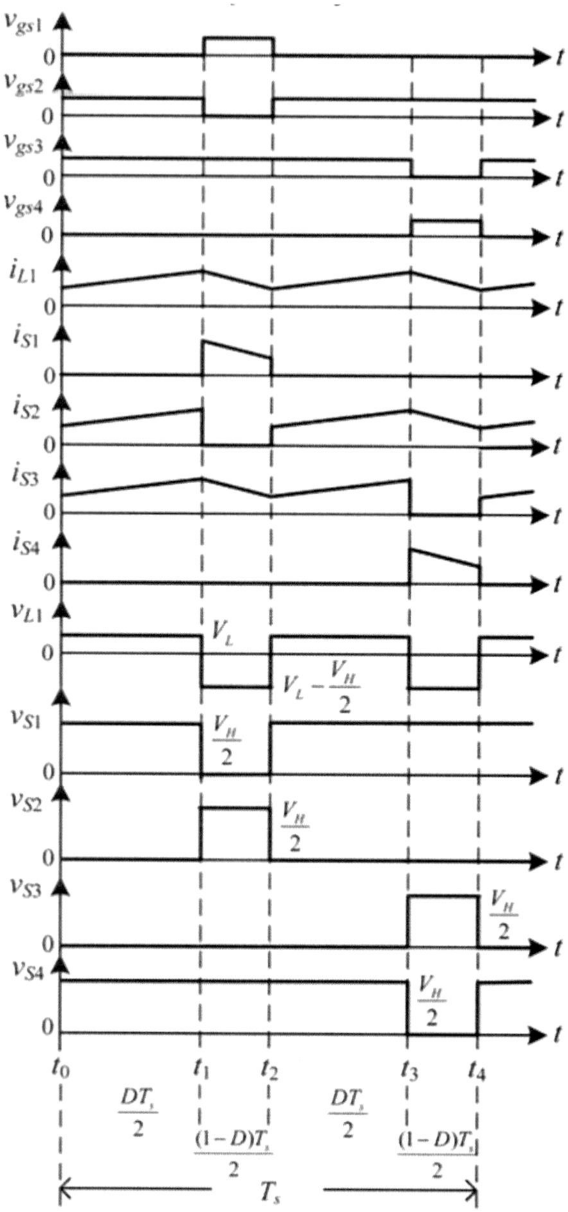

Fig. 7 Waveforms for step-up mode in CCM

The inductor current is achieved as

$$i_{L1}(t) = \frac{1}{L_1}\left(V_L - \frac{V_H}{2}\right)(t - t_1) + i_{L_1}(t_1) \tag{18}$$

Interval 3 (t_2–t_3): In this interval, operation is same as interval 1. Therefore, the inductor voltage is

$$V_{L_1} = V_L \tag{19}$$

The current through the inductor is achieved as

$$i_{L1}(t) = \frac{V_L}{L_1}(t - t_1) + i_{L1}(t_1) \tag{20}$$

Interval 4 (t_3–t_4): The devices S_2 and S_4 are ON and S_1 and S_3 are OFF. Meantime, the switch S_4 is employed for the active (synchronous) rectifier. The direction of current flow is shown in Fig. 6d below. During this period, the inductor starts releasing energy to capacitor C_{H2} along with supply voltage V_L and both the capacitors combined together to discharge for the load. Thus, the inductor voltage is

$$V_{L1} = V_L - \frac{V_H}{2} \tag{21}$$

The current through the inductor is achieved as

$$i_{L1}(t) = \frac{1}{L_1}\left(V_L - \frac{V_H}{2}\right)(t - t_3) + i_{L_1}(t_3) \tag{22}$$

By using volt-sec balance theory on the inductor L_1, one can obtain

$$\int_0^{DT_S/2} V_{L1}(interval1)dt + \int_0^{(1-D)T_S/2} V_{L1}(interval2)dt + \int_0^{DT_S/2} V_{L1}(interval3)dt + \int_0^{(1-D)T_S/2} V_{L1}(interval4)dt = 0 \tag{23}$$

By substituting Eqs. (9), (11), (13), and (15) into Eq. (17), the voltage gain is given by

$$\int_0^{DT_s/2}\left(V_L - \frac{V_H}{2}\right)dt + \int_0^{(1-D)T_s/2} V_L dt + \int_0^{DT_s/2}\left(V_L - \frac{V_H}{2}\right)dt + \int_0^{(1-D)T_s/2} V_L dt = 0 \tag{24}$$

$$\text{voltage gain } M_{su} = \frac{V_H}{V_L} = \frac{2}{1 - D} \tag{25}$$

The inductor current peak value is found to be

$$i_{L1Peak} - \frac{DT_s V_L}{2L_1} \tag{26}$$

From the boundary condition mode [8] by using amp-sec balance principle on capacitor C_{H1}, the subsequent equation can be achieved as

$$i_{C_{H1}} = \frac{(1/2)((1-D)T_s/2)i_{L_1\text{peak}} - i_H T_s}{T_s} = 0 \tag{27}$$

C. **Energy Management:** Here the solar PV system is the main power source to fulfill the power demand of the load, and the battery bank is utilized as auxiliary supply. If the generated power is more enough to meet the demand, then the excess energy has stored in inductor, meanwhile if the generated power is not enough then both solar PV and battery will carry the load [4].

3 Results and Discussions

The simulation model of proposed 3 kW ($V_{\text{mpp}} = 37.6$ V, $I_{\text{mpp}} = 8.06$A and series connected cells $N_s = 12$) stand-alone PV system with energy storage module is fabricated in MATLAB/Simulink model. The performance of BDC is estimated. The below Figs. 8 and 9 show the switching stresses during boost and buck operations of BDC, respectively. It clearly observed that S_2 has more stress than S_1.

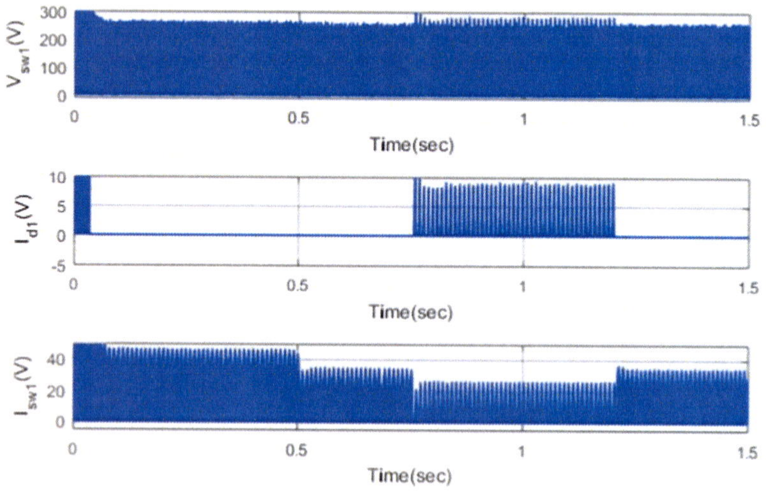

Fig. 8 S_1 stresses during boost mode operation

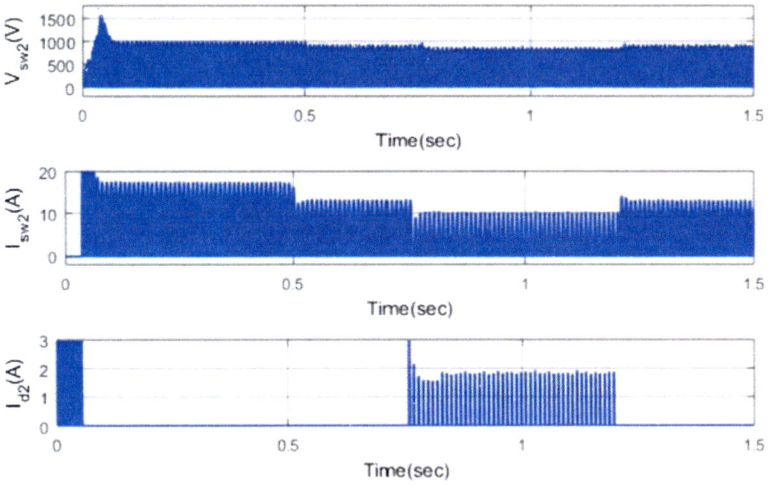

Fig. 9 S_2 stresses during buck mode operation

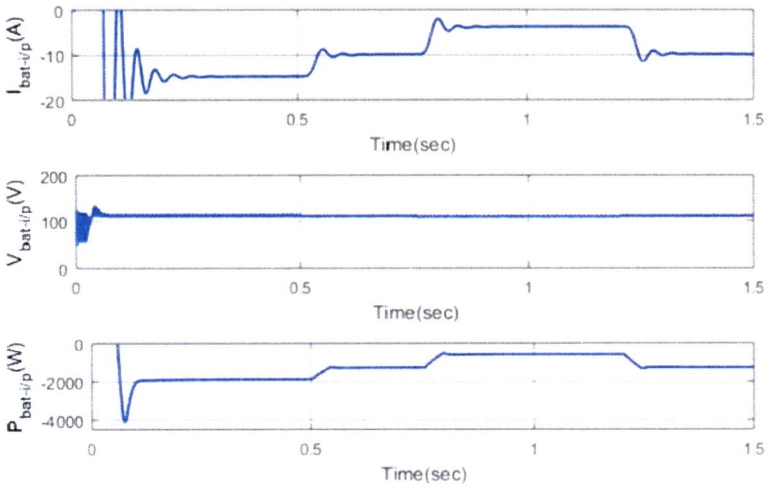

Fig. 10 Battery fed BDC input parameters

It is required to changing the battery voltage by varying the duty cycle to maintain constant voltage at DC link. Here BDC is the interface between battery and DC link. Further, the DC link voltage required to maintain 320 V, which is the input voltage to the stand-alone inverter. From the Figs. 10 and 11 during boost operation, the duty is 0.7 and during buck operation the duty is 0.3.

The input and output parameters of battery fed BDC is illustrated in above Figs. 10 and 11, respectively. Here the BDC input voltage is 100 V and the output voltage is

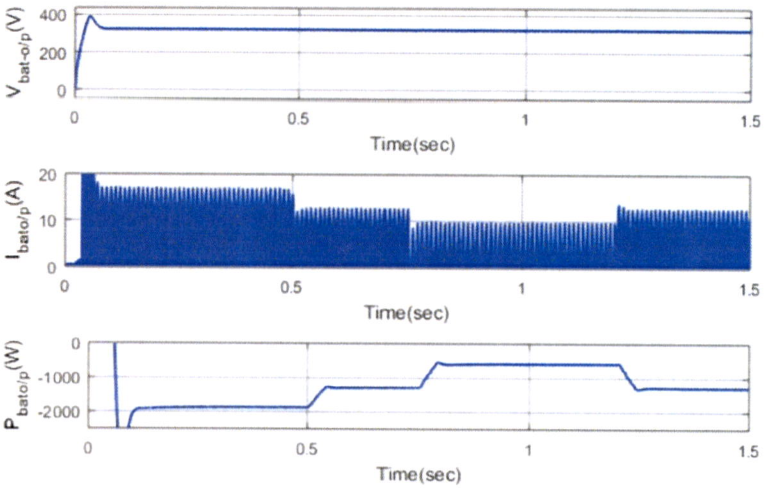

Fig. 11 Battery fed BDC output parameters

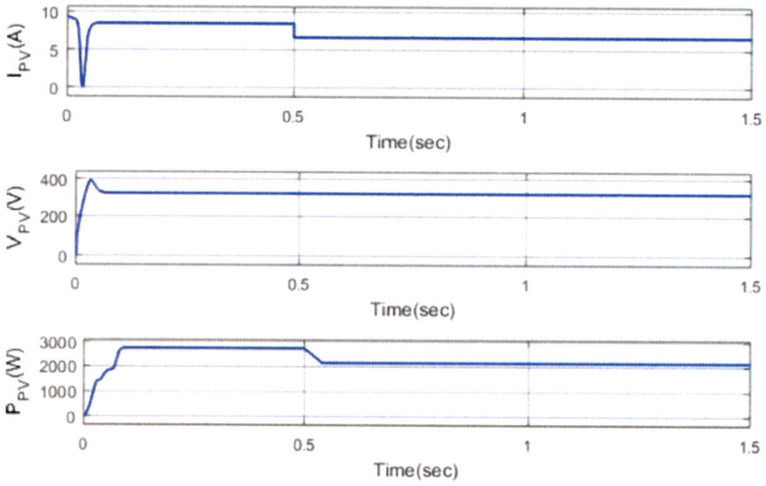

Fig. 12 PV panel parameters

around 320 V. It is necessary to maintain the PV module voltage and the BDC output voltage should be same (Figs. 12 and 13).

The output load current increases when the load power escalates from 0.75 s to 1.25 s and starts decreasing from that instant. The voltage appears across the load is 600 V peak-peak. In the above Fig. 14, the load is increased during the period 0.75 s and then decreased at 1.25 s, and at the same time the irradiance is also decreased. During this period, battery is feeding the load.

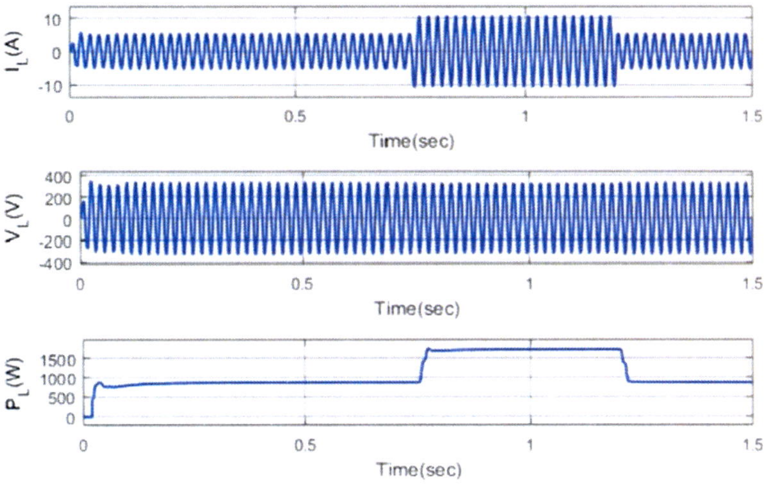

Fig. 13 Load parameters

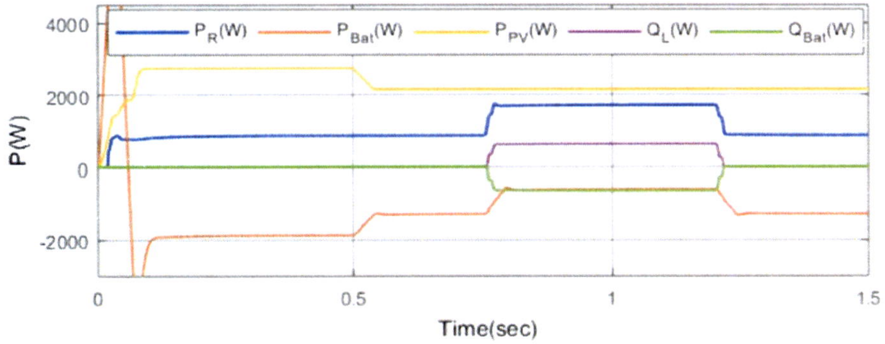

Fig. 14 Power supply management

4 Conclusion

The use of a single-stage power conditioning unit in a solar-powered battery-assisted RAPS has been successfully validated for remote area applications. The converter's critical performance is tested under a variety of environmental and load conditions. According to the findings, the proposed converter had a better dynamic response and good DC bus voltage regulation. Inverters successfully convert dc to AC, allowing better power quality to be fed to AC loads. The results confirmed the converter's feasibility and demonstrated that it outperforms existing solutions in terms of dynamic performance while feeding power to AC loads under both line and load varying conditions.

References

1. Mekhilef S (2008) Performance of grid connected inverter with maximum power point tracker and power factor control. Int J Power Electronics 1:49–62
2. Hairul NZ, Mekhilef S (2010) Comparison study of maximum power point tracker techniques for PV systems. In: International Middle East Power Systems Conference (MEPCON'10), 750–755
3. Trishan E, Patrick LC (2007) Comparison of photovoltaic array maximum power point tracking techniques. IEEE Trans Energy Convers 22:439–449
4. Nahidul HS, Norhafizan BA, Imtiaz AC, Zahari BT (2014) Modeling, control, and simulation of battery storage photovoltaic-wave energy hybrid renewable power generation systems for Island electrification in Malaysia. Hindawi Publishing Corporation 2014:1–21
5. Khanchai T, Krissanapong K, Sirichai T, Veerapol M (2004) FPGA-based islanding detection for grid connected inverter. IEEE Industrial Electronlcs Society, 1978–1982
6. Hiren P, Vivek A (2008) Control of a stand-alone inverter-based distributed generation source for voltage regulation and harmonic compensation. IEEE Trans Power Delivery 23:1113–1120
7. Malek R., Shuhui L, Saeed G (2017) Analysis and controller design for standalone VSIs in synchronous reference frame. IET Power Electronics, 1003–1012
8. Lin CC, Yang LS, Wu G, (2012) Study of a non-isolated bidirectional DC–DC converter. IET Power Electronics, 30–37
9. Hossein A, Rouzbeh RA, Sajad NR (2014) Non-isolated bidirectional DC–DC converter analysis and implementation. IET Power Electronics, 3033–3044
10. Duan RY, Lee JD (2011) High-efficiency bidirectional DC-DC converter with coupled inductor. IET Power Electronics, 115–123
11. Sangtaek H, Deepak D (2008) Bi-directional DC/DC converters for plug-in hybrid electric vehicle (PHEV) applications. IEEE, 784–789
12. Hongfei Wu, Kai S, Liqun C, Lei Z, Xing Y (2015) High step-up/step-down soft-switching bidirectional DC-DC converter with coupled-inductor and voltage matching control for energy storage systems. IEEE Trans Indus Elect
13. Mangu B, Fernandes BG (2014) Multi-input transformer coupled DC-DC converter for PV-wind based stand-alone single-phase power generating system. IEEE, 5288–5295
14. Yong-Won C, Woo-Jun C, Jung-Min K, Bong-Hwan K (2016) High-efficiency bidirectional DAB inverter using a novel hybrid modulation for stand-alone power generating system with low input voltage. IEEE Trans Power Electron 31:4138–4147
15. Sreekanth T, Narasamma N, Mahesh KM (2014) A single stage high gain buck-boost inverter with coupled inductor. In: IEEE International Conference on Power Electronics, Drives and Energy Systems (PEDES)
16. Tomar A, et al (2020) Machine learning, Advances in Computing, Renewable Energy and Communication, LNEE volume 768. Springer Nature, Berlin, 659 p. https://doi.org/10.1007/978-981-16-2354-7. ISBN 978-981-16-2354-7

Multistring SPPV Feeding Five-Level MLI: An Application to PV Grid Connected System

Dinanath Prasad and Devansh Bhatnagar

Abstract This script opens about the involvement of multistring solar-photovoltaic system (SPPV) in grid connected systems. Solar powered plants can be used to supply domestic needs as well as surplus can be fed to utility grid when not in use. Grid connected systems are controlled solar strings feeding the utility grid. The output is taken from different SPPV strings connected to independent placed DCBC (DC-to-DC boost) converters. The modified configuration of multistring SPPV increases the modularity and reduces the efforts made at first stage of DC-to-DC stepping up conversion stage. The output of DCBC converter is given to the common DC BUS. This BUS will be useful in feeding power to two individual H-bridges connected in cascade scheme of five-level cascaded H-bridge multilevel inverter (MLINV). However, current profile present in the grid is distorted by the harmonics produced by nonlinear switches used in power converters. Hence, we used MLINV topology with SPWMT for controlling gating pulse of switches. As there will be increase in the number of levels and switching frequency is also reduced eventually reduces the harmonic content and filter dimensions. We put in efforts to design and validate results of grid integration with use of multistring solar SPPV.

Keywords Multistring photovoltaic generation · Maximum power point tracking · Partial shadings · MLINV · Five-level cascaded H-bridge MLINV · Total harmonic distortion · Utility grid

1 Introduction

Impact of human driven global heating laid down by climatologist of IPCC and UN declared code RED due to highly intensifying change in Earth's climate. There is

---

D. Prasad (✉) · D. Bhatnagar
Ajay Kumar Garg Engineering College, Ghaziabad, Uttar Pradesh 201009, India
e-mail: prasaddinanath@akgec.ac.in

D. Prasad
Delhi Technological University, Rohini, Delhi 110042, India

A. Tomar et al. (eds.), *Proceedings of 3rd International Conference on Machine Learning, Advances in Computing, Renewable Energy and Communication*, Lecture Notes in Electrical Engineering 915, https://doi.org/10.1007/978-981-19-2828-4_39

extreme surge in Earth's temperature due to emissions and carbon footprint from fossil-based power generation. In later years, Government of India targets to achieve 450 GW RESYS (renewable energy systems) generation by 2030.

Present day development makes utilization and commercialization of solar aback due to huge investment and efficiency issues. With years of evolution, there come to advancement in improving the module efficiency at manufacturing end, and effective utilization at utility end either by modification in specific configuration or by enabling electrical trackers with improved results in terms of dynamic response [1–4]. According to authors, there are many configurations in which SPPV modules can be arranged for supplying power to grid. Like centralized SPPV string, single handedly feeding power to grid through power conditioning circuit [5, 6]. Main drawback of this scheme is system is unsusceptible to uncertainty of environmental situations like partial shading causing huge losses in output power generation due to the power mismatch between the several modules [7]. In order to overcome that, slight modification is inhibited in our scheme that is multistring SPPV MPPT configuration. This enables to have different SPPV strings with independently working MPPT connected each with DCBC converter for DC-to-DC conversion in its first stage [8]. The distributed form of DC-to-DC conversion enables the tracker to track the MPP of each string discretely, which makes power extraction increases even under shading or partial shading conditions [9]. It also provides flexibility to optimize the parameters and number of strings (i.e., modularity) and converters according to certain application level of power hence, we can observe the high-operational overall efficiency of system. This is most preferred configuration in medium power applications as it allows to have converter of different size and configuration for different string connected to common BUS [10, 11]. BUS is input to single-phase voltage source inverter here MLINV. Owing to low-total harmonic distortion, low-dV/dt stress, low-electromagnetic interference (EMIN), and higher efficiency, MLINVs are seen widely used in grid connected RESYS [12–14]. Besides that, it also reduces the size of inductor filter we used hence ultimate size and cost reduce [15]. There are innumerable approaches for extracting maximum power from SPPV nowadays intelligent and accurate methods are in craze. However, we selected traditional P&O owing to its ease of execution and less complex implementation [16, 17]. Owing to our motive of implying two stage grid connected system, we employed basic DCBC for the stage 1 conversion. Conventional MLINV started with three level such as the 0 V, Vdc, and -Vdc. However, we used five-level cascaded H-bridge MLINV which is able to generate multi-step output voltage levels [18–20]. They are 2 V, V, 0, − V, − 2 V, higher the number of voltage levels more improvement in the harmonic profile takes place. According to the latest research, further development is going for multiple micro-string inverters to be used for multiple SPPV strings which would increase the efficiency of system by 5–8%. In this work, we verified our idea through simulating grid connected multilevel inverter and verify its output voltage waveform, output current waveform henceforth total harmonic distortion (THD) profile and have improved harmonic profile as compared to work we are referring to which is attached in references [21]. Keeping the grid voltage and current sinusoidal and in phase by sinusoidal pulse width modulation (SPWMT)-based control strategy on

MLINV and LCL filter that act as coupler to feed inverter output to grid by removing high-frequency harmonics, thereby injecting high-sinusoidal current to grid [22, 23].

The simulation analysis of our work is containing novel approach for single-phase multistring grid integrating system supplied by a 20kWp capacity multistring SPPV for delivering400 volts single-phase grid and have THD analysis on grid current.

2 Methodology

Block Diagram of Dual Multistring Configuration:

The basic architecture of the SPPV grid connected system integrated with MLINV developed in this script for single-phase grid connected SPPV system is given as seen from the Fig. 2. It shows the development of four different parts. Namely, first DUO SPPV multistrings, secondly DCBC for steady output and operating SPPV string at maximum power point (MPP) with help of dedicated MPPT block installed along with it. As the length of the string increases, it increases the probability of string getting vulnerable by shading. DCBC helps to make steady power flow between string and MLINV and besides that helps in fostering up the voltage to a level somewhat reduces the modules connected in series. Thereby achieving optimum operation from such a topology, it shows circuit design of five-level cascaded H-bridge MLINV, and its control approach for proper switching required to produce five voltage levels is shown in Fig. 1. Besides this input current to grid is fed to PR current controller in order to make sure grid synchronization that is current in phase with voltage. The reference voltage at output of this controller will be the modulating wave compared with high-frequency carrier wave to carry out switching operation inverter. Eventually, calculating parameters for LCL in order to filter the current harmonics in current input to the grid, thereby integrating inverter with utility grid.

Graphical Development of Proposed Architecture:

3 Multistring SPPV Configuration:

In general, there can be as many strings independently working with respective MPPT, however, we used dual multistring SPPV that is also referred to as the dual multistring configuration in detail as shown in Fig. 1. Power extracted from multistring SPPV with capacity each of 20kWp using perturbation and observation technique (P&O) approach for operating each SPPV string at its respective, however, proposed dual multistring configuration with five-level inverter is shown in Fig. 1. There will be two different DCBC fed by individual SPPV string eventually connected to common DC BUS feeding H-bridges of MLINV. It is easily observed in the Fig. 2 SPPV string 1 and SPPV string 2 via their respective DCBC converters are feeding single-phase inverter. SPPV string parameter calculation: Several cells in series form a module and modules arranged in parallel forms string.

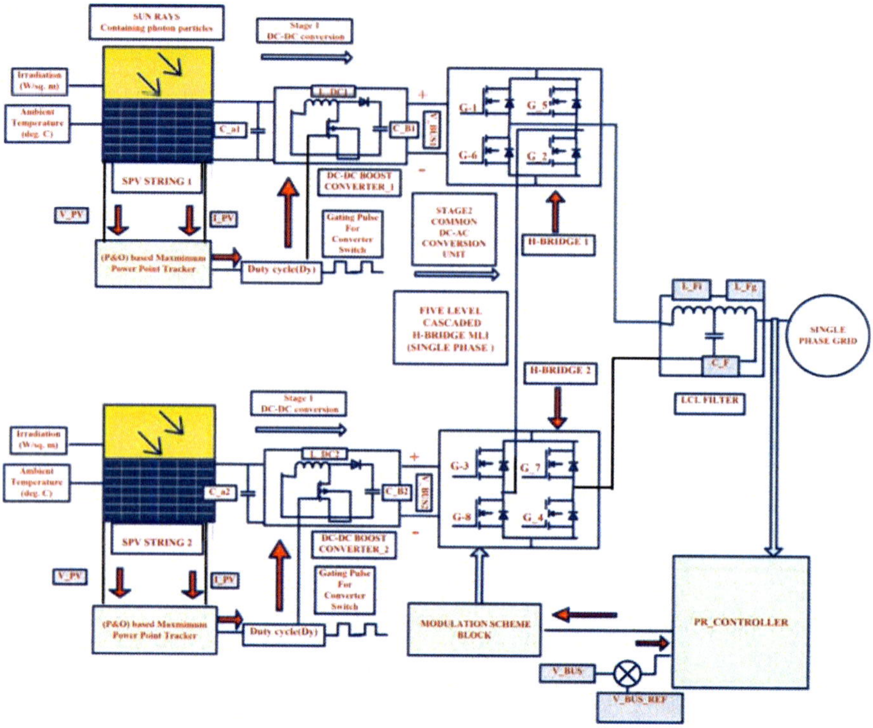

Fig. 1 Block diagram representation of the proposed architecture

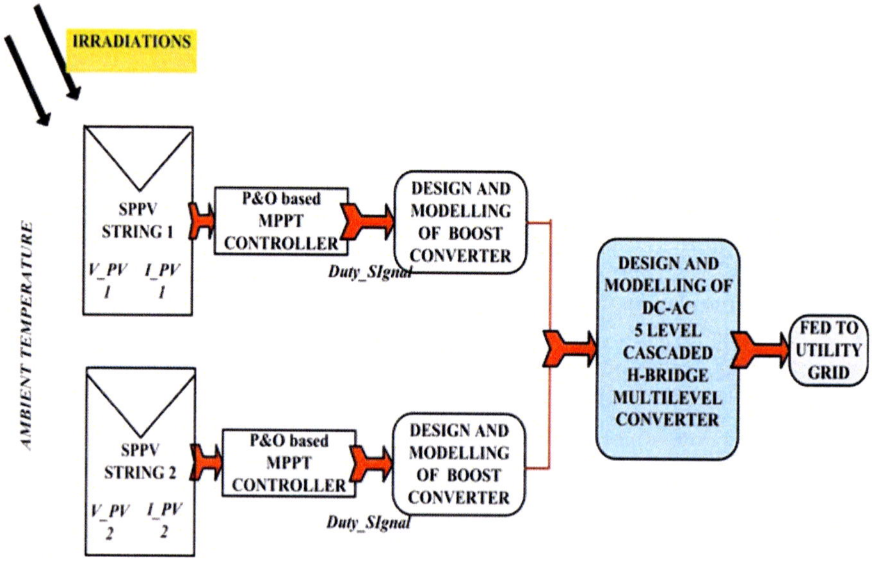

Fig. 2 Layout of proposed architecture

It is known that grid is supplied single-phase 400 V 50 Hz, and here, we are designing SPPV setup of 20 KW peak so number of modules in series and parallel can easily be calculated as shown below:

- No. of modules connected in series = 400 / 35.1 = 12
- Number of parallel modules is (20,000/400) / 7.69 = 6

Thus for the given SPPV array voltage at MPP *Vmpp* is (12 × 35.1) = 421.2 V; current at MPP *Impp* is (6 × 7.69) = 46.14 amperes; peak power = 19,434.168 W peak.

Number of modules per string can also be varied as per requirement to allow DCBC boost voltage up to the voltage level required at DC link BUS. This offers additive modularity to the system. Each SPPV string has its independent MPPT tracker along with individual dedicated DCBC installed further making it more tolerant to situations of unequal shading/ power mismatch conditions among modules. Thereby no or minimum chances of power loss due to shading which can count significantly if centralized string configuration is used. Furthermore, the advantage of multistring configuration is that the existent SPPV panel strings could be outstretched by adjoining new higher output panels without trading off with the overall SPPV string reliability and optimum energy yield. On top of that safety while installation and maintenance add to its supremacy. At times of failures defect or failure can easily be detected and that too precisely. Besides this wholesome functioning of system is not distressed by the breakdown of the single SPPV module.

Tracker here utilized the unidirectional DCBC to obtain MPP at reduced switching losses implementing P&O approach for impedance matching. A DCBC operates according to duty generated by MPPT tracker in such a way that converter makes impedance seen from SPPV output that is the input terminals of DCBC, matches with impedance at output terminals of DCBC. This will allow maximum power transfer from SPPV. Here comes with second stage of the system designed where DC input from DCBC BUS will be converter to the AC with help of MLINV. Further to it BUS output from the boost DCBC is fed to each of H-bridges of the five-level cascaded MLINV. To be precise two capacitors with same rating is used as DC BUS in order to isolate power source from H-bridge effective voltage seen at output of five-level MLINV is almost double the magnitude of voltage at DC link BUS. Minimum criteria of voltage level to be fed at H-bridge of five-level MLINV should be 115% that is 1.15 times that of voltage to be supplied at grid. In order to generate 5 levels at output, MLINV makes use of two H-bridges each containing four high-frequency switched device (IGBT). Since we know that use of nonlinear switches is responsible for harmonic or higher frequency components of sinusoidal wave. Thereby utilizing these switches in way to generate the output which contains lesser or reduced harmonics is the aim. Hence, we make it use of SPWMT switching scheme. Now before implementing this, we need to have grid synchronization with help of a PLL loop using PR current controller. This will maintain input current to grid in phase with of voltage owing to maintain power factor nearly unity. Many results and verifications are made by authors that unipolar modulation strategy is superior to bipolar modulation from the viewpoints of power loss and harmonics.

Table 1 Switched configuration of multi-stepped output voltage generated by 5 level MLINV

G1	G2	G3	G4	G5	G6	G7	G8	V
1	1	1	1	0	0	0	0	+
0	1	1	1	1	0	0	0	+
0	0	1	1	1	1	0	0	0
0	0	0	1	1	1	1	0	–
0	0	0	0	1	1	1	1	–

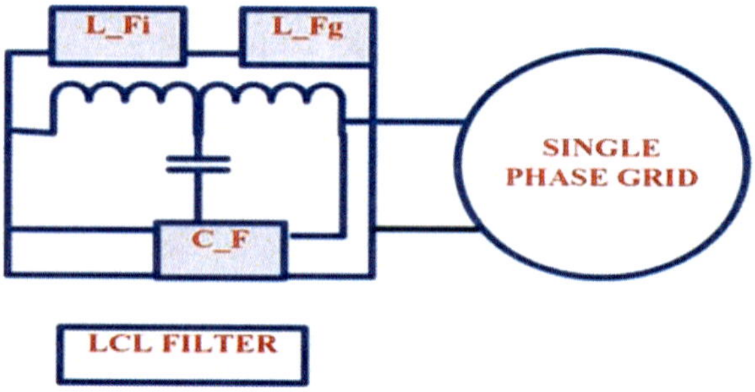

Fig. 3 Layout of LCL filter feeding grid

Hence, we are employing here unipolar SPWMT injection technique. Moreover, MLINV is able to achieve its power quality and efficiency of system requirements with the help of multi-stepped sinusoidal waveform. Forthwith to execute SPWMT higher frequency carrier signal generally triangular wave is compared with the low frequency (or frequency at which grid is found operating) sinusoid voltage wave, and accordingly, the switches will operate. Switch configuration and their operation as per the comparison are shown in Table 1. Output waveform of single-phase inverter is in the form of multi-stepped quasi-SINE wave with levels as Vdc, 2Vdc, 0, − 2Vdc, and − Vdc. It is seen that output waveform is seen in its best shape (that is good power quality) with higher number of levels in MLINV. Before the power is fed to the grid, it is passed through filter so as to decouple the resonance effect between inverter side and grid side filter inductance. Primarily selection of LCL filter is to reduce the high-frequency (or switching frequency) current harmonics supplied into the grid (Fig. 3).

Inverter side inductance calculated as 10 mH, grid side inductance as12mH, and filter side capacitance as 2.2 μF.

4 Simulation Results

In this section of script, we will discuss about the results of simulation obtained by executing the proposed simulation model in MATLAB 2018 version. It shows all the waveforms of SPPV string 1 and string 2 input as well as the output parameters.

In all each string is comprising 12 × 6 that is 12 modules in series and 6 modules in parallel giving maximum peak output of 20 kW peak ± 5%. SPPV input environmental conditions like solar irradiations are kept fixed that is 1000 W/sq. m., and the fixed ambient temperature 25C at which solar cell performs best is used in SIM (simulation). DCBC with independent trackers have been seen gracefully doing their job by making the output as steady firstly. Secondly, they implemented P&O-based MPPT approach for extracting maximum power from SPPV by impedance matching. PV_POWER output is shown with lesser variations or fluctuations in waveform. Average output voltage at BUS obtained is 410.6 V approx. which will be supplying as isolated dc source to H-bridge of MLINV. While modeling the layout, sample time steered as 10e-6 and solver configuration is performed at auto mode. After modeling and design four stages like SPPV, DCBC, DC-AC MLINV, LCL and integrating them in order to supply utility grid. Here, we took switching frequency for IGBT as, $F_{Sf} = 10$ kHz. Value of capacitance C_m is kept 100 μ F for steadiness in DCBC at boost input terminals. Inductor at input terminal od DCBC is taken as 25mH and capacitance at output of DCBC is 5800 μF. $R_{equivalent} = 0.00001$ ohms in series with inductor (Fig. 4).

Similarly, the inputs to the SPPV string 2 shown in system layout are simulated to get waveform of SPPV output along with the BUS output waveform is shown in Fig. 5.Which is as identical as waveform in SPPV string 1 shown in Fig. 4.

Here, 1 implies ON state and 0 implies OFF state.

In a similar fashion, we can design and achieve higher number of multilevel voltages with minimum harmonic distortions in the output. For a general multi-level inverter, if m is the number of H-bridge used in inverter, there are 2 m + 1 number average output voltage levels. Here, 5 level inverter is designed so equating 2 m + 1 = 5, and m is computed to be 2. This implies 2 H-bridges in cascaded fashion

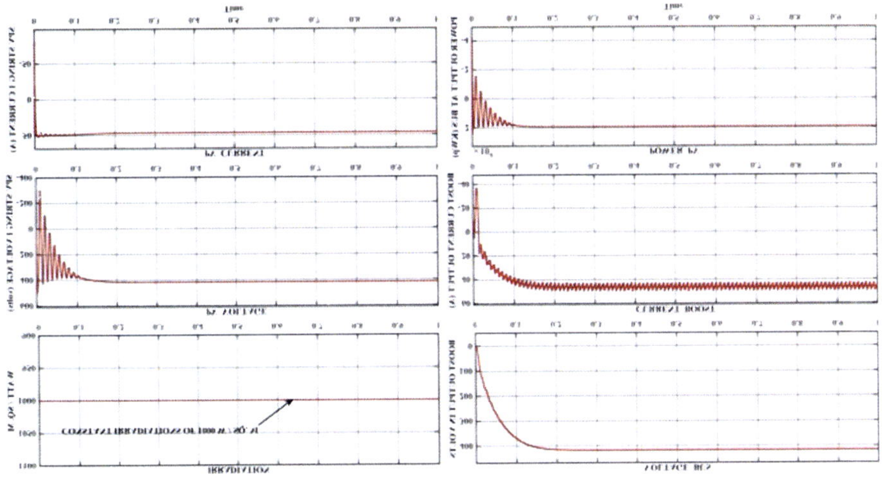

Fig. 4 SPPV string 1 input and output parameters along with DCBC output

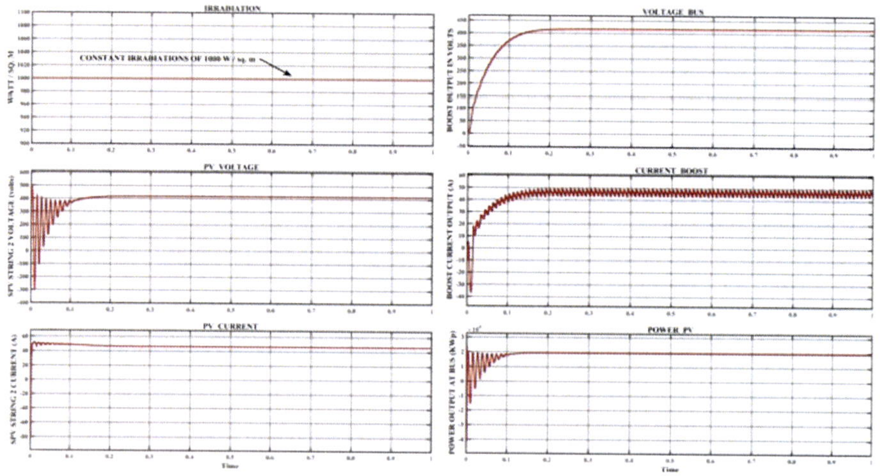

Fig. 5 SPPV string 2 input and output parameters along with DCBC output

is arranged and simulated for DC-to-AC conversion of DC obtained at the BUS. Each H-bridge is composed of four high-switched frequency switches that is IGBT. Voltage waveform consisting of five levels is shown in Fig. 6. This waveform is obtained by firing the appropriate switches at specified instants shown in Table 1.

As per the switch placed in H-bridge in Fig. 6, the following chronology will be followed for switch operation:

- If V mod > Vcarrier, then gate pulse 1 and 3 are high and switch 1 and 3 operate.
- If V mod < Vcarrier, then gate pulse 6 and 8 are high and switch 6 and 8 operate.
- If –V mod > Vcarrier, then gate pulse 5 and 7 are high and switch 5 and 7 operate.

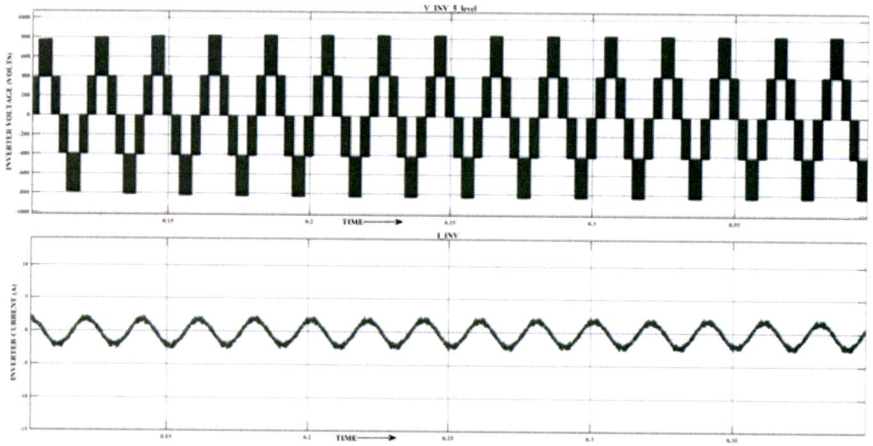

Fig. 6 MLINV output phase voltage containing five levels and its current waveform

- If –V mod < Vcarrier, then gate pulse 2 and 4 are high and switch 2 and 4 operate

The most feasible switching scheme is utilized here for medium and high-power applications as well. As they are able to generate voltage at output with lesser harmonics and lesser dv/dt stress. Also it draws the current with low-harmonic distortions. Here, we have got five levels of voltage in the form quasi-sine wave. Peak voltage of waveform is 800 V approximately as said that is twice that of voltage at the DC link. THD analysis done on inverter current gives 16.43% if observed from 0.2 cycle to 1 cycle which is further passed through filter in order to filter high-order frequency component from current (Fig. 7).

Here, THD analysis using FFT tool showing THD % to be 3.06% which is below 5% refer Fig. 8, as per IEEE standards. Hence, we can say we successfully executed proposed architecture and injected the current of good power quality in the utility using employed multistring approach. In fact, if we will implement FFT tool analysis

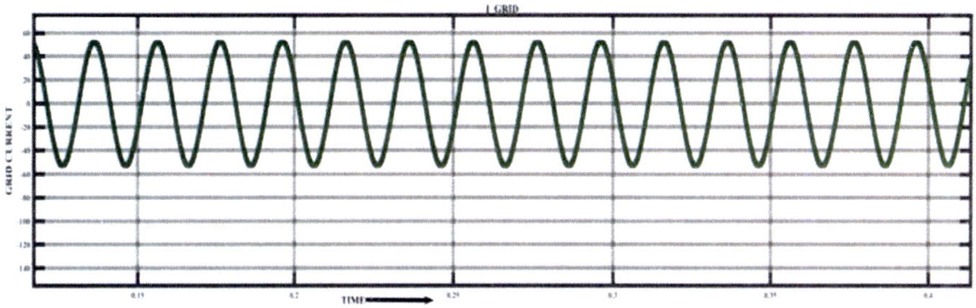

Fig. 7 Grid current waveform injected through five-level MLINV LCL filter

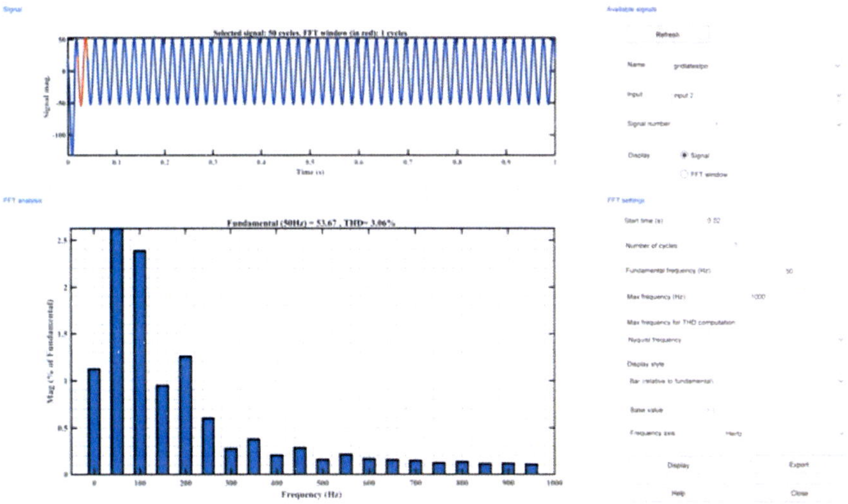

Fig. 8 THD variation through FFT analysis tool in MATLAB

for THD % in conventional or a 3 level inverter, it will come out of larger than THD generation in five level hence proving its use in system.

5 Conclusion

Our simulation work is designed and modeled in four different parts in MATLAB 2018A. They are multistring SPPV and their dedicated DCBC with MPPT installed, five-level cascaded MLINV, LCL, and ultimately the grid integration.

The multistring approach is executed so far seems advantageous in terms of dynamic as well as static performance as discussed in script. Multistring is seen independently working with its DCBC clubbed with MPPT tracker. Thereby, it is seen more susceptible to power mismatch in case of shading of cells. Five-level cascaded H-bridge MLINV is seen producing five voltage levels effectively with lesser harmonic distortions as comparison with conventional inverter. The SPWMT control scheme for switching of inverter switches is explained with the help of a table. Multistring strategy lower downs the filter requirements to some extent and effective for transformer-less grid connected solar inverters. Also, novel proposition of single-phase multistring grid integrating system supplied by a 20kWp capacity multistring SPPV for delivering 400 V single-phase grid integration is successfully performed, and results are justified with waveforms. Moreover, harmonics have been analyzed with the help of fast Fourier transform (FFT) tool in MATLAB. THD coming out to be 16.43% at MLINV output before filter application and eventually 3.06% after applying LCL filter which is less than the 5% as per IEEE standard for harmonics that is IEEE 519–2014. Hence, we can say simulation of proposed system is able to achieve its goals in terms of extracting maximum power point from each SPPV string during its first stage and meeting the grid requirements in its second stage with overall improved efficiency.

In respect to coming future, there are various developments in field of power electronics and solar power extraction in terms of its reliability and power density. It paves the way for easy integration of renewable energy systems. Alongside, it will help researchers to design new tropology of multistring multilevel inverters employed in grid connected systems with increased efficiency of converters and ease of access.

References

1. Jain V, Keshri JP, Tiwari H, Pankaj (2022) Five-level single-phase converter using SiC with reduced switched voltage stress. In: Bansal RC, Agarwal A, Jadoun VK (eds) Advances in energy technology. Lecture notes in electrical engineering, vol 766. Springer, Singapore. https://doi.org/10.1007/978-981-16-1476-7_31
2. Bughneda A, Salem M, Richelli A, Ishak D, Alatai S (2021) Review of multilevel inverters for PV energy system applications. Energies 14, 1585. https://doi.org/10.3390/en14061585

3. Ali Khajehoddin S, Bakhshai A, Jain P (2008) A novel topology and control strategy for maximum power point trackers and multi-string grid-connected PV inverters. In: 2008 Twenty-Third Annual IEEE Applied Power Electronics Conference and Exposition, pp 173–178. https://doi.org/10.1109/APEC.2008.4522718
4. Yelaverthi DB, Das SP (2014) Dual multi-string PV topology fed three level grid connected inverter. In: 2014 IEEE Applied Power Electronics Conference and Exposition—APEC 2014, pp 1704–1710. https://doi.org/10.1109/APEC.2014.6803535.X. Liang (2017) Emerging power quality challenges due to integration of renewable energy sources. IEEE Trans Industry Appl 53(2):855–866, March–April
5. Rahim NA, Selvaraj J (2010) Multistring five-level inverter With novel PWM control scheme for PV application. IEEE Trans Industrial Electronics 57(6):2111–2123. https://doi.org/10.1109/TIE.2009.2034683
6. Rahim NA, Selvaraj J (2011) A novel multi-string five-level PWM inverter for photovoltaic application. IEEE Int Electric Machines & Drives Conference (IEMDC) 2011:510–514. https://doi.org/10.1109/IEMDC.2011.5994650
7. Rao AM, Sriniva V, Srividya B (2018) Seven-level single phase inverter for multistring photovoltaic applications. In: 2020 IEEE International Conference on Power Electronics, Smart Grid and Renewable Energy (PESGRE2020), 2020, pp 1–6. https://doi.org/10.1109/PESGRE45664.2020.9070361. Motahhir S, El Ghzizal A, Sebti S, et al (2018) Modeling of photovoltaic system with modified incremental conductance algorithm for fast changes of irradiance. Int J Photoenergy 2018, 13 p
8. Shah N (2018) Multilevel inverter based single-stage grid connected photovoltaic system using cascaded two-level inverter. In: 2018 IEEE International Conference on Power Electronics, Drives and Energy Systems (PEDES), pp 1–6. https://doi.org/10.1109/PEDES.2018.8707445
9. Katkamwar SS, Doifode VR (2016) Cascaded H-bridge multilevel PV inverter with MPPT for grid connected application. International Conference on Energy Efficient Technologies for Sustainability (ICEETS) 2016:641–646. https://doi.org/10.1109/ICEETS.2016.7583832
10. Samarth MP, Kadwane SG (2015) Single phase grid connected reduced switched multilevel inverter for photovoltaic system. In: 2015 IEEE Power, Communication and Information Technology Conference (PCITC), pp 136–141. https://doi.org/10.1109/PCITC.2015.7438148
11. Azad ML, Das S, Kumar Sadhu P, Satpati B, Gupta A, Arvind P (2017) P&O algorithm based MPPT technique for solar PV system under different weather conditions. In: 2017 International Conference on Circuit, Power and Computing Technologies (ICCPCT), pp 1–5. https://doi.org/10.1109/ICCPCT.2017.8074225
12. Paul Raj MM, Meenakshi SS (2016) Cascaded H-bridge five-level inverter for grid-connected photovoltaic system using proportional-integral controller. Measurement Cont 49(1):33–41. https://doi.org/10.1177/0020294016629175
13. Saxena H, Singh A, Rai JN (2019) Design and Analysis of different PLLs as load compensation techniques in 1-Ø grid-tied PV system. Int J Electronics. https://doi.org/10.1080/00207217.2019.1600745
14. Jain D, Kalla UK (2016) Design and analysis of LCL filter for interconnection with grid connected PV system. In: 2016 IEEE 7th Power India International Conference (PIICON), pp 1–6. https://doi.org/10.1109/POWERI.2016.8077457
15. Iqbal A, et al (2021) Intelligent data-analytics for condition monitoring: Smart grid applications. Elsevier, 268 p. https://www.sciencedirect.com/book/9780323855105/intelligent-data-analytics-for-condition-monitoring. ISBN: 978–0–323–85511–2
16. Tomar A, et al (2020) Machine learning, advances in computing, renewable energy and communication, LNEE volume 768. Springer Nature, Berlin, 659 p. https://doi.org/10.1007/978-981-16-2354-7. ISBN: 978–981–16–2354–7
17. Iqbal A, et al (2020) Renewable power for sustainable growth, LNEE volume 723. Springer Nature, Berlin, 2020, 805 p. https://doi.org/10.1007/978-981-33-4080-0. ISBN: 978–981–33–4082–4
18. Waseem Ahmad Md, et al (2022) Intelligent data-analytics for power and energy systems, LNEE volume 802. Springer Nature, Berlin. 641 p. https://doi.org/10.1007/978-981-16-6081-8. ISBN: 978–981–16–6081–8

19. Iqbal A, et al (2020) *Soft computing in condition monitoring and diagnostics of electrical and mechanical systems.* Springer Nature, Berlin, 496 p. https://doi.org/10.1007/978-981-15-1532-3. ISBN: 978–981–15–1532–3
20. Iqbal A, et al (2020) Meta heuristic and evolutionary computation: algorithms and applications. Springer Nature, Berlin, 949 p. https://doi.org/10.1007/978-981-15-7571-6. ISBN: 978–981–15–7571–6
21. Jafar A, et al (2021) AI and machine learning paradigms for health monitoring system: intelligent data analytics, SBD, volume 86. Springer Nature, Berlin, 513 p. https://doi.org/10.1007/978-981-33-4412-9. ISBN: 978–981–33–4412–9
22. Srivastava S, et al (2019) Applications of artificial intelligence techniques in engineering, SIGMA 2018, vol. 1, AISC volume 698. Springer Nature, 643 p. https://doi.org/10.1007/978-981-13-1819-1. ISBN: 978–981–13–1818–4
23. Srivastava S, et al (2019) Applications of artificial intelligence techniques in engineering, SIGMA 2018, vol. 2, AISC volume 697. Springer Nature, 647 pages. https://doi.org/10.1007/978-981-13-1822-1. ISBN: 978–981–13–1821–4

Energy Management System: A Review on Ruling and Reckoning Opportunities to Save Energy

Amisha Srivastava, M. Rizwan, and Rinchin W. Mosobi

Abstract The transition of conventional electric grid to smart grid is characterized with several concerns and challenges. Efficient, sustainable, and smart management of energy are one such issues. Energy management system combining hardware and software is intended for optimized energy consumption, reduced cost, and environmentally friendly solutions. It comprises collection of computerized tools integrated with home automation, buildings, advanced metering, electric vehicle, and demand side management to monitor and optimize the system performance. Energy management system communicates and interacts with both customers and energy devices to help adapt the energy consumption to the available energy supply. This paper reviews various aspects of energy management viz. home energy management system, building energy management system, advanced metering infrastructure, electric vehicle, and demand side energy management system. Such energy management system forms an interface between energy suppliers and consumers, their integration with Internet of things, machine learning/artificial neural network provide intelligent control and predictive action. Efficient use of energy at grid level and end-user level is achieved with advanced metering infrastructure. With the advancement of multilevel energy management system, its application in electric vehicles is also incorporated; owing to the major significance of energy management system in smart grid as well, this paper deals with all the important advancements and research in the same field. The comprehensive review of energy management system in various aspects is discussed, and their challenges and future perspective are also highlighted.

Keywords Energy management system · Smart grid · Machine learning · Advanced metering infrastructure · Demand response · Internet of Things

A. Srivastava (✉) · M. Rizwan · R. W. Mosobi
Department of Electrical Engineering, Delhi Technological University, Delhi 110042, India
e-mail: amishas.speed16@gmail.com

M. Rizwan
e-mail: rizwan@dce.ac.in

A. Tomar et al. (eds.), *Proceedings of 3rd International Conference on Machine Learning, Advances in Computing, Renewable Energy and Communication*, Lecture Notes in Electrical Engineering 915, https://doi.org/10.1007/978-981-19-2828-4_40

1 Introduction

The blooming smart technology has improved the living standard of people in most part of the world and with that the energy consumption pattern has drastically changed. Furthermore, the estimated world's population by the end of 2050 is to be around 9.8 billion [1]. Such surge in population escalates the energy consumption which may globally rise to 25% by 2040 [2]. With the escalating dependence of human life style on energy, the need for technological advancement with innovative solution for sustainable energy consumption becomes significant. Researchers have now shifted their concern toward devising tools that minimize energy loss and use it proficiently. Energy management system (EMS) is one such tool and is defined as the combination of all the hardware and software that are collectively used to minimize energy wastage and provide a sustainable solution to energy saving.

Smart grid and EMS go hand in hand. Multiple features of smart grid like bi-directional communication between producer and consumer, transparent energy transmission system, eco-friendly environment, etc., are achieved to a great extent using EMS. The smart home energy management system (SHEMS) and **b**uilding energy management system (BEMS) are the emerging areas in energy management. Many software and devices have been designed to make smart homes and reduce electricity bills in big industries. The most important component of HEMS and BEMS is the advanced metering infrastructure (AMI). AMI has been adopted by a large number of countries worldwide. Smart meters unlike manual meters provide real-time monitoring and also facilitate bi-directional communication for consumers.

Electric vehicles (EVs) have gain significant importance in recent years and their role in smart grid cannot be neglected. EVs are true solution for reducing greenhouse gases. This paper also discusses the impact of plug-in EVs on EMS and how their charging is managed via HEMS. Studies reveal that researchers are developing interests toward demand response (DR) program. DR is primarily concerned with adjusting demand according to supply instead of adjusting supply according to demand. It allows consumers to actively participate in energy saving and is cost effective.

The paper is categorized into four sections. This section briefs the introduction; Sect. 2 presents the literature review in implementing EMS in different fields like SHEM, role of AMI, BEMS, and grid side EMS like demand response. Section 3 describes various challenges faced and possible solutions to it. And finally, a conclusion is deduced in Sect. 4.

2 Literature Review

EMS is designed with compound techno-economic objectives that are implemented at various levels. These objectives include minimizing electricity bills, maintenance cost, power losses, maintaining stability, handling frequency, and voltage deviation

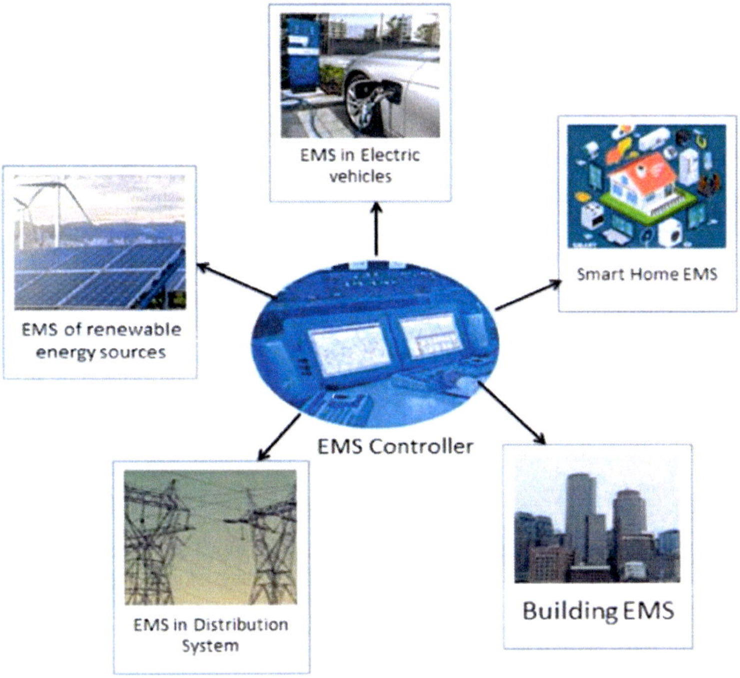

Fig. 1 Various aspects of EMS

at grid level and at the same time maximizing consumer's relaxation. Figure 1 shows various aspects of EMS in smart grid. The studies done in the field of EMS are reviewed in the further progression of paper.

2.1 Smart Home Energy Management System

Households play a significant role in increasing energy demand. One-third of the total energy mandate comes from the residential segment. The main idea behind smart homes was to design an environment where everything including generation and storage is decided, monitored, and controlled automatically with little or zero human interference. Using home area network, every digital device is interlinked with each other and can be remotely controlled making it easier and economic for user. A lot of progress is already been made toward SHEMS. SHEMS is a demand response tool that shifts and curtails demand to improve the energy consumption and production profile of a house according to electricity price and consumer comfort. The HEMS can communicate with household devices and the utility, as needed, and receive external information to improve the energy consumption and production schedule of household devices [3]. Loads in homes are categorized into controllable/scheduled loads and uncontrollable/unscheduled loads. SHEMS makes use of controllable loads such as washing machines, refrigerators, air conditioners, etc. such that appliance scheduling can be done and optimization of energy can be implemented. Inclusion of renewable energy resources with EMS has opened way for new prospects toward eco-friendly environment.

In [4], a multi-objective-mixed integer nonlinear programming (MO-MINLP) model is being discussed which aims at optimally using energy and at the same time taking care of user's comfort level in thermal and electric zones. It also deals with scheduling various loads according to a definite algorithm in order to save power. Under different user constrictions, the simulation results could provide successful reduction in energy use and optimal task scheduling. Another such work is presented in latest study [5] in which ICT is used to introduce an automatic home management system. The two main algorithms used in the system are power limit management (PLM) and smart electrical task scheduling (SETS). PLM takes care of overloading and SETS is having heuristic approach handles scheduling of loads. The software part of the system is implemented on IoT platform using Message Queuing Telemetry Transport (MQTT) protocol that transferences messages between various devices. The benefits of the system are also discussed using a case study of Gaza Strip which has restricted power sources.

Some important network protocols of smart homes include Zigbee, Z-wave, Thread, Bluetooth Mesh, and Wi-Fi. One of the concerning factor in homes is consumption of unnecessary power by some appliances such as standby power. In [6], a Zigbee communication module is used to design a wireless power strip constructing a low cost and low power networks based on IEEE 802.15.4 standard in order to reduce this standby power and the results have shown substantial decline in power consumption. The literature in [7] describes the effective switching of loads between renewable sources and grid energy using artificial neural network and machine learning algorithm, support vector machine (SVM), thereby concluding superiority of SVM over ANN. These studies reveal that EMS together with artificial intelligence can provide wonderful solutions to optimize energy and household loads can be smartly managed using SHEMS.

2.2 *Building Energy Management System*

EMS has touched almost all the sectors including industrial, residential, and academic. Building's energy consumption is a matter of great concern, and hence, managing it efficiently is the need of the hour. BEMS architecture mainly consists of communication system, interfacing technology, and sensing technology. Several environmental factors like temperature, humidity, air quality, luminance, etc., play a significant role in design of an effective BEMS which can manage almost all the considerations of building like heating, cooling, ventilation, security, alarm systems, and all such. In [8], an efficient air conditioning system is designed that operates based on the occupancy of the building and is cost effective too. It automatically adjusts the number of air conditioner units that are required to be operated at a time sensing the presence of people. Results say that it can save up to 22% of electricity bill. Similarly in [9], an EMS is designed for smart meters in residential building using fuzzy logic and implemented on microcontroller MSP430G2553. It saved energy consumption of the day by 7%, and peak demand was reduced by 34%.

In June 2011, ISO 50001 standard was created by International Organization for Standardization (ISO) that postulates the requirements for instigating an EMS in any type of organization irrespective of its size, segment, or topographical location. The ultimate aim is to reduce electricity bills, save energy, and control greenhouse effect. [10] Presents a detailed study of various aspects of ISO 50001 implementation and required tools for energy management.

2.3 Advanced Metering Interface

AMI forms the basis of EMS architecture and mainstay of smart grid. AMI is further made up of smart meters, communication network, meter data acquisition system, and gateways.

The electro-mechanical meters have been in existence since long time now but owing to their major disadvantages like collecting data door to door which may be erroneous or unidirectional communication, and emphasis is now shifted to smart energy meters (SEMs). SEMs provide an interactive interface, self-healing, or erroneous data and are multi-functional. They operate in a real-time environment and facilitate two-way communication between energy providers and consumers [11]. AMI allows remote-control of meter data and thus any customized change during peak loading hours or off-peak loading hours can be done in order to save energy and power. In [12], a system is designed using KEIL software used to write an 8051 microcontroller program that merges GSM with AMI. It basically alarms users when their pre-paid balance for meters is too low and thus overloading of appliances can be controlled. Another interesting work is presented in [13]. In this a numerical optimization technique, differential evolution algorithm is used for automatic load scheduling in SEM and simulation was implemented on MATLAB platform which ultimately minimizes energy consumption. A test site in New Delhi was selected for the purpose, and the results showed that approximately 19.42% power was saved using DE Algorithm.

The robust communication protocol used in AMI is mainly described by three networks, wide area network (WAN), neighborhood area network (NAN), and home area network (HAN). These networks allow communication among digital devices within a home or neighborhood. Meter data acquisition system performs periodic evaluation of data collected from SEMs, and accordingly logics are defined to sterilize the data (Fig. 2).

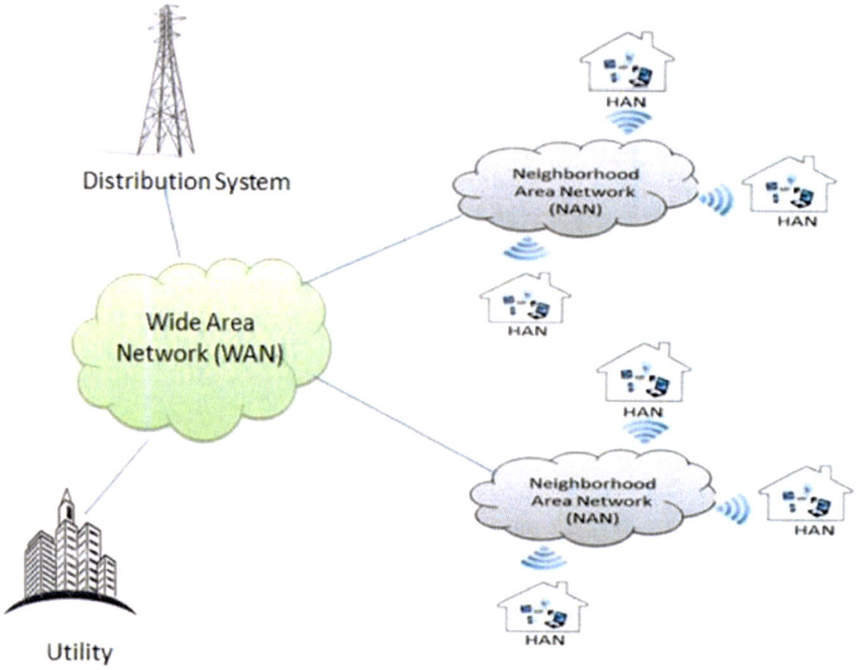

Fig. 2 Communication protocol in AMI

2.4 EMS in Electric Vehicles

Introduction of EVs in smart grid has not only controlled pollution but also plays a significant role in power system optimization with their two ways operation, i.e., vehicle to grid (V2G) in which they supply power to grid and grid to vehicle (G2V) in which power is supplied by the grid. Studies reveal that with the use of EVs, CO2, and NOX emission will be reduced [14]. Several studies have been done to analyze the impact of EVs charging and discharging in distribution system. Strong control strategies for EV charging stations reducing load burden on main grid are presented in [15]. A strong control is required in EVs which is achieved using EMS. [16] Describes the control strategies for diverse power sources used in hybrid electric vehicles. BEMS can help reduce electricity demands from residential and industrial sectors during peak loading hours, and this energy can be used to charge EVs thus maintaining a balance between generation and consumption and simultaneously improving power quality.

2.5 Demand Side Energy Management System

Owing to two-way communication in SG, DSM allows users to enthusiastically participate in energy saving and conserving renewables by monitoring and altering their power consumption plans according to the dynamics. This is also called demand response program (DRP). Residential DRP is classified into incentive-based and price-based DRP, and the details are given in [17].

Deep learning has become one of the popular methods for forecasting and solving complex problems with great ease. One such model depicting the overall behavior of demand response is presented in [18]. A multilevel deep learning model with multi-stage ensemble has been proposed for power forecasting at appliance level, and the performance of the algorithm is assessed on GREEND and UK-DALE, datasets that are openly obtainable. The presented model is robust, accurate, and assures proper implementation of demand response programs.

Also in a country like India which is dominated by small-sized and medium-sized buildings, Building Energy Management Open Source Software (BEMOSS) is a platform that supports implementation of DR [19–21]. Since it is an open source platform, it can be implemented via multiple protocols. It has also facilitated integration of IoT devices to monitor and regulate the data and simultaneously check energy consumption. Ultimately, DRP is encouraging consumers to become prosumers and increases reliability of grid.

3 Challenges in the Way of EMS

Although implementation of EMS in distribution system has numerous benefits. However, there are some limitations that have to be taken care of in the future designs. Some of the challenges identified and anticipated solutions are listed below.

Privacy issues are of utmost concern [22]. Large data collection may result in power thefts and cyber-attacks which are a frequent problem in AMI. More enhanced developments like advanced algorithms for cryptography are needed to maintain the security aspect of EMS. Literature in [23] presents a detailed analysis of energy savings algorithms implemented so far which can further be increased considering real-time scenarios and developing prediction based models. Another challenge faced is maintenance. EMS is composed of sensing and controlling technologies that require frequent updating and maintenance in order to check system's performance which becomes a tedious task [24]. Several complex architectures of EMS make it difficult for user to implement. More simplified designs are expected in the future with the use of artificial intelligence.

The major challenge that comes for DRPs is the establishment of appropriate control strategies and reliable market frameworks for its optimal implementation [25]. It is suggested to consider aggregated demands from various sources in order to form a novel modeling approach. Also, at times acquiring and monitoring data may be expensive which further restricts the fulfillment of the purpose of EMS. For this, use of distributed energy resources is a resilient way out. They are cost effective and simultaneously meet sustainability goals.

It is expected that the future works in the field of energy management system would consider these challenges and a more socio-economic and techno-economic developments can be seen.

4 Conclusion

Extensive research and development are being carried out to make EMS more reliable and secure. This paper provides a wide-ranging review of EMS application in various domain such as smart home energy management system (SHEM), building energy management system (BEMS), advanced metering infrastructure (AMI), and grid side demand response EMS to give readers an idea of its vast applications. Further, the several challenges encountered are also discussed. In addition, few of the recent advancement for future EMS is given below.

Merging artificial intelligence techniques and big data analytics with EMS is an area to be explored so that zero human interaction can be achieved with reduced complexity. Time of use (TOU) metering can be implemented where the utility firms charge the customers based on their energy consumption arrays during peak, off-peak and mid-peak hours. TOU tariffs encourage customer to shift their consumption to off-peaks times and thus balance the demand. This would lessen the strain on the grid. Future efforts should be aimed at creating regional or local energy hubs for centrally collaborating energy carriers and analyzing load curves of residential or industrial sectors for optimum energy saving.

Acknowledgement The proposed research was carried out as a part of sponsored research project funded by SERB, vide File No. CRG/2018/000291.

References

1. United Nations Department of Economic and Social Affairs. https://www.un.org/development/desa/en/news/population/world-population-prospects-2020
2. World Energy Outlook Report 2020. https://www.iea.org/reports/world-energy-outlook-2020
3. Marc B, Hamidreza Z, Arman Kiani B, Antony S (2014) Residential energy management using a two-horizon algorithm. IEEE Trans Smart Grid 5(4), 1712–1723. https://doi.org/10.1109/ISGTEurope.2012.6465896
4. Anvari-Moghaddam A, Monsef H, Rahimi-Kian A (2015) Optimal smart home energy management considering energy saving and a comfortable lifestyle. IEEE Trans Smart Grid 6(1):324–332. https://doi.org/10.1109/TSG.2014.2349352

5. Hamouda YEM, Dwedar SJI (2020) Optimally automated home management for smart grid system using sensor networks: Gaza Strip as a case study. Technol Econ Smart Grids Sustain Energy 5(16). https://doi.org/10.1007/s40866-020-00089-1
6. Sunghoi P, Myeong-in C, Byeongkwan K, Sehyun P (2013) Design and implementation of smart energy management system for reducing power consumption using ZigBee wireless communication module. Procedia Computer Sci 19, 662–668. https://doi.org/10.1016/j.procs.2013.06.088
7. Prakash NK, Vadana PD (2017) Machine learning based residential energy management system. In: 2017 IEEE International Conference on Computational Intelligence and Computing Research (ICCIC), pp 1–4. https://doi.org/10.1109/ICCIC.2017.8524383
8. Anas MH, Ibrahim AA, Khaled WA, Hatim HA, Talal HA (2021) Design of a smart air conditioning controller based on the occupancy of a building. Int J Electrical Electronic Engineering Telecommun 10(3):160–167. https://doi.org/10.18178/ijeetc.10.3.160-167
9. Chojecki A, Rodak M, Ambroziak A, Borkowski P (2019) Energy management system for residential buildings based on fuzzy logic: design and implementation in Smart Meter. IET Smart Grid 3. https://doi.org/10.1049/iet-stg.2019.0005
10. Fiedler T, Mircea P-M (2012) Energy management systems according to the ISO 50001 standard—Challenges and benefits, 1–4. https://doi.org/10.1109/ICATE.2012.6403411
11. Chakraborty AK, Sharma N (2016) Advanced metering infrastructure: technology and challenges. IEEE/PES Trans Distribution Conf Exposition (T&D) 2016:1–5. https://doi.org/10.1109/TDC.2016.7520076
12. Kavithakumari KS, Paul PP, CatherineAmalaPriya E (2017) Advance metering infrastructure for smart grid using GSM. Third International Conference on Science Technology Engineering and Management (ICONSTEM) 2017:619–622. https://doi.org/10.1109/ICONSTEM.2017.8261396
13. Vikram A, Karna D, Kumar A, Rizwan M (2020) An analytical approach of integrating automated load scheduling to a smart energy meter using differential evolution algorithm. In: IOP Conference Series: Materials Science and Engineering, 946. https://doi.org/10.1088/1757-899X/946/1/012007
14. Aruna P, Vasan PV (2019) Review on energy management system of electric vehicles. In: 2019 2nd International Conference on Power and Embedded Drive Control (ICPEDC), pp 371–374. https://doi.org/10.1109/ICPEDC47771.2019.9036689
15. Deshmukh RR, Ballal MS (2018) An energy management scheme for grid connected EVs charging stations. In: 2018 International Conference on Power, Instrumentation, Control and Computing (PICC), pp 1–6. https://doi.org/10.1109/PICC.2018.8384741
16. Goswami I, Suhag S (2020) Energy management in electric hybrid vehicle with diverse power sources. IEEE Students Conference on Engineering & Systems (SCES) 2020:1–6. https://doi.org/10.1109/SCES50439.2020.9236696
17. Shareef H, Ahmed MS, Mohamed A, Al Hassan E (2018) Review on home energy management system considering demand responses, smart technologies, and intelligent controllers. IEEE Access 6, 24498–24509. https://doi.org/10.1109/ACCESS.2018.2831917
18. Sharda S, Singh M, Sharma K (2021) A complete consumer behaviour learning model for real-time demand response implementation in smart grid. Appl Intell. https://doi.org/10.1007/s10489-021-02501-4
19. Khamphanchai W, et al (2014) Conceptual architecture of building energy management open source software (BEMOSS). In: IEEE PES Innovative Smart Grid Technologies, Europe, pp 1–6. https://doi.org/10.1109/ISGTEurope.2014.7028784
20. Pipattanasomporn M, Kuzlu M, Khamphanchai W, Saha A, Rathinavel K, Rahman S (2015) BEMOSS: An agent platform to facilitate grid-interactive building operation with IoT devices. IEEE Innovative Smart Grid Technologies - Asia (ISGT ASIA) 2015:1–6. https://doi.org/10.1109/ISGT-Asia.2015.7387018
21. Tomar A, et al (2020) Machine learning. In: Advances in Computing, Renewable Energy and Communication, LNEE volume 768. Springer Nature, Berlin, 659 p. https://doi.org/10.1007/978-981-16-2354-7. ISBN: 978-981-16-2354-7

22. Hussain S, El-Bayeh CZ, Lai C, Eicker U (2021) Multi-level energy management systems toward a smarter grid: a review. IEEE Access 9:71994–72016. https://doi.org/10.1109/ACCESS.2021.3078082
23. Al-Ghaili AM, Kasim H, Al-Hada NM, Jørgensen BN, Othman M, Wang J (2021) Energy management systems and strategies in buildings sector: a scoping review. IEEE Access 9, 63790–63813 https://doi.org/10.1109/ACCESS.2021.3075485
24. Badar AQH, Anvari-Moghaddam A (2020) Smart home energy management system—A review. Advances in Building Energy Res, 1–26. https://doi.org/10.1080/17512549.2020.1806925
25. Alam MS, Arefifar SA (2019) Energy management in power distribution systems: review, classification, limitations and challenges. IEEE Access 7:92979–93001. https://doi.org/10.1109/ACCESS.2019.2927303

Forest Fire Detection and Prevention System Using IoT

Malti Gautam Singh, Sharini Rithigaa, Deepa Raj, and Sharmila

Abstract In India, over the past 10 years, forest fires are uncontrollable which causes damages to a forest ecosystem, natural ecosystem, and biodiversity. An increase in population and change in climate causes the decline of Indian forests. The main objective of the proposed system is to design low cost and effective detection and prevention system using GSM modules. This proposed system can easily detect forest fire at the initial stage. In the monitoring, the hardware which we use is GSM modem, MQ-2 sensor, and the Flame sensor for transferring the data and also some devices. We will also use an MQ-2 sensor which will sense the gases like methane, CO2, and smoke, etc., and a Flame sensor which will sense the forest fire and will send the data to the control unit. We will also use an LCD screen that will display the data then an alert will be sent to us. Instead of using a PC, we use an LCD screen due to its small size and portability which will make this system of low cost. And we used a rechargeable battery as a power supply in our project, and for providing charging to the battery, we used a solar panel. And, for the prevention method, we provide a water motor pump that is connected to a relay, where the relay will provide a set of input terminals for single or multiple control signals. This system provides us a remote, safe, accurate detection, and prevention from fire and smoke. So, this is the most effective system for monitoring forest fire that occurs in remote places with the help of the GSM module.

Keywords Fire detection · Alert · Internet of Things · Sensor · Monitor

M. G. Singh · Sharmila (✉)
Krishna Engineering College, Ghaziabad, India
e-mail: sharmila1ece@gmail.com

S. Rithigaa
Accenture Solution Pvt. Ltd, Bengaluru, India

D. Raj
System Engineer, Infosys Limited, Ghaziabad, India

A. Tomar et al. (eds.), *Proceedings of 3rd International Conference on Machine Learning, Advances in Computing, Renewable Energy and Communication*, Lecture Notes in Electrical Engineering 915, https://doi.org/10.1007/978-981-19-2828-4_41

1 Introduction

In India, forest fire occurrence is wide. Most of the occurrence of Indian forest fires is due to human involvement and natural. To prevent the spread of forest fire in initials, a good detection and prevention system is required to protect the forest ecosystem, natural ecosystem, and biodiversity so that we can control fire spreading on time and can also, prevent it by using the prevention method. The various elements which stop the authorities to act during the forest fire are communication problems, transportation facilities, weather conditions, etc. So, to stop the spreading of forest fire, we proposed a system which will help the officials and forest authorities. This project is based on a detection and prevention system with the help of a GSM module and some other devices. In this system, we provide several detectors are placed on every tree in the forest area which will help the forest authorities and officials to detect forest fire and smoke. We also placed sensors like MQ-2 which will sense other gases like methane, CO_2, and smoke, and the Flame sensor will sense forest fire which will sense the data and send it to the control unit. And for sending SMS messages, we will use the GSM modem which is used for wireless transmission. In this system, for transmission, we used a control unit to store the data and also we used an LCD screen that will display the data, then an alert will be sent to us and also we used buzzer which will beep during the occurrence of forest fire and LEDs will also be turned ON. And for supply, we use a rechargeable battery that is connected to the solar panel, where the solar panel will recharge the battery so that it can be used whenever it is required. Also, instead of using a PC, we use an LCD screen due to its small size and portability which will make this system of low cost.

The main influence of this paper is to propose efficient fire detection for forest using IoT technologies. Section 2 summarizes the literature survey of the proposed system. Section 3 discusses the proposed method to implement the system. Section 4 concludes the paper.

2 Related Work

A Global System for Mobile Communication (GSMs) and radio frequency (RF) modules are used to detect and surveillance system for forest fire detection. It is used to detect the forest fire and also surveillance it with the help of GSM and RF modules. In their proposed system, hardware has server, nodes, and head. This technology contains a Global System for Mobile Communication model, fire, sensor to detect fire, and RF system for the transmitting of data and also used an antenna to receive the data at the reception side. IoT-based forest fire detection system [1–6] is IOT-based technology used for forest fire detection. In their system, they assembled a fire finder utilizing NodeMCU which is interfaced with a temperature sensor, LCD, and smoke sensor, where the temperature sensor is used to sense the temperature and smoke sensor used to sense the smoke in the environment. They attached the buzzer to the Arduino for the alert.

The IoT-based fire detection system [7–9] is done to improve the security system and also increase the protection in a forest area. Their project aims to design a simple wireless and low-cost protection system and also to provide an alert to the control room so it can be controlled at the initial. This technology contains a temperature sensor with Arduino to detect fire and temperature in the forest environment. They connect the DHT-11 sensor to the NodeMCU board.

The wireless sensor networks (WSNs) are used for fire detection [3] which is done for the detection of forest fire in early stage and also takes preventive measures to protect the forest. In this, they used two sensors that are smoke and a fire sensor for the automatic fire detection system. They also use the GSM module and RF module to alert the officials and authorities (Fig. 1).

The main objective of the proposed method is quick, reliable, detection, and localization of fire. In this, due to the installed sensors, monitoring and detection are done and to alert the authorities and official message alerts are used.

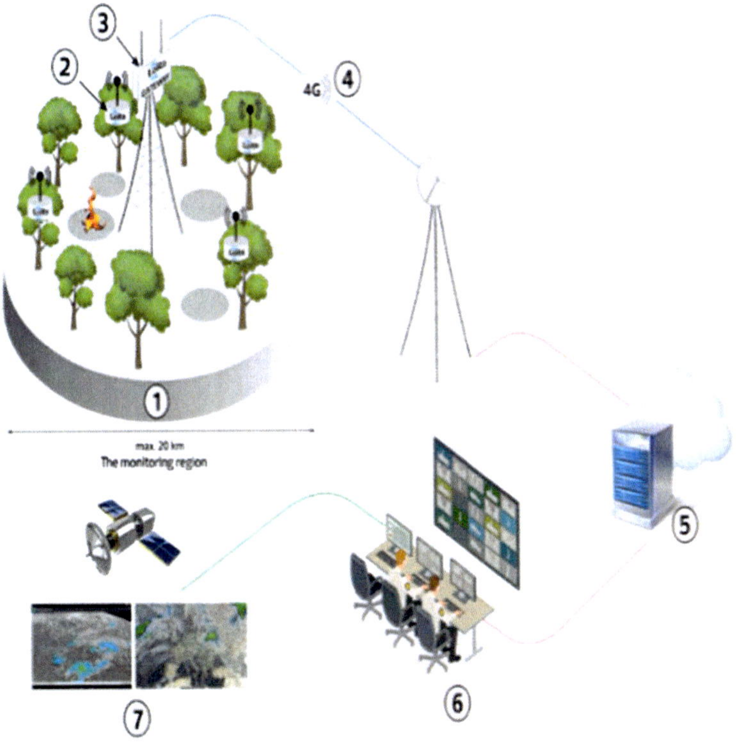

Fig. 1 Wireless sensor network

3 Proposed Method

To control forest fires on time, it is required to provide a system that is essential to the officials and forest authorities to prevent the spread of forest fire in beginning. A good detection and prevention system by using GSM modules and other components is required to protect the forest ecosystem, natural ecosystem, and biodiversity so that we can control fire spreading on time. In the sensing unit, we use several sensors to sense the environment in a forest area. The purpose of the study is to provide a detection and monitoring system by using GSM and RF modules with the help of real-time methods and to stop the forest fire at the beginning stage to protect biodiversity and also to detect poaching, deforestation, and natural disasters like landslides, etc. The all above data will be sent to the officials and forest authorities so that they can manage to take appropriate decisions at the beginning stage.

3.1 Methodology

In this proposed system, we design an IC sensor that will help us to sense the fire and smoke in the forest area. This system having a control unit that helps in determining the intensity of fire and smoke which will later show the rate of fire and smoke on an LCD screen and sends an SMS message with the help of a GSM modem. Our system comprises three important phases: detecting, communication, and processing unit.

For the detecting unit, the sensor senses the changes in the surrounding environment, a sensor such as, i.e., MQ-2 sensor which has very high sensitivity towards propane, methane, LPG, CO_2, and hydrogen. These sensors help to sense a fire in the forest. By communication, we will receive the message or call on the registered mobile number when the input of the fire sensor or MQ-2 sensor is high and simultaneously. Our prevention system ON which switches the water motor ON to put out the fire or flame. Now, Fig. 2 shows the overall block diagram of this system.

3.2 System Design

For the system designing of this project, we divide it into two system design parts such as software and hardware parts. In hardware system design, we have done the hardware part which consists of four units in which various components are used such as MQ-2 gas sensor, AT89C52 microcontroller, encoder, decoder, RF module, and GSM module, etc. In software system design, we perform the programming and initialized it in the control unit.

1. **Hardware System Design**—To design the hardware system, the four units are considered such as sensing unit, control unit, and processing unit.

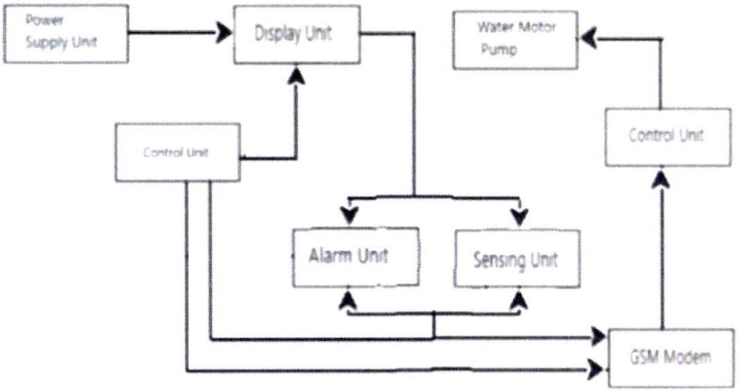

Fig. 2 Block diagram of the proposed system

2. **Software System Design—**In the software system design, we initialized and program the control unit.

3.3 Hardware System Design

Sensing Unit. In this unit, there are both sensors which are fire sensor and MQ-2 sensor. But the sensor works together in detecting the flame or fire when there is either smoke or flame is detected in the forest. Then, it sends the information of flame and smoke to the control unit and a red led glows as well as a piezoelectric buzzer starts to beep to alert the authorities. Figure 3 shows the block diagram of sensing unit.

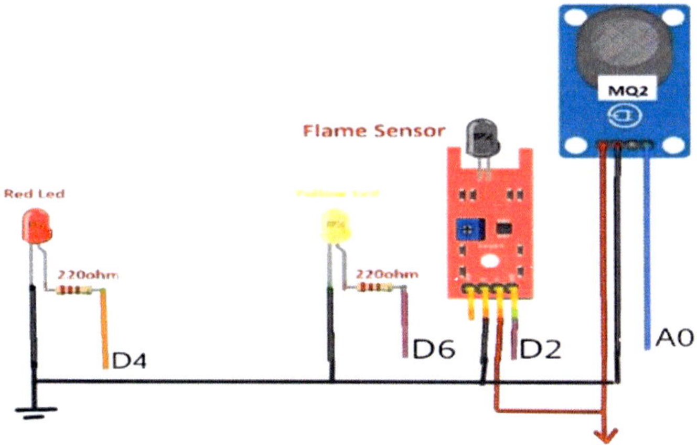

Fig. 3 Circuit diagram of the sensing unit

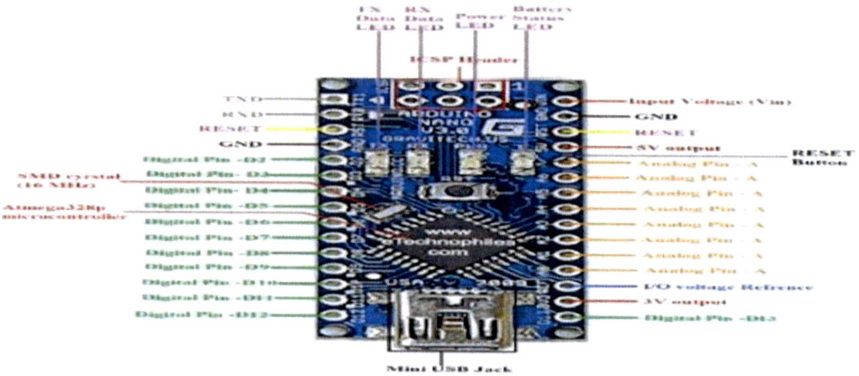

Fig. 4 Schematic diagram of the sensing unit

Control Unit (Arduino Board and Relay). Arduino Nano board has ATmeag328 microcontroller. It is in a different package, either more or less it has some functions of the Arduino Duemilanove. For Arduino Nano, a mini USB cable is required instead of a standard one and it lacks a DC power jack. Figure 4 shows the Arduino Nano schematic diagram.

Control Unit (Arduino Board and Relay). It is in a different package, either more or less, and it has some functions of the Arduino Duemilanove. For Arduino Nano, a mini USB cable is required instead of a standard one and it lacks a DC power jack. Figure 4 shows the Arduino Nano schematic diagram.

Relay. It is a switch that is operated electrically. It has a set of input terminals that are required to control single or multiple signals, it is also a set of contact terminals which control a circuit through self-regulating small power signals. It is used for many reasons one of them is that we use a relay for high voltage or high current devices. While to power controls we use low voltage which will further energize the relay. Figure 5 shows the relay pin diagram. Figure 6 shows proposed system block diagram.

3.4 Software System Design

In the software system design, we initialized and program the control unit. Below shows the steps of initializing the control unit and programmed it:

- In the first step, we will monitor either there is any sense of smoke or flame in the forest with the help of sensors.
- In the second step, if there is smoke and flame in the environment, then the transmission will take place from the control unit, as the sensors will inform the control unit to take action.

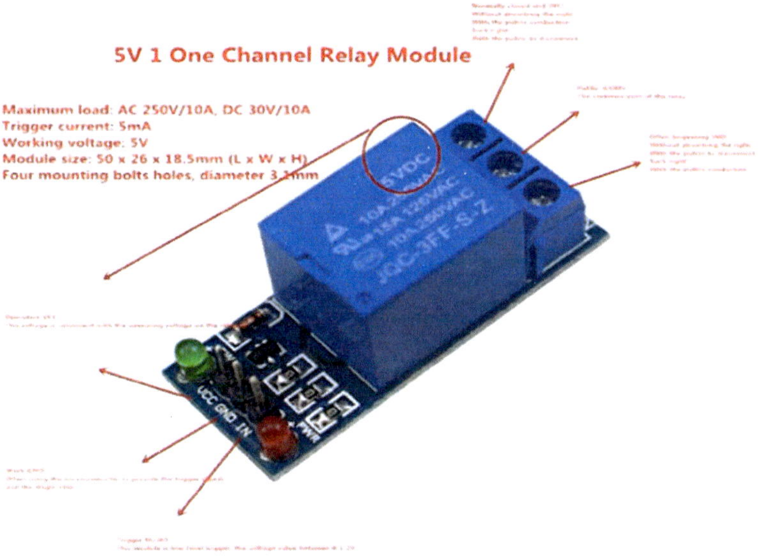

Fig. 5 Relay pin diagram

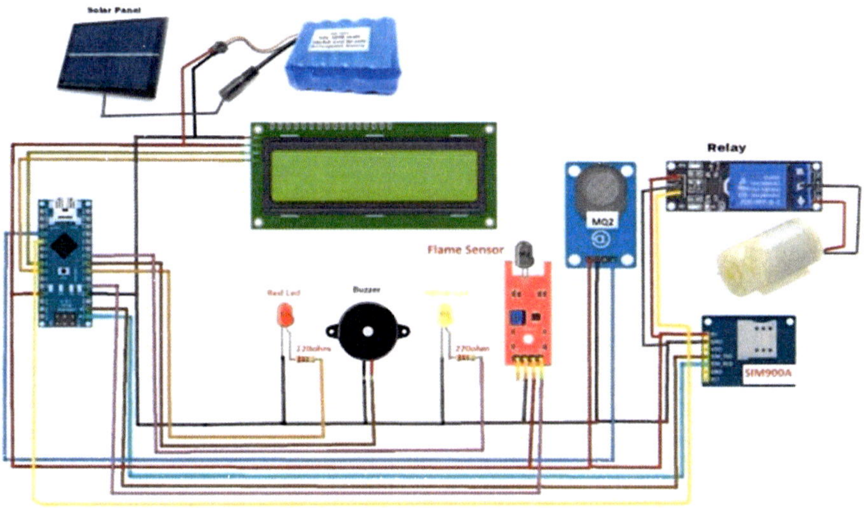

Fig. 6 Proposed system block diagram

- In the third step, after the transmission is done, reception will take place immediately with the help of LCD, buzzer, LEDs, and GSM module.

- In the fourth step, the Arduino Nano will display the message on a liquid crystal display and turns on the piezoelectric buzzer to make the sound to alert the officials and also sends a message to the official mobile number.
- At last, in the fifth step, the relay will start the water motor to put out the fire and works as a fire extinguisher.

4 Conclusion

Nowadays, monitoring is done by computerized which is very essential and useful for the officials and authorities of the forest. Instead of suppressing fire in later stages, it is easy to suppress it in the early stages. This system provides the detection and monitoring of forest fire at the early stages and it also sends message alerts to the forest authorities. The feature of this project is low cost, effective, flexible, wireless system, and power efficient. To block the fire to spread further, it gives instantaneous information to authorities.

References

1. Ransing RS, Rajput M (2015) Smart home for elderly care, based on wireless sensor network. In: 2015 International conference on nascent technologies in the engineering field
2. Herutomo A, et al (2015) Forest fire detection system reliability test using wireless sensor network and OpenMTC communication platform. In: 2015 3rd international conference on information and communication technology (ICoICT). IEEE, 2015
3. Bento AC (2018) IoT: NodeMCU 12e X Arduino Uno, results of an experimental and comparative survey. Int J 6(1)
4. Othman MF, Shazali K (2012) Wireless sensor network applications: a study in environment monitoring system. Proc Eng 41:1204–1210
5. Li GH, Zhao J, Wang Z (2006) Research on forest fire detection based on wireless sensor network. In: Proceedings of the 6th world congress on intelligent control and automation, June 21–23, 2006, Dalian, China
6. Tomar A, et al (2020) Machine learning, advances in computing, renewable energy and communication, Vol 768. Springer Nature, Berlin, LNEE, p 659. doi: https://doi.org/10.1007/978-981-16-2354-7. ISBN 978–981–16–2354–7
7. EUFOFINET Project Detection Synthesis of Good Practices Zvolen, SlovakiaNational Forest Centre
8. Vikram R, Sinha D, De D, Das AK (2020) EEFFL: energy-efficient data forwarding for forest fire detection using localization technique in wireless sensor network, Springer. Dampage U, Bandaranayake L, Wanasinghe R, et al (2022) Forest fire detection system using wireless sensor networks and machine learning. Sci Rep 12:46. https://doi.org/10.1038/s41598-021-03882-9
9. Iqbal A, et al (2021) Intelligent data-analytics for condition monitoring: smart grid applications. Elsevier, p 268. https://www.sciencedirect.com/book/9780323855105/intelligent-data-analytics-for-condition-monitoring. ISBN:978–0–323–85511–2

Sustainable IT and Energy Management for Metropolis Substructure Expansion

Sonali Vyas, Shaurya Gupta, and Deepkshikha Bhargava

Abstract In today's world, the societal development is very much in progress keeping in view rapid expansion in the field of science and technology. While talking about sustainable IT ecology, the inter collaboration among physical infrastructure and IT aims at enhancing conservational efficiency along with optimizing the cost in terms of development and maintenance. For integration of above, two verticals generally five dynamics are involved which involve ecology design, ascendable reserves micro grids, persistent intuiting and information detection. This approach could be applied to varied verticals for attaining sustainability in an urban infrastructure. The sustainability approach is enhanced in terms of social, economic and environmental factors in a metropolitan scenario keeping the economic growth of the metropolitan in consideration.

Keywords Metropolis infrastructure · Sustainability · Information technology · Ecology

1 Introduction

Populace progression in current scenario necessitates the need of new-fangled cohort of metropolises in developing financial prudence around the whole world. This physical structure will involve substantial quantities of resources besides energy for structure creation, procedures and disposal of waste resources. Nowadays, the biosphere faces intimidations of environment transformation besides developing shortage of

S. Vyas · S. Gupta (✉) · D. Bhargava
School of Computer Science, University of Petroleum and Energy Studies, Dehradun, India
e-mail: shaurya.gupta@ddn.upes.ac.in

S. Vyas
e-mail: svyas@ddn.upes.ac.in

D. Bhargava
e-mail: deepshikhabhargava@gmail.com

A. Tomar et al. (eds.), *Proceedings of 3rd International Conference on Machine Learning, Advances in Computing, Renewable Energy and Communication*, Lecture Notes in Electrical Engineering 915, https://doi.org/10.1007/978-981-19-2828-4_42

water, sustenance and energy, which in turn demands suitable adoptions of least material and energy design apart from procedures for next-level organizations. Contrasting to preceding cohort of metropolises constructed through the industrialized era, the subsequent cohort will be having access to expertise from Information era. However, the necessity for info besides the aptitude to interconnect has remained backbones of metropolises; though they are now embedded with info and communication expertise. However, investigators and city developers are continuously working on concepts of intelligent metropolises and smart region [1], along with the incorporation of computing and recognizing expertise with physical ecosphere [2, 3]. A succinct structure for accomplishing such incorporation of IT into physical substructure is deficient at varied points. This paper discusses a framework: which is suitable to sustainable IT ecosystems, apart from other substitutes to conventional physical substructures, which may be extra sustainable because of its incorporated design and organization with respect to IT. The structure of buildings plays a very important role while considering its effects on natural environment. In recent times, the concept of ecology buildings along with the concept of ecology metropolis is taken into consideration for the ultimate objective of sustainability. These constructions are erected keeping in mind the natural conservationism of atmosphere and surroundings. These structures should follow the local eco-cycle principles keeping in mind the topography, climate of the region. The design of such buildings, which include landscape, aeriation, water supply and drainage illumination, should follow strict norms and procedures as per the ecosystem of the region.

2 Life-Cycle Design of Ecosystem

A perilous task while creating sustainable IT ecologies ensures that selections, which are made during, design period does not end into unfavorable behavior through the life-cycle. Life-cycle valuation delivers a very beneficial method in these kinds of design for environment. The concept of Life-cycle assessment is being discussed in [4–6] and it has been effectively functional in the past in numerous areas. Customary Life-cycle assessment methods are overall intricate in building comprehensive records of unit processes plus merchandises curving through the complete life-cycle—starting with abstraction of raw materials besides industrialized manufacture processes over delivery and merchandizing operations. While considering aggregation of all kinds of inventories via proxy indicators [7] like mass, energy, which is calculated in terms of procedures and merchandises. Consistent or recognized impact factors are functional applied over the inventory for attaining the environmental effect [8, 9]. Numerous shelf software posts [10–13] are accessible for executing life-cycle assessment for educational and residential purposes also. Regrettably, mostly life-cycle assessment methodologies are fixated on product- or system-scale application, which includes almost a thousand constituent's standard calculation practices. For creating sustainable IT ecologies, better progressive design tools and techniques are required, which permits a higher grade of hybridization for life-cycle assessment implementation. Tools provides better authentication, speediness apart from scalability in terms of design methodologies. Hybridization may comprise an amalgamation of reorganized procedures [14], monetary input–output methods generally in case of supply chain management [15], outmoded life-cycle assessment, object-based modeling [16], and other methods, which includes Design of Experiment techniques [17] apart from innovations in computational life-cycle assessment methods [18]. Once effectively incorporated, any toolkit enables originators with a multi-stage layered methodology allowing a preliminary tapering of design space into distinct subsections of constraints, wherein the resolution is progressively increased for sensitive substances because of the expansion of design space. Considering the systematic plus computational intricacy of these kinds of methods, which beforehand prohibited its extensive disposition, but now with the help of IT infrastructure—comprising software design procedures, besides hardware infrastructure—which permits scrutiny plus strategy consideration with the help of product systems. These technologies help in generation of collection of domain-specific procedures plus metrics followed by an incorporated method for ecosystem scaling and achieving functional design, which is applicable in terms of information technology for municipal substructures. Authors [19] have discussed varied aspects of sustainability. Informality sustainability of Ghana cities is being discoursed by authors [20].

3 Ascendable Microgrids of Resources

While taking resources disbursed while creation of the physical arrangement, sustainable IT ecology should be able to provide high operation within any provisions of transferring populace and keeping in mind the available resource accessibility patterns. A responsive set-up is always essentially required for efficiently familiarizing its variations throughout its prolonged operative life-cycle. As and when the populace progress occurs in urban centers, including mega metropolis, scalability always becomes a key characteristic of substructures or infrastructures [21, 22]. While taking consideration of metropolitan structures wherein safety, configurability, besides consistency forms critical individualities, which includes power supply and distribution [23] besides public safety. Achieving the anticipated safety, excellence, dependability and obtainability for urban infrastructures is feasible by enabling cohesive supply-side plus demand-side administration via a system of resource micro grids. These kinds of micro grids comprise numerous distinct sources which are interrelated in providing an accessible reserve pool, which is subject to reduction (supply-side management) as it provides suppleness plus a higher customizability in terms of fulfilling user requirements (demand management). The rise of power engendering systems apart from other ascendable skills provides innovative possibilities for generating resolutions improving dependability besides scalability of supply-side structures in a metropolis scenario. The characteristics of physical layer is composed

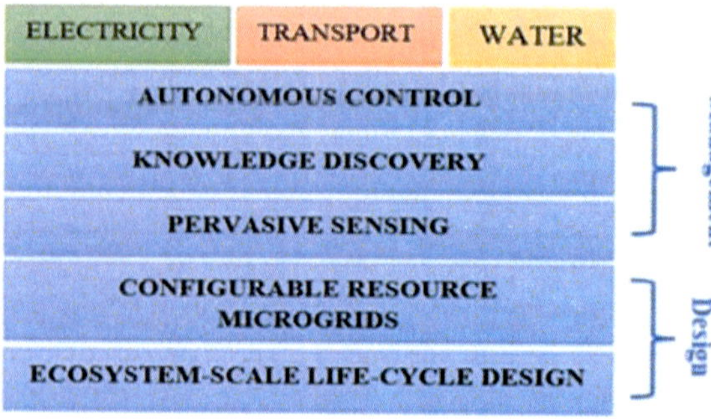

Fig. 1 Framework of IT ecosystem in city 2.0

of micro grids and hubs, which frequently delivers information. Figure 1 displays requirement-centered donating in terms of funds utilizing quantity.

Guidelines, which are driven by human collaboration and computerization, are quite helpful in guiding the resource administration across varied verticals. For an instance, a grid of regional water stowage tanks is storing water and the supply can be controlled from these tanks, which is completely based on traditional configurations besides supervised state. The existing central dissemination system, municipal water dissemination can ingest near to 1.5 MWh of electricity disseminated for micro grids for hydro usage. Similarly, the same fundamental can be used for distribution of varied possessions and facilities for example power transport, etc. The datasets involving the indicators for sustainable energy development is being discussed and is available on the following link: https://www.kaggle.com/natashalondon/sustainable-energy-for-all. Varied indicators like access to clean fuels, clean technology, electricity consumption, energy consumption and optimization are discoursed in the above dataset available at kaggle.com.

4 Ubiquitous Sensing

The current progresses in terms of design of semiconductor and circuit has made disposition of nodes particularly within municipal spaces. Nodes, which are deployed habitually, form a strong network which is wireless network. The introduction of technologies like system-on-chip, which have permitted for microcontrollers chip amalgamation, memory, radio transceivers, antennae altogether [24]. The implementation of low power design besides cautious planning besides energy planning for usage through system expedites varied procedures with the help of energy ingathering [25]. The grid-powered nodes are quite helpful in the conservation of network energy. Ubiquitous detecting has numerous applications through metropolitan regions. Abundant beneficial applications can be considered like smart transportation systems [26, 27], managing air quality [28], home safety [29, 30], flood controller, apart from monitoring of engineering assemblies. Additionally, prospective in terms of nanoscale sensors increases the probability of its installation in industrialized or manufacturing substructures. These kinds of sensors may be installed in the pavement at intersections for the purpose of transportation sensing plus control or it might be inoculated into constructions, channels and flyovers to measure stress; observing flow and seepage; or entrenched in the casing or isolation of electronic chains and cables for the purpose of monitoring and controlling control usage. Sensors are involved in recognition of events, which are of concern to owners or manufacturers of substructures. Numerous events may require instantaneous corrective action. They might be unpretentious or unswervingly sensed, like failure of a component. Another determination of detection is supplying data for tools that promotes management. The ultimate purpose of detecting calls for the enactment of autonomic mechanism systems. The size of data produced by a collection of sensors in WSN is extremely large and requires an ascendable architectural design for assembling in addition to amassing data though distributing it to numerous users. Thus, ubiquitous sensing necessitates a software stack by which the sensor data is accessible to consumers. About some of circumstances are fixated for sensors like exactness and accurateness in terms of range, dimension rates or bandwidths. Numerous applications necessitate timely distribution of sensor data. Real-time data feeds are predominantly significant in terms of independent control, besides visualizing consoles whereas offline data excavating depends on past documentations.

5 Conception and Information Exploration

Knowledge discovery denotes statistics, data excavating and machine learning practices, which alter and analyze data gathered from different sources as well as ubiquitous detecting networks. IT ecology creates enormous capacities of data associated to physical and operative state, comprising conservational sensor data for instance temperature, operative structures and hardware unit devices, i.e., utilizing values besides consumer request. Generally, these practices are functionally useful to real-time data streams, archived data, which may result in the following:

- Event and anomaly recognition, judgment and extrapolation
- Causativeness inference
- Patterns and associations detection
- Computerized Regulator Models
- Computerized forecasting and programming
- Summarization and visual conception of operating state besides metrics via dashboards.

These practices are extensively applied to numerous mechanisms of current urban substructures, comprising transference, power, water, telecommunication and networking. Taking an instance, smart transportation systems [31] depend on knowledge discovery methods for attaining economical, extraordinary maintainable and ecological systems such that they assist in agent-based independent plus disseminated traffic control, commuter traffic flow forecast, insinuation of human behavior identification of crash patterns which will help in better design implementation in future, apart from computerized resource distribution and planning. Correspondingly, when considering case of telecommunication and interacting systems, information detection procedures are widely used for applications comprising of resource distribution grounded on mandate, invasion recognition discovery and identification of fiascos [32], besides consumer association and administration behavior prediction [33]. While considering case of power systems, varied tasks use knowledge discovery practices for the purpose of expansion and approximating of power substructure postponements, connecting of future evaluation and minimum demands, upcoming charges associated to generation, energy ingestion, conveyance and communication of influence; fault analysis. Existing effort involves the role of information technology including ubiquitous detecting and information discovery methods, which can perhaps be characteristically deployed crossways dissimilar substructure verticals to acquire a more united assessment of metropolitan situation. Additional invention essential in the conception of a supportable IT ecology provided by amalgamation in terms of sustainability metrics into knowledge discovery practices.

6 Urban Energy Managing System

UEMS involves a collection of premeditated and applied procedures intended at accomplishing sustainable, effective usage of energy and fuel resources. In the upcoming years, it is premeditated that UEMS must be applied for all structures and edifices immaterial being private or secretarial. However, it has been implemented only for structure with a budget factor like infirmaries, universities, academies, playschools and cloistered buildings. UEMS is grounded on the extensive usage of information technologies including the concept of digital revolution. The ultimate objective of UEMS involves a likelihood for implementation in the budget policy for state goals in area of energy efficiency. Accomplishing the objective comprises of both administrative and practical errands in the urban administration, which includes the following:

- Multifaceted bookkeeping of each disbursed energy resources.
- Transmitting of data when energy consumption is taken in consideration for a certain and specific resource besides it inculcates engineering arrangements in terms of constructions.
- Automation in terms of monitoring process enactment of provincial, community energy-efficient programs.
- Operational and reliable foretelling in terms of energy costs.
- Discovery in terms of additional energy fatalities besides anticipation of disasters situations.

- Checking of eminence considerations acquiescence besides control of varied modes of supply of energy resources.
- Checking all the speculation agendas, which are aimed at improvement in terms of energy efficacy.

Firstly, practical data relating to state inclusion constructions besides engineering systems is augmented to the overall system for the purpose of evaluation of energy efficiency. As In Future, buildings may compulsorily require accreditation in terms of energy efficiency [34], as it provides adequate conviction and consistency in terms of certification as the certification is subject to inspection and observing by autonomous experts. According to energy expenditure curve, while associating normative indicators, which are cantered on mechanical aspects of the building, some of the energy-efficient events are differentiated on basis of asset attractiveness besides financial viability. Consequently, the functionality of system is designated in terms of:

- Physical or computerized information amassing.
- Development of set of buildings in terms of energy transformation.
- Conception of approximation of efficacy of energy resource consumption for taking appropriate administration conclusions.
- Comparison of energy consumption data among cities and rural area.

Implementation of urban energy observing system helps in energy saving in terms of renewable and non-renewable sources of energy [35]. Reserves are required for modernizing warming systems including heat generating base stations besides attachments of varied components for a building structure. However, it becomes quite essential in terms of preparing a rough draft of all tasks, which enables the assessment of successful employment of UEMS apart from providing translucent observing results. UEMS permits execution of dynamic examination of acquiescence to ethics and guidelines, in terms of nation's energy policy [36]. Apart from that, it aims at energy efficiency with the help of varied projects projected at energy conservation apart from monitoring messages, which in turn help system users to take operative administration resolutions besides activities in terms of improvement of energy efficacy [37]. System should inculcate all sorts of energy resources plus apart from the by-products produced from transformation of energy resources [38].

7 Conclusion and Future Scope

The aptitude of assimilating of IT and Energy considering their sustainability into account considering the physical substructures delivers an élite chance regarding advancement in terms of the policy and administration of physical substructures. The manuscript discusses the concept of sustainable IT ecologies ensuring need-centered administration of resources at both supply and demand end plus the implementation concept of Urban Energy Management System with the help of resource micro grids. Future work will involve more enhancement and improvement in context of substructures for urban metropolis. Varied verticals in case of metropolis like eco-park, eco-friendly buildings or substructures, battery operated transport system must be taken into consideration in future scope of study.

References

1. Steventon A, Wright S (eds) (2010) Intelligent spaces: the application of pervasive ICT. Springer Science & Business Media
2. Stankovic JA, Lee I, Mok A, Rajkumar R (2005) Opportunities and obligations for physical computing systems. Computer 38(11):23–31
3. Estrin D, Culler D, Pister K, Sukhatme G (2002) Connecting the physical world with pervasive networks. IEEE Pervasive Comput 1(1):59–69
4. Finkbeiner M, Inaba A, Tan R, Christiansen K, Klüppel HJ (2006) The new international standards for life cycle assessment: ISO 14040 and ISO 14044. Int J Life Cycle Assess 11(2):80–85
5. Chen L, Liu M, Huang J (2009) GB/T 24040-2008 environmental management-life cycle assessment-principles and framework national standard understanding. Stand Sci 2
6. Tomar A et al (2020) Machine learning, advances in computing, renewable energy and communication, Vol 768. Springer Nature, Berlin, LNEE, p 659. doi:https://doi.org/10.1007/978-981-16-2354-7. ISBN 978–981–16–2354–7
7. Potekhin I, Mischenko V, Mottaeva A, Zheltenkov A (2018) Evaluation of possibility to increasing sustainability of high-rise buildings through use university intellectual property. In: E3S Web of conferences, Vol 33. EDP Sciences, p 03020
8. Usta P, Arıcı A, Evci A, Kepenek E (2017) Sustainability of traditional buildings located in rural area. Periodicals Eng Nat Sci (PEN) 5(2)
9. Berardi U (2015) Sustainability assessments of buildings, communities, and cities. In: Assessing and measuring environmental impact and sustainability. Butterworth-Heinemann, pp 497–545
10. Bernheim A (2003) How green is green? Developing a process for determining sustainability when planning campuses and academic buildings. Plan High Educ 31(3):99–110
11. Grazieschi G, Asdrubali F, Guattari C (2020) Neighbourhood sustainability: state of the art, critical review and space-temporal analysis. Sustain Cities Soc 102477
12. Karaca F, Guney M, Kumisbek A, Kaskina D, Tokbolat S (2020) A new stakeholder opinion-based rapid sustainability assessment method (RSAM) for existing residential buildings. Sustain Cities Soc 60:102155
13. Alaloul WS, Musarat MA (2020) Impact of zero energy building: sustainability perspective. In: Sustainable sewage sludge management and resource efficiency. IntechOpen
14. Al-Sakkaf A, Zayed T, Bagchi A, Mahmoud S, Pickup D (2020) Development of a sustainability-rating tool for heritage buildings: future implications. Smart Sustain Built Environ
15. Walker AM, Vermeulen WJ, Simboli A, Raggi A (2020) Sustainability assessment in circular inter-firm networks: an integrated framework of industrial ecology and circular supply chain management approaches. J Clean Prod 125457
16. Braulio-Gonzalo M, Bovea MD (2020) Relationship between green public procurement criteria and sustainability assessment tools applied to office buildings. Environ Impact Assess Rev 81:106310
17. Mao D, Ma Q, Zhou BB (2020) Sustainability of human–environment systems through the lens of landscape
18. Lazaretti K, Giotto OT, Sehnem S, Bencke FF (2019) Building sustainability and innovation in organizations. Benchm Int J

19. Song X, Feng Q, Xia F, Li X, Scheffran J (2021) Impacts of changing urban land-use structure on sustainable city growth in China: a population-density dynamics perspective. Habitat Int 107:102296
20. Azunre GA, Amponsah O, Takyi SA, Mensah H (2021) Informality-sustainable city nexus: the place of informality in advancing sustainable Ghanaian cities. Sustain Cities Soc 67:102707
21. Mahmoud S, Zayed T, Fahmy M (2019) Development of sustainability assessment tool for existing buildings. Sustain Cities Soc 44:99–119
22. Roostaie S, Nawari N, Kibert CJ (2019) Sustainability and resilience: a review of definitions, relationships, and their integration into a combined building assessment framework. Build Environ 154:132–144
23. Zarghami E, Fatourehchi D, Karamloo M (2019) Establishing a region-based rating system for multi-family residential buildings in Iran: a holistic approach to sustainability. Sustain Cities Soc 50:101631
24. Aquilino F, Della Corte FG, Merenda M, Zito F (2008) Fully integrated wireless temperature sensor with on-chip antenna. In: Sensors, 2008 IEEE, pp 760–763
25. Mathuna CO, O'Donnell T, Martinez-Catala RV, Rohan J, O'Flynn B (2008) Energy scavenging for long-term deployable wireless sensor networks. Talanta 75(3):613–623
26. Banerjee T, Bose S, Chakraborty A, Samadder T, Kumar B, Rana TK (2017) Self-driving cars: a peep into the future. In: 2017 8th annual industrial automation and electromechanical engineering conference (IEMECON). IEEE, pp 33–38
27. Karpis O (2013) Wireless sensor networks in intelligent transportation systems. Int J Mod Eng Res (IJMER) 3(2)
28. Woo JH, Kim H, Lim SB, Kim JJ, Lee J, Ryoo R, Kim H (2010) Air scope: a micro-scale urban air quality management system. In: International conference on algorithms and architectures for parallel processing. Springer, pp 520–527
29. Yu J, Chun W, Nejat G, Noel E, Tang KW (2007) Self-powered wireless sensor balls for homeland security. In: ASME international mechanical engineering congress and exposition, Vol 43033, pp 963–972
30. Mills KL (2007) A brief survey of self-organization in wireless sensor networks. Wirel Commun Mob Comput 7(7):823–834
31. Dietzel S, Kargl F, Heijenk G, Schaub F (2011) Modeling in-network aggregation in VANETs. IEEE Commun Mag 49(11):142–148
32. Akkaya K, Younis M (2005) A survey on routing protocols for wireless sensor networks. Ad Hoc Netw 3(3):325–349
33. Andriamampianina L, Ravat F, Song J, Vallès-Parlangeau N (2020) A generic modelling to capture the temporal evolution in graphs. In: EDA, pp 19
34. Trejo-Perea M, Moreno GR, Castañeda-Miranda A, Vargas-Vazquez D, Carrillo-Serrano RV, Herrera-Ruiz G (2013) Development of a real time energy-monitoring platform user-friendly for buildings. Proc Technol 7:238–247
35. Suciu G, Necula L, Iosu R, Usurelu T, Ceaparu M (2019) IoT and cloud-based energy monitoring and simulation platform. In: 2019 11th international symposium on advanced topics in electrical engineering (ATEE), IEEE, pp 1–4
36. Yazdanie M, Orehounig K (2021) Advancing urban energy system planning and modeling approaches gaps and solutions in perspective. Renew Sustain Energy Rev 137:110607
37. Nik VM, Perera ATD, Chen D (2021) Towards climate resilient urban energy systems: a review. Nat Sci Rev 8(3):nwaa134
38. Gupta S, Vyas S (2021) IoT in green engineering transformation for smart cities. Smart IoT Res Indus 121

Optimal and Efficient Consumption of Distributed Generation Using Soft Computing Techniques

Gagandeep Kaur, Sobiya Mukhtar, and Naveen Kumar Sharma

Abstract Energy is compulsory for economic enlargement, enlightening the excellence of life and increasing opportunities for green power growth. Generation of power is possible through several sources and the chief sources covering the major generation are fossil fuels. With passing time the availability of fossil fuels is decreasing, though it is a key source of GHG emissions, generation of electricity is transferred toward green power like renewable energy sources. The applications of distributed generation (DG) models are rising in current smart grids and power devices. A significant issue with DG's is their particular optimal area and dimension within the circulation system. The optimization problem is articulated and explained using the Optimal Power Flow (OPF) technique that studies all transmission restraints and power flow limits. In this paper involves the use of Particle Swarm Optimization (PSO) in conjunction with Genetic Algorithms (GA) suggested for sizing and optimal placement of DG models to be able to decrease network blockage. The IEEE 30-Bus system test program has been used to illustrate the potency of the recommended comparison with Artificial Neural Network (ANN). The proposed approach better optimized the cost, power loss and LMP due to its global and local forecast.

Keywords Index.dispersed generation · Distributed generation resources · Load flow

1 Introduction

Global energy demand is increasing at such a rapid pace that conventional generation alone will be unable to meet it in the near future. Energy is a necessary component of any society playing a key role in the progress. Energy demand is predicted to rise from 7.5 billion in 2017 to 8.2 billion by 2030, demand for energy is expected to rise at a faster rate over the same time period. The worldwide demand for energy has risen as a result of industrialization and population growth. Furthermore, worldwide

G. Kaur (✉) · S. Mukhtar · N. K. Sharma
I. K. Gujral Punjab Technical University Jalandhar, Main Campus Kapurthala, Kapurthala, India
e-mail: gaganpitk@gmail.com

A. Tomar et al. (eds.), *Proceedings of 3rd International Conference on Machine Learning, Advances in Computing, Renewable Energy and Communication*, Lecture Notes in Electrical Engineering 915, https://doi.org/10.1007/978-981-19-2828-4_43

environmental challenges, as well as continuous advances in green energy technology, give new avenues for utilizing renewable energy instruments. The deregulation of electricity segment has led to multiple openings for the development of new technologies. In the aggressive power market, internationally abundant methods have been experimented to contract with the overcrowding on transmission lines. In the competitive electricity power market generation, transmission, and distribution are self-regulating activities. Generators compete with one another to offer power to various clients. The key benefits of the deregulation comprise economical electrical energy, effective capacity development planning, cost minimization, choice, and improved service. For exploitation of social welfare Location Marginal Price (LMP) aims for assigning DGs. LMP is the langrangian multipliers linked with the vigorous power flow equations for every bus. LMP build communal vital information for market players to enlarge their approaches toward quoting of bids. LMP at the exacting location is the marginal cost to supply a supplementary increment of authority lacking any violations of limits. Most situations include DG owners making decisions on the technological foundation of generation, such as solar, wind, and fuel cell, as well as bus numbers for interconnection, because distribution system operators (DSOs) have only a limited amount of influence over the same. The deregulation method focuses on improving system quality, increasing service levels, and developing a competitive market. The approach to power restructuring plan is really significant. In India, the process of electrical reform is still ongoing. The terminology "Dispersed Generation" dictates generation of small scales electric units nearer to the location of consumption.

Energy production from RES can play a significant role in providing a consistent source of electricity in rural regions, which also contributes to meeting the targeted radiation reduction objective. In an energy market that has been liberalized, one of the System Operator's primary responsibilities is to permit power transactions and match the required demand-supply balance. Ancillary Services, which provide power transfer while preserving dependability and the required degree of quality and safety, are critical to maintaining this balance. As a result, Ancillary Services have become a major issue since they are essential for a power system's reliable and secure operation. Sharma et al. [1] a PV, hydro power, battery, conventional generator, and power electronic converter green microgrid system. It identified the finest location for arrangement of DG and optimum size approaches for addressing OPF issues are very vulnerable to beginning points and typically converge locally to optimum solutions [2]. Hadded et al. [3] suggested an innovative approach for improving the quality and reliability of DG systems in the presence of diverse disruptions or operational circumstances. It is critical to confirm that the accessible DG are used as efficiently as possible. A great impact of losses occurred in transmission system has been observed with optimal locations and capacities of DGs [4, 5]. In these papers [6, 7] the popularly known meta-heuristic technique is utilized, in which voltage profile improvement and minimization of losses have been done in distribution network using PSO technique. For [8] minimization of power losses considering the different loading has been done by sizing and placement of DG using GA technique. Bindeshwar Singh, et al. [9] presented the improvement and effects of the losses in system, profile of voltage and price by employing DGs of various sizes in the distribution network. The suggested method has been evaluated for the IEEE-14 bus distribution test system. Rahul et al. [10] presented a novel synchronized Volt/VAr design which can be utilized in Allocation Managing System (AMS). PSO finds application if it is compatible with combinatorial variables, sturdy in solving non-linear optimization problems and its simple execution. Sriramulu et al. [11] carried out the work in order to minimize the voltage deviation and loss reduction in the radial distribution system by optimizing the location and size of DG units. The most suitable size was found using a PSO algorithm. A generalized optimum model for deciding the location and assessment of wind power output in a novel power market with a double auction [12, 13]. A soft computing was utilized to randomly and identifies the optimal locations of candidate buses for the appointment of DG units in a radial distribution system. A novel method for the most favorable allotment of ancillary services in deregulated power sector [14, 15]. Huddar et al. [16] explained that in order to maximize the DG placement profits, it is important to do more studies in this field. In Ref. [17] used a sensitivity analysis for the identification of best possible candidate positions for DG and location of capacitor and used technique named heuristic arc mounting to find optimum capability of DG in the network. A proposed model for appropriate positioning and DGs size to decrease network crowding and to reduce LMP differences among different buses [18]. Kazemi et al. [19] presented an algorithm for exact DG installation in distribution system, and object is to reduce the losses occurring in network and agreement of a suitable profile of voltage. The technical effect of DG on voltage regulation which is associated with a typical distribution system [20–23]. Computer prototypes were constructed along with simulation and implementation of control systems to investigate the relationship of DG with voltage regulators and capacitors. To demonstrate the aforementioned relationship, test cases were presented.

Many nations have achieved grid parity with renewable energy, and numerous strategies have been presented to achieve 100 percent use of green energy sources by 2050. The amount and location of DGs in a network have been determined by using hybrid optimization techniques; Genetic Algorithm and Particle Swarm Optimization and Artificial Neural Network (ANN). The objective of this research is to manage the congestion in network which further leads to exceeding to thermal limits and initiation of multiple problems. The hybrid optimization has also shown the promising results as compare to ANN. The DGs of various sizes and power factor have been integrated alternatively to select the range of DG size. The proposed optimization techniques have been tested on standard test system. The relevance of DG models is elevating in current smart grids and power strategy. A momentous affair with DGs is their scrupulous proper area and dimension surrounded by the circulation system.

2 Proposed Methodology and Optimization Framework

The issue with optimum distribution and size of DG units consists of a number of parameters. To achieve maximum benefit of DG units in the system networks prior definition of the constraints related to its objective, functions, and procedures are required [3, 13].

(a) Objective Function: The problem objective can be categorized into single and multiple objectives for proper location and sizing of DG. This novel single objective function might be in form of minimization of system power deficiency (real and reactive loss), enhancement in levels of voltage at buses, maximization of DG size, minimization of generation cost and minimization of voltage eccentricities. For formation of multi-objective functions the single objective functions are combined using weighting factors. For the sizing and position of DG in optimal manner is possible with the minimization of power loss in distribution system. The total real power loss in a distribution system is given by:

$$P_{\text{losses}} = \sum_{t=1}^{N} \left| I_i^2 \right| R_i \tag{1}$$

where

I_{th} = magnitude of current in each branch; Rth = ith branch system resistance;

The loss minimization may be represented using the N bus distribution system as follows:

$$\text{minimize } P_l = \sum_{i}^{N} \sum_{j}^{N} \left[\alpha_{ij} \left(P_i P_j + Q_i Q_j \right) + \beta_{ij} \left(Q_i P_j - P_i Q_j \right) \right] \tag{2}$$

$$\alpha_{ij} = \frac{R_{ij}}{V_i V_j} \cos\left(\delta_i - \delta_j \right) \tag{3}$$

$$\beta_{ij} = \frac{R_{ij}}{V_i V_j} \sin\left(\delta_i - \delta_j \right) \tag{4}$$

$$Z_{ij} = R_{ij} + X_{ij} \tag{5}$$

Here, R_{ij} = resistance of the line between buses i and j;
Z_{ij} = line impedance between bus i and bus j;
V_i = voltage magnitude on the ith bus;
V_j = scale of voltage at jth bus;
δ_i = angle of voltage at the bus i;
δ_j = voltage angle at bus j;
P_i, P_j and Q_i, Q_j = injection of active and reactive power at bus i and j, respectably;

(b) Constraints: For equality constraint at each bus, the demand and supply must be met for satisfaction of load balance Eq. (6):

$$\sum_{i=1}^{n} P_{\mathrm{DG}i} \leq \sum_{i=1}^{n} P_{\mathrm{DG}i} \tag{6}$$

For in-equality constraint at each bus, upper and lower voltage bounds should be fixed as voltage limits as in Eq. (7).

$$\begin{gathered} |V_i|^{\min} \leq |V_i| \leq |V_i|^{\max} \\ 0.95_{p.u} \leq V_i \leq 1.05_{p.u} \end{gathered} \tag{7}$$

where $V_{\min}$ is the minimum voltage on the bus and $V_{\max}$ is the maximum voltage on the bus.

(c) Types of DG units: Generally based on the terminal characteristics in form of capability to deliver real and reactive power, DGs are categorized into four main types;

1. DGs for injection of only real power
2. DGs for injection of only reactive power
3. DGs for both actual and reactive power injection
4. DGs for injection of real power but taking reactive power.

The applications these DGs are: type 1 DGs are based on integration into main grid such as micro-turbines, photovoltaic, and fuel cells. In type 2, there is placement of synchronous compensator like in gas turbines. In type 3, synchronous machines act as DG for applications in cogeneration and gas turbines. In type 4, DG is largely induction generator having application in wind farms.

(d) Cost Criteria: It plays a considerable position in deciding the capability of DGs. Contemplation of costs for investment, costs for operation, capacity and durability indexes that differentiate the DGs from former conservative reparation equipments. The potential of power generation (reactive) is an important trait that depends on selection done for type of DG and adopted technology for that system. The number of DGs with dissimilar technologies of DGs can be diffused for several applications such as gas, wind, combined heat and power (CHP), photovoltaic, and fuel cell units.

Now the cost function is proposed for DGs is:

$$C(\rho) = (a\mathrm{DG})\rho^2 + (b\mathrm{DG})\rho + c\mathrm{DG}$$

where a_{DG}, b_{DG}, and c_{DG} are coefficients for this function.

(e) Congestion Criteria: The excess loading of power lines and huge divergence in LMP on buses is an indication of congestion in distribution networks. The calculation for LMP is done as follows:

$$\mathrm{LMP}_i = \lambda + \lambda \mathrm{L}_i + \lambda C_i$$

where $\lambda L_i = \lambda \frac{\partial PL}{\partial P_i}$, $\lambda C_i = \sum_N^{j=1} \mu L_{ij} \frac{\partial P_{ij}}{\partial P}$.

In LMP equations, at reference bus the marginal energy component is presented by λ and its identical for all buses in network, marginal loss shown by λL_i and congestion is by λC_i. Accordingly, the LMP on each and every bus differs on level of congestion, its position and corresponding losses. Hence, the key indicator for congestion management is LMP criteria. As the LMP differences flanked by buses are reduced then it should be taken as optimized congestion.

For optimal placement of DGs the proposed technique has been tested and analyzed on IEEE 30 bus system, which is consisted of 6 generators, 24 load buses, and 41 transmission lines. 283.40 MW is the total active power of load and 126.200 MVAR is the total reactive power of load. The total real power generation is 300.928 MW and the total reactive power generation is 141.21 MVAR. The total real line losses are 17.528 MW as well as total reactive line losses are 68.888 MVAR. The total congestion cost is 123.20 ($/MWh). In this Fig. 1 shows real and reactive loss at 0.8 power factor with and without different fixed DG.

Case 1: The proposed methodology using ANN optimization approach is executed in IEEE 30 bus system (Fig. 2).

Step by step procedure is as shown below:

1. Start the load/power
2. Initialize the load flow
3. Allocate the generators and evaluate the cost
4. Process the ANN for prediction
5. If the output of ANN has optimized neurons

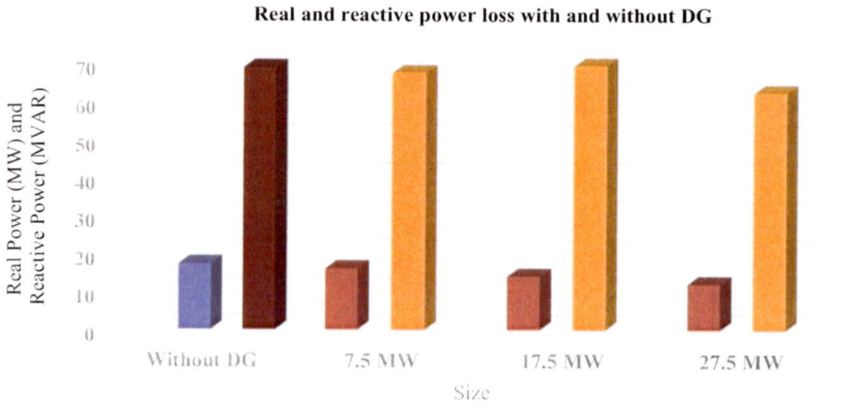

Fig. 1 Loss, both real and reactive, with and without DG

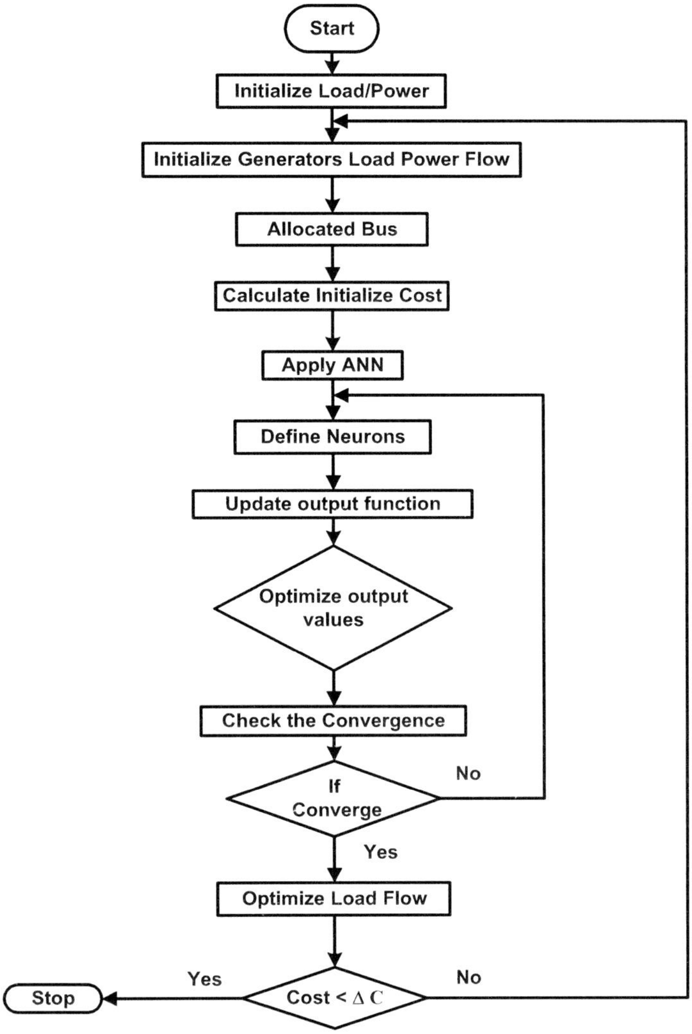

Fig. 2 Flow chart for ANN optimization approach

6. Check the system convergence. If system converges, so check the cost features otherwise go to initialize the neurons and repeat step 5
7. If the system cost function is less than ΔC so the system stops (Table 1).

Case 2: The proposed methodology using PSO-GA optimization approach is executed in IEEE 30-bus system.

Step by step procedure shown below:

1. Start the load/power
2. Initialize the generator load power

Table 1 DG size and location optimal in case 1

Sr. No.	Parameters	Optimal values
1	DG size	10.23 MW
2	LMP	120 $/MWh
3	Loss	15.3422 MW
4	Bus location	9

3. Allocate the generators and analyze the cost
4. Apply the PSO for optimization
5. If optimization by PSO is done then verify convergence, otherwise GA start its optimization with the below steps

    (a) Initialization of chromosomes.
    (b) Cross over among chromosomes.
    (c) Accept roulette selection.
    (d) Confirm optimization status. If it is optimized then follow convergence, otherwise check either loop is running a waiting the objective to achieve.

6. Confirm the status of convergence. If it converges then confirm the cost features otherwise again start initialization of particles and repeat step 5
7. If the cost feature satisfy, less than ΔC then stop.

The proposed optimization approach is implemented in modified IEEE 30 bus system. In Fig. 3 shows the flow chart of suggested algorithm. In Table 2 presents the optimal values for IEEE-30 bus system. Figure 4 presents the LMP comparison for case 1 and case 2.

3 Conclusion

Green energy footprint in distribution power networks can be increased, help to meet rising energy demands while also lowering fossil fuel usage and giving economic and technological benefits. DG can play a vital role in India's strategy to not only add more capacity, but also to boost energy security, solve environmental concerns, and lead the enormous renewable energy market. Optimal location of various DG in deregulated power sector, which can either be operated in grid-connected or islanded mode, helps to satisfy the local demand. In the competitive power market environment, DG is an operative method to accomplish presentation, procedure and control of the distribution system. This work clearly presents the valuable approach for consideration of a number of vital points to be noted with optimal positioning and sizing of DG in IEEE 30-bus system. Congestion management has been turned out to be a chronological come up to lessen the congestion problems. In this cursory, to assuage congestion, innovative suggestions and features are engaged to the performance of the power system to generate the proficient methods by the use of contemporary technologies.

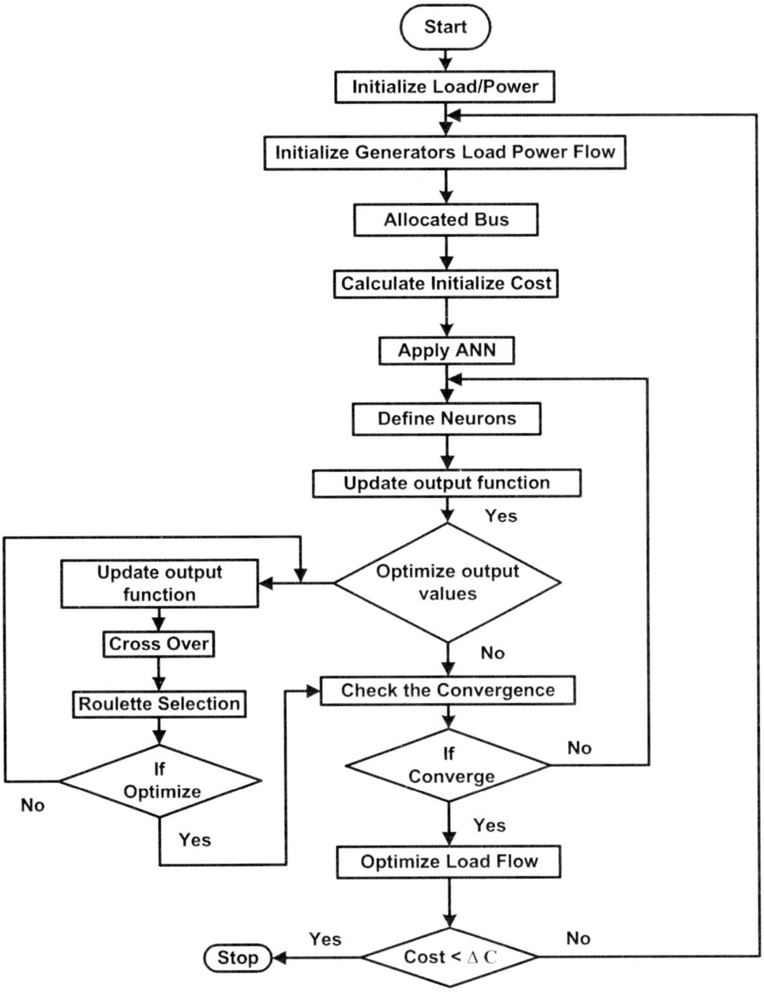

Fig. 3 Flow chart for PSO-GA optimization approach

Table 2 DG size and positioning optimal in case 2

Sr. No.	Parameters	Optimal values
5	DG size	7.23 MW
6	LMP	90 $/MWh
7	Loss	12.333 MW
8	Bus location	24

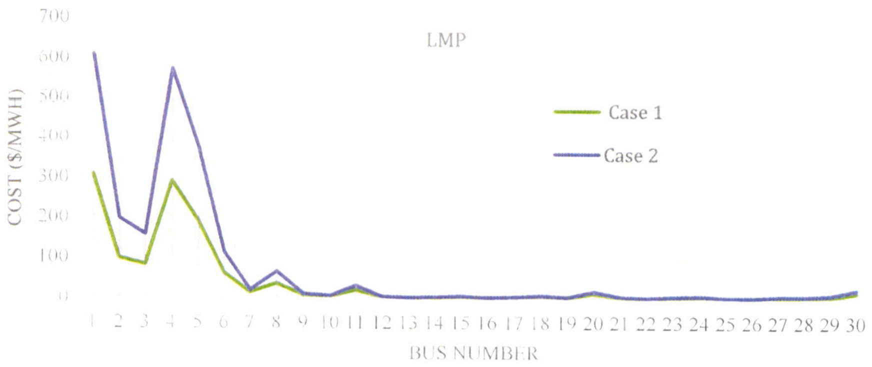

Fig. 4 Case 1 and case 2 locational marginal pricing

Optimization methods are acting for prevailing role in congestion management. The settlement and sizing of DGs in optimal manner are prepared to supervise the congestion in a deregulated power system by implementing on modified IEEE 30 bus test system. The results likewise indicate that DG devices form a successful tool in dropping the loading of transmission lines. The proposed methodology presents the reduced amount of cost as indicated by the power factor.

References

1. Sharma NK, Sharma S, Sood YR, Maheshwari A, Banshwar A (2022) Optimal planning of green hybrid microgrid in power industry. Mach Learn Adv Comput Renew Energy Commun LNEE 768
2. Maheshwari A, Sood YR, Sharma S, Sharma NK (2022) Optimal power flow and its vindication in deregulated power sector. Mach Learn Adv Comput Renew Energy Commun LNEE 768
3. Haddad RJ, Guha B, Kalaani Y, El-Shahat A (2018) Smart distributed generation systems using artificial neural network-based event classification. IEEE Power Energy Technol Syst J 5(2):18–26
4. Kaur J, Banshwar A, Sharma NK, Sood YR, Shrivastava R (2018) Strategic utilization of resources in a microgrid in an uncertain electricity market. IEEE international conference on power energy, environment and intelligent control, pp 220–225
5. Tooryan F, Collins ER (2018) Optimum size and placement of distributed generators in microgrid based on reliability concept. IEEE Power Energy Conf Illinois 1–6
6. Prakash DB, Lakshminarayana C (2016) Multiple DG placements in distribution system for power loss reduction using PSO algorithm. Proc Technol 25:785–792
7. Fanish R, Bhadoriya J (2007) Optimal placement of multi DG in 33 bus system using PSO, Vol 3297, 2320–3765
8. Singh D, Singh D, Verma K (2008) GA based optimal sizing and placement of distributed generation for loss minimization. KES J 12:147–156
9. Singh B, Gyanish BJ (2018) Impact assessment of DG in distribution systems from minimization of total real power loss viewpoint by using optimal power flow algorithms. Energy Rep 4:407–417

10. Kumar RA, Devriese G, Srivastava AK (2018) Voltage and reactive power control to maximize the energy savings in power distribution system with wind energy. IEEE Trans Ind Appl 54(1):656–664
11. Sriramulu M, Rahul MR (2016) Optimal placing and sizing of DG in a distribution system for voltage stability improvement. In: *2016* International conference on electrical, electronics, and optimization techniques (ICEEOT), pp 1469–1475
12. Sharma NK, Sood YR (2012) Reduction in subsidy for solar power as distributed electricity generation in Indian future competitive power market. Int J Renew Sustain Energy 6(1):13123
13. Sharma NK, Sood YR (2014) Strategy for optimal location and rating of wind power generator with maximization of social welfare in double auction competitive power market. Int J Renew Sustain Energy 6(1):13123
14. Banshwar A, Sharma NK, Sood YR, Srivastava R (2015) Determination of optimal capacity of pumped storage plant by efficient management of renewable energy sources. In: 2015 IEEE students conference on engineering and systems (SCES), pp 1–5
15. Banshwar A, Sharma NK, Sood YR, Shrivastava R (2016) A sequential optimization approach for competitive procurement of energy and Ancillary Services. In: 2016 international conference on electrical power and energy systems (ICEPES), pp 498–503
16. Huddar S, Kantharaj B, Mohan KR, Patil SB, Magadum R (2014) Optimal location and sizing of DG using fuzzy logic. Int J Eng Res Technol (IJERT) 3(02):2278–3181
17. Mughal SN, Sood YR, Jarial RK (2021) Design and optimization of photovoltaic system with a week ahead power forecast using autoregressive artificial neural networks. Mater Today: Proc
18. Sharma NK, Tiwari PK, Sood YR (2012) Current status, policies and future perspectives of Indian power sector moving towards deregulation. IEEE Stud Conf Electr Electron Comp Sci 2012:1–6
19. Kazemi A, Sadeghi M (2009) Distributed generation allocation for loss reduction and voltage improvement. In: 2009 IEEE Asia-pacific power and energy engineering conference, pp 1–6
20. Pokharel P, Poudel L (2019) Impact of distributed generation in distribution network losses and voltage profile. Int J Eng Appl Sci (IJEAS) 6(10):13–20
21. Kaur G, Sharma NK, Kaur J, Bajaj M, Zawbaa HM, Turky RA, Kamel S (2021) Prospects of biogas and evaluation of unseen livestock-based resource potential as distributed generation in India. Ain Shams Eng J 101657
22. Iqbal A et al (2021) Intelligent data-analytics for condition monitoring: smart grid applications. Elsevier, p 268. https://www.sciencedirect.com/book/9780323855105/intelligent-data-analytics-for-condition-monitoring. ISBN: 978–0–323–85511–2
23. Tomar A et al (2020) Machine learning, advances in computing, renewable energy and communication, Vol 768. Springer, Berlin, LNEE, p 659. doi:https://doi.org/10.1007/978-981-16-2354-7. ISBN 978-981-16-2354-7

Dataset Classification Using Various Machine Learning Algorithms

Muskaan Nagpal and Richa Mishra

Abstract The extensively used CIFAR-10 dataset, which is commonly used to analyze image classification technique. The analysis of an efficient classification approach for the CIFAR10 dataset using a convolutional neural network, dense neural networks, and support vector machines is presented in the research. Convolutional neural network (CNN) model was enhanced with learnable filters, pooling, and padding layers, and the dataset was transformed into two dimensions for support vector machine (SVM) model using picture preprocessing in Python.

Keywords Convolutional neural networks · Cifar10 · Image classification · Support vector machine

1 Introduction

Many people in all walks of life use image classification. It pervades every aspect of the social, economic, and corporate spheres of influence, all over the world. The demand for more precise, specific classification enhances the need for machine learning algorithm changes, adaptations, and breakthroughs. Many picture classifying models have been developed to aid in the resolution of the most pressing issue of identification accuracy. Image classification is a key subject in the field of computer vision, having a wide range of practical applications [1–17].

Image classification tasks have been successfully handled by employing convolutional neural networks. It is a form of a deep, feed—forward network which has gotten a lot of interest from the research industry and community after showing promise in tasks including natural language processing, object recognition, transfer learning, speech recognition, and signal processing. Whereas, in the land covering imaging

M. Nagpal (✉) · R. Mishra
Computer Science Department, Amity University, Noida, Uttar Pradesh, India
e-mail: Muskaannagpal24@gmail.com

R. Mishra
e-mail: richas.phoenix@gmail.com

A. Tomar et al. (eds.), *Proceedings of 3rd International Conference on Machine Learning, Advances in Computing, Renewable Energy and Communication*, Lecture Notes in Electrical Engineering 915, https://doi.org/10.1007/978-981-19-2828-4_44

community, support vector machines are a relatively recent supervised classification tool. They have their origins in Statistical Learning Theory and have acquired popularity as a result of their resilience, precision, and efficiency, even when only a small training sample is used. SVMs are binary classifiers by nature, but they can be used to handle the many classification tasks that are frequent in remote sensing applications.

1.1 Real Time Applications

There are various real time applications of data classification such as:

Malware Detection: On the basis of comparable traits of similar malware, a multinomial classification can be used to classify new/emerging malware. Malware categorization can be done using machine learning methods like Naive Bayes, KNN, and tree-based models.

Credit card Fraud Detection: For credit card fraud detection, a binary classification model can be employed, in which historical transaction data of a client is evaluated using machine learning methods such as Naive Bayes and KNN.

Web text classification: Web text is classified or tagged based on pre-determined categories learned from previous data.

Product Classification: A multinomial classification can be used to group products supplied by multiple shops into the same categories, regardless of the categories that the retailers have given to the products.

2 Literature Review

In paper [1] already trained CNN models with training images are utilized to extract features from many levels, mixing features from various layers in provided CNN models. According to paper [2] the detection accuracy is improved by combining the learned low and high frequency features and processing the composite feature mapping. The paper [3] proposes a revolutionary picture classification system that combines two exceptional classifiers, the CNN and eXtreme Gradient Boosting. By using CNN as a trainable feature, the given CNN-XGBoost model produces more exact results. In [4] aims at reducing testing error by using a deep convolutional network with sparsity, although photos are not sparse, they can be conceived of as sparse by adding padding. The paper [5] integrates a new method for picture recognition which is Biomimetic Pattern Recognition with CNNs. BPR can address some of the drawbacks of classical pattern recognition by performing class detection using a union of geometrical cover sets via a high dimensional feature set. In [6] a new Tensorflow-based, specialized GPU-accelerated simulation framework for approximation hardware that allows approximate DNN inference and retraining is proposed.. Paper [7] investigates the performance of the DNN classifier under four different forms of distortions and proposes two methods for reducing image distortion by fine-tuning and retraining via noisy images. The research proposed in paper [8] emphasized on increasing the accuracy of support vector machine for image classification problems by selecting the proper kernel function and appropriate approach for parameter selection. The paper [9] presented a multi label SVM model for active learning by incorporating two selection methods: Mean Max Loss s and Max Loss. Paper [10] encourages the selection of samples that are not currently part of the current training set based on three strategies: Euclidean distances and Parzen window in the spatial domain and idea of spatial entropy. The paper [11] enhances the efficiency of CNN algorithm on image classification by PCAPool which is a Principal Component Analysis (PCA)-based pooling approach. The sample matrix is formed by arranging all row vectors in the matrix by sliding the pooling window. The PCA algorithm then extracts all eigenvectors from the sample matrix to create the eigenvector matrix, which right multiplies the sample matrix to produce the principal component matrix. Paper [12] proposed a modified Generative Adversarial Networks. The generative adversarial network is based on the standard generative adversarial network. By building numerous generation models and incorporating a cooperation mechanism, the generation models can learn from one another and develop together in the training process, improving the model's ability to fit real data and improving classification quality. The paper [13] proposed PCANet is a basic deep learning foundation for image classification that uses PCA instead of stochastic gradient descent to learn the filter banks in each layer. With only a few parameters and no backpropagation process, it performs well for image classification applications. The paper [14] focuses on loss function design so it proposes a model to increase the probability difference between the correct and incorrect classes, competing ratio loss calculates the posterior probability ratio between the correct and competing incorrect classes. The paper [15] highlights the need of mathematical proofs to create image categorization efficacy metrics and test their utility with real-world and hypothetical cases.

3 Problem Statement

Image classification on cifar10 dataset is usually performed using dense neural networks or convolutional neural networks which provides a good accuracy rate. Whereas in the proposed project image classification is done using support vector machine which generally is a normal binary classification problem. Image classification using dense neural networks is a reliable approach than using support vector machines, as image classification using support vector machine is little problematic approach as it is slower than CNN and DNN, plus the input data should be in two dimensions which increases the number of rows and columns in case of image classification problems. Support vector machines can yield a good accuracy score if some compression methods are applied for example using Principal Component Analysis.

4 Methodology

The project involves use of three machine learning methods: Convolutional neural networks (CNN), dense neural networks (DNN) and support vector machines (SVM).

A. Working With Dense Neural Networks

Deep neural networks have a hierarchical network of neurons that resembles the human brain. Based on the information received, the neurons send the signal to other neurons. The output will be passed if the value is larger than threshold value, otherwise it will be ignored. The data is transferred to the input layer, which produces output for the next layer, and so on, until it reaches the output layer, which makes a probability-based prediction of yes or no. A layer is made up of numerous neurons, each with its own function termed the Activation Function. They serve as a conduit for signal transmission to the next linked neuron. The input influences the output of the following neuron, while the last output layer is influenced by the weight. The weights are assigned at random at first, but when the network is trained iteratively, the weights are tuned to ensure that the network generates accurate predictions (Fig. 1).

B. Working With Convolutional Neural Networks

To detect objects, identify faces, and so on, convolutional neural networks use picture categorization. It consists of neurons, weights and biases. Neurons receives inputs and it calculates the weighted total, which is further passed to activation function.

The CNN network, involves many layers like the input layer then the convolutional layer which is the basic fundamental unit, it handles computational hard work in the network. Filters are tiny blocks that handles entire data/images they are used as sliding windows over the data. The next is the activation layer, this

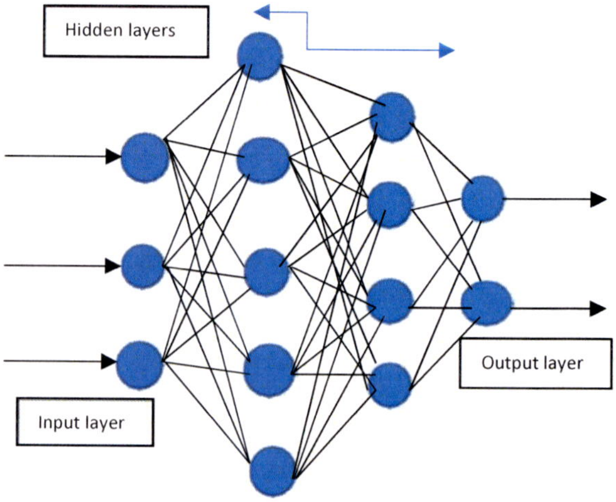

Fig. 1 Working of dense neural network

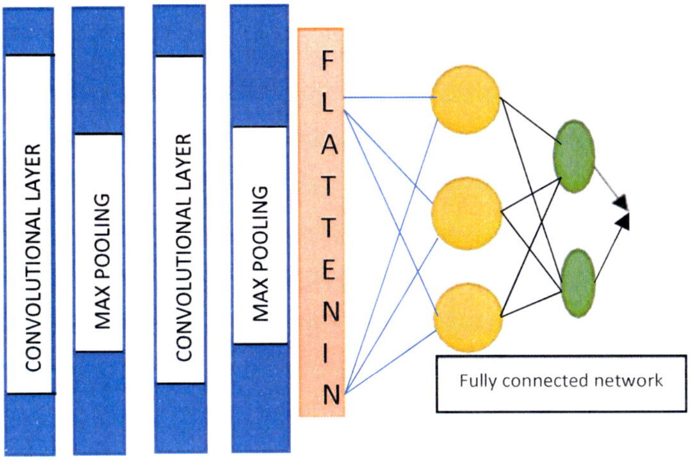

Fig. 2 Working of convolutional neural network

uses an activation function for instance rectified linear unit (RELU) to increase the non-linearity of the network. Then is the pooling layer which enables feature sampling. Pooling is performed to each layer in a 3-D volume.

Then last is the fully connected network, which includes the last step called flattening. The whole feature map is transformed into differentiable columns which are further passed to the network for calculations (Fig. 2).

C. Working With Support Vector Machine

Support vector machine (SVM) is a linear model used for regression and classification issues. It tackles linear as well as nonlinear problems, so that's why it is useful for various applications. Basic concept of SVM is that it divides the dataset into classes by incorporating a hyperplane. The SVM technique finds the data point from each class which are closest from the line. Then distance between lines and support vectors is calculated using these points, which are known as support vectors. The distance computed is called margin. The ideal hyperplane is the one for which the margin is largest (Fig. 3).

5 Results

The dataset contains 60,000 32 $\times$ 32 color photos divided into ten classes, each with 6000 pictures. The complete set of photos is separated into 2 sets: training data of 50,000 images and testing data of 10,000 images. Training model for CNN or DNN uses 10 epochs each on 50,000 images, whereas for SVM only 17% of the data of used for training that is 3000 images and for testing 1000 images were used. After training graphs were plotted between accuracy vs epochs and loss vs

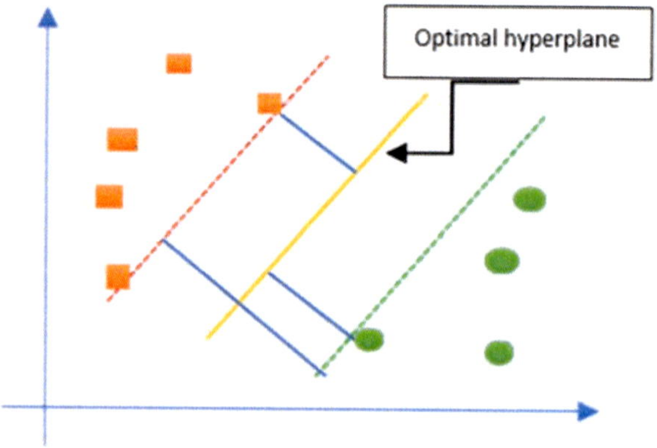

Fig. 3 Working of support vector machines

epochs for DNN and CNN both, whereas for SVM three different kernels were tested upon.

A. **Accuracy by using Dense Neural Network**

Above figures namely, Figs. 4 and 5 represent the accuracy and loss in the dataset, respectively, when dense neural network is applied on the training set containing 50,000 images. The model achieved 50% accuracy in 10 epochs and the loss also reduced as the number of epochs were increased.

B. **Accuracy by using Convolutional Neural Networks**

The two graphs shown in Figs. 6 and 7 represent accuracy and loss in the dataset when convolutional neural network is applied on training set containing

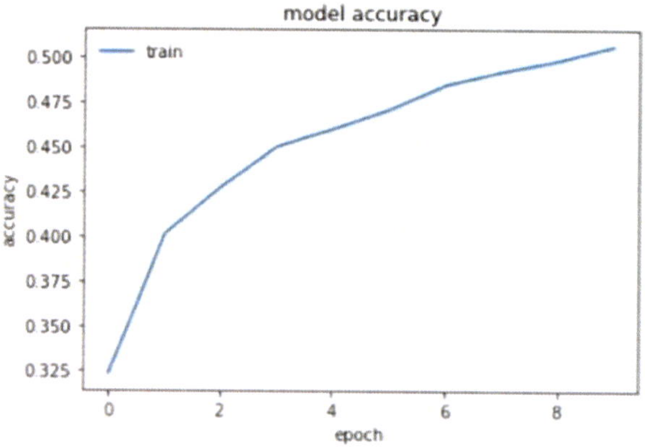

Fig. 4 Accuracy versus epoch on training data

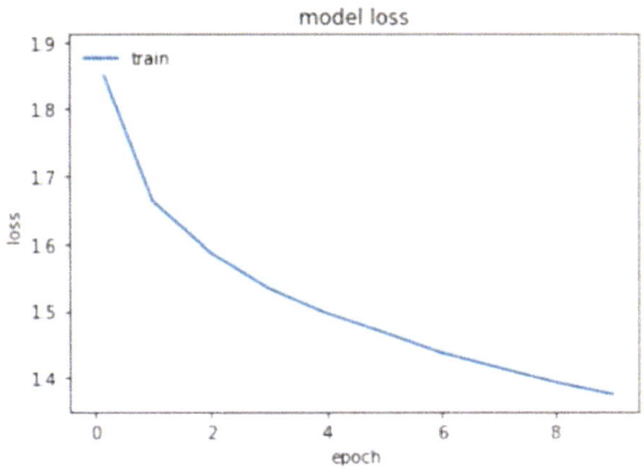

Fig. 5 Loss versus epoch on training data

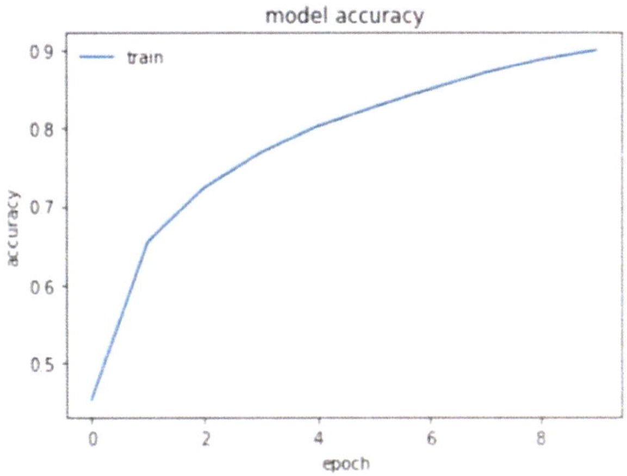

Fig. 6 Accuracy versus epoch on training data

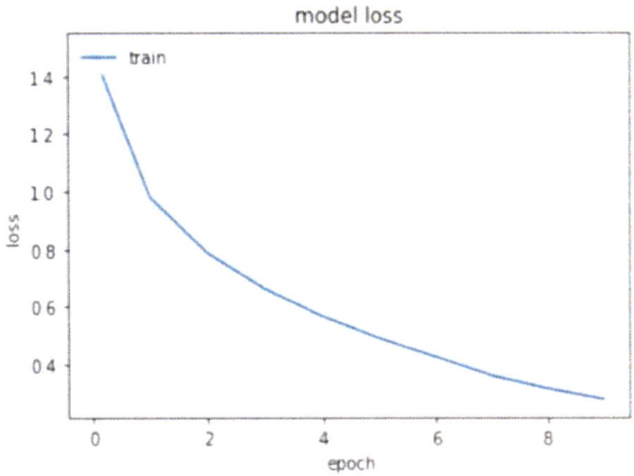

Fig. 7 Loss versus epoch on training data

50,000 images. The training model almost reached 90% accuracy with 0.4 loss by performing 10 epochs on the model.

C. **Accuracy by using different SVM Kernel**

Figure 8 is the plot between accuracy and values of C for training and testing data using SVM with linear kernel. The red line show cases the training data, whereas the orange line show cases testing data. The accuracy on training data is 100% and on testing data it is 27.9% as inferred from the graph.

Figure 9 is the plot between accuracy and values of C for training and testing data using SVM with polynomial kernel. The red line show cases training data, whereas the orange line show cases the testing data. The accuracy on training data is 80% and on testing data it is 26% as inferred from the graph.

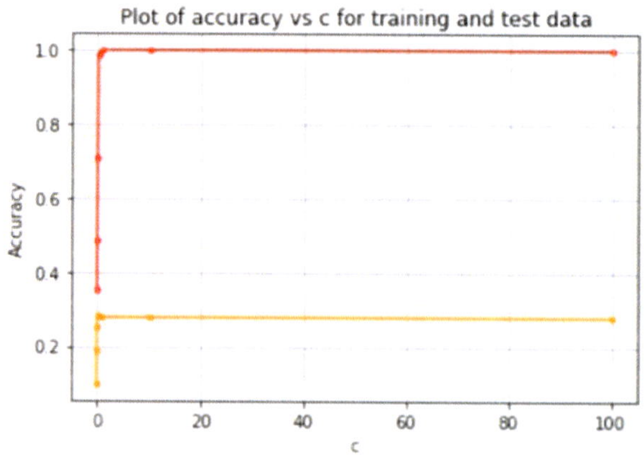

Fig. 8 Using linear kernel

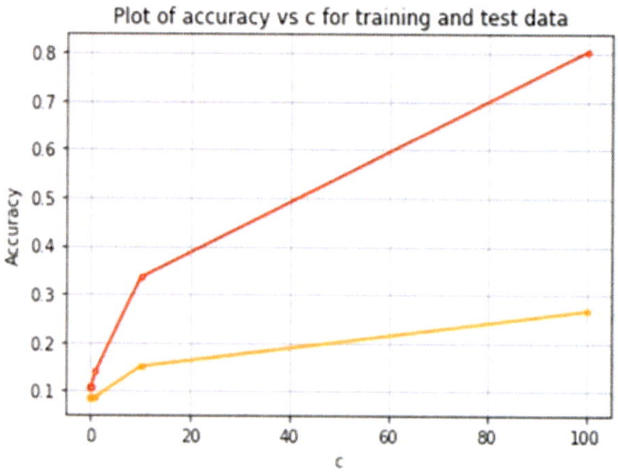

Fig. 9 Using polynomial kernel

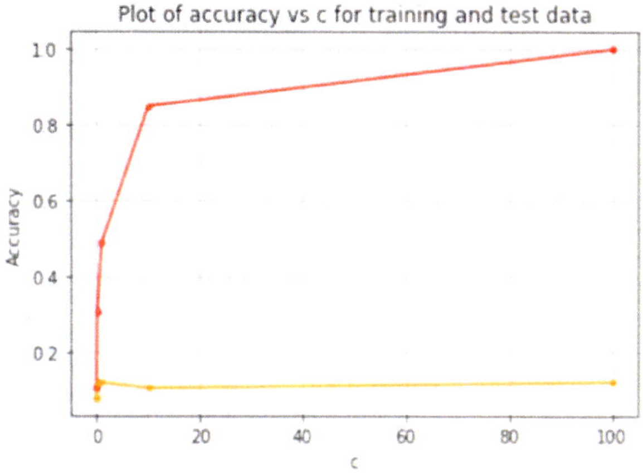

Fig. 10 Using RBF kernel

Lastly Fig. 10 is the plot between accuracy and values of C for training and testing data using SVM with radial basis function (RBF) kernel. The red line show cases the training data whereas, the orange line show cases testing data. The accuracy on training data is 100% and on testing data it is 11% as inferred from the graph.

6 Discussion and Conclusion

6.1 Challenges Faced

The major challenge countered during the implementation of various algorithms on CIFAR10 dataset was related to the time taken by the algorithms for training as the size of the dataset is huge. As we know that to enhance the accuracy of a model while using CNN the number of epochs should be manipulated, but this is a challenging task as applying 100 or 1000 epochs on a dataset containing 60,000 images as it would takes hours. On the other hand while using SVM on larger datasets it takes days for training the dataset,even though SVM is a good algorithm for classification but for image classification for 10 classes and with 60,000 images it is a tedious task, also coding time for image processing and classification is more than that with CNN or DNN.

6.2 Future Scope

By 2023, the Data Classification Market is expected to increase by INR 117,203.63245 million. In future machine learning will flourish in many fields such as medical and finance. Obtaining and arranging massive amounts of data for the computer to learn from. Extending the model's capabilities by deploying it to make analytical predictions or feeding it with new types of data. Therefore the present algorithms should be manipulated and enhanced in such way that it takes less training time and produces a good accuracy.

6.3 Conclusion

The dataset was trained by three machine learning methods dense neural networks (DNN), convolutional neural networks (CNN) and support vector machines (SVM). The results state that convolution neural networks is the best approach as it gives almost 90% accuracy, whereas dense neural networks gives 50% accuracy and support vector machines performs pretty good on training data when linear kernel was used but poorly on testing data. The accuracy via support vector machines can be further increased using Principal Component Analysis (PCA).

References

1. Yim J, Ju J, Jung H, Kim J (2017) Image classification using convolutional networks with multi-stage features. In: Robot intelligence technology and application, Department Electrical Engineering, KAIST, Korea, 2017
2. Williams T, Li R (2016) Advanced image classification using wavelets and convolutional neural networks. In: Proceedings of the 15th IEEE international conference on machine learning and applications
3. Ren X, Guo H, Li S, Wang S, Li J (2017) Novel image classification method with CNN-XGBoost model. In: International workshop on digital watermarking. Department Cyber Space. Security, University of Shanghai Jiao Tong
4. Graham B (2014) Spatially-sparse convolutional neural network. Department of Statistics, University of Warwick, United Kingdom
5. Zhou L, Li Q, Huo G, Zhou Y (2017) Image classification using biometric pattern recognition with convolutional neural networks features. University of Hohai, Hohai
6. Parra CDL, Guntoro A, Kumar A (2020) ProxSim: GPU-based simulation framework for cross-layer approximate DNN optimization. Design, automation and test in Europe conference and exhibition
7. Zhou Y, Song S, Cheung N (2017) On classification of distorted images with deep convolutional neural networks. In: IEEE international conference on acoustics, speech and signal processing (ICASSP)
8. Arya MA, Bedi SS (2018) Survey on SVM and their application in image classification. Department of CSIT University. MJP Rohilkhand
9. Li X, Wang L, Sung E (2004) Multilabel SVM active learning for image classification. In: International conference on image processing
10. Pasolli E, Melgani F, Tuia D, Pacifici F, Emery WJ (2014) SVM active learning approach for image classification using spatial information. In: Conference on IEEE transactions on geoscience and remote sensing
11. Zhao B, Dong X, Guo Y, Jia X, Huang Y (2021) PCA dimensionality reduction method for image classification. Neural Process Lett

12. Zhao Z, Li R (2021) Modified generative adversarial networks for image classification. Evolution Intell
13. Alotaibi, Mubarakah & Wilson, Richard. (2021). Multi-layer PCA Network for Image Classification.
14. Zhang K, Guo Y, Wang X, Chang D, Zhao Z, Ma Z, Han T (2021) Competing ratio loss for discriminative multi-class image classification. Neurocomput
15. Guofan S, Lina T, Hao Z (2021) Introducing image classification efficacies. IEEE Access
16. Jafar A, et al (2021) AI and machine learning paradigms for health monitoring system: intelligent data analytics, Vol 86. Springer Nature, Berlin, SBD, p 513. doi:https://doi.org/10.1007/978-981-33-4412-9. ISBN 978-981-33-4412-9
17. Tomar A, et al (2020) Machine learning, advances in computing, renewable energy and communication, Vol 768. Springer Nature, Berlin, LNEE, 2020, p 659. doi: https://doi.org/10.1007/978-981-16-2354-7. ISBN 978-981-16-2354-7

A Generic Review on Anomaly Detection

Nishu Bansal and Swimpy Pahuja

Abstract Detection of anomalies in the data has now emerged as a hot topic among diverse areas for researchers, especially in the cyber security field. Also, it is one of the important aspects of risk management strategies. A broad range of strategies for protecting against these anomalies have been developed in the literature, however certain gaps remain, necessitating the creation of robust detection systems. This article provides a compendium of some anomaly detection techniques explored by the researchers.

Keywords Anomaly detection · Clustering · Anomaly · Intrusion detection

1 Introduction

The problem of finding diversions from the majority of data present in the dataset from its expected behavior is termed Anomaly Detection. In other words, it is the unusual behavior detection of any application or system in consideration. In many application areas such as scam detection, impingement detection, healthcare, and others, this divergence from a typical code of behavior is termed as oddity, anomalies, outliers, delusion, strident values, irregularities, or defilement. The abnormalities caused due to these anomalies in the data may result in serious consequences which makes anomaly detection very important and crucial factors to be taken into consideration. For example—A CT image that is aberrant may suggest the existence of cancerous tumors [1]. Inaccuracies in credit card payment details, for example, might result in credit card or information theft, resulting in financial losses, or anomalies in spacecraft sensor data might indicate a problem with a spacecraft hardware, which, if not detected and resolved in a timely manner, could lead to mass destruction. These

N. Bansal (✉)
Ajay Kumar Garg Engineering College, Ghaziabad, India
e-mail: bansalnishu@akgec.ac.in

S. Pahuja
SOET, CMR University, Bangalore, India
e-mail: swimpy.p@cmr.edu.in

A. Tomar et al. (eds.), *Proceedings of 3rd International Conference on Machine Learning, Advances in Computing, Renewable Energy and Communication*, Lecture Notes in Electrical Engineering 915, https://doi.org/10.1007/978-981-19-2828-4_45

few examples make clear that the detection and timely resolution of anomalies is very much important to avoid serious consequences.

The irregular patterns or trends in data that do not adhere to a standard defined term are known as anomalies. The anomalies in Fig. 1. are depicted in a two-dimensional space in which the bulk of the data is contained in two areas, N1 and N2, which are referred to as normal regions. Data points that are significantly away from these zones are referred to be anomalies, such as points A1 and points in region A2. Anomalies can be classified as Contextual Anomalies, Point Anomalies, or Collective Anomalies. In the detection industry, there are several open-source point anomaly detection approaches accessible, including [2]. A data snippet abnormal in one scenario but not in another is termed as Contextual Anomaly. A collective anomaly arises when a group of linked information instances deviates from the overall information set. Individual data occurrences in a collective anomaly are not unusual in and of itself, but their occurrence as a group is extraordinary. In medical terms, anomalies are classified into Major anomalies as well as Minor anomalies. Major anomalies harm the health of the patient while the latter has no serious consequences on the health of the patient but minute consequences on the appearance of the patient [3]. Csabay et al., for example, addressed one of the primary abnormalities, genetically inherited deformities of the central nervous system (CNS), and the researchers also revealed several diagnostic tools or methodologies used to detect such defects in [4].

In the cyber security arena, for example, false alert rates, diverse kinds of data, and intractable integration of physical and cyber dynamics are key problems for anomaly detection. To address these challenges, Ferragut et. al. proposed a probabilistic and unsupervised approach based on a reduced-order model for automated anomaly detection for energy grid systems [5].

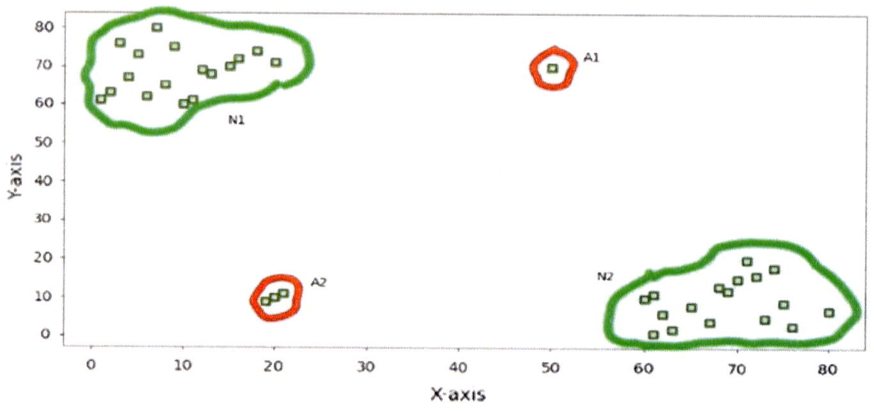

Fig. 1 Example of anomalies in the 2-D dataset

2 A Broad View of Anomaly Detection

This portion throws light on the domain's richness and emphasizes the necessity for a diverse set of anomaly detection methods. Anomaly detection involves various components which are summarized below:

2.1 *Input Data*

Any method for detecting anomalies is dependent on the normality of the model. An event, pattern, object, case, sample, entity, or observation is a collection of data elements known as input. An attributes set, also known as features, variables, and dimensions, can be used to represent each data point.

2.2 *Data Labeling*

A data instance's labels specify whether the data item is legitimate or problematic. It is worth noting that acquiring reliable and representative tagged data for a variety of operations might be fairly costly. Because human expert labeling is common, obtaining the labeled training data set requires a significant amount of effort. Anomaly detection is separated into three types depending upon the data label's availability.

2.2.1 Supervised Anomaly Detection

It is a technique in which training is accomplished in supervised mode with the use of a coaching information set including tagged examples for both traditional and anomalous categories for model creation. Any previously unknown data object is contrasted to the model to identify the group to which it corresponds. To discover irregularities in video objects, Georgescu [6] used supervised detection approach.

2.2.2 Semi-Supervised Anomaly Detection

When using this technique, it is assumed that the training set for the standard data would consist entirely of tagged examples. They are more usable than guided procedures since they no longer require labels for the unusual category. The authors of [7] suggested an approach for large dimension datasets based on a deep auto encoder and nearest neighbor graphs, which resulted in decreased computing cost and increased detection accuracy. PLELog, a new practical log-based technique proposed by Yang et al. [8], outperforms current log-based anomaly detection techniques.

2.2.3 Unsupervised Anomaly Detection

The strategies that function unattended without the need of training data are most common now-a-days. This set of strategies assumes that common occurrences outnumber anomalies in testing set by a factor of ten. If this presumption is incorrect, then techniques in this category have a large rate of false alarms and low detection rates. Authors in [9] used an unsupervised clustering technique to identify network anomalies using four classifiers: naive Bayes, rule induction, decision tree, and closest neighbor, and discovered that distance-based outlier detection worked better. To find unsupervised anomalies in Big Data, Maleki et al. [10] proposed utilizing Enhanced LSTM Auto Encoders (ELSTMAE). Barua et al. [11] presented a real-time system based on Hierarchical Temporal Memory (HTM).

2.3 Outcome of Anomaly Detection

A major facet of any detection algorithm is the way irregularities are reported. Anomaly detection system's results often fall into one of two groups namely, scores and labels. Anomaly scoring algorithms provide a ranked list, allowing the analyst to pick either the leading handful anomalies or anomalies based on some cutoff criteria. For example, the researchers in [12] compute the Loaded Anomaly Score using a distance-based anomaly score metric. The analyst uses anomaly detection algorithms based on scoring to choose the most significant anomalies based on a domain-specific threshold.

3 Anomaly Detection Techniques

Researchers have used various techniques for anomaly detection as depicted Fig. 2 below.

3.1 Classification-Based Anomaly Detection Techniques

There are two steps to the classification-based technical testing and training procedure. The classifier is taught using accessible labeled entities in the training module. Because a training set of correctly recognized observations is provided, it is also known as supervised learning. For accessible data, an algorithm is utilized to create a classifier which distinguishes between anomalous and non-anomaly data. Various classification techniques used for Anomaly Detection:

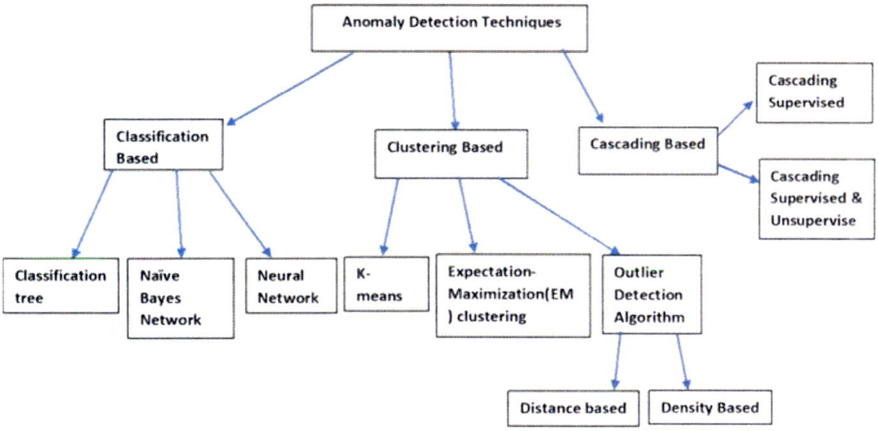

Fig. 2 Nomenclature of anomaly detection techniques

3.1.1 Classification Tree

It is also known as a call tree or a prediction model. The interior nodes represent a check property, each branch represents the analysis results, and the end terminals or leaflets provide the class label category in a tree-style graph with a flow graph structure. ID3, as well as the statistical classifier C4.5, are the most fundamental and extensively used algorithmic rules for the classification tree. Top-down and bottom-up pruning strategies are used to create these trees. Top-down tree architectures include ID3 and C4.5 [13].

3.1.2 Naïve Bayes Network

There are various examples of applied mathematics correlations or causative links between system variables. It may be hard to clearly describe the stochastic linkages between these variables. In other words, the system's historical data is restricted to the fact that certain variables may be modified by others. To take use of this intrinsic link between the explanatory variables of a drag, a stochastic network model known as Naive Bayesian Networks (NB) is commonly utilized [14]. The decision tree has a higher accuracy than the Bayesian network, but it takes less time to compute. As a nutshell, utilizing NB models will be cost effective whenever the sample sizes is really large. For anomaly detection, Manimurugan [15] employed naive Bayes and PCA.

3.1.3 Neural Networks

It is a network of connected units mimicking the human thought processes. Each node is connected to a large number of alternative nodes in nearby layers through a weighted connection. Both supervised and unsupervised learning may be accomplished using neural networks. Furthermore, the user chooses a set of hidden units based on the amount of nodes inside each hidden unit. In a variety of applications, MLP neural networks have proven to be more productive than other computational learning approaches currently available, generating more accurate results. Murugesan et al. [16] reported an anomaly detection approach for video surveillance that obtained an accuracy of 98.56%. Similarly, Yuan-Cheng Lai [17] deployed MLP to produce parameters for anomaly detection automatically.

3.2 Cluster-Based Anomaly Detection Technique

Clustering is a technique for organizing related data elements into bunches. Although clustering is typically an unsupervised method, semi-supervised clustering has recently gained popularity. For Anomaly Detection, the following clustering approaches are used:

3.2.1 K-Means

It is a form of unsupervised learning that is utilized when the data is unlabeled. This technique's purpose is to find a pattern in the data. To differentiate between anomalous and normal clusters, the centroid feature weights are utilized. Researchers in [18] employed the K-means clustering approach to discover black holes and misdirection nodes in the network.

3.2.2 Expectation–Maximization (EM) Clustering

It is an iterative approach for allocating a data point to a group based on the cluster mean. It employs the expectation-maximization method. The data point is not assigned to a specific group; rather, it is assigned based on a weight reflecting the likelihood of participation. This gives high precision results; however, the approach suffers from a sluggish convergence difficulty in some instances.

3.2.3 Outlier Detection Algorithms

It is a method for finding data anomalies which do not follow the overall trend. An outlier is a specified item that is substantially distinct from other objects with specific attribute values. Based on the efficiency and accuracy required, we may choose any of the outlier identification techniques to deal with the anomaly detection problem. There are two approaches available in outlier detection:

Distance-Based Approach

It is a non-parametric method that uses the concept of nearest neighbor algorithm which states that the greater the value of a node from its neighboring data node, the more likely node will act as an outlier. Authors in [9] used this detection method for detecting unknown network intrusions. Sugiyama et al. also used this approach for outlier detection in the data mining field [19].

Density-Based Approach

It is a parametric method based on the worldwide distribution of a given collection of data points that is mostly utilized for detecting collective anomalies. The denser section of the hyperplane is regarded as normal, whereas the thin part is seen as abnormal. The authors of [20] suggested a new robust metric called commute distance that combines distance-based and density-based methods to detect abnormalities in network traffic. However, both of these techniques proved to be adequate for detecting contextual anomalies.

3.3 Cascading-Based Anomaly Detection Techniques

To obtain a high detection rate, different advantages of varying detection mechanisms can be combined in a way like:

3.3.1 Cascading Supervised Techniques

Several categorization methods are combined to improve accuracy. A decision tree and naive Bayes algorithm combination was presented. This hybrid technique obtained 99 percent accuracy upon that KDD cup dataset. It concentrated on the ID3 method and the Naive Bayes theorem (NB) classifier's performance [31].

3.3.2 Combining Supervised and Unsupervised Techniques

The employment of unsupervised methodologies greatly increased the efficiency of the abnormality detection rate, hence enhancing the supervised algorithm's effectiveness [32–34]. The authors of [32] achieved 98 percent accuracy by combining k-means and ID3 techniques for identifying abnormal and ordinary behaviors in ARP data. Its goal is to see how effective strategy-assisted machine learning, is at detecting intrusions. As a result, hybrid approaches outperform traditional approaches by combining numerous strategies and overcoming their flaws, resulting in higher anomaly detection accuracy. Table 1 lists some of the strategies proposed in the literature, as well as their benefits and drawbacks.

4 Conclusion

In today's fast-paced environment, a novel problem or assault emerges every day, necessitating the development of various methods for detecting aberrant behavior. This paper gives an overview of different anomaly detection strategies that have been presented in recent years. This overview assists the reader in gaining a fundamental understanding of anomaly detection and the strategies available to identify it. It has been discovered that the fusion or combining of pre-existing algorithms enables better and more accurate anomaly detection, and this has proven to be a popular study area in recent years. For example, there are numerous new techniques within decision tree modification such as Genetic Algorithm, and SVM that may be modified or coupled to give new pathways toward better detection in various application areas.

Table 1 Summary of some anomaly detection techniques

Reference	Technique used	Advantages	Disadvantages	Dataset used	Accuracy
Gunawansyah et. al. [21]	Genetic algorithm, ANN	Good accuracy with different datasets	Increased computation time Overfitting problem	Local dataset	84.6%
Min-Ki Lee et. al. [22]	SV regression	Reduced RMSE by an average of 45.44% Good computation time (root mean square error)	It does not take into consideration meteorological elements	Dataset of Korean meteorological department, 2014	89.2% (for rainfall)
Soumya et. al. [23]	Combination of both supervised and unsupervised techniques	Outstanding results in the situation of anomalous behavior detection in sensor data	Might not have been applicable in a variety of situations	sensory data used	close to 100%
Grigorev et. al. [24]	Adaptive suppression	Can be applied to any dataset for which similarity metrics exist Good for parallel programming	Missing distance metrics in the dataset Large number of computational resources used	QMUL junction dataset and UCSD Anomaly detection dataset	74.6% on QMUL and 70% on UCSD
Sarita Chaudhary et. al. [25]	Gaussian mixture model (GMM)	Automatic detection without human intervention	Object overlapping not considered	Created own dataset with 45 videos of 3 activities	90%
Bamaqa et. al. [26]	Cortical learning algorithm (CLA) model	It could also identify space-time irregularities in deterministic and chaotic domains, and it can operate with real-time data	A functional framework is required to satisfy the needs of a crowd control application in terms of early anomaly detection	crowd dataset	Coordinate-CLA Model has accuracy of 92.58%

(continued)

Table 1 (continued)

Reference	Technique used	Advantages	Disadvantages	Dataset used	Accuracy
Raihan Ul Islam1 et. al. [27]	Belief rule-based association rule (BRBAR)	It can be used in various application areas like disaster management, surveillance and environmental monitoring	More investigation is required for choosing the benchmarked data	Temperature and rainfall sensor dataset of Chittagong in Bangladesh	99% (for rainfall and temperature data)
Joshi et. al. [28]	Machine learning and deep learning techniques	Best accuracy with naive Bayes approach	Some techniques are not considered	2020 Signal processing cup image dataset	92.079%
Kurniawan et. al. [29]	Weka open tool	Used power consumption, battery temperature and network traffic for anomaly detection in Android	Few parameters are taken into consideration	Android applications (malware samples from android malware genome project)	85.6%
Salman et. al. [30]	Linear regression, random forest algorithm	Enables detection as well as categorization of different attacks Applicable in multi-cloud setups	Inability to categorize some attacks	UNSW dataset	99% (for detection) and 93.6% (for categorization)

References

1. Ripan RC, Sarker IH, Hossain SMM, Anwar MM, Nowrozy R, Hoque MM, Furhad MH (2021) A data-driven heart disease prediction model through K-means clustering-based anomaly detection. SN Comp Sci 2(2):1–12
2. Fisch A, Eckley IA, Fearnhead P (2018) A linear time method for the detection of point and collective anomalies. arXiv preprint arXiv:1806.01947
3. Hennekam RC, Biesecker LG, Allanson JE, Hall JG, Opitz JM, Temple IK (2013) Elements of morphology: general terms for congenital anomalies. Am J Med Genet Part A 161(11):2726–2733
4. Csabay L, Szabo I, Papp C, Tóth-Pál ERNÔ, Papp Z (1998) Central nervous system anomalies. Annals New York Acad Sci 847(1):21–45
5. Ferragut EM, Laska J, Czejdo B, Melin A (2013) Addressing the challenges of anomaly detection for cyber physical energy grid systems. In: Proceedings of the eighth annual cyber security

and information intelligence research workshop, pp 1–4
6. Georgescu MI, Barbalau A, Ionescu RT, Khan FS, Popescu M, Shah M (2021) Anomaly detection in video via self-supervised and multi-task learning. In: Proceedings of the IEEE/CVF conference on computer vision and pattern recognition, pp 12742–12752
7. Song H, Jiang Z, Men A, Yang B (2017) A hybrid semi-supervised anomaly detection model for high-dimensional data. Comput Intell Neurosci
8. Yang L, Chen J, Wang Z, Wang W, Jiang J, Dong X, Zhang W (2021) Semi-supervised log-based anomaly detection via probabilistic label estimation. In: 2021 IEEE/ACM 43rd international conference on software engineering (ICSE). IEEE, pp 1448–1460
9. Syarif I, Prugel-Bennett A, Wills G (2012) Unsupervised clustering approach for network anomaly detection. In: International conference on networked digital technologies. Springer, pp 135–145
10. Maleki S, Maleki S, Jennings NR (2021) Unsupervised anomaly detection with LSTM autoencoders using statistical data-filtering. Appl Soft Comput 108:107443
11. Barua A, Muthirayan D, Khargonekar PP, Al Faruque MA (2020) Hierarchical temporal memory based machine learning for real-time, unsupervised anomaly detection in smart grid: WiP abstract. In: 2020 ACM/IEEE 11th international conference on cyber-physical systems (ICCPS). IEEE, pp 188–189
12. Ghoting A, Otey ME, Parthasarathy S (2004) Loaded: link-based outlier and anomaly detection in evolving data sets. In: Fourth IEEE international conference on data mining (ICDM'04). IEEE, pp 387–390
13. Wu SY, Yen E (2009) Data mining-based intrusion detectors. Expert Syst Appl 36(3):5605–5612
14. Tsai CF, Hsu YF, Lin CY, Lin WY (2009) Intrusion detection by machine learning: a review. Expert Syst Appl 36(10):11994–12000
15. Manimurugan S (2021) IoT-fog-cloud model for anomaly detection using improved Naïve Bayes and principal component analysis. J Amb Intell Human Comp 1–10
16. Murugesan M, Thilagamani S (2020) Efficient anomaly detection in surveillance videos based on multi layer perception recurrent neural network. Microprocess Microsyst 79:103303
17. Lai YC, Zhou KZ, Lin SR, Lo NW (2019) Flow-based anomaly detection using multilayer perceptron in software defined networks. In: 2019 42nd international convention on information and communication technology, electronics and microelectronics (MIPRO). IEEE, pp 1154–1158
18. Wazid M, Das AK (2016) An efficient hybrid anomaly detection scheme using K-means clustering for wireless sensor networks. Wireless Pers Commun 90(4):1971–2000
19. Sugiyama M, Borgwardt K (2013) Rapid distance-based outlier detection via sampling. Adv Neural Inf Process Syst 26:467–475
20. Khoa NLD, Babaie T, Chawla S, Zaidi Z (2010) Network anomaly detection using a commute distance based approach. In: 2010 IEEE international conference on data mining workshops, IEEE, pp 943–950
21. Liong, T. H. (2017, May). Prediction and anomaly detection of rainfall using evolving neural network to support planting calender in soreang (Bandung). In *2017 5th International Conference on Information and Communication Technology (ICoIC7)* (pp. 1–6). IEEE.
22. Lee MK, Moon SH, Yoon Y, Kim YH, Moon BR (2018) Detecting anomalies in meteorological data using support vector regression. Adv Meteorol
23. Shaw S, Joshi K, Pathak A, Thyagarajan AK, Vidya G, Hemal Shah R, Alex JSR (2022) Anomaly detection in drones with machine learning algorithms. In: Futuristic communication and network technologies: select proceedings of VICFCNT 2020. Springer, pp 433–441
24. Grigorev A, Severiukhina O, Derevitskii I (2019) Anomaly detection using adaptive suppression. Proc Comp Sci 156:274–282
25. Chaudhary S, Khan MA, Bhatnagar C (2018) Multiple anomalous activity detection in videos. Proc Comp Sci 125:336–345
26. Bamaqa A, Sedky M, Bosakowski T, Bastaki BB (2020) Anomaly detection using hierarchical temporal memory (HTM) in crowd management. In: Proceedings of the 2020 4th international conference on cloud and big data computing, pp 37–42

27. Islam RU, Hossain MS, Andersson K (2018) A novel anomaly detection algorithm for sensor data under uncertainty. Soft Comput 22(5):1623–1639
28. Joshi K, Vidya G, Shaw S, Thyagarajan AK, Pathak A, Shah RH, Alex JSR (2022) Anomaly detection in drone-captured images using machine learning techniques and deep learning architectures. In: Futuristic communication and network technologies. Springer, pp 783–791
29. Kurniawan H, Rosmansyah Y, Dabarsyah B (2015) Android anomaly detection system using machine learning classification. In: 2015 international conference on electrical engineering and informatics (ICEEI). IEEE, pp 288–293
30. Salman T, Bhamare D, Erbad A, Jain R, Samaka M (2017) Machine learning for anomaly detection and categorization in multi-cloud environments. In: 2017 IEEE 4th international conference on cyber security and cloud computing (CS cloud). IEEE, pp 97–103
31. Farid DM, Harbi N, Rahman MZ (2010) Combining naive bayes and decision tree for adaptive intrusion detection. arXiv preprint arXiv:1005.4496
32. Yasami Y, Mozaffari SP (2010) A novel unsupervised classification approach for network anomaly detection by k-means clustering and ID3 decision tree learning methods. J Supercomput 53(1):231–245
33. Iqbal A, et al (2021) Intelligent data-analytics for condition monitoring: smart grid applications. Elsevier, p 268. https://www.sciencedirect.com/book/9780323855105/intelligent-data-analytics-for-condition-monitoring. ISBN: 978-0-323-85511-2
34. Tomar A, et al (2020) Machine learning, advances in computing, renewable energy and communication, Vol 768. Springer, LNEE, p 659. doi:https://doi.org/10.1007/978-981-16-2354-7. ISBN 978-981-16-2354-7

Optimal Sizing of a Grid-Connected Biomass/Biogas/PV System for Rural Electrification

Bandana Sharma, M. Rizwan, and P. Anand

Abstract Electrical energy demand is skyrocketing as a result of fast industrialization, new technology inventions, rising household consumption, and globalization. The energy generated from traditional resources is growing more expensive due to rising fossil fuel prices and depletion, and it also has a negative impact on the environment. According to this viewpoint, countries must place as much emphasis as possible on renewable energy-based power generation in order to gain access to affordable energy and achieve their sustainable development goals while also combating global warming. However, renewable energy sources provide fluctuating power, which is not ideal for the utility grid. Thus, to alleviate the stated problem, adopting hybrid renewable energy systems is a viable choice to provide access to clean, reliable, and inexpensive energy. Therefore, the present work conducts a feasibility assessment and size optimization of a grid-connected solar photovoltaic/biomass/biogas/battery-based hybrid system for a village in India for supplying continuous energy at minimal cost. The particle swarm optimization technique has been applied for reducing the cost of the developed hybrid system.

Keywords Photovoltaic (PV) · Renewable energy system (RES) · Cost of energy (COE) · Particle swarm optimization (PSO) · Greenhouse gas (GHG) emissions · Hybrid renewable energy sources

1 Introduction

In the recent years, due to the depletion of conventional sources, rising fuel prices, and increasing ecological concerns, the use of RES for the production of energy has grown significantly. As a result, the utilization of RES is becoming progressively more

B. Sharma (✉) · M. Rizwan
Department of Electrical Engineering, Delhi Technological University, Delhi, India
e-mail: bandana.elec@gmail.com

P. Anand
Department of Electronics and Communication Engineering, Bhagat Phool Singh Mahila Vishwavidyalaya, Khanpur Kalan, Sonipat, India

A. Tomar et al. (eds.), *Proceedings of 3rd International Conference on Machine Learning, Advances in Computing, Renewable Energy and Communication*, Lecture Notes in Electrical Engineering 915, https://doi.org/10.1007/978-981-19-2828-4_46

popular, especially in rural or isolated regions. A hybrid renewable energy system (HRES) that includes several RES with storage can offer cost-effective and steady power [1, 2]. An HRES, however, is used to accommodate the nonlinear response and unpredictable nature of RES and demand profiles, allowing for efficient and affordable power across the system. To get the best performance from the HRES, each component must be optimally sized for maximum efficiency at the lowest possible cost [2–6]. Several studies have looked into the feasibility of deploying HRESs during the last few years. Bahramara et al. carried out the research for both COE and GHG emissions [7]. Erdinc et al. used genetic algorithm (GA) to handle a variety of complex techno-economic challenges for HRES [8]. Using the multi-objective PSO, Borhanazad et al. conducted research to find an optimal configuration for HRES [9]. Nadjemia et al. developed a new discrete cuckoo search method for the best size of PV/wind/battery for a grid-connected HRES. [10]

Zebarjadi and Askarzadeh used harmony search (HS) optimization to determine the size of a grid-connected photovoltaic (PV) power plant for six households in Iran. The research found that installing PV systems might be cost-effective in future if energy prices rise [11]. To minimize the CoE and maximize dependability, Mohamed et al. presented a PSO method for optimum design of PV/wind HRES with a grid. In terms of cost reductions, a comparison of the optimized HRES to the actual electric grid was conducted and found to be more appropriate [12]. Anand et al. determined the optimal size of a grid-connected HRES for rural regions in Haryana, India using the PSO method. The optimum design for the research was determined to be a grid-connected HRES. The ideal size of the HRES grid was determined using hourly simulations [13]. Anand et al. also find out the optimal size of HRES components using HS. A grid/PV/wind/biomass(BM)/biogas(BG)/battery(G_PV_Biomass_Biogas_Battery) HRES was found to be more optimal than the other configurations [14].

Above all, the goal of this study is to develop and size a G_PV_Biomass_Biogas_Battery HRES for rural locations in Agra, Uttar Pradesh, India using the PSO technique. This hybrid system will ensure the cluster of villages receives uninterrupted power. The obtained results from PSO have been compared with HOMER and found to be the most desirable in terms of least NPC and COE.

2 Description of Proposed Site

The feasibility assessment based on existing RES is the first step in designing and developing a site-specific HRES. Availability of data for the main RES is thus gathered and provided in the successor sections at the appropriate place. The study area includes a cluster of villages located in the District of Agra, Uttar Pradesh, India. The solar energy and wind energy data are obtained for the selected area from NASA meteorological data [15]. From the available resource of wind, it is found that a sufficient amount of wind is not available in this area which can produce energy. Therefore, wind is not taken for modeling HRES. There is a total BM production of

795 tons/year from agricultural leftovers and a daily BG production of 239 m^3/day in this region. On average, the research region receives 5.267kWh/m2/day of solar radiation.

3 Setup for Proposed HRES and Modeling

The grid-connected BM, BG, PV, batteries, and a utility grid are all part of the proposed system and depict in Fig. 1. Accurate source and load modeling are essential for efficient HRES system design. HRES is therefore designed in this study using the usage of stochastic and time-varying probabilistic models. In the forthcoming section, the HRES system is briefly described in terms of its mathematical modeling.

3.1 PV System

The PV power (p_o^{pv}) can be calculated by using Eq. 1.

$$p_o^{pv} = R_{solar} \times \eta^{pv}(t) \times A^{PV} \tag{1}$$

where R_{solar} is solar radiation, A^{PV} represents PV modules area, η^{pv} is PV module efficiency

$$\eta^{pv}(t) = \eta_{ref} \times \eta_{mppt} \times [1 - \beta \times (T_{amb}(t) - T_{ref}) - \beta$$

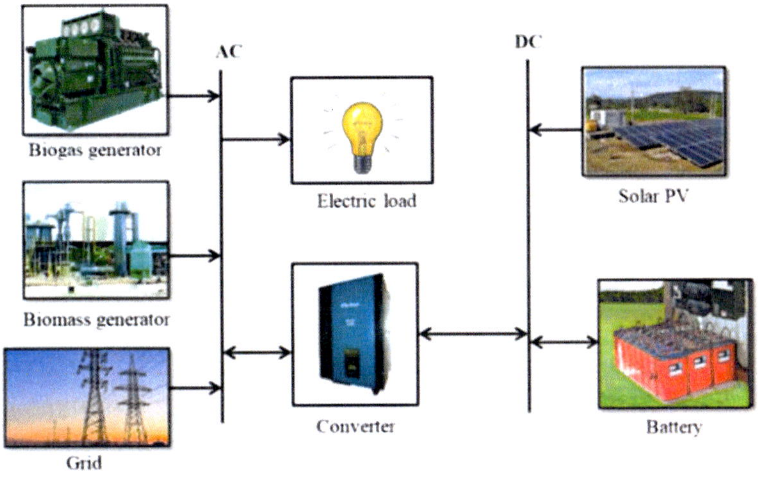

Fig. 1 Setup for proposed HRES

$$\times R_{\text{solar}} \times \left(\frac{T_{n_{\text{cell}}} - 20}{800}\right) \times \left(1 - \eta_{\text{ref}} \times \eta_{\text{mppt}}\right)\Bigg] \tag{2}$$

where $T_{n_{\text{cell}}}$ is cell temperature, η_{ref} is an efficiency, T_{amb} is the ambient temperature, T_{ref} is reference temperature.

3.2 BG System

The power of the BG system $P_{\text{b\_g}}(t)$. can be calculated by using Eq. (3).

$$P_{\text{b\_g}}(t) = \left(q_{\text{b\_g}} \times f_{\text{b\_g}} \times n_{\text{b\_g}}\right)/\left(860 \times h_{\text{b\_g}}\right) \tag{3}$$

where $q_{\text{b\_g}}$ is biogas/day, $f_{\text{b\_g}}$ is calorific value,$n_{\text{b\_g}}$ is conversion efficiency, $h_{\text{b\_g}}$ is of BG operational hours/day. Energy generated (EG) by biogas $e_{\text{g}}(t)$ can be calculated by equation Eq. (4).

$$e_{\text{g}}(t) = P_{\text{b\_g}}(t) \times \Delta t \tag{4}$$

3.3 BM System

The BM output $(p_{\text{m}}(t))$ can be given by Eq. (5) as follows

$$p_{\text{m}}(t) = (q_{\text{am}} \times f_{\text{m}} \times n_{\text{m}} \times 1000)/(365 \times 860 \times h_{\text{m}}) \tag{5}$$

where q_{am} is biomass (tons/year). f_{m} is calorific value, n_{m} is conversion efficiency, h_{m} is BM operational hours /day. EG by BM $e_{\text{m}}(t)$ can be estimated by Eq. (6)

$$e_{\text{m}}(t) = p_{\text{m}}(t) \times \Delta t \tag{6}$$

3.4 Battery System

In the charging stage, capacity of battery may be determine as

$$\begin{aligned} e_{\text{b}}(t) = (1 - \gamma) \times e_{\text{b}}(\tau - 1) + \epsilon_{\text{x\_m}}(\tau) + e_{\text{x\_g}}(\tau) \\ + \epsilon_{\text{x\_pv}}(\tau) \times \eta_{\text{charging}} \end{aligned} \tag{7}$$

where γ—Self-discharging rate of battery,$e_b(\tau)$ is energy stored in battery, $e_{xpv}(\tau)$ and $e_{xm}(\tau)$ $e_{xg}(\tau)$ are extra energy by PV, wind, BM, and BG system and $\eta_{charging}$ is charging efficiency.

In the discharging stage, capacity of battery may be determine as

$$e_b(\tau) = (1 - \gamma) \times e_b(\tau - 1) - \frac{e_{df}(\tau)}{\eta_{inverter} \times \eta_{discharging}} \tag{8}$$

where $\eta_{discharging}$, $\eta_{inverter}$ are battery discharging and inverter efficiency, respectively.

4 PSO Method

As a result of bird congregating behavior, PSO is a stochastic-based optimization approach. An initial population of random solutions, termed particles, is used to identify the best solution by iteratively improving the generation process. In PSO, a particle is represented by a vector with m decision variables. There are b particles with a velocity $u_n(b)$. To achieve a better position $y_n(b+1)$, each particle uses its memorto fly across the search space with an enhanced velocity (b). Information about the particle's past experiences (p_{bn}) and information about the group's past experiences (g_{bn}) comprise the memory information [5, 16]. Particle velocity is changed in each iteration, as follows:

$$u_{n(b+1)} = w \times u_n(b) + k_1 \times s_1(p_{bn}(b) - y_n(b)) + k_2 \times s_2(g_{bn}(b) - y_n(b))$$
$$y_n(b+1) = u_n(b+1) + y(b)$$

w is inertia weight component; k_1 and k_2 are learning coefficients; s_1 and s_2 ranging from 0 to 1.

5 Result and Discussion

This study concentrated on the development of cost-effective G_PV_BM_BG_battery HRES using the PSO approach. The MATLAB environment is used to build the PSO algorithm. Further, a comparison of results obtained from PSO with HOMER is carried out and found the most optimal solution. Table 1 provides technical and pricing information for different components that were used to obtain the optimization findings. In addition, Fig. 2 illustrates the load profile. For one year, hourly simulations for the selected model were run in MATLAB using the PSO approach. The findings obtained after hourly simulation for the chosen model are listed in Table 2.

Table 1 Various costs of different components of HRES [16]

RES	Capacity consider	Capital price ($)	O and M price (%)	Salvage value	Fuel price ($/ton)
PV	235 W	192	2	10	–
BM	1 KW	1033	5	30	15
BG	1 KW	660	5	30	8

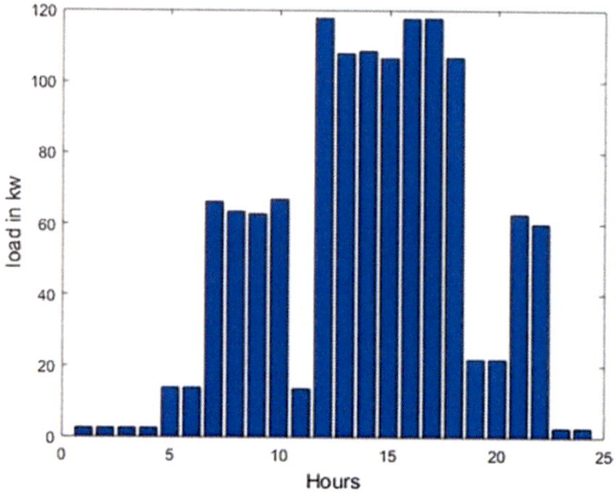

Fig. 2 Load profile of study area

Table 2 The results of PSO and HOMER

Technique	PV (KW)	BM (KW)	BG (KW)	Battery (kWh)	NPC($)	CoE ($/kWh)
PSO	372	60	50	36	$451,050	0.08769
HOMER	280	80	60	48	67,500	0.0997

Further, the results obtained from PSO are compared with HOMER and shown in Table 2 that indicates that the PSO is found to be more accurate and economical. The proposed HRES for selected area has an optimal size of 372 kW PV, 60 kW BM, 50 kW BG, and battery bank storage of 36 kWh. Total NPC and CoE of the proposed HRES are $\$4.5 \times 10^5$ and $0.08769/kWh, respectively. Further, the share of RES in annual electricity generation is presented in Figs. 3 and 4 shows the contribution of different components in NPC of the HRES. Further, the PSO convergence curve for NPC is shown in Fig. 5.

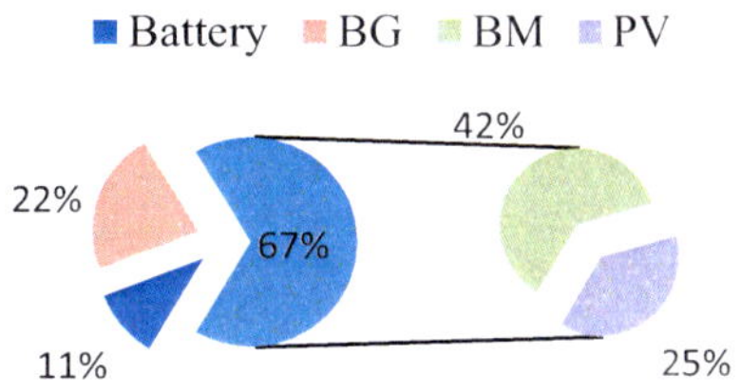

Fig. 3 Share of components of HRES in terms of NPC

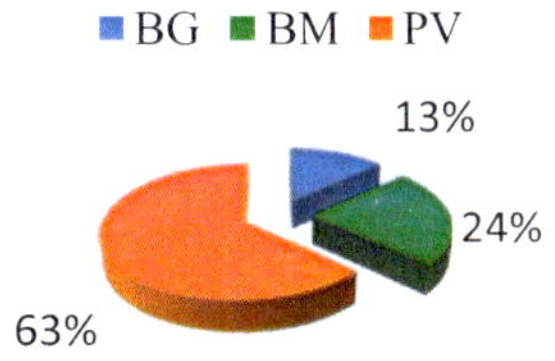

Fig. 4 Share of RES components in annual power generation

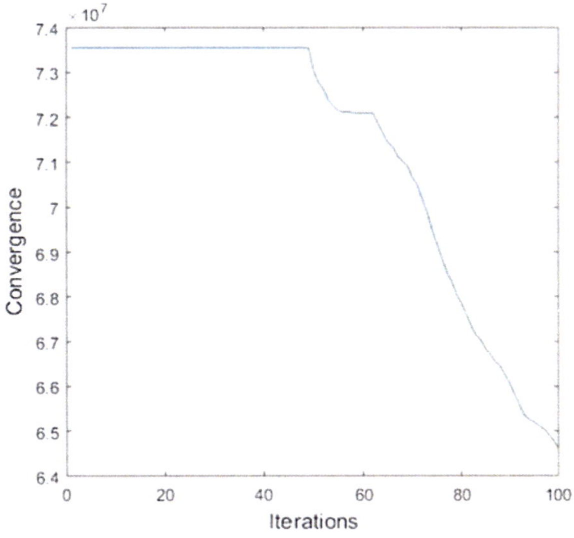

Fig. 5 Convergence curve of PSO

6 Conclusion

The current work conducted a feasibility assessment and size optimization of a grid/PV/BM/BG HRES for supplying continuous energy to a village located in India using the PSO technique. The findings obtained from PSO are also compared to HOMER and are shown to be more accurate and cost-effective. The proposed HRES has a total NPC and CoE of $\$4.5 \times 10^5$ and $0.08769 kWh, respectively. The suggested system would supply the given area with steady, clean, and affordable energy. People in the community will greatly benefitted from this research because they will have access to a continuous and cheap supply that is available 24 × 7. Therefore, these rural areas may get benefit from the successful implementation of this mission in terms of improved lifestyles, improved healthcare facilities, and modern methods of education. Such studies may also be carried out in various distant areas.

References

1. Anand P, Bath SK, Rizwan M (2016) Feasibility analysis of solar-biomass based standalone hybrid system for remote area. Am J Electr Power Energy Syst 5(6):99–108
2. Chauhan A, Khan MT, Srivastava A, Saini RP (2019) Performance optimization of a grid-connected PV/biomass-based hybrid energy system using BBO algorithm. In: Mishra S, Sood Y, Tomar A (eds) Applications of computing, automation and wireless systems in electrical engineering. Lecture Notes in Electrical Engineering, 553. Springer, Singapore
3. Eteiba MB, Barakat S, Samy MM, Wahba WI (2018) Optimization of an off-grid PV/biomass hybrid system with different battery technologies. Sustain Cities Soc 40:713–727
4. Kharrich M, Mohammed OH, Kamel S, Selim A, Sultan HM, Akherraz M, Jurado F (2020) Development and implementation of a novel optimization algorithm for reliable and economic grid-independent hybrid power system. Appl Sci 10(18):6604–6604
5. Kusakana K (2015) Feasibility analysis of river off-grid hydrokinetic systems with pumped hydro storage in rural applications. Energy Convers Manage 96:352–362
6. Li Q, Loy-Benitez J, Nam K, Hwangbo S, Rashidi J, Yoo C (2019) Sustainable and reliable design of reverse osmosis desalination with hybrid renewable energy systems through supply chain forecasting using recurrent neural networks. Energy 178:277–292
7. Bahramara S, Parsa Moghaddam M, Haghifam MR (2016) Optimal planning of hybrid renewable energy systems using HOMER: a review. Renew Sustain Energy Rev 62:609–620
8. Erdinc O, Uzunoglu M (2012) Optimum design of hybrid renewable energy systems: overview of different approaches. Renew Sustain Energy Rev 16(3):1412–1425
9. Borhanazad H, Mekhilef S, Ganapathy VG, Modiri-Delshad M, Mirtaheri A (2014) Optimization of micro-grid system using MOPSO. Renew Energy 71:295–306
10. Nadjemi O, Nacer T, Hamidat A, Salhi H (2017) Optimal hybrid PV/wind energy system sizing: application of cuckoo search algorithm for Algerian dairy farms. Renew Sustain Energy Rev 70(5):1352–1365
11. Zebarjadi M, Askarzadeh A (2016) Optimization of a reliable grid-connected PV-based power plant with/without energy storage system by a heuristic approach. Sol Energy 125(3):1221–1221
12. Mohamed MA, Eltamaly AM, Alolah AI (2017) Swarm intelligence-based optimization of grid-dependent hybrid renewable energy systems. Renew Sustain Energy Rev 77:515–524
13. Anand P, Bath SK, Rizwan M (2019) Renewable energy based hybrid model for rural electrification. Int J Energy Technol Policy 15(1):86–113
14. Anand P, Bath SK, Rizwan M (2020) Size optimisation of RES-based grid connected hybrid power system using harmony search algorithm. Int J Energy Technol Policy 16(3):238–276
15. https://www.nrel.gov/gis/solar.html access on 20/8/2021
16. Tomar A, et al (2020) Machine learning, advances in computing, renewable energy and communication, Vol 768. Springer Nature, Berlin, LNEE, 2020, p 659. doi:https://doi.org/10.1007/978-981-16-2354-7. ISBN 978-981-16-2354-7

Comparative Analysis of Baseline Models, Ensemble Models, and Deep Models for Prediction of Graduate Admission

Meet Vasani, Smit Patel, and Jasleen Kaur

Abstract As there is a rise in number of academic institutes all over the world, shortlisting the universities to apply has become a tedious task for aspiring graduate student. There are many crucial factors involved while submitting the application in any university. The cost involved while applying for any university is one of them. If student profile gets rejected, this may lead to wastage of time and money both. In order to handle this problem, we developed a machine learning-based admission prediction system. This proposed prediction system considers various parameters such as English language score, university rank, statement of purpose, letter of recommendation, cumulative grade point average, and research experience for predicting the chances of admission in a university using ensemble model. Total of 13 machine learning-based prediction models were trained and tested. These 13 models were divided into three categories: baseline models, ensemble models, and deep models. Different performance parameters such as root mean squared error, mean squared logarithmic error, and R2 were used. Execution time for each model was also recorded. In category wise comparison, ensemble models outperformed baseline models and deep models. While in total of 13 models, multiple linear regression and ridge regression outperformed all other models (in terms of high-R2 score, less root mean square error, and less execution time) followed by gradient boosting regression and extra tree regression.

Keywords Baseline models · Deep models · Ensemble models · Graduate admission · Prediction · Regression and ridge regression

1 Introduction

The world's job and business sectors are growing rapidly and persistently, in terms of skillset, and knowledge. Young individuals who want to succeed are always looking for higher education/degree to improve their skills and ability. In the last decade,

M. Vasani · S. Patel · J. Kaur (✉)
School of Engineering, P P Savani University, Kosamba, Gujarat, India
e-mail: sidhurukku@yahoo.com

A. Tomar et al. (eds.), *Proceedings of 3rd International Conference on Machine Learning, Advances in Computing, Renewable Energy and Communication*, Lecture Notes in Electrical Engineering 915, https://doi.org/10.1007/978-981-19-2828-4_47

there is an increase in number of individuals who are opting for higher education from esteemed foreign universities [1–3]. When it comes to taking decision about selection of universities, specific preparation is necessary. Various factors such as English language score, English language specific test, and consulting the experts for guidance are predominant factors in admission process. The Admission Committee of the respectful Universities accept/reject a fellow's application on the basis of profile which comes under their criterion. If profile gets rejected, the entire resources become a complete waste, additionally the students will be demotivated and will have lost their crucial time to prepare for the worst case.

This fact motivated us to carry out this research work to predict the possibility of acceptance and rejection of application of student based on various factors which include GRE score, TOEFL score, undergraduate university ranking, SOP, LOR, CGPA, and research experience using machine learning and deep learning. We proposed a deep learning-based methodology which helps students to predict their chances of admission in any esteemed university in the world.

Rest of paper is organized as follows. Section 2 presents related work carried out in proposed direction. Section 3 is presented by complete methodology followed for implementation of this work. Section 4 forecasts detailed results and analysis followed by Sect. 5 with concluding remarks.

2 Previous Work

This section provides a brief glimpse about machine learning application in various domains.

Mohan S. Acharya et al. presented a technique using various regression models such as linear regression, decision tree, random forest, and support vector regression [1]. D. A. Chithra Apporva et al. present a machine learning-based model using linear regression, K-nearest neighbors, random forest, and ridge regression and linear regression performs well [2]. P. Janani et al. use decision tree algorithm-based technique to predict the possibility of admission in particular university on basis of input scores of students [3]. Sara Aljasmi et al. present a technique to find chances of getting accepted in university using K-nearest neighbors, random forest, multi-layer perceptron, and multiple linear regression [4]. Kanadpriya Basu et al. developed a system using data of liberal arts college in California which classifies whether a particular student will take admission in university or not [5].

A. J. Alvero et al. developed a college admission system which will be able to predict gender and household income of student from the narrative essay submitted for admission [6]. Sushruta Mishra et al. proposed a machine learning technique by which institute can enhance the quality of student's admission using K-means clustering [7]. Simon Fong et al. applied a hybrid model of neural network and decision tree classifier for predicting student's likelihood of getting admission in university [8]. Abdul Hamid M. Ragab et al. presented college admission system using two cascaded hybrid recommenders, first recommender allocates study tracks for preliminary year students, and second recommender allocates the specific school for students who breezed through the preliminary year tests effectively [9]. Shital Girase et al. developed a recommender system that comprehends data seeker's need and likewise produces recommendation through basic interface [10]. Sashank Sridhar et al. developed a system which could utilize information identified with past candidates of different colleges and their admit or reject status [11]. V. Raghavendran et al. presented a system based on investigating the dataset of various colleges for predicting the opportunity of being admitted in university using multiple linear regression, decision tree regression, polynomial regression, and random forest regression [12].

3 Methodology

Architecture of proposed methodology is depicted in Fig. 1. Proposed system consists of two main phases: phase 1 and phase 2. The detail description of phase 1 and phase 2 is as follows:

3.1 Phase 1

It consists of following sub-phases.

3.1.1 Data Collection and Understanding the Dataset

For this research work, we have utilized admission dataset from Kaggle [13].

It contains various attributes:

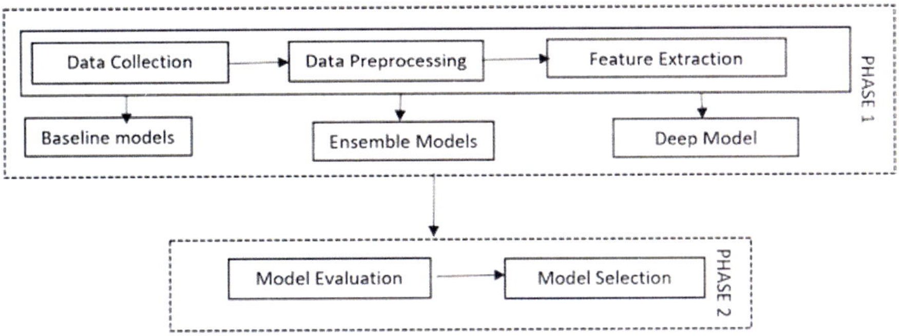

Fig. 1 Architecture of admission prediction system

Table 1 Dataset description

Total records	Attributes	Types of attributes
501	7	Simple attributes
501	1	Derived attribute

GRE Score, TOEFL Score, University Rating, SOP, LOR, CGPA, Research and Chance of Admit (Table 1).

3.1.2 Data Pre-processing

All the specified attributes were passed through various pre-processing phases [14, 15]. All the attributes were scaled. Scaling is a procedure which will level up all the values of the whole dataset, so that the model does not give dominance to a particular feature [14]. In this research work, scaling is carried out with the help of min–max scalar [14].

3.1.3 Feature Extraction

For this research work, all the attributes were considered as features except the first column (i.e., serial no.) because it does not add up as a feature. All features were divided into input and output features.

3.1.4 Model Training

In total, 13 models were trained and tested on this dataset. These models were divided into three categories: baseline models, ensemble models, and deep models. The detail of each category is as follows:

1. Baseline Models: It includes multiple linear regression, support vector machine, decision tree regression, K-nearest neighbor, lasso regression, ridge regression, elastic net regression [16–19].
2. Ensemble Learning Models: It includes random forest regression, gradient boosting regression, AdaBoost regression, and extra tree regression [20, 21].
3. Deep Learning Models: It includes neural network and multi-layer perceptron [22, 23].

3.2 *Phase 2*

Performance evaluation of all the prediction model is carried out using RMSE, MSLE, and R^2 score. Final model selection is carried out using Akaike information criteria (AIC), time, and performance parameters.

4 Results

This section provides the result and analysis of the entire experimentation. Table 2 presents the results of baseline models, ensemble models, and deep learning models, respectively.

Table 2 Performance evaluation of baseline models, ensemble models, and deep learning models

	Model	RMSE value	MSLE value	R^2 score	Time (in ms)
Baseline learning models	Multiple linear regression	0.003634	0.001382	0.825631	0.004246
	Ridge regression	0.003642	0.001389	0.82524	0.004984
	K-neighbors regression	0.004315	0.001655	0.792956	0.005667
	SVM	0.005061	0.001853	0.75718	0.008969
	Decision tree regression	0.007739	0.002977	0.628668	0.006028
	Lasso regression	0.020903	0.00738	−0.00293	0.004321
	Elastic net regression	0.020903	0.00738	−0.00293	0.005145
	Average	0.003634	0.001382	0.825631	0.004246
Ensemble learning models	Gradient boosting	0.004107	0.001574	0.802956	0.055536
	Extra trees regression	0.004161	0.001587	0.800365	0.14483
	Random forest	0.004193	0.001613	0.798815	0.030146
	Ada boosting	0.005082	0.001892	0.756177	0.08085
	Average	0.004386	0.001666	0.789578	0.077841
Deep learning models	Neural network	0.004328	0.001649	0.792346	1.9771
	Multi-layer perceptron regressor	0.0084	0.003216	0.59697	0.090733
	Average	0.006364	0.002433	0.694658	1.033917

From Table 2, it can be observed that, with RMSE value of 0.003634 and R^2 value of 0.825631, multiple linear regression performed better as compared to other baseline algorithms. Time taken to predict the result was reported to be 0.004246 ms for multiple linear regression algorithm. For multiple linear regression, independent variables are GRE score, TOEFL score, university ranking, statement of purpose, letter of recommendation, cumulative grade point average, and research experience, whereas dependent variable is chance of admit. Average RMSE value for baseline models was reported to be 0.009457, and average R^2 value was reported to be 0.546258.

In second experimentation, different ensemble models were trained and tested for prediction task. From Table 2, it can be observed that with lowest RMSE value and highest R^2 value, gradient boosting turns out to be the best among ensemble learning methods. RMSE value and R^2 value for gradient boosting were reported to be 0.004107 and 0.802956, respectively. Time taken to perform prediction using gradient boosting was 0.055536 ms. Parameters values taken into consideration for gradient boosting are number of random states with value 42. Average RMSE, MSLE, and R^2 value for ensemble learning models were 0.004386, 0.001666, and 0.789578, respectively.

In third experimentation, two different deep models were trained and tested. From Table 2, it can be observed that with lowest RMSE value and highest R^2 value, neural network turns out to be the best among deep methods. RMSE value and R^2 value for neural network were reported to be 0.004328 and 0.792346, respectively. Time taken to perform prediction using gradient boosting was 1.9771 ms. Parameters values of neural network are number of units (neurons) is 11, activation is ReLU, input_dim is 7, and epochs are 50. Average RMSE, MSLE, and R^2 value for deep learning models were 0.006364, 0.002433, and 0.694658, respectively.

For comparative analysis of models based on time parameter, different performance parameters were plotted against time as depicted in Figs. 2 and 3. With high-R^2 score and less execution time, multiple linear regression and ridge regression performed best followed by gradient boosting regression and extra tree regression. With less RMSE value and R^2 value, multiple linear regression performed better followed by gradient boosting regression and extra tree regression.

4.1 Comparative Analysis of Baseline Models, Ensemble Models, and Deep Models

For comparative analysis of these three models, average values of different performance parameters were considered. Table 3 depicted the results obtained in three different model categories.

From Figs. 4 and 5, it can be observed that while comparing models on basis of categories (baseline models, ensemble models, and deep models), ensemble models

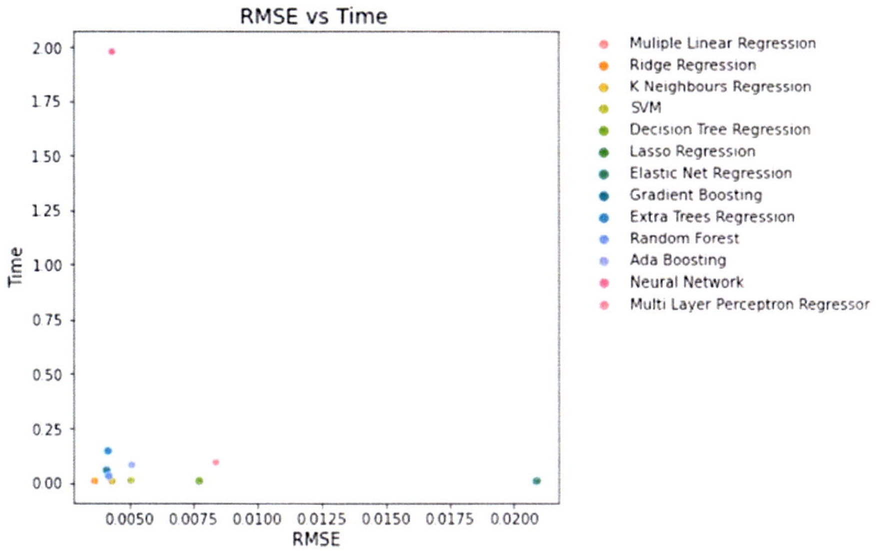

Fig. 2 Algorithms comparison with RMSE values and execution time

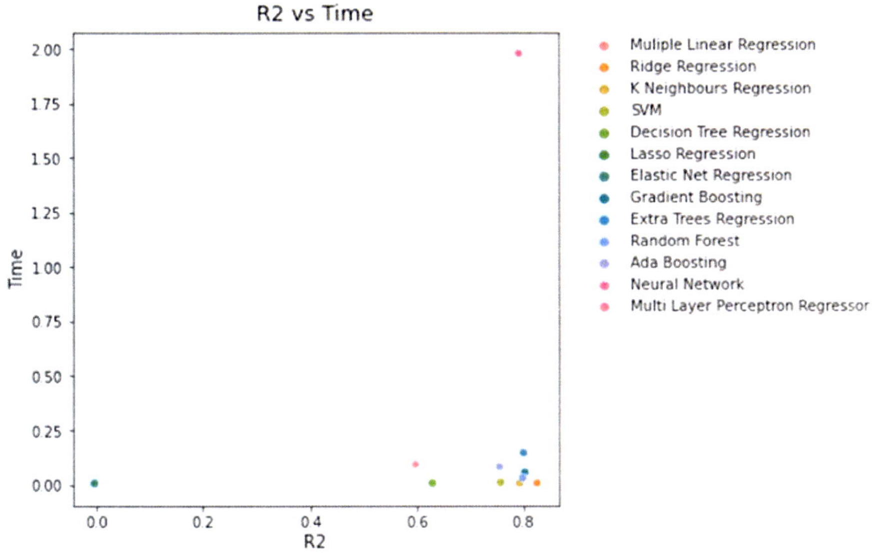

Fig. 3 Algorithms comparison with R^2 scores and execution time

Table 3 Comparative analysis of three model categories

Model category	RMSE value	MSLE value	R^2 score	Time (in ms)
Baseline models	0.009457	0.003431	0.546258	0.005623
Ensemble models	0.004386	0.001666	0.789578	0.077841
Deep models	0.006364	0.002433	0.694658	1.033917

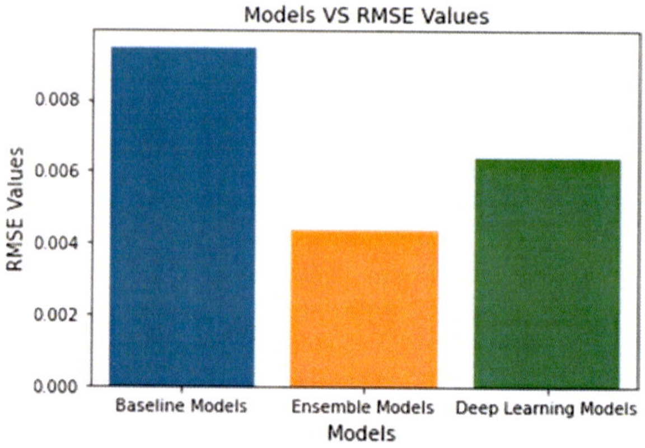

Fig. 4 Models comparison with RMSE values

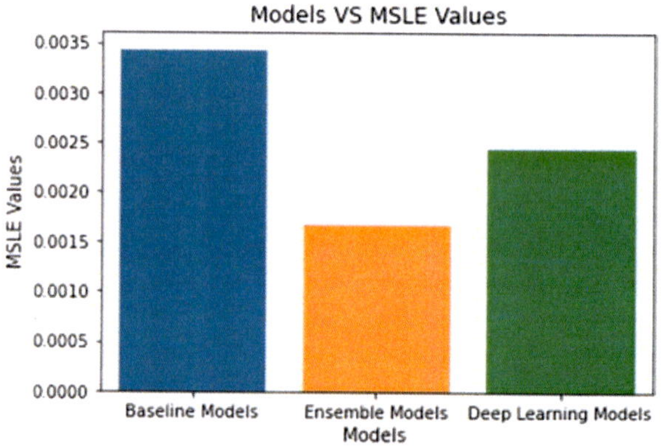

Fig. 5 Models comparison with MSLE values

performed best with less average root mean squared error and mean squared logarithmic error values compared to other two. From Fig. 6, it can be observed that, with high average R^2 score, ensemble models outperformed baseline and deep models. As depicted in Fig. 7, on basis of execution time, baseline models reported the lowest value.

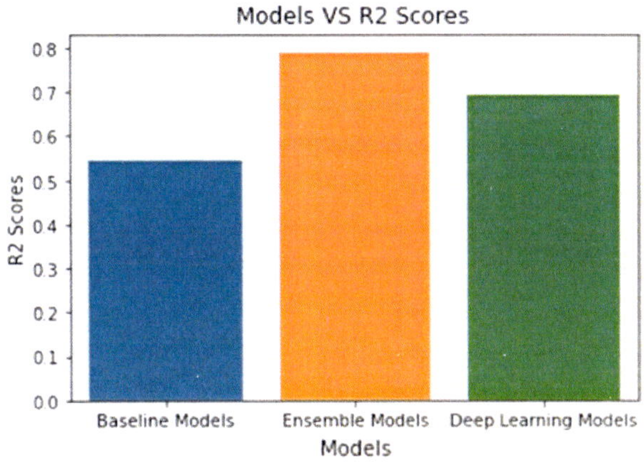

Fig. 6 Models comparison with R^2 scores

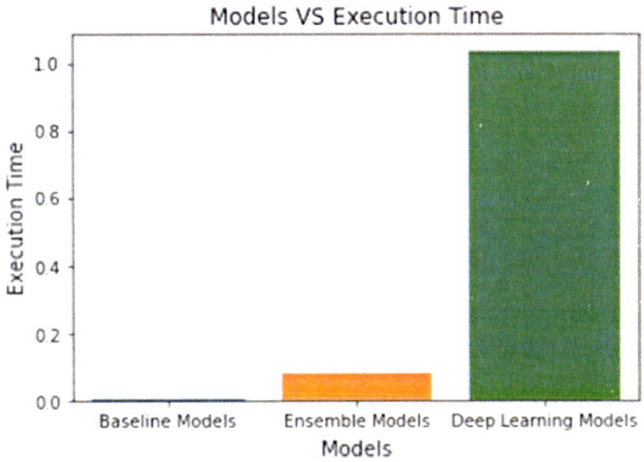

Fig. 7 Models comparison with execution time

5 Conclusion

In order to speed up the admission process and save the cost involved in admission process, we proposed a machine learning-based approach for predict the chances of admission. This proposed system takes into consideration various parameters for prediction task. Total of 13 models were trained and tested on admission dataset. For comparative analysis purpose, these models were divided into three categories: baseline, ensemble, and deep models. Root mean squared error, mean squared logarithmic error, R^2, and time parameters were used for evaluation of model. Multiple linear regression, gradient boosting, and neural network outperformed all other models in each category. From all 13 models, multiple linear regression and ridge regression outperformed all other models, followed by gradient boosting and extra tree regression. RMSE value and R^2 value were reported to 0.003634 and 0.825631, respectively, for multiple linear regression. From three model categories, ensemble models performed better as compared to baseline and deep models with lowest RMSE score and highest R^2 value.

References

1. Acharya MS, Armaan A, Antony AS (2019) A comparison of regression models for prediction of graduate admissions. In: 2019 international conference on computational intelligence in data science (ICCIDS)
2. Chithra Apoorva DA, Chandu Nath M, Rohith P, Bindu Shree S, Swaroop S (2020) Prediction for university admission using machine learning. Int J Rec Technol Eng (IJRTE) 8(6). ISSN: 2277-3878
3. Janani P, Hema Priya V, Monisha Priya S (2020) Prediction of MS graduate admissions using decision tree algorithm. Int J Sci Res (IJSR) 9(3). ISSN: 2319-7064
4. Aljasmi S, Nassif AB, Shahin I, Elnagar A (2020) Graduate admission prediction using machine learning. Int J Comp Commun 14. ISSN: 2074-1294
5. Basu K, Basu T, Buckmire R, Lal N (2019) Predictive models of student college commitment decisions using machine learning
6. Alvero AJ, Arthurs N, Antonio AL, Domingue BW, Gebre-Medhin B, Giebel S, Stevens ML (2020) AI and holistic review: informing human reading in college admissions. In: Proceedings of the AAAI/ACM conference on AI, ethics, and society (AIES '20). Association for Computing Machinery, New York, NY, USA, pp 200–206
7. Mishra S, Sahoo S, Mishra BK, Satapathy S (2016) A quality based automated admission system for educational domain. In: 2016 international conference on signal processing, communication, power and embedded system (SCOPES), pp 221–223
8. Fong S, Biuk-Aghai RP (2009) An automated university admission recommender system for secondary school students. In: The 6th international conference on information technology and applications
9. Ragab AHM, Mashat AFS, Khedra AM (2014) Design and implementation of a hybrid recommender system for predicting college admission. Int J Comp Inf Syst Indust Manage Appl 6:35–44. ISSN 2150-7988
10. Girase S, Powar V, Mukhopadhyay D (2017) A user-friendly college recommending system using user-profiling and matrix factorization technique. In: 2017 international conference on computing communication and automation (ICCCA), pp 1–5
11. Sridhar S, Mootha S, Kolagati S (2020) University admission prediction system using stacked ensemble learning. In: 2020 advanced computing and communication technologies for high performance applications (ACCTHPA), pp 162–167
12. Raghavendran CV, Pavan Venkata Vamsi C, Veerraju T, Veluri RK (2021) Predicting student admissions rate into university using machine learning models. Mach Intell Soft Comput Adv Intell Syst Comput 1280. Springer, Singapore
13. https://www.kaggle.com/mohansacharya/graduate-admissions?select=Admission_Predict_Ver1.1.csv
14. Jamal Muhammad Ali P, Hassan Faraj R (2014) Data normalization and standardization: a technical report. Mach Learn Rep 1(1)
15. Tomar A et al (2020) Machine learning, advances in computing, renewable energy and communication, Vol 768. Springer Nature, Berlin, LNEE, p 659. doi:https://doi.org/10.1007/978-981-16-2354-7. ISBN 978-981-16-2354-7
16. Naga Satish G, Raghavendran CV, Sugnana Rao MD, Srinivasulu C (2019) House price prediction using machine learning. Int J Innov Technol Expl Eng (IJITEE) 8(9). ISSN: 2278-3075
17. Frizzarin M, Gormley IC, Berry DP, Murphy TB, Casa A, Lynch A, McParland S (2021) Predicting cow milk quality traits from routinely available milk spectra using statistical machine learning methods. J Dairy Sci. ISSN 0022-0302

18. Dawoodi HH, Patil MP (2021) Rainfall prediction in North Maharashtra Region using support vector machine. Turkish J Comp Math Educ 12(7):1501–1505
19. Wan Fairos WY, Syerina Azlin MN, Wan Faizah WY, Norafefah MS (2019) Supervised data mining approach for predicting student performance. Indonesian J Electr Eng Comp Sci 16(3):1584–1592. ISSN: 2502-4752
20. Deokar A (2021) Using visceral adipose tissue measurements to build classification models for gestational diabetes mellitus. Paper presented at 2020 Northeast Section Meeting, Online. https://doi.org/10.18260/1-2-020-36266
21. Lee S, Chung JY (2019) The machine learning-based dropout early warning system for improving the performance of dropout prediction. Appl Sci 9:3093
22. Mishra S, Vijay Neurkar S, Patil R, Petkar S (2021) Heart disease prediction system. Int J Eng Appl Phys (IJEAP) 1(2). ISSN: 2737-8071
23. Abu Zohair LM (2019) Prediction of student's performance by modelling small dataset size. Int J Educ Technol High Educ 16:27

Smart Door System for Maintaining Social Distancing and Controlling Spread of Infectious Disease

Priyanka Malhotra, Aditya Dubey, and Rahul Goel

Abstract COVID-19, also known as coronavirus, has spread throughout the world and changed all facets of lives drastically. Though the vaccines are being distributed, the situation is still miserable in many countries. To slow down the spread of this virus, social distancing is required to be maintained. This paper proposes a smart door system that helps in fast temperature scanning of people and ensures social distancing at public places. The proposed system also automatically opens or closes the doors at public places and switches the appliances on the basis of the number of people present. The system design is based on Internet of things (IoT) and is embedded using Arduino Uno. This is an efficient and low-cost model to control the spread of infectious disease.

Keywords COVID-19 · IoT · Contactless door · Sanitization · Energy management · Temperature monitoring

1 Introduction

The novel COVID-19 is a highly infectious disease causing respiratory illness due to SARS-CoV-2 virus. The survival time [1] of this virus and its mutated versions on different surfaces is not certain, but like other viruses, it spread through direct or indirect transmission. The virus can survive on a surface for a long period of time depending on various factors. The coronavirus symptoms have shown reduction with the invention of vaccines, but the ability of vaccines to prevent the transmission of virus has been unclear [2]. The pandemic called COVID-19 had lead to the research for technology that can be used to control its spread. The U.S. base center for disease control and prevention (US CDC) [3] has launched a self-checker application on cloud platform to help the patient find an appropriate healthcare service through

P. Malhotra (✉) · A. Dubey · R. Goel
Chitkara University Institute of Engineering and Technology, Chitkara University, Punjab, India
e-mail: priyanka.malhotra@chitkara.edu.in

A. Tomar et al. (eds.), *Proceedings of 3rd International Conference on Machine Learning, Advances in Computing, Renewable Energy and Communication*, Lecture Notes in Electrical Engineering 915, https://doi.org/10.1007/978-981-19-2828-4_48

a set of questionnaires. But there are many COVID cases with no symptoms or asymptomatic carriers.

The COVID-19 cases are increasing exponentially, and many new variants [4] had being registered. The majority of infected individuals have mild to severe symptoms. Most common symptoms are cough, fever, tiredness, among which only fever is measurable parameter. The people with these symptoms must not visit public places or crowded areas. At public places so as to stop the spread of infection, the body temperature is checked, and the person is sanitized. A person on duty at the entrance of a bank, ATM, grocery store, etc., checks the body temperature of the people. The people with the body temperature within the safe range (97–99 °F) are allowed to enter inside after hand sanitization. These public places can also get crowded many times, so the person on duty also needs to ask people to wait outside. This kind of practice puts the person on the duty at a great risk of getting exposed to the infection related to this virus.

The cases of infection can be reduced by utilizing IoT-based technologies at the public places. To effectively curb COVID-19 spread, innovations and novel strategies should be implemented. The focus should involve the use of smart technology to minimize the transmission of the virus. The present research work proposes a prototype smart door system that can check the people at the entrance of public place and help in maintaining social distance. The proposed model can be employed to identify the infected person and automatically sanitize all the people entering a public place. It is hoped that the present work will help in cubing COVID-19-related infection at public places. The idea of the proposed model has been patented under intellectual property, India [5]. The paper contributes in the following way:

1. A prototype model based on IoT has been proposed for automatic identification of an infected person at public places.
2. The proposed model uses sensors to open and close doors at public places and sanitize the people on entry.

Section 2 discusses the background for working on the proposed idea. Section 3 discusses the workflow for proposed model. Section 4 explains the design and operation of proposed model. At the end, section 5 concludes the paper.

2 Background

The COVID-19-related infection begins with human interaction or by touching the infected surfaces. The solution to the problem may be the automation of the process. But automation may not be applicable everywhere. As related to public places, there are certain operations which have a high risk of infection involved. At these places, security personnel outside the premises have infrared thermometers to check the body temperature of visitors. The security guard also asks them to disinfect themselves by sanitizers, and sometimes even records the count of people entering inside to limit the number of people so as to maintain social distance. The person on duty is at great risk of catching infection.

Internet of things or IoT [6]-based systems can help in this regard. IoT systems include sensors for detection, processing units, and other technologies to accumulate data from different devices over the Internet. In [7], IoT -ased architecture has been proposed to control the spread of COVID-19. The architecture utilized smart sensors to identify the infected person and maintain social distance. The system has been connected using cloud computing to make decisions based on the real-time data. In [8], the authors employed IoT-based system for monitoring the health of a person so as to prevent the spread of the disease, and the authors proposed portable framework to display the heartbeat, temperature, and other health parameters of a person. A health monitoring system based on IoT has also been proposed for patients [9, 10].

IoT has been employed to provide various solutions for automatic door control. In [11], the author proposed a system designed using ARM-11 and Raspberry Pi-3 board to control the door of the home. Authors in [12] used IoT and the smartphone communication to open or close a door remotely. IoT has also been employed in [13] to control the industrial building lighting.

In the present work, a prototype model based on IoT has been proposed for curbing the spread of COVID-19-related infection at public places. The model can sense the visitor's body temperature automatically. It has built in mechanism to disinfect the people. It keeps an accurate record of count of people at the public place/premises. The model also has a mechanical system to automatically open/close the door depending on the count of people inside the hall/building. The proposed model also caters to the need for a spontaneous way to turn on the essential appliances according to the presence of people.

3 Proposed Model

The proposed model for curbing COVID-19 spread is based on IoT. It provides an automatic scanning and sanitization of the visitors at the public places/offices. The proposed model also gives the count of the number of person entering a building/hall. The working of the proposed model under three different conditions is as given below:

(a) **Person enters the room**: The operation of the proposed model on entry of a person inside the room/building is depicted in Fig. 1a. The sensor-based system installed outside the room senses the movement of a person. On sensing the presence of a person, the smart contactless infrared thermometer senses the temperature of human body. If the body temperature falls between 36 °C to 38 °C, the sanitization machine will get activated, and it will disinfect the person. The model also utilizes a mechanism for head count in the room. If the number of person present in the room is within the permissible limit, the room door opens and LED becomes 'green' else it remains 'red' and does not allow person to enter the room. On entry of a person, the counter is incremented by one. The proposed model also has the mechanism to automatically switch on the lights and fans in the room during the presence of people and automatically switches off the light and fans if there is no one inside.

(b) **Person exits the room**: The operation of proposed model when a person exits the room is depicted in the flowchart of Fig. 1b. The sensor-based system installed inside the room senses the movement of a person. On sensing the presence of person near the door, the LED glows 'red' else it remains 'green'. The person counter is decremented by one if the person exits the room. The

a

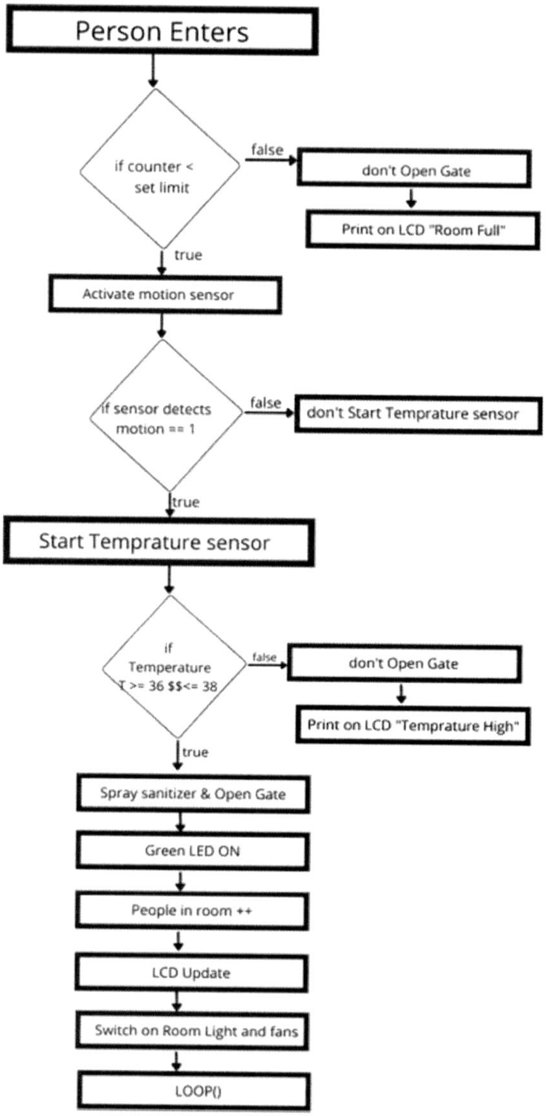

Fig. 1 **a** Flowchart representing the process on person entry followed when a person enters the room. **b** Flowchart representing the process followed when a person exits the room. **c** Flowchart representing the process followed when the number of person present in room reaches the permissible limit

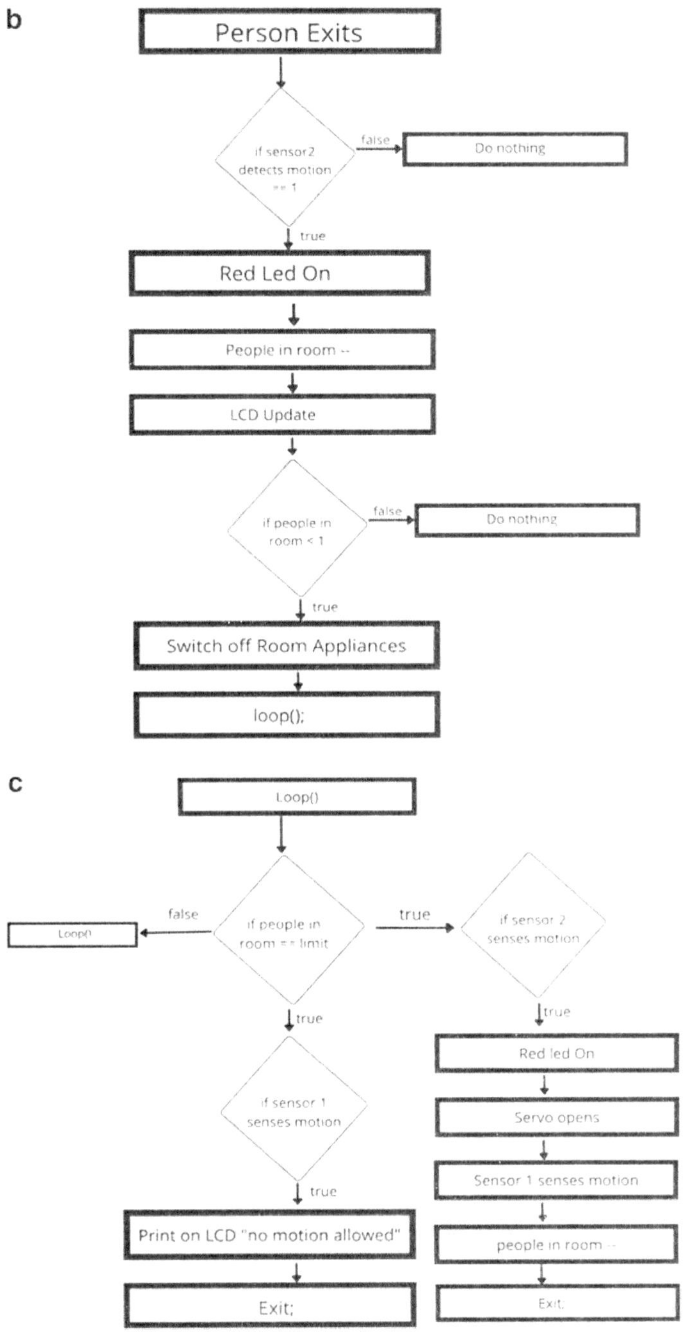

Fig. 1 (continued)

proposed model also automatically switches off the lights and fans in the room on absence of people.

(c) **Room full**: The operation of the proposed model when the room is full is depicted in the flowchart of Fig. 1c. If the head count in the room reaches the defined permissible limit, then the system disallows any further entry of people. Under such conditions, the model displays 'no motion allowed'. The model allows the entry and opens the door if the person count is below the permissible limits.

4 Proposed Model Design

The proposed model design for curbing the spread of COVID-19 is aimed at eliminating the interaction of security personnel and automatically controlling the entry and sanitization of people at public places. The representation of proposed model is shown in Fig. 2 in the form of a block diagram.

4.1 Important Components

The important components used in the proposed model are explained below:

(a) **Arduino Uno**: Arduino Uno [14] is a board with ATmega328P microprocessor. It has 14 I/O pins (digital), 6 input pins (analog), 16 MHz resonator, connector power, and USB as shown in Fig. 3. Arduino Uno has been widely used in

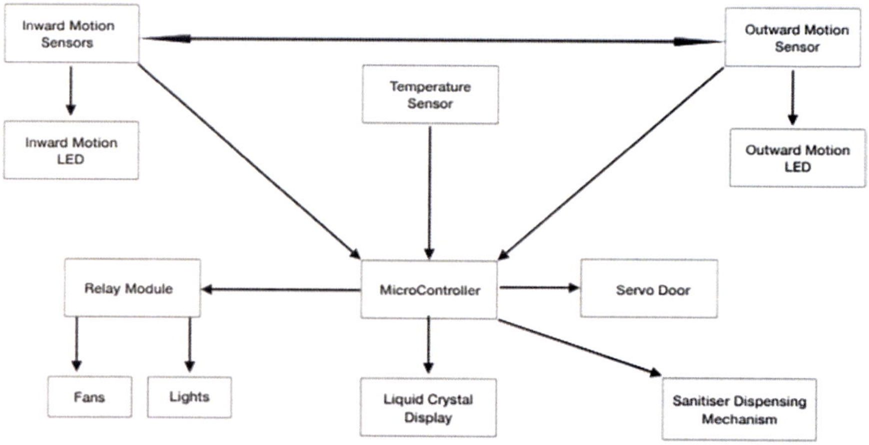

Fig. 2 Block diagram representing the proposed model

Fig. 3 Arduino Uno board [16]

many IoT-based products [15]. The microcontroller is interfaced with sensors to obtain related sensor reading data.

(b) **MLX90614**: It is an infrared-based contactless thermometer. MLX90614 thermometer has a built in low-noise amplifier, powerful DSP unit with 17-bit ADC which results in great accuracy and resolution of this small sensor [17] as shown in Fig. 4.

(c) **Ultrasonic sensor**: It calculates the distance between the target object and the sensor [19]. It has an emitter (head) of the sensor which emits ultrasonic sound waves and one receiver which receives those pulses. The sensor finds out the distance from a target by measuring time lapse between the transmission of the pulse and its reception (see Fig. 5).

(d) **Servo Motor**: A servo motor makes the shaft move at various angles between 0° to 180°. The speed of rotation of the motor can also be varied according to the need. It uses a closed loop mechanism for control. It works on digital signals [21].

(e) **Liquid Crystal Display (16*2)**: Liquid crystal display is a 16 pin device. Its working principle is that it blocks rather than dissipates light. Its brightness

Fig. 4 MLX90614 temperature sensor [18]

Fig. 5 Ultrasonic sensor [20]

is adjustable, and one can use it to display characters, special characters, and numerals.

(f) **Relay Module**: Relay module is an important sensor in many electronic IoT-based projects, it acts like a switch. It is used to open and close different circuits electrically. It works on the concept of electromagnetism. It uses a 5 V supply of power. It can also be provided using solar cells [22, 23].

4.2 Operation of the Proposed Model

The proposed model helps in maintaining social distancing in a very intuitive way. The circuit diagram of the proposed model is shown in Fig. 6. The diagram shows the placement of Arduino Uno, ultrasonic sensors, temperature sensor, servo motor, relay, and display system.

Firstly, MLX90614 IR contactless temperature sensor measures the temperature of the person entering the gate or premises of a building. The recorded temperature is fed to Arduino Uno to check whether the person's temperature lies in the healthy range (31.0–35.6 °C) or not. If the temperature lies within the desired range, the

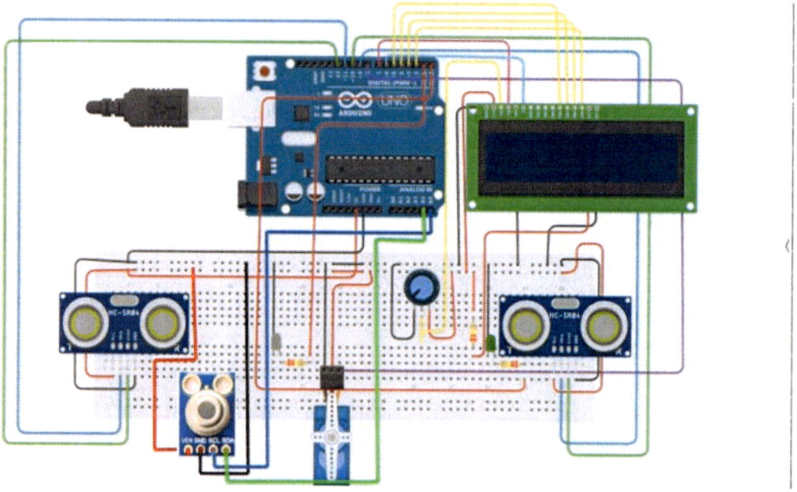

Fig. 6 Circuit diagram of proposed model

person is allowed to enter the gate. Figure 7 shows a model of the project displaying the temperature of the person. After temperature scanning, each person is sanitized before entry to the gate.

To automate the contactless opening and closing of the door, the proposed system is equipped with a digital servo motor. Arduino Uno has 14 digital pins, 6 of which provide PWM output, the digital servo motor is connected to one of them, and the system is programmed in such a way that it will operate once the In_US or Out_US is activated, and the body temperature of the visitor is within normal range. The servo here is rotated from 0 to 90° which is just an enactment of an opening or closing door, and then, it closes once the visitor walks in.

For keeping an accurate record of the number of person entering or exiting the room, a single door bi-directional person counter using two ultrasonic sensors placed around a feet apart is used. When a person enters the room, the first ultrasonic sensor senses the motion and is activated, and this interference is noted as referred to In_US and stored; afterward, the time is subtracted from the millis() function of the Uno, which gives us a virtual timestamp of the person starting to enter the room (referred as time_I). When the person has traveled distance of one foot, the second ultrasonic sensor gets activated and the same process of recording the virtual timestamp (time_II) from ultrasonic II (referred as Out_US) is done. The difference between time_I and time_II is calculated, and if it is positive and lower than or equal to 500 ms (ample time for crossing one foot distance), then it will be considered as the person has entered the room. Similarly if the Out_US is activated first and then In_US is activated afterward, this implies that the time_I will be lodged before time_II, therefore the difference in time_II from time_I will be negative, and if it is again smaller than the accepted time quantum of 500 ms, this will make the Uno decide that the person is exiting.

According to the head count present in the room, the lights, fans, and other appliances are switched on or off. A signal is sent to the relay which acts as a circuit bridge

Fig. 7 Proposed system displaying temperature

handling 220–230 V to 50 Hz AC supply to the connected appliances and turns the device on or off.

5 Conclusion and Future Scope

The proposed IoT-based smart door system provides detection of the symptoms of COVID-19 or other viral infection and controls the outbreak of infection. The smart door system facilitates testing the symptoms like fever or high temperature and helps in maintaining social distancing. An automatic door opening system helps in contactless operation. The model also has a sanitization system which will include sprinklers to sanitize a person before entering inside. The proposed model will also save electricity as the electrical appliances are switched on/off in accordance with the head count present in the room. It is a low cost and efficient model for controlling the spread of infectious diseases. The future scope of the proposed model includes implementation of such systems at malls, banks, doctor's clinic, and other public places. The future model may include a face recognition system to authenticate the person before entry as well as keep the record of the same with the duration of stay of the person.

References

1. Herman J, Biegel B, Huang L (2021) Inactivation times from 290 to 315 nm UVB in sunlight for SARS coronaviruses CoV and CoV-2 using OMI satellite data for the sunlit earth. Air Qual Atmos Health 14(2):217–233
2. Bartsch SM, O'Shea KJ, Ferguson MC, Bottazzi ME, Wedlock PT, Strych U, McKinnell JA, Siegmund SS, Cox SN, Hotez PJ, Lee BY (2020) Vaccine efficacy needed for a COVID-19 coronavirus vaccine to prevent or stop an epidemic as the sole intervention. Am J Prev Med 59(4):493–503. https://doi.org/10.1016/j.amepre.2020.06.011
3. https://www.cdc.gov/coronavirus/2019-ncov/symptoms-testing/coronavirus-self-checker.html
4. Li Q, Wu J, Nie J, Zhang L, Hao H, Liu S, Zhao C, Zhang Q, Liu H, Nie L, Qin H, Wang M, Lu Q, Li X, Sun Q, Liu J, Zhang L, Li X, Huang W, Wang Y (2020) The impact of mutations in SARS-CoV-2 spike on viral infectivity and antigenicity. Int J Health Pol Manage. Cell 182(5):1284–1294.e9. doi:https://doi.org/10.1016/j.cell.2020.07.012
5. Goel R, Dubey A, Malhotra P (2021) Inventors; system and method for controlling spread of infectious disease. Patent File Number – 202111029009, filed on 29 June, 2021
6. Belkhiri H, Messai A, Belaoued M, Haider F (2021) Security in the internet of things: recent challenges and solutions. In: Bououden S, Chadli M, Ziani S, Zelinka I (eds) Proceedings of the 4th international conference on electrical engineering and control applications. ICEECA 2019. Lecture Notes in Electrical Engineering, vol 682. Springer, Singapore. https://doi.org/10.1007/978-981-15-6403-1_79
7. Kumar K, Kumar N, Shah R (2020) Role of IoT to avoid spreading of COVID-19. Int J Intell Netw 1:32–35. https://doi.org/doi.org/10.1016/j.ijin.2020.05.002
8. Valsalan P, Baomar TAB, Baabood AHO (2020) IoT based health monitoring system. J Crit Rev 7(4):739–743

9. Nishitha KR, Vittal Bhat M (2022) IoT-based patient health monitoring system using STM32F103C8T6. In: PS, Prabhu NKS (eds) Advances in renewable energy and electric vehicles. Lecture Notes in Electrical Engineering, vol 767. Springer, Singapore. https://doi.org/10.1007/978-981-16-1642-6_20
10. Zareb M, Bakhti B, Bouzid Y, Kadourbenkada H, Bouzgou K, Nouibat W (2021) Novel smart air quality monitoring system based on UAV quadrotor. In: Bououden S, Chadli M, Ziani S, Zelinka I (eds) Proceedings of the 4th international conference on electrical engineering and control applications. ICEECA 2019. Lecture Notes in Electrical Engineering, vol 682. Springer, Singapore. https://doi.org/10.1007/978-981-15-6403-1_30
11. Quadri SAI, Sathish P (2017) IoT based home automation and surveillance system. Int Conf Intell Comput Contr Syst (ICICCS) 2017:861–866. https://doi.org/10.1109/ICCONS.2017.8250586
12. Jeong JI (2016) A study on the IoT based smart door lock system. In: Information science and applications (ICISA) 2016. Springer, Singapore, pp 1307–1318
13. Minoli D, Occhiogrosso B (2020) IoT-driven advances in commercial and industrial building lighting. In: Butun I (eds) Industrial IoT. Springer, Cham. https://doi.org/10.1007/978-3-030-42500-5_3
14. Badamasi YA(2014) The working principle of an Arduino. In: 11th international conference on electronics, computer and computation (ICECCO), 2014, pp 1–4, doi: https://doi.org/10.1109/ICECCO.2014.6997578
15. Lalbihari Barik (2019) IoT based temperature and humidity controlling using Arduino and raspberry Pi. Int J Adv Comp Sci Appl (IJACSA) 10(9). http://dx.doi.org/https://doi.org/10.14569/IJACSA.2019.0100966
16. https://d2rormqr1qwzpz.cloudfront.net/photos/2013/06/12/48912-arduinouno_r3_front.jpg
17. Jin G, Zhang X, Fan W, Liu Y, He P (2015) Design of non-contact infra-red thermometer based on the sensor of MLX90614. Open Autom Contr Syst J 7:8–20. https://doi.org/10.2174/1874444301507010008
18. https://miliohm.com/wp-content/uploads/2018/03/MLX90614_2.jpg
19. Carullo A, Parvis M (2001) An ultrasonic sensor for distance measurement in automotive applications. Sensors J IEEE 1:143–147. https://doi.org/10.1109/JSEN.2001.936931
20. https://free-electronic.com/wp-content/uploads/2019/07/Ultrasonic-Sensor-HC-SR04.png
21. van de Straete HJ, Degezelle P, De SJ, Belmans RJM (1998) Servo motor selection criterion for mechatronic applications. IEEE/ASME Trans Mechatron 3(1):43–50. doi:https://doi.org/10.1109/3516.662867
22. Madan J, Pandey R, Sharma R (2020) Device simulation of 17.3% efficient lead-free all-perovskite tandem solar cell. Sol Energy 197:212–221
23. Tomar A, et al (2020) Machine learning, advances in computing, renewable energy and communication, Vol 768. Springer Nature, Berlin, LNEE, p 659. doi:https://doi.org/10.1007/978-981-16-2354-7. ISBN 978-981-16-2354-7

A Survey of the Workload Forecasting Methods in Cloud Computing

Archana Yadav, Shivam Kushwaha, Jyoti Gupta, Deepika Saxena, and Ashutosh Kumar Singh

Abstract A few years ago, cloud computing had widely altered the way of computation and storage. It is quite challenging for cloud service providers to maintain the required quality of service (QoS) standard without violating a service level (SLA) agreement. To improve the performance of the cloud, accurate workload prediction plays a vital role. Cloud computing promises scalability, on-demand service, and virtualization to improve the quality of service (QoS) offered to end-users. In this paper, we present a comprehensive survey of workload prediction approaches in cloud environments. It also highlights the emerging challenges like resource wastage, excess power consumption, quality of service violations, etc.

Keywords Cloud computing · Workload prediction · Scalability · Classification

1 Introduction

Cloud computing continues as a model for IT service delivery at an astonishing speed driven by a wide range of interactive and interactive features. It is widely used for computation and storage over the Internet for various domains like marketing, business, education, banking, entertainment, etc. Instead of storing files on a portable hard drive or local storage device, cloud-based storage makes it easier to store

A. Yadav (✉) · S. Kushwaha · J. Gupta · D. Saxena · A. K. Singh
National Institute of Technology, Kurukshetra, Haryana, India
e-mail: archana_52010001@nitkkr.ac.in

S. Kushwaha
e-mail: shivam_52010007@nitkkr.ac.in

J. Gupta
e-mail: jyoti_52010049@nitkkr.ac.in

D. Saxena
e-mail: 13deepikasaxena@gmail.com

A. K. Singh
e-mail: ashutosh@nitkkr.ac.in

A. Tomar et al. (eds.), *Proceedings of 3rd International Conference on Machine Learning, Advances in Computing, Renewable Energy and Communication*, Lecture Notes in Electrical Engineering 915, https://doi.org/10.1007/978-981-19-2828-4_49

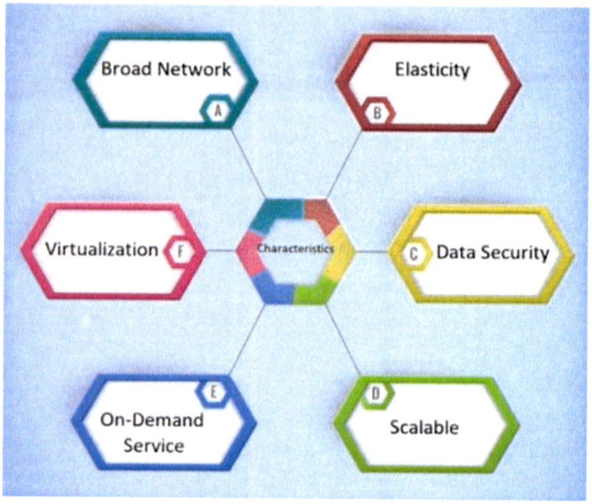

Fig. 1 Characteristics of cloud computing

them in a remote database [1–17]. The popularity and usage of cloud computing are increasing day by day; cloud-based applications make people highly dependent on it for their day-to-day activities. It offers several advantages compared to traditional computer models along the way including reduced costs and increased flexibility. Cloud computing is a much-needed delivery of IT resources to the Internet at pay-as-you-go rates. Cloud is present at remote locations rather than local servers. It serves several benefits like broad network, elasticity, data security, scalable, on-demand service, virtualization, etc. Figure 1 illustrates the various types of characteristics of cloud computing. We can process and access data and applications from servers via the Internet. It can also help save our planet by providing a complete computing environment. Cloud computing can be described as a new computer-style where powerful and easy-to-use resources are offered as an Internet service. Cloud computing will redesign information technology (IT) processes and the IT market and is also a powerful computer infrastructure for many applications. The purpose of using the cloud is to allow users to benefit from all of these technologies, without the need for in-depth information about each of them. The cloud aims to reduce costs and help users focus on their core business instead of being blocked by IT barriers. Cloud computing creates privacy concerns because a service provider can access data in the cloud at any time. It may alter the information by mistake or intentionally or delete the information. The main motivation is to reduce the error in workload prediction by extracting knowledge from previous existing models so that we could identify current as well as predict future workloads in the cloud environment.

This document presents an in-depth investigation of the technical predictions of the project, their strategies used, and motivations to conduct them. The main contributions of this paper are as follows:

- We are describing what is cloud, what is cloud computing, and how it works.
- The benefits and highlighting emerging challenges of cloud computing.

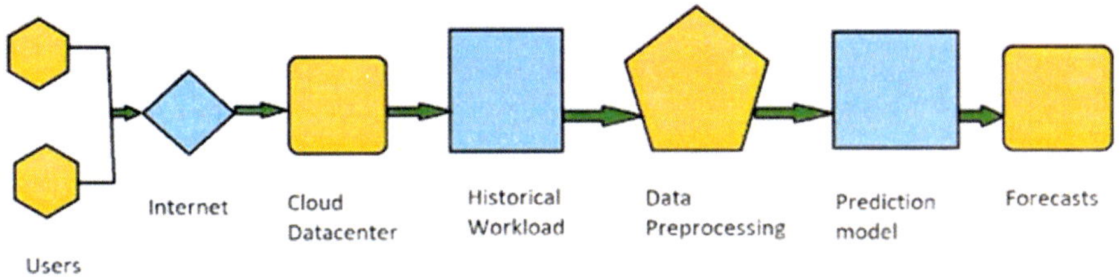

Fig. 2 General workload forecasting model

- What is workload prediction including a survey of various existing workload prediction models.
- Comparison among various existing workload prediction models in a tabular form.

The rest of the paper is organized as follows: Sect. 2 discusses related activities. Comparison of different approaches is presented in Sect. 3, then we have provided emerging challenges in Sect. 4, and finally, we end up with a conclusion in Sect. 5.

Workload Prediction

Forecasting takes information available in the present and uses it to predict the future. Workload prediction is a technique that is used to improve efficiency and reduce the operational cost of the cloud. Application management and resource management are the two very basic requirements that motivate prediction. Figure 2 illustrates the general scenario of workload forecasting at the cloud data center. Servers receive millions of requests sent by users. Applications are processed there as well all requests that come at a certain time of speculation are compiled as historical data, which is later used to predict future work. Historical data is collected and processed in advance for standardization. Then, normal data transferred to a forecasting system to predict future workload.

2 Related Work

2.1 Neural Network-Based Approaches

Saxena et al. [1] have proposed a model of network load variability to predict the average load between consecutive forecast periods. The variable load prediction model shows the predictable demand for demand in cloud data centers. This data is used to train and evaluate data in the neural network predictive system. The AADE learning algorithm then trains the neural network. If the predicted workload is almost equal to the actual workload, then these data are sent to trained workload prediction models, otherwise, it is again sent to the AADE learning algorithm.

Lu et al. [2] have proposed a model of a predictive load forecast to predict the future value of work for cloud conservationists. The projection model for the proposed novel, called RVL BPNN, is based on the BP neural network algorithm and predicts a much larger future load in terms of how to use the internal relationships between the incoming workloads. The test results show that the proposed RVL BPNN model is gaining clarity and efficiency.

Saxena et al. [3] have proposed the provision of energy-efficient resources and a distribution framework to predict the changing needs of future applications. Predicting the use of multiple resources simultaneously, a predictive model for the online multi-resource feed-forward neural network (OM-FNN) is proposed. The tri-adaptive differential evolution (TaDE) learning algorithm is designed for optimal use of OM-FNN predictions. Simultaneously with future applications, automated VM deployment based on the integration of specific resource requirements and reduced VM allocation enables PMs.

Kumar et al. [4] have developed a workload prediction model using a neural network and a different evolution algorithm. In the author's way, using a structure containing n-p-q neurons where n, p and q are the number of neurons in the sequence of input, hidden, and output.

A workload forecasting framework based on a neural network model with supervised learning technique is proposed in [5]. To upgrade the network learning process, a BiPhase adaptive differential evolution (BaDE) learning algorithm is introduced to train the model. The BaDE learning algorithm is used to train the proposed framework for predicting the network load (WFNN) function. To train the neural predictive network, historical data is actively provided by the cloud data keeper. The proposed activity is measured by two real-world stocks, Saskatchewan HTTP and NASA.

Bi et al. [6] have presented an integrated forecasting system, provided by sound filtering and data representation, known as Savitzky-Golay and wavelet supported by stochastic configuration networks (SGW-SCNs), to indicate the amount of workload in the future.

2.2 *Heuristic Approaches*

Kumar et al. [7] have proposed a performance management system (SDWF) that identifies the practice of predicting errors by incorporating the latest forecasts and using it to increase the accuracy of other predictions. The model uses a healing method based on dark conditions in training neurons. This paper offers to create a way to predict the burden of learning from past mistakes and deal with them appropriately. It also develops one of the meta-heuristic optimization algorithm proposed by black-hole scenarios to achieve the most accurate network tools.

Nguyen et al. [8] have established a predictive model for a series of novel timelines, based on an advanced reading machine. Due to uncertainty in the weights of installing over-the-counter learning equipment requires a large number of hidden neurons to achieve better results, for this reason, the model weight is increased. Winning the issue is a new war of attrition based on opposition to extreme learning machines, choosing the right input tools, and hidden bias. So in this model, two algorithms are used for extreme learning machines (ELM) and opposition warfare (OTWO). ELM was raised to overcome the regression of media-based approaches in a single hidden layer of supplying neural networks through two phases.

Zhu et al. [9] have raised the cache memory network (LSTM) encoder network. The model has two components: the LSTM-based network encoder–decoder network. First, the data is encrypted. In context, the vector then selects the code to produce the middle prediction result of the output layer. Finally, the output layer displays the predictive values of the workload. The result was satisfactory in predicting mixed loading in the computer environment.

2.3 Recurrent Neural Networks-Based Approaches

Gao et al. [10] have compared the making of predictable methods of independent artwork. They suggest how to make predictions for a while before a predetermined time point to allow for sufficient time for work planning based on the predicted workload. Continuing to improve the accuracy of the forecast, they introduced a combined method of predicting the load, which begins to combine all the tasks into several stages and trains the forecast model for each phase in sequence.

Kumar et al. [11] have analyzed and compared the accuracy of the assumptions for different machine learning algorithms aimed at predicting the load of server logs. The proposed speculation model for the comparison study was used using linear regression (LR), K-nearest neighbors (KNN), support vector machine (SVM), ARMA, ARIMA, and support vector regression (SVR) for web applications to select the appropriate algorithm depending on the characteristics of each load.

Singh et al. [12] have developed a dynamic predictive model using linear regulation, ARIMA, and supported vector regression for web applications. In addition to the cloud-based information discussed in web applications with a new weather module called technocrat load forecaster. In the cloud environment, the start of the VM takes 5–10 min, so the task of predicting work requires a solution in the first step to help prepare the required VM for the incoming workload.

2.4 Hybrid Approaches

Chandy et al. [13] have proposed machine learning methods to manage resource allocation to cloud computing for large data processing systems, the simulation of the proposed model using network simulator two enables better performance and resource utilization at a higher cost, time, power, and memory usage. A random forest algorithm used uninstall features that use the bootstrap reset process and upgrade the resolution tree for all visual effects.

Saxena et al. [14] have proposed predictive resource management models in cloud environments. In the conceptual framework for cloud resource management, m users request different applications to be executed at a data center. Each application had a specific resource requirement that is to be fulfilled by the data center. The workload forecasting accuracy and virtual machine resource prediction should be increased to ignore SLA violations.

Kumar et al. [15] have proposed a test of six different ways to predict the real-world tracking of web servers and cloud created. All analyzes were performed three times as three different tasks were used to measure the deviation of the predictions, i.e., a three-step guessing error, means a complete error, and a root means a double error.

Kumar et al. [16] compared the predictive accuracy of the various machine learning algorithms proposed to predict server load uploads. Line reversal is often the basic strategy used in statistical analysis where all the attributes included in the expectations are numeric.

3 Comparison Table

Author, year	Algorithm	Dataset	Programming language/tool	Advantage	Disadvantage
Saxena et al. (2020)	Auto-adaptive learning based	NASA and Saskatchewan HTTP traces	Python 3	Optimal prediction accuracy	Hardware dependence
Bi et al. (2018)	Integrated machine learning (SGW-SCN)	Google workload datasets	Savitzky-Golay filter	Better forecasting performance	It is never completely accurate
Kumar et al. (2020)	Self-directed learning based	HTTP weblogs from three different World Wide Web server	MATLAB R2017a	Provides more accurate weights of the network	It is never completely accurate
Nguyen et al. (2020)	Extreme learning machine and enhanced tug of war optimization	The Internet traffic (in megabyte (from EU) and in bytes (from the UK))	Opposition-based tug of war optimization (OTWO)	Good accuracy	Overfitting problem
Zhu et al. (2019)	Attention-based LSTM encoder–decoder network	Alibaba and dinda workload traces dataset	ARIMA, PSR + EA-GMDH	Controls prediction accuracy	Require large amounts of memory bandwidth

(continued)

(continued)

Author, year	Algorithm	Dataset	Programming language/tool	Advantage	Disadvantage
Jayakumar et al.	Self-optimized generic workload prediction framework (load dynamics)	Azure and LCG workloads	The inference and training were executed on a 16-core Intel Xeon platinum 8153 CPU	The prediction error is very low	Dependency on long short-term memory (LSTM)
Hu et al. (2016)	Elastic mechanism	CPU workload time series	Time series approach, Kalman filter model	Easy scalability	Data loss
Singh et al. (2018)	Technocrat ARIMA and SVR model	ClarkNet and NASA	MAE, MSE, RMSE, and MAPE	Improves resource provisioning oscillation	Workload is nonlinear
Qiu et al. (2016)	Deep learning	Workload datasets	EWMA model	Good performance	Require a large amount of data
Kumar et al. (2018)	Artificial neural network and adaptive differential evolution	NASA and Saskatchewan servers' HTTP traces	MATLAB	Fast evolution	Hardware dependence
Kumar et al. (2017)	Long short-term memory recurrent neural network (LSTM-RNN)	HTTP traces of, Calgary server, and Saskatchewan server	Python along with Keras library	Remembers information for a long time	Prone to overfitting
Lu et al. (2016)	Random variable learning rate back-propagation neural network (RVLBPNN)	Publicly available google workload traces	MATLAB 7.14	High accuracy	Less secure
Linma et al.	Query-based model	Three real-world database traces	MySQL and PostgreSQL	Improves system performance	Require more powerful hardware

4 Emerging Challenges

Cloud computing has placed many challenges in different aspects of data handling. Some of these challenges are:

(1) Decreased performance and overuse—Performance is an important factor when considering cloud-based solutions. If cloud performance is unsatisfactory, it can drive users away and reduce profits. Even a slight delay in loading an application or web page can lead to a significant decrease in the percentage of users.
(2) Adaptability—The word "adaptability" means the quality of being able to adjust to new conditions. The prediction model should be adaptable to behavior changes.
(3) Security and privacy—Data security is a major problem when switching to cloud computing. The information held by the user or organization is sensitive and confidential. Cloud security issues include identity theft, data breach, malware infection, and many more that ultimately reduce trust among users of your apps
(4) Workload fluctuation—Evaluate your workload and determine a pattern of factors that influence workload.
(5) Accurate workload prediction—Accuracy is a key factor in the forecasting of workloads and existing methods remain to produce 100% accurate results.

5 Conclusion

Providing a variety of remote services and services to its customers is a key objective of the cloud computing paradigm. Problems such as under-provision or over-provision could be caused by the error of predicting that the past reduces the lead to SLA violations and cloud performance, and this ultimately leads to the problem of resource wastage. Various load forecasting schemes are provided in the literature relating to the lack of accurate workload forecasting based on job history data and handling issues such as Slashdot results and workload fluctuations. This paper begins by introducing the basic concepts and challenges to the burden forecasting systems. After that, they passed the research on the proposed methods of forecasting the load and described their main contribution, and used an algorithm to make predictions. A comparison table is included in this paper to compare algorithms, data sets, etc.

References

1. Saxena D, Singh AK, et al (2020) Auto-adaptive learning-based workload forecasting in dynamic cloud environment
2. Lu Y, Panneerselvam J, Liu L, Wu Y, et al (2016) RVLBPNN: a workload forecasting model for smart cloud computing
3. Saxena D, Singh AK, et al (2020) A proactive auto-scaling and energy-efficient VM allocation framework using online multi-resource neural network for cloud data center
4. Kumar J, Singh AK, et al (2018) Workload prediction in cloud using artificial neural network and adaptive differential evolution

5. Kumar J, Saxena D, Singh AK, et al (2020) BiPhase adaptive learning-based neural network model for cloud datacenter workload forecasting
6. Bi J, Yuan H, Zhang L, Zhang J, et al (2018) SGW-SCN: an integrated machine learning approach for workload forecasting in geo-distributed cloud data centers
7. Kumar J, Singh AK, Buyya R, et al (2020) Self-directed learning-based workload forecasting model for cloud resource management
8. Nguyen T, Hoang B, Nyuyen G, Nguyen BM, et al (2020) A new workload prediction model using extreme learning machine and enhanced tug of war optimization
9. Zhu Y, Zhang W, Chen Y, Gao H, et al (2019) A novel approach to workload prediction using attention-based LSTM encoder-decoder network in cloud environment
10. Gao J, Wang H, Shen H, et al (2020) Machine learning-based workload prediction in cloud computing
11. Kumar K, Gangadhara Rao K, Bulla S, Venkateswarulu D, et al (2021) Forecasting of cloud computing services workload using machine learning
12. Singh P, Gupta P, Jyoti K, et al (2018) TASM: technocrat ARIMA and SVR model for workload prediction of web applications in the cloud
13. Chandy A, et al (2019) Smart resource usage prediction using cloud computing for massive data processing system
14. Saxena D, Singh AK, et al (2021) Workload forecasting and research management models based on machine learning for cloud computing environment
15. Jitendra Kumar, Ashutosh Kumar Singh, et al.Performance Assessment Of Time Series Forecasting Models For Cloud Datacenter Network Workload Prediction.2020
16. Kumar K, Gangadhara Rao K, Bulla S, Venkateswarulu D, et al (2021) Forecasting of cloud computing services workload using machine learing
17. Tomar A, et al (2020) Machine learning, advances in computing, renewable energy, and communication, Vol 768. Springer Nature, Berlin, LNEE. p 659. DOI: https://doi.org/10.1007/978-981-16-2354-7. ISBN 978-981-16-2354-7

Designing Architecture for Container-As-A-Service (CaaS) in Cloud Computing Environment: A Review

Anshita Malviya and Rajendra Kumar Dwivedi

Abstract Nowadays, every company is trying to become a tech company as we have seen that software is consuming the entire world. So, how the company approaches software development and serves their customers show how efficient their employees are or how agile their organization is. Therefore, it is of utmost importance for the companies to make correct decision about the architectural choices early in the software development life cycle. It has been observed that containers are taking place of virtual machines in the IT industry because they do not require to incorporate guest operating system in each instance like virtual machines, and they simply support the property and resources of host OS. This paper presents a survey on containers, and its architecture and how containers are better than virtual machines for software development.

Keywords Cloud computing · Containers · Virtual machines · Microservice architecture · Monolithic architecture · Container orchestration

1 Introduction

In the past decades, we have seen that global economy is changing and moving toward services rather than manufacturing. Cloud computing has a greatest advantage in the service industry, and it also benefits the business computing to a new standard. The on-demand distribution of IT assets over the Internet with pre-pay pricing is cloud computing. It is used to transform the IT industry into a utility. Lowered IT costs, improved agility, easy scalability, and cost effectiveness are some of the assets of cloud computing [1-32].

Cloud containers are packages of software which consists of all the dependencies like libraries needed to run software services and particular versions of programming language runtimes. It can run anywhere such as programmer's personal laptops,

A. Malviya (✉) · R. K. Dwivedi
Department of Information Technology and Computer Application, MMMUT Gorakhpur, Gorakhpur, India
e-mail: anshitamalviya2010@gmail.com

A. Tomar et al. (eds.), *Proceedings of 3rd International Conference on Machine Learning, Advances in Computing, Renewable Energy and Communication*, Lecture Notes in Electrical Engineering 915, https://doi.org/10.1007/978-981-19-2828-4_50

Table 1 Containers versus virtual machines

Containers	Virtual machines
It virtualizes operating systems	It virtualizes hardware
Many workload run on a single OS instance	Runs many OS instances
Lightweight than VMs	Heavyweight than containers

public cloud, or private datacenters as it virtualizes the operating system. Separation of responsibilities, workload portability, application isolation, and low overhead are some of the advantages of containerization.

Table 1 depicts the comparison between containers and virtual machines. There are different services in cloud computing but the one which is gaining popularity these days is the container-as-a-service (CaaS). It is a cloud-based service or can say is a cloud service model which helps the users to organize, manage containers, applications, and clusters. It also helps the IT departments and software developers to upload, run, and manage containers with the help of container-based virtualization or application programming interface (API) or Web portal interface. CaaS depends on the use of containers unlike PaaS. It is similar to the other cloud computing services in which the user can pay only for the resources they use. It is a subgroup of infrastructure-as-a-service. Portability, high efficient and cost cuttings, security, speed, scaling, and streamlined developments are some of the advantages of CaaS [30, 31].

The architecture of virtual machine and container is shown in Figs. 1 and 2. The architecture which helps to package the software and its dependencies in an isolated component called container to run consistently in any environment is known as containerized architecture. Unlike the traditional software deployment, containers are easily portable which helps the software to move from one environment to the other without errors and incompatibilities. In traditional virtualized architecture, containers are like virtual machines but they are lightweight and are faster to setup. We have seen that to manage microservice applications containerization is really helpful. There are some building blocks of containerized applications such as Container Engines, Container Orchestrators, and managed Kubernetes Services. The benefits of containerized architecture are lower costs, scalability, instant replication, flexible routing, resilience, full portability, OS independent, lightweight, and fast deployment.

The traditional way of developing the applications is the monolithic architecture. The application made by monolithic architecture is built as a single and indivisible component. It consists of a large codebase and lacks modularity. If we want to update or change something, the entire codebase has to be accessed, thus have to make updation on the whole stack at once.

Less cross-cutting concerns, easier debugging, simple to deploy and test, simple to develop are some of the strengths of monolithic architecture. The major weakness of it is to make changes. This drawback is overcome by the microservice architecture which builds the applications as a collection of smaller independent units. Each unit performs a separate service. Independent components, easier understanding,

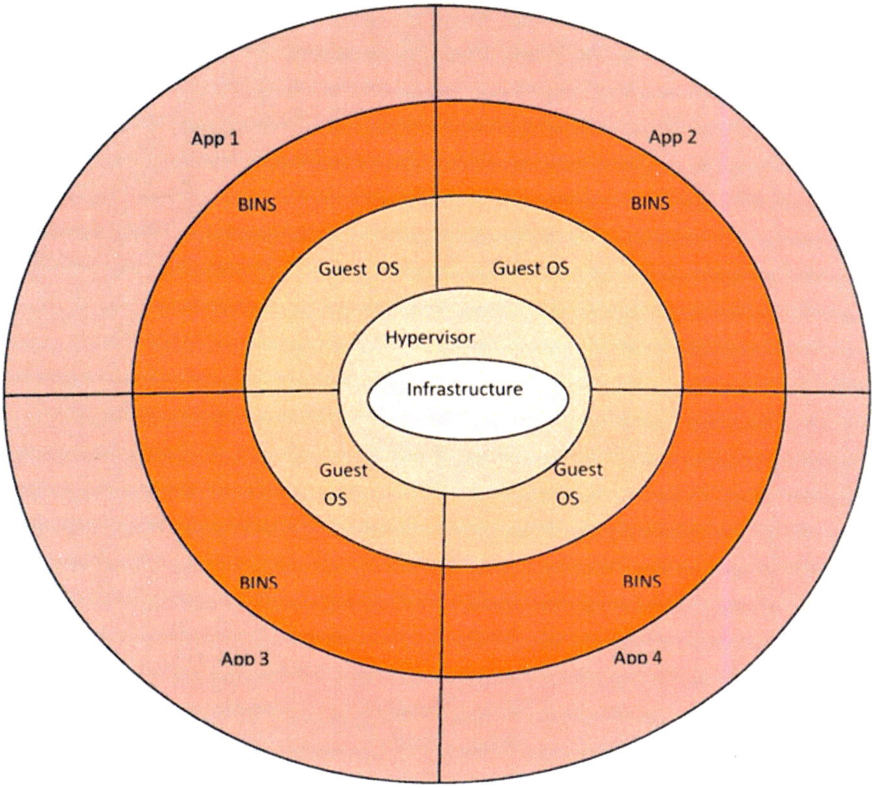

Fig. 1 Architecture of virtual machine

better scalability, flexibility in choosing the technology, higher level of agility are the advantages of microservice architecture.

Figure 3 presents the lifecycle management of a single container. The lifecycle of container consists of four elements, namely driver which is used to map container to kernel specific control, runtime engine which is used for lifecycle and resource control and state monitoring, repository which is used to version control like interface and also external service and container image which is used to store archive of file system tree.

The benefits of deploying applications through containers instead of virtual machine are that they are lightweight, flexible, and several applications can run on a single host. The demand of this technology is increasing day by day as it provides lower system costs. The main significance of this work is to give an overall understanding of this technology to the beginners for exploring and making this field as their research area.

The rest of the paper is organized as follows. Types of container are explained in Sect. 2. Section 3 describes container orchestration. Section 4 discusses about some important issues in designing containers. Section 5 presents a review on container

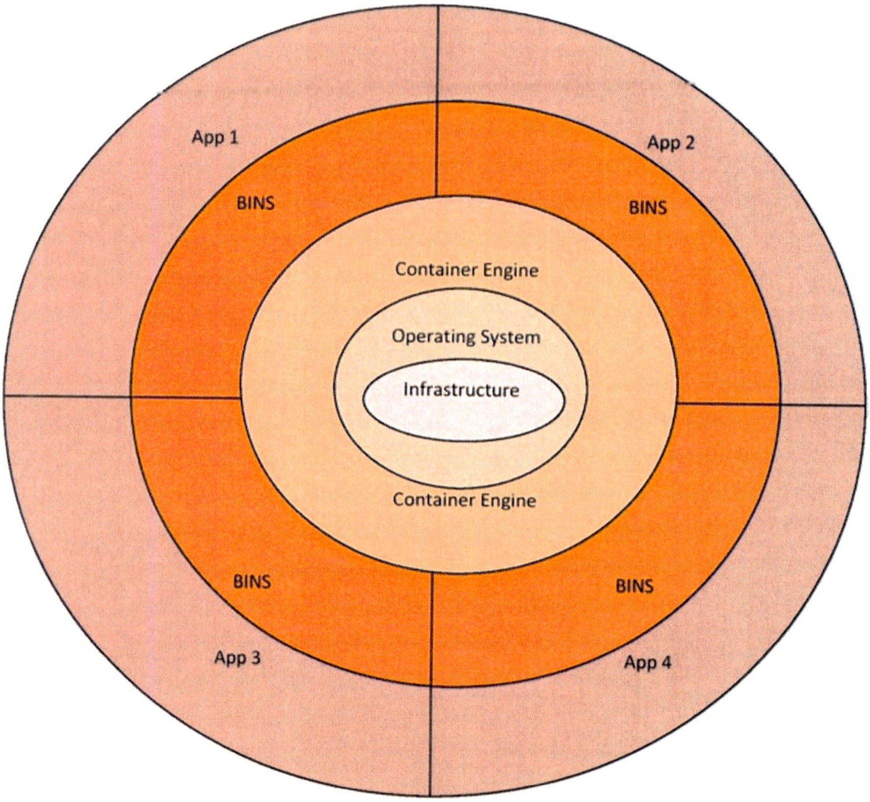

Fig. 2 Architecture of container

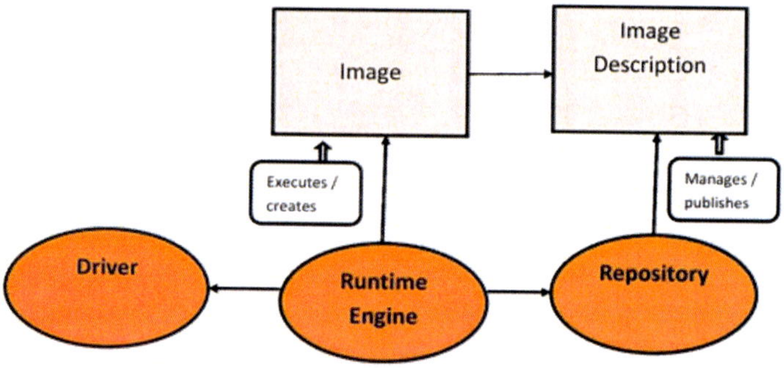

Fig. 3 Lifecycle of container

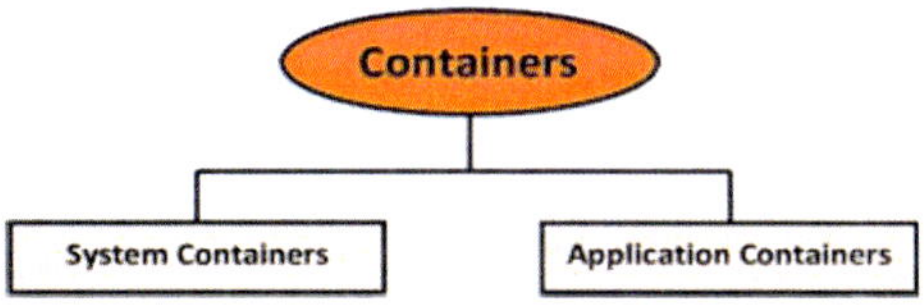

Fig. 4 Container types

architecture. Section 6 provides the major findings of the survey. Section 7 concludes this survey.

2 Types of Containers

Figure 4 shows that containers are of two types—system containers and application containers. System containers are similar to virtual machines as they are the oldest types of containers. They are used to run multiple processes. System containers are stateful and used for the development of monolithic applications. They are used to organize tools, architectures, and configurations implemented for virtual machines. The different applications of system containers are LXC/LXD, OpenVZ, BSD jails, etc. LXC is an open-source project which is used to provide independent application environment similar to virtual machines. It does not involve the overhead of running their own kernel. Unix process model is followed by LXC. OpenVZ permits the servers to operate many independent OS instances known as virtual environments. It is similar to LXC.

Application containers are new type of containers. They operate only a single process. They are stateless and horizontally scalable. Rigid and transient infrastructures are suitable for application containers. Different implementations of application containers are Dockers, CRI-O, etc. The most popular container platform is Docker. The usage of Linux containers is enabled by Docker. It is a tool used for the creation, deployment, and operation of software applications by containers. It has been embraced by almost all the IT industries. CRI-O is an implementation of Kubernetes and is an open-source tool. The major aim of CRI-O is to take the place of Docker. It is used to support runC and Kata containers.

3 Container Orchestration

It is the process of programming and organizing automatically individual containers for application development which uses microservice architecture. The orchestration of containers becomes necessary when around 900 containers and 500 applications are working. The container's lifecycle is managed by it. Resource allocation among containers, removing containers, allocation of resources among containers, balancing of load, scaling, availability and redundancy of containers, monitoring containers, and its hosts are all controlled, managed, and automated by container orchestration.

Kubernetes and Docker Swarm are the tools for container orchestration. These tools get to know about how to manage between containers and storing of logs with the help of configuration files. The deployment of containers and determining their hosts are done by these tools. When the hosts of the containers are determined, then these tools use to manage the lifecycle of the containers according to already known specifications. Any environment that runs containers is suitable for these tools for working. Docker machine, Docker swarm, and Docker compose are the orchestration tools for Docker whereas failed containers automatic rescheduling, load balancing, exposure of network ports to systems, automatic deployment, replication of containers, and online scale are the tools for Kubernetes.

4 Important Issues in Designing Container

Lack of education and experience is the most common matter influencing the orchestration and deployment of containers. This issue has raised because till now people do not known much about container technology. They need significant training on this technology, know about its advantages and disadvantages, and how containers are taking place of virtual machines in the IT industries. We have dearth of trained Docker, Kubernetes, and other tool professionals. We do not have experienced engineers in this technology.

As we know that world is changing fast, so it is expected from people to learn about this technology and reform the problem. Around the application development, there is a major need to automate tooling. The tools used for container orchestration and containerization are still in the state of development which is the lacking feature for this technology. There is a need of tenacious storage and networking in this area. When data move from one site to another, it has been seen that organizations have the inability to store it. It gets complicated when various containers interact with one another. Scalability is also a matter of concern in this technology. We need to focus more on the organization of the environment for the security of the container image.

Security is also one of the burning issues during the entire lifecycle of containers and container orchestration systems. We have seen that many developments have been made in this system, but security controls are straggling behind. The potential susceptibility of the container platforms should be completely understood by the DevOps team working on it. It has been observed that orchestration platforms are a considerable surface for attacks. The problem of security arises because of the highly zestful nature of orchestration systems. The containers should have its own machine identity to solve this problem. These are some of the major issues in the design and development of container technology.

5 Literature Survey

This section discusses a detailed review on containers and containerized architecture, advantages of containers over virtual machines, and different performance parameters. It has been observed that since few years' researchers have shown their great interest in this area.

Singh et al. [1] showed that any application made using microservice approach and deployed using Docker containers lowers the time and effort for deployment and continuous integration of the application. They have also compared the development and performance of applications using microservice and monolithic approaches and concluded that microservice application with less response time and higher throughput, and also microservice-based applications are used to achieve higher scalability because of its granular design. Jmeter was used by them for comparing both the approaches. HAProxy, Consul, Jenkins, GIT Repository, and Docker Registry are the tools used. They have concluded that containers are the best catalyst in comparison with VMs for the applications developed using microservice approach.

Hussain et al. [2] proposed a placement architecture for containers on virtual machine and at the same time taking into account the placement of VMs on the physical machines (PMs). Best fit, max fit, and ant colony optimization algorithms were used for the placement of containers on VMs as well as VMs on PMs. They concluded that their proposed architecture is a multitier scheduler that targeted in maximizing the utilization of active PMs and instantiated VMs while minimizing the number of active VMs and PMs.

Kang et al. [3] have explored different opportunities and challenges by the use of containers and considering the microservice design approach for managing and operating cloud infrastructure services. They discussed that operational efficiency improved when they took OpenStack as a case study while containerizing cloud infrastructure services and combined with the microservice approach architecture. They also found that to access the entire benefits of containerization and microservice approach, it is not sufficient to simply run the container services.

Pahl et al. [4] have conducted a survey of 46 chosen studies. They found that the management of container clusters is necessary, and the careful organization of the construction and deployment is a central concern. They revealed that container technologies research is still in a formative stage. They showed that there is a lack of tool support to imbrute and facilitate the management and orchestration in clustered cloud architectures. They observed that there is an impact of containers on both development and deployment aspects positively. They demonstrated that containers have better resource efficiency than VMs.

Casalicchio [5] discussed that the container technology is being adopted for internal usage as well as for commercial offering. The author told that the cloud technology is still not in the maturity stage, and many more researches are need to be performed. The author has discussed about the research problems in automatic container orchestration which includes monitoring synchronization between design and execution environments, performance modeling, resource management at run time and many more.

Durairajan et al. [6] discussed various architectures used to build applications such as traditional IT architecture (3 tier monolithic model), client-server model, service-oriented architecture (SOA), and the micro service architecture (MSA). The authors concluded that the corporations should employ loosely coupled microservices architecture for multiple smaller applications which helps in avoiding the complexity and interdependency. They also told that when an application is built using microservice architecture, it involves a lot of design changes, and the things are done in diverse ways.

Arango et al. [7] presented the comparison of the performance of container-based virtualization tools such as Docker, LXC, and singularity against bare metal. The authors concluded many results such as for HPC implementation, singularity containers are more efficient than Docker or LXC, LXC containers are efficient in network point of view, and many more.

Amaral et al. [8] discussed several advantages of containers and server virtualization in industries. The authors explored the related processes per container (RPPC) and found that it is suitable for microservice implementations. They conducted experiments for performance evaluation between two models for implementing RPPC and found that nested container approach is efficient than master slave model.

Espe et al. [9] evaluated two frequently accessed runtimes contained (industry standard) and CRI-O on runC and gVisor which are open container initiative runtimes. The evaluation tool used is touchstone was developed. The authors concluded runC as the best open container initiative runtimes for the starting option in any use-case.

Zheng et al. [10] discussed several advantages and disadvantages of microservice approach. They told that the most eminent advantage of microservice architecture is that tangled applications can be divided into series of independent service units which can be developed, tested, deployed, and expanded easily. The authors also discussed the advantages of traditional single architecture model but concluded that microservice architecture is more efficient.

Bernstein [11] told that container technology is an eye catching and emerging technology in cloud computing infrastructure. The authors discussed about Docker containers, abstractions on top of VMs and containers, and also about open-source cluster manager for Docker containers.

Namiot et al. [12] presented an overview on microservice architecture. The authors have discussed about the benefits and drawbacks of this approach. One of the major drawbacks of this approach is that for developing any distributed system, it requires additional complexity, but this architectural model is advantageous over monolithic architectural model. They have also discussed about the usage of microservice architecture in M2M applications.

Yu et al. [13] proposed a reference architecture model based on microservice. Their model consists of basic architectural components and elementary units. They suggested IT professionals to study their reference architectural model so it could help them developing IT transformation architecture solutions. They believed that their model would help the IT industries to use their microservice relevant technologies efficiently without confusion.

Leitner et al. [14] proposed a CostHat network model which is useful for cost sensitivity analysis. This model is used for microservice-based applications as well as for the applications that are executed on top of AWS Lambda services. The authors concluded that their proposed model warns the developers about costly code changes.

Khazaei et al. [14] developed an analytical model (microservice platform) for providing analysis of the performance of microservice platforms. The authors studied different performance indicators during experiments such as cluster size, utilization, rejection probability, and response time. They concluded that at large-scale provisioning performance of microservice platform can be studied.

Piraghaj et al. [15] discussed about the challenges faced by cloud providers. One of the major issues is to minimize power consumption of data centers. The authors also discussed about the gaining popularity of containers. They proposed a framework for consolidation on virtual machines and modeled the power optimization problem in CaaS environment. They conducted four times the simulation experiment and evaluated energy consumption of data center for selecting containers for migration, selecting destinations, and triggering migrations using correlation-aware placement algorithm and other placement algorithm.

Pahl [16] told about containerization which is a lightweight virtualization response and applicability of containers for platform-as-a-service clouds. The author also discussed about different container models and the need to use applications through distributed multicloud platforms.

Felter et al. [17] explored the performance comparison of both virtual machine and Linux containers by using hypervisor KVM and container manager Docker. The results showed that containers outperform the virtual machines in nearly all the cases. They discussed about the implementation of their results in future cloud architectures.

Taibi et al. [18] discussed about the decomposition process which is used to migrate a system from monolithic to microservice. They reduced the decomposition process by proposing a 6 step framework. They also validated their process by using industrial project. They told that their proposed framework would help the companies to solve issues in software development.

Jaramillo et al. [19] described about the microservice architecture which are used to lessen the time of application development. The authors also described about the Docker which is a troublesome technology in application development and its distribution but it is very useful and efficient while implementing microservice architecture.

Barik et al. [20] told about cloud computing and its variety of services which are efficient to overcome several problems. The authors discussed about network virtualization and their impact on virtual machines. They explored about the hosting of software applications using containers in cloud environments. They did the performance comparison of containers and virtual machines on various parameters and found that containers overcome the virtual machines but they have some disadvantages in the area of security.

Naik [21] presented the software development process on multiple clouds by using simulation of constructing a virtual system of systems which is based on Docker swarm, VirtualBox, Mac operating system X, and many more. This virtual system of systems can also be created by Amazon Web services, OpenStack, Microsoft Azure, etc. The authors concluded that such type of system could be constructed by using any cloud platform but require a reliable account on those clouds with valid subscription.

Silva et al. [22] presented an overview of container technology which is suitable for virtualization. They told that container technology dynamics have not been popular yet. They also mentioned about two types of technology used in virtualization of hardware which are container and hypervisor. They conducted an organized and systematic literature review on container technology and its future trends in research.

Han et al. [25] proposed control architecture to access container image and discussed about its significance. The authors told that the protection of container image could be done easily without any manipulation done by the uses. They concluded that the security of the container platform can be achieved by their proposed architecture.

Aksakalli et al. [26] discussed about the application deployment approaches and microservice communication platforms. They told that these two are very essential for the applications which are based on the microservice architecture.

Liu et al. [27] presented an overview of microservice architecture for development of applications. They also compared both monolithic and microservice architecture and also told about various challenges of microservice architecture.

Zhou et al. [28] told about the significance of containerization technology and its adoption in high-performance computing technology. They proposed a hybrid architecture consisting of both HPC cluster and cloud cluster and described the implementation of a platform named torque operator.

Vayghan et al. [29] discussed about the microservice architecture and container orchestration platform. They proposed HA state controller which works with Kubernetes and helps in replication of the state of application.

6 Major Findings of the Survey

On the basis of the survey, we have seen that many researchers concluded that microservice architecture is better than monolithic architecture for the development of applications. Container technology is the emerging technology of present time. In terms of resource efficiency, CPU usage and scalability containers are better than virtual machines. Table 2 shows various comparison of researches done on containers and its architecture.

It has been found from the survey that the coming generation would focus mostly on application virtualization rather than hardware virtualization in cloud computing. We have also found that the major issues in designing and development of containers

Table 2 Comparison of various researches on container architecture

References	Author(s)	Tools used	Architecture proposed	Comparisons of performance of containers and virtual machines	Remarks
[1]	Singh and Peddoju (2017)	HAProxy, Consul, Jenkins, GIT Repository, and Docker registry	Microservice architecture	Yes	Microservice approach is efficient
[2]	Hussain et al. (2019)	MATLAB	Multitier scheduler	No	In terms of VM and PM utilization, ACO is effective
[3]	Kang et al. (2016)	Implemented on Linux kernel	Microservice architecture	Yes	If research problems solved, container technology would be greatly adopted in cloud infrastructure management
[4]	Pahl et al. (2016)	No tool used	None	Yes	Containers have better resource efficiency than virtual machines
[5]	Casalicchio (2010)	No tool used	None	Yes	Discussed automatic container orchestration problem
[6]	Durairajan et al. [6]	Lamp	Microservice architecture	No	Application built using microservice architecture involves a lot of design issues

(continued)

Table 2 (continued)

References	Author(s)	Tools used	Architecture proposed	Comparisons of performance of containers and virtual machines	Remarks
[7]	Arango et al. (2017)	NAMD-benchmark	None	No	Performance comparison of Docker, LXC, and singularity against bare metal
[8]	Amaral et al. (2015)	CPU-benchmarking tool	Microservice architecture	Yes	Nested container model is more efficient than master slave model for implementing microservice architecture
[9]	Espe et al. (2020)	Touchstone	None	Yes	Containers are better in terms of CPU usage, memory latency, and scalability aspects
[10]	Zheng and Wei (2018)	Play Web framework, Angular.js, Netty servers	Microservice architecture	No	Microservice architecture is suitable for project development
[11]	Bernstein (2014)	No tool used	No architecture proposed	Yes	Cloud container is the emerging technology of present time
[12]	Namiot and Sneps-Sneppe (2014)	No tool used	Microservice architecture	No	Overview on microservice architecture

(continued)

Table 2 (continued)

References	Author(s)	Tools used	Architecture proposed	Comparisons of performance of containers and virtual machines	Remarks
[13]	Yu et al. (2016)	No tool used	Microservice-based reference architectural model	No	Proposed reference architectural model for IT enterprise architecture
[14]	Leitner et al. (2016)	Eclipse IDE	CostHat model based on microservice architecture	No	Proposed model is used for cost sensitivity analysis
[14]	Khazaei et al. (2016)	Amazon cloud	Microservice architecture	No	Performance of microservice platform is studied

are the lack of knowledge about container technology among people and the security control.

7 Conclusions and Future Directions

A brief description about containers and virtual machines is presented in this paper. Furthermore, container-as-a-service and containerized architecture have been discussed. We found that containers have taken the place of virtual machines in the software industries. Microservice architecture is more efficient than monolithic architecture for developing applications.

As future work, we could do the survey of more research papers on container architectures. We have also seen that there is a need of development of performance models and energy consumption models for container systems and adaptation models for container orchestration. There is a need of well-organized device sharing among containers.

References

1. Singh V, Peddoju S (2019) Container-based microservice architecture for cloud applications. In: International conference on computing, communication and automation. IEEE
2. Hussain M, Mousa M, Alqarni M (2019) A placement architecture for a container as a service (CaaS) in a cloud environment. In: Journal of cloud computing: advances, systems and applications. Springer
3. Kang H, Le M, Tao S (2016) Container and microservice driven design for cloud infrastructure DevOps. In: International conference on cloud engineering. IEEE
4. Pahl C, Brogi A, Soldani J, Jamshidi P, Cloud container: a state-of-art review. IEEE
5. Casalicchio E, Automatic orchestration of containers: problem definition and research challenges. ACM
6. Durairajan S, Vinod V, Cloud Micro Services Architectures for portable workloads using Container Technologies and Standards. In: International journal of applied engineering research, Vol 12. ISSN 0973-4562
7. Arango C, Dernat R, Sanabria J (2017) Performance evaluation of container based virtualization for high performance computing environments
8. Amaral M, Polo J, Carrera D, Mohomed I, Unuvar M, Steinder M (2015) Performance evaluation of microservice architectures using containers. In: 14th international symposium on network computing and applications. IEEE
9. Espe L, Jindal A, Podolskiy V, Gerndt M (2020) Performance evaluation of container runtimes. In: 10th international conference on cloud computing and services science. SCITEPRESS CLOSER
10. Zheng L, Wei B (2018) Application of microservice architecture in Cloud environment project development. EDP Sciences MATEC web of conferences 189, 03023-MEAMT 2018
11. Bernstein D (2014) Containers and cloud: from LXC to docker to kubernets. Cloud Comput
12. Namiot D, Sneps-Sneppe M (2014) On microservices architecture. Int J Open Inf Technol 2(9). ISSN: 2307-8162
13. Yu Y, Silveria H, Sundaram M (2016) A microservice based reference architecture model in the context of enterprise architecture. In: Advanced information management communicates, electronic and automation control conference (IMCEC). IEEE
14. Leitner P, Cito J, Stockli E (2016) Modelling and managing deployment costs of microservice based cloud applications. In: 9th international conference on utility and cloud computing. IEEE/ACM
15. Khazaei H, Barna C, Beigi-Mohammadi N, Litoiu M (2016) Efficiency analysis of provisioning microservices. In: 8th international conference on cloud computing technology and science. IEEE
16. Piraghaj SF, Dastjerdi AV, Calheiros RN, Buyya R (2015) A framework and algorithm for energy efficient container consolidation in cloud data centers. Res Gate
17. Pahl C (2015) Containerization and PaaS cloud. In: Cloud computing magazine IN PRESS—-accepted for publication, IEEE 6 May 2015
18. Felter W, Ferreira A, Rajamony R, Rubio J (2015) An updated performance comparison of virtual machines and Linux containers. In: 978-1-4799-1957-4/15/$31.00. IEEE
19. Taibi D, Systa K (2019) From monolithic systems to microservices: a decomposition framework based on process mining. In: 9th international conference on cloud computing and services science (CLOSER 2019), pp 153–164. ISBN: 978-989-758-365-0. SCITEPRESS – Science and Technology Publications, Lda
20. Jaramillo D, Smart R, Nguyen DV (2016) Leveraging microservices architecture by using Docker technology. In: 978-1-5090-2246-5/16/$31.00. IEEE
21. Barik RK, Lenka RK, Rao KR, Ghose D (2016) Performance analysis of virtual machines and containers in cloud computing. In: International conference on computing, communication and automation. IEEE
22. Naink N (2016) Building a virtual system of systems using docker swarm in multiple clouds. In: 978-1-5090-0793-6/16/$31.00. Crown

23. Silva VG, Kirikova M, Alksnis G (2018) Containers for virtualization: an overview. J Appl Comp Syst 23(1):21–27. doi:https://doi.org/10.2478/acss-2018-0003
24. Ernst D, Bermbach D, Tai S (2016) Understanding the container ecosystem: a taxonomy of building blocks for container lifecycle and cluster management. https://ise.tu-berlin.de/fileadmin/fg308/publications/2016/Ernst_woC_container-taxonomy
25. Bellishree P, Deepamala N (2020) A survey on docker containers and its use cases. Int Res J Eng Technol (IRJET) 07(07). www.irjet.net. ISSN: 2395-0072, e-ISSN: 2395-0056
26. Han S, Lee H, Lee S, Kim S, Jan W (2020) Container image access control architecture to protect applications. IEEE Access 8:162012–162021. doi:10.1109
27. Aksakalli IK, Celik T, Can AB, Tekinerdogan B (2021) Deployment and communication patterns in microservice architecture: a systematic literature review. J Syst Softw 180:1–25. https://doi.org/10.1016/j.jss2021.111014
28. Liu G, Huang B, Liang Z, Qin M, Zhou H, Li Z (2020) Microservices: architecture, container, and challenges. In: 2020 IEEE 20th international conference on software quality, reliability and security companion (QRS-C), pp 629–635. doi:https://doi.org/10.1109/QRS-C51114.2020.0017.ISBN: 978-1-7281-8915-4
29. Zhou N, Georgiou Y, Pospieszny M, Zhong L, Zhou H, Niethammer C, Pejak B, Marko O (2021) Container orchestration on HPC systems through kubernetes. J Cloud Comput Adv Syst Appl 10:16. https://doi.org/10.1186/s13677-021-00231-z
30. Vayghan LA, Saied MA, Toeroe M, Khendek F (2021) A kubernetes controller for managing the availability of elastic microservice based stateful applications. J Syst Softw 175:110924. https://doi.org/10.1016/j.jss.2021.110924
31. Dwivedi RK (2012) From grid computing to cloud computing and security issues in cloud computing. Technia-Int J Comput Sci Commun Technol IJCSCT 5(1):805–809. ISSN: 0974-3375
32. Tomar A, et al (2020) Machine learning, advances in computing, renewable energy and communication, Vol 768. Springer Nature, Berlin, LNEE, p 659. doi:https://doi.org/10.1007/978-981-16-2354-7. ISBN 978-981-16-2354-7

Performance Analysis of Distributed Power Flow Controller Employing Genetic-Based Fuzzy Logic Controller in Photovoltaic-Wind Hybrid System

V. Sowmya Sree and C. Srinivasa Rao

Abstract The deployment of hybrid non-conventional energy sources has become inevitable due to decrease in natural resources, global warming, growing energy costs, and expanding electrical consumption are all factors to consider. Because solar and wind energy are both free and environmentally beneficial, they are regarded as the finest options for remote (or rural) electricity. The combination of solar power and wind power is a reliable source of energy creating a constant energy flow by avoiding the fluctuations. But this hybrid system gives rise to complications related to power system stability. Most of the industrial loads are controlled by power electronic converters that are sensitive to power system disturbances. Hence, the power quality issues diminution is more focused in recent times as it is vital in power supply industry. A number of FACTS devices such as power semiconductor devices are developed to overcome the above power quality issues. Distributed power flow controller (DPFC), which is emerged from unified power flow controller, is considered as the best reliable device among the others. The DC link is the key distinction between these devices. In the case of DPFC, the DC connection that links both converters does not exist. This paper presents the performance analysis of DPFC when operating with and without genetic algorithm-based fuzzy logic controller, and the simulation results are validated using MATLAB/Simulink software.

Keywords Distributed power flow controller (DPFC) · Fuzzy logic controller (FLC) · Genetic fuzzy logic controller (GFLC) · Power quality · Solar and wind system

---

V. S. Sree (✉)
Research Scholar, JNTUA, Ananthapuramu, Andhra Pradesh, India
e-mail: sowmya.sree14@gmail.com

Assistant Professor, GPCET, Kurnool, Andhra Pradesh 518452, India

C. S. Rao
GPCET, Kurnool, Andhra Pradesh 518452, India

A. Tomar et al. (eds.), *Proceedings of 3rd International Conference on Machine Learning, Advances in Computing, Renewable Energy and Communication*, Lecture Notes in Electrical Engineering 915, https://doi.org/10.1007/978-981-19-2828-4_51

1 Introduction

Nowadays, the renewable energy-based power systems play a major role replacing the conventional sources-based power systems. The key use of renewable energy resources is that they are environment-friendly and does not produce any greenhouse gases during generation of electricity. Solar energy and wind energy are considered to be rapid developing sources globally. Solar power is proportional to solar irradiation, whereas wind power is proportional to wind speed. But the photovoltaic and wind energy sources create frequency and voltage oscillations in power system due to its intermittent nature [1–4]. Therefore, the renewable energy sources inject a fluctuating power into the grid.

Power quality is the major problem faced by an integrated microgrid system [5, 6]. The different types of power quality issues in a microgrid are caused by sudden changes in load, under voltage load shedding, sudden loss of resource, faults at load side, and during islanding. The major source of poor power quality in the network is voltage unbalance [4, 7]. The use of FACTS controllers is considered to be the best solution for power system stability problems especially steady-state control [6, 8]. The expansion of FACTS devices started with increasing capabilities of power electronic components [9–11]. Figure 1 shows the classification of power quality issues. In this work, the authors projected a new device called distributed power flow controller (DPFC), an enhanced version of UPFC for improving operation of power system network [12]. The absence of DC link between the converters is the major difference between DPFC and UPFC. In DPFC, the third frequency harmonic component acts as link between both converters for dynamic power trading [13].

In this view, the projected dynamic system is presented in Sect. 2. Section 3 deals with various control strategies of distributed power flow controller (DPFC), involving FLC and genetic-based FLC. The detailed presentation of simulation is presented in Sects. 4 and 5 is where the work's conclusions are provided.

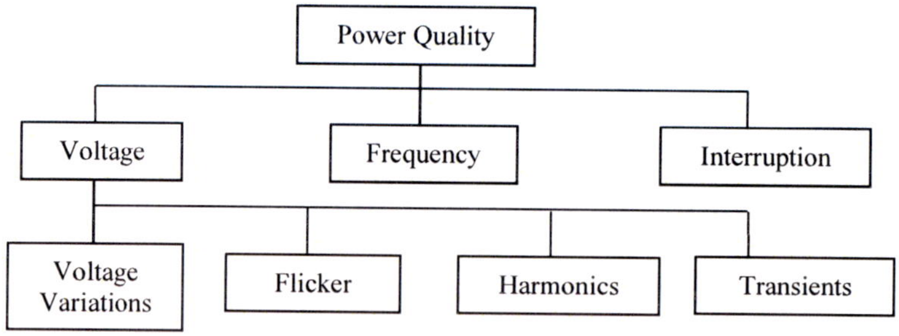

Fig. 1 Classification of issues in power quality

2 Proposed Dynamic System

Figure 2 shows the suggested dynamic system block diagram. Whenever a problem occurs at the PCC, it affects both output and grid power. So, a special device called DPFC is utilized to improve power quality concerns under load [8, 14–17]. In this study, DPFC and genetic-based fuzzy logic controllers are employed to regulate converters.

2.1 Concept of Wind Energy Conversion System

The essential components of a wind energy alteration system are the turbine, permanent magnet synchronous generator, and power electronic converter [2].

2.1.1 Mathematical Design of Wind Turbine

Wind turbine does the conversion of kinetic energy obtained from wind into mechanical power [9]. The power of wind turbine is expressed as

$$P_m = \frac{1}{2}\rho C_p A_r V_w^3 \tag{1}$$

$$C_p(\lambda, \beta) = 0.73\left(\frac{151}{\lambda_i} - 0.58\beta - 0.002\beta^{2.14} - 13.2\right)e^{\frac{18.4}{\lambda_i}} \tag{2}$$

$$\text{Where } \lambda_i = \frac{1}{\frac{1}{\lambda - 0.02\beta} - \frac{0.003}{\beta^3 + 1}} \tag{3}$$

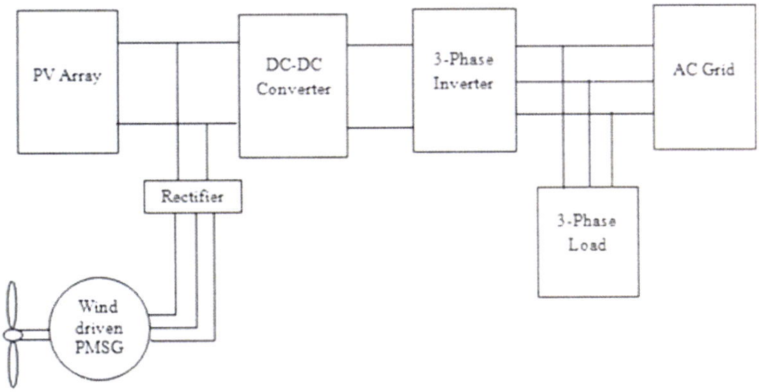

Fig. 2 Photovoltaic-wind hybrid system block diagram

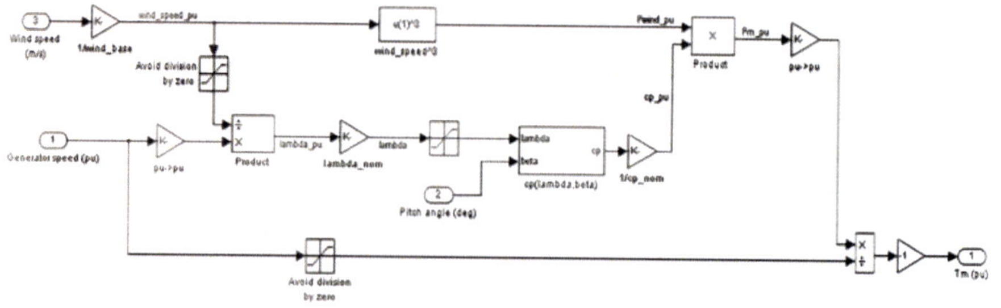

Fig. 3 MATLAB/Simulink model of wind turbine

$$\text{and TSR}(\lambda) = \frac{\omega_r R_r}{V_w} \tag{4}$$

Cp is assumed to be 0.59 according to Betz's Law, and the rotor pitch angle is considered to remain constant. Performance coefficients (Cp) vary from 0.2 to 0.4 in practise. Figure 3 shows the wind turbine system's MATLAB/Simulink diagram.

2.1.2 Permanent Magnet Synchronous Generator Modeling

In order to calculate the dynamic model of the PMSG with quadrature of 90° between the d-axis and q-axis with regard to the direction of rotation, the 2-phase synchronous reference frame is utilized, as seen in Fig. 4. Equations (5–7) represent the dynamics of PMSG in a frame of synchronous reference. Assume reference frame speed w.r.t. generator as ωe [9].

$$V_{gd} = R_{sg} i_{gd} + L_{sg} \frac{di_{gd}}{dt} - \omega_e L_{sg} i_{gq} \tag{5}$$

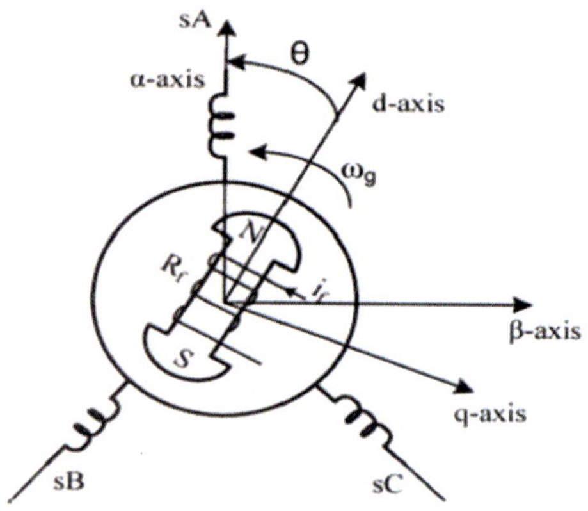

Fig. 4 d-q axes of synchronous machine

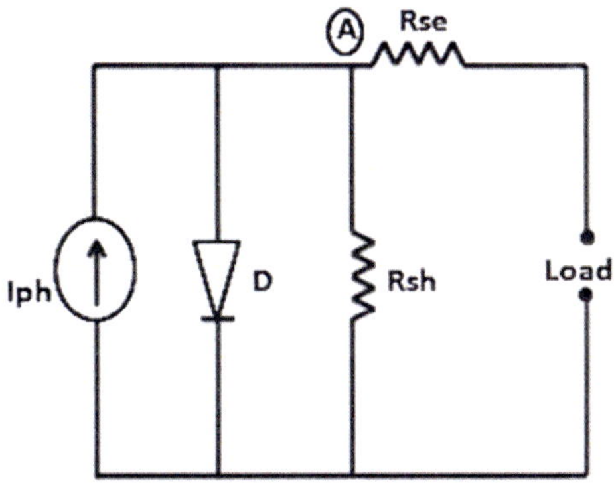

Fig. 5 Equivalent circuit diagram of solar PV module

$$V_{gq} = R_{sq} i_{gq} + L_{sg} \frac{di_{gq}}{dt} + \omega_e (L_{sg} i_{gd} + \lambda_m) \tag{6}$$

Electromagnetic torque is given as

$$T_e = \frac{3}{2} \frac{p}{2} \lambda_m i_{gq} \tag{7}$$

2.2 *Concept of Solar-PV System*

The analogous circuit of a solar PV module is seen in Fig. 5. It transforms solar energy into electrical energy. A typical PV cell is shown through a current source coupled parallely to a diode [2]. From Kirchhoff's current law,

$$I_{ph} = I_d + I_{RP} + I \tag{8}$$

$$I = I_{ph} - (I_{RP} + I_d) \tag{9}$$

2.3 *Distributed Power Flow Controller*

DPFC consists of shunt and distributed static series compensators for compensating active-reactive powers provided negative and zero sequence components of currents as well [12]. Shunt and series converters can be placed independently in UPFC with higher flexibility due to the absence of DC link. In DPFC, the use of multiple 1-phase converters in contrary to single 3-phase converter of large size reduces the

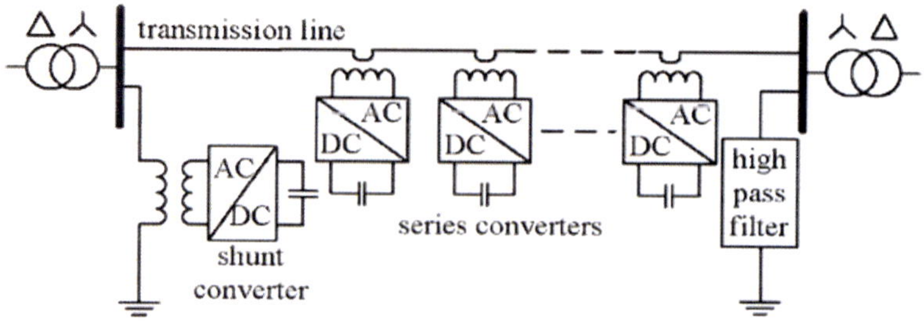

Fig. 6 Internal circuit of distributed power flow controller

components rating and provides high reliability because of redundancy [13]. Figure 6 shows the internal circuit diagram of DPFC [18, 19]. The converter control scheme is employed at the respective terminals of the converter and is called local controllers. The centralized control coordinates the individual converter controllers in addition to the control of various parameters of DPFC. DPFC uses central, series, and shunt controllers. The central controller is responsible for signaling other controllers. Series and shunt controllers compensate current and voltage harmonics [20]. DPFC management solutions have a significant role in decreasing power quality concerns which include harmonics, sag, and swell. The controllers are to be designed such that they not only compensate for voltage and harmonics but also capable of detecting and investigating faults occurring in the system. In addition, they also provide inverter gate pulses and draw negative harmonic currents.

3 Control Strategies of DPFC

In this paper, DPFC that incorporates fuzzy logic and genetic-based fuzzy logic controllers is installed to reduce problems such as harmonics, sags, and swells. The performance of the system incorporating DPFC is observed using MATLAB/Simulink software.

3.1 DPFC with Fuzzy Logic Controller

The FLC is dependent on theory which involves fuzzy set and human cognitive processing. Figure 7 depicts the FLC structure with three core blocks: fuzzification, rule-base interfacing, and defuzzification. Modeling FLC uses error and variation in error signals [6, 10]. Seven fuzzy sets are investigated in this report, namely NGL, NGM, NGS, ZER, PSS, PSM, and PLG. For simplicity, FLC uses triangle membership functions (Fig. 8). Defuzzification is conducted using the centroid approach. Mamdani fuzzy interface is employed for fuzzy logic blocks arrangement providing

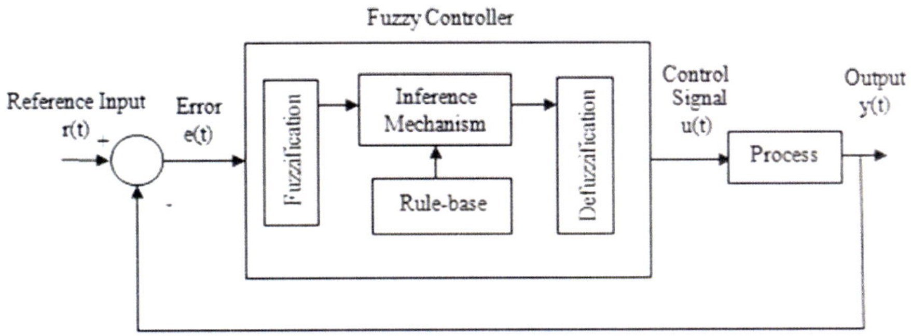

Fig. 7 Schematic diagram of FLC

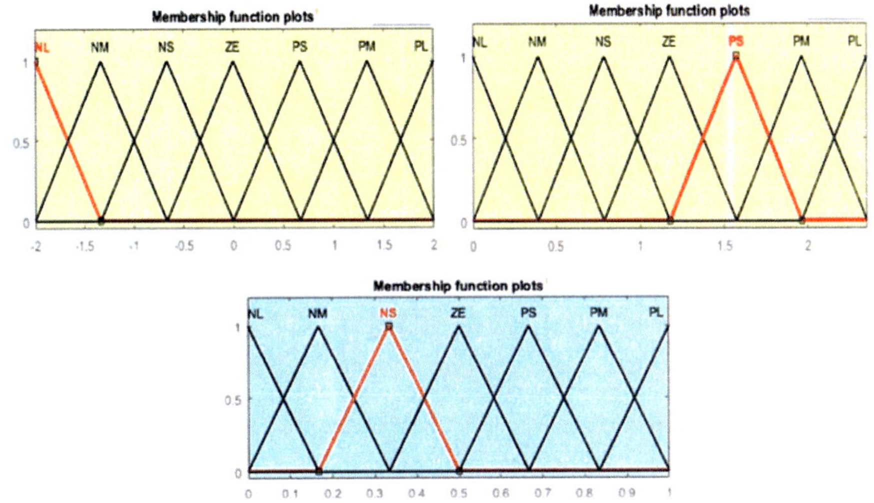

Fig. 8 Input and output membership functions

the system output to reach desired value. The voltage is evaluated to a fixed value in DPFC, and difference is given to FLC to regulate the shunt controller [11]. Inverter pulses are generated by comparing the reference value to the filter current. The fuzzy control rule design integrates input variables to output model attributes. Table 1 shows how to create FLC.

3.2 *DPFC with Genetic-Based Fuzzy Logic Controller*

GA can be applied to enhance the effect of FLC by tuning it. Optimization of both rule-base (RB) and data-base (DB) of FLC was proposed by Lee and Takagi using

Table 1 Rule-table of FLC

e Δe	NGL	NGM	NGS	ZER	PSS	PSM	PSL
NGL	NGL	NGL	NGL	NGL	NGM	NGM	ZER
NGM	NGL	NGL	NGL	NGM	NGM	ZER	PSS
NGS	NGL	NGL	NGM	NGS	ZER	PSS	PSM
ZER	NGL	NGM	NGS	ZER	PSS	PSM	PSL
PSS	NGM	NGS	ZER	PSS	PSM	PLG	PLG
PSM	NGS	ZER	PSS	PSM	PLG	PLG	PLG
PLG	ZER	PSS	PSM	PLG	PLG	PLG	PLG

triangular-shaped membership functions and binary-coded GA. FLC can be designed manually or automatically using GA. In manual design of FLC, GA tunes knowledge-base (KB) using set of training scenarios [14, 17]. FLC determines the outputs for a set of inputs within accuracy limit after GA-based tuning is done. Optimization using genetic algorithm is based on natural selection process. The population of individual solutions is developed repeatedly in this algorithm. The selection of individuals from current population is random for producing next generation children [15, 16]. In this paper, an initial population of binary-coded GA is created at random. Linear mapping rule is used to obtain the real values of input and output variables of each string. Then, absolute value of deviation in prediction is calculated for all the training scenarios, and fitness of each string of GA population is calculated. The population is GA-strings is then modified by carrying out various operations such as reproduction, crossover, and mutation [8, 17, 21]. Figure 9 shows the flowchart of proposed GA-FLC. The triangular membership function required for FLC is given by Equation (10) [8, 16, 17]. The fuzzification and defuzzification processes are being performed using this triangular membership functions.

$$\mu(x) = \max\left[\min\left(\frac{x - x1}{x2 - x1}, \frac{x3 - x}{x3 - x2}\right), 0\right] \tag{10}$$

4 Simulation Results and Discussions

This work presents with the performance of DPFC device employing FLC and GA-based FLC. MATLAB/Simulink software is used for developing these models. At first, the proposed PV-wind hybrid system is developed without DPFC device and connected to the grid. Later, the DPFC device with FLC and GA-based FLC is deployed in the hybrid system. In this paper, simulation result of DPFC device with different controllers is observed at load point. At different time instants, various power quality issues are introduced at system's load point.

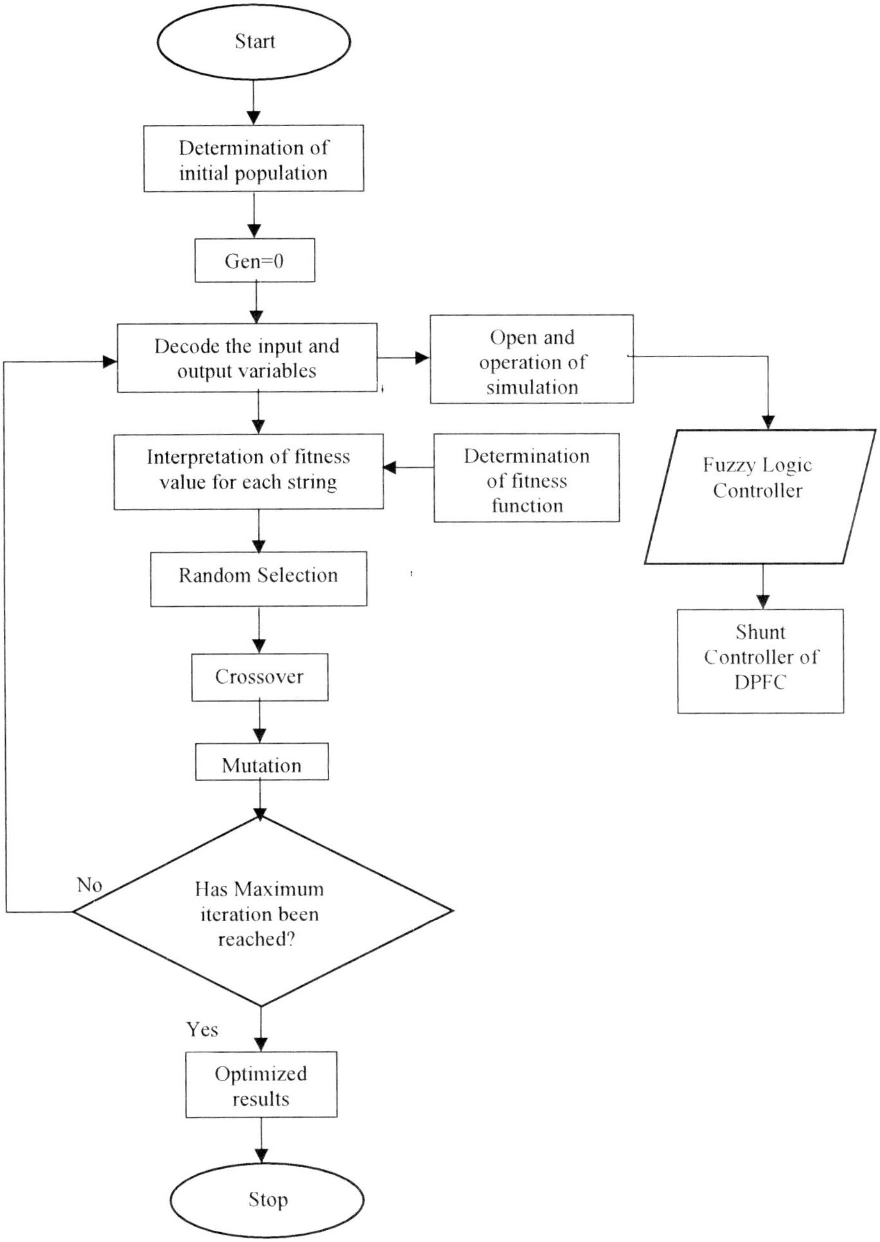

Fig. 9 Flowchart of proposed GA-FLC

4.1 Simulation Results of PV/Wind Hybrid System Without DPFC Device

The results of proposed PV/wind hybrid system without DPFC is shown. Figures. 10 and 11 show the current and voltage waveforms at different load points. Figure 12 depicts the power generated at load point. In this system, voltage sag and swell are created in between 0.2 to 0.4 sec and 0.6 to 0.8 sec, respectively, as shown in Fig. 12. Load voltage harmonic spectra are being shown graphically in Fig. 13. At t = 0.2sec (voltage sag harmonic) and t = 0.6sec (voltage swell harmonic), the percentage of THD is 6.26 and 4.17, respectively.

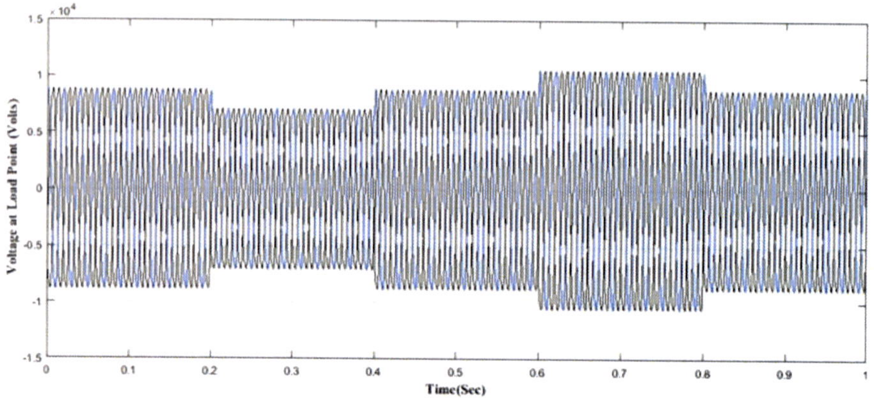

Fig. 10 Waveform of voltage at load point before DPFC connected

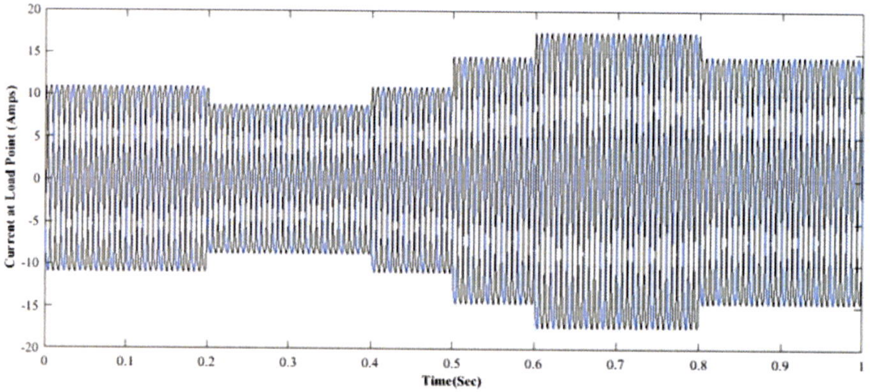

Fig. 11 Waveform of current at load point before DPFC connected

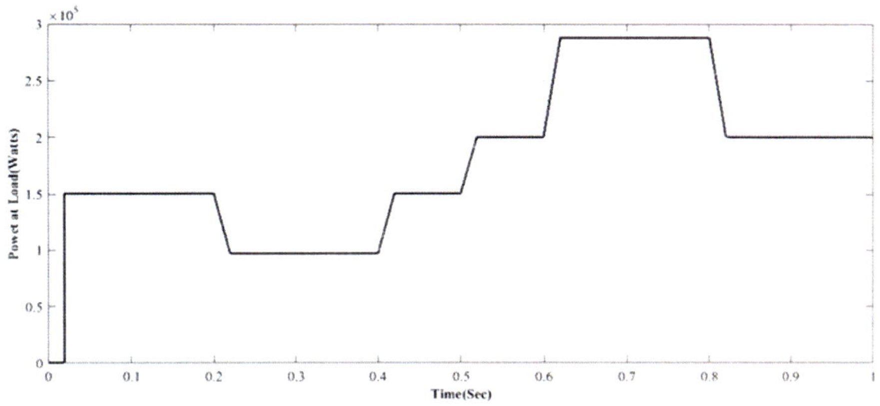

Fig. 12 Active power at load point before DPFC connected

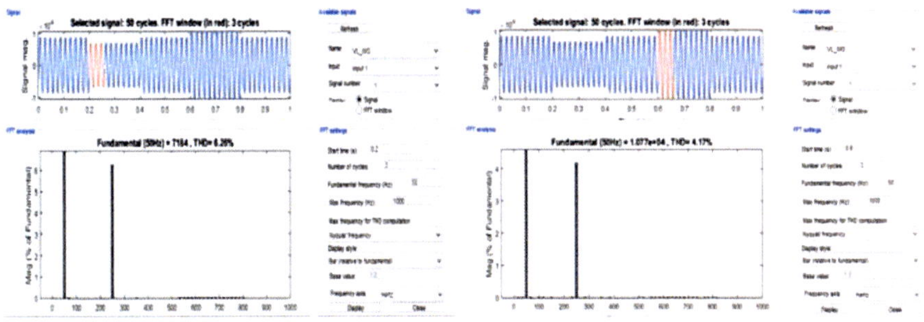

Fig. 13 Load voltage harmonic spectra at t = 0.2 s (for sag condition) and at t = 0.6 s (for swell condition)

4.2 *Simulation Results DPFC Device with Fuzzy Logic Controller*

It is noticeable from Figs. 14 and 15 that the DPFC device with FLC has effectively compensated for said sag and swell that occurred due to disturbances in the system amid evaluation. Figure 16 depicts the load's necessary active power, i.e., up to t = 0.5 sec, it is 150KW, and after t = 0.5 sec, it is 200 KW. Load voltage harmonic spectra are included in Fig. 17, correspondingly. At t = 0.2 sec, the percent THD is 0.31; at t = 0.6 sec, the percent THD is 0.30.

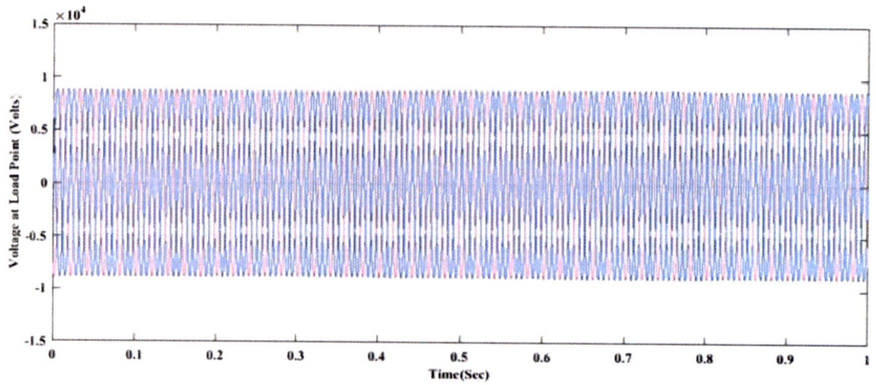

Fig. 14 Waveform of voltage at load point for DPFC with FLC

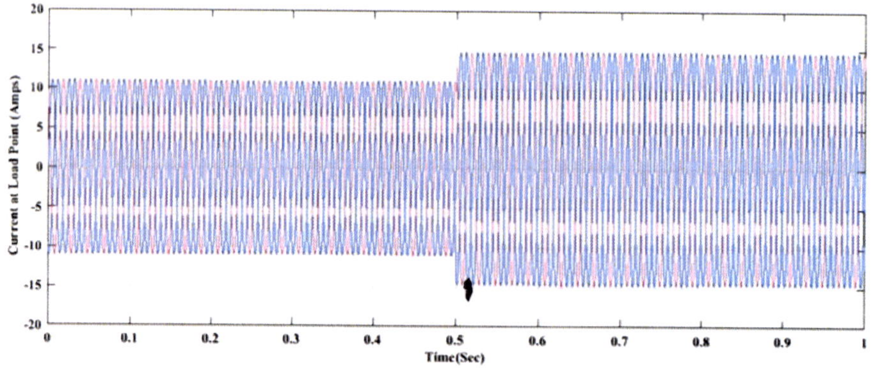

Fig. 15 Waveform of current at load point for DPFC with FLC

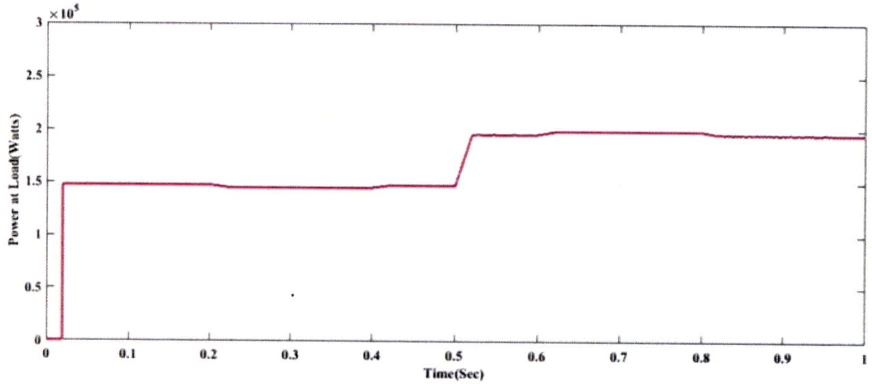

Fig. 16 Active power at load point for DPFC with FLC

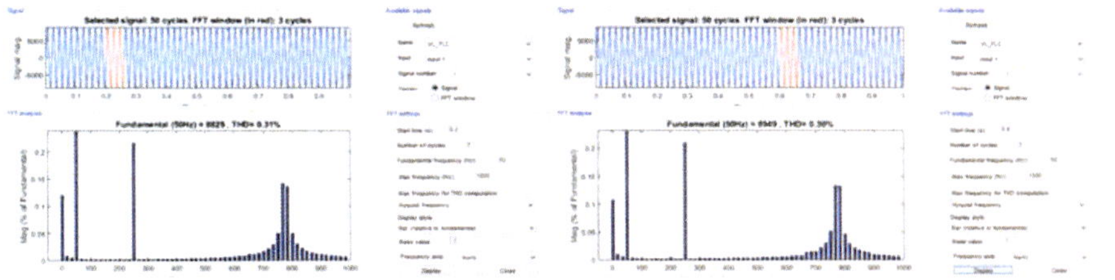

Fig. 17 Load voltage harmonic spectra at t = 0.2 s (for sag condition) and at t = 0.6 s (for swell condition)

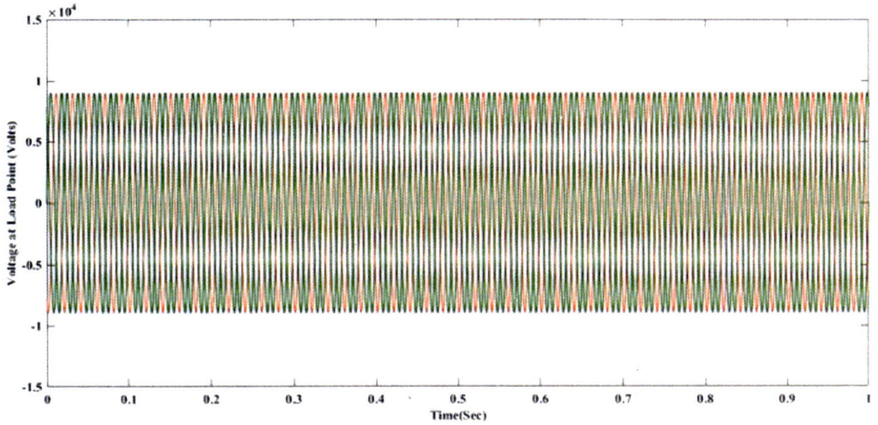

Fig. 18 Waveform of voltage at load point for DPFC with GFLC

4.3 *Simulation Results DPFC Device with Genetic-Based Fuzzy Logic Controller*

The currents and voltages at the load area could be seen in Figs. 18 and 19. As shown in Fig. 20, the load demands 150 KW of active power before to time zero, but requires 200 KW once that time has passed. At various points in time, harmonic spectra of load voltage are depicted in Fig. 21 article, correspondingly t = 0.2sec (voltage sag harmonic) and t = 0.6sec (voltage swell harmonic) have THD percentages of 0.03 and 0.03, respectively.

4.4 *Comparative Analysis*

See (Table 2 and Fig. 22).

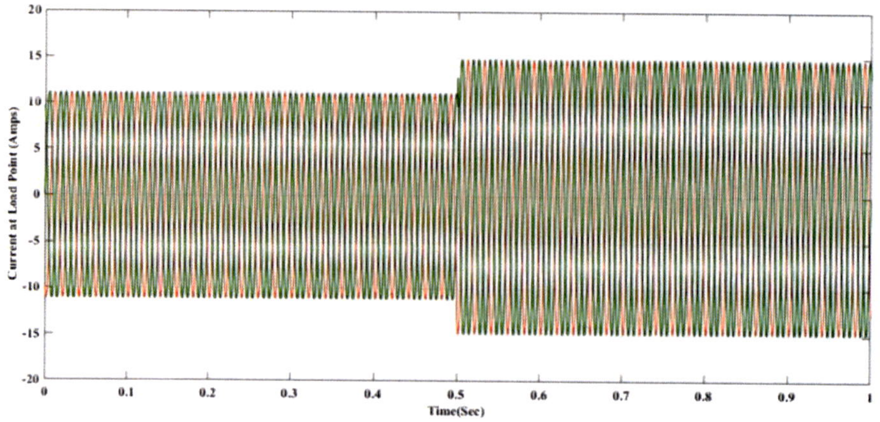

Fig. 19 Waveform of current at load point for DPFC with GFLC

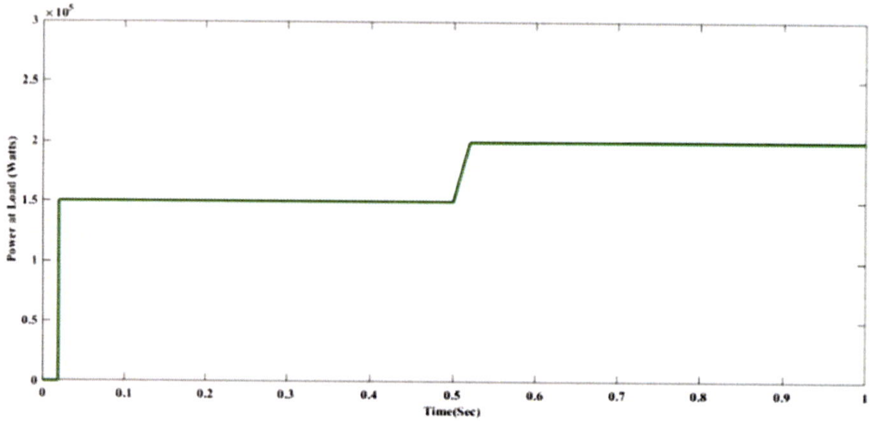

Fig. 20 Load active power for DPFC with GFLC

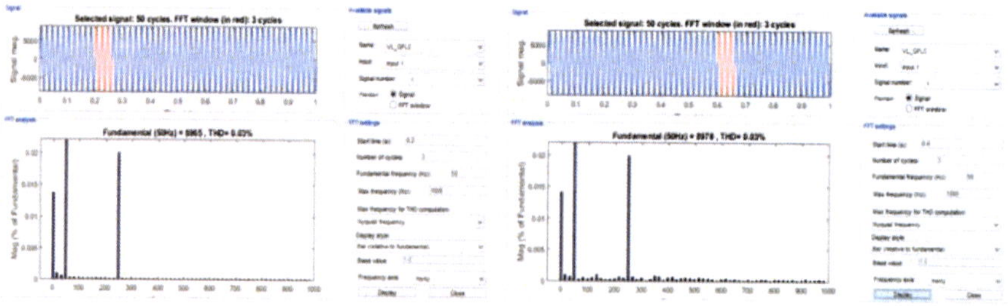

Fig. 21 Load voltage harmonic spectra at t = 0.2 s (for sag condition) and at t = 0.6 s (for swell condition)

Table 2 % THD values of load voltage waveforms

%THD	Without DPFC	DPFC with FLC	DPFC with genetic-based FLC
At = 0.2 s, %THD for Vload	6.26	0.31	0.03
At = 0.6 s, %THD for Vload	4.17	0.30	0.03

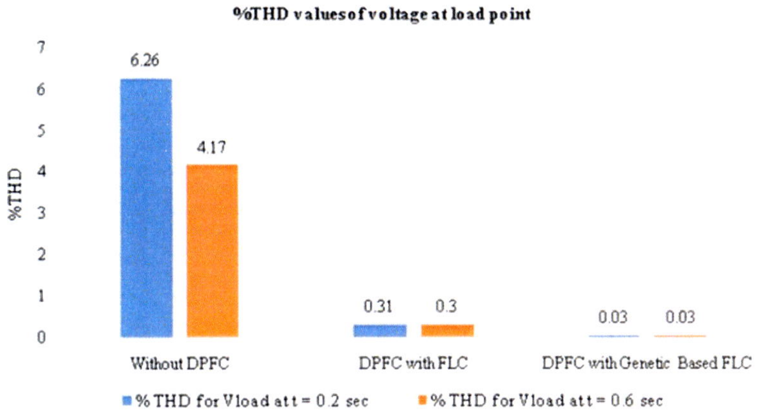

Fig. 22 %THD values of voltage at load point

5 Conclusion

The DPFC device is employed to reduce concerns involving sags and swells. DPFC has a comparable structure to UPFC and may affect system parameters. The DPFC has three control loops: central, shunt, and series. The system under investigation is a PV-wind hybrid. Swells and sags near the load approximate dynamic performance. The performance of a DPFC device was studied using internal control mechanisms using fuzzy logic and genetic-based fuzzy logic. In addition, harmonic content is assessed at 0.2–0.6 s intervals. The simulation findings indicate that both controllers fully offset the sag and swell harmonics; however, the genetic-based fuzzy logic controller has higher compensation capabilities and lessens harmonic distortion.

References

1. Serban R. Teodorescu JM, Guerrero, Marinescu C (2009) Modeling of an autonomous micro-grid for renewable energy sources integration. In: 2009 35th annual conference of IEEE industrial electronics, 3–5 Nov, pp 4311–4316. https://doi.org/10.1109/IECON.2009.5414923.
2. Prakash SL, Arutchelvi M, Jesudaiyan AS (2016) Autonomous PV-array excited wind-driven induction generator for off-grid application in India. IEEE J Emerging Selected Topics Power Electron 4(4):1259–1269. https://doi.org/10.1109/JESTPE.2016.2579678

3. Pota HR, Hossain MJ, Mahmud MA, Gadh R, Bansal RC (2014) Islanded operation of microgrids with inverter connected renewable energy resources. IEEE PES General Meeting Washington DC, 27–31 July 2014. https://doi.org/10.3182/20140824-6-ZA-1003.01091
4. Lavanya V, Senthil Kumar N (2018) A review: control strategies for power quality improvement in microgrid. Int J Renew Energ Res 8(1):149–165
5. Mandi RP, Yaragatti UR (2016) Power quality issues in electrical distribution system and industries. Asian J Eng Technol Innov Spec Conf Issue 3:64–69
6. Alsammak AN, Mohammed HA (2021) Power quality improvement using fuzzy logic controller based unified power flow controller (UPFC). Indonesian J Electr Eng Comput Sci 21(1). https://doi.org/10.11591/ijeecs.v21.i1.pp1-9
7. ul Hassan S, ul Abideen Z, Izhar T (2017) Advanced control techniques for micro-grids power quality improvement. In: 2017 Asian conference on energy, power and transportation electrification (ACEPT), 24–26 Oct, pp 1–6. https://doi.org/10.1109/ACEPT.2017.8168564
8. Elyaalaoui K, Labbadi M, Ouassaid M, Cherkaoui M (2021) Optimal fractional order based on fuzzy control scheme for wind farm voltage control with reactive power compensation. Math Probl Eng. https://doi.org/10.1155/2021/5559242
9. Reddy PGR, Vijaya Kumar M (2015) Analysis of wind energy conversion system employing DFIG with SPWM and SVPWM type converters. J Electr Eng (JEE) 15(4):95–106
10. Raut A, Raut SS (2019) Review: different technology for distributed power flow controller. Int Res J Eng Technol (IRJET) 6(3):6803–6809
11. Dash PK, Morris S, Mishra S (2014) Design of non-linear variable gain fuzzy controller for FACTS devices. IEEE Trans Control Syst Technol 12(3)
12. Yuan Z, de Haan SW, Ferreira JB, Cvoric D (2010) A facts device: distributed power flow controller (DPFC). IEEE Trans Power Electron 25(10):2564–2572. https://doi.org/10.1109/TPEL.2010.2050494
13. Bahamani AK, Reddy GMS, Ganesh V (2017) Comparative of performance for UPFC with DPFC. In: 2017 international conference on electrical, instrumentation and communication engineering (ICEICE2017). https://doi.org/10.1109/ICEICE.2017.8191843.
14. Pratihar DK (2013) Soft computing: fundamentals and applications. Alpha Science International Ltd. First Edition–2013
15. Cam E, Gorel G, Mamur H (2017) Use of the genetic algorithm-based fuzzy logic controller for load-frequency control in a two area interconnected power system. Appl Sci 7(308):1–22
16. Abdelouahed T, Zidi SA, Fadi A (2018) Performance evaluation of fuzzy-logic controller applied to a UPFC transmission system. In: 2018 Electrotehnica Electronica Automatica (EEA) vol 66, no 1, pp 122–131, ISSN 1582-5175
17. Duvvuru R, Rajeswaran N, Sanjeeva Rao T (2019) Performance of distributed power flow controller in transmission system based on fuzzy logic controller. Int J Recent Technol Eng (IJRTE) ISSN: 2277–3878, 8(3): 2039–2043. https://doi.org/10.35940/ijrte.C4530.098319.
18. Yuan Z, de Haan SW, Ferreira B (2009) DPFC control during shunt converter failure. In: 2009 IEEE energy conversion congress and exposition, pp 20–24. https://doi.org/10.1109/ECCE.2009.5316070.
19. Rao DN, Varma PS (2019) Enhancing the performance of DPFC with different control techniques. Int J Innov Technol Exploring Eng (IJITEE) ISSN: 2278–3075 8(6): pp 1002–1007
20. Naidu RP, Meikandasivam S (2020) Power quality enhancement in a grid-connected hybrid system with coordinated PQ theory and fractional order PID controller in DPFC. Sustain Energ Grids Netw 21:100317 Elsevier Ltd. https://doi.org/10.1016/j.segan.2020.100317.
21. Tomar A et al (2020) Machine learning, advances in computing, renewable energy and communication, Springer Nature, Berlin, LNEE volume 768, pp 659 https://doi.org/10.1007/978-981-16-2354-7. (ISBN 978–981–16–2354–7)

A Strategic Approach to Model the Machine-to-Machine Communication of Industrial IoT System for MQTT Protocol with a Case Study

Shital Pawar and Suhas Patil

Abstract The Industrial IoT systems are now becoming more complex and ubiquitous as they are integrating large number of physical devices, and virtual services are expanding over the architecture layers by using various connection patterns. Hence, it has been proved that the design and research of these systems is quite challenging. So, the researchers are now using the relevant modeling and simulation platforms to generate the precise representation of such systems. The heart of Industrial IoT is the programmable logic control (PLC) which is handling the logic of automation. The PLCs have been expanded to synchronize with current demands. One of the standards IEC 61499 is trying to follow such a progression and used for modeling the distributed industrial automation solutions. In this paper, we have modeled machine-to-machine communication of Industrial IoT system for MQTT protocol with one case study. To estimate the performance of the IoT applications, we have also represented the delay model for M2M communication of IoT application.

Keywords Industrial IoT (IIOT) · Programmable logic controller (PLC) · MQTT · Machine-to-machine communication (M2M communication) · IEC 61499

1 Introduction

Internet of Things is now extensively used in variety of industries and applications with great attention on machine-to-machine (M2M) communication [1]. For proposed system, we refer the Industrial IoT application scenario where device sensors will be represented by programmable logic controller (PLC) logic for one module [2, 3]. IoT technology is currently trying to get control on physical devices which connected through the Internet. Here, we are considering the automated industrial system which allows the user to control the machines, efficiently [4]. For presenting such a system, here we have used PLC loads to demonstrate the machines of industry and a motor is used to represent the industrial motor. The microcontroller

S. Pawar (✉) · S. Patil
Bharati Vidyapeeth (Deemed to be University) College of Engineering, Pune, India
e-mail: shital.pawar@bharatividyapeeth.edu; shitalp16@gmail.com

A. Tomar et al. (eds.), *Proceedings of 3rd International Conference on Machine Learning, Advances in Computing, Renewable Energy and Communication*, Lecture Notes in Electrical Engineering 915, https://doi.org/10.1007/978-981-19-2828-4_52

loads and operates the motors as per timer. In the actual industrial setup, the system state is actually displayed on the LCD display. For simulation purpose, we have used Ethernet control of PLC [5]

Thus, we can automate industrial units/components using PLC design for industry IoT automation. Refer Fig. 1 for industrial representation [6]. The "regulator" shown in figure for temperature control sensor called "Thermostat" is further modeled for MQTT protocol which is discussed in further section [7]. The components of above units are explained below:

Transformer: It is used to increase and decrease the current and voltage levels in the electrical circuit. In Fig. 1, it is used to raise the voltage level to decrease losses of lines during their transmission [8, 9].

Rectifier: Further, a rectifier is used for alternating current (AC) to direct current (DC) conversion. This procedure is called as rectification. Rectifier is used as radio signal detector as well as power supply component [9].

Regulator: To avoid voltage fluctuation and damage control, regulator is used. Attacker can malfunction regulator to damage the whole system. Voltage regulator is used to maintain the voltage within specified range which is being tolerated by the various electrical appliances [10].

Wi-Fi Modem: Industrial wireless network is basically a group of various connected devices, and such networks can be utilized for machine-to-machine communication

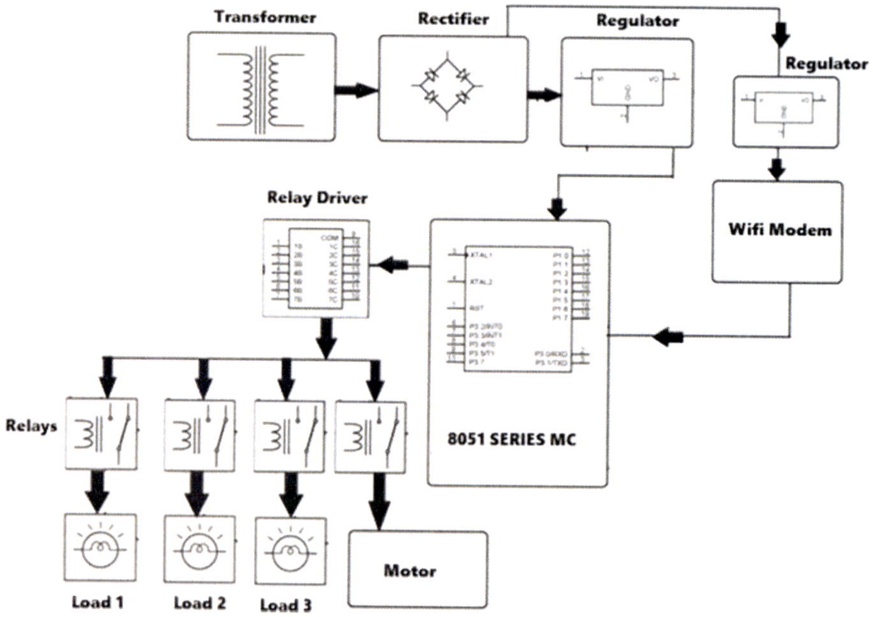

Fig. 1 Schematic representation of industrial automation

through the access point or bridge. Hence, for M2M communication demo, we will use Ethernet modeling with eclipse 4diac [10, 11].

Relay driver: It is an electromagnetic switch which will be utilized when we use the circuit of low voltage to switch a light bulb ON and OFF which is basically connected to the 220 V supply.

8051 Microcontroller: The usage of microcontroller is now increasing rapidly in various fields like mobile communications, industrial processing, electronics, medical applications, etc. In Fig. 1, the 8051 microcontroller is utilized for PLC communication. It is containing 40 pins integrated circuit. It also consists of four parallel 8-bit ports which can be programmable with PLC and addressable according to the requirements [12].

Relays: It acts as a switch which is generally used to turn ON and turn OFF the circuit. The relays can be operated through the magnetic force.

The relays can control the remote devices. They can provide automated control to the remote devices. Almost many industrial applications use relays for the productive working [13, 14].

Load 1, 2, 3: Can be any machine in unit like motors, boilers, mixers, conveyor belts or filters, coolers, etc. To make machine understand the microcontroller communication, PLC programming is required [15, 16].

PLC Programming: Programmable logic controller is one of the important components of industrial automation. PLC program contains the set of instructions which can be written in text or graphical form.

Ladder logic is the simplest language used to program the PLCs. It is a simplest form to create logic expressions for PLC through which we can easily automate the various machine task sequences. In earlier days, such automation was carried out through hardwired control system [17, 18].

2 Methodology for M2M Modeling with MQTT Protocol

Start with defining the various devices which are available in the system. We can use the System Configuration Editor for customizing the hardware required for automation, which can be opened just by doing selection in system configuration tree node which is displayed in System Explorer. By doing drag and drop the resource, device and type of segment will be inserted from the palette. Various IEC 61499 resources can be configured on IEC 61499 devices. We can add resources directly to devices.

Device and Resource arguments are mentioned precisely in each Device/Resource View. The most significant arguments are the IP address and the device management interface port because they are used for the connection between Engineering tool and the Devices [19, 20]. We can connect them just by clicking on one device and dragging it toward another one. Each segment has two sides, left one is used for moving; whereas, right one is used for connections. We can connect the devices by clicking on second row of segment which is device type [21, 22].

In Fig. 2, one resource named as EMB_RES (Embedded Resource) now added to the device. This device will run on IP: PORT which is represented by MGR_ID localhost: 61499. For this case, it will listen to port number 61499 and run locally. Here, we have given device name as M2M_Run. We can distinguish the different devices with the help of device name. We have named EMB_RES as M2M_test.

A. Configure the Resource

By clicking on M2M_run resource displayed in System Configuration, the resource editor will get open. In Fig. 3, the START function block is the default starter function block of M2M_test.

It will trigger the COLD event when PLC is set to START state and if whenever PLC will set to STOP state it will trigger STOP. It will trigger the WARM event whenever the MQTT requests by PLC changes from state STOP back to state RUN. We have connected WARM as well as COLD to the E_CYCLE. START.

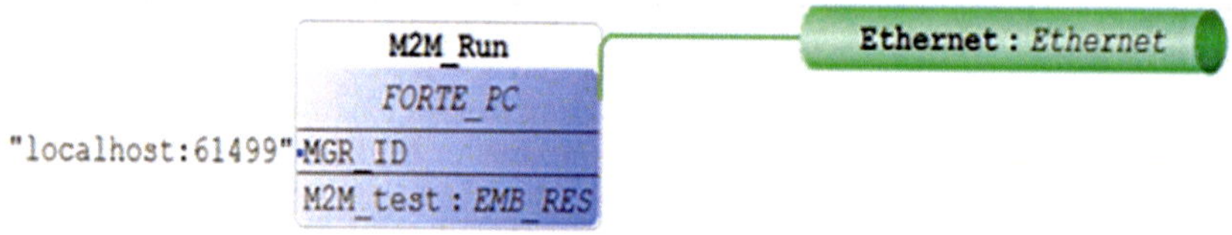

Fig. 2 Configuration of automation hardware

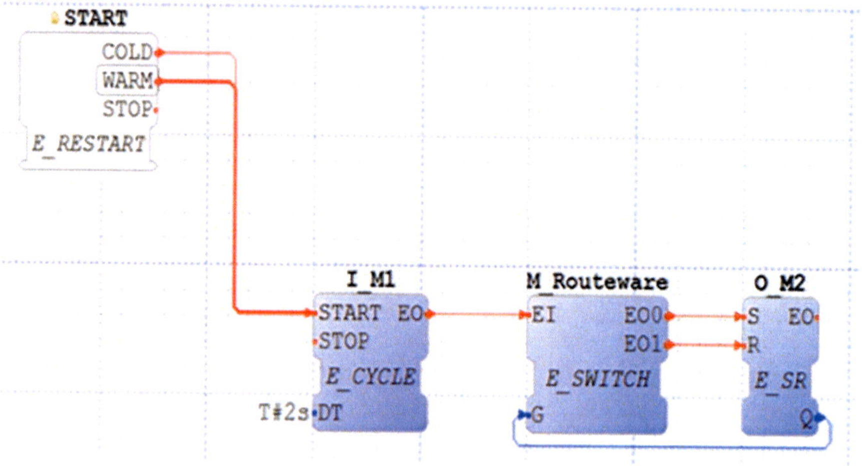

Fig. 3 Configuring the resource

Now, consider the event is arriving to E_CYCLE.START at every 2 s the output of an event will be triggered at E_CYCLE.EO. When G is zero, the output of the event will be triggered at EO0 of E_SWITCH. Else it will be triggered at E01 of E_SWITCH. If this event arrives at S of E_SR, then the value of Q will become 1 and it will becomes 1 when event arrives at R of E_SR (Fig. 3). The deployment of application is shown in Fig. 4.

B. Monitoring the Application

We have observed the output of our application by using 4diac's monitoring functionality.

1. Click on debug perspective button.

2. Go to the system explorer and then right click on System and select Monitor System (Fig. 5).

To watch variables, you can:

1. Right Click on Variable and Select Watch/Right Click on Functional Block and Select Watch.

2. The event shows the number of times the event is triggered, and data displays the present value. We have seen the alternate true and false values. When controller output is connected, a sensor of device would now process request at 2 s intervals (Fig. 6).

C. **Add another Device**

Fig. 4 Deploying application to 4diac FORTE

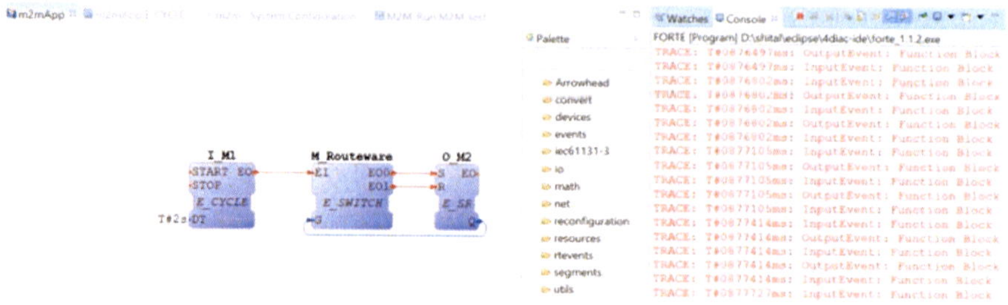

Fig. 5 Monitoring the application

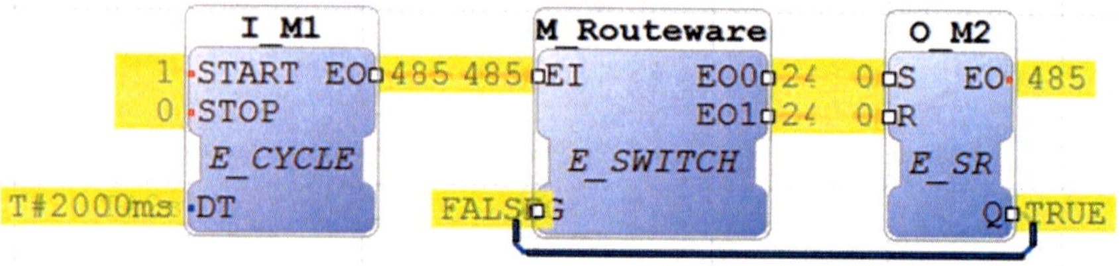

Fig. 6 Monitoring the data and events through watches

We have added another device to M2M system from the palette of System Configuration editor through drag and drop action. Now the counter will run on this new device. We have renamed this device as counter. The devices must be run on different local hosts, so we have set local host to 61500. Here we have set the name of resource as **Counter_RES** (Fig. 7).

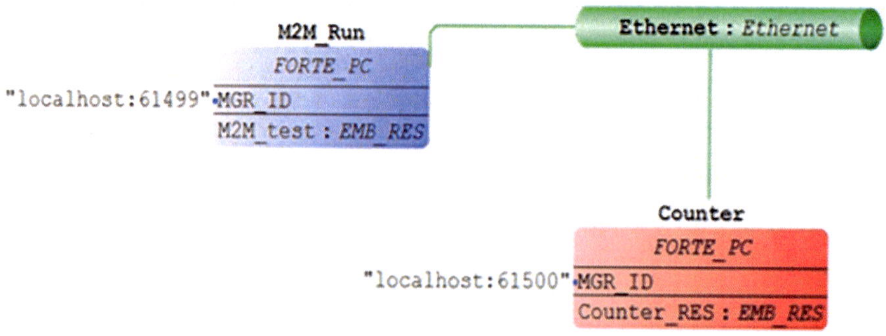

Fig. 7 Addition of another device

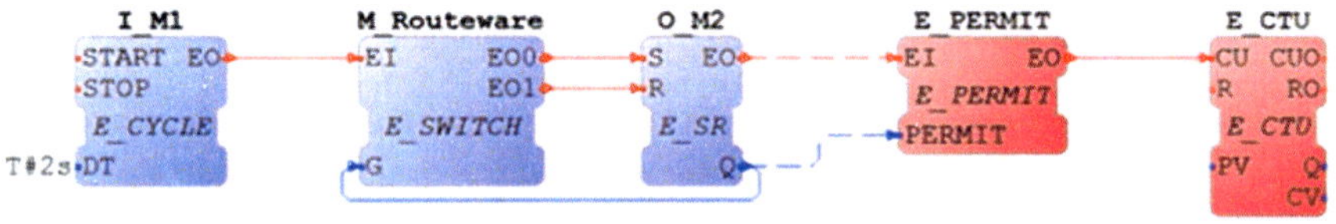

Fig. 8 Addition of two functional blocks E_PERMIT and E_CTU and connection between functional blocks

D. Set up the Sensor Blinking Part of the Application

We have done the changes in M2M_test Resource Editor. We have inserted the PUBLISH_1 Function Block from the palette and done the connections for it. It will be better to directly connect the INIT event to the START event. In Fig. 8, the value of QI is set to 1 and for PUBLISH_COUNT Function Block, the value of ID input data is set to 239.0.0.1:61000.

E. Setting up counting part of an application

We have renamed Counter as Counter RES resource editor. We have inserted SUBSCRIBE_1 Function block from palette, done the connections for it and then it is initialized. In Fig. 9, the value of QI is set to 1 and for SUBSCRIBE_COUNT Function Block, the value of ID input data is RES resource editor.from palette, done the connections for it and set to 239.0.0.1:61000 (Figs. 10 and 11).

F. Background of the Communication Function Blocks

Here PUBLISH_X Function Block is sending the messages through the network and received by an SUBSCRIBE_X Function Block. At each time the message is as per the ID input whenever Request (REQ) is triggered. The value of X in PUBLISH_X

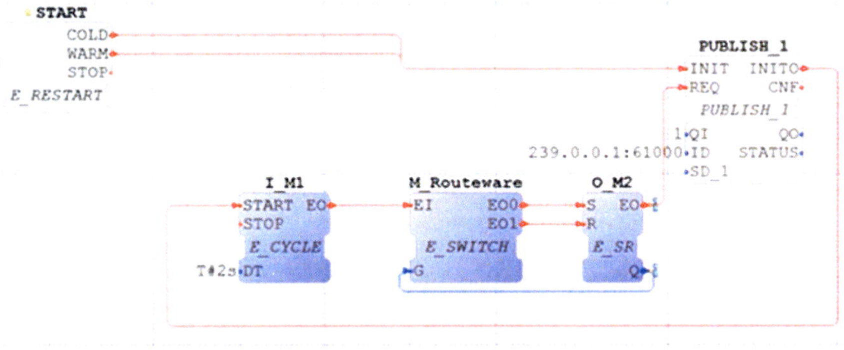

Fig. 9 Setup for sensor blinking part of an application

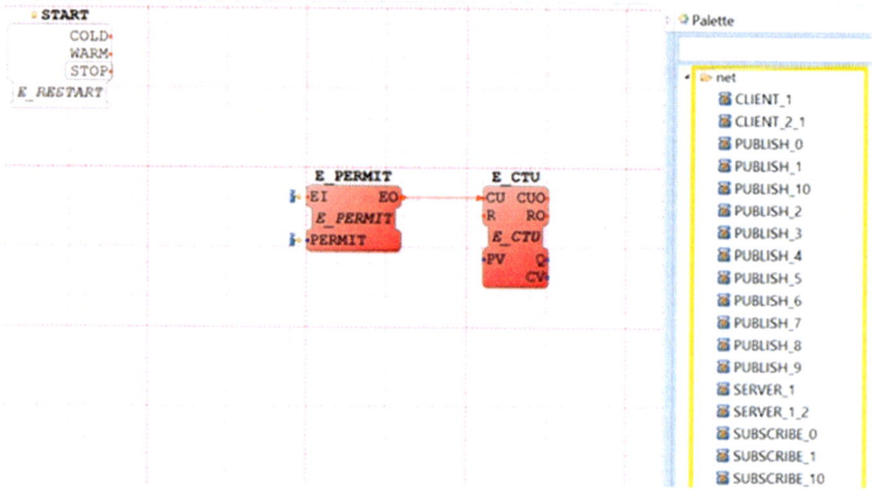

Fig. 10 Initialization of counting part of an application

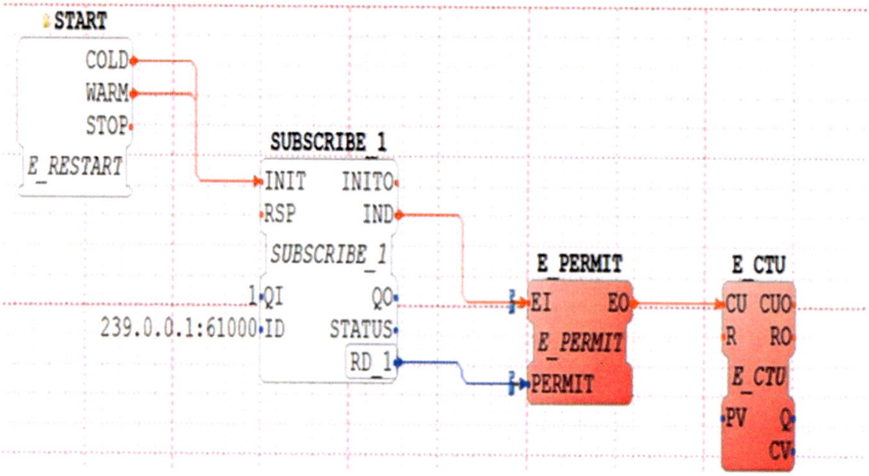

Fig. 11 Setup for the counting part of an application

specifies the number of data items which we wish to deliver in the messages. Here, we are sending only one data item, so we have represented it in figure as PUBLISH_1. We can specify the network protocol by specifying the value ID input. The ID value determines the IP: PORT pair. We have used IEC 61499 Compliance Profile for Feasibility Demonstrations which specifies that this is the UDP multicast address in a specific address range. We have selected address 239.0.0.1 and the port 61000 will do nothing with the localhost: 61499 or localhost: 61500 which are used for the device configuration. First one is a UDP multicast address which is used for

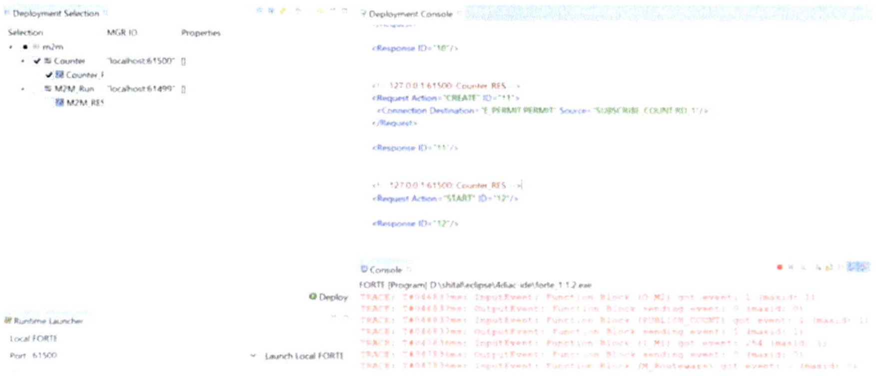

Fig. 12 Deployment output

sending messages between the devices and the latter is TCP connections for deploying applications to device and manage them. We must have to take care of not to use the same UDP socket for different PUBLISH_X/SUBSCRIBE_X pairs.52.

G. Deployment

Figure 12 represents the deployment of an application.

3 Results and Discussion

A. Testing

The following diagram represents the test results of M2M communication along with watch variable values (Fig. 13). In the Figs. 14 and 15, we have represented the test results of an application with publish and subscribe count.

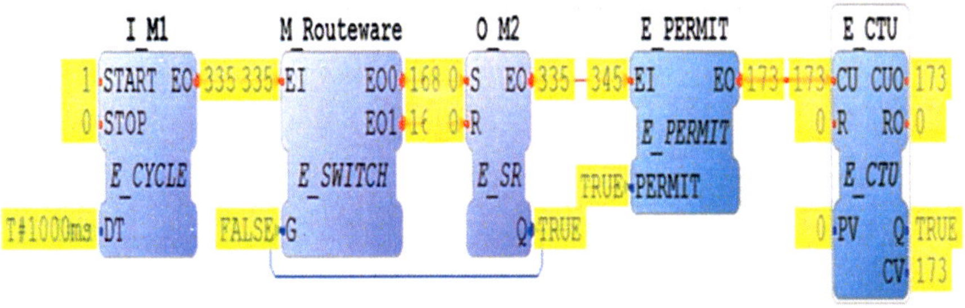

Fig. 13 Testing output with watch variable values

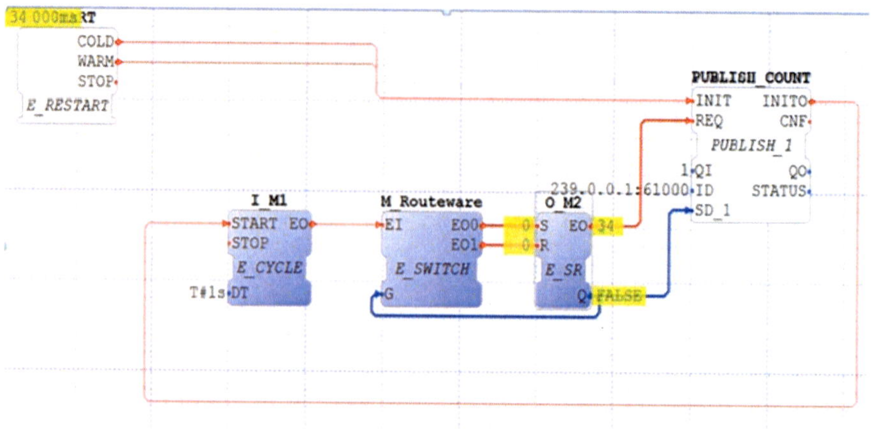

Fig. 14 Testing of an application with PUBLISH_COUNT

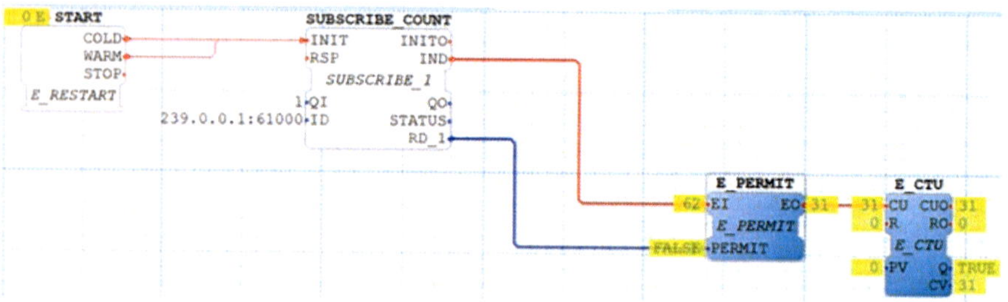

Fig. 15 Testing of an application with SUBCRIBE_COUNT

B. Delay Model

The delay model is presented for IIoT with consideration of request departure, request arrival, and malfunctioning delay against the pre-determined expected arrival period for each normal system operating time. Here we have assumed the maximum delay as 60 seconds. The Table 1 justifies the assumptions and represents the calculated delay (Fig. 16).

Table 1 Delay calculation

Request	Malfunction request delay (MRD) (ms)	Request departure delay (RDD) (ms)	Expected request arrival period (ERA) (ms)	Calculated delay (ms)
1	3	2	3	0.25
2	1	3	2	0.13
3	2	3	4	0.33
4	3	2	2	0.16

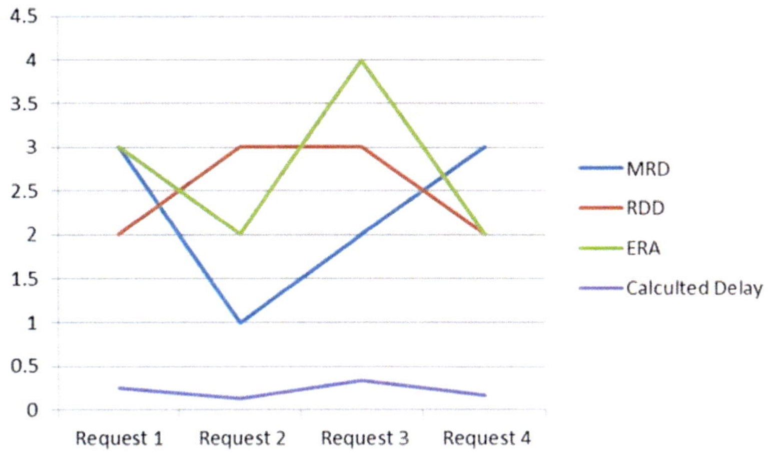

Fig. 16 Calculated delay

$$\text{Request Delay } (D_r) = \frac{(\text{Malfunctioning Request Delay} + \text{Request Departure Delay})}{\left(\frac{60}{\text{Expected Request Arrival Period}}\right)} \quad (1)$$

$$D_r = \frac{(R_{md} + R_d)}{\left(\frac{60}{R_a}\right)} \quad (2)$$

4 Conclusion

The PLC is, basically, a small computer which can be programmed to handle the machine automation in industry. It consists of inputs and outputs for running the various systems. In general, buttons and sensors generate the inputs, while the output controls motors. For proposed MQTT demo, we defined the devices which are present in our system. An automation hardware is modeled with the System Configuration Editor. The Device, Resource and Segment types are added for simulation. For illustration, we used "Thermostat" control application module. The "timer" as a sensor will trigger the event which can be passed through MQTT communication. MQTT protocol uses the Publish –Subscribe model for communication. So, we have represented the publish-subscribe count for M2M communication of MQTT protocol. At security level, the "delay" in signaling can be risky for any industrial setup. Hence, there is need to calculate the delay in M2M communication of Industrial IoT (IIOT) system. So, we have presented the delay model to calculate the delay in M2M communication for security concern. Here, machine-to-machine (M2M) communication is demonstrated with IEC 61499standard along with Eclipse 4diac as per product quality model defined in ISO/IEC 25010. We can use this modeling approach to model and simulate complex and distributed industrial applications. We can refer this M2M communication model for quality of service (QoS) evaluation of MQTT protocol.

References

1. Radanliev P et al (2019) Definition of Internet of Things (IoT) cyber risk discussion on a transformation roadmap for standardisation of regulations risk maturity strategy design and impact assessment. *arXiv* preprint arXiv:1903.12084
2. da Cruz MAA et al (2018) A reference model for internet of things middleware. IEEE Internet Things J. https://doi.org/10.1109/JIOT.2018.2796561
3. Iraji S et al (2017) Recent advances in M2M communications and Internet of Things (IoT). Int J Wireless Inf Networks 24:240–242. https://doi.org/10.1007/s10776-017-0362-3
4. Pawar S, Patil S (2019) Development of QoS evaluation algorithm for MQTT protocol with reference to threat model. Int J Eng Adv Technol (IJEAT) 8(6):1557–1562
5. Lutu A et al (2020) Where things roam: uncovering cellular IoT/M2M connectivity. In: Proceedings of the ACM internet measurement conference, October 2020, pp 147–161. https://doi.org/10.1145/3419394.3423661
6. Pawar S, Patil S (2021) A novel approach for enhancement of security through evaluation of quality-of-service parameters in industrial internet of things. Int Conf on Intelligent Technologies(CONIT) 2021:1–6.https://doi.org/10.1109/CONIT51480.2021.9498340
7. Jalali MS et al (2019) The internet of things promises new benefits and risks: a systematic analysis of adoption dynamics of IoT products. IEEE Secur Priv 17(2):39–48
8. Husseinet AH et al (2019) Internet of Things (IOT): research challenges and future applications. Int J Adv Comput Sci Appl (IJACSA) 10(6)
9. Qureshi TN et al (2019) Enhanced robustness strategy for IoT in smart cities based on data driven approach. Workshops of the international conference on advanced information networking and applications. Springer, Cham
10. Kuzminykh I, Carlsson A (2018) Analysis of assets for threat risk model in avatar-oriented IoT architecture. Internet of things, smart spaces, and next generation networks and systems. Springer, Cham, pp 52–63
11. Chauhan YK, Ratan R Study on placement of sensors for readings accuracy level enhancement in greenhouse. In: Mishra S, Sood Y, Tomar A (eds) Applications of computing, automation and wireless systems in electrical engineering. Lecture notes in electrical engineering, vol 553. Springer, Singapore. https://doi.org/10.1007/978-981-13-6772-4_22
12. Aazam M et al (2018) IoT resource estimation challenges and modeling in fog. Fog computing in the Internet of Things. Springer, Cham, 17–31
13. Xu T, Wendt JB, Potkonjak M (2014) Security of IoT systems: design challenges and opportunities. In: 2014 Proceedings of the 2014 IEEE/ACM international conference on computer-aided design. IEEE Press
14. Moraes P, Reale R, Martins J (2018) A publish/subscribe QoS-aware framework for massive IoT traffic orchestration. arXiv preprint arXiv:1806.03157
15. Lin J et al (2017) A Survey on internet of things: architecture, enabling technologies, security and privacy, and applications. IEEE Internet Things J
16. Lei Y, Fengyu Z, Yugang W, Xianfeng Y, Yang Z, Zhumin C (2016) Design of a cloud robotics visual platform. In: 2016 sixth international conference on instrumentation measurement, computer, communication and control (IMCCC)
17. Chin WL, Lin YH, Chen HH (2016) A framework of machine to machine authentication in smart grid: a two-layer approach. IEEE Commun Mag

18. Rachikidi EE (2017) Modeling and placement optimization of compound service in a converged infrastructure of cloud computing and internet of things. Université Paris-Saclay, Evry
19. Shakya AK, Ramola A, Pokhariya HS, Kandwal A (2019) Fusion of IoT and machine learning approach to prevent confidential data from digital crimes and cyber mugging for covert transmission. In: Mishra S, Sood Y, Tomar A (eds) Applications of computing, automation and wireless systems in electrical engineering. Lecture notes in electrical engineering, vol 553. Springer, Singapore. https://doi.org/10.1007/978-981-13-6772-4_49
20. Verma PK, Verma R, Prakash A, Agrawal A, Naik K, Tripathi R, Alsabaan M, Khalifa T, Abdelkader T, Abogharaf A (2016) Machine-to-machine (M2M) communications: a survey. J Netw Comput Appl vol 66 https://doi.org/10.1016/j.jnca.2016.02.016
21. Qin J, Li Z, Wang R et al (2021) Industrial Internet of Learning (IIoL): IIoT based pervasive knowledge network for LPWAN—concept, framework and case studies. CCF Trans Pervasive Comp Interact 3:25–39. https://doi.org/10.1007/s42486-020-00050-2
22. Tomar A et al (2020) Machine learning, advances in computing, renewable energy and communication. Springer Nature, Berlin, LNEE vol 768, pp 659 https://doi.org/10.1007/978-981-16-2354-7. ISBN 978–981–16–2354–7

Bidirectional DC-DC Converter as a Better Alternative for Charging and Discharging of Battery Backup in a MPPT-Controlled Solar PV System

Prashant, Amurt Prakash, Shivam Kumar, Kaushik Dubey, and Prashant Kumar Pandey

Abstract Energy consumption is increasing all over the world, and a large percentage of this need is now being met by the usage of fossil fuels, which are rapidly diminishing and are also harmful to the environment. Thus, in order to maintain these resources and reduce the strain on the ecosystem, there is a rapid transition to the use of renewable energy to meet energy demand and to provide a sustainable environment. Energy conversion and management using solar energy has developed in recent decades, and it has proven to be a superior alternative as an alternate source of energy for reducing environmental pollutants. However, solar energy has its own set of restrictions. As a result, power converters are being used to revamp the dependability and flexibility of the solar photovoltaic system. In this research, a bidirectional DC-DC converter scheme that efficiently meets all of the requirements of a power converter in a solar photovoltaic system is suggested and MATLAB Simulink has been used to implement the proposed technique. The non-isolated topology of this converter using MPPT mechanism has been designed and simulated using Simulink showing charging and discharging of battery backup as well as allying it with DC plus AC load, which can then be used efficiently to satisfy load requirements.

Keywords Solar photovoltaic · Bidirectional DC-DC converter (BDC) · MPPT · Battery backup · Simulink · Double PI control

Prashant · A. Prakash (✉) · S. Kumar · K. Dubey · P. K. Pandey
Department of Electrical Engineering, J.S.S Academy of Technical Education, Noida, U.P., India
e-mail: amurtprakash@gmail.com

Prashant
e-mail: prashant27@jssaten.ac.in

S. Kumar
e-mail: shivamalig2209@gmail.com

K. Dubey
e-mail: kaushikjss10@gmail.com

P. K. Pandey
e-mail: pp038552@gmail.com

A. Tomar et al. (eds.), *Proceedings of 3rd International Conference on Machine Learning, Advances in Computing, Renewable Energy and Communication*, Lecture Notes in Electrical Engineering 915, https://doi.org/10.1007/978-981-19-2828-4_53

1 Introduction

The accelerated depletion of the fossil fuel and its unenviable effect on the ecosystem is the wakeup call for us to look for an alternate source of energy. Amongst all the alternatives the solar energy conversion is the most favored one due to its cleanliness, easy operation, availability in major area and low maintenance cost. Solar photovoltaic system looks promising, but it has its own complications and limitations as it is affected by conditions such as availability of solar radiation, temperature and other weather conditions [1]. The MPPT method is utilized to optimize the solar PV system's output power [2]. In current work, incremental conductance methodology has been implemented for MPPT functionality in a solar PV system to keep track of the solar system's maximum operational point.

As we aware that a battery energy storage system (BES) is required to provide consistent power delivery to the load and to build a more efficient and dependable solar PV system [3]. In a solar PV system, power converters are utilized to operate as a power interface between different components of the solar PV system and to keep the power flowing in the battery energy storage (BES) system. A non-isolated architecture of bidirectional DC-DC power converter with bidirectional flow of power between the load and BES has been implemented in this research [4–6]. When there is excess solar power available during the day, BES will be charged with power flowing from the solar panel to BES via the bidirectional DC-DC converter (**BDC**), and BES will be discharged with power flowing from BES to the load via the BDC when there is insufficient power available across the load [6, 7].

A unidirectional buck or buck-boost converter is used in a traditional solar PV hybrid system for charging and discharging the battery backup in various modes of operation during the day. In our proposed method, using a BDC, we were able to accomplish buck and boost converter functionalities. This work models and implements a non-isolated topology of BDC in Simulink for charging and discharging battery backups, as well as integrating it with DC and AC loads, so that it may be utilized efficiently and effectively to satisfy load needs.

2 MPPT Functionality

MPPT increases the solar PV system's reliability and efficiency by dynamically adjusting the solar PV system's output power to the greatest power peak at that particular circumstance [2]. In current study, incremental conductance method of MPPT has been implemented which is more efficient and popular hill climbing method of MPPT.

2.1 Incremental Conductance (INC)

The authors in paper [2] give the concept of the incremental conductance of MPPT. The value of incremental conductance is zero at the highest power point. The total of the instantaneous and incremental conductance is negative on the right-hand side of the maximum power point, whereas it is positive on the left-hand side. The INC method, which is based on this notion, examines a PV generator's immediate and incremental conductance to determine the duty cycle of the power converter linked [2] (Fig. 1).

3 Bidirectional DC-DC Converter

Due to rapid advancement in the field of power electronics, the bidirectional DC-DC is getting popular these days. The bidirectional DC-DC converter provides a better

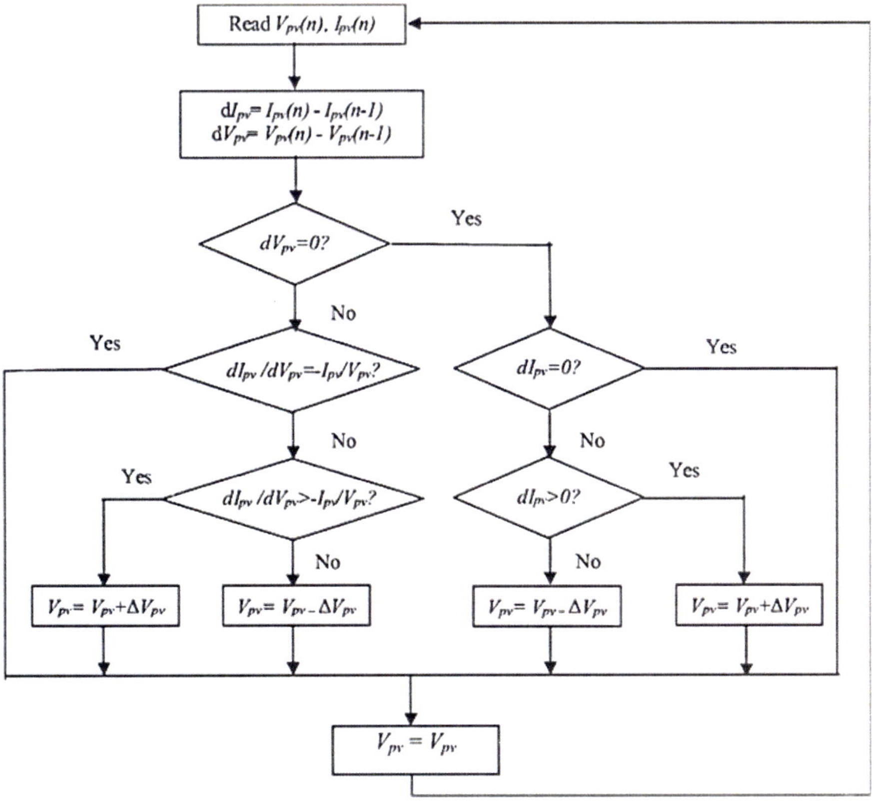

Fig. 1 Flowchart of algorithm of incremental conductance [3, 4, 8]

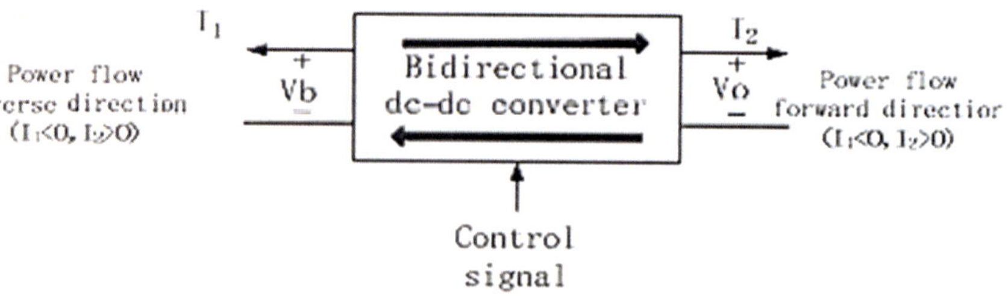

Fig. 2 Illustrative diagram of BDC [4]

flexible option as a power interface between the main and auxiliary power supply in EVs, solar PV system, hybrid electric vehicles, etc. [9–11]. The two-way flow of power is the key attribute of the BDC which gives its edge over other conventional power converters (Fig. 2).

3.1 Configuration of Non-isolated Bidirectional DC-DC Converter (BDC)

The BDC can mainly be categorized into two types: non-isolated and isolated type. As stated earlier in this study, non-isolated topology of BDC has been designed and tested. A parallel combination of diode and controlled switches is utilized for bidirectional power flow. Bidirectional buck-boost converters are basically anti-parallel integrations of buck and boost converters, with switches regulated according to the needed power direction.

3.2 Modes of Operation

Bidirectional DC-DC converters (BDC) can operate in two modes for bidirectional power flow:

(a) **Buck Mode**. When BDC operates in buck mode, the voltage magnitude is stepped down. During this mode of operation, switch s1 and diode D2 go into conduction mode depending on the duty cycle, while switch s2 and diode D1 remain off (refer to Fig. 3).

(b) **Boost Mode**. In boost mode, the output voltage magnitude of a BDC is increased. During this mode of operation, switch s2 and diode D1 go into conduction mode depending on the duty cycle, while switch s1 and diode D2 remain off (refer to Fig. 3).

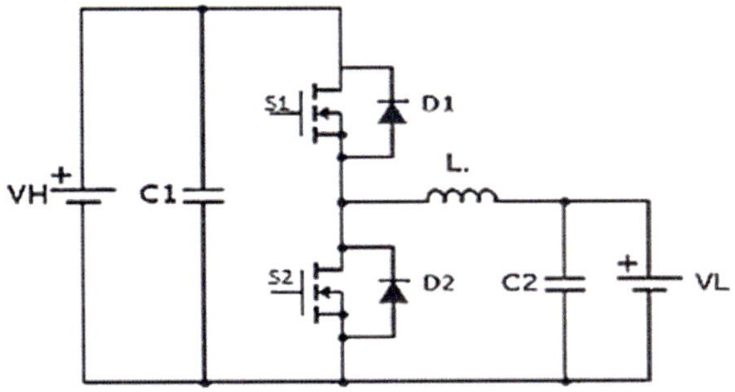

Fig. 3 Bidirectional DC-DC converter [6, 7]

4 Bidirectional DC-DC Converter Integrated with Solar PV System

In conventional solar PV hybrid system, a unidirectional buck or buck-boost converter is used for charging the battery backup when excess solar power is available. When sufficient power is not available from the solar PV modules, then battery backup discharges via. separate boost converter, as shown in Fig. 4.

Whereas, in the proposed technique, we have implemented buck and boost converters functionality using a single BDC. For starters, a unidirectional DC-DC converter acts as a power link between solar PV panels and the DC bus. The duty cycle of this converter is so well controlled that it acts as a maximum power point tracker, allowing the system to get the most power from solar PV arrays. A bidirectional DC-DC converter then serves as a power link between the solar PV array and the battery system for charging and discharging. The full configuration is depicted in Fig. 5 as a generic block diagram.

5 Controlling of Bidirectional Converter

There are various control techniques that may be used with a BDC, some of which are discussed by the authors in papers [13, 14].

Two loop control scheme of PI controllers is implemented in this work, using two autonomous PI controllers, one for DC bus voltage control and the other for current control. This method starts by comparing the DC bus voltage to a reference value, which is the intended voltage across the DC bus. The voltage PI controller processes the error signal provided by the voltage comparison and generates the reference current value. The instantaneous battery/inductor current is then compared to the reference current generated in the first loop of the PI control, and an error signal is sent to the current PI controller. The PI controller then provides a control signal that is sent to the PWM DC-DC generator, which generates gating pulses to control the bidirectional DC-DC converter's switches [4, 13] (Fig. 6).

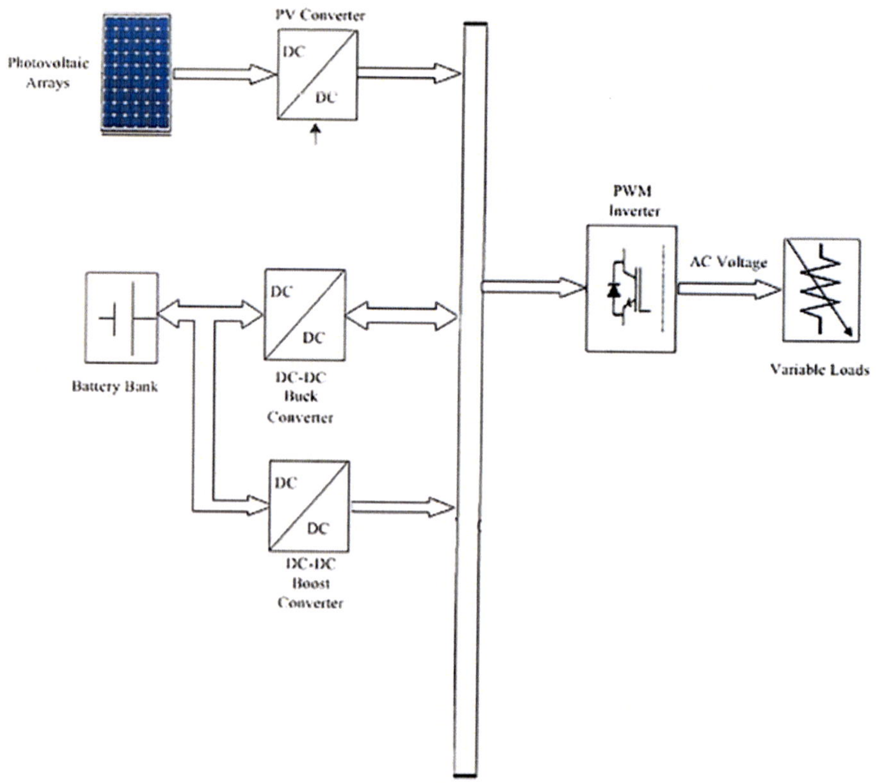

Fig. 4 Conventional technique of converters employed in solar PV system [6]

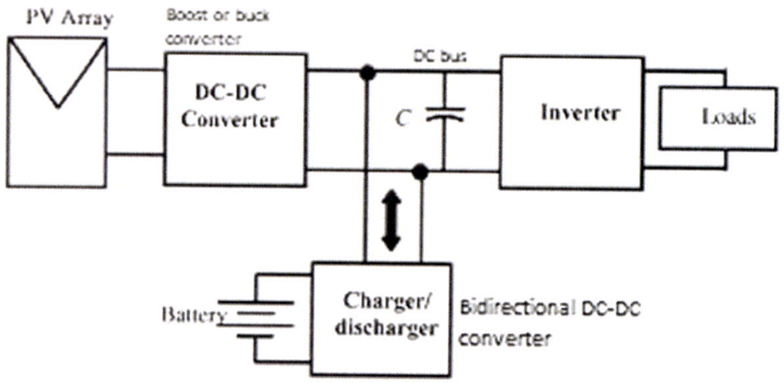

Fig. 5 Block diagram representation of the proposed system [4–7, 12]

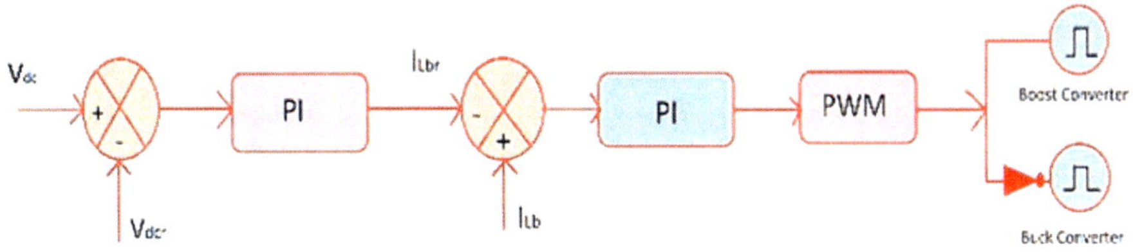

Fig. 6 Double loop PI control scheme

6 Modeling of Parameters

The non-isolated topology of BDC is essentially an anti-parallel connection of buck and boost converters. So, for deciding the parameter value, the following equations are considered [12, 15].

6.1 Basic Parameters of Boost Converter

(a) *Duty cycle*: The duty cycle of switching for the converter is usually calculated based on the specified output and input voltage.

$$D = 1 - \frac{v_{in} \times \eta}{v_{Out}} \tag{1}$$

(b) *Inductor*: Inductor value based on required voltage levels at the input and output side and switching frequency is given in Eq. (2).

$$L = \frac{V_{in} \times (V_{Out} - V_{in})}{\Delta I_L \times V_{Out} \times f_s} \tag{2}$$

(c) *Output capacitor:* The best choice of capacitor for this type of converter is a low ESR capacitor. The Eq. (3) tells how to decide capacitor value based on the required levels.

$$C_0 = \frac{I_{Out} \times D}{f_S \times \Delta v_{Out}} \tag{3}$$

6.2 Basic Parameters of Buck Converter

(a) *Duty cycle*: The duty cycle of switching is based on the required output voltage and input voltage.

$$D = \frac{v_0}{v_{in} \times \eta} \tag{4}$$

(b) *Inductor*: Inductor value based on required voltage levels at the input and output side and switching frequency is given in Eq. (5).

$$L = \frac{v_{Out} \times (v_{in} - v_{Out})}{\Delta I_L \times f_s \times v_{in}} \tag{5}$$

(c) *Capacitor*: Capacitor value based on required voltage levels at the input and output side and switching frequency is given in Eq. (6).

$$C_0 = \frac{\Delta I_L}{8 \times f_s \times \Delta v_{Out}} \tag{6}$$

V_{out}	*Desired output voltage.*
V_{in}	*Voltage on input side.*
f_s	*Switching frequency.*
ΔI_L	*Ripple current of inductor (assumed 10%-20%).*
η	*Efficiency of the converter.*
Δv_{Out}	*Voltage ripple (assumed 10%).*

When the bidirectional converter is in buck mode, the DC bus voltage is the input voltage, and the battery side charging voltage is the output voltage. During boost operation of the bidirectional converter, the battery terminal voltage will be the input voltage, and the DC bus voltage will be the output voltage.

7 Simulation and Results

7.1 System Description Under Study

In this research work, a boost converter controlled by incremental conductance MPPT algorithm has been used which extracts and feds maximum power to the DC load from the solar PV arrays. For bidirectional power flow across the auxiliary power supply, a non-isolated BDC has been designed. Depending on the voltage magnitude at the DC bus and availability of power from PV arrays, the power flow is controlled in the bidirectional converter [13, 16].

7.2 Simulink Model

MATLAB Simulink platform is used for the modeling of the system. Figure 7 depicts the said Simulink model of MPPT-controlled solar PV system with a bidirectional converter (Figs. 7 and 8) (Table 1).

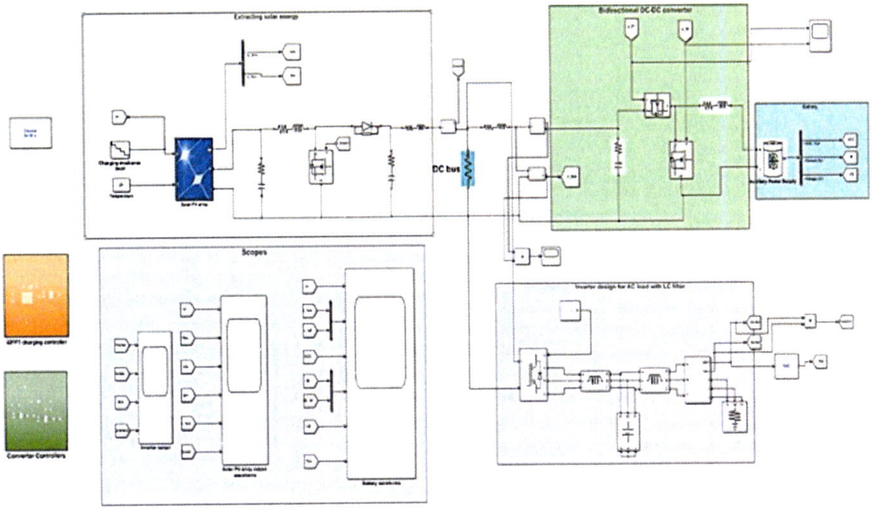

Fig. 7 Bidirectional converter integrated with solar PV array for battery charging and discharging

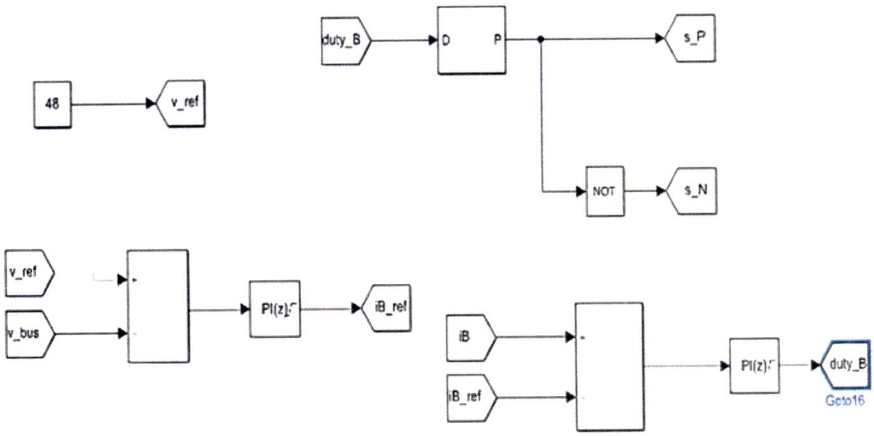

Fig. 8 Double loop PI control logic circuit for the bidirectional converter

Table 1 Simulation parameters

Duty cycle	0.5
F_s (switching frequency)	10,000 Hz
DC load power rating	384 Watts
DC bus rated voltage	48 V
Battery nominal voltage	24 V
Inductor value	1.2 mH
Inductor ripple current	10%
Capacitor (LCL filter)	6.67 μF
Inductor (LCL filter) (L_1=L_2)	**3.80mH**

7.3 Simulation Results

For checking the working of the model with this new bidirectional converter topology, the irradiance level was made to change at a regular time interval, and the corresponding waveforms are perceived.

It can be observed from Figs. 9, 10, 11, 12, 13 and 14 that when the solar PV array receives adequate irradiance, the module generates extra power, which is then used to charge the auxiliary battery backup via the BDC, and a constant DC bus voltage is maintained. When irradiance falls below a specific level and the DC load's power requirement cannot be met by the solar PV system, the BDC switches to boost mode and transfers the deficient power from the battery backup. As depicted in Figs. 10 and 11 when power output of solar PV array is not sufficient, battery soc% starts

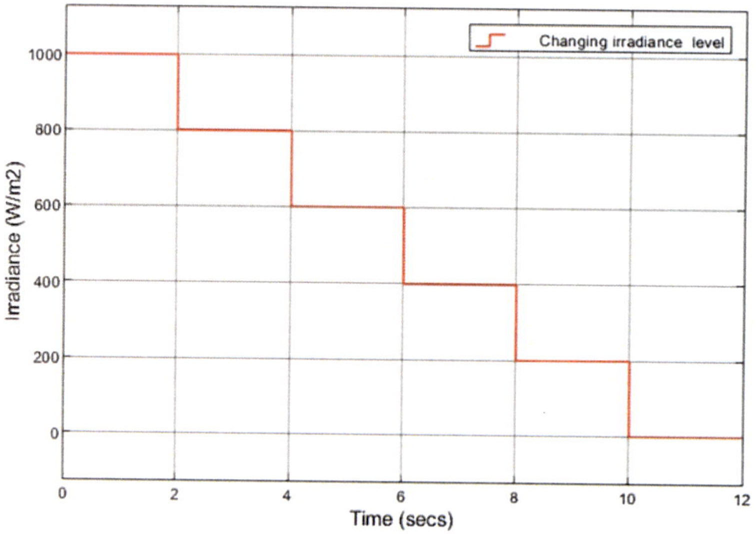

Fig. 9 Change in irradiance level with respect to time

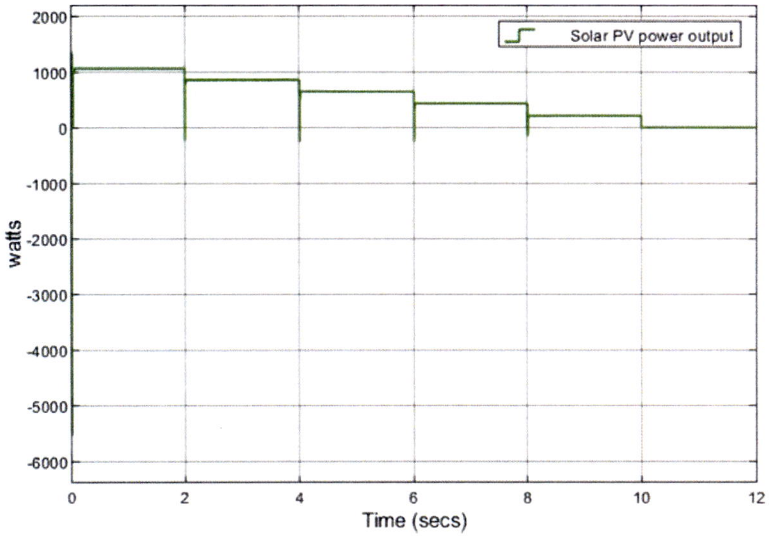

Fig. 10 Power output from solar PV module

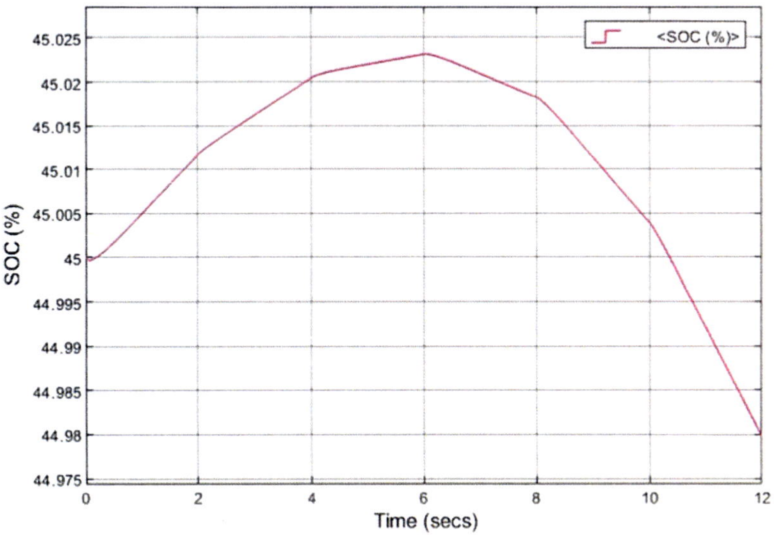

Fig. 11 Battery state of charge (SOC in %)

falling, i.e., battery is getting discharged.[1] When no power is available from the solar PV array, then entire power demand of the DC load is getting satisfied by the

[1] Constant current method of charging and discharging of battery is implemented in the proposed study.

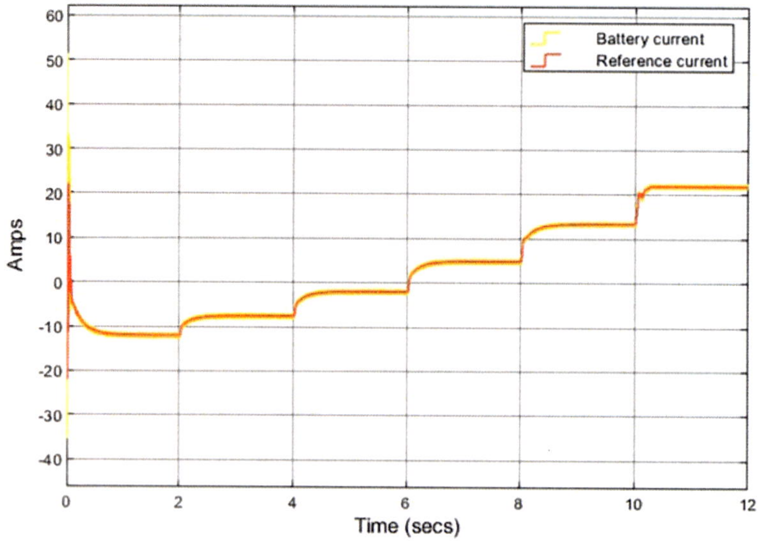

Fig. 12 Battery current and reference current generated by the controller

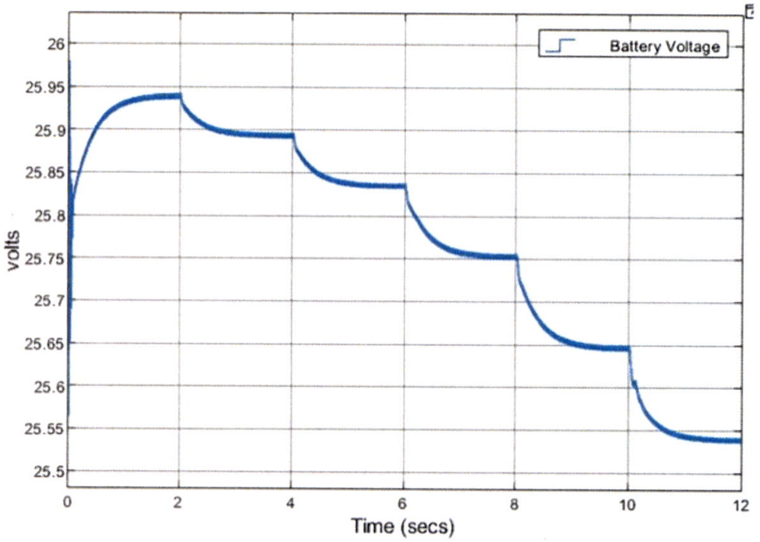

Fig. 13 Battery voltage

battery backup, as can be seen as the rapid decline in battery soc %. Throughout this processes, the voltage across the DC bus is maintained 48 V as shown in Fig. 14.[2]

[2] To obtain these results, the controlling of bidirectional converter is done by double loop PI control which is explained in the Sect. 5 of this paper.

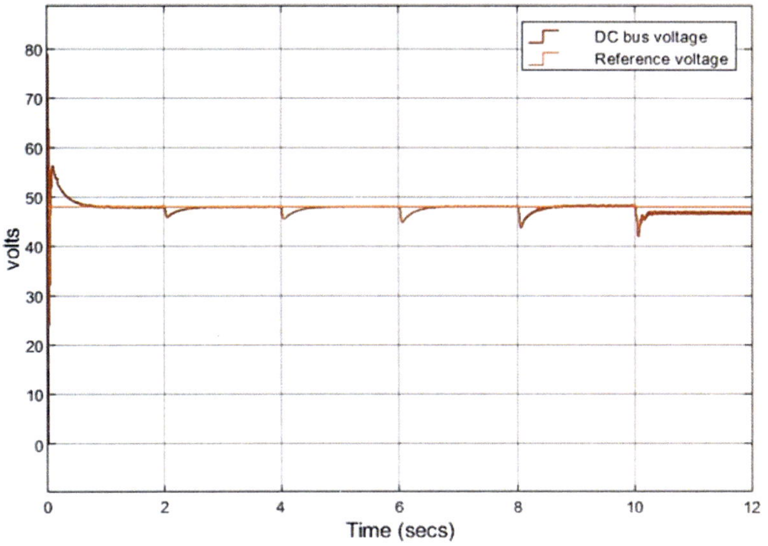

Fig. 14 DC bus voltage

AC Output

In this research work, integration of inverter at the DC bus is done for showing feasibility of the system for AC load. The output of DC bus is fed to a single phase to three phase inbuilt inverter from MATLAB Simulink library, in which corresponding gating signal is generated by level 2 PWM generator with internal generation of reference signal. The output of inverter is then fed to a LCL filter (author in paper [12] has discussed designing of LCL filter) for smoothing of three-phase AC waveform. Then, three-phase voltage and current output are observed which are shown in Figs. 15, 16 and 17 which shows the effectiveness of the proposed methodology.

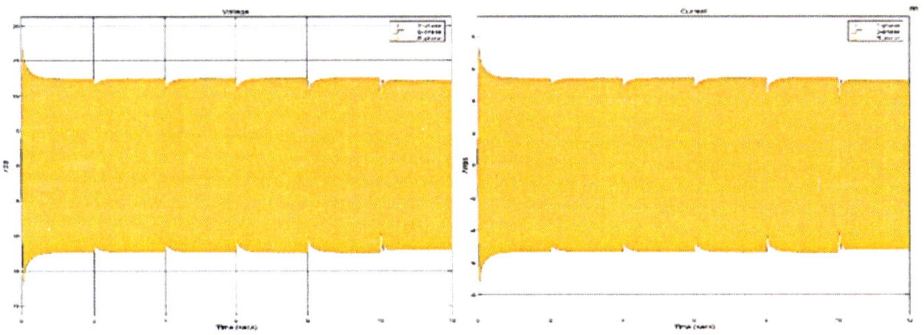

Fig. 15 Output of the inverter

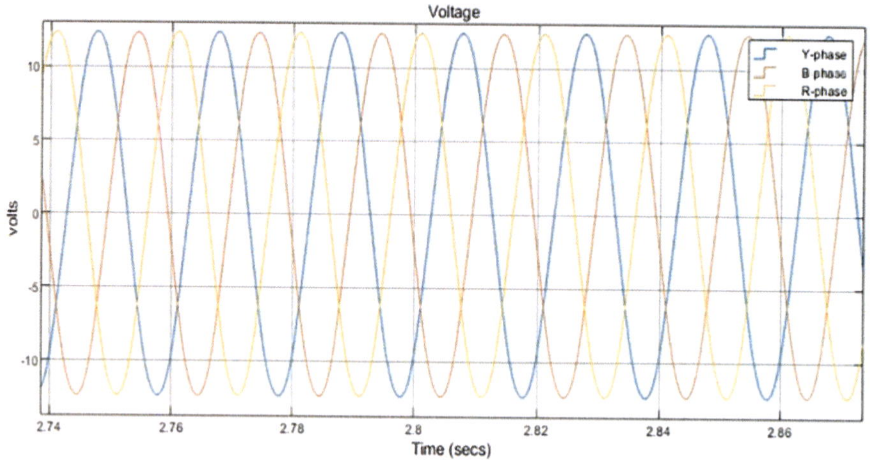

Fig. 16 AC output voltage (magnified view)

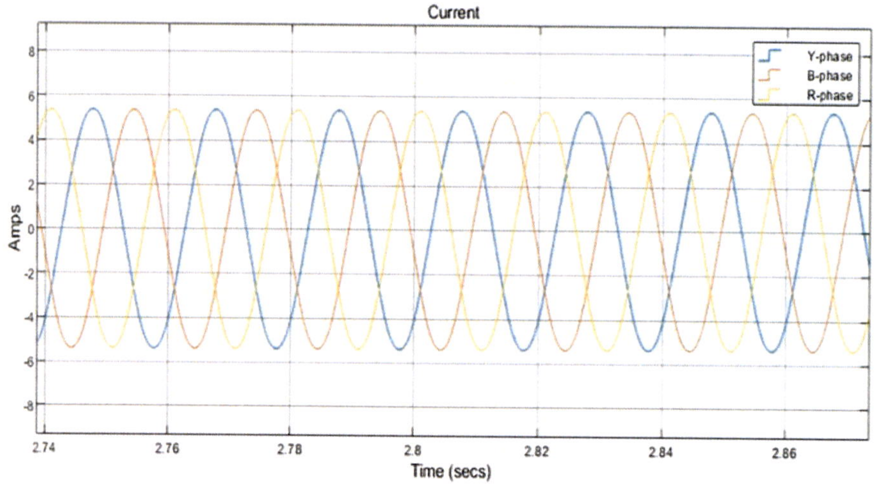

Fig. 17 AC output current (magnified view)

8 Conclusion

A viable alternative strategy for battery charging employing a non-isolated bidirectional converter connected with a solar PV system is proposed in this paper. From the study and test results, it can be concluded that bidirectional converter can work as an alternative for the charging and discharging of the auxiliary power supply. It enables us to perform both buck and boost operation from a single converter and therefore is more efficient and cost-effective in implementation. In this work, it is also found that bidirectional converter can not only be used for DC load, but it is also compatible with AC load which makes it eligible for implementation in the expanding solar PV power plants synchronized with AC grid. The efficiency and dependability of a solar PV system are improved by using a bidirectional converter. The bidirectional converter is available in a variety of topologies, allowing it to work with a wide range of voltages.

Further research work can also be done on the topologies of the converter to improve charging and discharging of battery using various soft computing techniques like genetic algorithm, artificial neural network, fuzzy logic, etc.

References

1. Venkatramanan D, John V (2019) Dynamic modeling and analysis of buck converter based solar PV charge controller for improved MPPT performance. IEEE Trans Ind Appl 55(6):6234–6246. https://doi.org/10.1109/TIA.2019.2937856
2. Elgendy MA, Zahawi B, Atkinson DJ (2013) Assessment of the incremental conductance maximum power point tracking algorithm. IEEE Trans Sustain Energ 4(1):108–117. https://doi.org/10.1109/TSTE.2012.2202698
3. Ranaweera I, Midtgard OM (2016) Optimization of operational cost for a grid-supporting PV system with battery storage. Elsevier Renew Energ 88:262–272
4. Xu X, Zheng C, Hu C, Lu Y, Wang Q (2016) Design of bi-directional DC-DC converter. In: 2016 IEEE 11th conference on industrial electronics and applications (ICIEA), pp 2283–2287. https://doi.org/10.1109/ICIEA.2016.7603972.
5. Tomar PS, Srivastava M, Verma AK (2020) An improved current-fed bidirectional DC–DC converter for reconfigurable split battery in EVs. IEEE Trans Ind Appl 56(6):6957–6967. https://doi.org/10.1109/TIA.2020.3024165
6. Tytelmaier K, Husev O, Veligorskyi O, Yershov R (2016) A review of non-isolated bidirectional dc-dc converters for energy storage systems. In: 2016 II international young scientists forum on applied physics and engineering (YSF), pp 22–28. https://doi.org/10.1109/YSF.2016.7753752
7. Caricchi F, Crescimbini F, Noia G, Pirolo D (1994) Experimental study of a bidirectional DC-DC converter for the DC link voltage control and the regenerative braking in PM motor drives devoted to electrical vehicles. In: 1994 proceedings of IEEE applied power electronics conference and exposition—ASPEC'94 vol 1, pp 381–386. https://doi.org/10.1109/APEC.1994.316373.
8. Deopare H, Deshpande A (2015) Modeling and simulation of incremental conductance maximum power point tracking. Int Conf Energ Syst Appl 2015:501–505. https://doi.org/10.1109/ICESA.2015.7503400
9. Onar OC, Kobayashi J, Erb DC, Khaligh A (2012) A bidirectional high-power-quality grid interface with a novel bidirectional non inverted buck-boost converter for PHEVs. IEEE Trans Veh Technol 61(5):2018–2032. https://doi.org/10.1109/TVT.2012.2192459
10. Jadhav S, Devdas N, Nisar S, Bajpai V (2018) Bidirectional DC-DC converter in solar PV system for battery charging application. Int Conf Smart City Emerg Technol (ICSCET) 2018:1–4. https://doi.org/10.1109/ICSCET.2018.8537391
11. Jin K, Yang M, Ruan X, Xu M (2010) Three-level bidirectional converter for fuel-cell/battery hybrid power system. IEEE Trans Industr Electron 57(6):1976–1986. https://doi.org/10.1109/TIE.2009.2031197
12. Solatialkaran D, Zare F, Saha TK, Sharma R (2020) A novel approach in filter design for grid-connected inverters used in renewable energy systems. IEEE Trans Sustain Energ 11(1):154–164. https://doi.org/10.1109/TSTE.2018.2887079

13. Bhagiya RD, Patel RM (2019) PWM based double loop PI control of a bidirectional DC-DC converter in a standalone PV/battery DC power system. In: 2019 IEEE 16th India council international conference (INDICON), pp 1–4. https://doi.org/10.1109/INDICON47234.2019.9028974.
14. Waghmare T, Chaturvedi P (2020) Study of bidirectional dc-dc converter: control schemes and switching techniques. In: 2020 IEEE first international conference on smart technologies for power, energy and control (STPEC), pp 1–6. https://doi.org/10.1109/STPEC49749.2020.9297713.
15. Rashid MH Power electronics circuits, devices and applications. Pearson publication, 3rd edn. ISBN: 978-93-325-1844-5
16. Islam MA, Merabet A, Beguenane R, Ibrahim H (2013) Modeling solar photovoltaic cell and simulated performance analysis of a 250W PV module. IEEE Electr Power Energ Conf 2013:1–6. https://doi.org/10.1109/EPEC.2013.6802959

Stress and Performance in NGOs: The Mediating Role of Social Intelligence and Ergonomics

Tanushree Sanwal and Puja Sareen

Abstract Stress, most popularly known as strain, tension and pressure, is the experience when we believe we cannot cope up effectively with the contemporary environment. Stress is basically used in physical science which means the pressure of one object on another object. We cannot eliminate stress from our life as people feel stress since they have no control over it. Therefore, it happens time and again. Basically, stress damages both the health of a human being and economy of business enterprises when it comes to industry. In nongovernmental organizations, stress acts as neutral motivating force. However, if the stress is beyond excessive intake, it hampers productivity too. The performance of an employee decreases, and the growth of the industry or NGOs slows down. Hence, it is the primary concern in industrial arena to arrest all the stress contributing factors. Social intelligence as a trait many times helps to reduce stress in a person. This paper is an endeavor to examine the relationship between stress, organizational performance and social intelligence. The paper also analyzes the significance of ergonomics in prevention and management of stress. Ergonomics interventions at the workplace are becoming more and more critical to employee performance and their health especially in nongovernmental sectors.

Keywords Social intelligence · Performance · Stress · Ergonomics · NGOs

1 Introduction

The word 'stress' has now become a very common word in today's world especially during pandemic. The Oxford Dictionary defines it as 'a state of affair which involves demand on physical or mental energy'. Stress is psychological condition or

T. Sanwal (✉) · P. Sareen
Amity University, Noida 201313, India
e-mail: tanushreesanwal12@gmail.com

P. Sareen
e-mail: psareen@amity.edu

A. Tomar et al. (eds.), *Proceedings of 3rd International Conference on Machine Learning, Advances in Computing, Renewable Energy and Communication*, Lecture Notes in Electrical Engineering 915, https://doi.org/10.1007/978-981-19-2828-4_54

circumstance which is not always adverse, but it definitely disturbs a person's whole physiological system leading to many health issues.

Stress is a biological as well as psychological term which was first coined in the 1930s. It is an outcome of our feelings when we react to some events in an extra ordinary manner. Many times, we meet these tough situations with strength and stamina. We are more focused and are highly alert [1], but sometimes stress is also the result of the failure of our body to respond properly to the physical and emotional threats of environment [2].

The symptom of stress includes high adrenaline secretion. Exhaustion, irritation, muscular tension and lack of concentration are also few of its coping symptoms. Stress symptoms include a state of alarm, short-term resistance and a lot of physiological reactions such as headache, anxiety, high blood pressure and elevated heart rate.

Stress is a part of everyone's life, be it personal or professional. Stress in professional life is dynamic because an employee is confronted with both opportunities as well as constraints. An employee is often confused of what he or she desires in their working life. The outcome or result perceived by him or her is generally uncertain and important. Stress is highest in those employees, who recognize that it is doubtful whether they will win or lose. It is lowest in those people at work, who think that winning or losing is not all an important outcome. Sharma [3] Work stress is generally recognized worldwide as a major issue and challenge to a worker's health and which might affect the productivity of any organization or NGOs very badly (ILO [4, 5]).

1.1 Stressors

The incidents that incite stress are called stressors. They cover a lot of situations that directly or indirectly affects a person. Job stressors are very common in today's era especially during pandemic. Research taken up by an industrial body Assocham in 2007 had revealed that 200 employees in India in higher positions as engineers, architectures, managers, quality controllers, editors and copy writers brave stress and take home its undesirable effects as adverse health issues. The ill impact of stress on physical and mental health has become a part of our existence. Leaders in commercial and government sectors such as construction, shipping, banks, health care, IT experts, export and import houses, electronic and print media, courier firms, small-scale enterprises, retail and card franchising organizations are no longer immune to the stress of their jobs [6].

Stressors are caused by both external as well as internal factors. Both personal factors related to individual and environmental agents are prime cause of stress. Conflicts, feeling of guilt and fear of inadequacy are other factors leading to stress. Perception of individual is also at times cause of stress in individual. Environmental stressors are conditions which are external and beyond our control. External stressors, according to Bhagat [7], might seriously disrupt the work routine. Structure issues, management's use of authority, boredom, a lack of opportunities for promotion, enormous duties, confusing demands from leaders, value conflicts and unrealistic workloads are all major sources of stress in NGOs. A worker's personal life, which includes his or her family, friends, health and financial circumstances, can also cause stress and have an impact on his or her professional performance.

1.2 Organizational Performance

The relative potency of an individual's involvement in a corporation is referred to as 'organizational performance.' Employees with excellent performance are less likely to quit for another job and are more likely to execute at greater levels while staying in the same employment, which is vital in today's society. The quality of related functions or processes carried out in companies is referred to as 'work performance' [8]. Performance is also a measure or indication that indicates an organization's level of achievement and progress [9]. In pandemic scenarios, an employee's excellent performance becomes an essential prerequisite for generating better results in an organization, particularly in the non-profit sector. When employees are not under stress, they perform better. For such employees, finding a work–life balance is critical. When their personal and professional lives are stress-free, they may perform well. Furthermore, social intelligence has been used as a method to improve human performance. People who are socially intelligent in the workplace perform better and accomplish more. They aid in the achievement of an organization's goals and objectives [10–13]. In today's world, social intelligence is the buzzword. A person who is socially intelligent is intelligent and works in accordance with the current situation. He or she is diplomatic and has a good recall. He or she is also self-assured and cooperative, both of which are necessary for great performance in any organization or NGO. As a trait, social intelligence aids in the reduction of stress in people, which in turn improves organizational performance, particularly in non-profits. Many studies have shown that a socially aware manager or leader can help employees perform better in the workplace by lowering stress [12]. Employees' social and emotional intelligence has also served as a stress-relieving factor as well as a facilitator of good health [5]. Employees who are no longer stressed out enjoy their jobs and perform better in all situations, whether pandemic or new normal.

There are various issues at work that cause stress and lower organizational performance, such as physical and mental aspects.

Stress has a number of effects on a person's physique in addition to lowering organizational effectiveness. According to Williams [14], excessive stress levels can manifest themselves in a variety of ways, including mood swings, exhaustion, anxiety, outbursts, sadness and so on.

1.3 The Effects of Work Stress

Stress affects different people in different ways. When people are in stress, they cannot maintain a balance between work and personal life. At the same time, they get themselves engaged in unhealthy activities, and their immunity decreases. When in stressed employees may be:

- Distressed and irritable
- Lose concentration
- Have difficulty in thinking logically and making decisions
- Low motivation
- Depressed
- Anxious.

The organizations or NGOs which do not get the best from their stressed workers are at a high risk of low productivity and growth. This overall deteriorates the image of the company and leads to:

- Increased absenteeism
- Lack of engagement and commitment
- Increasing employee turnover
- Low productivity
- Enhanced poor working practices and accident rates
- Increased complaints from clients and customers.

On the whole we can say the symptoms of stressed employees are many like adverse effect on performance, psychological disorders, physical disorders, etc.

1.4 Stress-Related Hazards in NGOs

There are many stress related negative outcomes when it comes to working with non-governmental sectors like employees feel heavy work load with extended working hours or extensive travel or when meeting the expectations of the caregivers or stakeholders. Maintaining work–life balance gives lot of stress to employees.

1.5 Stress Management in NGOs

Stress is a pervasive part of any workplace in the industry or NGOs. It is bound to be happened and comes in to the behavior of employee be it physical or mental. Stress is also related to customers. Researchers using machine learning and algorithms have predicted high stress in customers [15]. Therefore, it becomes quite pertinent for managers to deal with it. Hence, the relation of stress with performance and social intelligence needs to be linked properly [16].

1.6 Relationship of Stress and Organizational Performance

Stress and organizational performance are linked positively as well as negatively. An employee in any NGO will not perform properly until some amount of stress is built on him or her. However, if the amount of stress exceeds or work load increases, the employee starts showing negative results in his or her performance. The productivity decreases, and overall performance and productivity of an organization decreases.

There are three levels of stress in any employee:

1. Extremely high stress
2. Moderate stress
3. Extremely low stress.

Figure 1 depicts the relation of stress with performance of an employee. The low-level and extremely high-level stress as shown calm and distress, respectively, shows deterioration in performance. However, in case of moderate level of stress, depicted as 'eustress' shows the highest level of performance in an employee. The extreme of high level and low level of stress have negative impact on employee performance. It reduces the productivity of an employee along with his or her health. Thus, proper management of stress must take place in the organizations and NGOs to boost performance and reduce health hazards.

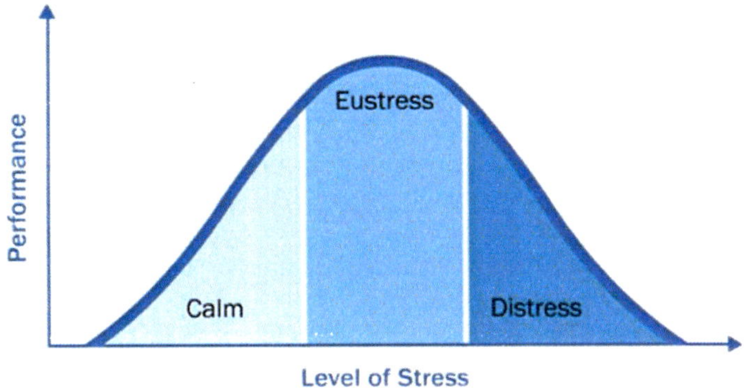

Fig. 1 Relation of stress with performance of an employee

1.7 Relationship of Stress and Social Intelligence

Studies have revealed that socially intelligent people can overcome stress and live a healthy life [5, 17]. Stress and social intelligence have a unique relation and both affect the health. Stress generally has ill effect on health but people who have social intelligence have a better physical and mental health [9]. They generally excel in a better way in life than others. Hence, we can say that social intelligence as a trait helps to manage stress in individuals specially working in NGOs.

Besides social intelligence other possible and viable methods through which stress could be arrested in NGOs are as follows:

(a) **Identify unhealthy stress in organization**

The moment it is understood that effect of stress on performance, it becomes imperative to identify the causes of unhealthy stress [18]. It is quite difficult for managers to identify the level of stress because people behave in a different manner.

(b) **Create Organizational Climate supportive of employee**

Generally, the bureaucracies are developed through formal, inflexible and impersonal climate. Therefore, the stress bounds to be happened. Hence, the organizational environment may be made less formal and more supportive to employee in their needs.

(c) **Job to be made more interested**

The management if tries to make the job interesting as far as possible, (without tempering and hampering the original nature of work) it can also prevent the development of stressor at workplace.

(d) **Design and create career counseling**

The management must give proper counseling on career growth to the employees.

(e) **Encourage and help employees handling stress**

The stressors are quite common in circumstances where employees are interacting with different situations. In such cases, either the behavior of employee is insufficient or inappropriate to deal with the situation. [16, 19]. Hence, organizations specially NGOs must take steps to overcome such conditions of stress in the employees.

1.8 Stress Prevention and Management Through Ergonomics

The word ergonomics is made up of two Greek words 'Ergon and Nomos' which means work and natural law, respectively. The combined meaning of ergonomics is the science of work and an individual relationship to that work. The meaning given in Oxford dictionary is the study of people's efficiency in their working environment. In today's world, an employee spends major portion of time at workplace, keeping his health always in risk. Hence, today there are not only specialist for advising about the right posture and exercise but also highly specialist furniture and accessories that are available in the market for preventing repetitive ergonomics injuries and promoting health. Ergonomics is also known as human factor engineering. Using proper posture and body mechanic, good placement of computer equipment comfortable height of tables is all aspect of ergonomics. These ergonomically designed structures help to reduce stress in employees which in turn enhances their health and performance in any kind of organization.

2 Enhance Organizational Performance and Social Intelligence Through Ergonomics

Better output of employees and organizational growth can be achieved through ergonomics. Eco design principles help in better output [20]. Moreover, when the workplace is designed properly, a person is able to show his caliber and intelligence in a better manner. He or she becomes more creative and cooperative. Following are some of the ways through which enhanced performance and socially intelligent behavior can be achieved ergonomically:

- Correction in physical environment at workplace
- Handling of back injuries
- Behavior-based safety
- Regular health checkup camp
- Yoga classes
- Provide gym at workplace.

3 Conclusion

From the above review of literature, it is very clear that stress plays an important role in our life. Healthy body and healthy mind are both needed for us. Both of this can be achieved in any NGOs when we design our work station according to the principles of ergonomics. Ergonomics helps us to live stress-free life which in turn boosts employee performance in an organization. Productivity of the organization can also be increased by hiring socially intelligent behavior. Social intelligence is an important parameter which helps us to reduce our stressors and work smartly. Our skills of social intelligence like patience, cooperativeness, confidence, etc., can also be enhanced when we follow the principles of ergonomics. Thus, we can say ergonomics is an important aspect which will help us to prevent and manage stress. It will also help us to enhance performance in any organization or NGOs. Keeping these in mind the researcher has proposed the following model, as shown in Fig. 2 after extensive literature survey:

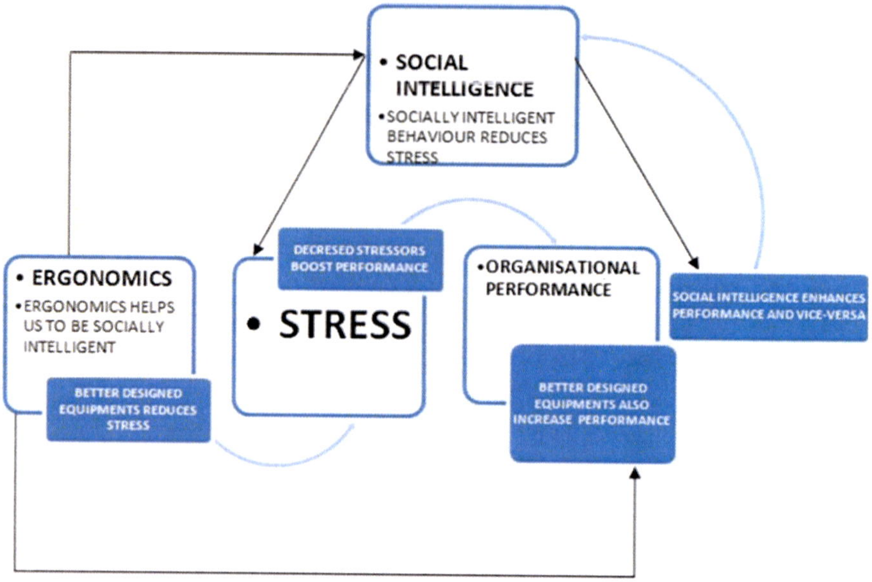

Fig. 2 Model

4 Discussion

The workplace is venue for employee passing substantial time with due application of all the energy and knowledge; therefore, the pressure of carrier growth on employee and getting maximum output by employer are two major concerns. Sometime the clashes of these two concerns to prove on the criteria given by employer create strain and stress in employee. This is the time for employee either to go away with unwanted elements of workplace or get minimum impact and survive. But it is not as easy as said or written. Stress damages both the health of a human being and economy of business enterprises when it comes to NGOs. Hence, it is the primary concern in nongovernmental sectors to arrest all the stress contributing factors. The cool temperament, the balance state of mind and body and awareness about the path adopted can pave the way. Ergonomics too has lot of importance in stress management. Ergonomics interventions at the workplace are becoming more and more critical to employee health. These interventions are boosting performance of an employee and making them more socially intelligent. To achieve the state of sound mind in sound body, it is imperative to take measures such as yoga, gym, exercise and participation in sports. The physical environment is also required to be made and framed to make the employee more and more comfortable at work. The carrier counseling for employee is another way of handling it. But the dynamic nature of industrial arena adds new visible and invisible stressors to be handled at work place. Hence, it is on the part of both employee and employer to find out ways and means either to eradicate or nullify the impact of stressors as far as possible.

References

1. Teens Health (2010) Available at http://kidshealth.org/teen/your_mind/emotions/stress.html Accessed on 18 Apr 2010
2. Hans S (1956) The stress of life. McGraw-Hill, New York
3. Sharma RK (2010) Industrial psychology, Kedar Nath Ram Nath, Merrut
4. International Labour Organization (ILO) (1992) Preventing stress at work. conditions of work digest, 11, International Labour Office, Geneva
5. Landa JMA, López-Zafra E, Martos MPB, del Carmen Aguilar-Luzon M (2008) The relationship between emotional intelligence, occupational stress and health in nurses: a questionnaire survey. Int J Nurs Stud 45(6):888–901
6. Dance with shadows, Work Stress in India (2010), available at http://www.dancewithshadows.com/society/work-stress.asp Accessed on 18 Apr 2010
7. Bhagat RS (1983) Effects of stressful life events on individual performance effectiveness and work adjustment processes within organizational settings: a research model. Acad Manag Rev 8(4):660–671
8. Neely AD, Adams C, Kennerly M (2002) The performance primes: the scorecard for measuring and managing stakeholder relationship, Financial Times/Prentice Hall, London
9. Hooda D (2009) Social intelligence as a predictor of positive psychological. Health 35(1):143–150
10. Albrecht K (2009) Social intelligence: the new science of success. Pers Excellence 10(12):5
11. Emmerling RJ, Boyatzis RE (2012) Emotional and social intelligence competencies: cross cultural implications. Cross Cultural Manage: An Int J 19(1):4–18
12. Goleman D (2006) Social intelligence: the new science of social relationships. Bantam Books, New York
13. Riggio RE, Reichard RJ (2008) The emotional and social intelligences of effective leadership: an emotional and social skill approach. J Manag Psychol 23(2):169–185
14. Williams JC (1982) Human behaviour in organizations (Cincinnati: South-Western), 212–13
15. Srinivas KNR, Manikanta KSS, Jacob TP, Nagarajan, G, Pravin A (2020) Customer stress prediction in telecom industries using machine learning. In: International conference on emerging trends and advances in electrical engineering and renewable energy, pp 491–498. Springer, Singapore
16. Certo SC, Certo ST (2008) Modern management, PHI learning, New Delhi
17. Goleman D, Boyatzis R (2008) Social intelligence and the biology of leadership. Harv Bus Rev 86(9):74–81
18. Smereka CM (1990) Outwitting, controlling stress for a healthier life style. Healthc Fin Manage 44:70–75
19. Tomar A et al (2020) Machine learning, advances in computing, renewable energy and communication, Springer Nature, Berlin, LNEE vol 768, 659 p https://doi.org/10.1007/978-981-16-2354-7 (ISBN 978–981–16–2354–7)
20. Lhopital V, Bordignon M (2017) Overview of eco-design applications on various types of electronic product development. In: International conference on eco-design in electrical engineering, pp 3–7. Springer, Cham

Imputing Missing Data in Electronic Health Records

Aashi Pathak, Shivani Batra, and Himanshi Chaudhary

Abstract *Introduction*: Internet of Medical Thing is an amalgamation of IoT devices with health sensors devices. Together these devices are a blessing to humans. IoMT has helped humans a lot as so many health aspects like cholesterol level, creatinine ratio, urea, blood sugar level, and many more are known before worsening the health of people. IoMT devices produce data, on which medical decisions are taken. This data is crucial and must complete. In data, there can be missing values or corrupted data. This paper is dealing with missing values that must be imputed. *Objective*: To impute the missing values, the diabetes dataset produced from the IoMT devices. So, to impute the data, we have used two categories of imputation techniques: statistical techniques—zero, mean, median, and mode; machine learning algorithms—SVM and KNN. Accuracy is evaluated for all these above-mentioned techniques, that is what percent of data is correctly imputed with these mentioned. *Methods*: We compared the above-mentioned algorithm on the electronic healthcare dataset. Analysis of the most suited algorithm on EHR data is done. *Results*: After considering EHR data, 1000 rows and eight attributes with 10%, i.e., 800 missing values, and after applying imputation techniques, we got the result as when compared with the above-stated algorithms, KNN has given the best accuracy in imputing missing data. *Conclusion*: KNN produced the best results by imputing missing values by 72.87% in EHR data. The future scope of this paper says that accuracy can be increased more if the value of K in KNN algorithm is changed to another number.

Keywords Electronic health record · Database · Data · Missing value · Imputation · Zero · Statistical methods · KNN imputation

1 Introduction

As EHRs were created to record and enhance patient care and expedite billing, rather than as research resources, there are major obstacles in using this data to learn more

A. Pathak · S. Batra (✉) · H. Chaudhary
KIET Group of Institutions, Delhi-NCR, Ghaziabad, India
e-mail: pathakaashi2@gmail.com

A. Tomar et al. (eds.), *Proceedings of 3rd International Conference on Machine Learning, Advances in Computing, Renewable Energy and Communication*, Lecture Notes in Electrical Engineering 915, https://doi.org/10.1007/978-981-19-2828-4_55

Fig. 1 IoT layer architecture

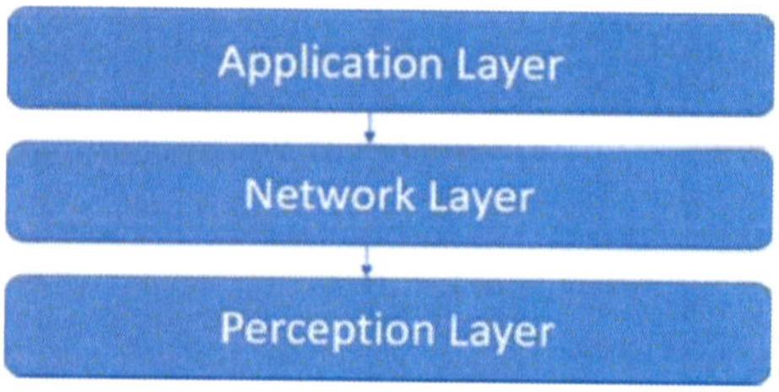

about human health. It is critical to create techniques for dealing with missing data as EHR data becomes more widely used as a source of phenotypic information for biomedical research [1]. IoT in the medical field is often referred to as the Internet of Medical Things (IoMT). The Internet of Medical Things (IoMT) integrates Internet of Things (IoT) technology with healthcare services to just provide real time, remote patient monitoring, and treatment. Because most communication protocols are not created particularly for the demands of connected medical devices, it is necessary to categorize the various IoT communication technologies in terms of security [1]. IoMT describes that how almost all the devices that are used in the medical field are integrated and connected over the Internet. Perception layer, network layer, and application layer are the three levels that make up the Internet of Things, from bottom to top [2] as shown in Fig. 1. The perception layer, which comprises sensors for perceiving and receiving information about the environment, oversees data collecting among them. It recognizes a variety of physical features as well as other intelligent objects in the environment [3]. The network layer takes care of the communication of one layer to other one [3] and is also responsible for transmitting signals from sensors to the cloudlets. Finally, in the application layer, analytics and diagnosis process are performed [3]. The data is provided by the application layer. Application-specific services are provided to the user.

However, the data that we get has high chances of having some missing data because of reasons such as faulty sensor, malware attack, any human error, any network drift, any bug in the software pipeline, etc. Anything that results in the missing data should get. Missing values must be handled with care and precision accurately [4].

2 Related Work

The purest form of electronic clinical data acquired in a medical facility, hospital, clinic, or practice at the point of treatment. Electronic health records (EHRs) and electronic medical records (EMRs) are two terms for the same thing.

Even though hospitals and ambulatory care centers are rapidly embracing EHRs, the majority of user-level usage is still focused on billing [5]. ECG, temperature, electromyography (EMG) muscle activity, respiration rate, sweating, and blood glucose levels will all be measured via wearable sensors. Those sensors must be powered by tiny and low-energy batteries. The batteries are supposed to function continuously without needing to be recharged or replaced. Most data gathered in IoT is the time series data [6]. Data loss is a regular occurrence in the IoT for a variety of reasons, including poor network connectivity, synchronization problems, faulty sensors, and other forms of device failure [2]. The data with missing values is removed from the initial partial dataset to make it a full dataset in case deletion is necessary [7]. The way how we replace or estimate the missing data is known as imputation. This problem of the missing data of data science can be broadly classified in three categories [8].

- **Missing values/data at random (MAR)**: Missing at random (MAR) is a more realistic assumption for anesthetic research. The data is deemed MAR when the likelihood of missing responses is established by the collection of actual responses rather than the specific missing values expected to be created [9].
- **Missing values/data completely at random (MCAR)**: In this the missing values appear at random, implying that the data is missing regardless of seen or unobserved data [7]. Missing entirely at random occurs when the chance of missing data is unrelated to either the specific value that should be reached or the set of observed responses (MCAR) [10].
- **Missing values/data not at random (MNAR)**: If the record characters do not match those of MCAR or MAR, they are categorized as missing not at random, but rather on a case-by-case basis (MNAR) [11]. In this case, the only way to get an unbiased estimate of the parameters is to model the missing data. Deleting the value that is missing is known as marginalization [12]. The other way is to estimate the missing value which is known as imputation. There are two ways how we can impute the missing values
- **Single-valued imputation**: In single value imputation, missing data is replaced with a single value, such as the mean score of the complete instances in the study sample (i.e., mean imputation) [13]. Single imputation involves a little less computation for the imputation of the missing data. Single imputation approaches, in general, does not count for the uncertainty of missing data, resulting in estimates with excessively tiny standard errors [14]. For example, forward the previous value to the missing position, mean imputation, or many other statistical imputation methods. Also, we have machine learning approaches like KNN, SVM, etc.
- **Multiple-valued imputation**: It begins by sampling from an imputation model to input values for missing data numerous times. If factors that determine unresponsiveness are not included as the conditioning variables, related imputed values are marked as "satisfied," and users are free to avoid the imputations. Because a single unique value may be computed using more easy methods, this methodology is not intended for inventing different values [15]. For multiple imputation, we can use ensemble methods [16].

3 Dealing with Medical Data

The steps that are required to be followed for the process of imputation by ant method are mentioned in Fig. 2. These steps show the overall process of dealing with the data.

As shown in Fig. 2, the first or preliminary step is the data collection step, in which it gets decides that which particular data is to be worked on, followed by the next step which is data preprocessing in which basic steps are taken, which is then followed by the data imputation step in which various methods are used to replace or impute the missing values and the last step is to evaluate the imputed value. Now when we have the data with us it depends if the data is extremely critical or moderately critical or low critical. Some other ways, we must impute the missing data are as follows.

- Impute the missing value with blank, NAN, and placeholder.
- Dropping the rows with null values.

These ways are one of the easiest ways to replace or estimate the missing value, but sometimes the replacements made from these methods are much different from the original missing data, and the values does not match with the actual missing data which many times causes erroneous dataset. Apart from these methods, there are many other approaches as well like Jordan neural network, genetic algorithm as a form of wrapper method, and many others, also we have statistical methods to impute the missing data. Statistical techniques that appropriately manage missing data can assist to reduce biasness and increase estimate precision [17].

Few of the statistical methods are zero imputation, mean imputation, median imputation, mode imputation, standard deviation imputation, etc. Some commonly used machine learning techniques are KNN imputation, SVM imputation, regression, ensemble methods, etc.

1. **Zero Imputation**: In this, the missing value is replaced or imputed by zero (0) [17]. This is represented by the following function. This is an easy and simple way to compensate for the dataset that is a nearly non-critical [12].

$$Z(I(a,b)) = \begin{cases} 0 \text{ if } I(a,b) \text{is the missing value} \\ I(a,b), \text{ otherwise} \end{cases} \tag{1}$$

where Z represents the zero imputation,
a is the tuple, and b being the attribute.
I is the instance of (a, b).

2. **Mean Imputation**: Imputes the missing value with the average of the attribute in the entire dataset

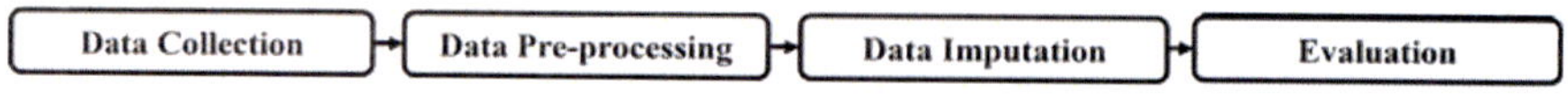

Fig. 2 Data flow architecture

$$A(I(I,f)) = \begin{cases} \dfrac{\sum_{k=1}^{|d|} I(i, f) \text{if (i,f)is the missing value}}{|D|} \\ I(i, f) \text{otherwise} \end{cases} \quad (2)$$

where A is the mean imputation, I is the instance at ith tuple and fth attribute, and d is the entire dataset.

3. **Median Imputation**: One of the simple imputation methods includes median imputation. Median is the method in which the middlemost value of a particular attribute gets imputed at the missing position of the dataset [18].
4. **Mode Imputation**: One of single imputation strategy is by substituting the missing values by the mode. Mode is the most frequently used value among all the values of that attribute in the entire dataset [19].
5. **Impute the missing values with SVM**: Support vector machines (SVM) is used for prediction of missing values by using classification based on the statistical learning theory [17].

    a. **Linear SVM**: For the cases, when we have linearly separable datasets that is classifying the dataset using a straight line.
    b. **Nonlinear SVM**: When we have the data that cannot be classified by a straight line, and the data is nonlinearly separable.

6. **Impute the missing values with KNN**: This K-nearest neighbor (KNN) algorithm is a robust way that is used to impute the missing values in the dataset. This same method can be used on EHR data. KNN imputation is also often regarded as effective replacement of some traditional imputation methods. To use KNN imputation, Python is considered best. K-nearest neighbor imputation (KNNI) splits the entire dataset into two [12]. One dataset consists of incomplete data and the other one is the complete data, i.e., without missing values. This algorithm finds the k-nearest node using few distance methods like the [20] Euclidean distance, Manhattan distance and there are several other ways to find the distance between the missing data node and k-nearest node. In this paper, we will discuss KNN using the Euclidean distance. The Euclidean distance is calculated by distance(p,q). This method of imputing EHR can be used over cloud as well [13].

$$\text{Distance (p, q)} = \sum_{i=1}^{n} \sqrt{p^2 - q^2} \quad (3)$$

where p = {p1, p2, p3...p_n} and q = {q1,q2, q3...q_n}.

The distance measure used to identify the k-nearest neighbors of a sample point determines the performance of the KNN algorithm. Basically, in KNN technique, the value that's missing checks for the complete K values that are neighbor to it in the dataset. Once the K number of neighbors are decided, you can apply any method to impute the missing value.

7. **Ensemble Methods**: The ensemble method has two types. Homogeneous ensemble is constructed from the same type of learners. Heterogeneous ensemble is used when we have different types of learners. The presentation of concepts of medical field is done in archetypes [21, 22].

4 Results

The dataset has been taken from Mendeley [23]. The language used to compute on this dataset is Python using Spyder. In the healthcare area, electronic, digital, and standard health data is extremely desirable [24]. This electronic medical data must be exchanged globally via semantic interoperable standards [25]. The database that has been taken into consideration is having all the values, to apply the techniques some imputations have been introduced on this numerical data. The diabetes dataset was created by extracting data from patients' files from the City Hospital's laboratory and entered it into a database. This dataset that is used has 1000 tuples and eight attributes. Initially, as the medical data that was chosen did not have missing data as it was a complete dataset but for experimental reasons, we had to introduce 10% missing data. Calculations have been done according to the 10% missing data.

Table 1 shows the results of all imputation strategies, with zero imputation yielding 11%, mean imputation yielding 41.25%, mode imputation yielding 31.54%, and median imputation yielding 28.67%. Apart from these four statistical imputation techniques, two machine learning approaches were used: SVM imputation yielding 59.38% accuracy and KNN imputation yielding 72.87% accuracy. This demonstrates that, of all the algorithms employed in this research, KNN is the most appropriate and produces the closest answers to the true data. Figure 3 is the graphical representation of the Table 1 that consists of the accuracies of all the statistical and machine learning approaches that are discussed in this paper top impute the missing values.

Table 1 Accuracy of techniques

Approach	Accuracy (%)
Zero imputation	11
Mean imputation	41.125
Mode imputation	31.54
Median imputation	28.67
SVM imputation	59.38
KNN imputation (k = 2)	72.87

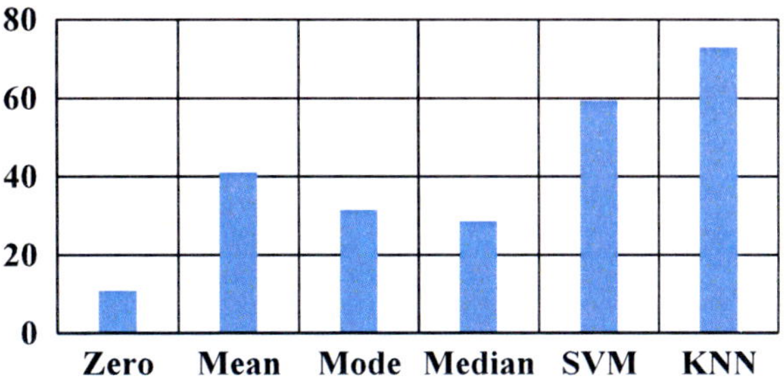

Fig. 3 Graphical representation of the accuracies

5 Conclusion

In this paper, we have discussed about the scenarios of how the medical data is collected, how it travels through various layers and how to deal with the data that has missing values. The accuracies of all the techniques have been mentioned in the chart and table. It is observed that machine learning approach is better than statistical approaches. We used six techniques to assess the correctness of imputation in this study; however, we lacked a sufficient decision-making mechanism to assist us in determining which strategy is the most appropriate. In the future, an algorithm in the future to give us an indication of which strategy will give us the best results for imputation.

References

1. Pigott TD (2001) A review of methods for missing data. Educ Res Eval 7(4):353–383
2. Huang SF, Cheng CH (2020) A safe-region imputation method for handling medical data with missing values. Symmetry 12(11):1792
3. Mera-Gaona M, Neumann U, Vargas-Canas R, López DM (2021) Evaluating the impact of multivariate imputation by MICE in feature selection. PLoS ONE 16(7):e0254720
4. Wood AM, White IR, Thompson SG (2004) Are missing outcome data adequately handled? A review of published randomized controlled trials in major medical journals. Clin Trials 1(4):368–376
5. Kim E, Rubinstein SM, Nead KT, Wojcieszynski AP, Gabriel PE, Warner JL (2019) The evolving use of electronic health records (EHR) for research. Semin Radiat Oncol 29(4):354–361. WB Saunders
6. Jakobsen JC, Gluud C, Wetterslev J, Winkel P (2017) When and how should multiple imputation be used for handling missing data in randomised clinical trials–a practical guide with flowcharts. BMC Med Res Methodol 17(1):1–10
7. Beaulieu-Jones BK, Lavage DR, Snyder JW, Moore JH, Pendergrass SA, Bauer CR (2017) Characterizing and managing missing structured data in electronic health records. bioRxiv, 167858.
8. Al-Milli N, Almobaideen W (2019) Hybrid neural network to impute missing data for IoT applications. In: 2019 IEEE Jordan international joint conference on electrical engineering and information technology (JEEIT).IEEE pp 121–125

9. Pedersen AB, Mikkelsen EM, Cronin-Fenton D, Kristensen NR, Pham TM, Pedersen L, Petersen I (2017) Missing data and multiple imputation in clinical epidemiological research. Clin Epidemiol 9:157
10. Koutras D, Stergiopoulos G, Dasaklis T, Kotzanikolaou P, Glynos D, Douligeris C (2020) Security in IoMT communications: a survey. Sensors 20(17):4828
11. Alsaber AR, Pan J, Al-Hurban A (2021) Handling complex missing data using random forest approach for an air quality monitoring dataset: a case study of Kuwait environmental data (2012 to 2018). Int J Environ Res Public Health 18(3):1333
12. Cheng C, Huang H (2019) A distance-threshold K-NN method for imputing medical data missing values. J Adv Comput Netw 7(1):13–17
13. Aleryani A, Wang W, De La Iglesia B (2020) Multiple imputation ensembles (MIE) for dealing with missing data. SN Comput Sci 1(3):1–20
14. Khan MMR, Arif RB, Siddique MAB, Oishe MR (2018) Study and observation of the variation of accuracies of KNN, SVM, LMNN, ENN algorithms on eleven different datasets from UCI machine learning repository. In: 2018 4th international conference on electrical engineering and information and communication technology (iCEEiCT). IEEE, pp 124–129
15. Karahalios A, Baglietto L, Carlin JB, English DR, Simpson JA (2012) A review of the reporting and handling of missing data in cohort studies with repeated assessment of exposure measures. BMC Med Res Methodol 12(1):1–10
16. Kang H (2013) The prevention and handling of the missing data. Korean J Anesthesiol 64(5):402
17. Khan SI, Hoque ASML (2020) SICE: an improved missing data imputation technique. J big data 7(1):1–21
18. Sethi P, Sarangi SR (2017) Internet of things: architectures, protocols, and applications. J Electr Comput Eng
19. Batra S, Sachdeva S, Bhalla S (2019) Generic data storage-based dynamic mobile app for standardised electronic health records database. Int J High Perform Comput Networking 15(1–2):91–105
20. Batra S, Sachdeva S, Bansal A, Bansal S (2018) Modeling sparse and evolving data. International conference on big data analytics, pp 204–214. Springer, Cham
21. Sathya M, Madhan S, Jayanthi K (2018) Internet of things (IoT) based health monitoring system and challenges. Int J Eng Technol 7(1.7):175–178
22. Tomar A et al. (2020) Machine learning, advances in computing, renewable energy and communication, Springer Nature, Berlin, LNEE 768, 659 pp https://doi.org/10.1007/978-981-16-2354-7. (ISBN 978–981–16–2354–7)
23. Rashid A (2020) Diabetes dataset. Mendeley Data, V1. https://doi.org/10.17632/wj9rwkp9c2.1
24. Batra S, Sachdeva S (2016) Organizing standardized electronic healthcare records data for mining. Health Policy Technol 5(3):226–242
25. Sachdeva S, Batra S, Bhalla S (2017) Evolving large scale healthcare applications using open standards. Health Policy Technol 6(4):410–425
26. Mary IPS, Arockiam L (2017). Imputing the missing data in IoT based on the spatial and temporal correlation. In: 2017 IEEE international conference on current trends in advanced computing (ICCTAC). IEEE (pp 1–4)

An Analytical Review of Machine Learning and Deep Learning Approaches for Document Forgery Detection

Garima Jaiswal, Arun Sharma, and Sumit Kumar Yadav

Abstract Document forgery cases are widespread and have flourished in the modern world of gadgets and technology. The manual approaches of document verification are time consuming and destructive, paving the way for efficient and automated document verification systems. In this paper, we reviewed the work done by the researchers to detect document forgery through source printer identification, verifying identity documents and signature. This study analyzes the machine learning and deep learning approaches, results, datasets, and limitations used to detect document forgery for each domain. This review can help the researchers, digital document forensic, and analysis communities to provide them with a view of various types of document forgeries and future research prospects.

Keywords Document forgery · Machine learning · Deep learning · Source printer identification · Identity documents · Signature verification

1 Introduction

With the growing evolution in the pervasive use of the Internet, smartphones, laptops, and other electronic gadgets have automated everyday tasks. Be it from buying groceries and fruits to making online transactions. This digitalization has made human life productive, faster, and relaxed by saving time, money, and effort. Despite all benefits, it has a flip side. It has also paved the way for many digital crimes and alterations. Among such crimes, document forgery is one of the most prevailing and destructive offenses.

G. Jaiswal (✉) · A. Sharma · S. K. Yadav
Indira Gandhi Delhi Technical University for Women, Delhi, India
e-mail: garima121@gmail.com

A. Sharma
e-mail: arunsharma@igdtuw.ac.in

S. K. Yadav
e-mail: sumit@igdtuw.ac.in

A. Tomar et al. (eds.), *Proceedings of 3rd International Conference on Machine Learning, Advances in Computing, Renewable Energy and Communication*, Lecture Notes in Electrical Engineering 915, https://doi.org/10.1007/978-981-19-2828-4_56

Document forgery is a money-making venture practiced in several parts of the world. It is done with the wrong intentions that generally involve creating, modifying, or emulating objects or documents. It can be of any form—simple forgery, freehand simulation, tracing, or electronic manipulation that involves signature forgery, falsifying or modifying printed or identity documents, emulating title or property certificates, paper forgeries, ink forgeries, and many more. Document forgery detection is a challenging task that imposes questions on the integrity, trust, and authenticity of the document [1]. Many approaches were examined to detect document forgeries such as anomaly detection, cryptographic hashing, digital signatures, optical character recognition (OCR), distance measure, bag-of-words, visual saliency, machine learning classifiers, and deep learning models.

Signatures are used in any form—offline or online as an essential way of authenticating the transactions. People signs checks, approve documents and contracts, validate credit card transactions, and verify activities through signatures [2]. Identity is the collection of unique features that relate to a particular person. These unique features can include the date of birth, fingerprints, ethnicity, birthmark, eye color, and many more. Identity authentication verifies these unique features and validates them against an actual and physical individual. Nowadays, the availability of printers has led to an intense usage of printed documents in our daily transactions. These printed documents are used in financial and personal identity documents [3]. As these types of documents' usage and their availability have increased, so is the growth of document forgery.

The present review is segregated into three sections. Section 2 elaborates the related work along with approach, experimentation, and results. Followed by the future research prospects and conclusion in Sects. 3 and 4.

2 Literature Review

This section discusses the papers related to document forgery and classifies them into three groups: source printer identification, verifying identity documents, and signature verification, as shown in Fig. 1. For each domain, the methods, approaches, dataset, results, and limitations are elaborated.

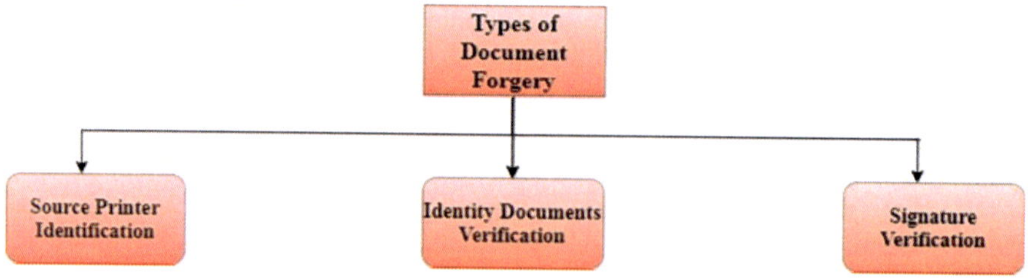

Fig. 1 Types of document forgery (reviewed in this study)

2.1 Source Printer Identification

Source printer identification is trending tremendously among researchers for document forgery analysis. Bibi et al. [3] presented an approach to identify the source printers to detect document forgery based on text. The authors extracted patches from the document images using textual features through SVM and CNN. A neural network cascading learning approach was explored in [4] to identify the source color laser printers. Adversarial training was deployed for halftone color decomposition, followed by ConvNet training. Authors in [5] examined the scanned and microscopic images for printer source identification by employing support vector machines (SVM) and CNN. Tsai et al. [6] extended the work in [5] by deploying seven-layered CNN to classify scanned text and document images and 13 layered CNN for detecting microscopic text and document images.

Ferreira et al. [7] merged the multiple forms of various data and inputted them to a CNN. To capture multiple forms of data, various preprocessing operations were implemented. The results depicted that the approach is robust to noisy data but examined only for cropped data. An individual classifier-based approach is proposed in [8] that outdo the usage of optical character recognition (OCR). The authors in [9] proposed multidirectional gray-level co-occurrence matrices (GLCM_MD) and multidirectional and multiscale gray-level co-occurrence matrices (GLCM_MDMS) approach. Tsai et al. [10] examined Japanese documents by capturing the essential features through gray-level co-occurrence matrix (GLCM), discrete wavelet transform (DWT), spatial filter, Wiener filter, and Gabor filter. They implemented support vector machines (SVM) for classification. The text dependency restricts its usage.

The authors in [11] utilized thirteen laser and seven inkjet printers. They successfully identified the inkjet printers with a classification accuracy of 93.57% compared to laser printers. Shang et al. [12] proposed a method to distinguish between the documents generated by different type of printers. Edge roughness was emphasized in [13] to differentiate the documents printed from laser and inkjet printers. Results illustrated that some preprocessing algorithms could be examined to segregate the text and images clearly (Table 1).

2.2 Identity Documents Verification

Identity documents are very crucial as they capture and authenticate the identity of an individual. With its excessive use, identity frauds are also increasing tremendously. Sirajudeen et al. [14] presented a method that utilized CNN on the MIDV-500 dataset to detect the forgeries done in identity documents —text, hologram, seal, and stamps. The authors in [15] examined pattern matching and various recognition methods to detect modifications done in identity documents by implementing combination of unsupervised (distance measure) and supervised (SVM) machine learning methods. In [16], an approach to detect a forgery in a remote ID verification system was

Table 1 Comparative study of the work done for printer identification approaches for detecting document forgery

Reference ID	Approach	Documents	Count of laser printers	Count of inkjet printers	Method	Limitations
Kim et al. [4]	Text dependent	-	8	-	ConvNet	High computational cost
Tsai et al. [5]			10	2	CNN, SVM	Feature-based SVM and deep learning approaches can be explored
Tsai et al. [6]		1200	8	4	CNN	Deep learning approaches can be explored
Ferreira et al. [7]		120	-	-	CNN	The approach can be examined for interpolated data
Joshi et al. [8]		-	-	-	Single classifier-based approach	The approach can be extended to make it text and language independent
Ferreira et al. [9]		120	10	-	GLCM_MD and GLCM_MDMS	The classifier and decision level fusion can be analyzed
Shang et al. [12]		-	10	6	WT and average gradient	Improvements can be made to enhance accuracy at lower resolution
Elkasrawi et al. [11]	Text independent	400	13	7	SVM	Segmentation algorithms can be examined to remove noise barriers
Bibi et al. [3]		1200	13	7	CNN, SVM	Comparison between text-dependent and text-independent approaches can be analyzed

(continued)

Table 1 (continued)

Reference ID	Approach	Documents	Count of laser printers	Count of inkjet printers	Method	Limitations
Gebhardt et al. [13]		1200	13	7	Anomaly detection	Preprocessing algorithms can be explored to differentiate between images and text

proposed. It uses deep learning for removing the background and machine learning classifiers for authenticating the documents.

Dlamini et al. [17] proposed a solution for detecting forgery in hardcopy documents comprising two steps—the generation process and the validation process. They merged the four approaches—cryptographic hashing, 2D barcodes, digital signatures, and OCR to validate the authenticity of documents.

Attivissimo et al. [18] initially captured the image's corners by utilizing the pixels of the examining document. The extracted image acts as an input for CNN, which is trained on the synthetic dataset. The visual saliency approach was used in [19] to detect identity documents. The saliency maps were computed by the Dahu distance, which requires little prior knowledge about the documents, and the images impose its usage in present scenarios. In [20], the approach follows three steps: pretreatment, segmentation, and recognition. To recognize the images, the author's utilized template matching for characters with fewer types, while for others, support vector machines (SVMs) was deployed. 90% accuracy was achieved for digital identification.

Table 2 sums up the work done by the researchers for detecting document forgery in identity documents. From the previous work done, it can be concluded that there is a dire need to design an extensive database of identity documents like passports, driving licenses, identity cards, etc., to provide a platform for the researchers to investigate. Moreover, a standard approach is required that applies to all types of identity documents.

2.3 Signature Verification

Nowadays, signatures are the primary means of authenticating daily transactions. Both the offline and online signatures are used pervasively [21–30]. This section elaborates on the various approaches used for offline and online signature verification. Alajrami et al. [21] proposed an approach for offline signature verification. They focused on learning features in a writer-independent way instead of hand-crafted features. The authors used CNN to capture the visual features from the signatures

Table 2 Comparative analysis for the work done in detecting document forgery for identity documents

Reference ID	Dataset	Approach	Results/limitations
Sirajudeen et al. [14]	MIDV-500	CNN	More feature detection algorithms and deep neural networks can be explored to enhance accuracy
Ghanmi et al. [15]	Private	Distance measure and SVM	Grid-3CD illumination invariance can be examined
Castelblanco et al. [16]	Colombian identity documents	Machine learning classifier	The approach can be evaluated to more extensive and other datasets
Dlamini et al. [17]	Private	Cryptographic hashing, 2D barcodes, digital signatures, and OCR	Requires evaluation for damaged documents
Attivissimo et al. [18]	Synthetic dataset	CNN	Error rates in the case of vertices detection can be minimized by further enhancing the proposed approach
Ngoc et al. [19]	Private	Visual saliency	Low contrast images may lead to poor results
Fang et al. [20]	Private	SVM and template matching	Requires accuracy enhancement in terms of Chinese character identification

regardless of the users. A hybrid approach of online and offline signature verification methods can be examined to append robustness to the system. The authors in [22] worked on CEDAR and MCYT-75 signature datasets. For feature encoding, bag-of-visual words (BoVW), and a vector of locally aggregated descriptors (VLAD) were employed. They focused on the usage of the cognitive process employed by forensic document examiners. The results depicted lower error rates for both datasets. In [23], CNN was implemented to verify the signatures. Moreover, to examine the efficiency of the proposed approach toward different languages, the model was evaluated for Chinese signatures.

For detecting document forgery in [24], the Persian signature dataset (UTSig) was used. The authors captured features using run-length distributions, slant distribution, entropy, histogram of gradients features (HoG), and geometric features. Engin et al. [25] worked to make the signature verification system writer independent. The signature representations extracted post to cleaning are evaluated against the threshold

Table 3 Comparative study of the work done for signature verification for detecting document forgery

Reference ID	Dataset	Approach	Limitations
Alajrami et al. [21]	GPDS-960, MCYT-75, CEDAR, and Brazilian PUC-PR	CNN	Online and offline signature verification methods can be concatenated
Okawa et al. [22]	CEDAR and MCYT-75	BoVW, VLAD, and cognitive process	Add more robustness to the system
Kao et al. [23]	ICDAR 2011 SigComp	DCNN	Did not perform well for Chinese characters
Ghanim et al. [24]	Persian offline signature (UTSig)	Bagging tree, random forest, and SVM	Deep learning approaches can be examined
Engin et al. [25]	Tobacco-800	Cycle GAN, and CNN	Unsupervised approaches can be examined
Yapıcı et al. [26]	GPDS	CNN	Explore feature extraction methods
Gumusbas et al. [27]	CEDAR	Capsule network	Explore various combinations of capsule networks

to determine its authenticity. Global equal error rate (EERglobal) was captured to evaluate the proposed approach.

Yapıcı et al. [26] proposed an offline signature verification system. They addressed the inter-personal variability in the offline signatures using the CNN architecture. The authors designed two models separately: writer dependent (WD) and other for writer independent (WI). The experimental results illustrated an accuracy of 62.5% for WD and 75% for WI. To remove the barrier of large datasets for deep learning models was addressed in [27]. The authors used capsule networks using fewer data with lower resolution. Results depicted that the proposed approach was capable of capturing the local features despite fewer data.

Table 3 illustrates the work done by the researchers for offline and online signature verification to detect document forgery. The researchers used various evaluation metrics to compute the results. From the work done so far, it can be viewed that CNN tends to enhance the results. To further enhance the results, deeper CNN models [28] and other deep learning approaches like GAN and LSTM can be explored.

3 Future Research Prospects

For future work, the below-mentioned areas can be explored:

- Supervised approaches require prior knowledge about the questioned documents for labeling them, which imposes a constraint on its usage for solving real-life problems. Unsupervised approaches can be examined to address this issue.
- Deep learning solutions have attained desired results in various domains. But the unavailability of publically available large datasets restricts its usage to solve document forgery problems. Capturing a large size dataset for various applications is an important future research area that needs to be addressed. More complex and advanced deep learning models like GAN, autoencoders, RNN, and transfer learning models may be examined to yield better results [28, 29].
- There may be variety of identity documents (passports, driving licenses, identity cards, etc.), legal documents, and many more. There is a dire need to design standard solutions to check all types of documents irrespective of text dependency. An algorithm may be designed that converts the dataset from one language to another to check its robustness and remove text dependency barriers.
- Ink mismatch detection and writer identification are well-known approaches for detecting document forgery. Researchers have achieved state-of-the-art results by using hyperspectral imaging to implement these approaches. Hyperspectral imaging is an emerging technique that captures the unique features from the images through their spectral signatures [28, 29]. Hyperspectral imaging can be explored to detect the authenticity of signatures, originality of identity documents, and printed documents for better results.

4 Conclusion

Document forgery is a very challenging task that involves creating, modifying, or emulating objects or documents. It can be of any form—simple forgery, freehand simulation, tracing, or electronic manipulation. Document forgery involves falsifying the documents in various ways like replicating the signatures, imitating the identity documents (passports, driving licenses, identity cards, etc.), modifying the printed documents, and many more. In this study, we reviewed the various work done by the researchers to inspect document forgery in the domains-mentioned above. Present paper captures the various machine learning and deep learning approaches implemented, results, datasets used, and limitations for the three different types of forgeries.

References

1. Srivastava S, Rastogi V, Jaiswal G, Sharma A (2022) Hyperspectral imaging in document forgery. in proceedings of data analytics and management, pp 105–115. Springer, Singapore
2. Jaiswal G, Sharma A, Yadav SK Critical insights into modern hyperspectral image applications through deep learning. Wiley interdisciplinary reviews: data mining and knowledge discovery, e1426

3. Bibi M, Hamid A, Moetesum M, Siddiqi I (2019) Document forgery detection using printer source identification—a text-independent approach. In: 2019 international conference on document analysis and recognition workshops (ICDARW). IEEE vol 8, pp 7–12
4. Kim DG, Hou JU, Lee HK (2019) Learning deep features for source color laser printer identification based on cascaded learning. Neurocomputing 365:219–228
5. Tsai MJ, Yuadi I, Tao YH (2018) Decision-theoretic model to identify printed sources. Multimedia Tools Appl 77(20):27543–27587
6. Tsai MJ, Tao YH, Yuadi I (2019) Deep learning for printed document source identification. Signal Process: Image Commun 70:184–198
7. Ferreira A, Bondi L, Baroffio L, Bestagini P, Huang J, Dos Santos JA, Rocha A (2017) Data-driven feature characterization techniques for laser printer attribution. IEEE Trans Inf Forensics Secur 12(8):1860–1873
8. Joshi S, Khanna N (2017) Single classifier-based passive system for source printer classification using local texture features. IEEE Trans Inf Forensics Secur 13(7):1603–1614
9. Ferreira A, Navarro LC, Pinheiro G, dos Santos JA, Rocha A (2015) Laser printer attribution: exploring new features and beyond. Forensic Sci Int 247:105–125
10. Tsai MJ, Hsu CL, Yin JS, Yuadi I (2015) Japanese character based printed source identification. In: 2015 IEEE international symposium on circuits and systems (ISCAS). IEEE, pp 2800–2803
11. Elkasrawi S, Shafait F (2014) Printer identification using supervised learning for document forgery detection. In: 2014 11th IAPR international workshop on document analysis systems. IEEE, pp 146–150
12. Shang S, Memon N, Kong X (2014) Detecting documents forged by printing and copying. EURASIP J Adv Signal Process 2014(1):1–13
13. Gebhardt J, Goldstein M, Shafait F, Dengel A (2013) Document authentication using printing technique features and unsupervised anomaly detection. In: 2013 12th international conference on document analysis and recognition. IEEE, pp 479–483
14. Sirajudeen M, Anitha R (2020) Forgery document detection in information management system using cognitive techniques. J Intell Fuzzy Syst (Preprint), 1–12
15. Nabil G, Awal AM (2018) A new descriptor for pattern matching: application to identity document verification. In: 2018 13th IAPR international workshop on document analysis systems (DAS), pp 375–380
16. Castelblanco A, Solano J, Lopez C, Rivera E, Tengana L, Ochoa M (2020) Machine learning techniques for identity document verification in uncontrolled environments: a case study. In: Mexican conference on pattern recognition, pp 271–281. Springer, Cham
17. Dlamini N, Mthethwa S, Barbour G (2018) Mitigating the challenge of hardcopy document forgery. In: 2018 international conference on advances in big data, computing and data communication systems (icABCD). IEEE, pp 1–6
18. Attivissimo F, Giaquinto N, Scarpetta M, Spadavecchia M (2019) An automatic reader of identity documents. In: 2019 IEEE international conference on systems, man and cybernetics (SMC). IEEE, pp 3525–3530
19. Ngoc MÔV, Fabrizio, J, Géraud T (2018) Saliency-based detection of identy documents captured by smartphones. In 2018 13th IAPR international workshop on document analysis systems (DAS) (pp 387–392). IEEE
20. Fang X, Fu X, Xu X (2017) ID card identification system based on image recognition. In: 2017 12th IEEE conference on industrial electronics and applications (ICIEA) (pp 1488–1492). IEEE
21. Alajrami E, Ashqar BA, Abu-Nasser BS, Khalil AJ, Musleh MM, Barhoom AM, Abu-Naser SS (2020) Handwritten signature verification using deep learning. Int J Acad Multi Res (IJAMR), 3(12).
22. Okawa M (2018) From BoVW to VLAD with KAZE features: offline signature verification considering cognitive processes of forensic experts. Pattern Recogn Lett 113:75–82
23. Kao HH, Wen CY (2020) An offline signature verification and forgery detection method based on a single known sample and an explainable deep learning approach. Appl Sci 10(11):3716

24. Ghanim TM, Nabil AM (2018) Offline signature verification and forgery detection approach. In 2018 13th international conference on computer engineering and systems (ICCES) (pp 293–298). IEEE
25. Engin D, Kantarcı A, Arslan S, Ekenel HK (2020) Offline Signature verification on real-world documents. In: Proceedings of the IEEE/CVF conference on computer vision and pattern recognition workshops, pp 808–809
26. Yapici MM, Tekerek A, Topaloglu N (2018) Convolutional neural network based offline signature verification application. In: 2018 international congress on big data, deep learning and fighting cyber terrorism (IBIGDELFT). IEEE
27. Gumusbas D, Yildirim T (2019) Offline signature identification and verification using capsule network. In: 2019 IEEE international symposium on innovations in intelligent systems and applications (INISTA) (pp 1–5). IEEE
28. Jaiswal G, Sharma A, Yadav SK (2022 Deep feature extraction for document forgery detection with convolutional. Comput Elect Eng 99:107770
29. Jaiswal G, Sharma A, Yadav SK (2019) Analytical approach for predicting dropouts in higher education. Int J of Inf Commun Technol Educ (IJICTE) 15(3):89–102
30. Anuradha T et al (2020) Machine Learning, Advances in Computing, Renewable Energy and Communication, (Springer Nature, Berlin, LNEE volume 768, 2020), 659 p. Doi: https://doi.org/10.1007/978-981-16-2354-7. (ISBN 978–981–16–2354–7)

Hybrid Model for Deepfake Detection

N. Rajesh, M. S. Prajwala, Nancy Kumari, Muhammad Rayyan, and A. C. Ramachandra

Abstract In recent years, there have been booming improvements in the fields of deep learning and artificial intelligence (AI). However, the growth also has resulted in a new age of realistic digital manipulation of videos known as deepfakes. With this new research, it is difficult to differentiate between real and manipulated media. The growing popularity of deepfakes and easy availability of open source creating tools have become a concern of privacy and a threat to society. This has led to the need to develop an accurate and efficient tool that can identify and stop the harm these media might cause. Recently, there has been a lot of exploration and research for identifying these synthetic media. Meanwhile, there is a lot of focus to find the best feature scales for detection and improving the existing algorithms. Features and attributes such as background comparison, eye blinking patterns, facial artifacts, and pose estimation are currently being used. The proposed Hybrid Model for Deepfake Detection (HMDD) which is aimed to achieve better performance in detecting.

Keywords Deep learning · Deepfakes · Convolutional neural network · Generative adversarial network · Face recognition · ResNet-50 · LSTM

1 Introduction

With growing advancements in technology and a breakthrough in the field of artificial intelligence, we are noticing an era where even digital images, videos, or audio recording can be manipulated and forged based on one's wishes. Hence, the idea of Deepfake was born. Deepfake is a method that is able to forge high-quality images and videos with utmost accuracy by relying on deep learning methods and algorithms. Deepfake can be referred to as simulated media where a person in an image or

N. Rajesh · N. Kumari · M. Rayyan · A. C. Ramachandra (✉)
Nitte Meenakshi Institute of Technology, Bangalore, Karnataka, India
e-mail: ramachandra.ac@nmit.ac.in

M. S. Prajwala
Enquero Global, Bangalore, Karnataka, India

A. Tomar et al. (eds.), *Proceedings of 3rd International Conference on Machine Learning, Advances in Computing, Renewable Energy and Communication*, Lecture Notes in Electrical Engineering 915, https://doi.org/10.1007/978-981-19-2828-4_57

video is swapped with another person. Deepfake can stitch anyone in the world into a video or photo they were never a part of, making them say or do things they never did. First known attempt of deepfake was made way back in 1865, where one of the iconic portraits of US President Abraham Lincoln was mixed with another politician [1] known as John Calhoun. After this attempt, the idea of manipulating images and videos gained popularity. It has expanded over Reddit to include synthetic media applications, where some of the online communities shared deepfake rendering celebrities, politicians, and other VIPs. Deepfakes have become more common and spreading misinformation is rampant nowadays. With constantly trained models, deepfake videos are created with exact match, in turn having a massive impact on the economy [2] and national security. Initial targets of deepfake videos were public figures such as celebrities and politicians as their images and videos are available over the Internet. However, according to reports from deeptrace, it was seen that the targets for manipulated images and videos are no longer limited to government leaders or actors. Therefore, deepfake has become one of the biggest worries for everyone, which contribute to factual relativism. It can be used to help public figures hide their immoral acts in the veil of deepfake and fake news, framing their actions as false. The technology has advanced and has potentially creating damage to the image of individuals, society, institutions, and democracies.

1.1 Deepfake Creation

To make a deepfake image or video, a model must be trained using real images or videos to understand how a "realistic" it would look like in various angles and under different lighting [3]. Then, the model is combined with a computer-graphic technique to superimpose a copy of someone's lookalike to create a fake image or a fake video. Considering the two most used deepfake creation models: Convolutional neural network (CNN) is a deep learning algorithm that utilizes images of various categories as input and detects differences between features which helps to differentiate these images from one another. CNN uses the predictions from different layers to produce a final output that presents a vector of probability scores to represent the likelihood that specifies the distinct features. This network comparatively requires less preprocessing and consists of two major components: convolutional tool layer and a fully connected layer. Generative adversarial network (GAN): GANs are a powerful algorithm in the deep learning domain that trains using unsupervised learning. GANs are combination of two machine learning models, namely generator and discriminator. The generator creates fake videos and makes the discriminator distinguish between real and fake videos. Each time the discriminator can detect or classify the input data to be fake, and it hints to the generator about how the data is detectable. As the discriminator gets better at detecting fake data, the generator gets better. Other methods were developed to create deepfake videos using deep learning algorithms such as recurrent neural network (RNN), long short-term memory (LSTM), and hybrid approaches. These models initially worked with frame-level features, extracted after processing the data. Hybrid approaches can be described as a combination of two algorithms aligned as a set of encoder–decoder developed for an image or video forgery.

2 Background Work

Younus et al. provide an abbreviate review [4] of the methodologies of a method comparison study that has been presented to assist, distinguish between real videos and fake ones which are uploaded daily on various Web sites across the Internet. Background comparison, temporal pattern analysis, eye blinking, facial artifacts, mesoscopic analysis, eye blinking, and pose estimation are some of those techniques which are efficient and fast methods. Nguyen et al. presented a survey of algorithms used to create deepfakes and, more importantly, methods proposed to detect deepfakes. By reviewing the background of deepfakes and state-of-the-art deepfake detection methods, this study provides [5] a comprehensive overview of Deepfake techniques and facilitates the development of new and more robust methods to deal with the increasingly challenging DeepFakes for both images and videos using shallow and deep classifiers. Chan et al. proposed an improved VGG network named NA-VGG to detect DeepFake face image, which was based on image noise and image augmentation. Firstly, to [6] learn the tampering artifacts that may not be seen in RGB channels, SRM filter layer is used to highlight the image noise features; secondly, the image noise map is augmented to weaken the face features. Finally, the augmented noise images are input into the network to train and judge whether the image is forged. The experiment carried out using the Celeb-DF dataset. Pan et al. proposed [7] two methods of deep learning approaches, namely Xception and MobileNet to automatically detect DeepFake videos. Models are known to select features that are common for each manipulated video using datasets from Face2Face, FaceSwap, and neural textures platforms.

In another study by Rana et al. [8] proposed a deep ensemble learning technique called DeepFake Stack was suggested. The possibility of using various deep learning models opened a new possibility of the use of blockchain technology to detect DeepFake. Also, a novel framework approach to detect Deepfakes by synthesizing three levels of features, named SDHF, was proposed by Liang et al. [9] a hierarchical framework using 2D CNN model for feature extraction and MBConv aggregator for extracting clip-level and video-level features are adaptive for comprehensive decisions. Jung et al. suggested detection of DeepFakes using different blinking patterns that depend on an individual's gender, age, and cognitive behavior. The study suggested that with the use of Deep Vision, abnormal blinking patterns can be used as a crucial feature for detection by comparing and analyzing the number of blinks, average blinking cycles, and duration of blinking. Observing blinking patterns while combining them with eyebrow matching as suggested by Nguyen et al. [10] accuracy of a detector can be drastically improved. Algorithms such as LightCNN, ResNet, DenseNet, SqueezeNet, and matching can be used to achieve the same. Moreover, one of the other most used features was understanding dynamic lip movement as suggested by Zhao et al. [11] using basic CNN to complex models such as SA-DTH net to extract information about subjects talking habits. This in turn would increase the chances to find DeepFake videos trained and created with utmost efficiency.

The most used approaches for detection of modified or DeepFake images and videos are with the help of face recognition, as suggested by Feng et al., and Patel et al. These proposed methods [12, 13] used full-face image of the image or video and find any tampering. Initially, faces are recognized; then, angles of the faces are judged, after which processing of those faces is done to recognize frame-level features that differentiate synthesized videos from real ones. As suggested by Kharbat et al. in one of the studies, frame-level features could be extracted using techniques [14] such as support vector machines, histogram of oriented gradient, oriented fast and rotated BRIEF, binary robust invariant scalable key-points, KAZE, speeded-up robust features, and features from accelerated segment test. To summarize the above context, it is observed that there are many methods that can be adapted for the detection of Deepfake videos.

3 Methodology of Proposed Model

The methodology used during prototyping of the proposed model for Deepfake detection has two major blocks such as preprocessing and Deepfake Detection is as shown in Fig. 1 along with other sub blocks as discussed subsequently. Dataset used for this model is collected from various sources such as Deepfake detection challenge dataset (Kaggle) [15], Celeb-DF dataset [16], and Face Forensics++ dataset [17, 18]. The data collected is preprocessed for further processing.

The steps followed in preprocessing are as shown in Fig. 2. A count of videos present in the dataset is noted and check for corrupt videos carried out. If any videos are found to be corrupt, the same is dropped. Then, the clean video dataset is taken and is split into frames. After frames are ready, a threshold is set to drop frames after a certain limit. This step is done to lower computational burden of the model. Since frames are ready, detecting frames with only a face using face recognition algorithm is carried out, then crop videos to have face-only frames. It is important on the GPU to drop all the frames without faces, to avoid model from processing unwanted frames, hence reduces computation burden. To have uniformity among

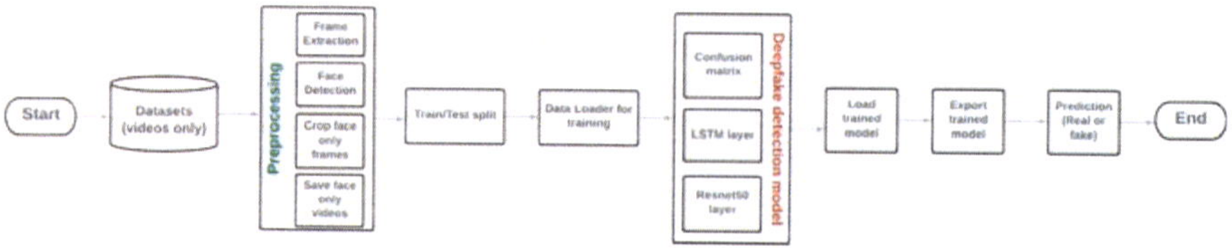

Fig. 1 Proposed model of HMDD

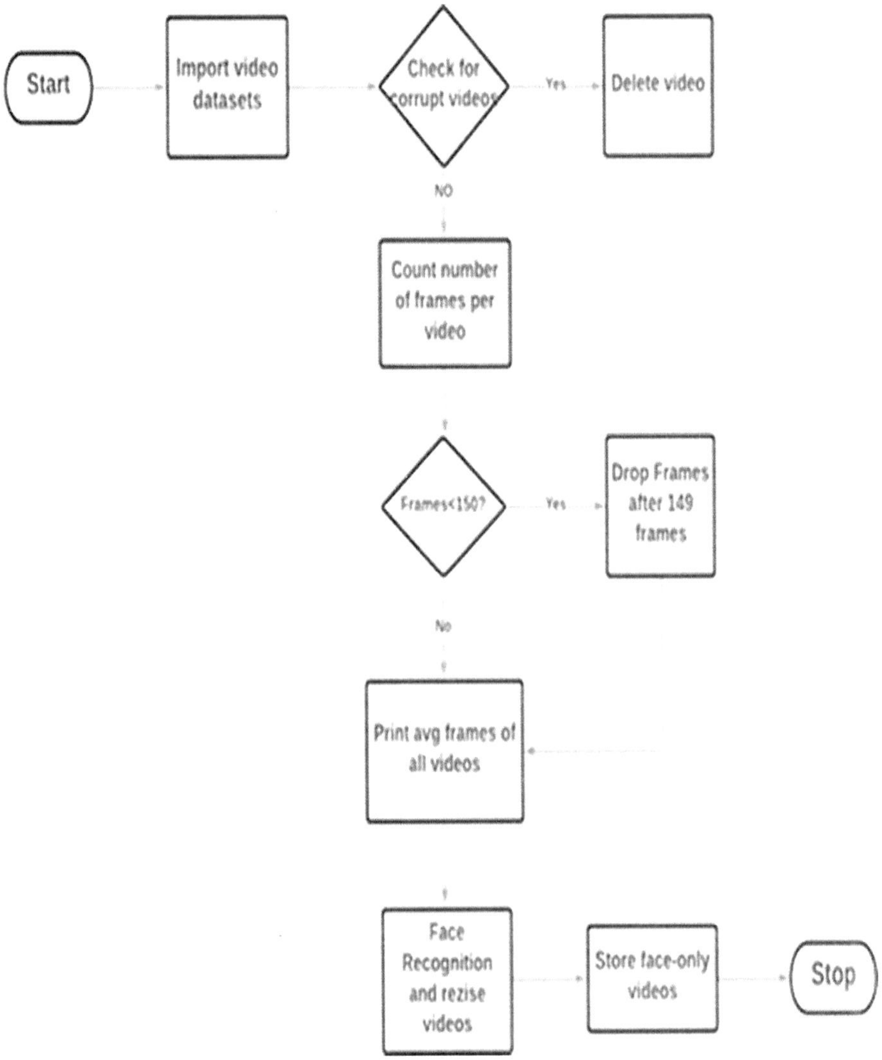

Fig. 2 Preprocessing of the images

all the processed videos, mean of all the frames are calculated, and new processed face-only dataset is created around the mean value of these frames.

The model is trained in a sequential ordered as shown in Fig. 3. Training dataset is used for training the model, i.e., the model uses the training dataset created after preprocessing to find patterns and learn them. The dataset available is bifurcated as training and test dataset: Test dataset is used for validating the performance of the model. The entire processed dataset is split in the ratio of 7:3 i.e., 70 and 30%, respectively.

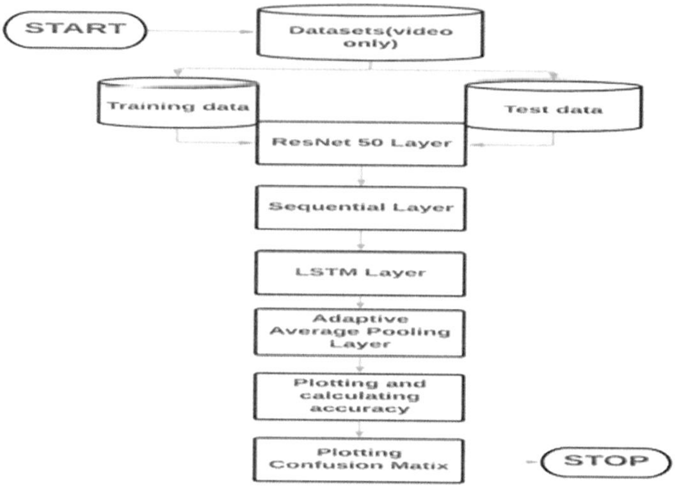

Fig. 3 Training model of the proposed model

First layer introduced in the model is ResNet-50 which is capable of learning 4-D features from the input dataset and hence is used for frame-level feature extraction, freezes all the existing layers, and updates only the parameters in the new layers. Then, a sequential layer is added, which refers to the way model is built in Keras. Next, a long short-term memory (LSTM) is added. This layer is used to reduce long-term dependency of the model; that is it decides which part of the features must be learnt and which can be ignored. For activation function, leaky rectified linear unit (ReLu) is used. Model then is given a linear layer, which is responsible for creating single layer feed forward network with a specific number of inputs. Then, an adaptiveAvgPool2d layer is applied, which provides a 2D adaptive average pooling over input signal composed of several input planes. A graph of epoch vs accuracy and a confusion matrix helps in understanding how well model has learnt as well as how many exceptions were learnt. Once the model attains desired accuracy, the trained model is saved for further classification with fresh data.

The classification process is as shown in Fig. 4. The videos which should be checked are resized to phase only frame and loaded to classification model. Since the model is trained to use facial features for classification, new input must be resized. After the video is in the desired format, using the concept of statistical estimation, i.e., frames are compared with each other for identification of any manipulation. Frames learnt are compared with new frames to detect an abnormal variation in facial features. Then, the frame used for classification is plotted using heatmaps, which gives the idea about the part of face model is looking into. Heatmaps are a way of finding the collinearity of the data and identify which rows or columns should or should not be included as part of results. Finally, the new video is classified into either real or fake.

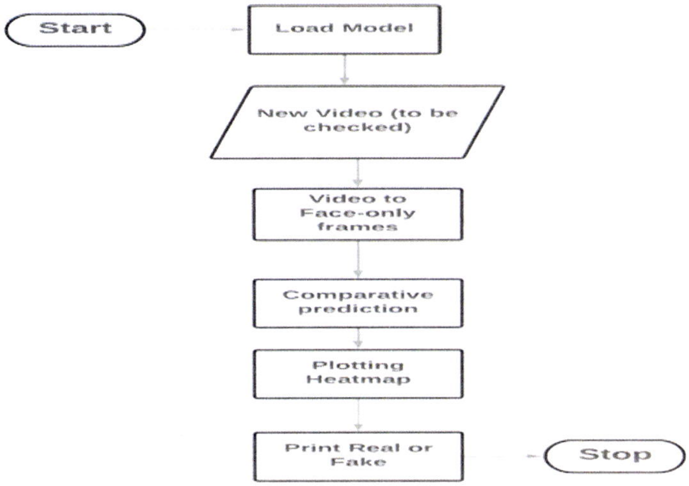

Fig. 4 Classification model for HMDD

4 Experimental Result and Analysis

The proposed model HMDD is tested for its efficiency at various levels, and it is analyzed for time, frame rate, and resolution in preprocessing stage. It is observed that the difference or fake identification was very difficult, and hence, time taken and frame rate reduced after preprocessing. The resolution is altered because of the resizing to face-only frames. Table 1 gives a comprehensive comparison before and after processing.

It is important to evaluate training models performance. For the proposed model, a graph, as shown in Fig. 5 of epoch versus accuracy, is plotted to compare the model's performance during its training and testing phase. The number of epochs, i.e., the number of times the entire dataset has been passed through the model, is 30. Along with a learning rate, which is a configurable parameter that controls the adaptivity

Table 1 Comparison between original and processed data

	Before processing	After processing
Time (seconds)	13	10
Frames/seconds	30	15
Resolution	1920 * 1080	112 * 112

Table 2 Tabulates the performance of proposed model, HMDD

Model (proposed)	Dataset	Method	Accuracy
HMDD	FF++, DFDC, Celeb-DF	Hybrid of ResNet and LSTM	86.5%

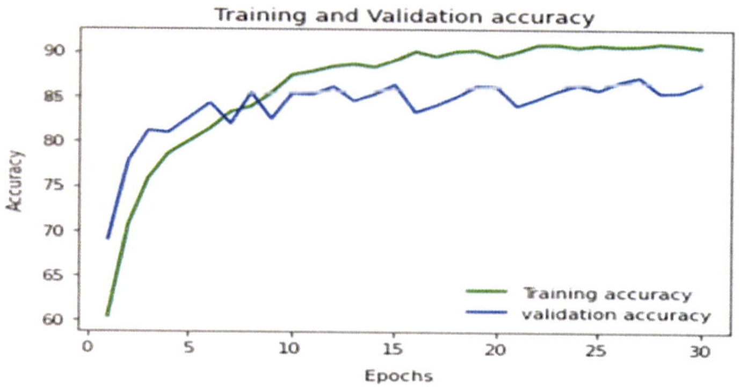

Fig. 5 Train-test accuracy graph

of the model, of 0.00001. The training accuracy is 91%, and testing accuracy is 86.469%.

Performance of the classification model is evaluated by using confusion matrix which is as shown in Fig. 6. A confusion matrix is a technique for summarizing the performance of an algorithm. Classification accuracy alone can mislead if an unequal number of observations are present in each class or if there are more than two classes in the dataset. Out of considered data set for analysis using HMDD, the model has identified 568 fake videos as fake which is considered as true positives and 544 real videos being predicted as real, and it is true negative. The false positive and false negative being fake predicted as real and real predicted as fake are 70 and 104, respectively.

Performance of the classification model is represented using heatmaps. The classification of the images considered as data set is done using heatmaps, which uses the features to examine the image considered. The heatmap image is as shown in Fig. 7 and is based on frame-level facial features. Selected features used by the model for classification are represented by red color and features ignored by blue color. Model

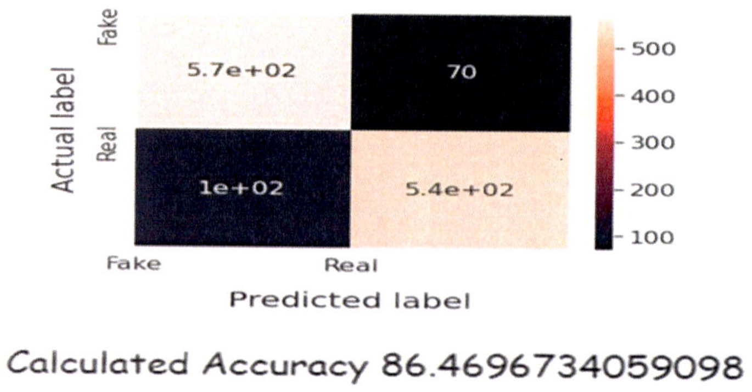

Fig. 6 Confusion matrix performance evaluation of the algorithm

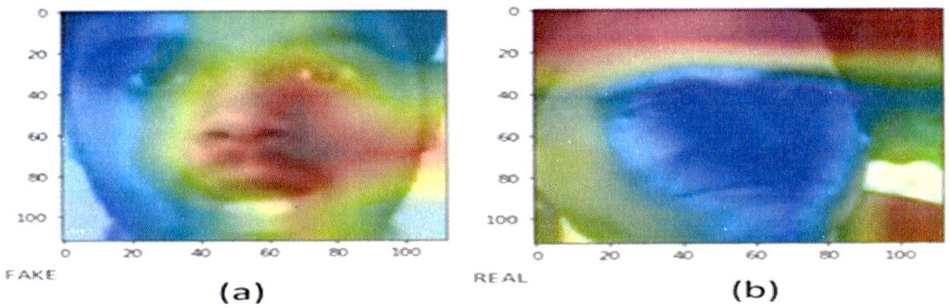

Fig. 7 Heatmap of fake video

is considering all the features of a face, even if it is at different angles, for identification of tampering or manipulation and is not restricted its references to any specific region of the face. In Fig. 7a, the video is classified as fake, features near subject's mouth are used for detection of manipulation in the video, and the other features are ignored. In Fig. 7b, the video is classified as real, and features near subject's forehead are used for identification of manipulation. This experimentation is carried out for the image existing in the standard dataset.

To further evaluate the performance of the model, a completely new video dataset was created with the help of several subjects and then introduced to the model. As seen in Fig. 8a, the video is classified as real and again features near subject's mouth is focused to represent the feature using heatmap. After which, the same video is manipulated digitally and then introduced to the model. It is observed from Fig. 8b that model classified the video to be fake and used features near the subject's mouth to identify the manipulation represented by heatmap. By these observations, it is understood that the model is performing well with any newly introduced video as well and is able to classify.

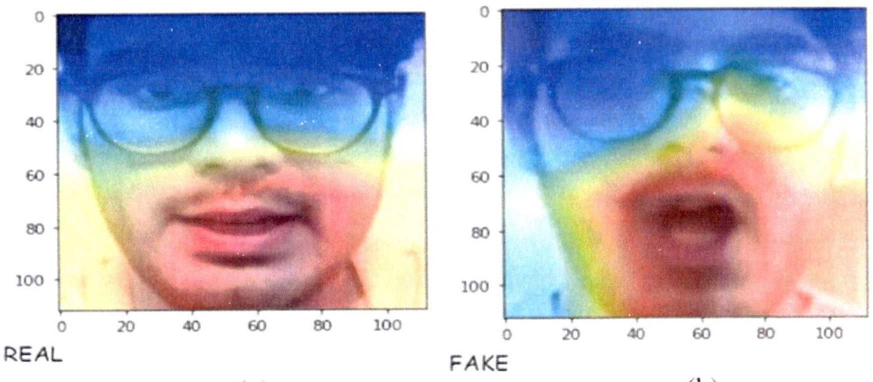

(a) (b)

Fig. 8 Heatmap of new real video

5 Conclusions

Deepfakes are realistic video forgeries that can make anyone say or do fictitious things on a recording. It is a potential threat to individuals, society, businesses, and democracy. By observing a manipulated video, individual cannot make judgment on its authenticity. Since these synthetic contents are flawless and not easy to identify, there is a need for a tool to distinguish fake from authentic videos. Detection of complex and precise forgeries has sparked exploration and research in this field. Different strategies, methods, and ideas are proposed and assessed to find efficient ways to deal with deepfake. Through proposed model, HMDD using simple deep learning algorithms, identification and pointing out frame-level manipulation of a video is possible. With each frame, the model is able to target significant facial features despite the angle the face is in the frame, using ResNet and prioritize target features using LSTM. Based on these insights, a comparative study between various inputs, it is established that the model uses all the facial features to interpret and comprehend tampering. Hence, it can be concluded that by combining datasets and with the help of simple hybrid model, it is possible to achieve better performance in identification of tampering with an efficiency of 86.469%.

References

1. Somers M (2020) Deepfakes, explained. MIT Management Sloan School
2. Jaiman A (2020) Deepfakes harms and threat modelling. Towards data science
3. Shen T, Liu R (2016) DeepFakes using generative adversarial networks (GAN). NoiseLab, UCSD.edu
4. Younus MA, Hasan TM (2020) Abbreviated view of deepfake video detection techniques. In: 2020 IEEE international conference on sustainable technology and development
5. Nguyen TT, Nguyen CM (2021) Deep learning for deepfake creation and detection: a survey AMC Comput Surv vol 7
6. Chang X, Wu J (2020) Deep fake face image detection based on improved VGG convolution neural network. In: 2020 Proceedings of the Chinese control conference
7. Pan D, Sun L (2020) Deepfake detection through deep learning. In: 2020 IEEE/ACM international conference on big data computing, applications and technologies
8. Rana MS, Sung AH (2020) Deepfake stack: a deep ensemble-based learning technique for deepfake detection. In: 2020 IEEE 7th international conference on cyber security and 7th international conference on edge computing and scalable cloud
9. Liang T, Chen P (2020) SDHF: spotting seefakes with hierarchical features. In: IEEE international conference on tool with artificial intelligence
10. Nguyen H, Derakhshani R (2020) Eyebrow recognition for identifying deepfake videos. In: 2020 international conference of the biometrics special interest group
11. Yang CZ, Ma J (2021) Preventing deepfake attacks on speaker authentication by dynamic lip movement analysis. IEEE Trans Inf Forensics Secur 16
12. Feng K, Wu J (2020) A Detect method for deepfake video based on full face recognition. In: 2020 IEEE international conference on information technology, big data and artificial intelligence
13. Patel M, Gupta A (2020) Trans-DF: a transfer learning based end-to-end deefake detector. In: 2020 IEEE 5th international conference on computing communication and automation

14. Kharbat FF, Elamsy T (2019) Image feature detectors for deepfake video detection. In: 2019 IEEE/ACS 16th international conference on computer system and application
15. DFDCdataset: www.kaggle.com/c/deepfake-detection-challenge/data
16. Celeb-DFdataset: https://github.com/yuezunli/celeb-deepfakeforensics
17. FaceForensics ++ dataset: https://github.com/ondyari/FaceForensics
18. Tomar A et al (2020) Machine learning, advances in computing, renewable energy and communication. Springer Nature, Berlin, LNEE vol 768, 659 p. https://doi.org/10.1007/978-981-16-2354-7. (ISBN 978-981-16-2354-7)

Web Usage Mining Cluster Simulation Using K-Means and Fuzzy Approach

Hardik A. Gangadwala and Ravi M. Gulati

Abstract The Internet continues to develop at an outstanding rate in both the scale and complexity of Web sites and is properly on its way to being the principal source of information and data. Due to this growth in scale and complexity of Internet, Internet site publishers are having so much trouble in appealing and associating Web users. To layout popular and appealing Web sites, publishers need to apprehend their customers' wishes. Consequently, inspecting users' behavior is a crucial part of Web site layout. Web usage mining (WUM) is the utility of mining methods to proxy server log repositories so as to determine the utilization patterns which need to investigate the Web user's surfing behavior. Clustering method is grouping a set of data points into several clusters. Clustering keeps related data points within the same cluster and keeps unrelated data points into distinct clusters. Paper presents comparative study of K-Means and Fuzzy C-Means methods. These methods offer adequate and precise investigation of proxy server log. The least distance between clusters is calculated with the aid of using the Euclidean distance equation.

Keywords Web usage log · K-Means algorithm · Fuzzy C-Means algorithm · Fuzzy cluster · Amount of gratification

1 Introduction

Due to the digitization and improvements in hardware and software, there are large numbers of data available in dataset. Considering a large number of data in dataset, it is beyond human being capacity to extract the required data without efficient data mining algorithms [1–21]. The massive degree of information, coupled with the

H. A. Gangadwala (✉)
Shri. Shambhubhai V. Patel College of Computer Science & Business Management, Veer Narmad South Gujarat University, Surat, India
e-mail: hardik.gangadwala@gmail.com

R. M. Gulati
Department of Computer Science, Veer Narmad South Gujarat University, Surat, India
e-mail: rmgulati@vnsgu.ac.in

A. Tomar et al. (eds.), *Proceedings of 3rd International Conference on Machine Learning, Advances in Computing, Renewable Energy and Communication*, Lecture Notes in Electrical Engineering 915, https://doi.org/10.1007/978-981-19-2828-4_58

necessity for effective data investigation systems has been described as a dataset huge but outcome less condition. It deals with the expertise within proxy server log file [2, 3]. Web usage mining is the detection of Web site visits patterns in the proxy servers [2, 3]. Web usage mining evaluates navigation of user interactions with a proxy server, such as database transactions at Web site or a group of associated Web sites. The motivation of Web usage mining analysis fuzzy models is to bifurcate Web user base on their interest. Keeping Web user into specific and various cluster depending on their degree of interest. Information gained from Web usage mining helps to improve Web site interface, predicting user next request and fruitful browsing.

The organization of the paper is as follows, Sect. 1 presents the introduction of research paper; Sect. 2 presents the essential steps of preprocessing of proxy server log file; Sect. 3 presents the experiments and result analysis work for the paper, and Sect. 4 concludes research and future work.

2 Pre-Processing of Proxy Server Log File

Web usage mining is the technique of making use of data mining strategies to find out page navigation patterns from proxy server log dataset as a way to apprehend and better serve the requests of customers. Web usage mining technique comprises three levels: (i) preprocessing, (ii) detection of probability patterns, and (iii) investigation of probability patterns. Pattern detection means finding common pattern in the proxy server log file. For its implementation, the data needs to be transformed within the preprocessing segment such that the result of the translation can be used as the input for algorithms. By using algorithms, resulting in outcomes of information through pattern detection for drawing conclusion. In Web usage mining, the input for the entire method is the proxy server log dataset. The data needs to be preprocessed on the way to have the best input for the Web mining and behavior pattern detecting fuzzy clustering algorithms. The distinct technique requires specific input form, as a result, the preprocessing section can offer three forms of output records. The frequent pattern identification phase requires simplest Web sites visited via a given Web user. In this situation, the order of the Web sites is irrelevant.

Table 1 presents fields and its description of VNSGU proxy server dataset and it does not denote performance metrics. We collected proxy server log from date 01/11/2017 to 31/03/2018. Available numbers of users are around 4139. Considering the dataset maximum request by a Web user 102,747. Based on it, we considered 5% records from the maximum request is 5137. We get 86 customers, with number more than 5137. Taken into consideration, customers visit number of Web sites are 10,486.

Table 1 Description of proxy server log fields

Field value	Description	Field value	Description
TID	Unique ID of the transaction	URL	URL being requested
URL Date	URL access date	Time Spend	Time duration spend on page
URL Time	URL access time	URL Type	Type of URL
IP	IP address of client	OS	Operating system of client
User ID	Unique user ID		

Table 2 Parameters of fuzzy C-Means algorithm

X	$X = \{x_1, x_2, \ldots, x_n\} \subseteq R_p$ is the dataset in the
P	p-dimensional vector space, total data point is denoted as p,
C	c denotes the total clusters with $2 \le c \le n-1$
V	$V = \{v_1, v_2, \ldots, v_c\}$ is the c centers or prototypes of the clusters,
V_i	v_i denotes the p-dimensional center of the cluster i, and
$d^2(x_j, v_i)$	d^2(xj, vi) denotes a length between cluster center v_i and object x_j
U	$U = \{\mu_{ij}\}$ denotes a fuzzy matrix with $u_{ij} = u_i(x_j)$ is the degree of gratification x_j in the ith cluster;
X_j	x_j is the jth of p-dimensional measured data point

Table 3 Results of cleaning and user identification

Source	VNSGU web sever	Considered users	86
Total records	3,016,859	Considered web site	10,486
Number of users	4302	Matrix	86 × 10,486
Total months	4 months	Algorithm	K-Means, FCM
Unique web site	37,859		
5% from highest request sender 1, 02,747			

2.1 Cluster

In this technique, a set of data items having relative attributes is kept within same group. With respect to fuzzy clustering approach, fuzzy partitioning of the data point where each data point is the grade of gratification for each cluster. Due to

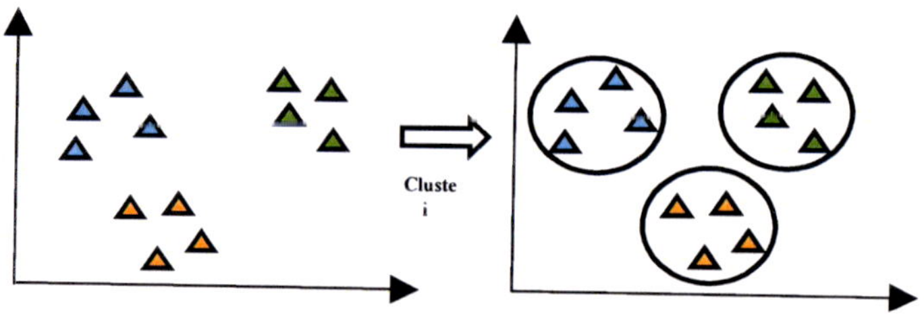

Fig. 1 Clustering

fuzzy clustering technique, clusters can increase into its original forms. Data points which does not belong to cluster have grade of gratification 0. In case the grade of gratification is nonzero, it represents the grade to which the data point to be a member of each cluster. Fuzzy clustering techniques are so much powerful to manage the outliers via imparting them very minor grade of gratification which is close to clusters. As a consequence, fuzzy clustering is better approach for managing original data point where vagueness, ambiguity, and imprecision [2].

2.2 *Cluster Analysis*

Cluster investigation has been doing an essential process in solving many issues for users like and dislike, Web user behavior investigation, and pattern detection. It provides a fuzzy cluster validity standard depending on a validity method which detects complete compactness and separateness fuzzy partitions. This fuzzy method depends on the dataset, geometric distance, distance among cluster centroids, and significantly on the fuzzy partition produced by fuzzy technique. The technique is statistically acceptable through its association to a precise crisp clustering validity method: The separation index, for which the circumstance of uniqueness has already been defined. The performance assessment of this validity method relates favorably to that of many others. Investigation of clustering is to detect clusters within the records. Cluster is a set of data points that are alike to each other. Likeness can be represented via distance equation, designated through Web users or specialists.

An excellent clustering technique can reduce the similarity between clusters and increase the similarity within cluster (Figs. 3 and 4)

2.2.1 K-Means Algorithm

K-Means cluster model is consider as base line model. The cluster centers are uniformly initialized in K-Means clustering. Then, for each data point x_i, the cluster v_j with the shortest distance to this data point is allocated. After the data points are allocated to clusters, the cluster centers are modified by taking the weighted average

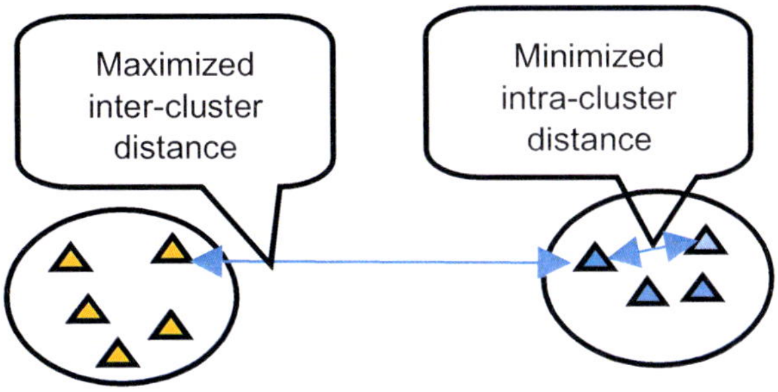

Fig. 2 Clustering distance

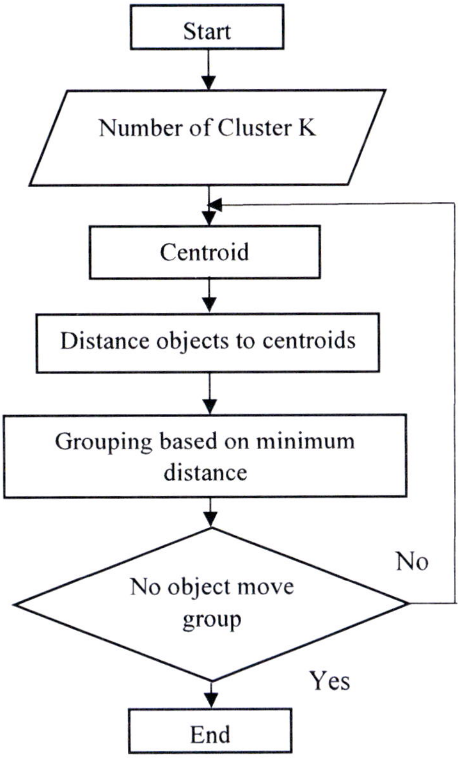

Fig. 3 Clustering flowchart

of all data points in that cluster. The cluster centers have been recalculated, resulting in a stronger cluster center set. The procedure is repeated until the cluster does not shift.

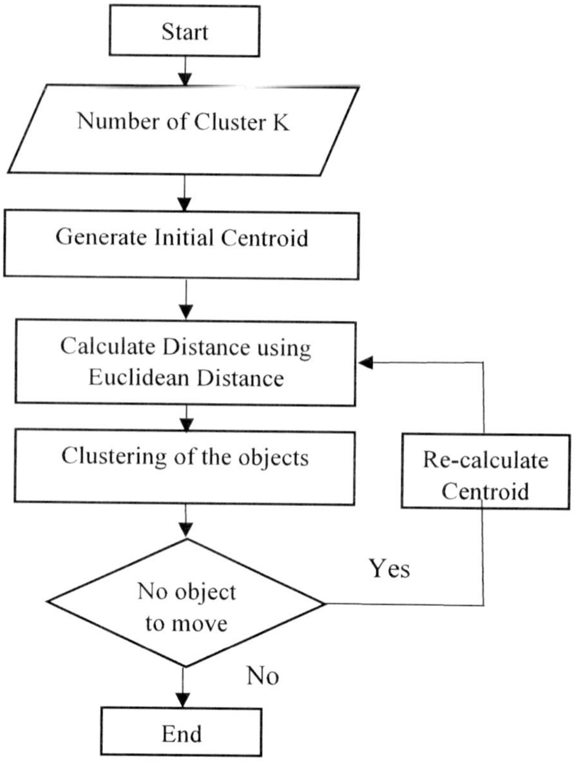

Fig. 4 K-means flowchart

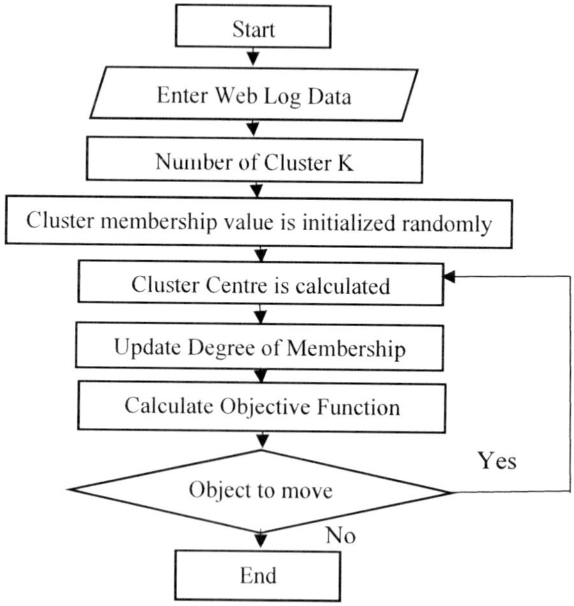

Fig. 5 Fuzzy C-Means flowchart

2.2.2 Algorithm: K-Means

The mean for each cluster's data points represents the cluster's nucleus as per K-Means technique for partitioning.

```
Input:
  K: required clusters.
  D: n data points available with data set.
Output: A set of K clusters.
Method:
Step 1. As the initial cluster center arbitrarily select K objects;
Step 2. Repeat
Step 3. (re)initialize every object to the cluster;
i.  Repeat:
ii. Compute the length for each centroids from a data point and the
data point which have least length from the centroid of a cluster
which is assigned to that specific cluster center. Compute the mean
for that cluster center.
iii. Edit the cluster mean to that centroid.
iv. Continue till last data point
Step 4. Continue till cluster values remain same;
```

2.2.3 Issue in K-Means

In K-Means, clustering technique is repeated until there may be no modification in cluster centers. K-Means method is effective in coping with the crisp data point which have perfect borders, but in reality, clusters have fuzzy borders and regularly overlying clusters. Usually, the natural data point consist of uncertainty, imprecision, and ambiguity [2].

2.2.4 Fuzzy Clustering Algorithm

Fuzzy C-Means (FCM) is fuzzy version of the K-Means method. **It's an innovative clustering technique which lets data point to belong to more than one clusters depending on their degree of interest**. Fuzzy C-Means contains fuzzy set an idea of incomplete gratification and can give outcome in the creation of overlying clusters. The fuzzy method estimates the cluster centers and initializes an amount of gratification to each data point between zero to one. This technique invented by Dunn in 1973 and it is upgraded by Bezdek in 1981. The fuzzy method is a repetitive clustering technique that find out an optimal C cluster (Table 2 and Fig. 5).

Algorithm for Proposed FCM [6]

```
Step 1: Assign
    Total clusters c = 2
    Number of patterns on the dataset (n)
    Fuzziness control (m, n), with (m, n) > 1
    Iteration number t = 0
    Stopping criteria ε = 0.001
```

```
    Initialize centroids (vi(0))
Step 2: Calculation
    For 1 ≤ k ≤ n
For 1 ≤ i ≤ c
Compute the distance d²(xk, vi) between the kth pattern
and ith the cluster center using Eqs. (1) or (2):
```

$$d(x_i, v_j) = \left| \sum_{k=1}^{n} x_k^i - v_k^j \right|^2 \tag{1}$$

$$d_{\text{cov}}(X_i, V_j) = \frac{d_{\text{cov}}(X_i, V_j)}{\sqrt{d_{\text{var}}(X_i) * d_{\text{var}}(V_j)}} \tag{2}$$

Edit the fuzzy membership degree $\mu_{ik}^{(t)}$ using (3)

$$\mu_{ik} = \left[\sum_{j=1}^{c} \left(\frac{d^2(x_k, v_i)}{d^2(x_k, v_j)} \right)^{\frac{2m}{(m-1)}} \right]^{-1} \tag{3}$$

Computer the cluster center $v^{(t)}$ using (4)

$$v_i = \frac{\sum_{k=1}^{n} \mu_{ik}^m x_k}{\sum_{k=1}^{n} \mu_{ik}^m}, \ 1 \le i \le c \tag{4}$$

```
Step 3: Stopping Criterion
      If max max_ik |v_ik^(t+1) − v_ik^t| < ε then stop.
    Else Increment iteration number t = t+1 and GoTo step
2.
```

3 Experiment and Result Analysis

For making comparison of the proposed technique, VNSGU proxy server dataset is chosen. The associated details with the dataset are as follows (Table 3)

By analyzing Table 4 and Fig. 6, Web user-based on their Web site visit. We have the following observation. Out of 86 users, almost 71% users are interested in the Cluster5. Out of 86 users, almost 25% users are interested in the Cluster1. Out of 86 users, remaining 3% users are interested in the Cluster2, the Cluster3, and the Cluster4. K-Means is speedy but less informative; whereas, limitation of FCM is prior requirement of the number of clusters, complex, and time-consuming but advantage of FCM is detailed, sturdy, and more explanatory. With K-Means algorithm, the user

Table 4 K-Means clustering algorithm user versus cluster

User	K-Means	User	K-Means	User	K-Means	User	K-Means	User	K-Means	User	K-Means
1	5	16	1	31	5	46	5	61	5	76	5
2	5	17	1	32	5	47	5	62	5	77	5
3	1	18	1	33	5	48	5	63	2	78	5
4	1	19	1	34	5	49	5	64	5	79	5
5	1	20	5	35	1	50	5	65	5	80	5
6	5	21	1	36	5	51	5	66	5	81	5
7	5	22	1	37	5	52	5	67	5	82	5
8	5	23	1	38	5	53	5	68	5	83	5
9	5	24	1	39	5	54	4	69	5	84	5
10	1	25	1	40	5	55	5	70	5	85	5
11	1	26	5	41	5	56	5	71	5	86	5
12	1	27	1	42	3	57	1	72	5		
13	1	28	5	43	5	58	5	73	5		
14	1	29	5	44	5	59	1	74	5		
15	1	30	5	45	5	60	5	75	5		

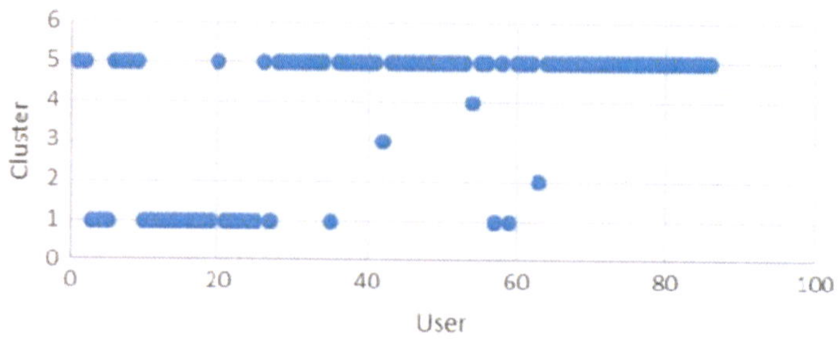

Fig. 6 K-Means clustering algorithm user versus cluster

can belong to any one cluster. We do not get much more information related to it. With FCM, we have received amount of gratification for user belonging to each cluster, relative level of interest of group of users in different clusters. We also get set of users visiting behavior with the help of FCM algorithm.

By analyzing Table 5 and Fig. 7, out of total 86 users, strikingly a type of group emerges out in the Cluster4 which comprises 27% users who share highest amount of gratification compared to which they have in remaining clusters. These 27% users are highly interested in the Cluster4 having average amount of gratification of 0.67—highest among all the clusters. This indicates significant number of users who are more interested in the Cluster4 than in any other clusters. Out of total 86 users, 73% users are more interested in the Cluster1 and the Cluster5 as compared with the Cluster4. Visiting pattern of all the users of the Cluster1 and the Cluster5 are similar but reverse in the Cluster4. All the users are rarely interested in the Cluster2 and Cluster3. 73% of the users are less interested in the Cluster4 as compared with the Cluster1 and the Cluster5 (Table 6).

By analyzing Figs. 8 and 9 and Table 5, we observed that the average amount of gratification of 73% of users in the Cluster1 and the Cluster5 is 0.369 and the same users have amount of gratification of 0.229 in the Cluster4. It represents interest gap which is only 0.14. 73% of the users are more likely and equally interested in the Cluster1 and the Cluster5 as compared with the Cluster4. By analyzing Fig. 9 and Table 7, we observed that average amount of gratification of 27% of users in the Cluster4 is 0.67 and in the Cluster1 and the Cluster5 is 0.17 represents interest gap of 0.47. 27% of the users is very less interested in the Cluster1 and the Cluster5. Least number of users are observed in the Cluster2 and the Cluster3 in both K-Means and FCM algorithm. By analyzing K-Means and FCM chart, we can say that only one user interested in Cluster4 as per K-Means chart and table but more users are interested in the Cluster4 as per FCM chart and table (Fig. 10).

With respect to Figs. 6 and 7, we have clear observation that all the users of the Cluster5 in K-Means algorithm are found equally interested in the Cluster1 and the Cluster5 of FCM algorithm.

We believe that the above results can be further improved if we use Fuzzy Possibilistic C-Means method to builds gratifications and possibilities concurrently. The

Table 5 Fuzzy C-Means clustering algorithm user amount of gratification versus cluster

User	Cluster1	Cluster2	Cluster3	Cluster4	Cluster5	User	Cluster1	Cluster2	Cluster3	Cluster4	Cluster5
5	0.061	0	0	0.877	0.061	73	0.366	0.001	0.003	0.264	0.366
18	0.066	0	0	0.868	0.066	44	0.367	0.001	0.002	0.263	0.367
13	0.091	0	0	0.817	0.091	66	0.367	0.001	0.003	0.263	0.367
21	0.094	0	0	0.811	0.094	6	0.369	0	0.002	0.26	0.369
22	0.094	0	0	0.811	0.094	39	0.369	0.001	0.003	0.258	0.369
19	0.101	0	0	0.798	0.101	33	0.369	0.001	0.003	0.257	0.369
23	0.111	0	0.001	0.777	0.111	28	0.369	0.001	0.003	0.257	0.369
12	0.119	0	0.001	0.762	0.119	60	0.372	0	0.002	0.253	0.372
25	0.123	0	0.001	0.754	0.123	70	0.373	0	0.002	0.252	0.373
24	0.141	0	0.001	0.717	0.141	72	0.374	0.001	0.003	0.248	0.374
4	0.156	0	0	0.689	0.156	32	0.376	0.001	0.003	0.245	0.376
3	0.159	0	0.001	0.681	0.159	30	0.382	0	0.001	0.234	0.382
16	0.189	0	0.002	0.619	0.189	74	0.385	0	0.002	0.228	0.385
15	0.19	0	0.002	0.617	0.19	75	0.387	0	0.001	0.224	0.387
10	0.202	0	0	0.595	0.202	82	0.387	0.001	0.003	0.223	0.387
36	0.223	0	0	0.554	0.223	62	0.394	0	0.001	0.21	0.394
27	0.234	0.001	0.003	0.528	0.234	61	0.397	0	0.001	0.204	0.397
57	0.247	0	0.002	0.504	0.247	79	0.398	0	0.001	0.202	0.398
26	0.253	0	0	0.494	0.253	83	0.401	0	0	0.198	0.401
59	0.26	0	0.002	0.479	0.26	69	0.401	0	0.001	0.197	0.401
17	0.265	0.001	0.005	0.464	0.265	48	0.402	0	0.002	0.194	0.402

(continued)

Table 5 (continued)

User	Cluster1	Cluster2	Cluster3	Cluster4	Cluster5	User	Cluster1	Cluster2	Cluster3	Cluster4	Cluster5
35	0.281	0.001	0.004	0.432	0.281	77	0.403	0	0.001	0.192	0.403
11	0.285	0.001	0.006	0.423	0.285	86	0.406	0	0.001	0.187	0.406
14	0.066	0	0	0.868	0.3	56	0.409	0	0.001	0.181	0.409
50	0.31	0	0.001	0.378	0.31	68	0.409	0	0.001	0.181	0.409
9	0.328	0.001	0.003	0.34	0.328	81	0.41	0	0.001	0.178	0.41
38	0.319	0.01	0.04	0.312	0.319	53	0.411	0	0.001	0.178	0.411
55	0.337	0.003	0.011	0.311	0.337	80	0.412	0	0.001	0.174	0.412
2	0.343	0	0.002	0.311	0.343	8	0.413	0	0.001	0.173	0.413
40	0.329	0.006	0.026	0.31	0.329	84	0.413	0	0.001	0.172	0.413
43	0.334	0.004	0.019	0.31	0.334	47	0.415	0	0.001	0.17	0.415
64	0.334	0.003	0.019	0.309	0.334	49	0.416	0	0.001	0.166	0.416
45	0.328	0.004	0.031	0.309	0.328	34	0.421	0	0.001	0.158	0.421
71	0.341	0.002	0.008	0.308	0.341	78	0.423	0	0.001	0.154	0.423
65	0.348	0.002	0.007	0.295	0.348	31	0.424	0	0.001	0.151	0.424
1	0.352	0	0.001	0.295	0.352	58	0.426	0	0.001	0.148	0.426
46	0.35	0.001	0.005	0.293	0.35	67	0.431	0	0	0.138	0.431
54	0.296	0.025	0.093	0.29	0.296	52	0.434	0	0.001	0.131	0.434
20	0.356	0	0.002	0.285	0.356	51	0.437	0	0.001	0.126	0.437
37	0.357	0.001	0.003	0.283	0.357	29	0.442	0	0	0.115	0.442
85	0.362	0	0.001	0.276	0.362	7	0.45	0	0	0.099	0.45
41	0.362	0.001	0.002	0.273	0.362	42	0	0	0.999	0	0
76	0.366	0.001	0.002	0.266	0.366	63	0	1	0	0	0

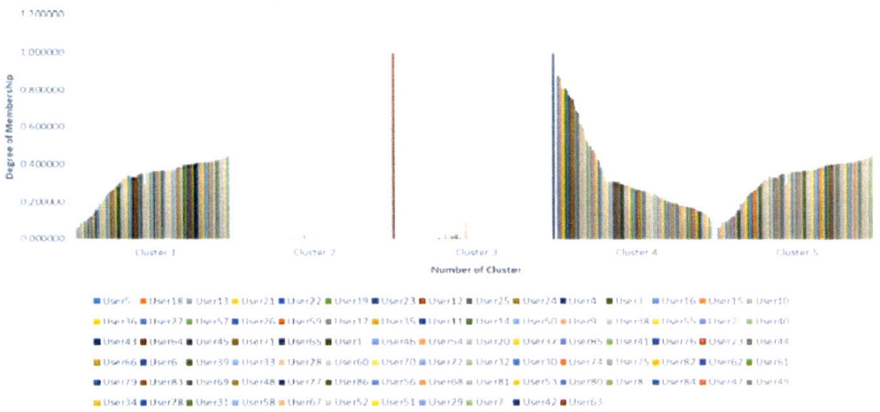

Fig. 7 Fuzzy C-Means algorithm user versus amount of gratification

Table 6 Average amount of gratification 73% user

Cluster1	Cluster2	Cluster3	Cluster4	Cluster5
0.368805	0.017356	0.021513	0.223522	0.368805

Fig. 8 Average amount of gratification 73%

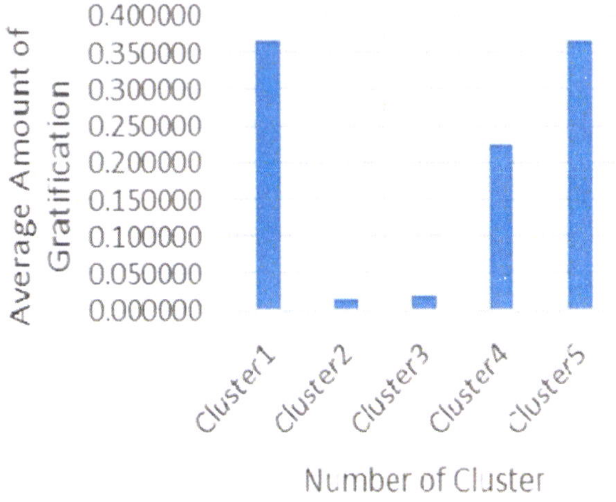

noise sensitivity and outliers issue of FCM are answered by FPCM, which solve the simultaneous clusters drawback of PCM.

Fig. 9 Average Amount of gratification 27%

Table 7 Average Amount of gratification 27% user

Cluster1	Cluster2	Cluster3	Cluster4	Cluster5
0.176919	0.000385	0.001719	0.644058	0.176919

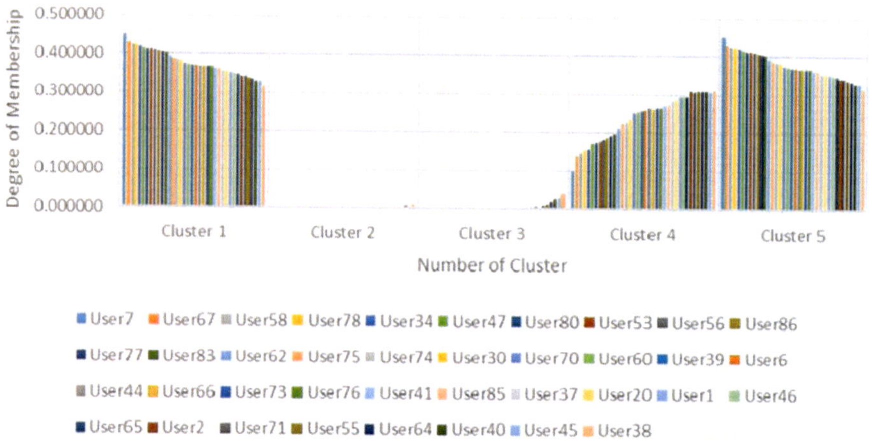

Fig. 10 Average amount of gratification cluster5

4 Conclusion

Data clustering performs an essential part in assembling the likely type of data point into a particular cluster. Cluster investigation objective at detecting out sets of related data points. Therefore, this assists to find out the identification of patterns and exciting associations in proxy server datasets. In case of Fuzzy C-Means, an expansion of the cluster examination which means the relationship of data points focusing to cluster through amount of gratification. A study of K-Means and fuzzy C-method uses proxy server log dataset to detect clusters. Clustering is curious, useful, and difficult concept. It has great capacity in area such as like item recognition and dataset cleaning and retrieval. But, it is possibly to take advantage of this capability only after detecting many visiting patterns. The K-Means and Fuzzy C-Means are popular clustering methods in the partition-based methods and they have been applicable in lots of application domain. It is very popular that each of the two methods has merits as well as demerits. K-Means is mentioned as rapid, robust, and modest to develop. K-Means is efficient in computational time complexity with its value. K-Means method can be a good choice for special clustering. K-Means is unable to handle overlapping clustering and nonlinear changes of issues of data point. This is to such an extent that portrayals of a particular dataset with Cartesian coordinates and polar coordinates which may also supply special clustering effects. K-Means fails to answer cluster noisy information and nonlinear datasets. Fuzzy C-Means offers pleasant outcome for overlapped situation within proxy log file. Similarly, it is better than K-Means strategy, contrasting to K-Means where the data point is the part to a single cluster. In Fuzzy C-Means, the data point is offered gratification to multiple clusters. Fuzzy C-Means, a priori requirement of the number of clusters. Fuzzy C-Means is little bit extra difficult and slower in execution, but effective approach for removal of noise. The present research has used the two methods—K-Means and Fuzzy C-Means. Fuzzy C-Means algorithm provides a more detailed view in comparison to K-Means algorithm.

References

1. Kumar GK, Chary TB, Premchand P (2013) A new and efficient K-Means clustering algorithm. Int J Adv Res Comput Sci Software Eng ISSN: 2277 128X 3(11)
2. Ansari Z, Azeem MF, Babu AV, Ahmed W (2011) A fuzzy clustering based approach for mining usage profiles from web log data. Int J Comput Sci Inf Secur (IJCSIS), ISSN 1947–5500, vol 9, no 6
3. Gupta A, Khandekar A (2016) Development of web log mining based on improved fuzzy C-Means clustering algorithm. Int J Sci Eng Technol Res (IJSETR) 5(3)
4. Gayathri K, Vasanthi D (2017) Brain tumor segmentation using K-Means clustering and fuzzy C-Means algorithms. Int J Sci Res Comput Sci Eng Inf Technol, 2017 IJSRCSEIT, 2(2) ISSN: 2456–3307
5. Huang YP, Bhalla K, Chu HC et al (2021) Wavelet K-Means clustering and fuzzy-based method for segmenting MRI images depicting parkinson's disease. Int J Fuzzy Syst 23:1600–1612
6. Shedthi BS, Shetty S, Siddappa M (2017) Implementation and comparison of K-Means and fuzzy C-Means algorithms for agriculture data. In: 2017 International Conference on Inventive Communication and Computational Technologies ICICCT
7. Kaur S, Singh C (2017) Use of fuzzy C-Means algorithm for web proxy server performance improvement. Int J Comput Sci Mobile Comput IJCSMC 6(4), pp 332–340, ISSN 2320 088X
8. Mishra H, Shuchi, Tripathi SP (2017) A comparative study of data clustering techniques. Int Res J Eng Technol (IRJET) 04(05):e-ISSN: 2395–0056, p-ISSN: 2395–0072
9. Bora DJ (2017) Performance comparison of K-Means algorithm and FCM algorithm with respect to color image segmentation. Int J Emerg Technol Adv Eng, ISSN 2250–2459, ISO 9001:2008 Certified Journal, vol 7(8)

10. Tseng JC (2017) Knowledge management based on fuzzy clustering algorithm with picard iteration. Int J Sci Eng Sci 1(12):32–37, ISSN: 2456–7361
11. Gayathri K, Vasanthi D (2017) Brain tumor segmentation using K-Means clustering and fuzzy C-Means algorithms. Int J Sci Res Comput Sci Eng Inf Technol IJSRCSEIT vol 2(2), ISSN: 2456–3307
12. Shi P (2017) An efficient approach for clustering web access patterns from web logs. Int J Adv Sci Technol
13. Agarwal A, Saxena A (2020) An approach for improving page search by clustering with reference to web log data. Int J Sci Technol Res 9(2) ISSN 2277–8616
14. Mining Projected on Improved Fuzzy C-Means Clustering Algorithm, © 2019 JETIR vol 6(3), ISSN-2349–5162
15. Wiharto ES (2020) The comparison of clustering algorithms K-Means and fuzzy C-Means for segmentation retinal blood vessels. Acta Inform Med 28(1):42–47
16. Zhang X, Wang H, Zhang Y et al (2021) Improved fuzzy clustering for image segmentation based on a low-rank prior. Comp Visual Media 7:513–528. https://doi.org/10.1007/s41095-021-0239-3
17. Rustam Z, Hartini S (2019) Classification of breast cancer using fast fuzzy clustering based on kernel. IOP Conf Series: Mater Sci Eng 546:052067. https://doi.org/10.1088/1757-899X/546/5/052067
18. Song Q, Wu C, Tian X et al (2021) A novel self-learning weighted fuzzy local information clustering algorithm integrating local and non-local spatial information for noise image segmentation. Appl Intell
19. Razavi SM, Kahani M, Paydar S (2021) Big data fuzzy C-means algorithm based on bee colony optimization using an Apache Hbase. J Big Data 8:64
20. Jahan M, Hasan M (2021) A robust fuzzy approach for gene expression data clustering. Soft Computing. 25 https://doi.org/10.1007/s00500-021-06397-7
21. Tomar A et al (2020) Machine learning, advances in computing, renewable energy and communication, Springer Nature, Berlin, LNEE vol 768, 659 p. https://doi.org/10.1007/978-981-16-2354-7. (ISBN 978–981–16–2354–7)

A Meta-Analysis of Student's Aptitude and Attitude Toward Learning Electronics Through Remote Experimentation

Deepti Prit Kaur, Shivani Malhotra, and Archana Mantri

Abstract For the courses based on fundamental electronics, the usual teaching involves delivery of theoretical content through lectures and tutorials and corresponding practical through hands-on sessions in the laboratory. This practice is based on "learning by doing" exercise to ensure maximum knowledge transfer among the learners. However, in certain situations, like one posed by the pandemic conditions worldwide, this approach may not be feasible for implementation, leading to disconnect between the theory and practice classes. To address this, this paper describes the meta-analysis of student's learning in the theoretical course on "basics of electronics engineering" integrated with virtual lab sessions through remote experimentation. The student attitude for the adopted methodology has also been analyzed on the basis of user interface, behavior of the virtual environment, visualization provided by the environment, and role of lab supervisor. For effective utilization of the virtual lab sessions with online teaching of theory course, the assessment strategy was also modified for evaluating the student's knowledge through a well-designed questionnaire based on integrated course curriculum. By comparing the end semester result of previous batch students who were administered traditional approach of theory class followed by physical lab sessions (control group) with those who were taught using integrated virtual laboratories (treatment group), it was found that the technical knowledge and overall grade for both the groups remain significantly the same, indicating effectiveness of the adopted methodology. Furthermore, the proposed solution supports technology enhanced learning (TEL) and offers flexibility in teaching through blended mode in an efficient manner for the institutions with constraints in the form of non-availability of appropriate infrastructure and number of instructors/faculty for real-time hands-on experimentation.

Keywords Engineering education · Blended learning · Virtual labs · Basic electronics · Educational technology

---

D. P. Kaur (✉) · S. Malhotra · A. Mantri
Chitkara University Institute of Engineering and Technology, Chitkara University, Punjab, India
e-mail: deeptiprit.kaur@chitkara.edu.in

A. Tomar et al. (eds.), *Proceedings of 3rd International Conference on Machine Learning, Advances in Computing, Renewable Energy and Communication*, Lecture Notes in Electrical Engineering 915, https://doi.org/10.1007/978-981-19-2828-4_59

1 Introduction

For education technology, there has been ongoing research to develop and define the actions for the teachers, so that student acquires appropriate practical knowledge relevant to respective technical concepts. During past few years, there has been a constant enhancement in instructional practices. Besides chalk-board teaching, considerable efforts have been taken to enhance the student's learning, by imbibing the use of technology in education with the use of computers and other multimedia material, simulations, games, smart phones, and immersive technology such as mixed reality (virtual or augmented) that can provide complete or partial immersion to its users [17]. When it comes to the discipline of electronics engineering, where conceptual models and analogies are equally important along with the presence of physical 3D objects; the approach of "learning by doing" becomes of paramount importance so that the theoretical concepts are supplemented with the practical experience and the attainment of learning outcome is maximized. This is because, the problem solving in electronics is closely related to physics concepts, which involves the visualization of complex spatial processes and manipulation of the graphs, diagrams, and concepts [19]. Engineering education provides comprehension of fundamentals through hands-on learning experiences, which eventually aids to the theoretical knowledge in the students. Researchers aim to improvise students' conceptual understanding through the use of new techniques, thereby, enhancing their learning and motivation. Previous studies [3, 8] have shown a relationship between the visualization skills of a student and his/her understanding for the complicated concepts based on science, technology, education, and mathematics. The emerging technologies in the field of engineering education have significantly contributed for refining the educator's pedagogy, enabling better inculcation of such developments into their curriculum, while imparting appropriate and relevant technical knowledge to their students. New trends such as games, massive open online courses (MOOCs) [13], immersive technologies (augmented and virtual reality) [2, 6, 7, 11], remote laboratories, [10, 14] etc., have been successfully used for educational purposes. More precisely, due to the widespread of novel Coronavirus, which eventually created a sudden shift of regular classrooms to online teaching–learning scenario, the use of aforementioned technologies in engineering education has been encouraged on a wider scale. This paper presents the results of investigation on students' attitude and aptitude after integrating the course on basic electronics engineering with virtual labs (also called as VLabs) [12] to carry out remote experimentation for improved understanding of concepts in freshmen students of undergraduate engineering courses. This study has been done to answer the following research questions:

Research Question 1: Is there any significant difference in student's learning of electronics fundamentals through integrated remote experimentation in comparison with hands-on physical laboratory experimentation?

Research Question 2: What is the student's response toward the use of adopted virtual system on the basis of user interface, behavior of the virtual environment, visualization provided by the environment, and role of lab supervisor?

This paper has been organized as follows.

Sect. 2—basic electronics and its importance for engineering undergraduate programs.

Sect. 3—virtual labs for remote experimentation.

Sect. 4—analysis of student's attitude and aptitude toward learning through virtual labs integrated with the theory course on basic electronics engineering.

Sect. 5—conclusion of the study and future scope for carrying out similar implementation for other courses in electronics engineering.

2 Importance of Teaching Basic Electronics Course for Engineering Undergraduate Programs

The course of basic electronics has been designed for freshmen students to make them familiar with electronics elements such as diodes and transistors; their functionality and applications in circuits such as rectifier, voltage regulator, and amplifiers. The course also introduces digital electronics for the students to perceive the concept of logic gates and integrated circuits in electronics. After learning this course, the students are expected to acquire an ability to explore physical systems by setting up experiments, collecting and analyzing data, identifying sources of uncertainty, and interpreting their results in terms of the fundamental principles and concepts of electronics.

Usually, for this course, the theory sessions are followed by lab sessions for real-time implementation of the theoretical concept in focus. As represented by Fig. 1, the integrated virtual lab sessions for the course covers the practical experiments alongside the respective theory topics in the blended mode [21], where the theory lecture delivery could be done through online platforms with practical experiments demonstrated and conducted via VLabs [5]. The contents of the course on basics of electronics engineering involve semiconductor diodes and applications, transistors and applications, introduction to digital electronics and logic circuits and integrated circuits such as logic gates, operational amplifier, timer, etc.

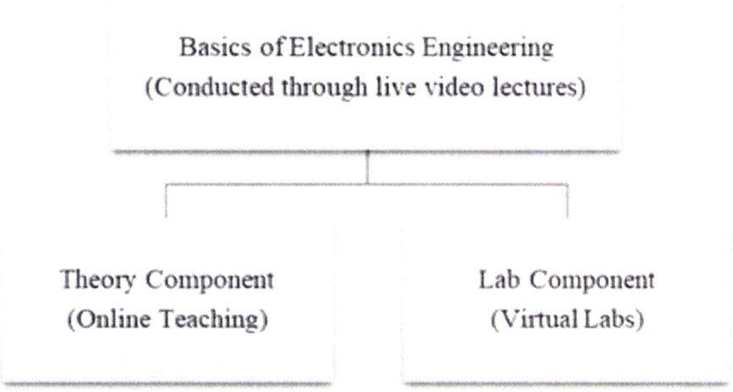

Fig. 1 Conduct of course on basics of electronics engineering

Table 1 Evaluation scheme for the course

Evaluation component	Type of component	No. of assessments	Weightage of component (%)
Component 1	Formative assessments	02	10
Component 2	Mid-semester sessional tests (STs)	02	30
Component 3	End term examination	01	60
Total		100%	

The evaluation strategy for the online conduct of this course has been represented in Table 1.

The technology is capable of enhancing the learning, if activities are smoothly integrated with teaching strategies. In order to achieve this for the course taught using virtual labs in addition to traditional teaching, the evaluation components 1 and 2 represent internal evaluation through formative assessment and mid-semester tests, and component 3 is external evaluation through end semester examination. For all these components, the questionnaire has been framed very carefully to test the theoretical as well as practical knowledge of the students. The final result (component 1 + component 2 + component 3) carries 100% weightage.

3 Virtual Labs for Remote Experimentation in Basic Electronics Courses

The virtual laboratories provide a platform for online hands-on experience for various fields of science and engineering and help to enhance the technical and laboratory skills of undergraduate science students and research scholars [16]. Many researchers have presented their work by utilizing virtual laboratories for engineering education course on electrical engineering [4, 18]. By enabling the remote experimentation for basic and advanced concepts of engineering, a complete learning system is made available to the learners at the places with constraints due to geographical area and lack of infrastructure. During pandemic period, virtual labs have emerged as a useful resource for delivery of lab-oriented theoretical courses by enabling the engineering undergraduate students to perform online experiments for observations and analysis of various engineering processes. In India, virtual labs is an initiative of Ministry of Education, under the National Mission on Education through information and communications technology (ICT). Various broad areas (such as electronics and communication engineering, computer science and engineering, electrical engineering, mechanical engineering, etc.) are offered by the participating institutes.

Figure 2 presents the simulation environment for connecting the circuit elements of a forward biased silicon diode, and Fig. 3 presents the corresponding V–I characteristics graph obtained after making the appropriate connections and observing the

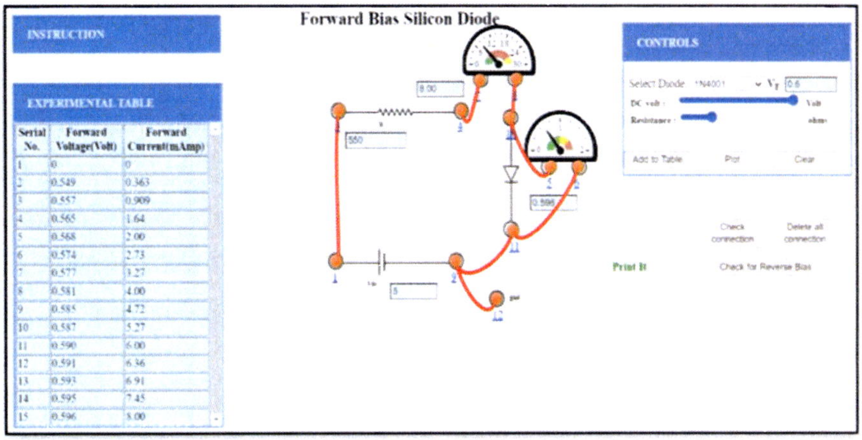

Fig. 2 Virtual labs simulation environment for a forward biased silicon diode

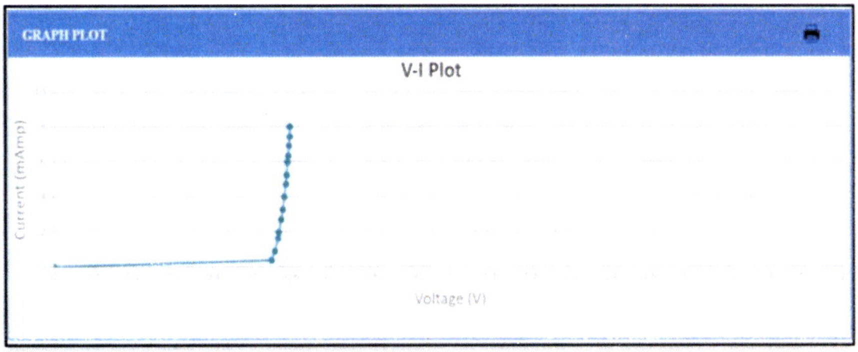

Fig. 3 Plotting of V–I characteristics graph on virtual Labs

output current corresponding to several values of input voltage varied through the controls provided on the interface.

For the present study, several experiments (such as PN junction diode, Zener diode, rectifier, transistor, logic gates, IC555 timer, etc.) were conducted through the VLabs for enhancing the theoretical knowledge of students in respective theoretical concepts.

4 Result Analysis and Discussion

This section presents the analysis of student's result and survey conducted to observe the aptitude and attitude of the students for the adopted methodology of virtual labs for imparting basic electronics education to engineering undergraduates. The two research questions as presented in Sect. 1 have been answered by this study.

Research Question 1: **Is there any significant difference in student's learning** for fundamentals of electronics through integrated remote experimentation in comparison with hands-on physical laboratory experimentation?

To answer this question, the end semester result of students learning through virtual labs for a theoretical course was compared with that of students learning through physical laboratories for hands-on experimentation for the same course. A sample of 30 students from each group was considered for analysis on the results of end term exam and final semester exam. Figure 4 presents the total sample base of 60 students. Table 2 clearly gives that the percentage average marks of the students of two groups are comparable with values of percentage average score as 85.35 for traditional teaching followed by hands-on physical lab experimentation, and 84.22 for students learning through online mode integrated with virtual laboratory experimentation.

Thus, research question 1 has been answered as there is no significant difference in student's learning for fundamentals of electronics through integrated remote experimentation in comparison with hands-on physical laboratory experimentation. However, there are certain factors, such as lack of physical contact with the device,

Fig. 4 Sample base

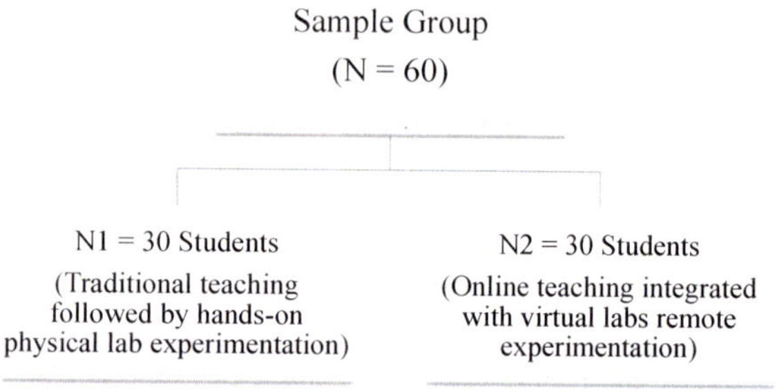

Table 2 Result analysis of two sample groups

Sample group	Mode of teaching	% Average score
N1	Traditional teaching followed by hands-on physical lab experimentation	85.35
N2	Online teaching integrated with virtual labs remote experimentation	84.22

lack of face-to-face interaction with teachers, lack of student attention, participation, etc., which may affect the outcome of online teaching [1, 9, 15, 20].

Research Question 2: What is the attitude of students toward the use of adopted virtual system on the basis of user interface, behavior of the virtual environment, visualization provided by the environment, and role of lab supervisor?

To answer this question, the students learning through remote experimentation were surveyed using a questionnaire based on the use of virtual labs integrated with theory course. A total of 30 participating students of group 2 responded as per the Tables 3 and 4, where Table 4 gives the number of responses of students on likert scale 5 (with 1-strong disagree, 2-disagree, 3-neutral, 4-agree, and 5-strong agree).

For the dichotomous questions based on yes/no answers, the student's response from Table 3 indicates that the adopted system was new for majority of the students (70%), as they had not worked using virtual experimentation before. Also, a greater number of students (87%) felt that virtual labs help for better understanding of concepts through remote experimentation.

Table 3 Student response for dichotomous questions

Sr. No	Statement	Yes	No
1	Have you ever used virtual mode of experimentation before being exposed to the same through the course on basics of electronics engineering?	9	21
2	Do you think virtual labs serve the purpose of imparting conceptual and practical knowledge?	26	4

Table 4 Student response for scaling questions

Sr. No	Statement	Average response on a scale of 5
1	The user interface of the platform was easy to handle as appropriate instructions were available for implementation of the experiment	4.1
2	The virtual experiments provided flexibility in terms of parameter modification in the circuit	3.97
3	Adequate visualization (in terms of output waveforms) was provided by the virtual experiments for comprehension of the theoretical concept in focus	4.17
4	The simulation results supported the theoretical concepts, hence relevant for the subject material	4
5	Faculty/instructor was helpful and there was adequate communication between teachers and students to carry out an experiment	4.53
6	On a scale 1–5, rate the understanding of experiments done in the basic electronics course (theory integrated with virtual labs)	4.07

The student's response from Table 4 indicates that the adopted system was well appreciated by the learners, as the average response for all the scaling questions is more than 4 on a scale of 5 (except for item number 2 in Table 4, which has response value of 3.97, which is nearly 4). Thus, the research question 2 is answered that the users show a positive attitude toward the use of adopted virtual system on the basis of user interface, behavior of the virtual environment, visualization provided by the environment, and role of lab supervisor.

5 Conclusion and Future Scope

This paper represented a comparative analysis of student's learning through the use of virtual labs with respect to the traditional lab-oriented teaching for a theory course in electronics engineering in order to enable the learners acquire the basic fundamentals via remote experimentation. The results indicated similar learning in engineering undergraduates on the basis of chosen approach (traditional hands-on laboratory or virtual simulations), thereby, providing an opportunity to inculcate hands-on practices among the students even in the absence of physical setup of laboratories. The main purpose of this study was to probe the student's behavioral analysis on usage of virtual tools available in support of technology enhanced learing (TEL). A blended learning approach on the basis of user interface, behavior of the virtual environment, visualization provided by the environment, and role of lab supervisor has been validated through this work, where the students learn through online simulation tools/platforms in addition to traditional ways of learning and were engaged in different activities for better utilization of time to enhance their knowledge. In addition, it is also useful with respect to the safety measures and saves cost of accidental damage of electronic components. Thus, remote experimentation provides an efficient solution to the problems raised due to non-availability of physical lab setup for hands-on practical sessions to support the theoretical concepts, and hence, enhance the student's understanding.

References

1. Azoulay U (2020, November). Global education coalition. UNESCO. https://en.unesco.org/covid19/educationresponse/globalcoalition
2. Cai X, Tu Y, He X (2018) An interactive augmented reality system based on LeapMotion and Metaio. In: Advanced multimedia and ubiquitous engineering. pp 237–244. Springer, Singapore
3. Contero M, Naya F, Company P, Saorin JL, Conesa J (2005) Improving visualization skills in engineering education. Computer graphics in education. Published by the IEEE computer society, pp 24–31
4. Hasan B, Al-Quorashy Y, Al-Mousa S, Al-Sahhaf Y, El-Abd M (2020) V-LAB–the virtual electric machines laboratory. In: 2020 IEEE global engineering education conference (EDUCON). IEEE, pp 72–77
5. https://www.vlab.co.in/

6. Jensen L, Konradsen F (2018) A review of the use of virtual reality head-mounted displays in education and training. Educ Inf Technol 23(4):1515–1529
7. Kaur DP, Mantri A, Horan B (2019) Design implications for adaptive augmented reality based interactive learning environment for improved concept comprehension in engineering paradigms. Interactive learning environments, 1–19
8. Kozhevnikov M, Thornton R (2006) Real-time data display, spatial visualization ability, and learning force and motion concepts. J Sci Educ Technol 15(1):111
9. Limpraptono FY, Faradisa IS (2016) Development of the remote instrumentation systems based on embedded web to support remote laboratory. In: 2015 Proceedings of second international conference on electrical systems, technology and information 2015 (ICESTI 2015). pp 537–543. Springer, Singapore
10. Limpraptono FY, Nurcahyo E, Faisol A (2021) The development of electronics telecommunication remote laboratory architecture based on mobile devices. Int J Online Biomed Eng (iJOE) 17(03):26
11. Lu A, Wong CS, Cheung RY, Im TS (2021) Supporting flipped and gamified learning with augmented reality in higher education. In: Frontiers in education. 6:110. Frontiers
12. Makarova I, Boyko A, Shubenkova K, Pashkevich A, Giniyatullin I (2019) Virtual laboratories: engineers' training for automotive industry. In: 2019 17th international conference on emerging elearning technologies and applications (ICETA). IEEE, pp 505–511
13. Martínez-Núñez M, Fidalgo-Blanco Á, Borrás-Gené O (2015) New challenges for the motivation and learning in engineering education using gamification in MOOC
14. Odeh S, Shanab SA, Anabtawi M, Hodrob R (2013) A remote engineering lab based on augmented reality for teaching electronics. Int J Online Eng (iJOE) 9(S5):61–67
15. Parmigiani D, Benigno V, Giusto M, Silvaggio C, Sperandio S (2020) E-inclusion: online special education in Italy during the Covid-19 pandemic. Technol Pedagogy Educ 1–14
16. Potkonjak V, Gardner M, Callaghan V, Mattila P, Guetl C, Petrović VM, Jovanović K (2016) Virtual laboratories for education in science, technology, and engineering: a review. Comput Educ. https://doi.org/10.1016/j.compedu.2016.02.002
17. Santos MEC, Chen A, Taketomi T, Yamamoto G, Miyazaki J, Kato H (2014) Augmented reality learning experiences: survey of prototype design and evaluation. IEEE Trans Learn Technol 7(1):38–56
18. Sarac VJ, Minovski DI, Stefanov GG (2020) Virtual laboratories-an innovative concept in teaching of electrical engineering. In: 2020 24th international conference on information technology (IT). IEEE, pp 1–4
19. Smith ME (2009) The correlation between a pre-engineering student's spatial ability and achievement in an elecetronics fundamentals course. All graduate theses and dissertations, p 254
20. Tomar A et al (2020) Machine learning, advances in computing, renewable energy and communication, Springer Nature, Berlin, LNEE vol 768, 659 p. https://doi.org/10.1007/978-981-16-2354-7. (ISBN 978–981–16–2354–7)
21. Yigit T, Koyun A, Yuksel AS, Cankaya IA (2014) Evaluation of blended learning approach in computer engineering education. Procedia Soc Behav Sci 141:807–812

Reduction of IoT Security Vulnerabilities Using Machine Learning Algorithm

Naveen and Uttam Sharma

Abstract The function of IoT-enabled spruce devices has increased to a big extent, it is estimated that in the last ten years, the use of spruce devices reached 20–21 billion, and this number will increase drastically in this decade. The use of smart devices makes human life easier as we are moving towards a connected world but we cannot shut our eyes towards the loopholes and vulnerabilities of IoT device security and network threats. The paper is lightening up some well-known and newly discovered security threats and we have provided a collective package of machine learning algorithms that possibly reduce the vulnerability to some extent. The concept which will work behind the proposed package of an algorithm is data which is being collected from several IoT-enabled device and main features which will play a significant role in the prediction about severity of attack and vulnerability are the type of attack, i.e. software, network, physical attack or encryption attack, the effect of the attack, time taken in the discovery of attack and the device or frame work or platform which is more vulnerable to attacks. The collected data need to be pre-processed using pre-process steps and stage, and then the algorithm implementation will give the prediction and using the predictions or we can say the set of rules the vulnerability can be detected and reduced.

Keywords Security threats · Vulnerability · Machine learning algorithms

1 Introduction

The paper aims to cover major security threats discovered in IoT-enabled smart environments around us and how machine learning algorithms can be given as a possible solution to overcome that vulnerability. The architecture design for various

---

Naveen (✉)
GL Bajaj Institute of Technology and Management, Greater Noida, India
e-mail: naveenbasoya@gmail.com

U. Sharma
Gautam Buddha University, Greater Noida, India

A. Tomar et al. (eds.), *Proceedings of 3rd International Conference on Machine Learning, Advances in Computing, Renewable Energy and Communication*, Lecture Notes in Electrical Engineering 915, https://doi.org/10.1007/978-981-19-2828-4_60

IoT devices does not follow a specific standard but every vendor has its framework and architectures but one thing is always common for every IoT device that its architecture always includes three layers including sensor network and application. The application of IoT in real life and its security vulnerabilities.

Sensor Devices

Heat Sensor The heat sensors are implanted in work place industries and living areas. These sensors are considered to be the most commonly used type of sensor as the range of this kind of sensor is used in thermostat devices for sensing temperature and heat, and this kind of sensor can serve from household for checking the boiling temperature to industry purpose where the sensitivity is much higher in the industry like furnaces plants. These sensor devices can be based on two logic like contact-based or no-contact type, as the name indicates the type will serve it's specific are of service like sensing liquids, gases or solid material and the no-contact type will serve in sensing the gases and liquids which produce radiant energy in form of infra-red radiation.

Example: LM35 Temperature sensor by robokart, W1206 Digital LED DC 12V + NTC Sensor.

Humidity Sensor This type of sensor works on the principle of sensing and measuring the dew drops in environments, and these are also known as humidity detectors. The working logic behind this kind of sensor is a measurement of the ratio of dew and absolute humidity by combining RH and temperature parameters. The sensor focuses on two elements: air and moisture. Example: DHT22 digital humidity sensor, DHT-11 humidity sensor.

Motion Sensor and Ultrasonic Sensors Motion sensors are based on a combination of various sensations like pressure, movements, vibration, density and several other parameters. Ultrasonic sensors are also known as object detection sensors or we can say distance measurement sensor, and the working principle behind the sensor is an emission of ultrasonic waves by which an estimation of the distance of the intended object is been calculated. The reflection of measurement can be in form of an echo pattern when the sound waves hit by an object, here the receiver and sender elements of the signal are known as a transducer which will send ultrasonic wave or high-frequency sound wave and receive back the echo pattern, and made the distance pattern estimation accordingly. Example: HC-SR04.

Security Vulnerability Associated with Sensors Sensor devices are prone to be a victim of network attackers in IoT environment, and the attacker usually find out the loop hole in the sensor layer and network layer. The attacker may attack via sensor layer or network layer and easily get access over a node and can delete and copy sensitive data, and there is a possible attacker can change the configurations and basic requirements by which monitoring intended to be done but with changed parameters, nobody will immediately know about attack and change in the system until a big issue has been discovered and till that a lot of data may be lost which will lead to big financial loss. Another possibility of attack is an attacker can overload the network with false notification and alerts which are meaningless but are intended to make the server busy enough that it cannot handle future requests and responses.

There are several applications by which attackers can easily analyse the traffic flowing in the network and internal communication between two devices or server-to-device communication.

Smart Authorization Devices

Smart locks and FRS systems: These devices are designed to provide security in a more authentic and customized way as compared to traditional lock systems and user identification systems which can be unlocked by duplicate keys or unauthorized person who may enter the traditional ledger. The smart lock device works on a cryptographic key and uses a wireless mechanism for connectivity to its authorized user. If the device is been accessed or damaged in any means, instant alert will be notified to the user, so he or she can get the message about possible mishappening. Here, in contrary to traditional lock, these locks do not depend on physical key for unlocking but it uses a smart phone for unlocking purpose with proper fingerprint verification, hence boost the security in every means. Example: J20-A series smart lock and YMI 70RB-A series smart locks.

Security Vulnerability Associated With Smart Authorization Devices

Smart tools are at risk to numerous types of attacks, unauthorized strikes software attacks, deprivation of information attacks and hardware attack. If we talk about smart locks, it depends on the key which is nothing but the connected smart phone with a legitimate user fingerprint or passes key, in basics there is no other way to access a smart lock, no tampering can be done, but what happens when the security of smart phone is being breached as there is a well-known loophole in smart phone security, an attacker can easily access the control over device if a user is not aware for security breaches or may not considering required access control over applications and accessibility. An attacker may extract credentials by which the alert and notification can be diverted to the false node instead of legitimate mode, in such case, the devise sensor is working fine but the information is not conveyed to the registered user so even if security has been compromised but nobody knows until the things are monitored carefully but it will take time. Smart locks use SSL to communicate which is an encrypted way of communication still using encryption attacks and software attacks the security can be violated. Voice assistance holds the vulnerability of storing and recoding users data like voice commands daily routine and other sensitive data about day-to-day life and several cases have been reported like the conversation of couple fight was been recorded and listened to by the company employee which was not legitimate at all, with the latest operating system, they gives the right to phone user for controlling the access and right been granted to an application installed but with granted rights still lots of application gain the access unknowingly to the end-user, and the details is set aside in the cache memory and cookie logs by app and this information and highly prone to be attacked and used by malicious user or attackers, they can easily use the voice commands, saved password of the user to create an action against user which will result in financial loss or some other harm to the user.

2 Literature Survey

As per the DoW, the original demonstrator should be: "moderately basic instances of essential action segments, for example, presence and area, methods of velocity furthermore, stance, and hand motions." The second era was to "join a few fundamental segments into more intricate exercises". At last, the third era was to "based upon a complex mix of essential parts and show situations propelled by and associated with reality application regions" For the original, the first arrangement was to put together the assessment concerning basic effectively repeatable tests. Not withstanding, after expounding conversation, the consortium has chosen to change that arrangement also, start the undertaking with an account of a huge, complex informational index that can be utilized for stage one, stage two, and to a huge degree stage 3. The choice has been founded on the accompanying contemplations:

Aiming at the investigation of dynamic, changing sensor designs the undertaking requires exceptionally multimodal informational collections. Recoding such informational indexes includes extensive exertion, even in little scale tests. Along these lines, when putting forth such an attempt it is attractive for guarantee that subsequent information can be utilized for more than one examination. AI strategies including directed learning, unaided education and fortification learning (RL) have been broadly try to better Internet security, for example, confirmation, way in command, against sticking discharge and malicious recognitions [1–15].

Supervised learning strategies, for example, uphold vector machine SVM, credulous Bayes and K-nearest neighbour (KNN), neural organization, profound neural organization (DNN) and arbitrary woods perhaps utilized on name of organizations business or application hints of IoT gadgets to fabricate characterization or relapse model [2]. For instance, IoT gadgets can utilize SVM to identify network interruption [2] and parodying assaults [5], apply KNN in the organization interruption [6] and malware [7] recognitions and use the neural organization to recognize network interruption [8] and DoS assaults [9] Credulous Bayes can be applied by IoT gadgets in the interruption identification [2]. The arbitrary backwoods classifier can be utilized to distinguish malware [7]. IoT gadgets with adequate calculation and remembrance assets could be use DNN to recognize ridiculing assaults [16].

Unsupervised learning does not need marked information in the directed learning and examines the closeness between the unlabelled information to bunch them into various gatherings [2]. For instance, IoT gadgets could be make use of multi-component relationship inquiry to accept DoS assaults [10].

Reinforcement learning procedures, for example, Q-learning, Dyna-Q, post-choice state (PDS) [17, 18] and profound Q-organization (DQN) [19] empower an IoT gadget to pick the safety convention just in spite of critical boundaries opposed to different assaults through experimentation [1]. Now instance, Q-learning is a model-free RL procedure that utilized to upgrade the exhibition about validation [1], hostile to sticking unload [3, 12, 13] and malicious location [4].

Scope for the proposal of machine learning algorithms: We have downloaded a data set named security vulnerability, the data set contains data from the year

2015–2019 about the security breaches detected over the year and it also highlights the type of security issue that occurred mostly in a particular year. In Fig. 1, it has a clear visualization of the type of vulnerability discovered in several years from the year 2015–2019. The vulnerabilities which are highlighted in the data set are denial of service attack, code execution mistakes, network overflow issue, memory vulnerability, SQL injection, XSS, directory traversal, bypassing of nodes and network, splitting of HTTP request/response, gain information, etc.

Figure 2 shows the comparison of the last three years, i.e. 2019, 2018 and 2017 which clearly shows that the rate of occurrence of DoS attack is increasing every year with a good proportion that indicates the breach is being identified by attackers but still the developers of devices and appliances has not focused on over coming this breach or the deployment of a possible solution is to the time taken, another possibility behind this the detection of such breaches is too slow or not that efficient, for proposing and for support of that proposal, we have also collected the data for vulnerabilities detected before the year 2015.

In Fig. 3, it shows majorly detected security vulnerability detected in those years so as shown in the diagram, the attacks which present in high proportion are overflow of severing, code exhaust issue and denial of service attack. The overflow of server or network and DoS was a major issue before the year 2015 as it ranges from 112–1574 to 894–7946, respectively. When we see the upcoming year in case of code execution attack the range of cases discovered increase till 2017 after that in 2018, there is a fall in several cases and in 2019, the number reduces to half. In the case of DoS attack, there is a fall in several cases after 2014 which remain in 2015 and 2016 but suddenly in the year 2017, there was a drastic increase in several attacks which keep growing in 2018 in 2019, there was a slight drop but the number of cases

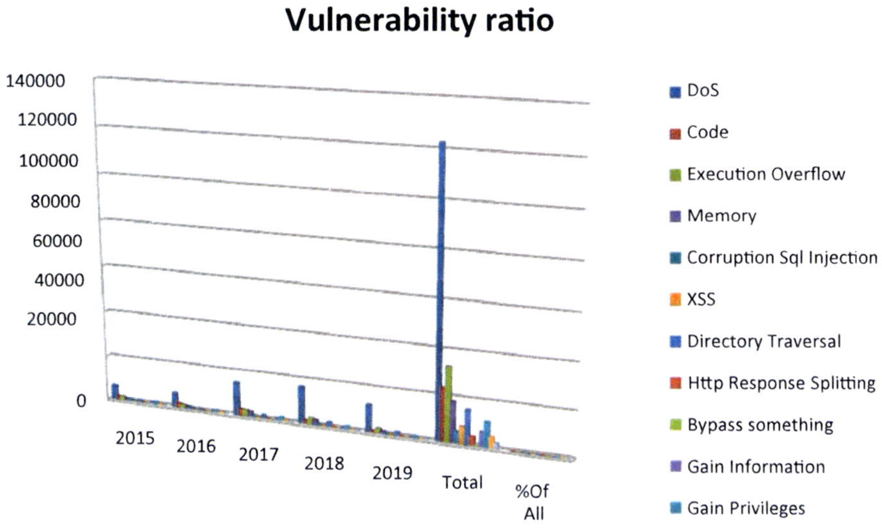

Fig. 1 Type of vulnerability detected over from year 2015–2019

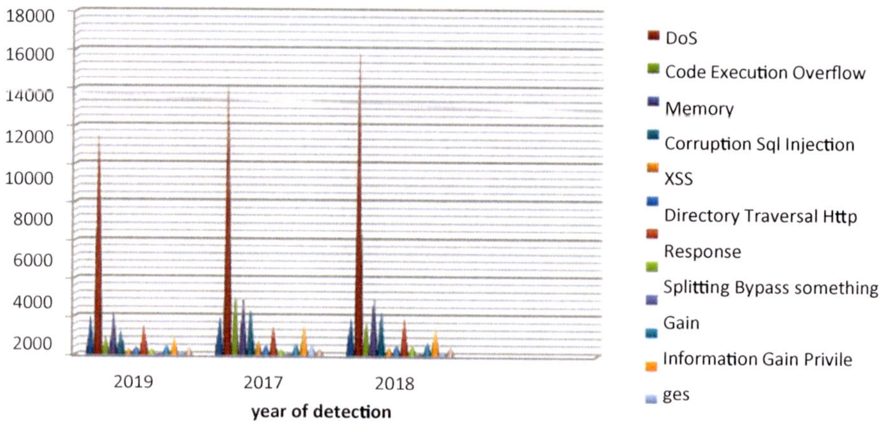

Fig. 2 Vulnerability detected in year 2019 in comparison with previous years

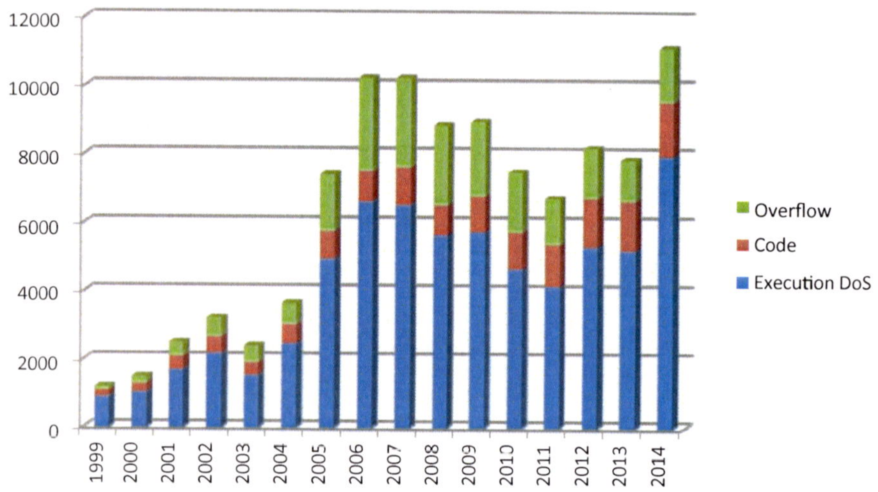

Fig. 3 Highest detected vulnerability before year 2015

still there. The whole scenario shows the weakness of the system for detection of vulnerability and in the address of these breaches, here we have proposed a package of machine learning algorithms which will act as the module in combination with their specific capabilities to detect the breach and problem in the existing framework so the solution for these can be proposed and employed in no time (Fig. 4).

Fig. 4 Visualization of classification error against DoS vulnerability

3 Module of Selected Algorithm and Their Capabilities

The algorithms we have selected for our proposed module are KNN, SVM and Naïve Bayes. The proposed module will be applied in a secure authentication framework where with capabilities of the algorithm, the like accuracy ration, missing value ratio, false-positive ratio, kappa statistics, false alarm rate, energy consumption rate and several other parameters will act as a key feature in concluding performance of device and detection of possible attacks and vulnerabilities in IoT devices. With the selected data set, we have applied some dimension reduction techniques for refining the data set for making it compatible for implementation on these algorithms and some of the output for prediction parameters are attached in Figs. 5, 6 and 7 they show the classification error rate ratio after implementation of algorithm module and the cluster assignment for focused vulnerabilities, i.e. DoS Attack, code execution and network/server over flow is shown in Figs. 7, 8 and 9.

With a low-dimension data set, the target parameters are calculated and measured for highly discovered vulnerabilities, and the prediction are visible in terms of visualization as well. Now the proposed module can be implemented on IoT devices with the required set-up in terms of software employments and with some hardware set-up. The target for this module to achieve authenticity is by labelling the detected breach vulnerability and gives the alert in terms of prediction so the delay which was present in earlier approaches can vanish and a smarter solution with smart prediction can be applied to upcoming IoT devices. The package is not limited to a new device but can also be added to the existing of software set-up of the device with a small upgrade like mechanism.

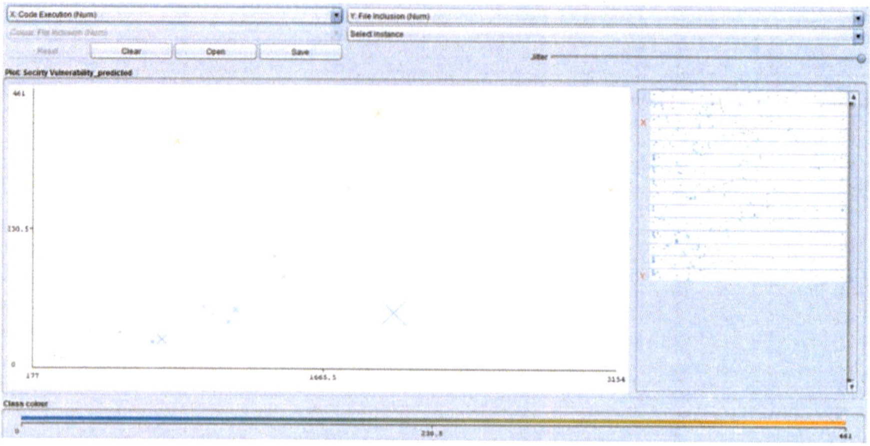

Fig. 5 Visualization of classification error against code execution vulnerability

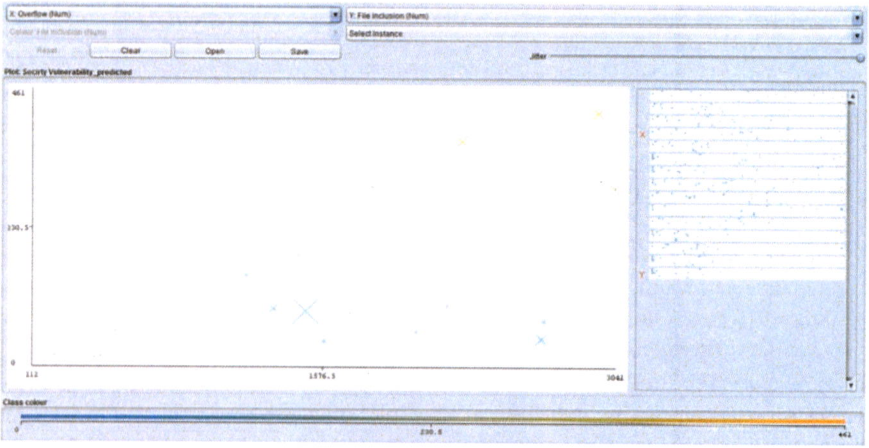

Fig. 6 Visualization of classification error against over flow vulnerability

4 Conclusion

The focus of the paper is to highlight the most commonly discovered security breach detected in the past year in IoT-enabled smart devices and appliances, and we have tried to address that's why a particular security breach has shown over year due to lack of detection system and slower deployment of the solution, here we have a focus on capability of machine learning algorithm, i.e. prediction and classification capabilities, these can be applied to detect the breach just by providing required data set which can easily be collected and maintained using smart device daily log file. As all IoT devices are connected and somehow use cloud computing services that mean

Fig. 7 Visualization of cluster assignment for code execution vulnerability

Fig. 8 Visualization of cluster assignment for over flow vulnerability

the data set is being prepared at the server side, with required modification, this data can be used to predict and classify the possible vulnerability in the software framework of the device or appliance. Proposed algorithm KNN, SVM and Naïve Bayes have their capabilities in outlier detection, missing value ratio, accuracy prediction, calculating false-positive ration and several other parameters for having accurate prediction about the performance of device if any point of time the performance is not addressed up to the mark, then the cross-check for vulnerability can be done immediately without wasting of time for a complete month or year stats, hence it will save time and effort in every possible way. About implementation of the proposed module, the only requirement is to have a well-formatted data set, with IoT devices

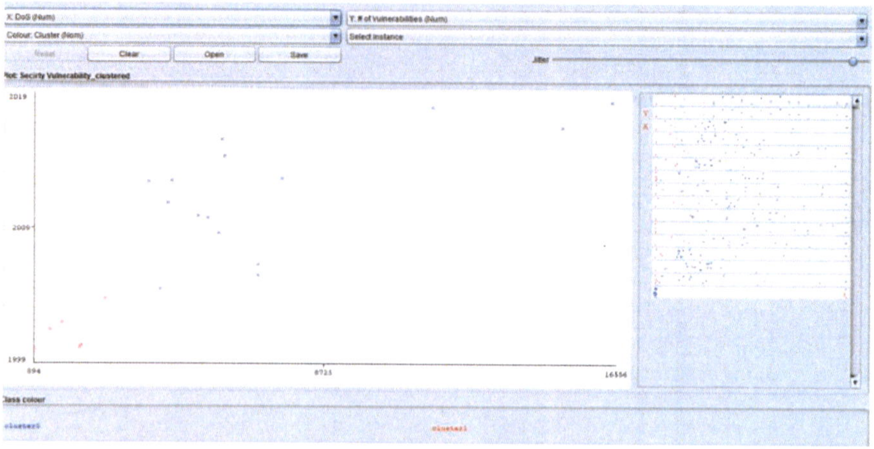

Fig. 9 Visualization of cluster assignment for DoS vulnerability

the data set of the daily log is already prepared in the existing framework so no additional burden will be there, just required modification and the data set is ready for prediction, hence for the deployment of this module no change in the framework is required just add the module at server side and all the prediction and classification will be done at server side only without any intervention and it make the existing IoT environment even smart era sit can detect the breaches and vulnerability by its own and rapid solution deployments and upgraded can be employed in no time once the breach has been identified.

References

1. Xiao L, Li Y, Liu G, Li Q, Zhuang W (2016) PHY-layers poofing detection with reinforcement learning in wireless networks. IEEE Trans Veh Technol 65(12):10037–10047
2. Alsheikh MA, Lin S, Niyato D, Tan HP (2014) Machine learning in wireless sensor networks: algorithms, strategies, and applications. IEEE Commun Surveys Tutorials 16(4):1996–2018
3. Xiao L, Xie C, Chen T, Dai H (2016) A mobile offloading game against smart attacks. IEEE Access 4:2281–2291
4. Xiao L, Li Y, Huang X, Du XJ (2017) Cloud-based malware detection game for mobile devices with offloading. IEEE Trans Mobile Comput 16(10):2742–2750
5. Ozay M, Esnaola I, Vural FT, Kulkarni SR, Poor HV (2015) Machine learning methods for attack detection in the smartgrid. IEEE Trans Neural Netw Learn Syst 27(8):1773–1786
6. Branch JW, Giannella C, Szymanski B, Wolff R, Kargupta H (2013) In-network outlier detection in wireless sensor networks. Knowl Inf Syst 34(1):23–54
7. Narudin FA, Feizollah A, Anuar NB, Gani A (2016) Evaluation of machine learning classifiers for mobile malware detection. Soft Comput 20(1):343–357
8. Buczak L, Guven E (2015) A survey of data mining and machine learning methods for cyber security intrusion detection. IEEE Commun Surveys Tutorials 18(2):1153–1176

9. Kulkarni RV, Venayagamoorthy GK (2009) Neural network based secure media access control protocol for wireless sensor networks. In: Proceedings international joint conference neural networks, pp 3437–3444, Atlanta, GA
10. Tan Z, Jamdagni A, He X, Nanda P, Liu RP (2013) A system for Denial-of-Service attack detection based on multivariate correlation analysis. IEEE Trans Parallel Distrib Syst 25(2):447–456
11. Xiao L, Yan Q, Lou W, Chen G, Hou YT (2013) Proximity-based security techniques for mobile users in wireless networks. IEEE Trans Inf Forensics Secur 8(12):2089–2100
12. Gwon Y, Dastangoo S, Fossa C, Kung HT (2013) Competing mobile network game: Embracing anti-jamming and jamming strategies with reinforcement learning. In: 2013 proceedings IEEE conference communication and network security (CNS), pp 28–36,National Harbor, MD
13. Aref MA, Jayaweera SK, Machuzak S (2017) Multi-agent reinforcement learning based cognitive anti jamming. In: 2017 Proceedings IEEE wireless communication and networking conference (WCNC), pp 1–6, San Francisco, CA
14. Li Y, Quevedo DE, Dey S, Shi L (2016) SINR-based DoS attack on remote state estimation: a game-theoretic approach. IEEE Trans Control Network Syst 4(3):632–642
15. Han G, Xiao L, Poor HV (2017) Two-dimensional anti-jamming communication based on deep reinforcement learning. In: 2017 IEEE international conference on acoustics, speech and signal processing, pp 2087–2091, New Orleans, LA
16. Shi C, Liu J, Liu H, Chen Y (2017) Smart user authentication through actuation of daily activities leveraging WiFi-enabled IoT. In: Proceedings ACM international symposium on mobile ad hoc networking and computing (MobiHoc), pp 1–10, Chennai, India
17. He X, Dai H, Ning P (2015) Improving learning and adaptation in security games by exploiting information asymmetry. In: 2015 IEEE Conference Computer Communication (INFOCOM), pp 1787–1795, Hongkong, China
18. Tomar A et al (2020) Machine learning, advances in computing, renewable energy and communication, Springer Nature, Berlin, LNEE volume 768, 659 p https://doi.org/10.1007/978-981-16-2354-7. (ISBN 978–981–16–2354–7)
19. Mnih V, Kavukcuoglu K, Silver D et al (2015) Human-level control through deep reinforcement learning. Nature 518(7540):529–533

A Novel Voltage Level-Up Shifter Design for Power Efficient Methods Using Dual Current Mirror Technique

Aylapogu PramodKumar, B. Kalivaraprasad, M. V. D. Prasad, A. Jayalakshmi, and S. Ravichand

Abstract The double supply voltage concept is extensively adoption in today's VLSI design to decrease power dissipation. Level shifters are helpful to convert small voltages to large voltages. As a low-power, efficient-energy level changer, the Wilson current mirror level shifter was presented. The recommended design was implemented using Cadence Virtuoso and 45 nm technology. At 1 MHz, an input voltage ranging from 0.5 V to 1 V was transformed with a power expenditure of 1.92 nW and a latency of 0.96 nS. In future, the proposed architecture might be implemented with fewer transistors, resulting in improved latency and power usage.

Keywords Twin-grid · Equipoise-converter · WCM · Energy utilization

Objectives
It is evaluated to provide a little setback time, effective output power. It is observed to build an effective circuit with short fixed power.

Motivation
Interfacing devices frequently employ level shifters. Other logics are necessary for interfacing. We switched to level shifters to prevent any additional embedded logic. Without any additional integrated logics, we can convert the logic levels.

A. PramodKumar (✉) · A. Jayalakshmi
Department of ECE, Vardhaman College of Engineering, Hyderabad, India
e-mail: pramodvce@gmail.com

A. Jayalakshmi
e-mail: a.jayalakshmi@vardhaman.org

B. Kalivaraprasad · M. V. D. Prasad
Department of ECE, KLU University, Vaddeswaram, Guntur, India
e-mail: baditakali@gmail.com

S. Ravichand
Department of ECE, Nalla Narasimha Reddy Institutions, Hyderabad, India
e-mail: ravichandusankuru1@gmail.com

A. Tomar et al. (eds.), *Proceedings of 3rd International Conference on Machine Learning, Advances in Computing, Renewable Energy and Communication*, Lecture Notes in Electrical Engineering 915, https://doi.org/10.1007/978-981-19-2828-4_61

Methodology
The Wilson current mirror concept is used to create the suggested model. We were able to determine the power and delay settings using this method. To balance the output impedance, the design might use a different circuit configuration. Here, the proposed design implemented by employing current mirror approach to boost the voltage levels from 0.5 to 1 V with minimum power and latency. We guaranteed that the outcomes will be more efficient than the current circuit.

Contribution
The primary goal of this work is to increase input voltage while reducing delay and power consumption in VLSI architectures.

1 Introduction

At a faster rate, the world is moving toward technology. When compared to the same technology now, electronic gadgets used to be somewhat huge. When the computer was first invented, it was so enormous that it took up a lot of space, but now, the same computer is so little that it can be taken around with ease. The improvements in the technology sector are the primary cause of this rapid change. Transistors played a crucial part in this shift in the electronic world. The size of transistors has shrunk in recent years, resulting in a reduction in the size of gadgets. The key area of interest is lowering the size of the transistor, and power dissipation is also diminished. There are several approaches for lowering the circuit's power usage. The power dissipation is diminished by lowering the supply voltage. Reduced supply voltage can result in a quadratic fall in power dissipation (CV2), but it also causes substantial latency. One of the most common strategies that makes use of deliver voltage scaling and has a big success amount in many-supply systems. In the more supply approach, there are double source voltages. The non-critical component of the circuit, known as VDDL, runs at a voltage lower than the supply voltage. The principal section of the circuit, i.e., VDD, works at nominal supply voltage and also recognized as VDDH. Using this technology, the circuit's dynamic power is lowered without affecting its performance.

The phrase "level shifter" refers to an interface circuit that changes signals between logic levels or voltage levels, providing interoperability between ICs with varying voltage requirements. The level shifter is most commonly used in different units or subunits operate at different voltages and speeds in digital circuits and SOC applications. It has the ability to act as a link between two voltage domains. In dual supply designs, shifters of voltage levels are employed to provide accurate voltage levels for consecutive digital sections.

The first conventional level conversion in dual VDD systems in which chip area was minimized by incorporating clustered voltage scaling (CVS) [1]. Delay, energy, and area penalties were focused while level conversion was done by CVS scheme. The use of a level converter on FF was tested to determine the network robustness versus supply, which is a critical element that distinguishes improved level proponent models. The multi-supply technique was used to create the level shifter, and the pull-up network was designed using the double-stage cross-couple approach [2]. An extra PMOS was linked in similar to the first stage of the pull-up structure to boost the pull-up network's potency. 90 nm CMOS technology was used to create the circuit. This circuit can change the voltage from 100 mV to 1 V. At 1 MHz. For a 200 mV source signal, the fixed dissipation power is 8.7 nW, the broadcast latency is 8.7 nW, and the entire force per conversion is just 77 fJ. The differential cascade voltage switch technology was used to construct the LS [3]. A split inverter was employed in the output stage to reduce power usage. In 180 nm CMOS technology, the design was realized. This concept was capable of converting a 100 mV source signal to 1.8 V. The propagation delay for a 400 mV source move to 1.8 V was 31.7 ns, the standard stationary potential was minor than 60 pW, and the vitality per changeover was 173 fJ. The level shifter's pull-up structure was created utilizing the regulated cross-couple (RCC) technology [4]. Using 180 nm CMOS technology, the post-layout simulation was delivered. A low voltage of 80 mV could be converted to 1.8 V using this equipoise changing. For a minimum/maximum supply voltage of 0.4/1.8 V and an input frequency of 1 MHz, the energy distraction and transmission delay are 123.1 nW and 23.7 ns, appropriately. A novel voltage equipoiser varying configuration that can execute together top and down level variable service has been developed. It employs transmission gates as well as a multiplexer [5], with the select line is the input voltage, while the input lines are the supply voltages, with the relax of the network serving as a level-up or level-down converter. The projected system can be employed in applications that require minimal power and fast speed. It can convert 0.4 to 1 V and can level falling from 1 V to 0.4 V. A groundbreaking voltage level changer using a WCM and CMOS logic gate architecture. It was created to be used in DVS applications [6]. "(a) Slighter vicinity in provisions of outline for auxiliary threshold voltage modification, (b) tiny power distraction in more than brink exercise, (c) reasonable grow and drop time lag in control space, (d) the operational range has no effect on transistor size or threshold voltage characteristics, and (e) bi-directional level changing activity were all factors in the development of the new level shifter". In this study, hybridization buffers, current mirrors, and delay circuits were employed to improve the level shifter's performance in terms of energy, latency, and duty cycle across a wide length of DVS functions. A network with no cross-coupled connections [7, 8], resulting in a delay reduction. The proposed level shifters are designed using transmission gates. The configured level shifter has a stable duty ratio and high-speed performance. It is ideal for broad I/O interface voltage applications in CMOS-based ultralow-power designs.

To reduce power consumption, a novel design of voltage level shifters was applied to Wilson mirror level shifters utilizing the altered Wilson current mirror circuit MTCMOS technology [9], however, leakage power dissipation is limited. Stack approach was used to overcome the problem of leaky power dissipation, and comparisons were done between deformed and non-deformed approach applied to altered Wilson current mirror-based level shifters. For power reduction in this circuit, forcing methods were used. Level shifter that has been proposed Wilson's current mirror hybrid buffer is modified utilizing both the forced PMOS approach and the non-forced PMOS approach. To build a unique LS, a altered Wilson current mirror hybrid buffer was employed [10]. To convert full-range and bi-directional levels, the recommended altered Wilson current mirror combination buffer level changing was constructed. The maximum operating voltage is specified by transistor technology as the standard supply voltage, although the nominal operating voltage may be deep under the threshold, shut to the least source energy of digital blocks. The circuit required a minimum input voltage 200 mV to function. The 65 nm technology was used to create the circuit. Half-latch-based methodology was used in conjunction with the present mirror technology to create a design [11]. As a current limiter, a single PMOS was used. This design would not operate if the input voltage was less than 100 millivolts. The broadcast latency be 18.4 nW, the energy utilization was 6.6 nW, also the strength for each conversion was 93.9 fJ for a 200 mV input voltage at 1 MHz.

The cross-coupled and current mirror approaches were combined to create a hybrid level shifter [12]. The split inverter was used to take the yield of the level shifter, which minimized the outflow current. This level changing might only exertion at a voltage of 330 mV. This level shifter does not perform properly for voltages below 330 mV. A device has a regulated current source and an inbuilt level shifter via a diode-associated transistor and an NMOS switch transistor in the input division. The output division is a CMOS inverter, and the ILS separated the output stage's gate terminals. Between the output inverter's gate terminals, a diode-connected level shifter was utilized, and simulations were run on 40 and 180 nm CMOS technology. The circuit required, just 4.2 fJ/transition when manufactured in 40-nm CMOS technology, with VDDL and VDDH of 0.35 V and 1.1 V, correspondingly. In 180 nm technology, this might work with an input voltage of 80 mV. The pass transistor was utilized to speed up the fall transition, and the reduced-swing buffer [14] was utilized to eliminate the large leakage current. The design could work in the sub-threshold range. The 65 nm CMOS technology was used to create the level shifter. These LS might changing a 300 mV signal to a 1 V output with a transmission time of 7.5 ns, outflow energy of 2.64 nW, and power per conversion of 123.8 fJ. The design took up 7.45 m^2 of floor space. A description of how a vibrant current generator [15] may be utilized to decrease stagnant energy dissipation by only operating when the logic levels of input and output do not match. The intensity of pull up/ down structure is kept in balance to execute level irregular excessive values of short input voltages that are under verge voltages. PVT corners were used to verify the level shifter's performance, and Monte Carlo simulations were used to estimate power. No fixed current flows among the supply rails because of this construction. Current mirror logic and DCVSL were worn to create the short-power voltage level changing. The power cutback approaches for turn deep submicron enter voltages to supposed output voltages were reported in an energy efficient sub-threshold level converter [16]. Multi-threshold devices, multiple supplies, level conversion delay, and extending minimum input voltages are all part of the model's implementation. This level shifter converts up to 1.2 V effectively and dependably. The supply voltage operating range is connecting 188 mV and 1.2 V, making it appropriate for together sub-threshold and DVS process. How whole energy distraction can be lowered utilizing dual approach: adaptive voltage level supply (AVLS) to boost supply potential and AVLG to raise ground potential.

Using the AVL technique, this research focuses on lowering active power, leakage power, and threshold voltage. This voltage level shifter's results confirm the low-power circuit features that may be employed in the design of low-power applications [17]. A revolutionary technique in which the primary transistors were body-tied was presented in interface sub and great threshold reconfigurable logic cells with ultra-small voltage level shifters. This method adds another link, resulting in an increase in energy use. The level shifter may be adjusted between 0.35 and 1.2 V. voltage level changers are divided into four categories, (a) voltage level shifter with twin supply, (b) voltage level changing with solitary supply, (c) voltage level shifter with pass-transistor semi latch, (d) electrical energy level changing with pre-charge circuit [18]. Several frequent misunderstandings have been uncovered and embittered. We describe an automatic layout approach that diminish energy usage by using two supply voltages.

Ishihara et al. [19] to save energy, the compact supply voltage is deflated in a clock tree. We adapted these strategies to a media processor chip by combining them. In this, they had described that the proficient on-to-off relation can be drastically improved by the use of to save electricity, the lower source voltage is also deflated in a timer tree. By linking these methodologies, we were able to apply them to a publishing processor chip. The united methodology compact power consumption by 47% in random-logic modules and 73% in the clock tree while preserving work. They stated that using this method can considerably improve the efficient on-to-off ratio [20]. Structures from the Schmitt trigger efficiently, reducing the gate output node leakage, thus stabilizing the output level. References [21–23], the multipliers are fully functioning at source voltages varied from 84 to 62 mV. The nominal supply voltage of 62 mV, 17.9 nW power consumption, and 5.2 kHz is measured with 8 × 8 bit multipliers.

2 Design Approach

Figure 1 shows the suggested diagram of a voltage level shifter. The suggested circuit is modeled using the Wilson current mirror level shifter approach. It has three branches: input, output, and voltage conversion. The source branch is made up of a CMOS inverter with little threshold PMOS and NMOS transistors. Current limiters, which are designed for voltage conversion, make up the voltage conversion branch. The current mirror is included in the pull-up network. A split inverter makes up the output branch. A high VDD is provided by the split inverter. The transistors' aspect ratio is displayed in Table 1. The MN1 transistor is triggered when the input A is HIGH, and the node Q1 is set to 0 (ground), leading the MP1 transistor to activate and Q3 to go low. Because the input is VDDL, MN2 is turned on, and Q2 is charged by turning off MP2, MP3, MP4, and MP5. MP6 and MN3 are switched on and off, correspondingly, because Q3 and Q1 are associated with ground. As a result, if the

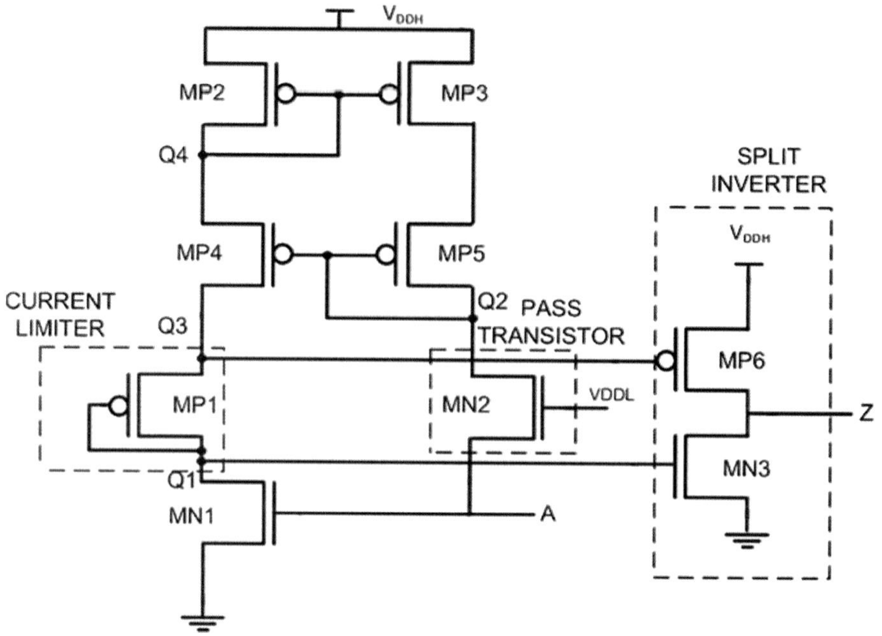

Fig. 1 Proposed level shifter design schematic

Table 1 Transistors aspect ratio

Equipment	(W/L) nm
MN2, MN3, MP2, MP3, MP4, MP5, MP6	240/180
MN1, MP8	300/180
MP1	450/180

output Z is set to HIGH. When the input A is small, the MN1 is turned off. Node Q2 receives the value of the input and reduces it. VDDL controls MN2, which is a pass transistor. Because the Q2 is at a low voltage, the MP2, MP3, MP4, and MP5 transistors are turned on. VDDH is charged with Q4, Q3, and Q1, leading the MP6 and MN3 to turn off and on, respectively. As a result, the output is grounded, resulting in a LOW signal.

Table 1 shows the design metrics for each transistor width and length in the proposed circuit.

3 Results and Discussion

Both 45 and 90 nm technology were used to create the proposed LS. The simulation results for 45 nm technology are shown here. On both circuits, various analyzes such as temperature study, frequency examination, and corner investigation were carried out. Temperature analysis was carried out for an source voltage of 0.5 V at a frequency of operation of 1 MHz by varying the temperature from −40 to 110 °C. Figure 2 depicts the voltage variation of the current design's input and output waveforms.

The supply voltage was changed from 1 to 0.6 V in steps of 0.1 V in 45 nm technology, and the variations in power and delay were observed as follows. The graphs are depicted in Fig. 3.

The temperature was changed from −40 to 110 °C, and the power and delay variations were observed as Fig. 4.

The frequency of input was changed from 1 to 100 MHz, and the power and delay were observed in the given Fig. 5.

The existing and proposed voltage was compared in 45 and 90 nm technology by performing supply voltage analysis by changing the supply voltage from 1.2 to 0.8 at 1 MHz frequency. The responses are represented in Fig. 6.

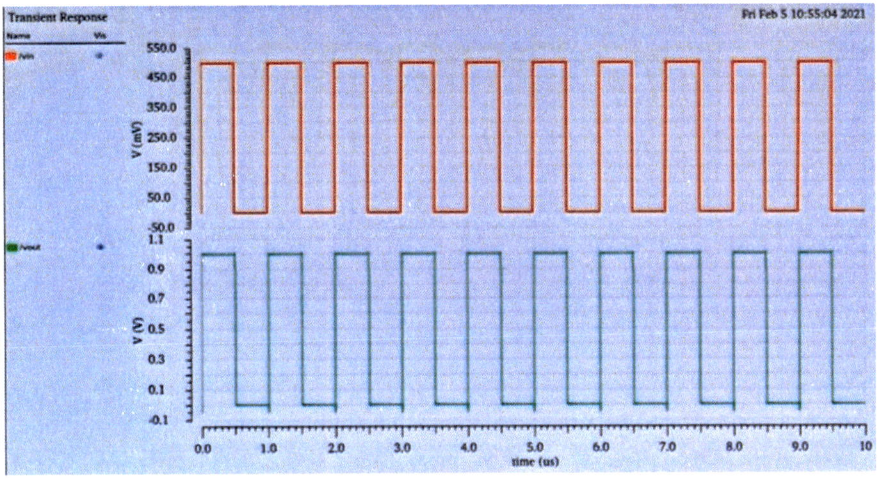

Fig. 2 Input and output wave forms of proposed design

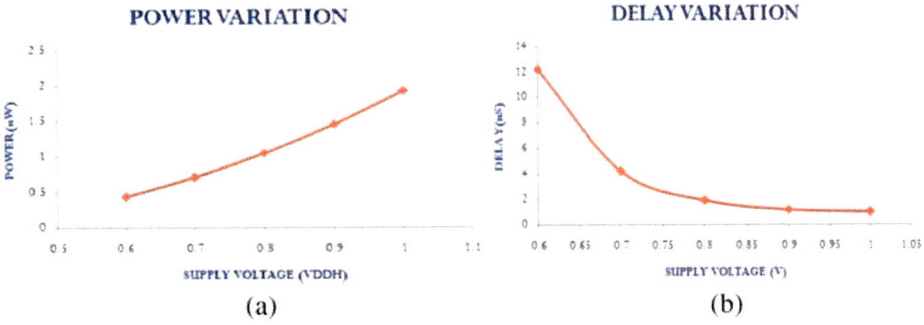

Fig. 3 **a**, **b** Voltage variations of power and delay

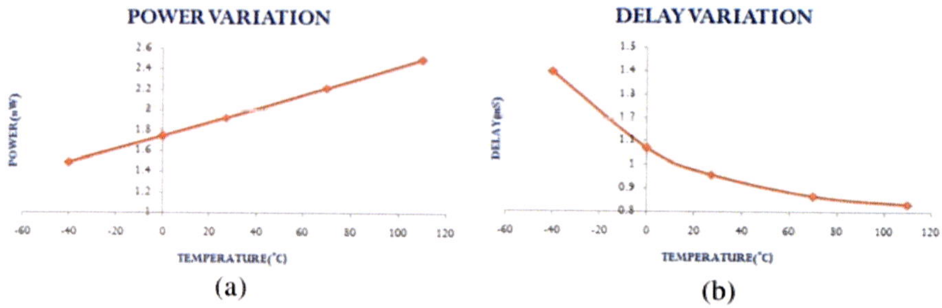

Fig. 4 **a**, **b** Temperature variations of power and delay from −40 to 110 °C

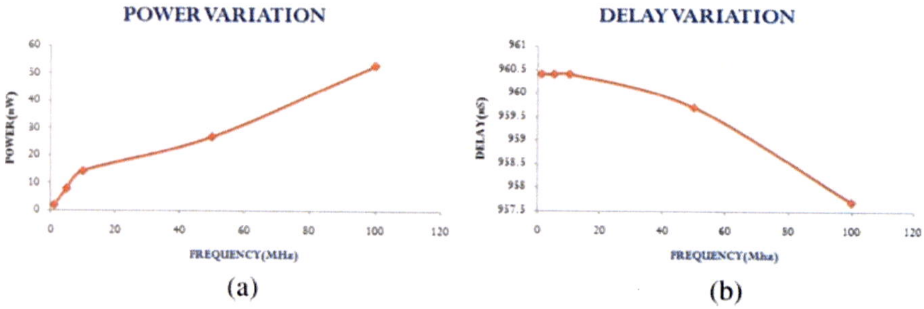

Fig. 5 **a**, **b** Power and delay variations from 1 to 100 MHz

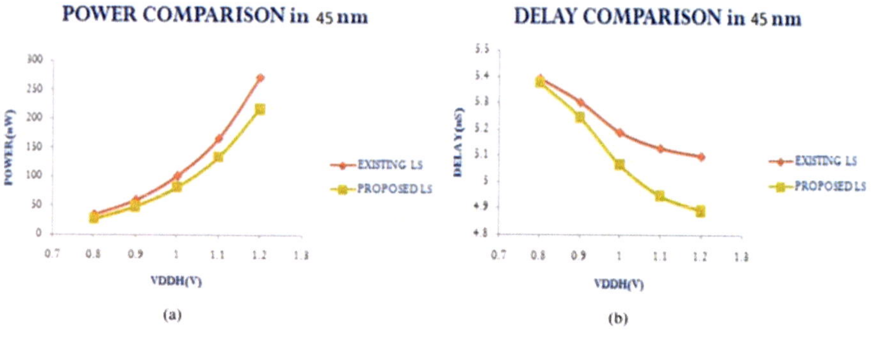

Fig. 6 **a** Energy and **b** Latency comparison of projected LS in 45 and 90 nm with change in the supply voltage

Temperature analysis was performed on existing and proposed designs by changing temperature from −40 to 110 °C in 45 and 90 nm technology. The obtained response is depicted in Fig. 7.

Frequency analysis was performed on both the designs by changing the frequency from 1 MHz to 100 MHz in 45 nm and 90 nm technology. Achieved graphs are shown in Fig. 8.

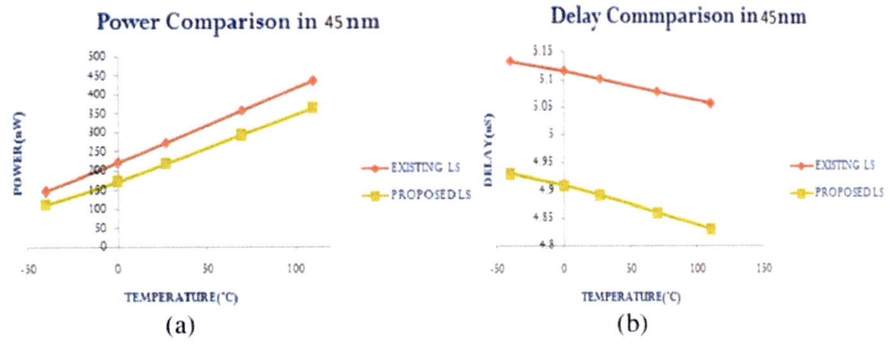

Fig. 7 **a** Energy and **b** delay comparison of proposed LS in 45 and 90 nm with change in the temperature

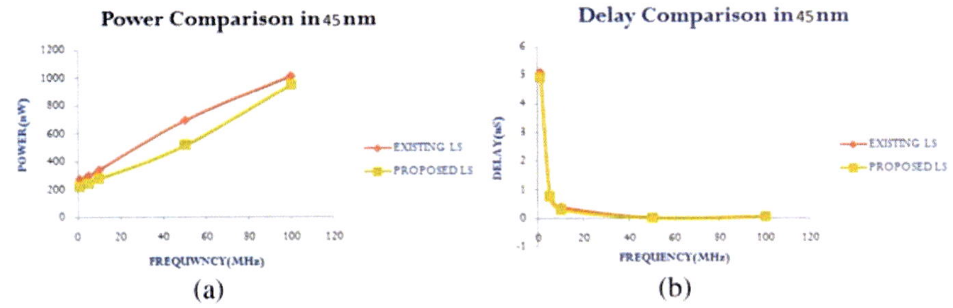

Fig. 8 **a** Energy and **b** Delay comparison of projected LS in 45 and 90 nm with change in the frequency

As the width of the transistor rises, so does the power consumption, and the latency of the transistor shrinkage as the channel length of the transistor shrinkage. This can be seen from the above figure as the temperature is increased, the power consumption gets increased rapidly, and the delay also tends to decrease in 45 nm compared to 90 nm. Throughout the investigation, the changes were tabulated by keeping the source train of pulse is 0.5 V, and the frequency of exploit is 1 MHz. The observations were graphically represented for the viewer's ease of comprehension. Figure 9 illustrates the fluctuation in characteristics when the frequency of operation was raised from 1 to 100 MHz for an input source of 0.5 V.

As can be seen, altering the transistor's size has a significant impact on when the frequency of operation is changed the parameters. Because energy dissipation is related to supply voltage, function frequency, and capacitance, increasing frequency leads to increased power. The increased in the channel length adds more effect, and the power at 45 nm has increased rapidly while compared to the 90 nm technology. The delay in 45 nm is negligible when compared to the delay in the 90 nm technology, which is due to the increased in the width of the gate. As the delay was low, it led the PDP of 40 nm technology to be lower compared to the 90 nm technology. The

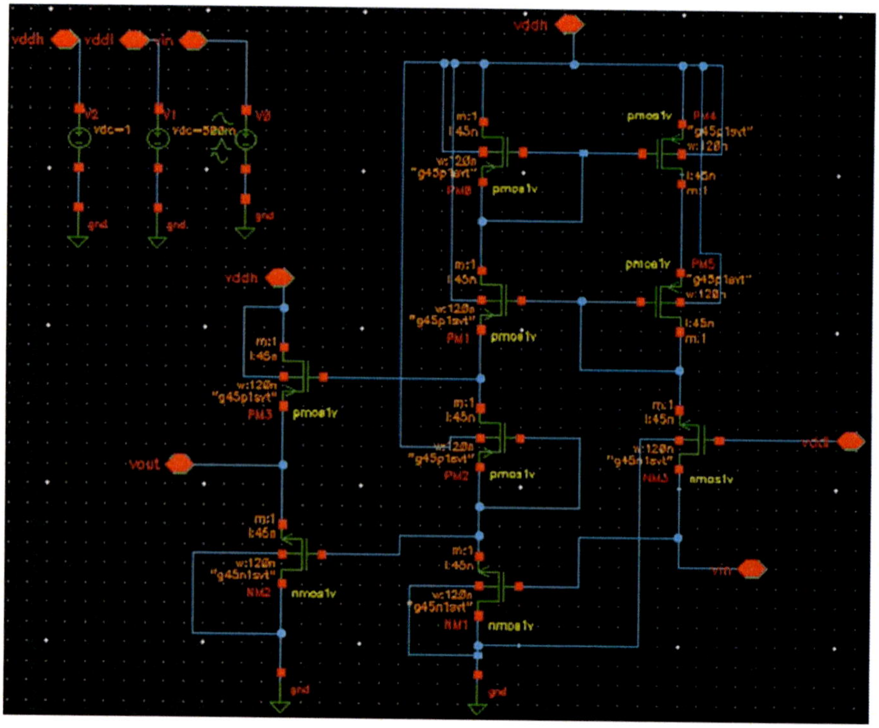

Fig. 9 Schematic plan of the projected design

power and delay variations of the suggested method are correlated with the existed methods, and it is represented in Table 2.

Table 2 Power and delay comparison with existed works

Level shifter design	Minimum VDDL (mV)	Power (nW)	Delay (nS)
Ref. [4]	200	107.58	7.5
Ref.[7]	200	123.1	23.7
Ref. [8]	350	272.6	5.1
Ref. [9])	100	6.6	18.4
Ref. [21]	100	8.7	16.6
Proposed design	100	1.92	0.96

4 Conclusion

Lower voltages are converted to high voltages using a level shifter. The level shifter proposed in this paper can convert voltages between 0.5 and 1 V. In 45 nm technology, the result was accomplished with the help of a cadence tool. At 45 nm, the projected idea may run at a reduced fixed energy of 3.26 pw. When compared to other existing designs, the proposed design reduced power by 20%. All simulation results, as promised, are well below the current design while the input voltage and the working frequency are the same.

Acknowledgements The reporters would like to convey their gratefulness to the VLSI group at Vardhaman College of Engineering in Hyderabad for allowing them to utilize the Cadence Virtuoso tool to generate the findings.

References

1. Alioto M (2012, January) Ultra-low power VLSI circuit design demystified and explained: a tutorial. IEEE Trans Circuits Syst—I: Regular Papers 59(1)
2. Usami K, Igarashi M, Minami F, Ishikawa T, Kanazawa M, Ichida M, Nogami K (1998, March) Automated low-power technique exploiting multiple supply voltages applied to a media processor. IEEE J Solid-State Circuits 33(3)
3. Lotze N, Manoli Y (2012, January) A 62 mV 0.13 μm CMOS standard-cell-based design technique using Schmitt-Trigger logic. IEEE J Solid-State Circuits 47(1)
4. Maghsoudlo E, Rezaei M, Sawan M, Gosselin B (2017) A high-speed and ultra low-power sub threshold signal level shifter. IEEE Trans Circuits Syst–I: Regular Papers 64:1164–1172
5. Lanuzza M, Corsonello P, Perri S (2014) Fast and wide range voltage conversion in multi supply voltage designs. IEEE Trans Very Large Scale Integr (VLSI) Syst
6. Lanuzza M, Crupi F, Rao S, De Rose R, Strangio S, Iannaccone G (2016) An ultra-low voltage energy efficient level shifter. IEEE Trans Very Large Scale Integr (VLSI) Syst
7. Kabirpour S, Jalali M (2018) A low-power and high-speed voltage level shifter based on a regulated cross-coupled pull-up network. IEEE Trans Circuits Syst II
8. Kabirpour S, Jalalil M (2019) A power – delay and area efficient voltage level shifter based on a reflected output Wilson current mirror level shifter. IEEE Trans Circuits Syst II Briefs
9. Lütkemeier S, Rückert U (2010, September) A sub threshold to above-threshold level shifter comprising a Wilson current mirror. IEEE Trans Circuits Syst—II: Exp Briefs 57(9):721–724
10. Luo S-C, Huang C-J, Chu Y-H (2014, June) A wide-range level shifter using a modified Wilson current mirror hybrid buffer. IEEE Trans Circuits Syst—I: Regular Papers 61(6)
11. Lotfi R, Saberi M, Rasool Hosseini S, Ahmadi-Mehr AR, Staszewsk RB (2018, February) Energy-efficient wide-range voltage level shifters reaching 4.2 fJ/transition. IEEE Solid-State Circuits Lett 1(2)
12. Le VL, Kim TT-H (2018) An area and energy efficient ultra-low voltage level shifter with pass transistor and reduced-swing output buffer in 65-nm CMOS. TCAS-II-02776
13. Hosseini SR, Saberi M, Lotfi R (2014, October) A low-power sub-threshold to above-threshold voltage level shifter. IEEE Trans Circuits Syst II, Exp Briefs 61(10):753–757
14. Luo S-C, Huang C-J, Chu YH (2014, May) A wide-range level shifter using a modified Wilson current mirror hybrid buffer. IEEE Trans Circuits Syst I, Reg Papers 61(6):1656–1665
15. Kesari P (2015, May) A novel energy efficient transmission gate voltage level shifter for multi VDD systems. Int J Sci Res Manage (IJSRM) 3:2263–2266

16. Srinivasulu G, Venkata Ramanaiah K, Padma Priya K (2014, May) Design of low power high speed level shifter. Int J Res Eng Technol (IJRET) 03:12–15
17. Rampuria P, Mishra DK (2016, June) Design of modified Wilson current mirror based level shifter. Int J Eng Sci Comput (IJESC) 6:8114–8117
18. Sharma R, Akashe S (2014, November) Analysis of low power reduction in voltage level shifter. In: International conference on innovative applications of computational intelligence on power, energy and controls with their impact on humanity (CIPECH14), pp 355–359
19. Ishihara F, Sheikh F, Nikolic B (2004, February) Level conversion for dual-supply systems. IEEE Trans Very Large Scale Integr (VLSI) Syst 12(2)
20. Wooters SN, Calhoun BH, Blalock TN (2010) An energy-efficient sub threshold level converter in 130-nm CMOS. IEEE Trans Circuits Syst-II, Exp Briefs 57(4):290–294
21. Chavan A, MacDonal E (2007) Ultra low voltage level shifter to interface sub and super threshold reconfigurable logic cells. IEEEC paper #1345 version12 updated Dec 17, 2007
22. Lanuzza M, Corsonello P, Perri S (2015) Fast and wide range voltage conversion in multi supply voltage designs. IEEE Trans Very Large Scale Integr (VLSI) Syst 23:388–391
23. Tomar A et al (2020) Machine learning, advances in computing, renewable energy and communication, LNEE vol 768. Springer Nature, Berlin, 659 p. https://doi.org/10.1007/978-981-16-2354-7. ISBN 978-981-16-2354-7

An Assessment of the Missing Data Imputation Techniques for COVID-19 Data

Aashi Pathak, Shivani Batra, and Vineet Sharma

Abstract In medical domain, the accuracy of the data supplied is critical. Missing values, on the other hand, are a typical occurrence in this sector for a variety of reasons. Most current science concentrates on establishing novel data imputation procedures, but more research on conducting a comprehensive review of existing algorithms is highly desired. Authors have evaluated the performance of four mostly adopted data imputation techniques, i.e., MICE, EM, mean, and KNN on a real-world dataset of COVID-19. KNN is an imputation approach that, according to the findings of the studies, is expected to be a good fit for dealing with missing data in the healthcare industry.

Keywords COVID-19 · Database · Missing value · Imputation · Mean · Statistical methods · MICE · KNN imputation

1 Introduction

The accuracy of input data is critical in healthcare profession that too in health-care surveillance systems. Well formatted healthcare data aids in generating accurate recommendations. The computer intends to handle episodes of comprehensive instances gathered from sensors so to conduct healthcare activities. Missing data are a usual issue caused by a variety of circumstances including equipment failures, incorrect measurements, data collection process limitations, or inaccurate sampling [1–15]. A missing data/value is described as an attribute that was never sampled or recorded in the entire dataset. The existence of missing values not only complicates data processing but also raises serious worries among experts. If there are more

A. Pathak · S. Batra (✉) · V. Sharma
KIET Group of Institutions, Delhi-NCR, Ghaziabad, India
e-mail: ms.shivani.batra@gmail.com

V. Sharma
e-mail: vineet.sharma@kiet.edu

A. Tomar et al. (eds.), *Proceedings of 3rd International Conference on Machine Learning, Advances in Computing, Renewable Energy and Communication*, Lecture Notes in Electrical Engineering 915, https://doi.org/10.1007/978-981-19-2828-4_62

than 5% missing samples, sophisticated handling strategies are necessary to improve accuracy [1, 2, 15].

To replace missing data with statistical prediction, a procedure known as "missing data imputation," many studies have been conducted, and a substantial amount of research has been published. However, the current study's major focus is on establishing novel imputation approaches, despite a dearth of research for a comprehensive valuation of prevailing techniques, particularly for medical data. Health/medical data can be analyzed, longitudinal, and complicated. As a result, researchers cannot treat healthcare data in the same way they treat other types of data.

Furthermore, details about each imputation methodology's performance can serve as suggestions for making the best methodological option in practice. This paper investigates four influential missing data imputation techniques on real healthcare datasets: EM, MICE, KNN, and mean imputation. The extract of this research is structured as follows. Most widely used missing data imputation techniques, algorithms as well as missing datasets, patterns, criteria, and evaluation along with data analytics methodologies, are discussed in Sect. 2. The findings of the experiments are provided in Sect. 3. The study concludes with the segment of conclusions in Sect. 4.

2 Methods

According to several studies, few most common missing data imputation methods for health care are regularized EM, MICE, KNN, and mean imputation. The experiment was carried out using many public online databases that include data of real-time healthcare systems. There are many features that are relevant to. The result has been computed by comparing the original values with missing value imputed in dataset.

2.1 Frequently Adopted Imputation Techniques

The imputation techniques have been chosen based on the projected usage, popularity, reference, extension in the healthcare data research community. Much research [3–5] identified EM, MICE, KNN, and mean as the frequently adopted imputation techniques.

- **Expectation Maximization (EM)**

It is a discourse for ensuring data coverage by echoing two phases till coverage is achieved [6] impute a value using other data (Expectation phase), and further, check if the value of data is the most likely (Maximization phase). Google Scholar has a total of 51,359 citations since its establishment. As a result, EM is one of the first successful techniques for imputation of missing values that leverage maximum likelihood as a strategy. Google Scholar indexed 5000 research papers in 2016 that used EM in healthcare applications.

- **Multivariate Imputation by Chained Equations (MICEs)**

It is a statistical method for responding with datasets that are incomplete [7]. MICE generates $M > 1$, but often $M \leqq 10$ whole datasets from the original data, each of which is evaluated independently before being integrated to yield a single set of complete conclusions. The application of this approach requires three steps: imputation, analysis, and pooling. As its introduction, this approach has received almost 15,000 citations in the literature, with over 8000 research using MICE in health domain according to Google Scholar.

- **K-Nearest Neighbors Imputation**

In a multi-dimensional space, KNN is every sample or independent entity with its nearest k-neighbors and further imputes missing value with a particular variable based on the average of k-neighbors value. Despite being acknowledged and contrasted in many of projects, KNN use in healthcare domain is still limited when compared to EM algorithms. Only, about 800 projects have used KNN to solve challenges in health care since its proposal.

- **Mean Imputation**

Mean is used to impute the missing value in this technique [8]. The sample size is preserved in this strategy, but the variability in the data is minimized. As a result, standard deviations and variance estimations are frequently underestimated. Mean imputation, on the other hand, is extensively utilized in academics because to its simplicity, especially when the amount of missing data are less. Only, about 6,490 projects have used mean imputation to deal with many issues related to medical data since its proposal.

2.2 Data Patterns

Based on the pattern, missing data are categorized as following [9].

- **Missing completely at random (MCAR)**

In MCAR, the missing values are dispersed in a random manner across all observations.

- **Missing at random (MAR)**

In MAR, the missing values are dispersed inside one or more sub-samples rather than randomly across observations,

- **Missing not at random (MNAR)**

If missing values aren't distributed evenly among observations or within one or more sub-samples, this is a problem. The comparison was done using the MCAR assumption.

The dataset's missing ratio is calculated as follows:

$$\alpha = \frac{\text{The number of Missing value}}{\text{The number of total values}} \quad (1)$$

3 Experiments and Results

In different statistical models, missing information has risen in popularity. Throughout data analysis, most researchers come across missing information. Missing information occurs for a variety of causes. Whenever a researcher utilizes a survey, for instance, respondents may decline to address certain queries leading to a shortage of availability or intimate details. As a result, in data analysis, researchers must appropriately impute the missing information. Results obtained from low-quality information are of inferior quality.

To increase the accuracy of the research, preprocessing is used [10]. In the event of missing information, an appropriate procedure must be used to impute the incomplete information before computing [11, 12]. Data are lost for a variety of causes, and investigators must establish what type of information is absent [4, 13].

The algorithms under consideration of this study were tested using data of eruption of COVID-19 disease in the United States [14]. This dataset contains missing value in an unclassifiable manner. As a result, the missing values must be imputed.

From the start of the epidemic (January 2020) until June 2021, this dataset consists statistics of each of 3142 American counties. The data comprise information regarding the number of verified COVID-19 instances and fatalities, and 46 additional demographic, regional, meteorological, epidemiology, and social variables that might impact the disease's transmission and consequences. This information comes from a variety of publicly accessible data sources and contains the everyday amount of COVID-19 confirmed incidence and mortality, and 46 other factors that could affect pandemic evolution, such as each country's population, regional, ecologic, vehicular, community, socioeconomic compliance, and system. Parameters having a missing data rate of greater than 50 percent were eliminated. We used mean, EM, MICE, and KNN to imputation. In all four imputation approaches, the information was divided into train and testing data using a tenfold approach. The efficiency of the systems was evaluated after they were run 5000 times. The model's efficiency was determined by the average accuracy. The results of all imputation approaches are shown in Fig. 1. Most reliable imputation technique for COVID-19 data seemed to be KNN in the trial.

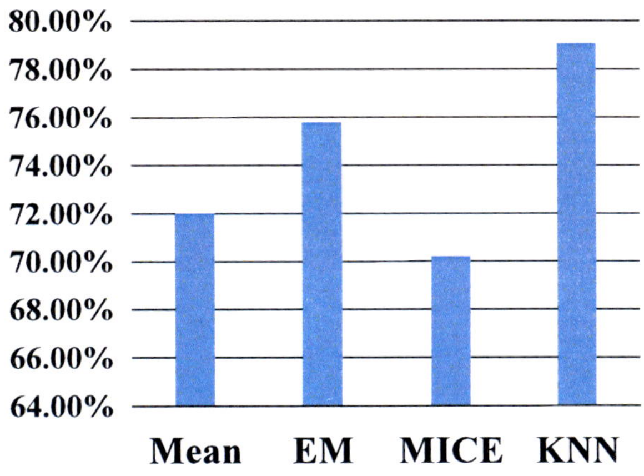

Fig. 1 Experimental results

4 Conclusion

This study compared four popular missing data imputation algorithms: expectation–maximization (EM), MICE, KNN imputation, and mean imputation based on the available data. As KNN proves to be the most robust and most efficient technique to impute the missing data in the domain of health care, it gives the highest accuracy on the dataset that has the missingness ranging from 2 to 82%. It has been validated by the experiment done through Python by taking the left and the right neighbors of the missing values. The best missing data imputation method, according to experimental results, is KNN.

References

1. Acuna E, Rodriguez C (2004) The treatment of missing values and its effect on classifier accuracy. In: Classification, clustering, and data mining applications. Springer, Berlin, Heidelberg, pp 639–647
2. Batra S, Sachdeva S (2016) Managing large-scale standardized electronic health records. In: Big data analytics. Springer, New Delhi, pp 201–219
3. García S, Luengo J, Herrera F (2016) Tutorial on practical tips of the most influential data preprocessing algorithms in data mining. Knowl-Based Syst 98:1–29
4. Allison PD (2001) Missing data. Sage
5. Cheema JR (2014) A review of missing data handling methods in education research. Rev Educ Res 84(4):487–508
6. Dempster AP, Laird NM, Rubin DB (1977) Maximum likelihood from incomplete data via the EM algorithm. J Roy Stat Soc: Ser B (Methodol) 39(1):1–22
7. Rubin DB (2004) Multiple imputation for nonresponse in surveys, vol 81. Wiley
8. Donders ART, Van Der Heijden GJ, Stijnen T, Moons KG (2006) A gentle introduction to imputation of missing values. J Clin Epidemiol 59(10):1087–1091
9. Little RJ, Rubin DB (2019) Statistical analysis with missing data, vol 793. Wiley

10. Batra S, Sachdeva S (2021) Pre-processing highly sparse and frequently evolving standardized electronic health records for mining. In: Handbook of research on disease prediction through data analytics and machine learning. IGI Global, pp 8–21
11. Deng W, Guo Y, Liu J, Li Y, Liu D, Zhu L (2019) A missing power data filling method based on improved random forest algorithm. Chin J Electr Eng 5(4):33–39
12. Sanjar K, Bekhzod O, Kim J, Paul A, Kim J (2020) Missing data imputation for geolocation-based price prediction using KNN–MCF method. ISPRS Int J Geo Inf 9(4):227
13. Rossi A, Pedreschi D, Clifton DA, Morelli D (2020) Error estimation of ultra-short heart rate variability parameters: effect of missing data caused by motion artifacts. Sensors 20(24):7122
14. Haratian A, Fazelinia H, Maleki Z, Ramazi P, Wang H, Lewis MA, Greiner R, Wishart D (2021) Dataset of COVID-19 outbreak and potential predictive features in the USA. Data Brief 38:107360
15. Tomar A et al (2020) Machine learning, advances in computing, renewable energy and communication, LNEE vol 768. Springer Nature, Berlin, 659 p. https://doi.org/10.1007/978-981-16-2354-7. ISBN 978-981-16-2354-7

Critical Insights into Machine Learning and Deep Learning Approaches for Personality Prediction

Shreya Biswas, Swati Bhat, Garima Jaiswal, and Arun Sharma

Abstract Personality prediction is used in many real-time applications such as job performance assessment, medical forums, psychology and many more. Over the years, various attempts have been made to predict personalities using different indicators. One such indicator is the Myers–Briggs Type Indicator (MBTI). This paper summarizes the most recent work done in predicting personality using MBTI from a time frame of 2010–2020. A total of 30 papers are reviewed. The paper gives a thorough literature review comparing and contrasting all these works based on parameters like dataset used, algorithm used, feature extraction method used and its limitations. Finally, giving a conclusion drawn based on these observations.

Keywords MBTI · Personality prediction · NLP · Machine learning · Deep learning

1 Introduction

The Myers-Briggs Type Indicator (MBTI) is a psychometric tool for analysing the personality of people on the basis of four categories each having a dichotomy. The four categories are: Extraversion/Introversion (E/I), Sensing/Intuition (S/N), Thinking/Feeling (T/F), Judging/Perception (J/P). Based on these four dichotomies, a person is assigned a four letter personality type (example, INTJ). So, there are a total of 16 such types [1–31].

S. Biswas · S. Bhat · G. Jaiswal (✉) · A. Sharma
Indira Gandhi Delhi Technical University for Women, New Delhi, India
e-mail: garima121@gmail.com

S. Biswas
e-mail: shreyab1999@gmail.com

S. Bhat
e-mail: swatibhat499@gmail.com

A. Sharma
e-mail: arunsharma@igdtuw.ac.in

A. Tomar et al. (eds.), *Proceedings of 3rd International Conference on Machine Learning, Advances in Computing, Renewable Energy and Communication*, Lecture Notes in Electrical Engineering 915, https://doi.org/10.1007/978-981-19-2828-4_63

Personality prediction is really useful as it has a lot of real-world applications. It has pretty much become a part of our daily lives. It helps in analysing people's behaviour and understanding the reason behind it. Some places where it is frequently used are medical forums which deal with psychology and recommendation systems for movies, products and services. Team formation is one particular area where predicting personalities is of great use. Mazni et al. [29] used personality traits to decide whether homogenous or heterogenous teams will be better for team performance. Poursafar et al. [22] aimed to find out the personality distribution in architect's offices and to scrutinize the effect of personality types on employee's productivity.

There are a lot of ways by which personalities can be judged, like through text, audio and visuals. In this paper, we will focus on personality prediction using text data or more precisely social media text data. Yamada et al. [10] have also used user behaviour or responses on textual data to extract information for personality prediction. They found that user behaviour is important for gathering information for users who post less tweets but the effect is not considerable when the user is quite active. Until recent times, most of the research work conducted on field was dependent on manual data collection which would not give much accurate results as data available was scarce and cumbersome to collect. The massive increase in the daily use of social networking platforms like Twitter and Facebook has made this type of classification necessary. People sharing their thoughts, ideas and emotions on these platforms has made collection and analysing of data for this task relatively easy. The evolution in technology has paved way more large-scale research work and development of better systems in the field. Multiple NLP techniques were used as we are dealing with text data.

There are two most popular ways of predicting personalities on text data, namely the MBTI and the Big Five, i.e., OCEAN classification (openness, conscientiousness, extraversion, agreeableness and neuroticism). A lot of research have already been done in both of them, especially in the last decade. We reviewed the work done explicitly related to MBTI in the past decade.

Many different datasets and algorithms have been used for this purpose. We will analyse each of them further.

2 Related Work

Logistic regression is one of the most commonly used machine learning algorithm. Plank and Hovy [1] had implemented logistic regression on a Twitter[1] dataset to classify MBTI personality types and genders. The authors of [2] used computational stylometry for gender and personality prediction from TWISTY. They used logistic regression and linear SVC along with binary features for character n-grams and word n-grams. For gender prediction, it performs better than both the majority and random baseline and better than weighted random baseline for personality prediction.

[1] Twitter: https://bitbucket.org/bplank/wassa2015/src/master/.

Gjurkovic et al. [3] worked on personality models and demographic data as they used PANDORA, a dataset of Reddit comments. They applied SVM, L-regularized (LR) and three-layer multilayer perceptron (MLP). Similar implementations can also be observed in the methodology proposed by Choudhary et al. [4] but using Kaggle[2] dataset. The authors have shown that using linear SVM instead of kernel SVM had a drop in model accuracy. The authors of [5] had implemented logistic regression and also a combination of feature models, thereby improving combined model accuracy of almost 69%. They used three feature models, namely bag of n-grams, part of speech and word vector for analysing language on Twitter. They concluded that for individual features, highest accuracy was given by word vectors, 0.651, and overall combination of all three feature models together gives the highest accuracy as, 0.661.

It can also be noticed that feature extraction plays a major role in improving model accuracy. In the work of Li et al. [6], the authors applied K-nearest neighbours on Kaggle dataset along with many feature extraction techniques like TF-IDF, word category and nuance. Choong and Varathan [7] also suggest using word-level features over character-level features due to higher prediction power and accuracy. In their attempt to get personality information from social forums, the authors used a number of classification models but LightGBM and SVM performed better than other models. Though the authors suggested using LightBGM over SVM as it has better robustness and chances of achieving convergence. Just like the accuracy can be increased by using or combining multiple feature models it can also improve our prediction system by combining data from multiple sources. The authors of [7] used data from personality cafe from Kaggle but they believed that the data may be insufficient and might have biases as it was accessed from a single source and by considering data from multiple sources they could have made the system better and more robust.

Bassignana et al. [8] used TF-IDF and n-words (lexical features), both word and character n-grams along with linear SVM model. The authors conducted tests on embedding-based representations and stylistic features. Cross-domain experiments were also conducted on PERSONALITY and TWISTY but the accuracies of in-domain experiments were always greater. Although the author concluded that there is no particular single best model, the lexical features gave the best result. Along with that, they also implemented weighted histograms cross validation and tenfold cross-validation. The results proved to be much better as it gave higher accuracies of 51.24% and 68.62%, respectively.

For personality prediction using MBTI, SVM seems to be the most commonly used machine learning algorithm so far. Bharadwaj et al. [9] and the authors of [10] had implemented SVM along with other algorithms, and it can be concluded that out of all the algorithms used, SVM gave the most promising results. Gupta et al.

[2] Kaggle: https://www.kaggle.com/datasnaek/mbti-type.

[11] had used NLP and SVM to peruse the dataset. They found that, in the original, only two attribute pairs, namely T/F and J/P had been balanced. From the results, it was concluded that for the under sampled dataset J/P has the highest accuracy, while for the original dataset T/F gave the best accuracy. Kumar et al. [12] found out that XGBoost performs better than linear SVM. They predicted personality traits on Twitter MBTI dataset with ensemble learning, count-based vectorization and GloVe word embedding technique. In this case, the dataset was unbalanced, so SMOTE was used to balance the classes using two sampling techniques—undersampling and oversampling. Even though simple language-based models work well as it employs a variegated set of features, the authors concluded that multimodal information fusion is better in this case than simple language-based models. Similar work has also been implemented by Khan et al. [13] and Amirhosseini and Kazemian [14], where XGBoost was used as the main algorithm.

It can be observed that the choice of dataset also affects the model's efficiency. Lukito et al. [15] have used Naive Bayes classifier on Twitter dataset in Bahasa Indonesian language. The dataset being in Indonesian language was a major drawback in an otherwise good model which could have provided better results with other datasets. The authors of [16] used Random forest for predicting personality type from the writing style of a person. The algorithm used was efficient but due to a limited dataset, the accuracy turned out to be not that great, with only 45.35% accuracy. Random forest algorithm was also seen to be used in the research of Abidin et al. [17] to analyse candidates for job assessment.

A relatively new and useful methodology is the use of pre-trained language models for the model training. Being pre-trained frees the user from the hassle of doing feature extraction. One such model is BERT. Mehta et al. [18] have used a BERT base with multilayer perceptron to predict personalities with psycholinguistic features. Also Keh and Chang [19] used BERT on personality cafe dataset on Kaggle. The prior one got a better accuracy than the latter due to using psycholinguistic feature extraction including Linguistic Inquiry and Word Count (LIWC), SenticNet, NRC emotion lexicon, valence, arousal and dominance (VAD) lexicon and readability. LIWC scores were used as data from Twitter by Raje and Singh [20]. They also used Artificial Neural Networks and Logistic Regression. Accuracies for both the models were found to be almost similar and vary between 51 and 59%. Classification into the Thinking and Feeling category achieved the highest accuracy. Ensembling methods like bagging, boosting and stacking were also quite efficient as seen in the findings of Das et al. [21]. Poursafar et al. [22] used linear regression in their work.

Deep learning algorithms proved to be really efficient than all other algorithms used so far. Cui et al. [23] performed duplication and reduction first on the skewed dataset of Kaggle and then used classification algorithms and deep learning framework of multi-layered LSTM. A deep learning encoder decoder framework was used. They concluded that regularized SVM performs better than baseline Naive Bayes models and deep learning outperforms SVM. Gjurkovic et al. [24] used multilayered perceptrons along with linguistic and user activity feature extraction on MBTI9K[3] Reddit dataset. Another useful and efficient deep learning algorithm is CNN. It performed well in the experiments of Pradhan et al. [25], where CNN is used along with max pooling on personality cafe questionnaire dataset.[4] The results of this experiment reduced the time and effort of answering 50–60 questions that too with high accuracy.

Brinks et al. [26] used multi-variate Bernoulli event model of Naive Bayes classifier. It was observed that the decisions the classifier made were dependent on the number of words provided. Multinomial event model solved the above problem but the performance of test data was not good. Feature set size was reduced to fix the high variance problem but the performance of the model did not improve much. It was concluded that natural language processing could give better results. Similarly, Gaussian Naive Bayes theorem was also used by the authors of [27] on stack overflow data. In Patel et al. [28], the authors observed that Naive Bayes with TF-IDF up sampling performs better than other algorithms.

Mazni et al. [29] had used rough set analysis to analyse team diversity and team performance. The authors of [30] in their work intended to apply extreme gradient boosting together with hyper parameters tuning. The authors concluded that XGBoost is an ideal algorithm when working on medium to large datasets. Comparative analysis of different works in predicting personalities on MBTI basis is elaborated in (Table 1).

3 Conclusion and Future Work

To conclude, we successfully reviewed 30 papers and we found limited and biased dataset as some of the major limitations. We can also conclude that the results vary largely depending upon the dataset we choose, the algorithm we implement and also the feature extraction involved as seen when XGBoost was used on Kaggle dataset the accuracy was better than when XGBoost was used on Twitter dataset.

XGBoost, SVM and logistic regression were some of the most used algorithms. Also deep learning algorithms proved to be really efficient than all other algorithms. In future works, the accuracy of the model can be improved by working on a large and unbiased dataset.

[3] Reddit: http://takelab.fer.hr/data/mbti.

[4] Personality Café Forum: https://figshare.com/articles/dataset/8k_MBTI_Dataset_From_Personality_Cafe/14587572.

Table 1 Comparative analysis of different works in predicting personalities on MBTI basis

Author	Dataset	Algorithm used	Feature extraction used	Evaluation metric: accuracy	Limitation and future work
Choudhary et al. [4]	Kaggle	Logistic regression	n-gram, TF-IDF	Accuracy 67.75%	Uses traditional machine learning algorithms, limited to those users who write comments
Bassignana et al. [8]	YouTube (from personality corpus and TWISTY)	Linear SVC	n-gram	Avg. (personality)-51.24% Avg.(TWISTY)-68.62%	Lexical features tend to overfit
Gottigundala et al. [16]	Kaggle	Random forest	TF-IDF, truncated SVD	Accuracy 45.35%	Limited dataset and questionnaire-based dataset could have saved time, algorithms like CNN can improve accuracy
Pradhan et al. [25]	Personality cafe forum	CNN	TF-IDF, max pooling	Accuracy 81.4%	Model is biased, website becomes slow with more traffic and chatbot can be created for future work
Mazni et al. [29]	Engineering students team	Rough set analysis	Not specified	Accuracy 71.86%	Uncertainties in finding dependencies between the different team attributes and team performance. Further work could be done on rule case validation and rule extraction
Yamada, Sasano and Takeda [10]	Twitter	SVM	BOW with SVD and DBOW	AUC E/I = 73.18%, S/N = 69.89%, T/F = 70.96%, J/P = 62.1%	Information could have been gathered from the replies and comments for users not part of experiment

(continued)

Table 1 (continued)

Author	Dataset	Algorithm used	Feature extraction used	Evaluation metric: accuracy	Limitation and future work
Wang et al. [5]	Twitter	Logistic regression	POS, BOW, n-gram and word vector	AUC-E/I = 69.1%, S/N = 65.3%, T/F = 68%, J/P = 61.9%	Not specified
Choong and Varathan [7]	Kaggle	Light GBM	Char-level TF-IDF, word-level TF-IDF and LIWC	AUCROC-88.86% for LightGBM and combo features	Limited and biased dataset. Inconsistent data could be produced as less information is given on the version of MBTI assessment used
Verhoeven et al. [2]	Twitter	Linear SVC	n-grams	F-measure E/I = 67.87%, S/N = 73.01%, T/F = 58.45%, J/P = 56.06%	Personality prediction is harder to perform. It is harder to predict S–N and J–P dimensions from linguistic signals
Gjurkovic et al. [3]	PANDORA	Linear/Logistic regression and NN	TF-IDF, count of words	Macro-averaged F1 score-Thinking = 0.739, Introverted = 0.654, Perceiving = 0.642, Intuitive = 0.606, Enneagram = 0.251, Gender = 0.904, Region = 0.626	Performance of deep learning baseline models is not so good. In future work, these gaps could be filled by advanced deep learning models and coherent user representations
Gjurkovic et al. [24]	MBTI 9K Reddit	Multilayer perceptron (MLP)	TF-IDF, LIWC, n-gram and LDA	F1 score E/I = 82.8%, S/N = 79.2%, F/T = 64.4%, J/P = 74%	Dataset could be improved by increasing size, other deep learning models can be used to improve accuracy
Plank and Hovy [1]	Twitter	Logistic regression	Stability selection, binary word n-gram	Accuracy E/I = 72.5%, T/F = 61.2, S/N = 77.5%, J/P = 55.4%	There is a lot of difference between corpus and general population personality types

(continued)

Table 1 (continued)

Author	Dataset	Algorithm used	Feature extraction used	Evaluation metric: accuracy	Limitation and future work
Li et al. [6]	Kaggle	K-nearest neighbours	TF-IDF, nuance, word category and weighted histograms	Accuracy E/I = 90%, S/N = 90%, F/T = 91.25%, J/P = 76.25%	Less importance is given to verbosity of people and binary pair combination could have improved performance
Bharadwaj et al. [9]	Kaggle	SVM	Emolex, LIWC, ConceptNet, TF-IDF and BOW	Accuracy E/I = 84.9%, S/N = 88.4%, T/F = 87%, J/P = 78.8%	Less weightage is given to gravity of the words, and difference in tone of voice can change meaning of the word/sentence
Kumar et al. [12]	Twitter	XGBoost	GloVe word embedding, count vectorization and SMOTE	Accuracy- E/I = 68.3%, S/N = 85%, T/F = 77.3%, J/P = 75.6%	Different interactions between people can be considered for future as a multi-layered graph
Khan et al. [13]	Kaggle	XGBoost	Count vectorization, TF-IDF and resampling	Accuracy E/I = 99.37%, S/N = 99.92%, F/T = 94.55%, J/P = 95.53%	Biased dataset, only English language text data, traditional machine learning models used, less weightage given to feature extraction
Amirhosseini et al. [14]	Kaggle	XGBoost	TF-IDF	Accuracy E/I = 78.17%, S/N = 86.06%, T/F = 71.78%, J/P = 65.7%	Limited dataset
Lukito et al. [15]	Twitter	Naive Bayes	n-gram	Accuracy- E/I = 80%, S/N = 60%, F/T = 60%, J/P = 60%	Limited dataset, lower accuracy due to only Bahasa Indonesian language corpus used
Abidin et al. [17]	Kaggle	Random forest	Word2Vec	Accuracy- E/I = 100%, S/N = 100%, F/T = 100%, J/P = 100%	Only twitter data was used. Only English data was considered. For future, XGBoost and deep learning can be used

(continued)

Table 1 (continued)

Author	Dataset	Algorithm used	Feature extraction used	Evaluation metric: accuracy	Limitation and future work
Mehta et al. [18]	Kaggle	BERT + MLP	LIWC, SenticNet, NRC emotion lexicon, VAD lexicon and readability	Accuracy E/I = 78.8%, S/N = 86.3%, F/T = 76.1%, J/P = 67.2%	Personality is not measured in continuous scores as it should have been in psychometric assessments, overestimation of model performance
Keh and Chang [19]	Kaggle	BERT	BERT pre-trained language model	Accuracy E/I = 75.83%, S/N = 74.41%, T/F = 75.75%, J/P = 71.9%	Limited dataset, a clean and large dataset could have been used, large BERT model could have been used
Raje and Singh [20]	Twitter	Logistic regression	LIWC	Accuracy E/I = 53.57%, S/N = 51.99%, T/F = 56.25%, J/P = 53.57%	Frequency of tweets and timing of different responses can be analysed to predict personalities of users
Das et al. [21]	Kaggle	Stacking ensemble	Word2Vec, TF-IDF	Accuracy E/I = 91.13%, S/N = 97.53%, T/F = 79.39%, J/P = 73.53%	For future, a comprehensive assessment can be carried out to determine the original objective behind the utility of phrases from user written information
Cui and Qi [23]	Kaggle	LSTM	BOW, POS, n-gram and capital letter count	Accuracy E/I = 89.51%, S/N = 89.85%, T/F = 69.1, J/P = 67.65%	Limited dataset for deep learning mechanisms using latest features of NLP and word embeddings can improve results

(continued)

Table 1 (continued)

Author	Dataset	Algorithm used	Feature extraction used	Evaluation metric: accuracy	Limitation and future work
Brinks et al. [26]	Twitter	Naive Bayes	TF-IDF	Accuracy- E/I = 63.9%, N/S = 74.6%, T/F = 60.8, P/J = 58.5%	Performance on training data is good but it does not give good results for testing data. Model also has a high variance problem which does not improve. Natural language techniques may give better results in further research
Patel et al. [28]	Kaggle	KNN, Naive Bayes, Logistic regression, RNN, and SVM	n-grams, BOW, TF-IDF, W2V and GloVe word embedding	E/I = 77.79%, N/S = 85.89%, T/F = 86%, P/J = 66.23%	The established system can be made **more efficient by** conducting more testing and feeding the system with a much more accurate dataset
Gupta et al. [11]	My personality forum	NLP and SVM	TF-IDF	Accuracy- Original dataset- T/F = 78.5% J/P = 67.9% Under Sampled dataset- E/I = 58.9% N/S = 32.9% T/F = 30.7% P/J = 60.2%	In future work, the accuracy can be enhanced by considering a balanced dataset of greater size. Adding more features can also be considered
Mushtaq et al. [30]	Kaggle	XGBoost, Naive Bayes, SVM and LSTM	TF-IDF	E/I = 89.01%, N/S = 85.96%, T/F = 84.19%, P/J = 85.42%	Applying some hyper parameter tuning techniques like increasing the number of iterations on a more balanced dataset or increasing the tree depth can result in generating optimal performance of the algorithm

References

1. Plank B, Hovy D (2015) Personality traits on twitter—or—how to get 1500 personality tests in a week. In: Proceedings of the 6th workshop on computational approaches to subjectivity, sentiment and social media analysis, pp 92–98
2. Verhoeven B, Daelemans W, Plank B (2016) Twisty: a multilingual twitter stylometry corpus for gender and personality profiling. In: Proceedings of the Tenth international conference on language resources and evaluation (LREC'16), pp 1632–1637
3. Gjurković M, Karan M, Vukojević I, Bošnjak M, Šnajder J (2020) Pandora talks: personality and demographics on reddit. arXiv preprint arXiv:2004.04460
4. Chaudhary S, Singh R, Hasan ST, Kaur I (2013) A comparative study of different classifiers for Myers-Brigg personality prediction model. Linguistic Anal 21
5. Wang Y (2015) Understanding personality through social media
6. Li C, Hancock M, Bowles B, Hancock O, Perg L, Brown P, Burrell A et al (2018) Feature extraction from social media posts for psychometric typing of participants. In: International conference on augmented cognition. Springer, Cham, pp 267–286
7. Choong EJ, Varathan KD (2021) Predicting judging-perceiving of Myers-Briggs Type Indicator (MBTI) in online social forum. PeerJ 9:e11382
8. Bassignana E, Nissim M, Patti V (2020) Personality: a novel youtube-based corpus for personality prediction in Italian. arXiv preprint arXiv:2011.05688
9. Bharadwaj S, Sridhar S, Choudhary R, Srinath R (2018) Persona traits identification based on Myers-Briggs type indicator (MBTI)-a text classification approach. In: 2018 International conference on advances in computing, communications and informatics (ICACCI). IEEE, New York, pp 1076–1082
10. Yamada K, Sasano R, Takeda K (2019) Incorporating textual information on user behavior for personality prediction. In: Proceedings of the 57th Annual meeting of the association for computational linguistics: student research workshop, pp 177–182
11. Gupta N, Madhavan A, Duvvuri D, Angeline R (2019) MBTI based personality prediction of a user based on their writing on social media. Int J Eng Adv Technol (IJEAT)
12. Kumar KNP, Gavrilova ML (2019) Personality traits classification on twitter. In: 2019 16th IEEE international conference on advanced video and signal based surveillance (AVSS). IEEE, New York, pp 1–8
13. Khan AS, Ahmad H, Asghar MZ, Saddozai FK, Arif A, Khalid HA (2020) Personality classification from online text using machine learning approach. Int J Adv Comput Sci Appl 11(3)
14. Amirhosseini MH, Kazemian H (2020) Machine learning approach to personality type prediction based on the Myers–Briggs type indicator®. Multimodal Technol Interact 4(1):9
15. Lukito LC, Erwin A, Purnama J, Danoekoesoemo W (2016) Social media user personality classification using computational linguistic. In: 2016 8th International conference on information technology and electrical engineering (ICITEE). IEEE, New York, pp 1–6
16. Gottigundala T (2020) Predicting personality type from writing style
17. Abidin NHZ, Remli MA, Ali NM, Eh Phon DN, Yusoff N, Adli HK, Busalim AH (2020) Improving intelligent personality prediction using Myers-Briggs type indicator and random forest classifier. Int J Adv Comput Sci Appl
18. Mehta Y, Fatehi S, Kazameini A, Stachl C, Cambria E, Eetemadi S (2020) Bottom-up and top-down: predicting personality with psycholinguistic and language model features. In: 2020 IEEE international conference on data mining (ICDM). IEEE, New York, pp 1184–1189
19. Keh SS, Cheng I (2019) Myers-Briggs personality classification and personality-specific language generation using pre-trained language models. arXiv preprint arXiv:1907.06333
20. Raje MS, Singh A (2016) Personality detection by analysis of twitter profiles. In: International conference on soft computing and pattern recognition. Springer, Cham, pp 667–675
21. Das K, Prajapati H (2020) Personality identification based on MBTI dimensions using natural language processing. Int J Creat Res Thoughts (IJCRT)

22. Poursafar Z, Devi NR, Rodrigues LR (2015) Evaluation of Myers-Briggs personality traits in offices and its effects on productivity of employees: an empirical study. Int J Curr Res Rev 7(21):53
23. Cui B, Qi C (2017) Survey analysis of machine learning methods for natural language processing for MBTI personality type prediction
24. Gjurković M, Šnajder J (2018) Reddit: a gold mine for personality prediction. In: Proceedings of the second workshop on computational modeling of people's opinions, personality, and emotions in social media, pp 87–97
25. Pradhan T, Bhansali R, Chandnani D, Pangaonkar A (2020) Analysis of personality traits using natural language processing and deep learning. In: 2020 Second international conference on inventive research in computing applications (ICIRCA). IEEE, New York, pp 457–461
26. Brinks D, White H (2012) Detection of Myers-Briggs type indicator via text based computer-mediated communication, pp 53–56
27. Katiyar S, Kumar S, Walia H (2020) Personality prediction from stack overflow by using Naïve Bayes theorem in data mining
28. Patel S, Nimje M, Shetty A, Kulkarni S (2020) Personality analysis using social media. Int J Eng Res Technol (IJERT)
29. Mazni O, Syed-Abdullah S-L, Hussin NM (2010) Analyzing personality types to predict team performance. In: 2010 International conference on science and social research (CSSR 2010). IEEE, New York, pp 624–628
30. Mushtaq Z, Ashraf S, Sabahat N (2020) Predicting MBTI personality type with K-means clustering and gradient boosting. In: 2020 IEEE 23rd International multitopic conference (INMIC). IEEE, New York, pp 1–5
31. Tomar A et al (2020) Machine learning, advances in computing, renewable energy and communication, LNEE vol 768. Springer Nature, Berlin, 659 p. https://doi.org/10.1007/978-981-16-2354-7. ISBN 978-981-16-2354-7

Copy-Move Image Forgery Detection Using SURF, SIFT, and KAZE

Anuj Rani and Ajit Jain

Abstract In this digital world, image forgery is the most common cause for any digital forgery. Nowadays, people accept digital images blindly, even without checking their integrity and authenticity. One of the critical facts in digital image forensics is that we need to keep high confidence in forgery detection algorithms. In this work, multiple image forger detection algorithms are evaluated to check their efficacy using their detection accuracy and execution time. Discrete wavelet transform (DWT) is used for image decomposition at the preprocessing stage. The objective of this work is to suggest a best copy-move forgery detection algorithm among SIFT, SURF, and KAZE. For modeling the code, MATLAB is used as a programming language. Using the SURF algorithm, the accuracy rate is 56.50% with average execution time 1.28 s, while it is 66.75% and 2.78 s using SIFT. The detection rate by using the KAZE algorithm is reported as 61.24% with execution time is 1.50 s.

Keywords SIFT · Copy-move · Image forgery · SURF · KAZE · DWT · Feature extraction

1 Introduction

The availability of digital images has been increased and found everywhere such as on the front of magazines, daily newspapers, courts, and across the web. We are exposed to their use throughout the day and often believe that "we trust what we see." Because of this, special techniques are required to identify traces of manipulation so that the authentication of digital data in domains such as multimedia communication, copyright management, jurisdiction, content analysis can be attained. With

A. Rani (✉) · A. Jain
Department of Computer Science, Banasthali Vidyapith, Tonk, Rajasthan, India
e-mail: anuanuj1989@gmail.com

A. Rani
Department of Computer Science, G L Bajaj Institute of Technology and Management, Greater Noida, India

A. Tomar et al. (eds.), *Proceedings of 3rd International Conference on Machine Learning, Advances in Computing, Renewable Energy and Communication*, Lecture Notes in Electrical Engineering 915, https://doi.org/10.1007/978-981-19-2828-4_64

(a) (b)

Fig. 1 Example of copy-move forgery

the use of computerized cameras, personal computers, and many easily available editing software, the manipulation of images is relatively common. However, there are many areas like medical imaging, advanced legal sciences, news-casting, intelligence services, criminal and forensic investigation, etc., where images are used as evidence. Therefore, image forgery detection is a major requirement [1]. Finding forgeries in copy-move-based manipulations is a tedious task as shown in Fig. 1a, b.

Multiple techniques have been implemented for identifying this problem. It is very important to find out that techniques we are implementing for identifying the copy-move forgery are reliable, robust, and able to find structural changes [2]. In this work, various methods are suggested to identify the image forgeries.

The classification of the article is done into five sections. Literature study is discussed in Sect. 2 followed by implementation in Sect. 3. Results are concluded in Sect. 4. Finally, conclusion with future work is presented in Sect. 5.

2 Literature Review

In this section, multiple methods are presented for understand recent works in this domain.

A. **Discrete Wavelet Transform (DWT)**

Discrete wavelet transform is mainly used in signal processing and image processing. It is a very effective tool that divides the different information into diverse segments. For image compression, DWT is handy because signal energy concentrates on particular wavelet coefficients. The signal will be passed through different scale frequencies to analyze the signal. The main highlight of DWT is multiscale representation capacity. 2D DWT works in the straightforward method by inserting transposition array in between two 1D DWT arrays. First, it divides the array vertically into two parts. The first part stores the average coefficients and second part stores the coefficients of detail. It repeats the same process for the column and with the help of output filter, four sub-bands will be produced [3].

B. **Singular Value Decomposition (SVD)**

In linear algebra, the SVD of any matrix is a factorization of that particular matrix into three matrices. SVD is a widely used technique for matrix calculations and investigation [4]. The primary purpose of SVD is to find the optimal set of segments which can predict the best outcome. In latent semantic analysis (LSA), SVD is used for preprocessing to text mining operations.

C. **KAZE**

It is a new multiscale 2D feature discovery and description algorithm in nonlinear scale spaces. Additive operator splitting (AOS) methods and variable conductance prolixity are used for nonlinear scale-space structure [5]. An image can induce a nonlinear space with the help of AOS and variable conductance. After that, it produces a maxes of the scale-regularized determinant of the Hessian response. In the final step, we get a scale and gyration steady descriptor of first-order image derivations after calculating the exposure of the key point.

D. **DCT**

Discrete cosine transform (DCT) is used to divide the image into various segments regarding image visual quality [6]. DCT separates images into different frequencies. In the quantization step, useless frequencies are discarded, which leads to lossy compression. DCT can detect tampered regions accurately [7]. Scanning the image from the upper left corner to the lower right corner by sliding the 2×2 block allows forgery detection. After that, for each block, DCT and quantized DCT have been calculated [8].

E. **SIFT**

Copy-move-based manipulation identification is the main actively investigated area in image forensics [9]. Using SIFT, features points can be extracted from an image. Using these features, the image can be classified into even and uneven regions with the help of a ratio of key points and total number of pixels [10]. The SIFT algorithm mainly contains the key point detection module and the descriptor generator module [11]. Using the SIFT algorithm, matching can be done between the initial image and the second image [12]. In [13], the hybrid technique has been proposed by combining FMT and SIFT.

Other relevant studies to understand the image forgery detection can be referred from [14–23].

F. **SURF**

A speeded-up strong component (SURF) is applied on forged images to separate features and afterward matched by utilizing Euclidean distance. The main ability of SURF is its fast computation of using box filters. It is widely used because of its high registration rate. Because of its convenience, it is mainly used for image features extraction and matching.

3 Image Forgery Detection Methodology

The forgery detection system model has been designed for copy-move images by identifying and matching key points in an image. The flowchart shown in Fig. 2 illustrates that how manipulation can be detected from an image using suggested approach.

The steps of the algorithm are explained below.

a. **Image conversion on RGB components**

The time–frequency of the signal is represented by using wavelet transformation. To analyze the non-stationary signals, wavelet transform used the multi-resolution analysis. When wavelet transformation gets to apply to an image, then four sub-images come from an original image. Wavelet transformation contains two filters, low-pass and high-pass filters. The process is shown in Fig. 3.

b. **Feature Extraction**

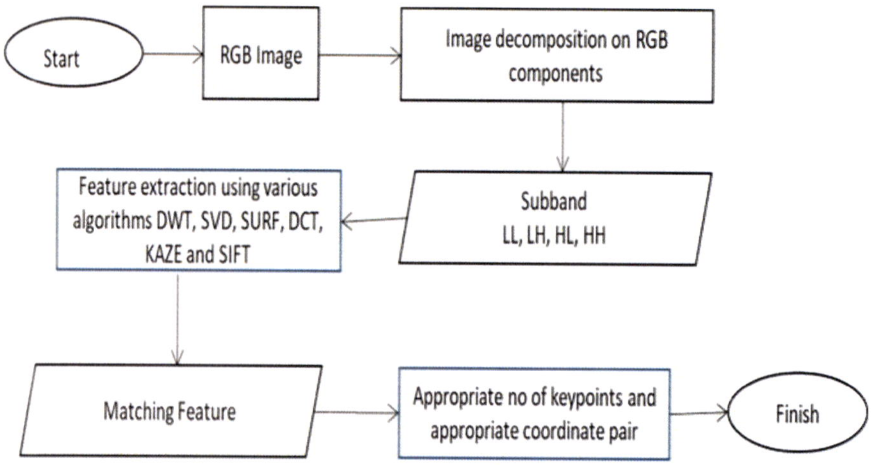

Fig. 2 Flowchart for image manipulation detection system

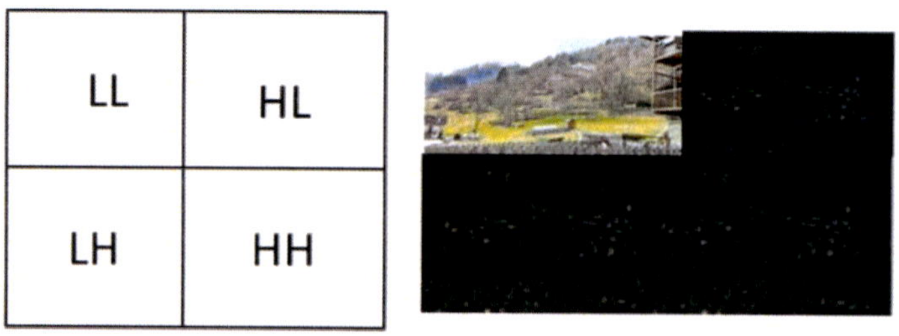

Fig. 3 Level 1 image transformation

Aims to extract features are to find the appropriate features from an image. In the feature extraction process, a feature is extracted from an image, and after that, this value of the feature is used in the next process. In this paper, various feature extraction algorithm is used, as shown below.

- Key point Bearings

  A very special process is required to deal with smooth regions to find accurate detection. A key point can be identified from an image, and these are the points whose value is fixed while we perform various transformations like scaling and many more. After identifying key point, edges have been selected by removing key point of the edges which contains low contrast points.
- Key point Descriptors Generation

  Histogram value and scale space in the given image will form the descriptor. The experiment was done by selecting 2×2 array histogram and eight orientation places, including 32 feature vector elements.

4 Results and Discussion

4.1 Preprocessing

In initial phase, we select an image to perform some analysis over it. After that, discrete wavelet transform (DWT) has been applied to decompose it. Moreover, after decomposition, we got the scaling function and wavelet coefficients. When DWT is applied to an image, its signal energy gets concentrated to a particular wavelet coefficient.

4.2 Method for Feature Extraction

In this phase, key points in image will be detected by using various methods like SIFT, KAZE, and SURF. After the decomposition process of the image, we get the LL sub-band, which is transformed into a grayscale image. The gray image features will be extracted using the SIFT, KAZE, SVD, and SURF algorithms. The existing library of MATLAB is used for feature extraction. From the image number and location of key point is obtained. Key point extracted by SIFT and SURF is: SIFT: 247, SURF: 66. Once the key points are obtained, we must check that key points are part of the forged image. Now, the next step is to match the appropriate features and calculate the falseness percentage in an image.

Table 1 Accuracy testing results

Test	Accuracy %			Highest accuracy algorithm
	SURF	SIFT	KAZE	
Test 1	69.05	72.43	63.29	SIFT
Test 2	48.39	54.89	51.59	SIFT
Test 3	56.65	69.48	61.49	SIFT
Test 4	52.12	70.23	68.59	SIFT

4.3 Research Result

The outcomes of the implementation are depicted in Table 1.

As we can see from Table 1, the accuracy of SIFT algorithm is higher than SURF and KAZE algorithms. Now, from Fig. 4, we can see the accuracy for implemented algorithms.

Now, the execution time of the algorithm can be seen in Table 2.

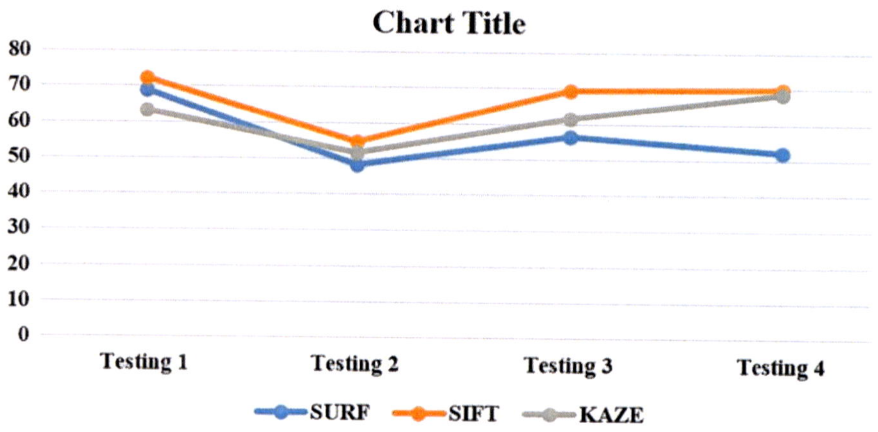

Fig. 4 Accuracy test results

Table 2 Algorithm execution time

Testing	Time (Second)			Algorithm with lowest execution time
	SURF	KAZE	SIFT	
Testing 1	1.42	1.56	2.79	SURF
Testing 2	0.98	2.34	2.84	SURF
Testing 3	1.48	1.45	2.93	SURF
Testing 4	1.24	0.67	2.58	SURF

5 Conclusion and Future Work

Digital image forgeries are nowadays a common task for which more scientific methods need to implement. In this work, three different methods are tested to identify the copy-move-based forgeries. Based on the results of the execution of the algorithm, the SURF algorithm we obtained an accuracy rate is 56.50%, and the average execution time is 1.28 s. Further, in the SIFT algorithm, the rate of copy-move detection is 66.75, and the average execution time is 2.78 s. Finally, the detection rate by using the KAZE algorithm is 61.24, and 1.50 s is the execution time. From the results, it is evident that SURF is outperforming. In the future work, the accuracies can be enhanced by making a hybrid approach that can identify different type of forgeries.

References

1. Birajdar GK, Mankar VH (2013) Digital image forgery detection using passive techniques: a survey. Digital Invest pp 226–245
2. Zandi M, Mahmoudi-Aznaveh A, Talebpour A (2016) Iterative copy-move forgery detection based on a new interest point detector. IEEE Trans Inf Forensics Security 11(11):1556–6013
3. Nuari R, Utami E, Raharjo S (2019) Comparison of scale invariant features transform and speed up robust feature for image forgery detection copy move. In: IEEE, 4th International conference on information technology, information systems and electrical engineering
4. Yusoff N, Alamro L (2019) Implementation of feature extraction algorithms for image tampering detection. Int J Adv Comput Res 9(43):197–211
5. Alcantarilla PF, Bartoli A, Davison AJ (2012) KAZE features. In: European conference on computer vision (ECCV)
6. Watson AB (1994) Image compression using the discrete cosine transform. Math J 4(1):81–88
7. Alkawaz MH, Sulong G, Saba T, Rehman A (2016) Detection of copy-move image forgery based on discrete cosine transform. Neural Comput Appl. https://doi.org/10.1007/s00521-016-2663-3
8. Mahmood T, Nawaz T, Irtaza A, Ashraf R, Shah M, Mahmood MT (2016) Copy-move forgery detection technique for forensic analysis in digital images. Math Probl Eng. https://doi.org/10.1155/2016/8713202
9. Liu Y, Guan Q, Zhao X (2017) Copy-move forgery detection based on convolutional kernel network. Multimedia Tools Appl. https://doi.org/10.1007/s11042-017-5374-6
10. Zheng J, Liu Y, Ren J, Zhu T, Yan Y, Yang H (2016) Fusion of block and keypoints based approaches for effective copy-move image forgery detection. Multidimensional Syst Sign Process, pp 989–1005
11. Sasikala N, Swathipriya V, Ashwini M, Preethi V, Pranavi A, Ranjith M (2020) Feature extraction of real-time image using SIFT algorithm. J Electr Eng Comput Sci 4(3)
12. Harihar S, Manjunath R (2020) SIFT image feature extraction for an efficient image registration. Int J Recent Technol Eng (IJRTE) 8(6)
13. Meena KB, Tyagi V (2019) A hybrid copy-move image forgery detection technique based on Fourier-Mellin and scale invariant feature transforms. Multimedia Tools Appl. https://doi.org/10.1007/s11042-019-08343-0
14. Kumar M, Aggarwal J, Rani A, Stephan T, Shankar A, Mirjalili S (2021) Secure video communication using firefly optimization and visual cryptography. Artif Intell Rev. https://doi.org/10.1007/s10462-021-10070-8

15. Rani A, Jain A, Kumar M (2021) Identification of copy-move and splicing based forgeries using advanced SURF and revised template matching. Multimedia Tools Appl 80(2021):23877–23898
16. Singh P, Diwakar M, Shankar A, Shree R, Kumar M (2021) A review on SAR image and its despeckling. Archiv Comput Methods Eng 28(2021):4633–4653
17. Aggarwa A, Kumar MAM, Sharma P, Alfarraj O, Deep V (2020) Principal component analysis, hidden Markov model, and artificial neural network inspired techniques to recognize faces. Concurrency Comput Pract Exp 33. https://doi.org/10.1002/cpe.6157
18. Kumar M, Rani A, Srivastava S (2019) Image forensics based on lighting estimation. Int J Image Graph 19(03):1950014
19. Kumar M, Srivastava S, Uddin N (2019) Forgery detection using multiple light sources for synthetic images. Aust J Forensic Sci 51(3):243–250
20. Kumar M, Srivastava S (2019) Image forgery detection based on physics and pixels: a study. Australian J Forensics Sci 51(2):119–134
21. Kumar M, Srivastava S (2019) Image authentication by assessing manipulations using illumination. Multimedia Tools Appl 78(2019):12451–12463
22. Kumar M, Srivastava S (2016) Identifying photo forgery using lighting elements. Indian J Sci Technol 9(48):105748
23. Tomar A et al (2020) Machine learning, advances in computing, renewable energy and communication, LNEE vol 768. Springer Nature, Berlin, 659 p. https://doi.org/10.1007/978-981-16-2354-7. ISBN 978-981-16-2354-7

IoT-Based Cloud Centric Framework for Weight Management Using Predictive Computing

Umang Kant and Vinod Kumar

Abstract Obesity is a grave and chronic condition that has affected a large number of people in recent years. It can have a negative impact on the body which can lead to other diseases such as diabetes mellitus, hypertension-blood pressure, heart related problems, and kidney problems amongst many others. Appropriate and immediate steps need to be taken by the affected persons at the early stages of the diseases, otherwise acute complications occur. In our proposed work, we have examined the way in which Internet of Things (IoT)-based cloud centric framework for weight management. This cloud centric proposed framework uses predictive computing to analyse the physical activities, as well as the dietary intake of the registered users. It also manages the alerts sent to the health inspectors and the predictions given by the system. The IoT-based development model consists of implementation, evaluation, feedback, and security layers each responsible for respective role. Separate clouds for users, mobile, and database are managed. An introduction of weight management, IoT, use of IoT in weight management, and predictive computing is discussed in Sect. 1. Section 2 explores the related work in the same domain which includes the related work done in the area of IoT-based Health Management frameworks, and Health Applications. A detailed description of proposed work is given in Sect. 3 which includes the proposed IoT-based development model, proposed user activity framework, and algorithm for the proposed framework. Section 4 deals with conclusion and future work.

Keywords eHealth · Brief interventions · Predictive computing · Business predictive analytics · IoT · Weight management

1 Introduction

Weight management is a practise of maintaining a long-term healthy lifestyle. It is also an approach to maintain a healthy body weight for long-term. Maintaining

U. Kant (✉) · V. Kumar
Delhi Technological University, Delhi, India
e-mail: umang.kant@gmail.com

A. Tomar et al. (eds.), *Proceedings of 3rd International Conference on Machine Learning, Advances in Computing, Renewable Energy and Communication*, Lecture Notes in Electrical Engineering 915, https://doi.org/10.1007/978-981-19-2828-4_65

a healthy lifestyle include much more efforts than just following a crash diet. It requires us to maintain a balance between eating healthy and including adequate physical activities to find equilibrium between energy intake and energy expenditure. An adequate weight management practise can be achieved by using effective weight management tools. To understand the process of weight management, we need to first understand what a healthy weight is. A healthy weight is such that depresses the risk of experiencing health problems [1–23]. Body Mass Index (BMI) and waist size are the first two features used to analyse if a person possess healthy weight. People usually associate healthy weight with attaining a certain number in scale or certain BMI. But the desired number on the scale is usually not the sign for having a healthy body. Following a crash diet to achieve certain number on scale does more damage to the person because dieting is done without having a proper knowledge about the body structure and other important factors which determine body weight. Instead of dieting, adopting healthy lifestyle works better and is long lasting. Body weight is only one aspect of health. Carrying some extra weight but eating healthy foods and being more active can help one lead healthy life by making one feel more energetic, by lowering health risks, by making one feel better and happy. One should aim for healthy weight as it is the best gift from us to ourselves. It helps prevent serious health concerns like heart related diseases, hypertension, diabetes mellitus, sleep apnea, etc. Brief intervention is another important component of weight management. Brief interventions are defined as intermediations that are focused on changing behaviour and are limited by time; often to a few minutes per session [2]. Figure 1 shows light on the components of these brief interventions. The main excuse of putting weight management process at the back seat by today's youth is the lack of time. And the facts prove that due to unrealistic work timings and work environment, people tend to ignore their health first. Be it mental health, emotional health or body health. Recent studies have shown that unrealistic work expectations from an employee by the employer and the advent of shift work has had a drastic and dramatic effect on threat of injury [3]. The results of these studies have shown that the risk of injury is 25–30% higher when working on a night shift as compared to working on a day shift [3]. They also present that working on 12 h shifts rather than on eight hours shifts increase the deterioration of sleep quality, quality of life, physical health in terms of metabolic rate and other bodily functions and cognitive functioning. The outline of health effects of anomalous work environment and shift work is discussed here in brief [1]. (1) reduction in quality of sleep, (2) reduction in quantity of sleep, (3) depression, anxiety, and heightened neurosis, (4) increased evidence of adverse cardiovascular effects, increased gastrointestinal ailments, (5) hormonal imbalance especially in females. Since the advent of Internet and android phone services, the time constraint can be minimized by taking advantage of various weight management applications. There are various weight management applications available on play store for free; people can make use of them to optimize better results.

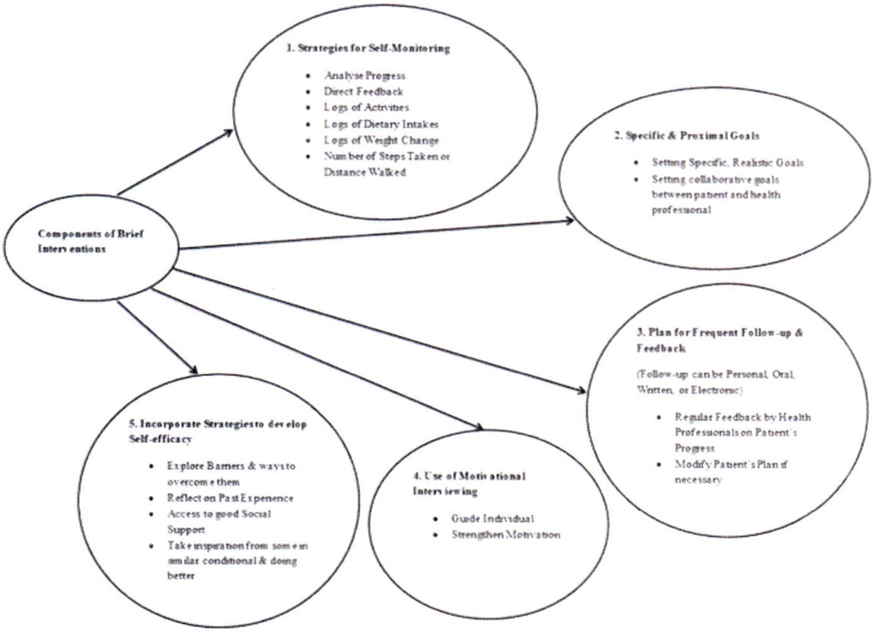

Fig. 1 Weight management: components of brief interventions

1.1 *IoT*

IoT technology covers a varied continuum of network products, systems, and sensors; also, it takes the advantage of current computing power, along with required electronic items and network interconnections. This technology is responsible for offering system capabilities which were not previously possible. Many definitions of IoT have been given by various organizations, one of the definitions (most applicable to the proposed work) given by IEEE Communications Magazine connects the IoT back to cloud facilities. The definition states that "The Internet of Things (IoT) represents an architecture in which everything has a depiction and existence on the Internet". Apart from the vast areas of application, IoT also has a huge impact on the business industry. We are discussing about the business domain as the proposed work also falls in the same category, as it would provide service to users. The three key advantages of IoT impacting day to day business include: (1) communication, (2) control, and (3) cost savings [6]. One another aspect of IoT service demand is the level of mobility. The level of agility is the extent to which devices and applications need to be itinerant [7]. Figure 2 shows the IoT service segmentation based on mobility. Internet of Things (IoT) connects every aspect of the world which we live in. IoT aims to connect various segments which make human lives better. IoT connects all possible objects to interact with each other over the Internet as now-a-days we all have Internet infrastructure in our devices. This in turn has provided a secure and comfortable life for humans.

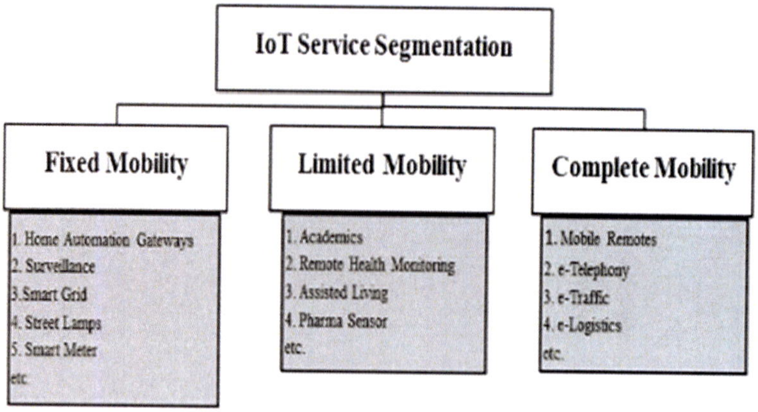

Fig. 2 IoT service segmentation applications w.r.t. mobility (cloud)

Every embedded computing device is exposed to Internet influence. Devices such as traffic lights, washing machines, microwave ovens, dish-washers, heart monitoring devices, have been launched with IoT technologies. IoT aims to provide complete automation to the devices used by humans in their day to day lives. Figure 3 gives the schematic representation of objects based on huge data available [10]. Here the application domains are identified based on the scale of impact of the generated data.

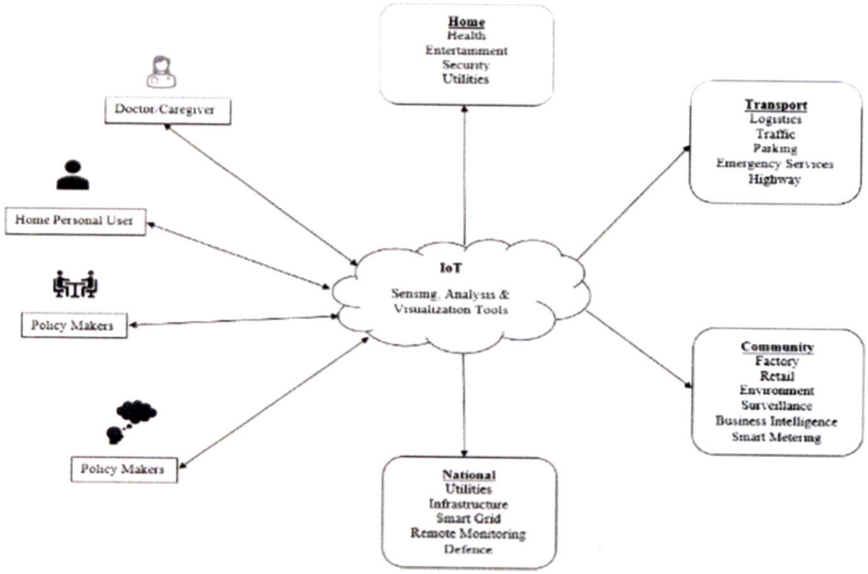

Fig. 3 IoT end users and application areas based on data [1, 10]

These generated data impacts users ranging from an individual or a wide group of people.

1.2 Use of IoT in Weight Management

In our work, we have selected the health monitoring aspect of IoT services, concentrating on weight management under remote health monitoring (Fig. 4). IoT has helped to provide service in the area of health along with other applications. This in turn has widened the access to health services and has improved the quality of health. As discussed in above section, due to lack to time, users can make use of their smart devices and install respective health applications to manage their weight. IoT and cloud services have made this farfetched idea, a reality. In recent times, the demand for health care services has exponentially increased. To address this huge demand, connected smart devices (smart phones, ipads, etc.) provide an extensive eHealth services to improve access and enable monitoring of chronic diseases or non-chronic diseases at remote locations [7]. This has major two benefits: one benefit being the time saving, and another benefit being improving the quality of care and life of patients without the need to having an enlarged infrastructure for healthcare system. Users can benefit from the service sitting at home without worrying about the time constraints and tackling the traffic conditions on the road. Figure 5 shows the general idea of eHealth service.

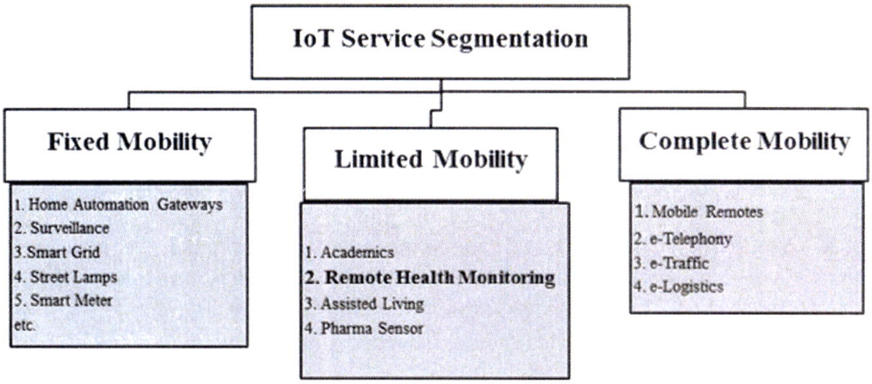

Fig. 4 Remote health monitoring (weight management using IoT)

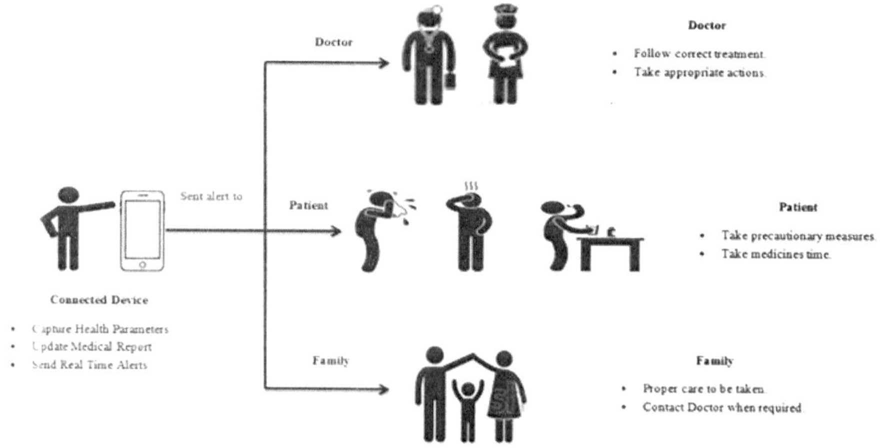

Fig. 5 General weight management process (eHealth)

1.3 Predictive Computing

Big data has brought about the era of predictive computing. This discipline uses various techniques and tools to analyse and predict the future results of various scenarios based on historical data. Predictive computing is a process that uses technology to analyse and predict the future. It allows companies to make better decisions and minimize risk. Predictive computing is a set of technologies that can predict future events and act on existing data. It can use machine learning and statistical data mining to analyse vast amounts of data. This concept is a vital component of any business intelligence system, which uses past events and forecasts to predict the future. Predictive Analytics is a field of data science that aims to predict future outcomes based on past data. With the help of sophisticated models and statistical techniques, it can produce predictions about the future behaviours and trends. Predictive computing draws its power from a variety of sources, including big data, machine learning, and statistical modelling. It uses this technology to identify and predict events and conditions that should happen at a specific time. With predictive analytics and computing, data can be analysed to identify patterns in the data that reveal potential risks and opportunities. This framework can then inform decisions regarding the procurement process. Predictive computing makes it possible to investigate the future and predict what's happening now. It can help retailers and airlines make better decisions and provide better prices. It can also help restaurants and hotels predict the number of guests during a given night. It aims to help organizations identify and stop various types of criminal behaviour prior to its inevitable damage.

2 Related Work

There is a huge potential for the adoption of Internet of Things (IoT) in the healthcare sector. To keep up with the increasing demand, various frameworks have been proposed to provide real time health monitoring of users. In Gupta et al. [11], have proposed a novel cloud centric architecture that enables the developers to perform predictive analysis on the activities of users in sustainable health centres. The architecture is based on the embedded sensor systems of the equipment. The cloud centric architecture is a conceptual concept that refers to the use of various cloud platforms such as public cloud, private cloud, and hybrid cloud. It is proposed to provide various advantages such as lower cost of ownership, better performance, and security. In Gupta et al. [12], have proposed a wearable Internet of Things (IoT)-based patient monitoring system. It uses a Wi-Fi module to monitor a person's heart rate and temperature. In Babu et al. [13], have introduced an approach to develop a smart health monitoring system that can support emergency medical services. This system can collect and integrate various data sources and provide interoperability of data. In Patil et al. [14], have presented a review on IoT-based smart health care system. Task deployment, collection and data analysis is simplified by the rapid growth of healthcare-related applications. This development has been made possible by the advances in sensor devices used in new applications of eHealth. Due to the advancement of sensor technology, mobile healthcare (also known as mHealth) becomes possible and brings innovative prospects for eHealth applications [11]. According to the data gathered by International Telecommunication Union, it reflects that there are approximately 6 billion mobile users worldwide, and most institutes, organizations and researchers hope to use this scope as a developing healthcare environment to provide healthcare services [15]. Wide-ranging selection of eHealth applications can be obtained for observing of blood sugar levels [16], blood pressure and hypertension [17], heart rate [18], physical workouts and activities [19], and over-weight factors [20] like regions. There are various healthcare applications available for various types of smartphones, tablets and iPads and other devices, called BlueBox [22], WIHMD [17], AppPoint, Heart-ToGo, Instant heart rate [21]. Although various types of applications can be found on the Internet, these applications lack security, privacy, reliability, efficiency, and acceptability. In addition, these apps are designed for devices like smartphones, and battery life is another issue. These apps constantly check the incoming data obtained by the sensors to obtain the user's heart and pulse-rate, electrocardiography, blood pressure, and blood sugar levels and other factors, and promptly notify healthcare employees in the event of an emergency.

3 Proposed Work

3.1 Proposed IoT-Based Development Model

After an intense literature review in the above sections, we can propose a framework specifically for weight management. In the above section, we have observed that most of the health management frameworks generate huge amount of data in the form of continues alerts sent to the users and health inspectors, which are not proving to be effective. Hence this shortcoming makes the framework over optimized. One more observation is the requirement of continuous connectivity of the smart devices for continuous monitoring and capturing of data. Continuous connectivity though necessary, is questionable as there are still many people not in the access of uninterrupted connectivity. In our research, we have tried to overcome the shortcomings by developing an IoT-based model shown in Fig. 6. This model consists of four layers: Implementation Layer, Evaluation Layer, Feedback Layer, and Security Layer. Each layer has a specific task associated with it.

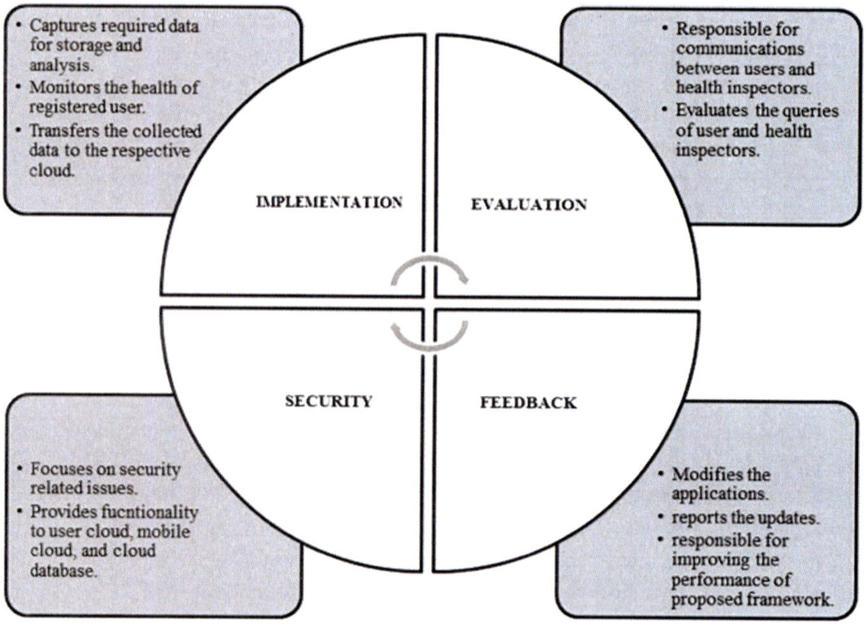

Fig. 6 IoT-based development model

3.2 *Proposed User Activity Framework*

The proposed framework is a user activity framework which keeps the user, as well as the health inspector or health care personal updated about the recent changes in the record. We have considered the user of this system to be a smart user (user who uses smart devices). The user in this case is the user interested in either losing weight or maintaining the current weight. Also, the user is considered to perform health related activities in the health centres (gyms, clubs, etc.). The health centres are considered to be equipped with required sensors, sufficient to capture the vital parameters of user (heart rate, calories burned, average speed on treadmill, distance travelled on treadmill, etc.). The framework updates the user's information once the user follows the suggested diet and completes the suggested workout session. A health inspector suggests and encourages users to follow the diet and workout instructions to be fit rather than being dependent on medications. Physical activities like brisk walking, swimming, running, yoga, aerobics, palates, etc. are recommended to the user depending upon the weight statistics and the physical condition of the respective user. The proposed framework (Fig. 7) monitor's the user's activities on regular intervals and maintains the updated captured details to the cloud. In case the suggested diet plan or workout plan is not being followed by the user, the same information is also updated in the cloud database. An alert regarding the same is sent to assigned health inspector, so that further communication regarding the status of the health can be established with the user. Regular check about the status of user's weight needs to be maintained by the system.

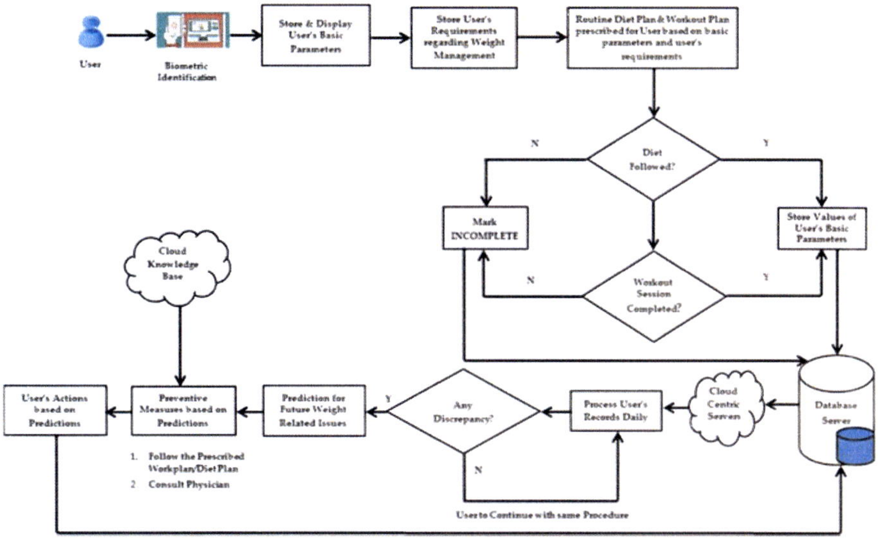

Fig. 7 Proposed framework for weight monitoring using predictive computing

The working of our proposed weight monitoring framework can be understood by Fig. 7. Once the user is registered with the system, basic parameters and the requirements are stored in the database for predictive analytics. For a good predictive weight monitoring system, a clear set of rules need to be defined so that best decisions can be derived. The cloud knowledge base stores the set of rules.

3.3 *Algorithm for Proposed Work*

Following is the algorithm for the proposed work, which ensures the efficient working of the framework.

Algorithm 1: Weight Monitoring using Predictive Computing

```
Procedure: Working of proposed framework
Generate(UserID)
Locate(UserID)
UserSession ← CollectBasicParameters()
if CollectBasicParameters(UserSession) < Threshold then
        AcceptableCategory(UserID)
        ContinueSameRoutine
elseif CollectBasicParameters(UserSession) > = Threshold then
        NeedAdviceCategory(UserID)
        UserStatus ← ConsultPhysician()
        UserLocation ← SmartLocation()
        ConductCloudServices(User ID, UserStatus, Userlocation)
        UpdateBasicParameters()
endif
endif
```

Algorithm 1 demonstrates the proposed framework for predictive-based weight monitoring. As discussed previously, the collected basic parameters of the user (CollectBasicParameters(UserSession)) is compared with a predefined threshold value. The further processing of the framework is decided on the basis of the comparisons between the user's basic parameters and the threshold value.

4 Conclusion and Future Work

In this competing era of technologies, hundreds of existing predictive computing and modelling techniques are used and new technologies are introduced everyday by researchers and practitioners. Researchers aim to address specific problems by making small changes in the available IoT-based predictive model, so that more accurate and efficient predictive models are produced. Hence predictive computing is a valued tool for enhancing efficiency in wide range of applications. Presented here is an IoT-based cloud centric framework for weight management using predictive computing. Considering the flexibility to support the fast changing and competing needs of the user, we have proposed a dedicated IoT-based development model and supporting framework and algorithm. The proposed framework and the supporting algorithm aim to work specially for weight management by considering the user's overall health related basic parameters, processing the collected data using cloud computing and then predicting future outcomes based on the previously collected data making use of predictive computing. In proposing the new framework, the associated challenges have been worked upon by considering the management of vast amount of data, its integrity and security. Further work is being worked upon the computation based on collected data, and other health related applications.

References

1. Singh M, Kant U, Gupta PK, Srivastava VM (2019) Cloud-based predictive intelligence and its security model. In: Predictive intelligence using big data and the internet of things. IGI Global, pp 128–143
2. Nourish by WebMD. What is a Healthy Weight? https://www.webmd.com/diet/obesity/healthy-weight#1
3. Jennings P, Cooney F, Hegarty M, Smith L (2018, November) Healthy weight for children (0–6 years) framework. http://www.hse.ie/eng/health/child/healthyeating/weightmanagement.pdf
4. Harrington JM, Health effects of shift work and extended hours of work. BMJ J Occupat Environ Med. https://oem.bmj.com/content/58/1/68
5. Internet Society (2015, October) The internet of things: an overview. https://www.internetsociety.org/wp-content/uploads/2017/08/ISOC-IoT-Overview-20151221-en.pdf
6. Lopez Research (2013, November) An introduction to internet of things (IoT). https://www.cisco.com/c/dam/en_us/solutions/trends/iot/introduction_to_IoT_november.pdf
7. Connected Living (2014, July) Understanding the internet of things (IoT). https://www.gsma.com/iot/wp-content/uploads/2014/08/cl_iot_wp_07_14.pdf
8. Alam JR, Sajid A, Talib R, Niaz M (2014) A review on the role of big data in business. Int J Comput Sci Mob Comput 3(4):446–453
9. Gupta PK, Maharaj BT, Malekian R (2016, October) A novel and IoT based cloud centric architecture to perform predictive analysis of user activities in sustainable health centres. Multimedia Tools Appl
10. Shivam G, Shivam K, Devesh PC (2017, March) IoT based patient health monitoring system. Int J Eng Technol (IRJET) 04(03)
11. Alcatel-Lucent (2021, March) The internet of things in healthcare. https://www.al-enterprise.com/-/media/assets/internet/documents/iot-for-healthcare-solutionbrief-en.pdf
12. Albesher AA (2019, February) IoT in health-care: recent advances in the development of smart cyber-physical ubiquitous environments. IJCSNS Int J Comput Sci Network Secur 19(2)
13. Information and Communication Technology (ICT) Statistics (2011) ITU. http://www.itu.int ITU-Dictindex.html. Accessed on 5 Mar 2016
14. Revathi R, Vijani S, Vijanu S (2015, April) Intelligent health care services based on IoT. Int J Sci Eng Res 6(4)
15. Lin Y, Ge Y, Li W, Rao W, Shen W (2014) A home mobile healthcare system for wheelchair users. In: Proceedings of 18th International conference on computer supported cooperative work in design (CSCWD). IEEE, New York, pp 609–614

16. Azumia Instant heart rate. http://www.azumio.com/s/instantheartrate/index.html. Accessed on 10 Mar 2016
17. Oresko JJ, Jin Z, Cheng J, Huang S, Sun Y, Duschl H, Cheng AC (2010) A wearable smartphone-based platform for real-time cardiovascular disease detection via electrocardiogram processing. IEEE Trans Inf Technol Biomed 14(3):734–740
18. Pasha M, Shah SMW (2018) Framework for E-Health systems in IoT-based environments. Wireless Commun Mob Comput
19. Farahani B, Firouzi F, Chang V, Badaroglu M, Constant N, Mankodiya K (2018) Towards fog-driven IoT eHealth: promises and challenges of IoT in medicine and healthcare. Futur Gener Comput Syst 78:659–676
20. Pirbhulal S, Wu W, Mukhopadhyay SC, Li G (2018, September) A medical-IoT based framework for eHealth care. In: 2018 International symposium in sensing and instrumentation in IoT Era (ISSI). IEEE, New York, pp 1–4
21. Semwal N, Mukherjee M, Raj C, Arif W (2019) An IoT based smart e-health care system. J Inf Optim Sci 40(8):1787–1800
22. Ullah K, Shah MA, Zhang S (2016, January) Effective ways to use internet of things in the field of medical and smart health care. In: 2016 International conference on intelligent systems engineering (ICISE). IEEE, New York, pp 372–379
23. Tomar A et al (2020) Machine learning, advances in computing, renewable energy and communication, LNEE vol 768. Springer Nature, Berlin, 659 p. https://doi.org/10.1007/978-981-16-2354-7. ISBN 978-981-16-2354-7

Efficient Cloud Orchestration Services in Computing

Ashish Tiwari, Sarvesh Kumar, Neeraj Baishwar, Sunil Kumar Vishwakarma, and Prabhishek Singh

Abstract In the buzz of technology, the IT Industry growing at a rapid speed with Cloud Orchestration. In the developing trends, cloud orchestration is becoming impressive and uncertain of dealing a large amount of information. The Cloud orchestration defines the proper arrangement and the coordination of various automated assignments for the proper flow of a process. These assignments fulfill the goal of Cloud Orchestration system. Cloud orchestration formulates the parameters like data storage capacity and capability, computing applications. The major areas where cloud orchestration is performing its goals are service workload, resource management, service level agreements, security, etc. The research is now trending with fuzzy and rough sets in cloud orchestration. In the innovation of cloud computing the important components are datacenters, cloud services, cloud parameters and users. There is a need of methodology by which the orchestration is being done. At last, the parameters used in cloud orchestration are being simulated using cloud simulators.

Keywords High-performance technology · Cloud organizations · Computing parameters · Resource management · Cloud services · Cloud organizations · Cloud providers · Ontology system · Cloud simulation

A. Tiwari · S. Kumar (✉) · N. Baishwar · S. K. Vishwakarma
BBD University, Lucknow, India
e-mail: Kr.sarvi91@gmail.com

A. Tiwari
e-mail: Ashish.fgiet@bbdu.ac.in

N. Baishwar
e-mail: baishwarn@bbdu.ac.in

S. K. Vishwakarma
e-mail: sunilvishwakarma83@bbdu.ac.in

P. Singh
Amity School of Engineering & Technology, Amity University, Noida, India

A. Tomar et al. (eds.), *Proceedings of 3rd International Conference on Machine Learning, Advances in Computing, Renewable Energy and Communication*, Lecture Notes in Electrical Engineering 915, https://doi.org/10.1007/978-981-19-2828-4_66

1 Introduction

According to the Research the term "cloud orchestration" tells that the computing is working with various architectures in parallel cloud automation devices. The end users of cloud computing form the database and the provider send the information into the network for the real time system. The middleware or the brokers are doing the communication for the end users. So, such type of system which is doing computing as automation processes, the term is cloud orchestration [1]. Cloud coordination apparatuses, regardless of whether local to the IaaS stage or outsider programming instruments, specify the assets, example types, IAM jobs, and so on that are required, and in addition the setup of those assets and the interconnections between them. In any case, each endeavor is in an alternate phase of actualizing the instruments and the rationality coordination infers. On the opposite end of the range, associations that coordinate computerization errands into standard however adaptable IT work processes under a solitary checking and arrangement programming interface are genuine Develops shops [1–4] (see Fig. 1).

Service arrangement is basic in the conveyance of cloud administrations consequently: Cloud administrations are expected to increase subjectively and powerfully, without requiring direct human mediation to do as such. Cloud administration conveyance incorporates satisfaction affirmation and charging. Cloud administrations conveyance involves work processes in different specialized and business spaces. Cloud organization innovation should work with heterogeneous frameworks, conceivably overhauling a worldwide cloud arrangement in various geological areas and with various suppliers. Many cloud orchestrator clients run public cloud and private organizations. Through our job-based admittance control and fine-grained multi-occupancy—engineers can approach their applications, while IT stays in charge of the assets and information [5–7].

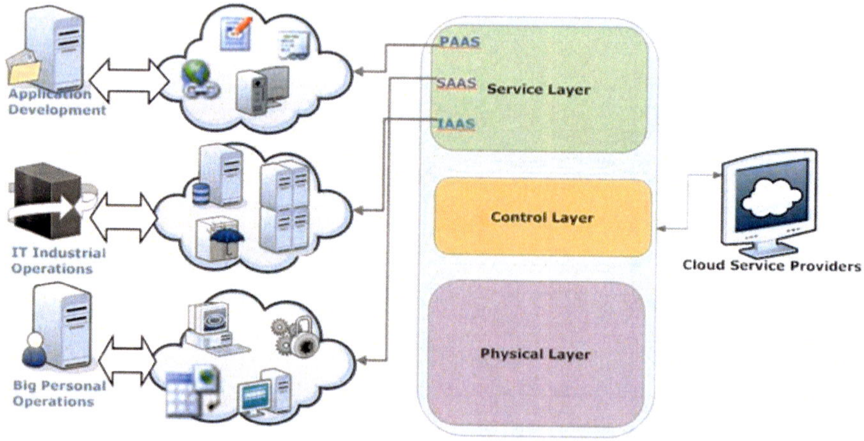

Fig. 1 Cloud system arrangement as per the NIST

2 Related Work

The arrangement is basic in the conveyance of cloud administrations hence: Cloud administrations are expected to scale-up self-assertively and powerfully, without requiring direct human intercession to do as such. Cloud benefits conveyance incorporates satisfaction confirmation and charging. Cloud administrations conveyance involves work processes in different specialized and business spaces [7]. While there were guarantees of Orchestration in the private cloud condition, in the previous year these guarantees are getting to be reality. What's fascinating in this space is that general society cloud players do not have an appealing Orchestration layer as they like to stay inside their own area. This has advanced the achievement of others to satisfy the guarantee of a half breed cloud while conveying in a private cloud [8–10] (Fig. 2).

An Orchestration layer varies from a Brokerage layer, although the two are regularly befuddled and have a few covers. The organization is centered on operational help administrations, incorporates propelled work process motor abilities, and incorporates the methods for administration union to drive institutionalization and acknowledge effectiveness. Fundamentally, coordination computerizes the robotization particularly, the request that assignments occur crosswise over particular machines, particularly where there are differing conditions [11].

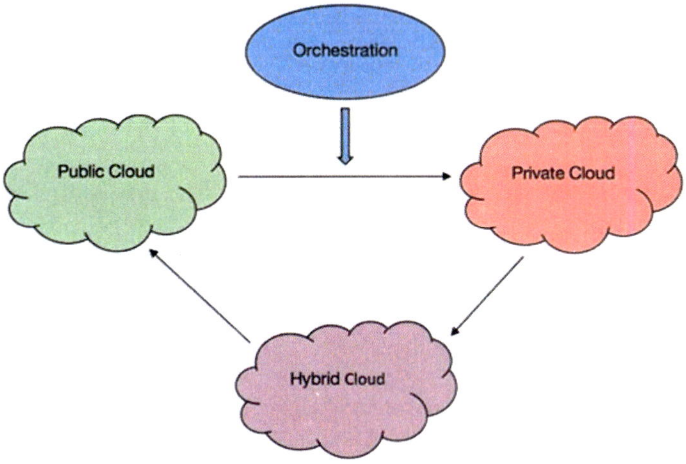

Fig. 2 Arrangement of cloud types

3 Research Challenges in Cloud Orchestration

The deftness managed by cloud foundation makes robotized organization always basic. Cloud merchants each have their own answer for provisioning VMs and administrations yet pass on the administration of the VM designs to fragile contents that are difficult to keep up with, or more awful—physically heated brilliant pictures [9–12].

4 Cloud Orchestration System

Orchestration package helps IT organizations standardize templates and enforce security practices. It is additionally an honest defense against VM sprawl—providing visibility into, and management over, cloud resources and, implicitly, costs. Because of the adapter platform oversees interactions of the many disparate parts of the applying stack, it will change the communication and connections from one work to alternative apps and users, further as guaranteed links square measure properly organized and maintained. Such merchandise sometimes embodies a web-based portal; thus, orchestration may be managed through one pane of glass. The centralized nature of associate orchestration platform permits directors to review and improve automation scripts [8–11, 13].

4.1 Cloud Orchestration Platform

The cloud service arrangement is important to a few IT associations and Develops adopters as the way of hustling the conveyance of administrations and cut back costs. Coordination upholds the conveyance of cloud assets to clients and closures clients, along with in an extremely self-administration model any place clients demand assets while not IT's contribution. In cutting edge associations, designers and line-of-business staff will between impart cloud coordination bundle as a self-administration component to send assets; chiefs will utilize it to follow the association's dependence on various IT contributions and deal with chargebacks [9–12]

Reference [17] is cited in the text but not provided in the reference list. Please provide the respective reference in the list or delete this citation.

4.2 Cloud Orchestration End Users

Many vendors offer cloud orchestrator products. When assessing cloud service arrangement items, it is suggested that overseers first guide the work processes of the applications in question. A portion of the other cloud organization merchants and their items incorporate the accompanying: [14–16]

- Amazon Web Services Cloud Formation
- Cisco Cloud Center
- CloudFX
- IBM Cloud Orchestrator
- Microsoft Cycle Computing.

5 Methodical Structure of Orchestration

An orchestration is the method involved with computerizing organization and redeployment of application parts so that intercomponent and underapplication joins are effectively settled and kept up with. Conveying application parts, including programming parts and data set parts; and creating network associations with take into consideration intercomponent correspondence and associations with clients and other applications [8–11].

5.1 Cloud Orchestration Tools

Most of the more full-highlighted cloud arrangement instruments fall into the DevOps item classification. DevOps depends on the rule that an application designer knows how the application should be facilitated and how the parts are associated [5]. Their most noteworthy disservice is that they portray the cycle and not the result. There are open source and business DevOps instruments accessible [7].

5.2 Cloud Orchestration Implementations and Results

Numerous merchants offer cloud orchestrator items. While assessing cloud organization items, it is prescribed that directors first guide the work processes of the applications included. In this research work the efficient time utilization if find out in the cloud services (Fig 3). In Fig. 4 the CPU utilization is being explained which provides the fast access of services. This progression will enable the chairman to imagine how confused the inside work process for the application is and how regularly data streams outside the arrangement of application segments.

By and large, cloud coordination devices or programming work likewise crosswise over open, private, and half breed mists, however the specifics of a given utilize case may support the highlights of one over another.

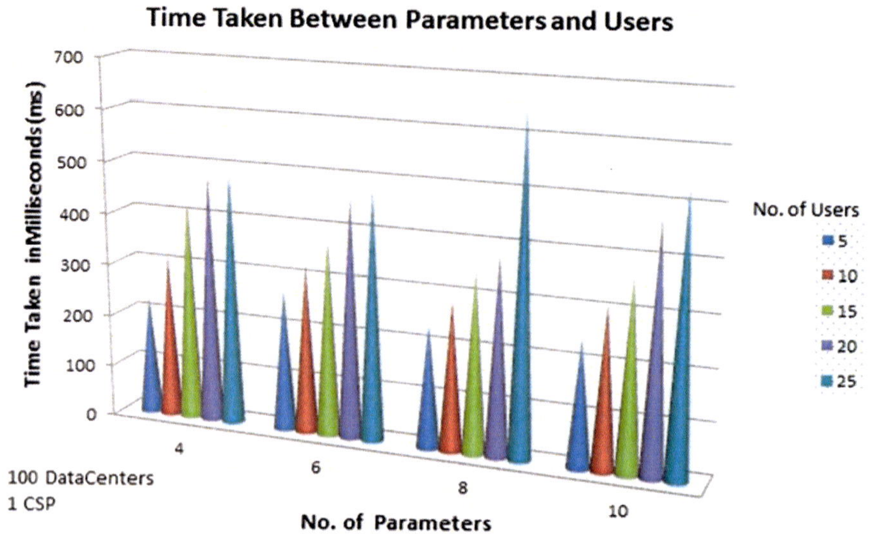

Fig. 3 Efficient time execution of services

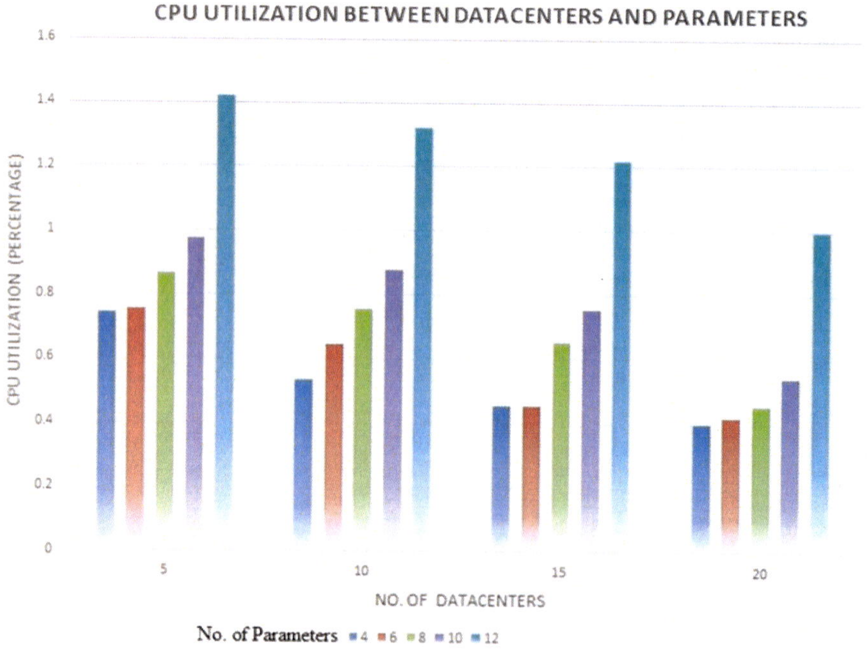

Fig. 4 Efficient CPU utilization for services

Different apparatuses and systems are accessible for coordination, every one of which are fitting for cases however for the implantation these products assume a critical job.

Conclusion and Future Scope

The research may be increased by using machine learning languages. The use of genetic algorithm makes the work stronger to get the performance of this research work. We will also try to work with the more IoT parameters for the system.

References

1. Buyya R, Rodriguez MA, Toosi AN, Park J (2018) Cost-efficient orchestration of containers in clouds: a vision, architectural elements, and future directions. arXiv preprint arXiv:1807.03578
2. Tiwari A, Sharma RM, Garg R (2020) Emerging ontology formulation of optimized internet of things (IOT) services with cloud computing. In: Pant M, Sharma TK, Verma OP, Singla R, Sikander A (eds) Soft computing: theories and applications, pp 31–52
3. Pawlak Z (2002) Rough set theory and its applications. J Telecommun Inf Technol 3
4. Tiwari A, Sharma RM (2016, August) Potent cloud services utilization with efficient revised rough set optimization service parameters. In: Proceedings of the international conference on advances in information communication technology & computing. ACM, New York, p 90
5. Manjumdar S (2012) Resource management on cloud computing, multifaceted problem and solutions in ACC
6. Diwakar M, Tripathi A, Joshi K, Memoria M, Singh P (2021) Latest trends on heart disease prediction using machine learning and image fusion. Mater Today: Proc 37:3213–3218
7. Singh P, Shree R (2016) Speckle noise: modelling and implementation. Int J Control Theory Appl 9(17):8717–8727
8. Chauhan H, Gupta D, Gupta S, Nayak SR, Shankar A, Singh P (2022) Framework for enhancing the traceability in supply chain using blockchain. J Interconnect Networks 2144008
9. Tyagi T, Gupta P, Singh P (2020, January) A hybrid multi-focus image fusion technique using SWT and PCA. In: 2020 10th international conference on cloud computing, data science & engineering (confluence). IEEE, New York, pp 491–497
10. Ghose S, Singh N, Singh P (2020, January) Image denoising using deep learning: convolutional neural network. In: 2020 10th International conference on cloud computing, data science & engineering (confluence). IEEE, New York, pp 511–517
11. Dhaka A, Singh P (2020, January) Comparative analysis of epidemic alert system using machine learning for dengue and Chikungunya. In: 2020 10th International conference on cloud computing, data science & engineering (confluence). IEEE, New York, pp 798–804
12. Sharma N, Gupta S, Mehta P, Cheng X, Shankar A, Singh P, Nayak SR (2022) Offline signature verification using deep neural network with application to computer vision. J Electron Imaging 31(4):041210
13. Singh P, Diwakar M, Chakraborty A, Jindal M, Tripathi A, Bajal E (2021, November) A non-conventional review on image fusion techniques. In: 2021 IEEE 8th Uttar Pradesh section international conference on electrical, electronics and computer engineering (UPCON). IEEE, New York, pp 1–7
14. Tiwari A, Tiwari AK, Saini HC, Sharma AK, Yadav AK (2013) A cloud computing using rough set theory for cloud service parameters through ontology in cloud simulator. In: ACITY-2013 Conference at Chennai, in CS and IT proceedings
15. Gopalakrishnan Nair TR, Sharma V (2012) A pragmatic scheduling approach for creating optimal priority of jobs with business values in cloud computing, ACC

16. Tomar A et al (2020) Machine learning, advances in computing, renewable energy and communication, LNEE vol 768. Springer Nature, Berlin, 659 p. https://doi.org/10.1007/978-981-16-2354-7. ISBN 978-981-16-2354-7

Efficient Data Aggregation in Wireless Sensor Network Cluster-Based Solution

Jainendra Singh and Zaheeruddin

Abstract By linking billions of devices capable of instantly sharing, receiving, and analyzing vast quantities of data to better satisfy business demands and improve decision-making, IoT systems have profoundly merged into our lives. Because data acquired by IoT devices are collected in real time, it can be analyzed and sent more quickly. With the introduction of software defined networking (SDN), network operators and users can now control and access network devices remotely while exploiting the network's global view. This paper proposes algorithms to make gathering and transmitting of data more energy efficient. Data aggregation is used to improve the transmitting efficiency by reducing transmission of redundant data to the sink. A clustering algorithm is also proposed which is based on residual energy of sensors and distance to sink. This proposed protocol, energy efficient cluster-based data aggregation (EECDA), will help to increase network lifetime by ensuring uniform distribution of energy among the nodes.

Keywords WSNs · Clustering · DB-SCAN · Steiner point · Rendezvous point

1 Introduction

A WSN is a group of sensors that are dispersed around the study area (ROI). Each sensor node can provide data to the transceiver to detect certain phenomena through sensing, such as automobile tracking, wildlife tracking, environmental monitoring, military, and so on [1–10]. The job of the transceiver is to collect data from the sensor node using the Internet. For data transmission, the transceiver connects to the network. The data from the surroundings are stored in the sensor node and forwarded to the base station via multi-hop communication. WSNs are data-gathering

J. Singh (✉) · Zaheeruddin
Department of Electrical Engineering, Jamia Millia Islamia, New Delhi 110025, India
e-mail: er.jainendra@gmail.com

Zaheeruddin
e-mail: zaheeruddin@jmi.ac.in

A. Tomar et al. (eds.), *Proceedings of 3rd International Conference on Machine Learning, Advances in Computing, Renewable Energy and Communication*, Lecture Notes in Electrical Engineering 915, https://doi.org/10.1007/978-981-19-2828-4_67

networks made up of small sensors dispersed around an area of interest. The application scenario in WSNs is often intervention-free and difficult due to geographical limits. WSNs use small sensors to collect data from their surroundings and transfer it to a sink. To forward data, single hop or multi-hop routing can be utilized. The network topology is set up such that neighboring nodes can choose a cluster head (static or dynamic) to assist with data forwarding based on the criteria. There are numerous clustering and cluster head selection options for the operation of WSNs.

The IoT represents the advancement of digitization in our lives, society, and economy. Using an amazing range of applications, large scale intercommunication is possible between objects and people. According to a report from Forbes [1, 2], McKinsey predicts the worth of the IoT market to be 581B American dollars by 2020, growing at a Compound Annual Growth Rate between 7 and 15%. The IoT aims at making our daily lives more sophisticated, flexible, and reachable to all objects and people around us. It derives as a three-dimensional approach of ANY-TIME, ANYPLACE, ANY-THING connection. Within these widespread interconnected networks, a huge amount of data is sensed, transformed, collected, transmitted, and processed, leading to the need for highly efficient and fast mechanisms for above mentioned processes. To handle the big data problem and its redundancy, many data aggregation techniques have been proposed and are utilized in real-time applications. On the Internet of things, billions of items, known as terminal devices, are discovered through various types of sensors and actuators that are interconnected via access networks such as WSNs. As a result, WSN technology is at the heart of IoT applications, where a plethora of sensor nodes monitor environmental variables such as temperature, humidity, and so on. One of the most essential prerequisites for the Internet of things is that things be made smart [2] via embedded intelligence techniques such as RFID. Users' and service providers' faith in the system, as well as management over the data shared over the network, are critical to the adoption of IoT in daily life. A key concern is the lack of standardization of IoT protocols. IoT is a simple idea with enormous possibilities that have the potential to change the old paradigm of outdated technology. We want to propose a resource data aggregation technique for WSNs systems that have become an essential part of our daily lives, through this work. It explains why the work is being done, what the goals are, what the backdrop is for the research, and how the study will be conducted. It also serves as a synopsis for the following portion of the article.

We start our problem by assuming that the WSNs operate in clustering rather than in a multi-hop fashion cluster head that collects data from member nodes. We proposed DB-SCAN clustering for WSNs; we provide the path planning using a Steiner point around those cluster heads' communication range. The mobile sink follows the data collection path, which seems more efficient than the existing proposals on many parameters [2, 3]. This section discusses the various factors that motivate the selection of this project for our dissertation. Firstly, IoT is swiftly and steadily becoming a major part of our lives, be it in the development of smart cities, smart home appliances, vehicles, environmental monitoring to control issues of global warming and pollution and in disaster management. Hence, efficient data aggregation becomes an important criterion to be considered for effective communication and fast delivery of services. Secondly, the massive amount of data which is generated and transferred within networks contains a prominent portion of data redundancy which leads to decreased network lifetime. Therefore, we are motivated to devise one such technique such that only the requisite data are sent to the sink node.

The rest of the paper has been organized into the following way as in the Sect. 2, the related work has been explained associated with the method of data collection using the mobile sink. In Sect. 3, the system model is presented, followed by our proposal in Sect. 4. In Sect. 5, our proposal's result and evaluation are given, followed by the conclusion in Sect. 6 substitutes. The background provides an insight into how WSNs and IoT systems have evolved through time, and they continue to do so. We present the objectives for our work and how we aim to make significant contributions to this novel model. At last, we provide a brief outline of the complete paper.

2 Related Works

A survey has been provided by [4] which explain the limitations of traditional networking in handling IoT requirements-of adequate control and access of connected devices in a flexible, scalable, efficient, seamless, and cost-effective manner. This is where SDN-based technologies come to the rescue. SDIoT is the new networking paradigm which has four major aspects: edge networking, access networking, core networking, and data networking. Data aggregation is an important aspect of SDN-based edge networking. According to [5], wireless sensor networks have revolutionized the way of sensing information. Energy conservation is the key issue in WSNs, since a large number of deployed sensor nodes are likely to transfer redundant data which increases communication cost and in turn decreases network lifetime. Energy conservation, query processing, and communication are major topics to be focused on. The big data paradigm and the integration of its major principles in the WSN technology are introduced by Boubiche [5]. The big data model can be a functioning resolution to collect the data so that it can further send to other receiver and at the same time, can be stored at respective place with the integration of WSN. In [5], authors have proposed WSN model which have 4 layers to simulate IoT ecosystem. This work intended to support systems which can work on low power as well on low cost. The different layers of communication are sensors, node, cluster hub, and cloud server. This framework has been tested for healthcare applications and can be further implemented in diverse IoT applications. Castro demonstrated an IoT-based WSN for monitoring environmental temperature and humidity, which was then used to product maintenance and pharmaceutical organizations. Following good feedback, expanding this framework across the whole hospital has become a viable option.

In WSNs scenario, data can be collected in two ways, first one is the direct, and second is known as rendezvous point (RP) [6]. The SMAC [7] is an early developed contention-based MAC protocol. The TDMA protocol forms the LEACH [8] protocol, one of the fundamental schedule-based MAC protocols is low-energy adaptive clustering hierarchy with deterministic cluster head selection (LEACH) [8] is a cluster-based scheduling protocol. The detailed description of scheduling in WSNs is documented in [9], and we will focus our related work on the mobile sink in WSNs only.

3 System Model

On the transmitter and receiver sides, the energy usage can be measured. Assume that $E_t(l, d)$ and $E_r(l)$ represent the consumed energy, respectively, by a transmitter and receiver when a message of 1-bit is broadcasted. This can be calculated Eqs. (1) and (2). A node's transceiver circuitry is given as

$$E_t(l, d) = E_{Cc}(l) + E_{am}(l, d) \tag{1}$$

$$E_r(l) = E_{Cc}(l) = l \times E_{Cc} \tag{2}$$

The sensors and network model are based on the following assumptions:

- The sink node is situated in the center of the sensing area.
- Location-aware item sensors are energy-constrained.
- Energy-aware and resource-adaptive item sensors are available.
- Symmetric propagation channel with stationary nodes and sink.

Here, the criteria to select of $E_t(l, d)$ at any case t depend on the distance d between receiver node p_j *and* transmitter p_i. It is represented in Eq. (3)

$$E_t(l, d) = \begin{cases} l \times E_{Cc} + l \times \varepsilon_{fs} \times d^2 & d < d_0 \\ l \times E_{Cc} + l \times \varepsilon_{mp} \times d^4 & d \geq d_0 \end{cases} \tag{3}$$

where ε_{fs} and ε_{mp} represent free-space and multi-path fading channel model, respectively.

Algorithm 1 CH Selection

Input: Sensors, model, r, X, Y
maxCH = p% of n
Average Energy $avgE = \frac{Total\ Energy}{Alive}$
CHcount = 0
for each node i in Network **do**

for each node j in Network **do**

D_{jj} = dist(i,j)

end for

end for

for each node i in Network **do**

D_{js}=dist(i,sink)

$\mathbf{D_{net=}}\sum D_{ij} + D_{is}$

end for

Sort D_{net}

for each node i in ID(D_{net}) **do**

if $Energy_i > 0 \quad \&\& Energy_i \geq avgE \quad \&\& \quad random < T(i)$ **then**

CH(CHcount+l)='C'

CHcount++

if $CHcount == maxCH$ **then**

break

end if

end if

end for

In this work, if the calculated distance d is more than the calculated threshold, then we have deployed multi-path fading channel and if not then free-space model. In order to achieve acceptable signal-to-noise ratio (SNR) while transferring a message of k-bits for a distance d the energy cost of transmission (ET_X).

4 Proposed Work

This section elaborates the prior requisite, which plays an essential role in our proposal. Our proposal is divided into two separate phases as setup and steady. In the first phase, the nodes which are deployed in the region initialize and start the neighbor discovery. Based on the set parameters, nodes select their leader node, which will be responsible for forwarding the data using multi-hop communication. This phase is called the network setup phase. There are plenty of centralizing, and distributed methods have been proposed for the selection of leader nodes.

LEACH works by randomly selecting cluster heads, which collect data from sensors in their vicinity and transfer it to the base station. The working model for LEACH as well as our proposal has shown in Fig. 1. Since data transmission dissipates energy, cluster heads are re-selected after each round. Operation of LEACH protocol

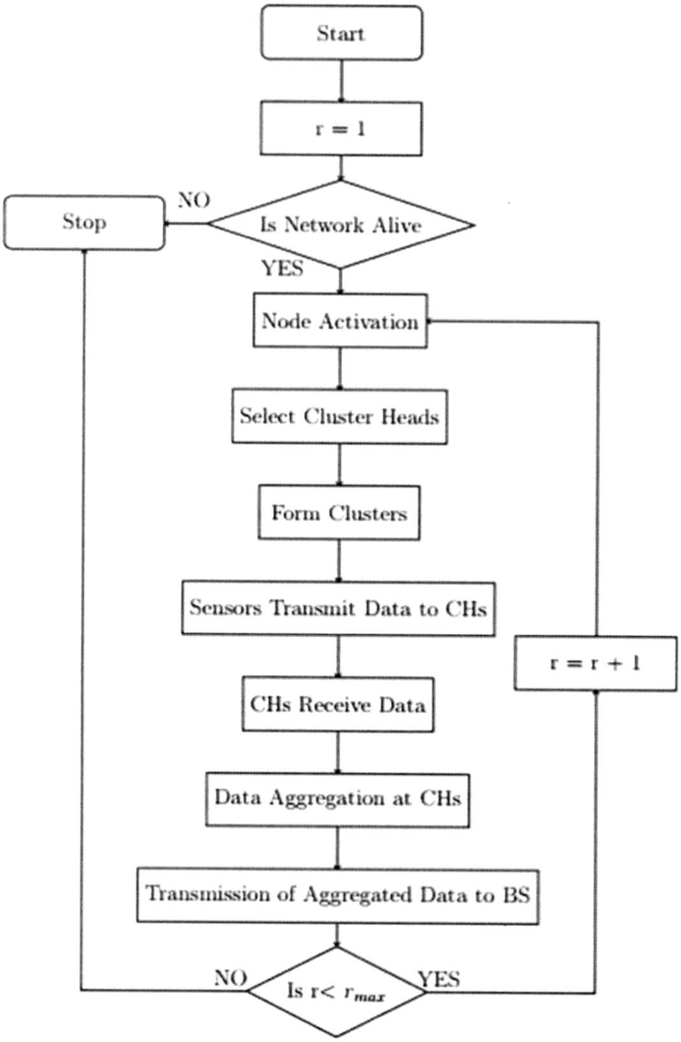

Fig. 1 Flow diagram

consists of several rounds, each of which is treated as a unit and every round consists of two phases:

- Set-up phase
- Steady phase

The following are the stages of the LEACH protocol: Initial setup during the setup phase, a predetermined group of nodes, *p*, is chosen as cluster heads using the following procedure. Every sensor node n selects a number *r* at random from 0 to 1.

Any node will be treated as a cluster head if the selected random integer is smaller than $T(n)$.

$$p(n) = \frac{p}{1 - p * (r * \bmod p^{-1}) \forall n \in G} \tag{4}$$

Algorithm 2 Data Aggregation

Input: Sender, DataPckt, CH
if CH.idxFill = = 0 **then**

CH.mlist(1) = DataPckt.Content
CH.freq(l) = l;
CH.idxFill = l;

end if
if CH.idxFill < CH.mlistlen **then**

if DataPekt.Content ∈ CH.mlist **then**

CH. freq (content_loc)++

end if

if DataPckt.Content ∉ CH.mlist **then**

CH. mlist (+ + idxFill) = DataPckt.Content
CH.freq(idxFill)++

end if
else

if DataPckt.Content ∈ CH.mlist **then**

CH.freq(content_loc)++

end if

if DataPckt.Content ∉ CH.mlist **then**

min_loc = Location(Min(CH.freq))
CH.mlist(min_loc) = DataPckt.Content
CH.freq(min_loc) = l

end if
end if

The r is a round number in this case. The chance of being chosen as CH is $frac1p$, where G denotes the set of nodes that were not cluster heads in the earlier rounds. The threshold is $T(n)$. Non-CH nodes get the cluster head advertising in the second phase and then submit a join request to the cluster head, indicating that they have joined the group underneath that base station. In the third phase, each of the selected

CHs generates a TDMA dynamically updated for their cluster's member nodes. After that, each node sends its data according to the timetable, and it was given. Phase of constancy cluster nodes transmit data to the central in constant phase. Each cluster's member sensors can only connect with the cluster head through a one hop transmission. Data are sent from the cluster head to the ground station. After a given amount of time has passed, the network returns to the initialization and begins a new round of CH selection. The stochastic cluster head selection in LEACH has the potential to create an energy imbalance in nodes, increasing the system's total energy waste. To ensure an equitable energy load distribution across the network, we add new variables depending on the node's remaining energy and its distance from the source and many other nodes. Our method's major aim is to avoid choosing CHs for nodes that have less residual energy and are farther away from the sink.

$$D_{\text{net}}(i) = \text{X}D_{ij} + D_{iSink} \tag{5}$$

We sort the D_{net} in ascending order to find the nodes with highest eligibility of being CH. The nodes which have residual energy larger than the average energy of the network are selected CHs. Also, to ensure that the nodes with similar above parameters can be selected CHs fairly, a randomness factor is introduced. We use the neighborhood concept to form the cluster, according to which a given node becomes a cluster member under a given CH provided that the CH is the closest node from the set of CHs which lie in its radio transmission range.

5 Result Analysis and Discussion

In this section, we will discuss the results of our simulation. The simulation is performed using the network package in the Python and MATLAB program. We have also assumed the following:

- All nodes have direct communication with the sink node. Nodes and sink items are not allowed to move around.
- In the first round, all nodes start with the same amount of energy.
- The energy used to transmit a packet from node 1 to node 2 is the same as the energy used to transmit a packet from node 2 to node 1.
- The sink node has a fixed position and theoretically infinite energy.
- The node dies when its energy is equal to 0.0 (Table 1).

The design model of the network has been depicted in Fig. 2 that have 100 similar sensor nodes who are arbitrarily dispersed in a 100 m * 100 m region. The base station is situated in the heart of the city. The nodes are thought to be stationary. Figure 2 depicts the network model at the simulation's first phase, when there are no dead nodes. The image depicts the entire process of cluster head selection and cluster formation. The blue lines reflect data sent from various cluster nodes to a designated cluster head. Now, the data aggregation has been executed by this

Table 1 Simulation parameter

Network X package based on Python simulator	
Number of node (n)	100
Area	(0,0) to (100,100)
P	0.1
sink.x,sink.y	(50,50)
E_0 ($J.node^-1$)	0.5
E_{TX} ($nJ.bit^{-1}$)	50
E_{RX} ($nJ.bit^{-1}$)	50
EDA (nJ)	5
$E_{fs}(nJbit^{-1}.m^-2)$	10
Emp ($pJbit1-1.m-4$)	0.0013
$Rmax$	500
Node Distribution	Random
DpacketLen (bits)	4000
HpacketLen (bits)	100

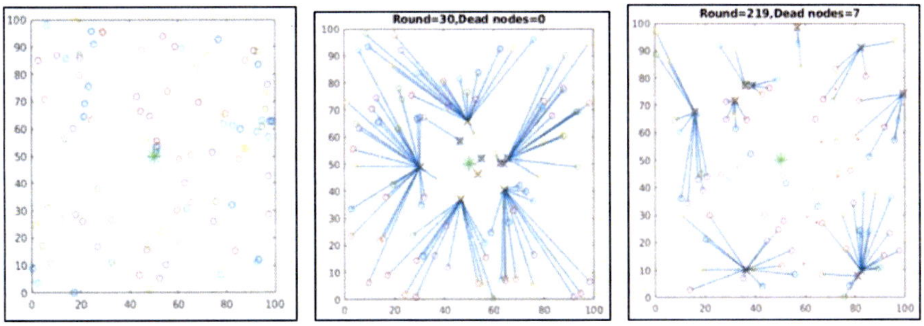

(a) Node deployment (b) Cluster formation and cluster head selection (c) Formation of dead nodes the final phase from DBSCAN clustering

Fig. 2 Cluster formation

cluster head before sending the aggregated data to the sink. Clustering is done in our proposed algorithm by considering the residual energy parameter, as well as the distance between each node and all other nodes, as well as the distance between each node and the sink, resulting in a consistent energy distribution and efficient cluster head selection and cluster formation. Figure 2 depicts the network model at the later phase of the simulation when a few nodes have died. In this figure, nodes farther from the sink form clusters and send their data to their respective cluster heads. But, nodes near the sink directly send their data to the sink as less energy would be required to send data to sink then to send data to cluster head in this case. The total energy of the model and the energy of individual nodes significantly

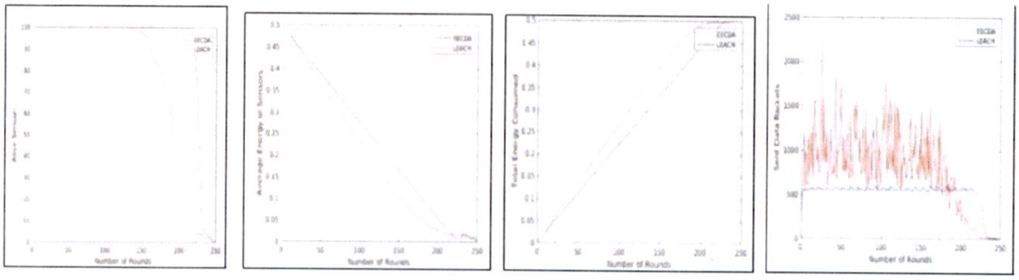

(a) Number of Alive sensors (b) Average energy of sensors per (c) Total energy consumed by (d) Number of data packet sent to per round sensor

Fig. 3 Network status after clustering

decrease in this phase and when the energy of the node becomes less than zero, then the node dies. The dead nodes are denoted by red dots. In the case of LEACH, the instability period is very high as the difference between the time when the first node dies, and the entire network dies is very large. But, in case of our proposed algorithm, the instability period is low due to load balancing in the network. This is very important because in many applications, when 50% of the nodes in a network die then the entire network becomes unserviceable. We have compared per round number of alive nodes between LEACH and EECDA that are illustrated in Fig. 3. A networks lifetime and stability are reflected by number of alive nodes. In LEACH, nodes begin to die after completion of 130 rounds whereas in case of EECDA, dead nodes start appearing after 220 rounds. All the nodes in both the protocols die at around 250 rounds. The instability period which is the difference between the time when all nodes in the network die and the first node dies are higher in case of LEACH than EECDA. This result shows that the proposed algorithm increases the network lifetime and stability of the network compared to LEACH. The energy is also uniformly distributed throughout the network as residual energy of the node is considered during cluster head selection instead of randomly selecting cluster heads. Distance of the node from the sink is also considered during cluster formation. In Fig. 2, the average energy of sensors per round is compared between LEACH and EECDA. Average energy of a sensor is the effective energy per sensor in a particular round. As we observed in Fig. 3a that in case of LEACH protocols, nodes start to die at around round 130 and at around 220 in case of EECDA protocol. The total energy of nodes in case of LEACH also decreases at a faster rate than EECDA. All nodes die at round 250 due to which average energy becomes 0. After round 200, there is a period when the average energy of sensors in case of LEACH is more than in case of EECDA. This occurs as a few nodes with high-energy remain alive in LEACH till the end due to non-uniform energy distribution. But in case of EECDA, there is uniform energy distribution due to which most of the nodes die in a very small time. Figure 3 shows the energy consumption of the nodes per round in the two protocols. In the graph, total energy consumed in case of LEACH increases at a faster rate as compared to EECDA. EECDA considers residual energy to calculate the next cluster head which improves the utilization of energy compared to LEACH which selects cluster heads randomly. Data aggregation at cluster head reduces the transmission cost of the network and reduces the total energy consumption. In Fig. 3, the number of data packets sent to sink per round is compared between LEACH and EECDA. In LEACH, the cluster head collects the data packet from sensors and sends the packets to sink in every round. But in case of EECDA, redundant data are removed using data aggregation, and only distinct data are sent to the sink in every round. This reduces the transmission cost of the network and helps to increase the network lifetime. In Fig. 3b, the average energy of sensors with respect to the number of sensors in plotted and the results are compared between LEACH and EECDA. In the proposed algorithm, the cluster heads are selected considering the residual energy of sensor nodes due to which energy is distributed uniformly in the network. The graph of average energy also decreases uniformly with increase in the number nodes. In case of LEACH, due to random cluster head selection and cluster formation, energy is not distributed uniformly and varies largely with change in the number of nodes. The figure also shows that average energy of the new algorithm is always more than that of leach due to efficient clustering and data aggregation used.

In example Fig. 3, the network status has shown once the clustering protocol forms and selects cluster head. In our case, we are performing a DB-SCAN clustering approach, which can be replaced with any clustering method. We can observe the network status without the range disk cluster head in Fig. 3. The most common path formation practice is to use the head node position to collect the data. The closed-loop uses a common approach to form the path considering all the head nodes shown in Fig. 2. In system, we are using the existing disk graph property of the head node to find the most suitable Steiner point as described in algorithm. 2. Our approach's advantage is that it minimizes travel time by opting for the RP far from the head node but near enough to collect data. In this way, we reduce the mobile sink's travel distance to any almost multiple of a head node with its disk radius value. It is a significant gain over the existing approach. The virtual division of a single position into four different places gives the locality-based path direction profitable for our cost optimization.

6 Conclusion

The paper proposed a cluster-based data aggregation protocol EECDA to prolong the lifetime of a sensor network which optimizes cluster head selection by considering distance and residual energy of the sensors. Clustering is one of the energy efficient techniques for extending the network lifetime. Clustering techniques organize nodes into clusters where the cluster heads can collect data from nodes in their respective cluster and transfer it to the base station. Further, we also introduced a data aggregation technique which aggregates the data at the cluster heads and then transfers the aggregated data to the base station. We compared the proposed method with the LEACH protocol. In LEACH, nodes that have been cluster heads cannot become cluster heads again for P rounds, where P is the desired percentage of cluster header elects the cluster head based on residual energy of the nodes and distance of the node from sink as well as other nodes in the network. We conclude the following results as sensors in LEACH protocol start to die in early rounds, whereas in EECDA, sensors remain alive for longer periods. The energy consumption in leach increases at a faster rate which reduces the overall network lifetime. The numbers of packets sent in each are more than in EECDA as it aggregates the data on the cluster head. As the number of nodes increases in the network, LEACH behaves very inefficiently with large variations in average energy of sensors which results in network instability, but EECDA maintains a uniform decrease in energy. The simulation results demonstrate that the proposed algorithm improves the network lifetime remarkably.

References

1. Chakrabarti A, Sabharwal A, Aazhang B (2006) Communication power optimization in a sensor network with a path-constrained mobile observer. ACM Trans Sensor Networks (TOSN) 2(3):297–324
2. Chakrabarti A, Sabharwal A, Aazhang B (2003) Using predictable observer mobility for power efficient design of sensor networks. In: Information processing in sensor networks. Springer, Berlin, pp 129–145
3. Xing G, Wang T, Jia W, Li M (2008) Rendezvous design algorithms for wireless sensor networks with a mobile base station. In: Proceedings of the 9th ACM international symposium on Mobile ad hoc networking and computing, pp 231–240
4. Shah RC, Roy S, Jain S, Brunette W (2003) Data mules: modeling and analysis of a three-tier architecture for sparse sensor networks. Ad Hoc Netw 1(2–3):215–233
5. Ma M, Yang Y (2008) Data gathering in wireless sensor networks with mobile collectors. In: 2008 IEEE international symposium on parallel and distributed processing. IEEE, New York, pp 1–9
6. Li X, Nayak A, Stojmenovic I (2010) Sink mobility in wireless sensor networks. Wireless Sensor and Actuator Networks 153
7. Van Hoesel L, Havinga P (2004) A lightweight medium access protocol (LMAC) for wireless sensor networks. In: 1st International workshop on networked sensing systems (INSS 2004)
8. Handy M, Haase M, Timmermann D (2002) Low energy adaptive clustering hierarchy with deterministic cluster-head selection. In: 4th International workshop on mobile and wireless communications network. IEEE, New York, pp 368–372
9. Sah DK, Amgoth T (2018) Parametric survey on cross-layer designs for wireless sensor networks. Comput Sci Rev 27:112–134
10. Tomar A et al (2020) Machine learning, advances in computing, renewable energy and communication, LNEE vol 768. Springer Nature, Berlin, 659 p. https://doi.org/10.1007/978-981-16-2354-7. ISBN 978-981-16-2354-7

Classification and Feature Extraction for Document Forgery Images

Rishabh Singh, Garima Jaiswal, Aditi Jain, and Arun Shrama

Abstract Hyperspectral images (HSIs) are captured over continuous wavebands. It includes the object's spatial as well as spectral information. Different materials have different reflection and absorption properties, which allows obtaining the underlying sample's physical structure and chemical composition. Therefore, hyperspectral imaging technology gives comprehensive information about the sample. This paper uses a hyperspectral imaging approach to detect document forgery in the documents through ink mismatch detection. The proposed approach utilizes PCA to handle multiple dimensions in HSIs. It captures the spectral features which act as an input to a convolutional neural network for image classification. The method is applied to the UWA Writing Ink HSI (WIHSI) database. The results are compared with the state-of-art results that prove the proposed approach's potential.

Keywords Hyperspectral imaging · Spectral · PCA · Document forgery · Deep learning

1 Introduction

Hyperspectral Imaging is an advanced technique based on spectroscopy. It collects images over a wide and continuous range of wavelengths for the same spatial area. It captures the spatial and spectral information from the object under analysis. It divides

R. Singh · A. Jain
Netaji Subhas Institute of Technology, University of Delhi, New Delhi, India
e-mail: rishabhs.ec.16@nsit.net.in

A. Jain
e-mail: aditijain0802@gmail.com

G. Jaiswal (✉) · A. Shrama
Indira Gandhi Delhi Technical University for Women, Delhi, India
e-mail: garima121@gmail.com

A. Shrama
e-mail: arunsharma@igdtuw.ac.in

A. Tomar et al. (eds.), *Proceedings of 3rd International Conference on Machine Learning, Advances in Computing, Renewable Energy and Communication*, Lecture Notes in Electrical Engineering 915, https://doi.org/10.1007/978-981-19-2828-4_68

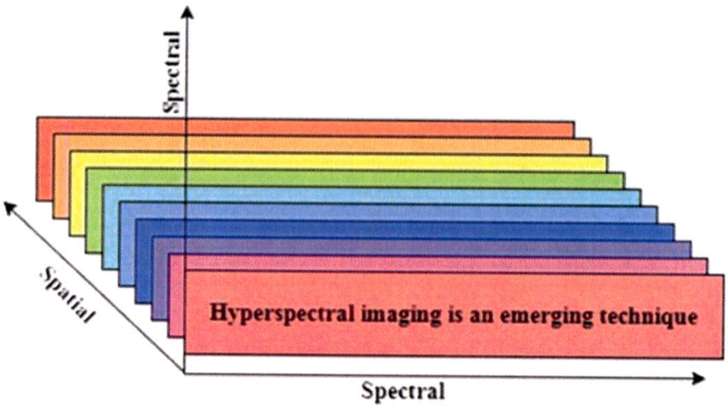

Fig. 1 Capturing Images at multiple bands

the spectrum into many more bands as illustrated in Fig. 1. It helps to determine the composition and assigns a unique spectral signature of the underlying material. It is being used in forensics, agriculture, archeology, security, medical diagnosis and surgery, remote sensing, and many more [1]. The unique spectral signature of each material in hyperspectral imaging (HSI) gives it tremendous power in identification and verification purposes [2].

Document forgery involves creating, falsifying, modifying the document with the wrong intention. In the present world of tools and techniques, detecting document forgery is a mysterious task that a naked eye cannot identify. In Document Analysis, the material's unique spectral signature helps us identify ink mismatches, the timeline of manuscripts, and recovering the degraded scripts. Cutting-edge technology like deep learning approaches has attained noticeable results in many domains [3].

In the present study, we utilized PCA for dimensionality reduction, and the extracted spectral features are fed to CNN to detect document forgery. The paper has been segregated into various sections. Section 2 illustrates the applications of hyperspectral imaging in various domains. Related work is elaborated in Sect. 3. The experimental setup and results are discussed in Sects. 4 and 5. Conclusion and Future work are depicted in Sect. 6.

2 Applications

Hyperspectral imaging is an emerging approach, gaining popularity for solving problems in many domains. This section highlights the application areas for hyperspectral imaging in the document analysis domain.

2.1 Document Analysis

Ink Mismatch Detection. The document forgery has always been of great concern and is frequently seen in fraudulent bank cheques, tampering with historical manuscripts and forensic evidence, and many more. If a document consists of more than one ink of similar color that may indicate a possible forgery. Currently, there are two approaches for identifying the ink mismatching-destructive and non-destructive. Non-destructive methods such as hyperspectral imaging are very useful in identifying the homogeneity of the ink.

Recovering Degraded Documents. The historical manuscripts, sometimes with time or due to some external factors, get degraded and unreadable. The unique spectral signature assigned to every element in an HSI can help to trace the text by applying image classification models [4].

Writer Identification. HSI is used to recognize the handwriting in a document. This approach helps to identify the writer or the owner of the document. It assists in identifying the modification and tampering done in the documents.

2.2 Supply Chain Management

Inventory Management. Supply chain management requires innovations and developments that may help it work in a more efficient and faster way in managing the inventory. The crucial role is to identify and also verify the articles. The problem usually comes in the verification of the article. It is a tedious task to separate articles manually and verify them. The customarily used cameras give us the RGB image, short of the subtle differences immune to naked eyes [5]. Here comes the beauty of HSIs, which allows capturing minute differences and the properties of the article's material. It allows us to exploit this and use it in varied applications.

One important application can be calculating the applicable charges on import and export according to the quality of the material. For example, there is a big difference in the import duty on varied fabrics. Still, they all may seem identical to a non-expert, or it usually becomes cumbersome to segregate the items manually. The materials appearing to be the same in the RGB camera should be charged differently according to the material's composition [6]. Here hyperspectral imaging can be used to facilitate knowledge of the properties of the material. Another practical application can be in the verification of the quality of the material. In businesses, there is a bulk sale and purchase of goods and ensuring the desired quality is their utmost priority. The use of HSI of the received articles will help to detect and match with the properties of the article promised.

Determine the quality of Food Products. The hyperspectral images of food products give essential spectral information about it. The commonly used colored image

(RGB) or seen by the naked eye only helps to determine the outer health [7]. While using hyperspectral images helps determine the complete health, i.e., inner and outer health [8]. It can be helpful in the management of the supply chain to separate unhealthy food like rotten vegetables or fruits. It can be used to effectively decide the product's cost based on their health.

3 Related Work

This paper focused on applying hyperspectral imaging for document forgery detection. We proposed to detect ink mismatch in HSIs to detect document forgery.

This section elaborates the related work for document forgery using machine learning and deep learning approaches in Table 1.

Table 1 Keypoints of the related work

S. No.	Paper ID	Approach	Limitations
1	Khan et al. [7]	JSBS for automatic mismatch detection for ink classification	Optimal band selection
2	Abbas et al. [8]	HySime for ink classification	HySime overestimates the number of inks
3	Khan et al. [9]	CNN with different architectures on spectral features	Spatial features unexplored
4	Khan et al. [10]	CNN with different architectures on spatial as well as spectral features	Uses supervised learning that requires prior knowledge
5	Islam et al. [11]	CNN on Spatio-Spectral features for writer identification	Unsupervised deep learning is not explored
6	Devassy et al. [12]	1-D-CNN, SAM and SID	Deeper neural networks unexplored
7	Qureshi et al. [13]	Literature Review of pattern recognition techniques	Handwriting classification unexplored
8	Devassy et al. [14]	Unsupervised clustering using t-SNE algorithm	Validation against other non-linear methods
9	Silva et al. [15]	PCA, MCR-ALS, and PLS-DA	Intersecting lines and identification of all the samples
10	Luo et al. [16]	Schwartz and P2P	Assumption on maximum number of inks mixed
11	Lian et al. [17]	Hyperspectral imager Nuance-Macro and software Nuance 1p46	Amount of ink applied and structure and surface of the paper substrate unexplored

4 Experimental Setup

This section elaborates the experimental setup for detecting document forgery using hyperspectral document images. We proposed a supervised neural network algorithm to detect document forgery in HSI that uses spectral information to classify the image's pixels (Fig. 2).

4.1 Database and Preprocessing

The WIHSI database [13] contains images of seven subjects. A single image in the database comprises five lines, all with same color (blue/black) but distinct ink. They are written in English by the subject. So, a total of 14 HSIs having 752 * 480 pixels, spanning across of 33 bands from 400 to 720 nm at a step of 10 nm, were captured [18]. The illumination in the images is non-uniform. Each hyperspectral image is exposed to preprocessing for further experimentation.

4.2 Preprocessing

We aim to process the data in such a way that it is ready for further experimentation. The first step is to separate each line to extract the background pixels via image thresholding. The global thresholding techniques fail to give satisfactory results as the image has non-uniform illumination. Sauvola's binarization method is used [9, 10] in this case as it threshold the image locally instead of globally. After this step, we have five hyperspectral data cubes, each containing an English phrase from every image along with their binary masks.

The objective of this work is to detect different inks in the same document with their unique spectral signature. To accomplish this, inks of the same subject were mixed in varying proportions [9, 10, 13]. No two different colored inks were mixed together, as it can be distinguished visually. Samples were generated using two, three, four, and five inks in equal and unequal proportions.

Fig. 2 Experimental setup

4.3 Proposed Approach

Principal Component Analysis is implemented before passing the image to the neural network to reduce the dimensions. With PCA, we can get rid of some of the features and map our dataset into a reduced subspace without losing essential information about the original dataset.

The objective is to extract the spatial features by preserving the spectral information of the HSI. The number of features to preserve has to be decided in this step. The number of principal components was varied from (3, 4, 5, 8, 9, 11, 13, 15, 17, 19, and 21). Its impact on accuracy was noted. It was concluded that with a count of 9 or above, the accuracy hardly changed. After the analysis, we selected 9 as the count.

Deep Learning is a cutting-edge technology that automatically captures features from a large dataset [19, 20]. A CNN consists of various layers: convolutional, activation, and pooling layers, followed by a connected layer that produces the output. The extracted spectral features were passed through a CNN model, as illustrated in Fig. 3 for classification.

5 Results

To analyze the proposed approach, we investigated the accuracy of ink mixing proportions for blue and black inks. The computed results are compared with the state-of-art approaches as illustrated in Table 2. The results depicted that black inks were challenging to identify compared to blue ink [19].

6 Future Prospects

The present study proposed a supervised deep learning-based method combined with PCA in HSIs for forgery detection. We evaluated the proposed approach by combining different inks in various proportions. The results clearly stated the effectiveness of the proposed approach. The present work may be extended using hybrid spectral and spatial features for forgery detection. Moreover, the supervised deep learning approach demands to know the count of the inks mixed in proportion in advance, which imposes a constraint for practical application. Unsupervised deep learning techniques may be examined to overcome this limitation.

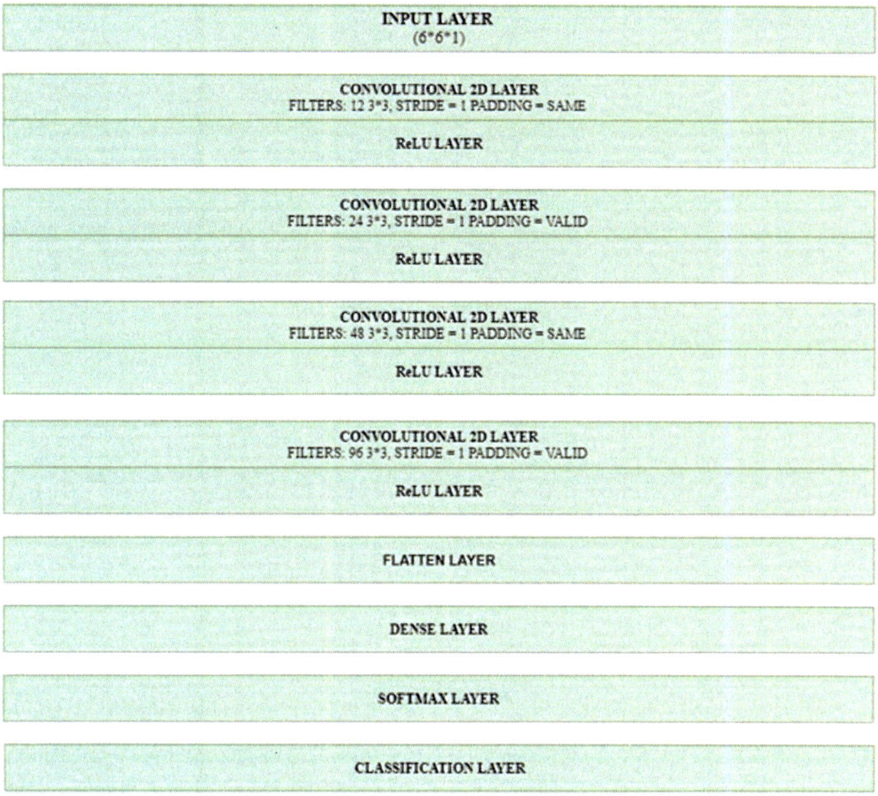

Fig. 3 CNN architecture

Table 2 Comparison of the proposed approach with state-of-art results

Citations	Approach	Average accuracy (%)		Maximum number of inks artificially mixed
		Blue inks	Black inks	
Proposed approach	Supervised deep learning	90.2	89.7	5
[13]	Feature selection	86.7	89.0	2
[14]	Unmixing	86.2	83.4	4
[21]	Unsupervised machine learning	89	82.3	2
[15]		86.7	81.9	2
[18]		85.6	81.4	2

References

1. Jaiswal G, Sharma A, Yadav SK (2021) Critical insights into modern hyperspectral image applications through deep learning. Wiley Interdisciplinary Reviews: Data Mining and Knowledge Discovery, e1426
2. Jaiswal G, Sharma A, Yadav SK (2022) Deep feature extraction for document forgery detection with convolutional autoencoders. Comput Electr Eng 99(107770)
3. Jaiswal G, Sharma A, Yadav SK (2019) Analytical approach for predicting dropouts in higher education. Int J Inf Commun Technol Educ (IJICTE) 15(3):89–102
4. Jyothsnaa S, Gandhe A, Deshpande A, Bodas S (2010, November) Automated inventory management and security surveillance system using image processing techniques. In: TENCON 2010–2010 IEEE Region 10 Conference. IEEE, New York, pp 2318–2321
5. Siche R, Vejarano R, Aredo V, Velasquez L, Saldana E, Quevedo R (2016) Evaluation of food quality and safety with hyperspectral imaging (HSI). Food Eng Rev 8(3):306–322
6. Lorente D, Blasco J, Serrano AJ, Soria-Olivas E, Aleixos N, Gomez-Sanchis J (2013) Comparison of ROC feature selection method for the detection of decay in citrus fruit using hyperspectral images. Food Bioprocess Technol 6(12):3613–3619
7. Khan Z, Shafait F, Mian A (2015) Automatic ink mismatch detection for forensic document analysis. Pattern Recogn 48(11):3615–3626
8. Abbas A, Khurshid K, Shafait F (2017, November) Towards automated ink mismatch detection in hyperspectral document images. In: 2017 14th IAPR International conference on document analysis and recognition (ICDAR), vol 1. IEEE, New York, pp 1229–1236
9. Khan MJ, Yousaf A, Abbas A, Khurshid K (2018) Deep learning for automated forgery detection in hyperspectral document images. J Electron Imaging 27(5):053001
10. Khan MJ, Khurshid K, Shafait F (2019, September) A spatio-spectral hybrid convolutional architecture for hyperspectral document authentication. In: 2019 International conference on document analysis and recognition (ICDAR). IEEE, New York, pp 1097–1102
11. Islam AU, Khan MJ, Khurshid K, Shafait F (2019, December) Hyperspectral image analysis for writer identification using deep learning. In: 2019 Digital image computing: techniques and applications (DICTA). IEEE, New York, pp 1–7
12. Devassy BM, George S (2019) Ink classification using convolutional neural network. NISK J 12:1–16
13. Qureshi R, Uzair M, Khurshid K, Yan H (2019) Hyperspectral document image processing: applications, challenges and future prospects. Pattern Recogn 90:12–22
14. MelitDevassy B, George S, Nussbaum P (2020) Unsupervised clustering of hyperspectral paper data using t-SNE. J Imaging 6(5):29
15. Silva CS, Pimentel MF, Honorato RS, Pasquini C, Prats-Montalbán JM, Ferrer A (2014) Near infrared hyperspectral imaging for forensic analysis of document forgery. Analyst 139(20):5176–5184
16. Luo Z, Shafait F, Mian A (2015, August) Localized forgery detection in hyperspectral document images. In: 2015 13th International conference on document analysis and recognition (ICDAR). IEEE, New York, pp 496–500
17. Lian Y, Liang L, Li B (2017) Hyperspectral imaging technology for revealing the original handwritings covered by the same inks. J Forensic Sci Med 3(4):210
18. Khan Z, Shafait F, Mian A (2013, August) Hyperspectral imaging for ink mismatch detection. In: 2013 12th International conference on document analysis and recognition. IEEE, New York, pp 877–881
19. Jaiswal G, Sharma A, Yadav SK (2021) Efficient ink mismatch detection using supervised approach. In: Singh M, Tyagi V, Gupta PK, Flusser J, Ören T, Sonawane VR (eds) Advances in computing and data sciences. ICACDS 2021. Communications in computer and information science, vol 1440. Springer, Cham. https://doi.org/10.1007/978-3-030-81462-5_65

20. Tomar A et al (2020) Machine learning, advances in computing, renewable energy and communication, LNEE vol 768. Springer Nature, Berlin, 659 p. https://doi.org/10.1007/978-981-16-2354-7. ISBN 978-981-16-2354-7
21. Khan Z, Shafait F, Mian A (2013, August) Hyperspectral imaging for ink mismatch detection. In: 2013 12th International conference on document analysis and recognition. IEEE, pp 877-881

A Greedy Jellyfish Search Optimization Algorithm

Jitendra Rajpurohit and Tarun K. Sharma

Abstract Jellyfish Search Optimizer (JSO) is one of the latest additions in the list of nature-inspired metaheuristic algorithms. This paper proposes a modified variant of JSO called Greedy Jellyfish Search Optimizer (GJSO). GJSO modifies the ocean current movement of JSO and adds greedy behaviour in the search process. The proposed variant has been tested on 20 diverse functions. The results are compared with two other peer algorithms. The results show superiority of the proposal over unimodal functions and comparative performance over multimodal functions.

Keywords Jellyfish Search Optimizer · Metaheuristic algorithms · Optimization

1 Introduction

Metaheuristic algorithms are set of heuristic search methods that are aimed at providing acceptable and near optimal solution to an optimization problem. Basic working of most of the metaheuristic algorithms follows similar kind of steps. They generate a set of search agents randomly in the search space. Value of the function to be optimized (objective function) is tested at the locations of these search agents. New locations are generated and tested using different methods in different algorithms. All such algorithms keep the location of the best agent stored and provides it as the output. There may be infinite points inside the search space, practically it is not possible to test all points by any metaheuristic algorithm. So, metaheuristic algorithms do not guarantee to provide the optimum solution, But, they almost always provide a near optimum and acceptable solution.

These algorithms do not use information of the search space, so are applicable on all optimization problems. Value of the objective function needs to be calculated

J. Rajpurohit (✉)
School of Computer Science, University of Petroleum and Energy Studies, Dehradun 248007, India
e-mail: jiten_rajpurohit@yahoo.com

T. K. Sharma
Shobhit University Gangoh, Saharanpur 247001, India

A. Tomar et al. (eds.), *Proceedings of 3rd International Conference on Machine Learning, Advances in Computing, Renewable Energy and Communication*, Lecture Notes in Electrical Engineering 915, https://doi.org/10.1007/978-981-19-2828-4_69

at each point being tested by the algorithm. This testing needs computing devices. That is one of the fundamental reasons that the popularity of these algorithms has increased along the improvement in computing technology.

To generate new locations to be tested, these algorithms take their inspiration from some existing optimization methods. Most of these algorithms take inspiration from nature. There are numerous natural species, methods and phenomena that have evolved over millions of years and are good source of mature search methods. Due to their inspiration from nature, these algorithms are often referred as nature-inspired optimization algorithms.

Researchers keep on developing new and more efficient algorithms and improve existing ones. In this paper, one of the latest algorithms, Jellyfish Search Optimization (JSO) algorithm has been modified for improvement. Efficiency and accuracy of the proposed method have been tested over a set of test functions, and the performance has been compared with other similar algorithms.

Remaining of this paper is structured in the following manner: Sect. 2 provides a brief review of literature pertaining to metaheuristic algorithms and their source of inspiration. Section 3 explains steps of working of JSO algorithm. Section 4 proposes the modified variant of the algorithm. Section 5 contains experimental results and comparison. Finally, Sect. 6 concludes the paper.

2 Literature Review

Development of some of the early algorithms includes simulated annealing [1], genetic algorithms [2] and differential evolution [3]. Some of the latest algorithms are discussed briefly below:

Grey wolf optimizer [4] is inspired from the herding behaviour of grey wolves. Each wolf belongs to one of the levels of hierarchy in the heard, namely alpha, beta, delta or omega. These hierarchies help in managing the herd. Further, hunting strategy of grey wolves is used to design search method of the algorithm. Hunting strategy includes chasing, encircling and attacking the prey.

Whale Optimization Algorithm [5] mimics the foraging behaviours of humpback whales. These whales live alone but often roam in groups. They use a special method for hunting called bubble-net feeding. Various steps of bubble-net feeding like encircling, bubble-net attacking, shrinking, encircling and spiral movement are used as methods to inspire searching in the algorithm.

Firefly Algorithm [6] models the search agents as fireflies that emit light due to a chemical reaction. Value of the objective function is denoted by the brightness of a firefly. A firefly is attracted towards a brighter firefly. Intensity of a firefly decreases with distance between two of them. A number of parameters are used to adjust the movement of fireflies in the search space.

Water Strider Algorithm [7] is an optimizer that is inspired by the unique characteristics of water strider bugs. Water striders are insects that are capable of living on water surface. Steps of the algorithm are inspired by various survival activities carried out by these bugs. The algorithm simulates activities like territory establishment, mating communication, feeding, dying, succession, etc.

Black Widow Optimization Algorithm [8] is inspired by the strange reproduction style of black widow, a spider species. After mating, the female black widow consumes the male. Even the offspring consumes the siblings and the mother itself. This cycle of elimination of weaker members is called cannibalism. The reproduction system leads to survival of fittest which is the basis of many other nature-inspired optimization algorithms. Steps of the algorithm include various activities of the reproduction process such as procreate, cannibalism and mutation.

3 Working of Artificial Jellyfish Search Optimizer (JSO)

JSO [9] is one of the latest metaheuristic nature-inspired algorithms. Experimental results show that it is one of the most efficient algorithms. In this algorithm, search agents are modelled as jellyfish in the ocean. The algorithm is inspired by the swimming and foraging behaviour of jellyfish swarm. The fundamental assumptions used in the algorithm are:

- A jellyfish either moves in the direction of the ocean current or inside the swarm.
- Jellyfish's movement in swarm can be either of type A (passive) or type B (active)
- Switching between ocean current/swarm movement and type A/type B motion is governed by a time control mechanism which is dependent on the iteration count.
- Jellyfish is attracted towards food-rich locations.

Following steps explain the complete process of the algorithm:

Initialization

Population of jellyfish can be initialized randomly or using any one of the chaotic maps. Initialization using logistic chaotic map provides more diverse population, hence can avoid trapping in local optima [10]. Population initialization using logistic chaotic map is done using Eq. (1).

$$X_{i+1} = \eta X_i(1 - X_i), \quad 0 \leq X_0 \leq 1 \tag{1}$$

where X_i is the logistic chaotic value of the ith jellyfish. X_0 is the initial value used, $X_0 \in (0, 1)$ and $X_0 \neq \{0.25, 0.50, 0.75, 1.00\}$. $\eta = 4$.

Movement along ocean current

In this movement, all jellyfish move along a vector that represents movement from mean location of the population to the global best jellyfish. New location of a jellyfish after movement along the ocean current is expressed in Eq. (2).

$$X_i(t + 1) = X_i(t) + rand(0, 1) \times \left(X^* - \beta \times rand(0, 1) \times \mu\right) \tag{2}$$

where X^* is the global best jellyfish; μ is the mean location of all jellyfish; β is distribution coefficient and $\beta > 0$. Best value of β obtained through sensitivity analysis is 3 [9].

Equation (2) denotes the movement of all jellyfish by a vector $\overrightarrow{\text{TREND}}$. Here, $\overrightarrow{\text{TREND}}$ is an indication of mean of all jellyfish subtracted from the global best.

$$\overrightarrow{\text{TREND}} = X^* - \beta \times rand(0, 1) \times \mu \tag{3}$$

where μ is the mean location of all jellyfish.

Movement inside the swarm

If not moving along the ocean current, then a jellyfish can move within the swarm. Swarm movements can be either of type A (passive) or type B (active).

In type A motion, surrounding locations of a jellyfish are explored. This movement can be given by Eq. (4).

$$X_i(t + 1) = X_i(t) + \gamma \times rand(0, 1) \times (U_b - L_b) \tag{4}$$

While in type B motion, a jellyfish moves either towards or away from a randomly selected other jellyfish from the swarm. Towards or away movement of the jellyfish depends upon the food quality of the location of the other jellyfish.

New location after type B movement can be described by Eq. (5).

$$X_i(t + 1) = X_i(t) + rand(0, 1) \times \overrightarrow{\text{Direction}} \tag{5}$$

where

$$\overrightarrow{\text{Direction}} = \begin{cases} X_j(t) - X_i(t), & \text{if } f(X_i) \geq f(X_i) \\ X_j(t) - X_i(t), & \text{Otherwise} \end{cases}$$

Time control mechanism

Selection of movement (ocean current or swarm movement/type A or type B) for jellyfish is controlled by time control mechanism. It uses stage of the execution of a run to control the movement. The selection between ocean current and swarm motion is done using the value of time control function ($c(t)$). Its value is calculated by Eq. (6).

$$c(t) = \left| \left(1 - \frac{t}{\text{Max}_{\text{itr}}} \right) \right| \times (2 \times rand(0, 1) - 1) \tag{6}$$

where t is the iteration count and Max_{itr} is the number of maximum iterations. Jellyfish follows ocean current if $c(t) > C_0$ and swarm motion otherwise. Here, C_0 is a constant and $C_0 \in (0, 1)$.

The same function $c(t)$ is used to select between type A and type B motion of jellyfish in a swarm. If $(1 - c(t)) > rand(0, 1))$, jellyfish exhibits type B motion and type A motion otherwise.

4 Proposed Greedy Jellyfish Search Optimizer (GJSO)

An efficient metaheuristic algorithm should observe a balance of exploration and exploitation. Exploration deals with expanding the search into newer locations for better objective function values. Exploitation means intense searching around existing locations to converge the search process. JSO also contains both these essential components. Ocean currents aim exploration and swarm movements target exploitation.

However, Eqs. (2) and (3) indicate a significant observation about the exploration component. All jellyfish are relocated in the same direction denoted by $\overrightarrow{\text{TREND}}$. This symmetry in the search process limits the randomness element and reduces the possibility of diverse search. Instead, the ocean current movement can be used to either enhance the diversity of the population by adding random direction or it can be used to speed-up convergence by applying a greedy approach. The proposed variant aims at improving convergence. It enables the jellyfish to move in different directions as well as improve the convergence process. Instead of using the same mean-based $\overrightarrow{\text{TREND}}$ vector, each jellyfish is relocated using a vector that is calculated using the distance between the jellyfish itself and the global best vector (X^*). As all jellyfish are moved towards available best vector, hence, it is called Greedy Jellyfish Search Optimizer (GJSO). So, for ocean current, Eq. (7) is used in place of Eq. (2).

$$X_i(t+1) = X_i(t) + rand(0,1) \times \left(X^* - \beta \times rand(0,1) \times X_i(t)\right) \tag{7}$$

As there is no common trend now, so Eq. (3) establishing $\overrightarrow{\text{TREND}}$ will not be applicable.

Complete working of the proposed variant is illustrated in Fig. 1. Figure 1a illustrates the major steps, while Fig. 1b explains the steps in an iteration.

5 Experimental Results and Comparison

Movement of all the vectors towards the global best in GJSO removes the symmetrical motion in the algorithm as well as it accelerates the convergence due to its greedy behaviour. Figure 2 shows the distribution of population while optimizing two-dimensional sphere function with optimum at (0, 0) with variable range [−100, 100]. The distribution is after five iterations of (a) JSO and (b) GJSO. Converging behaviour of the population can be observed in the figure.

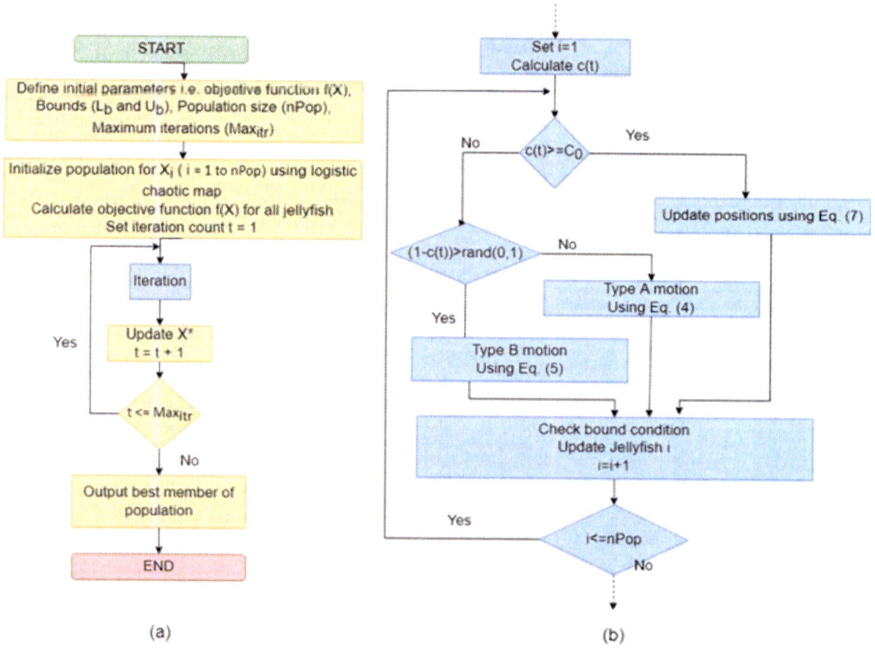

Fig. 1 **a** Major steps of GJSO **b** Steps inside an iteration of GJSO

The proposed variant has been tested on a testbed of 20 minimization functions. Table 1 shows the function details. Experimental setup and results are discussed further. F1 to F11 are unimodal and F12 to F20 are multimodal functions.

Performance of GJSO is compared with JSO and two other peer algorithms, namely grey wolf optimization [4] and Ant Lion Optimizer [11] for the same set of functions.

For all the experiments, population size has been kept 50. All algorithms were executed for 1000 iterations. Experiments were performed in MATLAB R2016a on an Intel(R) Core(TM) i7-8565U CPU @ 1.80 GHz Windows 10 machine. All the experiments are repeated 30 times to avoid any misrepresentation due to stochastic nature of the algorithms. Any number less than E−350 is considered as zero. For JSO and GJSO $C_0 = 0.5$, $\beta = 3$ and $\gamma = 0.1$. All algorithm specific parameters for GWO and ALO are adopted from their base papers [4] and [11], respectively.

Objective function values obtained by the algorithms are given in Table 2.

In Table 2, best and mean values of objective function are presented for JSO and GJSO for better comparison. While for GWO and ALO, mean values are given. Best performing algorithm for a function is considered the one with the best mean value. For functions F16, F17 and F18, there are more than one algorithms who perform equivalently good, in those cases all such algorithms are considered to be best performers. By this criterion, overall, JSO, GJSO, GWO and ALO are best performer for 7, 13, 2 and 4 functions, respectively. For unimodal functions, JSO, GJSO, GWO

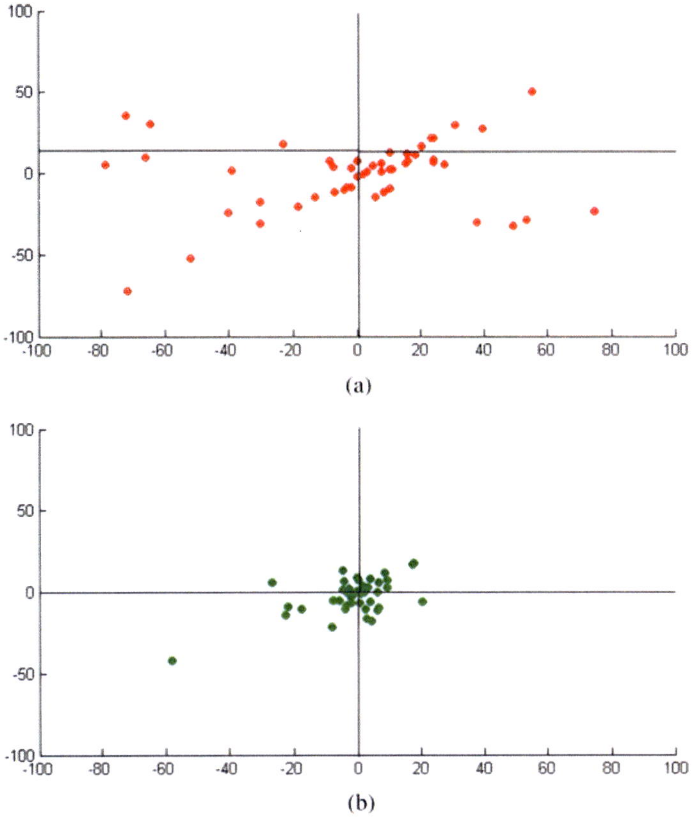

Fig. 2 Difference in convergence trend in **a** JSO and **b** GJSO after 5 iterations of optimizing two-dimensional sphere function with population in the range (−100, 100) and optimum at (0, 0)

and ALO are best performers for 2, 8, 1 and 0 functions, respectively. Similarly, for multimodal functions, JSO and GJSO are best performers for 5 functions each, while GWO and ALO are for 2 and 4 functions, respectively. These observations conclude that JSO and GJSO are the two best performing algorithms for this set of functions, and GJSO clearly outperforms all other algorithms. Similarly, for unimodal functions, GJSO is a clear winner. Although, for multimodal functions, JSO and GJSO are equivalent, and ALO also performs better. Infographics for the data is provided in Fig. 3.

The ocean current of JSO can either be utilized to diverse the population or to accelerate convergence. GJSO aims at converging faster by replacing uni-directional ocean current into greedy movements towards the global best. This modification increases exploitation. So, improved performance for unimodal functions and comparative performance for multimodal functions is expected.

Table 1 Details of functions used for testing of GJSO

Modality	Symbol	Function name	Range	Optimum value	Number of dimensions
Unimodal	F1	Sphere	[−100, 100]	0	30
	F2	SumSquares	[−10, 10]	0	30
	F3	Quatric	[−1.28, 1.28]	0	30
	F4	Matyas	[−10, 10]	0	2
	F5	Colville	[−10, 10]	0	4
	F6	Zakharov	[−5, 10]	0	10
	F7	Powell	[−4, 5]	0	24
	F8	Schwefel 2.22	[−10, 10]	0	30
	F9	Schwefel 1.2	[−100, 100]	0	30
	F10	Rosenbrock	[−30, 30]	0	30
	F11	Dixon-price	[−10, 10]	0	30
Multimodal	F12	Rastrigin	[−5.12, 5.12]	0	30
	F13	Schwefel	[−500, 500]	−12569.5	30
	F14	Michalewicz 5	[0, π]	−4.6877	5
	F15	Michalewicz 10	[0, π]	−9.6602	10
	F16	Shubert	[−10, 10]	−186.73	2
	F17	Shekel5	[0, 10]	−10.15	4
	F18	Shekel10	[0, 10]	−10.53	4
	F19	Powersum	[0, D]	0	4
	F20	Ackley	[−32, 32]	0	30

6 Conclusion

The paper proposed a greedy variant of Jellyfish Swarm Optimizer. The greedy behaviour of the variant is due to its tendency to converge towards the global best during ocean current movements. This tendency utilizes the symmetric ocean currents for exploitation of the search space. This results in faster convergence for unimodal functions. Although, the variant performs at par only for multimodal functions. In future, ocean currents are adjusted to balance exploitation, and exploration needs to be investigated.

Table 2 Objective function values obtained by the algorithms are given in Table 2

Function	Algorithm							
	JSO			GJSO			GWO	ALO
	best	mean	std	best	mean	std	Mean	mean
F1	7.17E−77	1.95E−66	8.72E−66	6.71E−171	**1.24E−161**	4.43E−161	1.70E−70	1.82E−07
F2	1.99E−61	1.79E−29	8.02E−29	8.54E−166	**3.49E−97**	1.56E−96	4.67e−72	6.01E−01
F3	3.56E−05	1.40E−04	9.79E−05	2.10E−05	**1.20E−04**	1.09E−04	1.10E−03	4.12E−01
F4	3.39E−179	5.89E−174	0.00E+00	9.66E−245	4.24E−233	0.00E+00	**1.07E−284**	4.82E−16
F5	3.71E−12	**1.80E−09**	3.72E−09	3.16E−13	5.63E−09	1.11E−08	1.33E+00	1.30E−03
F6	6.39E−41	1.06E−32	4.75E−32	3.45E−111	**2.73E−98**	1.15E−97	2.51E−89	1.96E−11
F7	1.40E−23	4.05E−10	1.14E−09	4.81E−161	**6.34E−92**	2.84E−91	1.31E−05	3.7E−01
F8	2.09E−36	2.09E−31	8.96E−31	1.20E−84	**8.02E−70**	3.58E−69	2.46E−41	8.18E−01
F9	9.63E−80	9.25E−72	3.95E−71	1.18E−172	**2.48E−162**	1.07E−161	8.59E−70	2.52E−01
F10	5.11E−06	**1.23E−02**	2.13E−02	4.88E−05	1.96E−02	4.67E−02	2.71E+01	9.48E+01
F11	1.02E−04	1.70E−03	1.90E−03	2.91E−04	**1.40E−03**	6.92E−04	6.66E−01	1.37E+00
F12	9.58E−11	**2.80E−03**	1.04E−02	0.00E+00	2.98E+00	9.19E+00	6.66E−01	8.45E+01
F13	−12,149.663	−10,245.2783	1028.626	−12,151.6772	**−11,773.2883**	1090.5321	−5299.4041	−4998.7103
F14	−4.6877	**−4.6697**	0.0361	−4.6877	−4.6143	0.0827	−4.4959	−3.7492
F15	−9.6552	**−9.5178**	0.0994	−9.6225	−9.4498	0.1708	−9.089	−6.8139
F16	−186.7309	**−186.7309**	2.351E−14	−186.7309	**−186.7309**	2.1626E−14	**−186.7309**	**−186.7309**
F17	−10.1532	**−10.1532**	2.0658E−11	−10.1532	**−10.1532**	2.5121E−15	−10.1528	**−10.1532**
F18	−10.5364	−10.5361	3.1303E−03	−10.5364	**−10.5364**	3.2856E−15	−10.5359	**−10.5364**
F19	2.00E−06	4.41E−04	8.33E−04	4.32E−06	3.48E−04	4.31E−04	0.0012399	**1.4695e−04**
F20	8.88E−16	3.55E−15	1.58E−15	8.88E−16	**2.13E−15**	1.74E−15	1.51E−14	2.73E+00

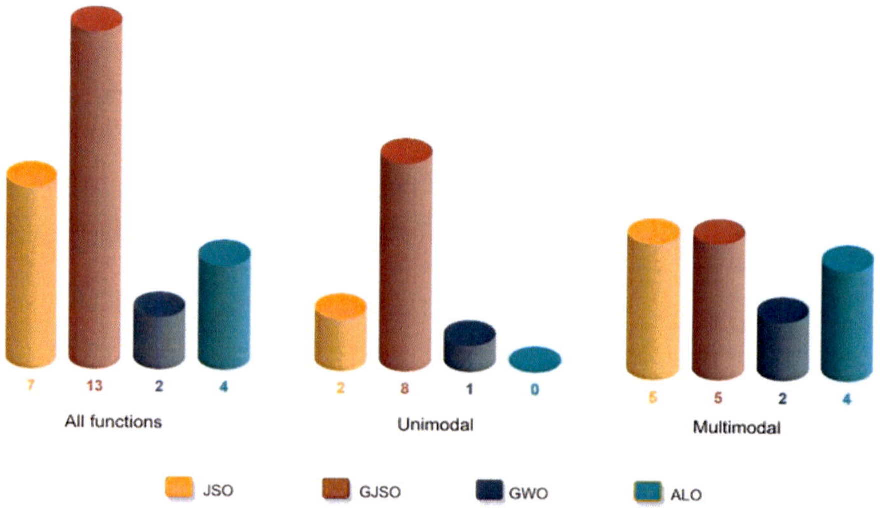

Fig. 3 Number of functions where JSO, GJSO, GWO and ALO performed best for all categories of test functions

References

1. Van Laarhoven PJ, Aarts EH (1987) Simulated annealing. In: Simulated annealing: theory and applications. Springer, Dordrecht, pp 7–15
2. Goldberg DE (1989) Genetic algorithms in search, optimization, and machine learning. Addison, Reading
3. Storn R, Price K (1997) Differential evolution–a simple and efficient heuristic for global optimization over continuous spaces. J Global Optim 11(4):341–359
4. Mirjalili S, Mirjalili SM, Lewis A (2014) Grey wolf optimizer. Adv Eng Softw 69:46–61
5. Mirjalili S, Lewis A (2016) The whale optimization algorithm. Adv Eng Softw 95:51–67
6. Yang XS (2010) Firefly algorithm, stochastic test functions and design optimisation. Int J Bio-inspired Comput 2(2):78–84
7. Kaveh A, Eslamlou AD (2020, June) Water strider algorithm: a new metaheuristic and applications. In: Structures, vol 25. Elsevier, Amsterdam, pp 520–541
8. Hayyolalam V, Kazem AAP (2020) Black widow optimization algorithm: a novel meta-heuristic approach for solving engineering optimization problems. Eng Appl Artif Intell 87:103249
9. Chou JS, Truong DN (2021) A novel metaheuristic optimizer inspired by behavior of jellyfish in ocean. Appl Math Comput 389:125535
10. Fister I Jr, Perc M, Kamal SM, Fister I (2015) A review of chaos-based firefly algorithms: perspectives and research challenges. Appl Math Comput 252:155–165
11. Mirjalili S (2015) The ant lion optimizer. Adv Eng Softw 83:80–98

Author Index

A. Tomar et al. (eds.), *Proceedings of 3rd International Conference on Machine Learning, Advances in Computing, Renewable Energy and Communication*, Lecture Notes in Electrical Engineering 915, https://doi.org/10.1007/978-981-19-2828-4